The MPEG-4 Book

ISBN 0-13-061621-4

9 790130 616219

90000

IMSC Press Multimedia Series

Integrated Media Systems Center

ANDREW TESCHER, Series Editor, *Compression Science Corporation*

Advisory Editors
LEONARDO CHIARIGLIONE, *CSELT*
TARIQ S. DURRANI, *University of Strathclyde*
JEFF GRALNICK, *E-splosion Consulting, LLC*
CHRYSOSTOMOS L. "MAX" NIKIAS, *University of Southern California*
ADAM C. POWELL III, *The Freedom Forum*

▶ Desktop Digital Video Production
  Frederic Jones

▶ Touch in Virtual Environments:
  Haptics and the Design of Interactive Systems
    Edited by Margaret L. McLaughlin,
    João P. Hespanha, and Gaurav S. Sukhatme

▶ The MPEG-4 Book
    Edited by Fernando Pereira and Touradj Ebrahimi

▶ Multimedia Fundamentals, Volume 1:
  Media Coding and Content Processing
    Ralf Steinmetz and Klara Nahrstedt

▶ Intelligent Systems for Video Analysis and Access Over
  the Internet
    Wensheng Zhou and C. C. Jay Kuo

Integrated Media Systems Center

The Integrated Media Systems Center (IMSC), a National Science Foundation Engineering Research Center in the University of Southern California's School of Engineering, is a preeminent multimedia and Internet research center. IMSC seeks to develop integrated media systems that dramatically transform the way we work, communicate, learn, teach, and entertain. In an integrated media system, advanced media technologies combine, deliver, and transform information in the form of images, video, audio, animation, graphics, text, and haptics (touch-related technologies). IMSC Press, in partnership with Prentice Hall, publishes cutting-edge research on multimedia and Internet topics. IMSC Press is part of IMSC's educational outreach program.

The MPEG-4 Book

Fernando Pereira
Touradj Ebrahimi
IMSC Press

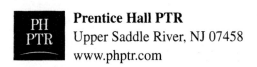

Prentice Hall PTR
Upper Saddle River, NJ 07458
www.phptr.com

Library of Congress Cataloging-in-Publication Data

Pereira, Fernando C. N.
   The MPEG-4 book / Fernando Pereira, Touradj Ebrahimi
      p. cm.-- (IMSC Press multimedia series)
   Includes bibliographical references and index.
   ISBN 0-13-061621-4
   1. MPEG (Video coding standard). 2. Video compression—Standards.
I. Ebrahimi, Touradj. II. Title. III. Prentice Hall IMSC multimedia series.

TK6680.5 .P47 2002
621.388—dc21

                                                2002070433

Editorial/production supervision: *BooksCraft, Inc., Indianapolis, IN*
Acquisitions editor: *Bernard Goodwin*
Editorial assistant: *Michelle Vincenti*
Marketing manager: *Dan DePasquale*
Manufacturing manager: *Alexis Heydt-Long*
Cover design director: *Jerry Votta*
Cover designer: *Anthony Gemmellaro*
Full-service production manager: *Anne Garcia*

© 2002 Pearson Education, Inc.
Publishing as Prentice Hall PTR
Upper Saddle River, New Jersey 07458

Prentice Hall books are widely used by corporations and government agencies for training, marketing, and resale.

For information regarding corporate and government bulk discounts please contact:
Corporate and Government Sales (800) 382-3419 or corpsales@pearsontechgroup.com

Printed in the United States of America
10 9 8 7 6 5 4 3

ISBN 0-13-061621-4

Pearson Education LTD.
Pearson Education Australia PTY, Limited
Pearson Education Singapore, Pte. Ltd.
Pearson Education North Asia Ltd.
Pearson Education Canada, Ltd.
Pearson Educación de Mexico, S.A. de C.V.
Pearson Education—Japan
Pearson Education Malaysia, Pte. Ltd.

Contents

Foreword xix

Preface xxiii

Abbreviations xxvii

1 Context, Objectives, and Process 1

    1.1 MPEG-4 Objectives 4

        1.1.1 Functionalities 5

        1.1.2 Requirements 7

        1.1.3 Tools 10

        1.1.4 Applications 12

    1.2 Formal Standardization Process 13

    1.3 MPEG Modus Operandi 15

        1.3.1 Mission 16

        1.3.2 Principles 17

        1.3.3 Standards Development Approach 18

    1.4 MPEG-4 Standard Organization 21

    1.5 MPEG-4 Schedule 25

    1.6 MPEG-4 Industry Forum 30

    1.7 Summary 31

    1.8 References 32

2 MPEG-4 Overview **37**

2.1 Design Goals 38

2.2 An End-to-End Walkthrough 40

2.3 Terminal Architecture 43

2.4 MPEG-4 Tools 47

    2.4.1 Systems Tools 47

    2.4.2 Visual Tools 48

    2.4.3 Audio Tools 50

    2.4.4 DMIF Tools 51

    2.4.5 Other MPEG-4 Tools 51

    2.4.6 Profiles and Levels 52

2.5 MPEG-4 and Other Multimedia Standards 52

2.6 MPEG-4 Applications 56

    2.6.1 Multimedia Portals 56

    2.6.2 Interactive Broadcasting 57

    2.6.3 Multimedia Conferencing and Communities 59

2.7 Summary 60

2.8 References 61

3 Object Description and Synchronization **65**

3.1 Object Descriptors: Entry Points to MPEG-4 Content ... 65

    3.1.1 Syntactic Description Language 66

    3.1.2 Object Description 70

    3.1.3 Stream Description 72

    3.1.4 Stream Relationship Description 76

    3.1.5 Content Complexity Description 77

    3.1.6 Streaming ODs 79

    3.1.7 Linking a Scene to Its Media Streams 82

    3.1.8 MPEG-4 Content Access Procedure 85

3.2 Semantic Description and Access Management 86

    3.2.1 Object Content Information: Meta Information About Objects ... 86

    3.2.2 Intellectual Property Management and Protection ... 89

3.3 Timing Model and Synchronization of Streams 90

    3.3.1 Modeling Time 90

    3.3.2 Time Stamps and Access Units 92

    3.3.3 Packetizing Streams: The Sync Layer 93

    3.3.4 Distributed Content, Time Bases, and OCR Streams ... 96

    3.3.5 Media Time 98

    3.3.6 System Decoder Model 98

3.4 Summary 100

3.5 References 101

4 BIFS: Scene Description **103**

4.1 Basics of BIFS 104

    4.1.1 Scene and Nodes 104

    4.1.2 Fields and ROUTEs 106

4.1.3 Node Types 108
4.1.4 Subscenes and Hyperlinks.................................... 109
4.1.5 Scene Changes... 109
4.1.6 Scene Rendering... 112
4.1.7 Binary Encoding.. 113
4.1.8 Quantization.. 114
4.2 Basic BIFS Features by Example 114
4.2.1 Trivial Scene... 114
4.2.2 Movie with Subtitles 117
4.2.3 Icons and Buttons.. 119
4.2.4 Slides and Transitions 121
4.2.5 Simple 3D Scene 124
4.2.6 Magnifying Glass.. 128
4.3 Advanced BIFS Features.. 129
4.3.1 Scripting.. 130
4.3.2 Encapsulation and Reuse............................... 131
4.3.3 Text Layout.. 134
4.4 A Peek Ahead on BIFS 136
4.4.1 Media Nodes 136
4.4.2 New Sensors 137
4.4.3 FlexTime 138
4.4.4 ServerCommand.. 139
4.5 Profiles.. 139
4.6 All BIFS Nodes 140
4.7 Summary.. 146
4.8 References 147

5 **MPEG-J: MPEG-4 and Java** **149**
5.1 MPEG-J Architecture 154
5.1.1 MPEGlets.. 156
5.1.2 Delivery... 158
5.1.3 Security... 159
5.2 MPEG-J APIs... 160
5.2.1 Terminal APIs 163
5.2.2 Scene APIs.. 164
5.2.3 Resource APIs... 173
5.2.4 Decoder APIs 176
5.2.5 Network APIs 178
5.2.6 Service Information and Section Filtering APIs 180
5.2.7 MPEG-J Profiles.. 180
5.3 Application Scenarios 181
5.3.1 Adaptive Rich Media Content for Wireless Devices.... 181
5.3.2 Enhanced Interactive Electronic Program Guide (EPG) 181
5.3.3 Enriched Interactive Digital Television.................. 182
5.3.4 Content Personalization.................................... 182

5.4 Reference Software 182
5.5 Summary 183
5.6 References 185

6 Extensible MPEG-4 Textual Format.. 187
6.1 Objectives 188
6.2 Cross-Standard Interoperability 189
6.3 XMT Two-Tier Architecture 190
6.4 XMT-Ω Format 193
    6.4.1 Reusing SMIL in XMT-Ω..194
    6.4.2 Extensible Media (xMedia) Objects 195
    6.4.3 Timing and Synchronization...198
    6.4.4 Time Manipulations 202
    6.4.5 Animation..203
    6.4.6 Spatial Layout 205
    6.4.7 XMT-Ω Examples...205
6.5 XMT-A Format...210
    6.5.1 Document Structure 210
    6.5.2 Timing 211
    6.5.3 Scene Description 212
    6.5.4 Object Descriptor Framework..219
    6.5.5 Deterministic Mapping of XMT-A 222
    6.5.6 Interoperability with X3D 223
6.6 Summary 224
6.7 References 224

7 Transporting and Storing MPEG-4 Content 227
7.1 Delivery Framework 229
    7.1.1 DMIF-Application Interface..231
    7.1.2 DMIF Network Interface...238
    7.1.3 Existent Signaling Protocols 246
7.2 FlexMux Tool 248
    7.2.1 Timing of a FlexMux Stream 251
7.3 MPEG-4 File Format...253
    7.3.1 Temporal Structure: Tracks and Streams, Time and Durations 254
    7.3.2 Physical Structure: Atoms and Containers, Offsets and Pointers........257
    7.3.3 MPEG-4 Systems Concepts in MP4..259
    7.3.4 MPEG-4 Track Types and Storage 259
    7.3.5 Hinting 260
    7.3.6 Atoms...262
    7.3.7 Random Access 265
    7.3.8 An MP4 Example 266
    7.3.9 Summary of MP4 268
7.4 Transporting MPEG-4 over MPEG-2..268
    7.4.1 Brief Introduction to MPEG-2 Systems..269
    7.4.2 Transport of MPEG-4 Elementary Streams over MPEG-2 Systems 274

7.4.3 Transport of MPEG-4 Scenes over MPEG-2 Systems 275
7.5 Transporting MPEG-4 over IP 280
    7.5.1 Brief Introduction to Streaming over IP 281
    7.5.2 Transport of Elementary Streams over IP..283
    7.5.3 Transport of SL-Packetized Streams over IP 285
    7.5.4 FlexMux Streams over IP and Timing Models 288
7.6 Summary..289
7.7 References 289

8 Natural Video Coding...**293**
8.1 General Overview 294
    8.1.1 Functionalities and Application Scenarios 294
    8.1.2 Basic Principles 295
8.2 Coding of Rectangular Video Objects..300
    8.2.1 Overview 300
    8.2.2 New Motion-Compensation Tools..303
    8.2.3 New Texture Coding Tools...311
8.3 Coding of Arbitrarily Shaped Video Objects..318
    8.3.1 Binary Shape Coding 318
    8.3.2 Gray-Level Shape Coding 323
    8.3.3 Coding of Boundary Macroblocks 324
8.4 Scalable Video Coding 330
    8.4.1 Spatial Scalability 332
    8.4.2 Temporal Scalability 337
    8.4.3 SNR Fine Granularity Scalability...340
8.5 Special Video Coding Tools..343
    8.5.1 Interlaced Coding 344
    8.5.2 Error-Resilient Coding...345
    8.5.3 Reduced Resolution Coding 355
    8.5.4 Sprite Coding...356
    8.5.5 Short Video Header Mode 359
    8.5.6 Texture Coding for High-Quality Applications....................................359
8.6 Visual Texture Coding 361
    8.6.1 VTC Tools 363
    8.6.2 Wavelet Coding..364
    8.6.3 Shape-Adaptive Wavelet Coding 368
    8.6.4 Spatial and Quality Scalability 369
    8.6.5 Bitstream Packetization 371
    8.6.6 Tiling..374
8.7 Summary..375
8.8 References 376

9 Visual SNHC Tools **383**
9.1 SNHC Overview...384
    9.1.1 VRML, X3D: Why Is SNHC Needed? 385
    9.1.2 SNHC Visual Tools...388

9.2 Face and Body Animation 389
  9.2.1 Overview 389
  9.2.2 Default Facial Expression and Body Posture 392
  9.2.3 FBA Object 393
  9.2.4 FDP and BDP Listing and Coding 396
  9.2.5 FAP and BAP Listing and Coding 403
  9.2.6 FBA and Text-to-Speech Interface 410
9.3 2D Mesh Coding 411
  9.3.1 2D Mesh Object 412
  9.3.2 Coding Scheme 413
  9.3.3 Example 420
9.4 3D Mesh Coding 421
  9.4.1 3D Mesh Object 422
  9.4.2 Coding Scheme 425
  9.4.3 Examples 437
9.5 View-Dependent Scalability 440
  9.5.1 View-Dependent Object 442
  9.5.2 Coding Scheme 444
  9.5.3 Example 446
9.6 Profiles and Levels 446
9.7 Summary 447
9.8 Acknowledgments 447
9.9 References 448

10 Speech Coding 451
10.1 Introduction to Speech Coding 452
10.2 Overview of MPEG-4 Speech Coders 454
10.3 MPEG-4 CELP Coding 455
  10.3.1 CELP Encoder 455
  10.3.2 CELP Decoder 456
  10.3.3 Parameter Decoding 457
  10.3.4 Multipulse Excitation 458
  10.3.5 Regular Pulse Excitation 459
  10.3.6 Scalability 460
  10.3.7 Silence Compression 461
10.4 MPEG-4 HVXC Coding 461
  10.4.1 HVXC Encoder 462
  10.4.2 HVXC Decoder 470
  10.4.3 Variable Bit-Rate Coding 474
10.5 Error Robustness 475
  10.5.1 Error-Resilient HVXC 477
  10.5.2 Error-Resilient CELP 481
10.6 Summary 482
10.7 References 483

11 General Audio Coding **487**

11.1 Introduction to Time/Frequency Audio Coding 488

11.2 MPEG-2 Advanced Audio Coding 490

    11.2.1 Coder Overview 491

    11.2.2 Gain Control 491

    11.2.3 Filterbank 493

    11.2.4 Quantization 495

    11.2.5 Noiseless Coding 497

    11.2.6 Temporal Noise Shaping 499

    11.2.7 Prediction 502

    11.2.8 Joint Stereo Coding 504

    11.2.9 Bitstream Multiplexing 507

    11.2.10 Other Aspects 508

11.3 MPEG-4 Additions to AAC 508

    11.3.1 Perceptual Noise Substitution 509

    11.3.2 Long-Term Prediction 511

    11.3.3 TwinVQ 513

    11.3.4 Low-Delay AAC (AAC-LD) 515

    11.3.5 Error Robustness 517

11.4 MPEG-4 Scalable Audio Coding 518

    11.4.1 Large-Step Scalable Audio Coding 519

    11.4.2 Bit-Sliced Arithmetic Coding 523

11.5 Introduction to Parametric Audio Coding 525

    11.5.1 Source and Perceptual Models 526

    11.5.2 Parametric Encoding and Decoding Concepts 528

11.6 MPEG-4 HILN Parametric Audio Coding 530

    11.6.1 HILN Parametric Audio Encoder 530

    11.6.2 HILN Bitstream Format and Parameter Coding 535

    11.6.3 HILN Parametric Audio Decoder 538

11.7 Summary 539

11.8 Acknowledgments 540

11.9 References 540

12 SNHC Audio and Audio Composition **545**

12.1 Synthetic-Natural Hybrid Coding of Audio 546

12.2 Structured Audio Coding 548

    12.2.1 Algorithmic Synthesis and Processing 549

    12.2.2 Wavetable Synthesis 552

    12.2.3 Interface Between Structured Audio Coding and AudioBIFS 554

    12.2.4 Structured Audio Applications 554

12.3 Text-to-Speech Interface 555

12.4 Audio Composition 557

    12.4.1 VRML Sound Model in BIFS 558

    12.4.2 Other AudioBIFS Nodes 560

    12.4.3 Enhanced Modeling of 3D Audio Scenes in MPEG-4 567

12.4.4 Advanced AudioBIFS for Enhanced Presentation of 3D
Sound Scenes..569
12.5 Summary 580
12.6 References 580

13 Profiling and Conformance: Approach and Overview............. 583
13.1 Profiling and Conformance: Goals and Principles.......................584
13.2 Profiling Policy and Version Management.................................588
13.3 Overview of Profiles in MPEG-4 592
13.3.1 Visual Profiling 592
13.3.2 Audio Profiling 602
13.3.3 Graphics Profiling...609
13.3.4 Scene Graph Profiling 613
13.3.5 Object Descriptor Profiling...619
13.3.6 MPEG-J Profiling 619
13.4 Summary 620
13.5 Acknowledgements...620
13.6 References 620

14 Implementing the Standard: The Reference Software............. 623
14.1 Reference Software Modules 625
14.2 Systems Reference Software 625
14.3 MPEG-4 Player Architecture..628
14.3.1 Player Structure 631
14.4 Scene Graph 632
14.4.1 Parsing the Scene Graph..633
14.4.2 Rendering the Scene Graph 638
14.5 PROTOs..639
14.6 Synchronization 643
14.6.1 Sync Layer 643
14.6.2 MediaStream Objects...644
14.7 Object Descriptors...646
14.7.1 Syntactic Description Language 646
14.7.2 Parsing the OD Stream 647
14.7.3 Execution of OD Objects...651
14.8 Plug-Ins 652
14.8.1 DMIF Plug-Ins..652
14.8.2 Decoder Plug-Ins 654
14.8.3 IPMP Plug-Ins 656
14.9 2D Compositor..656
14.9.1 Using the Core Framework 657
14.9.2 DEF and USE Handling..657
14.9.3 Rendering Optimization 658
14.9.4 Synchronization 659
14.10 3D Compositor..660
14.10.1 Differences Between the 2D and 3D Compositors 661

14.10.2 Using the Core Framework..662
14.10.3 Overview of the Key Classes...662
14.10.4 3D Rendering Process 662
14.10.5 Support for 2D Nodes..664
14.10.6 User Navigation 664
14.11 Summary...666
14.12 References 666

15 Video Testing for Validation 669
15.1 General Aspects 670
15.1.1 Selection of Test Material 671
15.1.2 Selection of Test Subjects...671
15.1.3 Laboratory Setup...672
15.1.4 Test Plan..672
15.1.5 Training Phase 673
15.2 Test Methods..673
15.2.1 Single Stimulus Method..674
15.2.2 Double Stimulus Impairment Scale Method 675
15.2.3 Double Stimulus Continuous Quality Scale Method......................676
15.2.4 Simultaneous Double Stimulus for Continuous Evaluation Method ...677
15.3 Error-Resilience Test 680
15.3.1 Test Conditions..681
15.3.2 Test Material 684
15.3.3 Test Method and Design 684
15.3.4 Data Analysis 686
15.3.5 Test Results 686
15.4 Content-Based Coding Test...688
15.4.1 Test Conditions..688
15.4.2 Test Material 689
15.4.3 Test Method and Design 689
15.4.4 Data Analysis 690
15.4.5 Test Results 690
15.5 Coding Efficiency for Low and Medium Bit-Rate Test............................691
15.5.1 Test Conditions..691
15.5.2 Test Material 692
15.5.3 Test Method and Design 692
15.5.4 Data Analysis 693
15.5.5 Test Results 693
15.6 Advanced Real-Time Simple Profile Test..694
15.6.1 Test Conditions..696
15.6.2 Test Material 699
15.6.3 Test Method and Design 700
15.6.4 Data Analysis 701
15.6.5 Test Results 702
15.7 Summary...704

15.8 References 705

16 Audio Testing for Validation 709
16.1 General Aspects...710
   16.1.1 Selection of Test Material 710
   16.1.2 Selection of Test Subjects................................711
   16.1.3 Laboratory Setup 712
   16.1.4 Test Plan 712
   16.1.5 Training Phase..712
16.2 Test Methods 713
   16.2.1 Absolute Category Rating Method.....................713
   16.2.2 Paired Comparison Method 714
   16.2.3 MUSHRA Method..715
16.3 Narrowband Digital Audio Broadcasting Test 717
   16.3.1 Test Conditions..718
   16.3.2 Test Material 719
   16.3.3 Assessment Method and Test Design................720
   16.3.4 Data Analysis...721
   16.3.5 Test Results 722
16.4 Audio on the Internet Test..724
   16.4.1 Test Conditions..724
   16.4.2 Test Material 725
   16.4.3 Assessment Method and Test Design................726
   16.4.4 Data Analysis...727
   16.4.5 Test Results 728
16.5 Speech Communication Test 730
   16.5.1 Test Conditions..730
   16.5.2 Test Material 732
   16.5.3 Assessment Method and Test Design................733
   16.5.4 Data Analysis...734
   16.5.5 Test Results 734
16.6 Version 2 Coding Efficiency Test 736
   16.6.1 Test Conditions..737
   16.6.2 Test Material 739
   16.6.3 Assessment Method and Test Design................740
   16.6.4 Data Analysis...740
   16.6.5 Test Results 740
16.7 Version 2 Error-Robustness Test 744
   16.7.1 Test Conditions..744
   16.7.2 Test Material 747
   16.7.3 Test Method and Experimental Design.............747
   16.7.4 Data Analysis...747
   16.7.5 Test Results 747
16.8 Summary 749

16.9 References 750

A Levels for Visual Profiles **753**
A.1 Video Buffering Verifier Mechanism 754
    A.1.1 Video Rate Buffer Verifier Definition 756
    A.1.2 Video Complexity Verifier Definition 760
    A.1.3 Video Reference Memory Verifier Definition 764
    A.1.4 Interaction Between the VBV, VCV, and VMV Models............ 766
A.2 Definition of Levels for Video Profiles................................. 767
A.3 Definition of Levels for Synthetic Profiles 767
    A.3.1 `Scalable Texture` Profile 775
    A.3.2 `Simple Face Animation` Profile 775
    A.3.3 `Simple FBA` Profile.. 775
    A.3.4 `Advanced Core` and `Advanced Scalable Texture` Profiles 776
A.4 Definition of Levels for Synthetic and Natural Hybrid Profiles..... 776
    A.4.1 `Basic Animated Texture` Profile 776
    A.4.2 `Hybrid` Profile... 778
A.5 References... 778

B Levels for Audio Profiles **781**
B.1 Complexity Units 781
B.2 Definition of Levels for Audio Profiles............................. 783
    B.2.1 `Main` Profile 784
    B.2.2 `Scalable` Profile... 784
    B.2.3 `Speech` Profile... 785
    B.2.4 `Synthetic` Profile.. 785
    B.2.5 `High-Quality Audio` Profile 786
    B.2.6 `Low-delay Audio` Profile 786
    B.2.7 `Natural Audio` Profile 786
    B.2.8 `Mobile Audio Internetworking` Profile....................... 786
B.3 References... 788

C Levels for Graphics Profiles **789**
C.1 `Simple 2D` Profile 789
C.2 `Simple 2D + Text` Profile... 790
C.3 `Core 2D` Profile 790
C.4 `Advanced 2D` Profile 793
C.5 References... 795

D Levels for Scene Graph Profiles **797**
D.1 `Simple 2D` Profile 797
D.2 `Audio` Profile.. 798
D.3 `3D Audio` Profile 801
D.4 `Basic 2D` Profile... 802
D.5 `Core 2D` Profile 802

D.6 Advanced 2D Profile..806
D.7 Main 2D Profile 809
D.8 References...813

E MPEG-J Code Samples.. **815**
E.1 Scene APIs 815
E.2 Resource and Decoder APIs 817
      E.2.1 Listener Class for Decoder Events...............................819
      E.2.2 Listener Class for Renderer Events.............................820
E.3 Network APIs 820
E.4 Section Filtering APIs 821

Index... **823**

Foreword

by Leonardo Chiariglione
MPEG-4 Convener

*C*ommunication standards have existed for centuries. Whereas in the past standards were generated by custom or by royal or republican edicts, with the progress of technology and the formation of standards bodies such as the ITU, a new process was put in place whereby innovations leading to new forms of communication could be developed in a particular environment and ratified by a standards body. This arrangement served the needs of all concerned parties for a long time. But those were times when innovation happened slowly and industries operated independently from one another.

MPEG came to the forefront at a time when it was becoming apparent that the old arrangement would no longer work. Too many similar developments happened in too many environments wherein each of the solutions developed could be applied reasonably well to all other environments. And this was happening at such a speed that the usual ways leading to standardization not only rendered the process ineffectual, because of the time required by traditional standardization, but also could exacerbate the impact of similar standards employed by different environments for similar applications. In contrast to this, MPEG sought the involvement of all players from all relevant industries and created a new process whereby the results of research could be fed into the standardization process before the industrial development.

Retrospectively, one could venture that MPEG-1, a process driven by industry's need to have a standard for interactive video on compact disc and digital audio broadcasting, has been a failure, because the specific products that were the targets of that standard have had scant or no success. The same can hardly be said of MPEG-2, because the product that was the target of the standard—digital television—has been incredibly successful. More than the success, or lack thereof, of the standards as originally conceived, what matters is the success of products that were enabled by the standards even though they were not considered, or were thought of as being of lower priority, as targets of the standard: Video CD, MP3, DVD, and 4:2:2 profile used in the television studio are the best known cases.

In order to understand the MPEG-4 project and its prospects, we must keep in mind the lesson provided by the first two MPEG standards. While working on the MPEG-2 standard, MPEG developers began wondering what the further needs of industry in digital audiovisual coding might be, given that the 1.5 Mbit/s range had been covered by MPEG-1 and the 5 Mbit/s range had been covered by MPEG-2, even as reference to bit rate had been—correctly— removed from MPEG-2. Developers sensed that there were latent needs for very low bit-rate coding that went beyond the simple moving-pictures-on-the-telephone-wire idea which was so common at that time.

It took more than two years to develop requirements for the new standard and three years to publish Version 1 of the standard. Eventually, MPEG-4 evolved into a comprehensive set of tools capable of satisfying, through the mechanism of profiles, multiple industry needs while providing a high level of interoperability. As a result, MPEG-4 provides solutions for the mobile environment, the Web, broadcast, rich virtual environments, and even very high bit-rate applications for the studio.

Clearly, people on the Internet are going to be major users of the MPEG-4 standard, but the marriage has not necessarily been decreed in heaven. The Internet has been built on the assumption that it should be an infrastructure accessible to everybody, and that multiple applications can be developed for it. Application standards, according to this assumption, are not needed because certain algorithms are implemented in any given piece of software, so that users can download the code, allowing them to decode the particular algorithm used to produce the bits of data of interest to them. Under these conditions is there still a need to develop a standard like MPEG-4?

In the early days of MPEG-4 development, this question used to be asked often; but today, with an ever-expanding use of MP3, it is easier to understand the benefits of a standard: A playback device is not necessarily connected to the network; it may be on a broadcast channel, or it may exist as a standalone or portable device. These devices can use many different CPUs, for which it might be too costly to develop playback software, or the hardware may use an ASIC for the audiovisual decoding that is not upgradeable or may have been designed to run specifically with the amount of RAM the standard algorithm requires. Setting aside other advantages, it is simply easier to have a common standard on which business opportunities can multiply, instead of struggling with incompatibilities between applications all over the place.

The world is sufficiently complex as it is without adding further confusion by blindly worshiping flexibility, particularly when one sees the mythical world of "many applications" turning out to be a world of a single application whose access and evolution is controlled by a single company.

In opposition to this model, MPEG (and MPEG-4 in particular) offers a standard that has been developed collaboratively. Access is open to anybody who wants to be in business, and its evolution is in the hands of technical representatives of the various players involved. MPEG-4 has even produced MPEG-4 reference software, through a process similar to open-source software that can be used by any given implementer.

Ability to access does not mean that access is free, however. For the last few decades, the world of digital audio and video has been the focus of research by thousands of companies and organizations. Some have worked for the sole purpose of increasing the collective knowledge of humankind; others have worked to create for themselves the opportunity for future revenues by securing patents for their inventions. Rightly, ISO does not allow consideration of patent issues in technical groups like MPEG, but this is no reason solutions cannot be found outside. A patent pool exists today for MPEG-2, and more are likely to be generated from the activity of the MPEG-4 Industry Forum.

This book was written by some of the most authoritative individuals playing key roles in developing the MPEG-4 standard. Although the book is not a substitute for the standard itself, it provides a general overview as well as insights into the critical elements of the standard and information pertinent to understanding MPEG-4 and its practical use.

Preface

The last decade has shown the quick growth of multimedia applications and services, with audiovisual information playing an increasingly important role. Today's existence of tens of millions of digital audiovisual content users and consumers is tightly linked to the maturity of such technological areas as video and audio compression and digital electronics and to the timely availability of appropriate audiovisual coding standards. These standards allow the industry to make major investments with confidence in new products and applications and users to experience easy consumption and exchange of content.

In this environment, the Moving Picture Experts Group (MPEG) is playing an important role, thanks to the standards it has been developing. After developing the MPEG-1 and MPEG-2 standards, which are omnipresent in diverse technological areas and markets (such as digital television, video recording, audio broadcasting, and audio players and recorders), MPEG decided to follow a more challenging approach, moving away from the traditional representation models by adopting a new model based on the explicit representation of objects in a scene. The new *object-based audiovisual representation model* is much more powerful in terms of functionalities that it can support. The flexibility of this new model not only opens new doors to existing multimedia applications and services, it also allows the creation of a wide range of new ones, offering novel capabilities to users that extend or redefine their relationship with audiovisual information.

The MPEG-4 standard is the first audiovisual coding standard that benefits from a representation model in which audiovisual information is represented in a sophisticated and powerful way that is close not only to the way we experience "objects" in the real world but also to the way digital content is created. In a way, MPEG-4 is the first digital audiovisual coding standard in which technology goes beyond a simple translation to the digital world of analog to exploit the full power of digital technologies.

With the MPEG-4 standard emerging as the next milestone in audiovisual representation, interested people worldwide are looking for reference texts that, while not providing the level of scrutiny of the standard itself, give a detailed overview of the technology standardized in MPEG-4. Because it takes advantage of many technologies, MPEG-4 may seem a large and complex standard to learn about. However, it has a clear structure that can be understood by interested people.

The purpose of this book is to explain the standard clearly, precisely, and completely without getting lost in the details. Although surely there will be other good references on MPEG-4, we tried hard to make this *the* reference by creating a book exclusively dedicated to MPEG-4, which addresses all parts of the standard, as timely and complete as possible, written and carefully reviewed by the foremost experts: those who designed and wrote the standard during many years of joint work, frustration, and satisfaction.

To help readers find complementary or more detailed information, the chapters include a large number of references. Some of these references are MPEG documents not readily available to the public. For access to these, first check the MPEG Web page at *mpeg.telecomitalialab.com*. Some of the most important MPEG documents are available from that site. If that does not work, contact the MPEG "Head of Delegation" from your country (check *www.iso.ch / addresse / address.html*), who should be able to help you get access to documents that were declared "publicly available" but still may be hard to obtain.

ORGANIZATION OF THE BOOK

The book is organized in three major parts: the introductory chapters, the standard specification chapters, and the complementary chapters.

The introductory chapters, Chapters 1 and 2, introduce the reader to the MPEG-4 standard. Chapter 1 presents the motivation, context, and objectives of the MPEG-4 standard and reviews the process followed by MPEG to arrive at its standards. Chapter 2 gives a short overview of the MPEG-4 standard, highlighting its design goals. It also describes the end-to-end creation, delivery, and consumption processes, and it explains the relation of MPEG-4 to other relevant standards and technologies. Lastly, it proposes three example applications.

The standard specification chapters describe and explain the MPEG-4 normative technology, as specified in the various parts of the standard. The first batch of these chapters addresses the technologies associated with the layers below the audiovisual coding layer. Chapter 3 addresses the means to manage and synchronize the potentially large numbers of elementary streams in an MPEG-4 presentation. Essential MPEG-4 concepts and tools such as object descriptors, the Sync layer, the system decoder model, and timing behavior are presented. Chapter 4 is dedicated to the MPEG-4 scene description format, a major innovation, supporting MPEG-4's object-based data representation model. It uses a number of examples to explain the BInary Format for Scenes, or BIFS format. Chapter 5 explains how it is possible to use the Java language to control features of an MPEG-4 player through the MPEG-J application engine. This chapter presents the MPEG-J architecture and describes the functions of an application engine. It also introduces the new Java APIs specific to MPEG-4 (Terminal, Scene, Resource, Decoder, and Network) that were designed to communicate with the MPEG-4 player. Chapter 6 presents the Extensible MPEG-4 Textual Format (XMT) framework, which consists of two levels of textual syntax and semantics: the XMT-A format, providing a one-to-one deterministic mapping to the MPEG-4 Systems binary representation, and the XMT-Ω format, providing a high-level abstraction of XMT-A to content authors so they can preserve the original semantic information. Chapter 7 describes the general approach and some specific mechanisms for the delivery of MPEG-4 presentations. It introduces the Delivery Multimedia Integration Framework (DMIF), which specifies the interfaces to mechanisms to transport MPEG-4 data, and describes two DMIF instances: MPEG-4 over MPEG-2 and MPEG-4 over IP. (Of course, MPEG-4 presentations can be delivered over other transport protocols as needed.) Finally, the chapter introduces the delivery-related tools included in the MPEG-4 Systems standard, notably the FlexMux tool and the MPEG-4 file format.

While Chapters 3–7 address technologies specified in MPEG-4 Part 1: Systems, Chapters 8–12 focus on media representation technologies specified in MPEG-4 Visual, Audio, and Systems as far as a few synthetic audio techniques are concerned. Chapter 8 introduces all the tools related to video and texture coding for rectangular and shaped objects, and it presents the tools for important functionalities such as error resilience and scalability. Chapter 9 presents the coding tools specified by MPEG-4 to support the representation of synthetic visual content. These tools address face and body animation, 2D and 3D mesh coding, and view-dependent scalability. Chapter 10 introduces the coding tools for natural speech. To address a large range of bit rate, quality, speech bandwidth, and other functionalities, MPEG-4 specifies two coding algorithms: CELP and HVXC. Chapter 11 addresses the general audio coding tools. Here, three coding algorithms are adopted to fulfill the requirements: an AAC-based algorithm with some extensions over MPEG-2 advanced audio coding (AAC); TwinVQ, which is a vector quantization algorithm suitable for very low bit rates; and HILN, which is a parametric coding algorithm providing additional functionalities. Chapter 12 presents the MPEG-4 audio synthetic-natural hybrid coding (SNHC) and composition and presentation tools. The main SNHC audio tools are structured audio and the text-to-speech interface. The audio composition and presentation tools are known as AudioBIFS and Advanced AudioBIFS.

Profiling and conformance are the major topics addressed in Chapter 13. Profiles and levels provide technical solutions for classes of applications with similar functional and operational requirements, allowing interoperability with reasonable complexity and cost. Moreover, they allow conformance to be tested, which is essential for determining if bitstreams and terminals are compliant.

Chapter 14 presents the concept of reference software in MPEG-4 and elaborates on the software architecture of the MPEG-4 Systems player, included in Part 5 of the MPEG-4 standard.

The complementary chapters (15 and 16) address the validation testing of the MPEG-4 video and audio technology. Although they do not cover MPEG-4 normative technology, they provide important information about the standard's performance from various points of view and for various potential applications.

ACKNOWLEDGMENTS

This book is the product of many years of work that resulted not only in thousands of pages of technical specification but also in many lifetime friendships among people all around the world.

The editors would like to thank all the MPEG members for the interesting and fruitful discussions, which not only substantially enriched their own technical knowledge but which, finally, motivated them to create this book. Special thanks go to the book reviewers, who spent many hours making suggestions for improvement, helping make the best book possible available for readers.

Finally, our thanks to the authors of the book who managed to dedicate substantial (and often personal) time in the midst of very busy lives to write their chapters. We sincerely hope that this book will turn out to be a useful tool in your multimedia related work, and that you will have many occasions to use it.

Enjoy the book!
Fernando Pereira, Touradj Ebrahimi

Abbreviations

3D	Three-Dimensional
3DMC	3D Mesh Coding
3GPP	3rd Generation Partnership Project
AAC	Advanced Audio Coding also Adaptive Arithmetic Coding
AAC LD	AAC Low Delay
AC	Arithmetic Codec also Alternative Component (in frequency representation)
ACR	Absolute Category Rating
ADPCM	Adaptive Differential Pulse Code Modulation
AFX	Animation Framework eXtension
AIR	Adaptive Intra Refresh
ALF	Application Level Fragmentation
AMD	Amendment
ANOVA	ANalysis Of VAriance
AOE	Applications and Operational Environments
API	Application Programming Interface
ASIC	Application-Specific Integrated Circuit
ATM	Asynchronous Transfer Mode

AU	Access Unit
AV	Audiovisual
AVC	Advanced Video Coding
BAB	Binary Alpha Block
BAC	Binary Arithmetic Coding
BAP	Body Animation Parameter
BB	subBand by subBand (in wavelet coefficients scanning modes)
BDP	Body Definition Parameter
BER	Bit Error Rate
BIFS	BInary Format for Scenes
BMLD	Binaural Masking Level Depression or Difference
BPP	Bit Per Pixel
BQ	Bi-level Quantizer (in wavelet coefficients quantization modes)
BSAC	Bit-Sliced Arithmetic Coding
B-VOP	Bidirectionally predictive-coded Video Object Plane
CAD	Computer Aided Design
CAE	Context-based Arithmetic Coding
CAT	Channel Association Tag
CBF	Compressed Binary Format (VRML)
CD	Committee Draft
CE	Core Experiment
CELP	Code Excited Linear Prediction
CGD	Computational Graceful Degradation
CIF	Common Intermediate Format
CIR	Cyclic Intra Refresh
CM	Composition Memory
CODAP	Component based Data Partitioning
COR	Technical Corrigendum
CPU	Central Processing Unit
CRC	Cyclic Redundancy Code
CS	Coefficients Scanning
CTS	Composition Time Stamp
CU	Composition Unit
DAB	Digital Audio Broadcasting
DAI	DMIF-Application Interface
DAM	Draft Amendment
DAVIC	Digital Audiovisual Council
dB	Decibel

DB	Decoding Buffer
DC	Direct Component (in frequency representation)
DCM	DC Marker
DCOR	Draft Technical Corrigendum
DCT	Discrete Cosine Transform
DECT	Digital Enhanced Cordless Telecommunications
DIS	Draft International Standard
DLL	Dynamic Link Library
DLS2	MIDI Downloadable Sounds 2 Format
DMIF	Delivery Multimedia Integration Framework
DNI	DMIF Network Interface
DOI	Digital Object Identifier
DP	Data Partitioning
DPCM	Differential Pulse Code Modulation
DPI	DMIF Plug-in Interface
DRC	Dynamic Range Control also Dynamic Resolution Conversion
DRM	Digital Rights Management also Digital Radio Mondiale
DSCQS	Double Stimulus Continuous Quality Scale
DSIS	Double Stimulus Impairment Scale
DTR	Draft Technical Report
DTS	Decoding Time Stamp
DV	Displacement Vector
DVB	Digital Video Broadcasting
DVD	Digital Versatile Disc
DWT	Discrete Wavelet Transform
EBU	European Broadcasting Union
EPG	Electronic Program Guide
ES	Elementary Stream
ESC	Error Sensitivity Category
ESI	Elementary Stream Interface
ETSI	European Telecommunications Standards Institute
EZW	Embedded Zero-tree Wavelet
FAP	Face Animation Parameter
FAPU	FAP Unit
FBA	Face and Body Animation
FCD	Final Committee Draft
FCR	FlexMux Clock Reference
FDAM	Final Draft Amendment

FDIS	Final Draft International Standard
FDP	Face Definition Parameter
FEC	Forward Error Correction
FFT	Fast Fourier Transformation
FGS	Fine Granularity Scalability
FGST	Fine Granularity Scalability Temporal scalable
FIG	FAP Interpolation Graph
FIR	Finite Impulse Response (in filter design)
FIT	FAP Interpolation Table
FMC	FlexMux Channel
FOV	Field Of View
FPDAM	Final Proposed Draft Amendment
FSS	Frequency Selective Switch
FTP	File Transfer Protocol (IETF)
GMC	Global Motion Compensation
GOB	Group Of Blocks
GPRS	General Packet Radio Service
GSM	Global System for Mobile Communications also Groupe Spécial Mobile
HCR	Huffman Codeword Reordering
HDL	Hardware Description Languages
HDTV	High Definition Television
HEC	Header Extension Code
HILN	Harmonic and Individual Lines plus Noise
HTML	HyperText Markup Language (W3C)
HTTP	HyperText Transfer Protocol (IETF)
HVXC	Harmonic Vector Excitation Coding
IDCT	Inverse Discrete Cosine Transform
IDFT	Inverse Discrete Fourier Transform
IEC	International Electrotechnical Commission
IETF	Internet Engineering Task Force
IFFT	Inverse Fast Fourier Transform
IGMP	Internet Group Management Protocol (IETF)
IMDCT	Inverse Modified Discrete Cosine Transform
I-MOP	Intra-Mesh Object Plane
IMT-2000	International Mobile Telecommunications-2000
IOD	Initial Object Descriptor
IP	Internet Protocol (IETF) also Intellectual Property
IPI	Intellectual Property Identification

IPMP	Intellectual Property Management and Protection
IS	International Standard
ISBN	International Standard Book Number
ISMN	International Standard Music Number
ISO	International Organization for Standardization
ITTF	Information Technology Task Force
ITU	International Telecommunication Union
ITU-R	International Telecommunication Union—Radio Standardization Sector
ITU-T	International Telecommunication Union— Telecommunications Standardization Sector
I-VOP	Intra-coded Video Object Plane
JAR	Java ARchiving format also Java Archive
JM	Joint Model
JND	Just Noticeable Difference
JPEG	Joint Photographic Experts Group
JTC 1	ISO/IEC Joint Technical Committee 1
JTPC	ISO/IEC Joint Technical Programming Committee
JVM	Java Virtual Machine
JVT	Joint Video Team
JWG	Joint Working Group
KBD	Kaiser-Bessel-Derived window
LAR	Logarithmic Area Ratio
LATM	Low-overhead MPEG-4 Audio Transport Multiplex
LAV	Largest Absolute Value
LPC	Linear Predictive Coding
LSF	Line Spectrum Frequency
LSP	Line Spectral Pair
LTP	Long Term Prediction
M4IF	MPEG-4 Industry Forum
MB	Macroblock
MDCT	Modified Discrete Cosine Transform
MDS	Multimedia Description Schemes
MHEG	Multimedia and Hypermedia information coding Expert Group
MIDI	Musical Instrument Digital Interface
MIME	Multipurpose Internet Mail Extensions (IETF)
MM	Motion Marker
MNRU	Modulated Noise Reference Unit

MOPS	Millions of Operations Per Second
MOS	Mean Opinion Score
MP4	MPEG-4 File Format
MPE	Multipulse Excitation
MPEG	Moving Picture Experts Group
MQ	Multiple Quantizer (in wavelet coefficients quantization modes)
MUSHRA	MultiStimulus test with Hidden Reference and Anchors
MV	Motion Vector
MZTE	Multiscale Zero-Tree Entropy
NADIB	Narrow Band Digital Broadcasting (consortium) now Digital Radio Mondiale (DRM)
NB	National Body
NBC	(MPEG-2) Non-Backward Compatible coding now Advanced Audio Coding (AAC)
NCT	Node Coding Table
NDT	Node Data Type
NEWPRED	New Prediction
NP	New Work Item Proposal
NSSD	Next Statistically Significant Difference
NTP	Network Time Protocol (IETF)
NTSC	National Television Standards Committee
OBMC	Overlapped Block Motion Compensation
OCI	Object Content Information
OCR	Object Clock Reference
OD	Object Descriptor
ODF	Object Descriptor Framework
OLA	OverLap/Add
OM	Optimization Model (MPEG-4)
OOP	Object Oriented Programming
OSI	Open System Interconnection
OTB	Object Time Base
OWG	Other Working Group
PAT	Program Association Table (MPEG-2)
PCR	Program Clock Reference (MPEG-2)
PCU	Processor Complexity Unit
PCW	Priority Codeword
PDA	Personal Digital Assistant
PDAM	Proposed Draft Amendment

PDF	Probability Density Function also Portable Document Format
PDTR	Proposed Draft Technical Report
PES	Packetized Elementary Streams (MPEG-2)
PEZW	Predictive Embedded Zero-tree Wavelet
PID	Packet Identifier (MPEG-2)
PLL	Phase Locked Loop
P-MOP	Predictive-coded Mesh Object Plane
PMT	Program Map Table (MPEG-2)
PNS	Perceptual Noise Substitution
PP	Parallelogram Prediction
PPD	Proposal Package Description
PQF	Polyphase Quadrature Filterbank
PS	Program Stream (MPEG-2)
PSD	Power Spectral Density
PSM	Program Stream Map (MPEG-2)
PSNR	Peak Signal to Noise Ratio
PSTN	Public Switched Telephone Networks
PT	Payload Type
P-VOP	Predictive-coded Video Object Plane
QCIF	Quarter Common Intermediate Format
QoS	Quality of Service
QP	Quantization Parameter
QSIF	Quarter Standard Image Format
RAM	Random Access Memory
RCU	RAM Complexity Unit
RM	Resynchronization Marker
RMS	Root Mean Square
RPE	Regular Pulse Excitation
RR VOP	Reduced Resolution VOP
RTP	Real-Time Transport Protocol (IETF)
RTSP	Real-Time Streaming Protocol (IETF)
RVLC	Reversible Variable Length Coding
S(GMC)-VOP	Sprite coded Video Object Plane with GMC sprite type
SA	Structured Audio
SA-DCT	Shape Adaptive Discrete Cosine Transform
SAOL	Structured Audio Orchestra Language
SASBF	Structured Audio Sample Bank Format
SASL	Structured Audio Score Language

SA-Wavelet	Shape Adaptive Wavelet
SBA	Segmented Binary Arithmetic coding
SC	Sprite Coding also Subcommittee
SCR	System Clock Reference (MPEG-2)
SDC	Sub-Division Coding
SDL	Syntactic Description Language
SDM	System Decoder Model
SDP	Session Description Protocol (IETF)
SDSCE	Simultaneous Double Stimulus for Continuous Evaluation
SE	Spectral Envelope (in speech coding)
SGML	Standard Generalized Markup Language
SI	Service Information (DVB)
SIF	Standard Image Format
SIT	Software Instrumentation Tool
SL	Synchronization Layer
SM	Simulation Model (MPEG-1)
SMIL	Synchronized Multimedia Integration Language (W3C)
SMPTE	Society of Motion Picture and Television Engineers
SNHC	Synthetic Natural Hybrid Coding
SPHIT	Set Partitioning in Hierarchical Threes
SPS	SL-Packetized Stream
SQ	Scalar Quantizer also Single Quantizer (in wavelet coefficients quantization modes)
SS	Single Stimulus
SSCQE	Single Stimulus Continuous Quality Evaluation
SSMR	Single Stimulus with Multiple Repetition
SSNCS	Single Stimulus Numerical Categorical Scale
SSR	Scalable Sampling Rate
STB	System Time Base
STD	System Target Decoder (MPEG-2)
SVG	Scalable Vector Graphics (W3C)
S-VOP	Sprite-coded Video Object Plane
SWG	Special Working Group
SYMM	Synchronized Multimedia working group (W3C)
T/F	Time/Frequency
TAT	Transmux Association Tag
TC	Technical Committee
TCP	Transmission Control Protocol (IETF)
TD	Tree-Depth (in wavelet coefficients scanning mode)

TDAC	Time Domain Aliasing Cancellation
TLSS	Tools for Large Step Scalability
TM	Test Model (MPEG-2)
TNS	Temporal Noise Shaping
TR	Technical Report
TS	Topological Surgery (in mesh coding) also Transport Stream (MPEG-2)
TTS	Text To Speech
TTSI	Text-To-Speech Interface
TwinVQ	Transform-domain Weighted Interleave Vector Quantization
UDP	User Datagram Protocol (IETF)
UEP	Unequal Error Protection
UML	Universal Modeling Language
UMTS	Universal Mobile Telecommunications System
URL	Universal Resource Locator (IETF)
UT	Uniform Triangulation (in mesh coding)
VBV	Video Rate Buffer Verifier
VCA	Vertex Clustering Array
VCV	Video Complexity Verifier
VDS	View Dependent Scalability
VHDL	Very High Speed Integrated Circuit Hardware Description Language
VLC	Variable Length Coding
VLD	Variable Length Decoding
VM	Verification Model
VMV	Video Memory Verifier
VO	Video Object
VOL	Video Object Layer
VOP	Video Object Plane
VPV	Video Presentation Verifier
VRML	Virtual Reality Modeling Language
VTC	Visual Texture Coding
VTR	Video Tape Recorder
VXC	Vector eXcitation Coding
W3C	World Wide Web Consortium
WD	Working Draft
WG	Working Group
WM	Working Model
X3D	eXtensible 3D

XM	eXperimentation Model (MPEG-7)
XML	eXtensible Markup Language
XMT	eXtensible MPEG-4 Textual Format
YM	sYstems Model (MPEG-21)
ZTC	Zero-Tree Coding
ZTE	Zero-Tree Entropy

Context, Objectives, and Process[1]

by Fernando Pereira

Keywords: audiovisual object-based representation, audiovisual objects, MPEG-4, standardization process

"What does it mean, to see? The plain man's answer (and Aristotle's, too) would be, to know what is where by looking. In other words, vision is the process of discovering from images what is present in the world, and where it is." [Marr82]

The first video coding standards and their underlying representation models mainly address the vision process by providing video representation in the form of a sequence of rectangular 2D frames, giving users a window to the real world: the television paradigm. However, the process of vision is often just the initial part of the task at hand, because typically humans need and want to see, to take actions after, to interact with the objects identified. A similar reasoning can be made regarding the process of hearing and the corresponding audio representation models [Pere99].

---

1. Some sections of this chapter are adapted from *Signal Processing: Image Communication*, 15 (4–5) (2000), Fernando Pereira, pp. 271–279, "MPEG-4: Why, What, How and When?", Copyright (2000), with permission from Elsevier Science.

Although the television paradigm dominated audiovisual communications for many years, the situation has been evolving quickly in terms of the ways audiovisual content is produced, delivered, and consumed [KoPC97]. Moreover, hardware and software are becoming increasingly powerful, with microelectronic technology providing new programmable processors, opening new frontiers to the representation technologies used and to the functionalities provided.

Producing content today is easier than ever before. Digital still cameras directly storing in JPEG format have hit the mass market. Together with the first digital video cameras directly recording in MPEG-1 format, this represents a major step for consumer acceptance of digital audiovisual acquisition and compression technology. This step transforms every individual into a potential content producer, capable of creating content that can be easily distributed and published on the Internet. In addition, more content is being synthetically produced—that is, computer generated—and integrated with natural material in truly hybrid audiovisual content. The various pieces of content, digitally encoded, can be successively reused without the quality losses typical of the previous analog processes.

Whereas audiovisual information, notably the visual portion, until recently was carried only over very few networks, the trend is now toward the generalization of visual information in every single network. Moreover, the increasing mobility in telecommunications is a major trend. Mobile connections will not be limited to voice; other types of data, including real-time media, are already emerging. Because mobile telephones are replaced every two to three years, new mobile devices can finally make the decade-long promise of audiovisual communications a reality. Notice that the need for visual communication is much stronger when people are not at home, and so have something to show besides the usual living room—for example, the nice beach where they are vacationing. This reinforces the relevance of audiovisual mobile communications.

The explosion of the World Wide Web and the acceptance of its interactive mode of operation have clearly shown that the traditional television paradigm will no longer suffice for audiovisual services. Users want to have access to audio and video as they now have access to text and graphics. This requires moving pictures and audio of acceptable quality at low bit rates on the Web, and Web-type interactivity with live content. It will be possible to activate relationships between users (in a potentially virtual world) through hyperlinking—the Web paradigm—and to experience interactive immersion in natural and virtual environments—the games paradigm.

As many of the emerging audiovisual applications demanded interworking, the need to develop an open and timely international standard addressing the needs mentioned above became evident. In 1993, MPEG (Moving Picture Experts Group) [MPEG] launched the MPEG-4 project, later formally called *Coding of Audio-Visual Objects*, to address (among other things) the requirements associated with the new applications resulting from these trends [N1177, N4319, N4505].

The need for any standard comes from an essential requirement relevant for all applications involving communication between two or more parts: *interoperability*. Interoperability is thus the requirement expressing the user's dream of exchanging any type of information without any technical barriers, in the simplest way. Without a standard way to perform some of the operations involved in the communication process and to structure the data exchanged, easy interoperability between terminals would be impossible. Having said that, it is clear that a standard shall specify the minimum number of tools needed to guarantee interoperability (because it is important that as many as possible non-normative technical zones exist), to allow the incorporation of technical advances, and thus to increase the lifetime of the standard, as well as to stimulate industrial technical competition. The existence of a standard also has important economic implications, because it allows the sharing of costs and investments and the acceleration of applications deployment.

MPEG has been responsible for the successful MPEG-1 (ISO/IEC 11172) and MPEG-2 (ISO/IEC 13818) standards, which have given rise to widely adopted commercial products and services, such as Video-CD, DVD, digital television, digital audio broadcasting (DAB), and MP3 (MPEG-1 Audio layer 3) players and recorders. The MPEG-4 standard (ISO/IEC 14496) is aimed at defining an audiovisual coding standard to address the emerging needs of the communication, interactive, and broadcasting service models as well as the needs of the mixed service models resulting from their convergence. The apparent convergence of the three traditionally separate application areas— communications, computing, and TV/film/entertainment—was evident in their cross-fertilization, with functionalities characteristic of each area increasingly emerging in the others (e.g., personal communications including video information or entertainment including interactive capabilities).

Following the previous successes—in fact, as a natural consequence of the vision underpinning MPEG-4—MPEG initiated in 1996 another standardization project addressing the problem of describing audiovisual content to allow the quick and efficient searching, processing, and filtering of various types of multimedia material: MPEG-7 (ISO/IEC 15938), officially called *Multimedia Content Description Interface* [N4509]. In fact, digital audiovisual information is more and more accessible to everyone, not only in terms of consumption but also in terms of production. But if it is much easier today to acquire, process, and distribute audiovisual content, it must be equally easy to access the available information, because huge amounts of audiovisual information are being generated all over the world every day. The need for a powerful way to quickly and efficiently identify, search, and filter various types of audiovisual content, by humans or machines (also using non-text-based technologies), directly follows from the urge to efficiently use the available audiovisual content and the difficulty of doing so. MPEG-7 will specify a standard way of describing various types of audiovisual information, including still pictures, video, speech, audio, graphics, 3D models, and synthetic audio, regardless of their representation format (e.g., analog or digital) and storage support (e.g., paper, film, or tape). In comparison with other available or emerging solutions for audiovisual content description, MPEG-7 can be mainly distinguished by (a) being general purpose, meaning its ability to describe content from many application environments; (b) its object-based representation model, meaning the capability of independently describing individual objects within a scene (be it MPEG-4 or any other format); (c) the integration of low-level and high-level features/descriptors into a single description framework, allowing it to combine the power of both types of descriptors; and (d) its extensibility, provided by the Description Definition Language, which allows MPEG-7 to keep growing, to be extended to new application areas, to answer newly emerging needs, and to integrate novel description tools [PeKo99]. The MPEG-7 standard was finalized in the summer of 2001.

Following the development of standards addressing more focused targets, MPEG acknowledged the lack of a big picture that described how the various elements building the infrastructure for the deployment of applications using multimedia content relate to each other, or even if there are missing standard specifications for some of these elements [N4333]. To address this problem, MPEG started the MPEG-21 project (first ISO/IEC 18034, now ISO/IEC 21000), formally called *Multimedia Framework*, with the aim of understanding if and how these various elements fit together, and to discuss which new standards might be required if gaps in the infrastructure exist. Once this work has been carried out, new standards will be developed for the missing elements with the involvement of other bodies, where appropriate; finally, the existing and novel standards will be integrated in the MPEG-21 multimedia framework. The MPEG-21 vision is thus to define a multimedia framework to enable transparent and augmented use of multimedia resources across a wide range of networks and devices used by different communities [N4333]. The MPEG-21 multimedia framework will identify and define the key elements needed to support the multimedia value and delivery chain, as well as the relationships between and the operations supported by them [N4511].

After briefly covering the context that motivated the birth of the MPEG-4 project, this chapter presents its major objectives in terms of functionalities, requirements, tools, and applications, as well as its organization and the sequence followed to achieve the goals defined. This chapter also addresses the MPEG modus operandi, notably its mission, principles, and specific approach to the development of standards. Finally, the objectives and working approach of the MPEG-4 Industry Forum will be presented.

1.1 MPEG-4 OBJECTIVES

Although MPEG discussions about projects beyond MPEG-2 began as early as May 1991, at the Paris MPEG meeting, it was not until September 1993 that the MPEG Applications and Operational Environments (AOE) subgroup was set up and met for the first time. The main task of this subgroup was to identify the applications and requirements relevant to the far-term, very low bit-rate coding solution to be developed by International Organization for Standardization (ISO)/MPEG as stated in the initial MPEG-4 project description [N271]. At the same time, the near-term hybrid coding solution being developed within the International Telecommunications Union-Telecommunications Standardization Sector (ITU-T) Low Bit-rate Coding (LBC) group started producing the first results (later, the ITU-T H.263 standard [H263]). It was then generally felt that those results were close to the best performance that could be obtained by block-based, hybrid, DCT/motion-compensation video coding schemes.

In July 1994, the Grimstadt MPEG meeting marked a major change in the direction of MPEG-4. Until that meeting, the main goal of MPEG-4 was to obtain a significantly better compression ratio than could be achieved by conventional coding techniques. Few people, however, believed it was possible, in the next five years, to make enough improvements over the LBC standard (H.263 and H.263+) to justify a new standard.[2] So the AOE subgroup was faced with the need to broaden the objectives of MPEG-4, believing that pure compression gains would not be enough to start a new MPEG standardization project. The subgroup then began an in-depth analysis of the audiovisual world trends, based on the convergence of the TV/film/entertainment, computing, and telecommunications worlds. The conclusion was that the emerging MPEG-4 coding standard should support new ways (notably content-based) of communicating, accessing, and manipulating digital audiovisual data.

1.1.1 Functionalities

Following this change of direction and the analysis made, the vision driving the MPEG-4 standard was explained through the eight new or improved functionalities described in the MPEG-4 Proposal Package Description (PPD), prepared by the time of the first MPEG-4 call for proposals in July 1995 [N998]. These eight functionalities came from an assessment of the functionalities that would be useful in future audiovisual applications, but which were not supported (or at least not well supported) by the available coding standards. The eight new or improved functionalities were clustered in three classes related to the three worlds—TV/film/entertainment, computing, and telecommunications—the convergence of which MPEG-4 wanted to address [N998, PeKo96]:

1. Content-based interactivity

☞ **Content-based multimedia data access tools:** MPEG-4 shall provide efficient data access and organization based on the audiovisual content. Access tools may be indexing, hyperlinking, querying, browsing, uploading, downloading, and deleting. Sample uses include content-based

---

2. At least under the acceptable complexity boundaries.

retrieval of information from online libraries and travel information databases.[3]

☞ **Content-based manipulation and bitstream editing:** MPEG-4 shall provide syntax and coding schemes to support content-based manipulation and bitstream editing without the need for transcoding. This means the user should be able to select one specific object in the scene/bitstream and change some of its characteristics. Sample uses include home movie production and editing, interactive home shopping, and the insertion of a sign language interpreter or subtitles.

☞ **Hybrid natural and synthetic data coding:** MPEG-4 shall support efficient methods for combining synthetic scenes with natural scenes (e.g., text and graphics overlays), the ability to code and manipulate natural and synthetic audio and visual data, and decoder-controllable methods of mixing synthetic data with ordinary video and audio (allowing for interactivity). For example, in virtual reality applications, animations and synthetic audio (e.g., MIDI) can be mixed with ordinary audio and video in games, and graphics can be rendered from different viewpoints.

☞ **Improved temporal random access:** MPEG-4 shall provide efficient methods to randomly access, within a limited time and with fine resolution, parts (e.g., frames or objects) from an audiovisual sequence. Example usage: audiovisual data can be randomly accessed from a remote terminal over limited-capacity media, a fast-forward can be performed on a single audiovisual object in the sequence.

2. Compression efficiency

☞ **Improved coding efficiency:** The growth of mobile networks creates an ongoing demand for improved coding efficiency; therefore, MPEG-4 set as its target providing subjectively better audiovisual quality compared to existing or other emerging standards (such as H.263), at comparable bit rates. Sample uses include efficient transmission of audiovisual data on low-bandwidth channels and efficient storage of audiovisual data on limited-capacity media, such as chip cards.

☞ **Coding of multiple concurrent data streams:** MPEG-4 shall provide the ability to efficiently code multiple views/soundtracks of a scene as well as sufficient synchronization between the resulting elementary streams.[4] For stereoscopic and multiview video applications, MPEG-4 shall include the ability to exploit redundancy in multiple views of the

---

3. Although this is one of the eight initial MPEG-4 functionalities, it relates more to the tools developed later in the context of the MPEG-7 standard [N4509].

4. This seems to be the only initial MPEG-4 functionality that has not been addressed at the end of 2001, as the companies participating in MPEG have shown no significant interest in it.

same scene, also permitting solutions that allow compatibility with normal video. Sample uses include multimedia entertainment (e.g., virtual reality games and 3D movies), training and flight simulations, multimedia presentations, and education.

3. Universal access

☞ **Robustness in error-prone environments:** Because universal accessibility implies access to applications over many wireless and wired networks and storage media, MPEG-4 shall provide an error robustness capability. Particularly, sufficient error robustness shall be provided for low bit-rate applications under severe error conditions. Sample uses include transmission from a database over a wireless network, communicating with a mobile terminal, and gathering audiovisual data from a remote location.

☞ **Content-based scalability:** MPEG-4 shall provide the ability to achieve scalability with a fine granularity in content, spatial resolution, temporal resolution, quality, and complexity. Content-scalability may imply the existence of a prioritization of the objects in the scene. Sample uses include user selection of decoded quality of individual objects in the scene and database browsing at different scales, resolutions, and qualities.

These functionalities were essential to shaping the MPEG-4 vision, balancing completely new functionalities with more traditional ones, and thus allowing a bridge from the past to the future not only in terms of functionalities but also in terms of tools and experts.

1.1.2 Requirements

Following the identification of the fundamental MPEG-4 functionalities, MPEG started a requirements development process, which has been continuously evolving since then. The requirements serve to drive the tools development process and assure that the right technology is being specified: Tools will be developed to fulfill the identified requirements, and no tools that do not address any requirement will be defined.

By the middle of 2001, the MPEG-4 requirements were structured as shown in Table 1.1. The requirements are organized in terms of major technical areas, which do not directly correspond either to Parts of the MPEG-4 standard or to MPEG working subgroups. Table 1.1 gives only a general flavor of the requirements (for the details, see [N4319]).

Within each category, the requirements are to be fulfilled by the set of tools selected for the standard, and not all requirements must be addressed by each individual tool. It is the right combination of tools that allows building the algorithms that can address the specific needs of a certain class of applications.

Table 1.1 MPEG-4 requirements [N4319]

Requirements for systems

Flexibility	Multipoint operation
Multiplexing of audio, visual, and other information	Object content information (OCI)
Composition of audio and visual objects	Video-related metadata
Application texture	Delay
Downloading	Configuration modes
User interaction	Priority of audiovisual objects
Media interworking	Dynamic resource management
Compatibility	Reference to associated MPEG-7 data
Robustness to information errors and losses	File format
Object-based bitstream manipulation and editing	Textual format

Requirements for natural video objects

Object-based representation	Object-based coding flexibility
Video content	Object-based scalability
Object-based bitstream manipulation and editing	Delay modes
Object-based random access	Formats
Object quality and fidelity	Bit-rate modes
Coding of multiple concurrent data streams	Complexity modes
Robustness to information errors and losses	Still images
	Tandem coding[*]

Requirements for synthetic video objects

Types of synthetic video objects	Text overlay
2D/3D mesh compression	Image and graphics overlay
Definition and animation parameter compression	View-dependent texture scalability
Texture mapping	Geometrical transformations
	Video object tracking

Requirements for natural audio objects

Object-based representation	Robustness to information errors and losses
Audio content	Delay modes
Object-based bitstream editing and manipulation	Complexity modes
Object-based scalability	Bit-rate modes
Object-based random access and user controls	Downmix[†]
Time scale change	Transcoding
Pitch change	Tandem coding
	Audio formats
	Improved coding efficiency

Table 1.1 MPEG-4 requirements [N4319] (Continued)

Requirements for synthetic audio objects

Low bit-rate speech	Text to speech
Synthetic speech data	Sound synthesis

Requirements for delivery multimedia integration format (DMIF)

Connectivity	End-to-end Quality of Service (QoS) management
Transparency	
Application service enablement	Network-based stream processing and management

Requirements for MPEG-J

Functional requirements	Byte code execution
Byte code delivery	Event mechanism
Authentication	

Requirements for multiuser environments

Scene graph representation	IPMP and sharing tools
Audiovisual objects and avatars representation	Application programming interfaces
Delivery and stream management	

Requirements for animation framework

Enhanced texture mapping	Reusability of scene graph nodes and animation streams
Animation support	
High-level shape representation	Persistence
	Compression of animated objects

Requirements for intellectual property management and protection (IPMP)

Identification of intellectual property	Intellectual property management and protection interfaces
Intellectual property management and protection hooks	

---

*Relates to the repeated encoding and decoding (cascading) of audio or video material, either with the same or different codecs.

†Relates to the ability to reduce the number of channels to a configuration with a lower number of channels for presentation purposes (e.g., listening to multichannel audio using stereophonic reproduction).

Although efficient compression was not the only first-priority requirement in MPEG-4 (as it had been for MPEG-1 and MPEG-2), it is clear that it has been a central requirement in MPEG-4 in the sense that, whatever the type of data that had to be represented by binary encoding (e.g., shape for video objects, facial animation parameters for 3D facial models, or even scene composition data), the target was always to reach the smallest number of bits for a certain level of quality. Although some people claim that efficient compression of data is not a must today because of the growing availability of bandwidth, MPEG always acknowledged that bandwidth resources (either for transmission or for storage) are still limited, and thus efficient compression is required. This is even truer for some relevant recent transmission cases, such as the Internet and mobile networks, where bandwidth limitations and efficient compression are major issues. However, in the context of MPEG-4, efficient compression must be balanced against other major requirements, such as those related to interactivity capabilities (which have a price in terms of compression efficiency compared with noninteractive representation schemes), if new functionalities not supported by frame-based coding schemes are to be provided.

As noted, the MPEG-4 requirements' development process has been evolving since the beginning of the standardization process; this evolution has been targeting the inclusion of additional requirements related to functionalities, which fit and complement well the MPEG-4 vision. Because the MPEG-4 standardization process is not yet finished, it cannot be said that all the requirements have been addressed by means of an MPEG-4 tool. However, it is possible to say that all of the not-yet-addressed requirements are either being worked on or should be removed in a short time if the industries do not show sufficient support to move to the technical development phase.

1.1.3 Tools

The major trends mentioned—notably the mounting presence of audiovisual media on all networks, increasing mobility, and growing interactivity—have driven, and continue to drive, the development of the MPEG-4 standard.

To address the identified functionalities and requirements [N4319], a set of tools was developed to perform the following functions [Pere99]:

☞ Efficiently represent a number of data types through media codecs

  ✗ Video from very low bit rates to very high-quality conditions

  ✗ Music and speech data for a very wide bit-rate range, from transparent music to very low bit-rate speech

  ✗ Generic dynamic 3D objects as well as specific objects such as human faces and bodies

  ✗ Synthetic speech and music, including support for 3D audio spaces

  ✗ Text and graphics

☞ Provide fine granularity scalability in the quality, temporal, and spatial dimensions

☞ Provide, in the encoding layer, resilience to residual errors for the various data types, especially under difficult channel conditions such as mobile ones

☞ Independently represent the various objects in the scene, allowing independent access for their manipulation and reuse

☞ Compose audio and visual (natural and synthetic) objects into one audiovisual scene in a synchronized way

☞ Describe the objects and the events in the scene

☞ Provide interaction and hyperlinking capabilities

☞ Manage and protect intellectual property on audiovisual content and algorithms, so that only authorized users have access to the content

☞ Provide a delivery-media-independent representation format, to transparently cross the borders of different delivery environments

The major difference with previous audiovisual coding standards, at the basis of the new functionalities, is the object-based audiovisual representation model that underpins MPEG-4 (see Figure 1.1). An object-based scene is built using individual independent objects that have relationships in space and time. This representation approach offers a number of advantages: First, different object types may have different suitable coded representations—for example, a synthetic moving head is best represented using animation parameters, whereas video benefits from a smart representation of pixel values. Second, it allows the harmonious integration of different types of data into one scene, notably with natural and synthetic origins—for example, an animated cartoon character in a real world environment, or a real person in a virtual studio set. Third, interacting with the objects and hyperlinking from them is

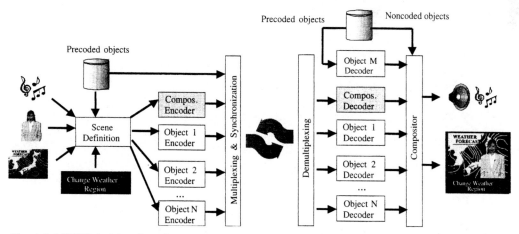

Fig. 1.1 MPEG-4 object-based representation architecture.

now feasible. There are many more advantages—such as the selective spending of bits, the easy reuse of content without transcoding, and the provision of sophisticated coding solutions for scalable content on the Internet—all of them resulting from the adoption of the object-based representation model.

The applications that benefit from the new concepts and functionalities are found in many (and very different) environments [N2724]. Therefore, MPEG-4 is constructed as a toolbox rather than a monolithic standard, using profiles that provide solutions for these different application settings (see Chapter 13). This means that, although MPEG-4 is a standard comprising a vast array of technologies, it is structured so that solutions are available at the measure of the needs. It is the task of each implementer to extract from the MPEG-4 standard the technical solutions adequate to his or her needs (likely a small subset of the standardized tools) by choosing the adequate profiling combination.

Because a standard is always a constraint on freedom, it is important to make it as minimally constraining as possible [Chia97]. To MPEG, this means a standard must offer maximum advantages by specifying the minimum necessary, allowing for competition and evolution of the technology in the *non-normative areas;* these non-normative areas correspond to the tools for which normative specification is not essential for interoperability. On the contrary, normative tools are those defined by the standard whose specification is essential for interoperability. For example, whereas video segmentation and rate control are non-normative tools, the decoding process needs to be normative. The strategy of specifying the minimum for maximum usability ensures that good use can be made of the continuous technical improvements in the relevant technical areas. The consequence is that better non-normative tools can always be used, even after the standard is finalized, and it is possible to rely on competition for obtaining ever-better results. In fact, it is through the non-normative tools that products will distinguish themselves, which only reinforces the importance of this type of tools.

1.1.4 Applications

MPEG-4 wants to address a wide range of applications, many of them completely new, as there are very new functionalities to take benefit from, and many others improved regarding those already available [N2724]. Unlike MPEG-2 where the *killer application* was digital television, in a first approach just understood as the digital translation of the rather old analog version, MPEG-4 does not target a major and exclusive killer application but opens many new frontiers. Playing with audiovisual scenes and creating, reusing, accessing, and consuming audiovisual content will become easier and more powerful. New and richer applications can be developed, for example, in enhanced broadcasting, remote surveillance, personal communications, games, mobile multimedia, and virtual environments. MPEG-4 allows services combining the traditionally different service models: broadcast, (online) interaction, and communication. As such, MPEG-4 addresses *convergence,* understood as the proliferation of audiovisual information in all kinds of services and on all types of (access) networks.

The MPEG-4 Applications document [N2724] describes application examples benefiting from the MPEG-4 technology that will serve as inspiration to the industry to create many more exciting applications. In this sense, MPEG-4 is a technical playground where many application constructions may be built by the manufacturers and service providers. The MPEG-4 Applications document suggests the following applications, using both audio and visual information or just one of them: broadcast, collaborative scene visualization, content-based storage and retrieval, digital amplitude modulation (AM) broadcasting, digital television set-top box, DVD, infotainment, mobile multimedia, real-time communications, streaming video on the Internet/Intranet, studio and television postproduction, surveillance, and virtual meetings.

1.2 FORMAL STANDARDIZATION PROCESS

MPEG (Moving Picture Experts Group) formally called *Coding of Moving Picture and Audio* is Working Group 11 (WG11) of Subcommittee 29 (Coding of Audio, Picture, Multimedia and Hypermedia Information) of the ISO/IEC (International Organization for Standardization/International Electrotechnical Commission) Joint Technical Committee 1 (JTC1) [MPEG]. JTC1's scope is standardization in the field of information technology: According to JTC1 [JTC1], information technology includes the specification, design, and development of systems and tools dealing with the capture, representation, processing, security, transfer, interchange, presentation, management, organization, storage, and retrieval of information.

Since the major objective of MPEG is to produce technical specifications, it is important to have an idea, even if brief, about the types of documents that can be produced by a group like MPEG. Through JTC1, MPEG can produce the types of documents presented in the following:[5]

☞ **International standards:** These are documents with the technical specification of the standard. After the first edition of a standard, new editions may be published, notably to incorporate in one document the previous edition and all the amendments and corrigenda meanwhile issued. Whenever a new edition for a certain Part of the standard includes new technology (and thus is not just a compilation of previously approved technology), it must be approved following a process similar to amendments.

---

5. For each type of document, the various successive stages, as well as the minimum time the voting period by national standardization bodies may take for each stage (this means the minimum time to take the promotion of a certain document from one stage to the next), are indicated.

The evolution process for an International Standard is as follows:

✗ New work item Proposal (NP): 3 months ballot (with comments)

✗ Working Draft (WD): no ballot

✗ Committee Draft (CD): 3 months ballot (with comments)

✗ Final Committee Draft (FCD): 4 months ballot (with comments)

✗ Final Draft International Standard (FDIS): 2 months binary (only yes/no) ballot (failing this ballot [no vote] implies going back to WD stage)

✗ International Standard (IS)

FDIS and IS documents are copyrighted by ISO.

☞ **Amendments:** Documents with technical additions or technical changes (but not corrections) to an International Standard; they are edited as delta documents to the standard they amend. Each amendment lists the status of all amendments and technical corrigenda to the current edition of the standard. Amendments are published as separate documents, which means the edition of the IS affected remains in print. When additions to a standard must be produced, it is decided if it is better to publish an amendment or a new edition of the IS that incorporates the additions. When a new edition is published, it includes a full revision of the previous edition, incorporating all the changes corresponding to the amendments and corrigenda meanwhile issued.

The evolution process for an Amendment is as follows:

✗ New work item Proposal (NP): 3 months ballot (with comments)

✗ Working Draft (WD): no ballot

✗ Proposed Draft Amendment (PDAM): 3 months ballot (with comments)

✗ Final Proposed Draft Amendment (FPDAM): 4 months ballot (with comments)

✗ Final Draft Amendment (FDAM): 2 months binary (only yes/no) ballot (failing this ballot [no vote] implies going back to WD stage)

✗ Amendment (AMD)

FDAM and AMD documents are copyrighted by ISO.

☞ **Corrigenda:** These are documents issued to correct technical defects in an IS (or an Amendment).[6] Technical corrigenda usually are not issued for the correction of a few editorial defects. In such cases, corrections can be incorporated in future technical corrigenda. Technical corrigenda are not issued for technical additions, which follow the Amendment procedure.

The evolution process for a corrigendum is as follows:

---

6. Notice that, formally speaking, an amendment is an IS, as it specifies new technology (and never corrections to previous technology, which is the task of corrigenda).

✗ Defect Report (DR): no ballot

✗ Draft Technical Corrigendum (DCOR): 3 months ballot (with comments)

✗ Technical Corrigendum (COR)

COR documents are copyrighted by ISO.

☞ **Technical reports:** These documents contain information of a different kind from that normally published as an IS, such as a model/framework, technical requirements and planning information, a testing criteria methodology, information obtained from surveys carried out among national bodies, information on work in other international bodies, or information on the state of the art regarding national body standards on a particular subject [JTC1].

The evolution process for a Technical Report is as follows:

✗ New Work Item Proposal (NP): 3 months ballot (with comments)

✗ Working Draft (WD): no ballot

✗ Proposed Draft Technical Report (PDTR): 3 months ballot (with comments)

✗ Draft Technical Report (DTR): 3 months ballot (with comments)

✗ Technical Report (TR)

DTR and TR documents are copyrighted by ISO.

For all ballots involving comments, MPEG produces a *Disposition of Comments* (DoC) document in which all comments made by national bodies are answered. During the MPEG-4 standard development process, MPEG used all the types of documents listed. Section 1.5 presents an overview of the many documents published in the context of the standardization process for the MPEG-4 standard.

1.3 MPEG MODUS OPERANDI

MPEG is open to experts duly accredited by an appropriate National Standards Body. On average, a meeting is attended by more than 300 experts representing more than 200 companies, universities, and research centers spanning all industry domains with a stake in digital audio, video, and multimedia. On average, more than 20 countries are represented. The group meets three to five times a year. Participants in MPEG-4 include broadcasters, equipment and software manufacturers, digital content creators and managers, telecommunication service providers, and publishers and intellectual property rights managers, as well as university researchers.

MPEG held its first meeting in May 1988 in Ottawa (Canada) and reached its 58th meeting in December 2001 in Pattaya (Thailand). A Convener chairs the MPEG activities: Dr. Leonardo Chiariglione has been the MPEG Convener since the start of MPEG.

Because of its size, an MPEG plenary meeting can be a difficult experience for a newcomer if he or she has no guidance. MPEG meetings are well known for their openness, lively discussions, and detailed dissection of issues. During the meetings, MPEG is organized in several subgroups, each one with a chairman, notably these [N4500]:

☞ **Specification development subgroups:** Systems, Video, Audio, Multimedia Description Schemes (MDS), and Synthetic Natural Hybrid Coding (SNHC)

☞ **Auxiliary subgroups:** Requirements, Test, Implementation Studies, and Liaison

To better coordinate the work between the MPEG plenary meetings, MPEG may decide to establish ad hoc groups (AHG). AHGs are established with mandate, membership, chairman, duration, and meeting schedule at the end of an MPEG meeting, for the sole purpose of continuing work between MPEG meetings. They are established by MPEG and report to it; the task of an AHG may cover just preparation of recommendations to be submitted to MPEG, with no decisions made. The duration of an AHG usually is limited to the period between two successive MPEG meetings; AHGs always cease to exist at the start of an MPEG meeting. Participation in AHGs is not restricted to the delegates present at the meeting.

1.3.1 Mission

MPEG's area of work is the "development of international standards for compression, decompression, processing, and coded representation of moving pictures, audio, and their combination, in order to satisfy a wide variety of applications." [N4500]. According to the MPEG *Terms of Reference* [N4500], the MPEG *Programme of Work* is as follows:

☞ Serve as responsible body within ISO/IEC for recommending a set of standards consistent with the area of work.

☞ Cooperate with other standardization bodies dealing with similar applications. For this, MPEG creates liaisons with other standardization bodies as well as with other relevant duly constituted organizations (e.g., industrial consortia), exchanging requirements and technical specifications with the aim of reaching the largest possible use of standards.

☞ Consider requirements for interworking with other applications, such as telecommunications and broadcasting, with other image coding algorithms defined by other SC29 Working Groups and with other picture and audio coding algorithms defined by other standardization bodies.

☞ Define methods for the subjective quality assessment of audio, moving pictures, and their combination for the purpose of the area of work.

☞ Assess characteristics of implementation technologies realizing coding algorithms of audio, moving pictures, and their combination.

☞ Assess characteristics of digital storage and other delivery media, targets of the standards developed by WG11.

☞ Develop standards for coding of moving pictures, audio, and their combination, taking into account the quality of coded media, effective implementation, and constraints from delivery media.

☞ Propose standards for the coded representation of moving picture information.

☞ Propose standards for the coded representation of audio information.

☞ Propose standards for the coded representation of information consisting of moving pictures and audio in combination.

☞ Propose standards for protocols associated with the coded representation of moving pictures, audio, and their combination.

1.3.2 Principles

Because the technological landscape changed from analog to digital, with all the associated implications, it was essential that standard makers acknowledged this change by modifying the way they create standards. Standards must offer interoperability across countries, services, and applications, and not just a system-driven approach by which the value of a standard is limited to a specific, vertically integrated system. This brings us to the toolbox approach by which a standard must provide a minimum set of relevant tools which (after they are assembled according to industry needs) provide the maximum interoperability at a minimum complexity—and very likely cost [Chia97]. The success of MPEG standards is based on this toolbox approach, bounded by the *one functionality, one tool* principle. In summary, MPEG wants to offer users interoperability and flexibility at the lowest complexity and cost.

To develop its standards, MPEG has been following a few principles. The most important of these principles, many of them very well known, are these [Chia97]:

☞ **Stick to the deadline:** Because of the importance of allowing industries to make serious planning and investments based on timely standards, MPEG rigorously follows the workplan set at the beginning of each standardization project. Never has an MPEG standard been delayed in reaching IS status compared to the planned dates.

☞ *A priori* **standardization:** To avoid becoming a standardization body endorsing industry-developed standards, MPEG identifies the maturity of technologies for standardization before industries make commitments. This approach allows the standard development process to be essentially technical and not biased by specific company interests.

☞ **Specify the minimum:** In order that a standard might be useful to several industries (and not especially tuned for any of them), will technically evolve (increasing its lifetime), and will keep space for industrial competition, it is essential that only the minimum set of tools essential for interoperability is specified.

☞ **Not systems but tools:** Because MPEG wants to specify standards that are useful for the various industries using the same technology, it is essential that tools rather than systems are specified, leaving the various industries the task of shaping those tools for their systems. To guide these industries, however, some major combinations of tools (profiling concept) are normatively specified so that functionality-driven and not application-driven combinations are adopted; this helps to increase interoperability and share costs.

☞ **One functionality, one tool:** To reach interoperability with an acceptable level of complexity justifying industry investments, it is essential that clear choices be made regarding the best tool to provide a certain capability. Past experience has shown that the *options* epidemic (often engaged to satisfy companies' wishes) prevents interoperability, many times preventing the standards from flying and thus destroying everybody's work.

☞ **Relocation of tools/algorithms:** So that the tools/algorithms are as useful as possible, not only must they be generic in the sense that they are not shaped to any specific application environment, it must also be possible to locate them at different positions in the final system, as different industries will likely position the same tool/algorithm differently (e.g., depending on the business model they follow).

☞ **Verification of the standard:** To check that a standard delivers what it should deliver, it must be evaluated and checked against the identified requirements in the same way a product is checked against the product specification. This verification process typically involves the performance of subjective or task-based tests.

These principles have guided the development of the MPEG-1, MPEG-2, MPEG-4, and MPEG-7 standards with the same success, and they will be applied to the MPEG-21 standard development process.

1.3.3 Standards Development Approach

In MPEG, any standardization project is the result of an exploration attitude by which MPEG proactively looks for the relevant problems to be addressed—for example, emerging applications, functionalities, or even technologies. This proactive exploration attitude allows the standards to be produced in a timely way (and not too late) when applications and technology are fully mature as, by that time, proprietary solutions will be conquering the markets. For this purpose,

MPEG holds open seminars to discuss relevant topics with people from industry and academia outside MPEG and creates ad hoc groups with the mandate to study the relevance in terms of standardization of specific topics.

After identifying a relevant area of interest, and in order to fulfill its objectives guided by the preceding principles, MPEG follows a standards development process with the following major steps [Chia97]:

1. **Applications:** Identify relevant applications using input from MPEG members.

2. **Functionalities:** Identify the functionalities needed by the applications.

3. **Requirements**: Describe the requirements following from the functionalities so that common requirements can be identified for different applications.

4. **Common requirements:** Identify which requirements are common across the areas of interest, and which are not common but still relevant.

5. **Specification:** Specify the tools supporting the requirements in three phases.

    i) **Call for proposals:** A public call for proposals is issued, asking all interested parties to submit technology that is relevant to the identified requirements and functionalities.

    ii) **Proposal evaluation:** The proposals are evaluated in a well-defined, adequate, and fair evaluation process, which is published with the call for proposals. The process can entail subjective testing, objective comparison, and evaluation by experts.

    iii) **Technical specification:** As a result of the evaluation, the technology best addressing the requirements is selected; this technology typically does not correspond to a single proposal but to the set of the best tools extracted from all proposals. This is the start of a collaborative process to draft and improve the standard. The collaboration includes the definition and improvement of a *working model*, which embodies early versions of the standard and can include non-normative tools to better and more completely test the normative tools. The working model evolves through comparing different alternative tools with those already in the working model, through the so-called *core experiments* (CE).

6. **Verification:** Verify that the tools developed can be used to assemble the target systems and to provide the desired functionalities with an adequate level of performance. This is done by means of *verification tests*. Until MPEG-4, the verification tests consisted of formal subjective tests aimed at evaluating the quality of either audio or video signals processed using specific MPEG algorithms; in MPEG-4, new types of tests have been performed [N999]. In order to obtain reliable and representative results, the tests are performed using optimized assessment methods and suitable panels of subjects.

This process is not rigid; some steps may be taken more than once and iterations are sometimes needed (as happened in MPEG-4). The time schedule, however, is always closely observed by MPEG. Although all decisions are made by consensus, the process keeps a fast pace, allowing MPEG to provide good technical solutions in a timely manner.

While the period until the proposals are evaluated is called the *competitive phase*, the period after the evaluation is the *collaborative phase*. During the collaborative phase, all the MPEG members collectively improve and complete the most promising tools identified at the proposals' evaluation. The collaborative phase is the major strength of the MPEG process, as hundreds of the top experts in the world, from many companies and universities, work together for a common goal. In this context, it does not come as a surprise that this superteam traditionally achieves excellent technical results, justifying the need for most companies to at least follow the process, if direct involvement is not possible.

Two working tools play a major role in the collaborative development phase that follows the initial competitive phase: the working model and core experiments [N1375]. In MPEG-1, the (video) working model was called Simulation Model (SM); in MPEG-2, the (video) working model was called Test Model (TM); and in MPEG-4, the various working models were called Verification Models (VMs).[7] In MPEG-4, there were independent VMs for the video, audio, synthetic and natural hybrid coding (SNHC), and systems developments. Regarding the MPEG-4 VMs and CEs, it is important to highlight a few points [Pere99].

1.3.3.1 Verification Models

A VM is a complete framework defined in text and with a corresponding software implementation, such that an experiment performed by multiple independent parties will produce essentially identical results. VMs are enabled to check the relative performance of different tools, as well as to improve the performance of selected tools. The MPEG-4 VMs were built after screening the proposals. The first VM (for each technical area, e.g., video, audio, and SNHC) was not the best proposal but a combination of the best tools, independent of the proposal to which they belonged. Each VM included normative and non-normative tools to create the *common framework* that allowed performing adequate evaluation and comparison of tools targeting the continuous improvement of the technology included in the VM. After the first VMs were established, new tools were brought to MPEG-4 and evaluated within the VMs following a core experiment procedure. The VMs evolved through versions as CEs verified the inclusion of new techniques or proved that included techniques should be substituted. At each VM version, only the best performing tools were part of the VM. If any part of a proposal was

---

7. In MPEG-7, the working model is called eXperimentation Model (XM), whereas in MPEG-21 the working model is called sYstems Model (YM).

selected for inclusion in the VM, the proposer had to provide the corresponding source code for integration into the VM software in the conditions specified by MPEG.

1.3.3.2 Core Experiments The improvement of the VMs started with a first set of CEs defined at the conclusion of proposal evaluation [N998]. The CE process allowed for multiple, independent, directly comparable experiments to determine whether a proposed tool had merit. Proposed tools targeted the substitution of a tool in one of the VMs or the direct inclusion in the VM to provide a new relevant functionality. Improvements and additions to the VMs were decided based on the results of CEs.

A CE must be completely and uniquely defined, so that the results are unambiguous. In addition to the specification of the tool to be evaluated, a CE also specifies the conditions to be used, again so the results can be compared. A CE is proposed by one or more MPEG experts and is accepted by consensus, providing that two or more independent experts agreed to perform the experiment.

It is important to realize that neither the text of the VMs nor any of the CEs ended (or will end) up in the standard itself, as they were just working tools to ease the development process. However the VMs' software is the basis for MPEG-4 Part 5: Reference Software [MPEG4-5] presented in the next section. Although it is not easy at this stage to tell how many CEs have been performed in MPEG-4, it is possible to state that they reached their goal by continuously improving and completing the technology to be included in the standard.

1.4 MPEG-4 STANDARD ORGANIZATION

The MPEG-4 requirements [N4319] have been addressed by a standard organized in several Parts, each one with multiple editions, amendments, and corrigenda. Amendments to each Part of the standard were also called *Versions* (the first edition itself is Version 1).[8] The list of documents issued as of July 2001 for the various Parts of the MPEG-4 standard is as follows:

☞ **Part 1: Systems.** This Part specifies scene description, multiplexing, synchronization, buffer management, and management and protection of intellectual property [MPEG4-1].

Version 1: First edition (14496-1:1999)

Version 2 (first amendment to first edition): Systems extensions (MPEG-4 file format, BIFS nodes)

---

8. Amendments to the second editions were sometimes called *extensions*.

Corrigendum 1 to first edition

Second edition of Part 1, including the 1999 edition, Amendment 1, and Corrigendum 1 (14496-1:2001)

Amendment 1 to second edition: Extended BIFS (Flextime)

Amendment 2 to second edition: Textual format (XMT)

Amendment 3 to second edition: Intellectual Property Management and Protection (IPMP) extension

Amendment 4 to second edition: Multiuser worlds and animation framework (AFX)

Amendment 5 to second edition: MP4 extensions (common with WG1/ JPEG)

Amendment 6 to second edition: MP4 extensions

Corrigendum 1 to second edition

Corrigendum 2 to second edition

☞ **Part 2: Visual.** This Part specifies the coded representation of natural and synthetic visual objects [MPEG4-2].

Version 1: First edition (14496-2:1999)

Version 2 (first amendment to first edition): Visual extensions

Version 3 (second amendment to first edition): 3D mesh profiles (withdrawn)

Corrigendum 1 to first edition

Corrigendum 2 to first edition

Second edition of Part 2, including the first edition, Amendment 1, and Corrigenda 1 and 2 (14496-2:2001)

Amendment 1 to second edition: Studio profiles

Amendment 2 to second edition: Streaming video profiles

Amendment 3 to second edition: New levels and tools

☞ **Part 3: Audio.** This Part specifies the coded representation of natural and synthetic audio objects [MPEG4-3].

Version 1: First edition (14496-3:1999)

Version 2 (first amendment to first edition): Audio extensions

Corrigendum 1 to first edition

Second edition of Part 3, including the first edition, Amendment 1, and Corrigendum 1 (14496-3:2001)

Corrigendum 1 to second edition

☞ **Part 4: Conformance Testing.** This Part defines conformance conditions for bitstreams and devices; it is used to test MPEG-4 implementations [MPEG4-4].

Version 1: First edition (14496-4:2000)

Version 2 (first amendment to first edition): Extensions to conformance testing

Corrigendum 1 to first edition

Second edition of Part 4, including the first edition, Amendment 1, and Corrigendum 1 (14496-4:2001)

Amendment 1 to second edition: Conformance extensions for studio and streaming video profiles and for Flextime

Corrigendum 1 to second edition

☞ **Part 5: Reference Software.** This Part includes software corresponding to most Parts of MPEG-4 (normative and non-normative tools); this means the VMs mentioned in Section 1.3.3.1. This software can be used for implementing compliant products as ISO waives the copyright of the code[9] [MPEG4-5].

Version 1: First edition (14496-5:1999)

Version 2 (first amendment to first edition): Reference software extensions

Corrigendum 1

Second edition of Part 5, including the first edition, Amendment 1, and Corrigendum 1 (14496-5:2001)

Amendment 1 to second edition: Reference software extensions for studio and streaming video profiles and for Flextime

☞ **Part 6: Delivery Multimedia Integration Framework (DMIF).** Part 6 defines a session protocol for the management of multimedia streaming over generic delivery technologies [MPEG4-6].

Version 1: First edition (14496-6:1999)

Version 2 (Amendment 1 to first edition): DMIF extensions

Corrigendum 1

Second edition of Part 6, including the first edition, Amendment 1, and Corrigendum 1 (14496-6:2000)

---

9. This does not mean that product developers do not have to license the patents involved.

☞ **Part 7: Optimized Visual Reference Software.** This Part includes
optimized software for visual tools such as fast motion estimation, fast
global motion estimation, and fast and robust sprite generation [N4554];
the optimized software is called *Optimization Model* (OM). Unlike the
other MPEG-4 standard Parts, this Part is a TR and not a standard
specification.
Version 1: First edition (14496-7:2001)

☞ **Part 8: Carriage of MPEG-4 Contents over IP Networks.** This Part
specifies the mapping of MPEG-4 content into several IP-based protocols
[N4427]; it is well known as *4on IP.*
Version 1: First edition (14496-8:2002)

☞ **Part 9: Reference Hardware Description.** Part 9 will include portable synthesizable/simulatable very high-speed integrated circuit hardware description language (VHDL) descriptions of MPEG-4 tools
[N4218].
Version 1: First edition (14496-9:2003)

☞ **Part 10: Advanced Video Coding (AVC).** This Part will specify video
syntax and coding tools[10] in the context of a joint project with ITU-T
SG16 [N4400], known as Joint Video Team (JVT); this activity used as a
starting point the available version of the H.26L video coding specification to address the identified requirements [N4466, N4508].
Version 1: First edition (14496-10:2003)

Although most parts are IS, Parts 7 and 9 are TRs with informative
value. Parts 1 to 3 as well as Parts 6, 8, and 10 specify the core MPEG-4 technology, whereas Parts 4, 5, 7, and 9 are supporting Parts. Parts 1, 2, 3, and 10
are delivery-independent, leaving to Parts 6 and 8 the task of dealing with the
idiosyncrasies of the delivery layer.

The major reason to develop the MPEG-4 standard in several rather
independent Parts (besides avoiding a single document with several thousand
pages) is to allow the various pieces of technology to be useful as stand-alones
and thus as much used as possible, even if in conjunction with proprietary
technologies. This has been the case, for example, for MPEG-2 Video, which
today is being used together with MPEG-2 Systems but not with MPEG-2
Audio in the context of the U.S. digital TV system. This means that within the
context of a certain standardization effort (e.g., MPEG-4), whenever a new
technological area is addressed that is different from the areas already
addressed (e.g., MPEG-4 on IP), an additional Part of the standard is created
to allow stand-alone use and a clearer organization of the tools specified by
the standard as a whole. However, although the various Parts may be used

---

10. The working model for this activity is called Joint Model (JM).

independently, they were developed to give optimal results when they are used together.

During the MPEG-4 development process, it was decided to issue successive versions of the several MPEG-4 Parts whenever new tools needed to be added to that Part of the standard. In this context, versions would serve to specify new tools, either offering new functionalities or bringing a significant improvement in terms of functionalities already supported. Formally speaking, the various versions (except the first) of a certain MPEG-4 Part correspond to amendments to that Part of the standard. So, whereas the IS for Systems (issued in 1999) is called Version 1, MPEG-4 Systems Version 2 corresponds to the first MPEG-4 Systems amendment, Version 3 to the Amendment 2 to Systems, and so on. It is important to note that new versions of a Part do not substitute or redefine tools specified in previous versions but simply add more tools. At each stage of specification, a certain Part of the MPEG-4 standard is the set of all the tools specified in all versions for that Part. In that sense, it is common to say that versions are backward-compatible, meaning that Version N may only add new tools and profiles to Version N-1 and not remove or redefine any tool or profile. This implies that existing terminals will always remain compliant, as profiles will not be changed in retrospect. The same reasoning applies to the amendments to the second and further editions of the various Parts of the standard.

1.5 MPEG-4 SCHEDULE

After the initial study phase, in which the major MPEG-4 objectives were identified, notably through the functionalities presented earlier, the first MPEG-4 call for proposals was issued in July 1995 [N998] and answers were received by September/October 1995. The call for proposals asked for relevant video and audio technology addressing the eight MPEG-4 functionalities as described in the MPEG-4 Proposal Package Description [N998]. The technology received was evaluated using subjective tests for complete algorithms and expert panels for single tools [N999]. In the case of algorithms addressing the eight MPEG-4 functionalities, three functionalities (one for each of the three classes) were selected as representative—content-based scalability, improved compression efficiency, and robustness in error-prone environments—and formal subjective tests were conducted for those. For the other five functionalities, proposals were evaluated by expert panels (in the same manner as tools), and the corresponding selected tools were thoroughly examined using the CE procedure.

The video subjective tests were performed in November 1995 at Hughes Aircraft Company, in Los Angeles; the audio subjective tests were performed in December 1995 at CCETT, Mitsubishi, NTT, and Sony. The video expert panels evaluation was performed in October 1995 and January 1996.

At this initial phase, MPEG decided to develop the Systems specification by means of a pure collaborative approach and thus no calls were issued for Systems tools.

After the evaluation of the technology received [M532], choices were made and the collaborative phase started with the most promising tools. In the course of developing the standard, additional calls were issued when insufficient technology was available within MPEG to meet the requirements—for example, for synthetic and hybrid coding tools in July 1996 [N1315], for identification and protection of content in April 1997 [N1714], and for an intermedia file format in October 1997 [N1919]. This is a typical solution when MPEG is missing some technology and there are good indications that the technology exists outside MPEG. Moreover, in order to check that the standardized technology is still among the best available, MPEG may issue calls for proposals to compare the already standardized tools with the most recent developments outside MPEG. This was the case with the calls for proposals on audio [N3992] and video [N4065] coding tools in March 2001. The results of the video tests [N4454] following the video call for proposals led to the creation of the joint project with ITU-T SG 16 to develop MPEG-4 Part 10 [N4400].

At the MPEG January 1996 meeting in Munich, the first MPEG-4 Video Verification Model (VM) was defined [N1172]. In this VM, and for the first time in a standardization process, a video scene was represented as a composition of arbitrarily shaped objects; each object was represented by a sequence of *Video Object Planes* (VOPs), which are the temporal instantiations of a video object at a certain moment. For this, the first MPEG-4 Video VM used ITU-T H.263 coding tools [H263] together with shape coding tools, following the results of the November 1995 MPEG-4 video subjective tests [M532].

A process similar to the one used for video was followed for audio, although with some initial delay due to the involvement of many audio experts in the advanced audio coding (AAC) MPEG-2 work [N1214].

Following this initial phase, the several MPEG-4 VMs evolved using the CE process. A new version of each of the MPEG-4 VMs has been issued at each MPEG meeting—for example, the Video VM was in its 18th version at the Pisa meeting in January 2001 [3908].

As highlighted in the previous section, the last step of the MPEG process is the verification of the technology in the standard aiming at testing the performance of the tools and demonstrating their potentialities. For MPEG-4, the verification step was performed through a set of verification tests addressing various parts of the standard. Many verification tests have already been performed for video and audio tools and profiles (see Chapters 15 and 16 on testing for validation).

For MPEG-4, the process highlighted in the previous sections translated to the time schedule presented in Table 1.2.

Table 1.2 MPEG-4 time schedule*

Date	Event
July 1995 (Tokyo)	Call for proposals on audio and video tools and algorithms [N998] Final version of the MPEG-4 evaluation document [N999]
November 1995	Subjective evaluation of video proposals
December 1995	Subjective evaluation of audio proposals
January 1996 (Munich)	Experts evaluation of video proposals First version of the MPEG-4 Video VM [N1172]
March 1996 (Florence)	First version of the MPEG-4 Audio VM [N1214]
July 1996 (Tampere)	Call for proposals on SNHC tools [N1315]
September 1996 (Chicago)	First version of the MPEG-4 SNHC VM [N1364] Call for proposals on synthetic audio [N1397]
November 1996 (Maceió)	WD, Parts 1, 2, 3, 5, and 6 Call for proposals on audio and video tools and algorithms [N1499]
April 1997 (Bristol)	Call for proposals on identification and protection of content in MPEG-4 [N1714]
October 1997 (Fribourg)	Call for proposals for an MPEG-4 intermedia format [N1919] CD, Parts 1, 2, 3, 5, and 6 WD, Part 4
March 1998 (Tokyo)	FCD, Parts 1, 2, 3, 5, and 6
October 1998 (Atlantic City)	FDIS, Parts 1, 2, 3, and 6
December 1998 (Rome)	CD, Part 4
March 1999 (Seoul)	Version 2, PDAM status, Parts 1, 2, 3, and 6
July 1999 (Vancouver)	Version 2, FPDAM status, Parts 1, 2, 3, and 6 FCD, Part 4 FDIS, Part 5 Version 2, PDAM status, Part 5
October 1999 (Melbourne)	COR 1, DCOR status, Parts 1, 2, and 6
December 1999 (Maui)	Call for proposals for an MPEG-4 textual format [N3157] Version 2, FDAM status, Parts 1, 2, 3, and 6 Amendment 1 to second edition, PDAM status, Part 1

Table 1.2 MPEG-4 time schedule* (Continued)

Date	Event
December 1999 (Maui) (cont)	FDIS, Part 4 Version 2, PDAM status, Part 4
March 2000 (Noordwijker-hout)	Call for proposals for a generic animation framework of synthetic objects [N3341] COR 1, COR status, Parts 1, 2, and 6 Amendment 1 to second edition, PDAM status, Part 2 Amendment 2 to second edition, PDAM status, Part 2 Version 2, FPDAM status, Part 5
May 2000 (Geneva)	Amendment 1 to second edition, FPDAM status, Part 1
July 2000 (Beijing)	Call for proposals for IPMP solutions [N3543] Call for proposals on multiusers worlds technology [N3574] Amendment 1 to second edition, FPDAM status, Part 2 Amendment 2 to second edition, FPDAM status, Part 2 COR 2, DCOR status, Part 2 COR 1, DCOR status, Part 3 Version 2, FPDAM status, Part 4 Version 2, FDAM status, Part 5 COR 1, DCOR status, Part 5
October 2000 (La Baule)	Call for proposals for new tools to further improve video coding efficiency [N3671] WG11 approval of second edition, Part 1 Amendment 1 to second edition, FDAM status, Part 1 Amendment 2 to second edition, PDAM status, Part 1 COR 1, DCOR status, Part 1 (second edition) COR 2, COR status, Part 2 WG11 approval of second edition, Part 6
January 2001 (Pisa)	Amendment 1 to second edition, FDAM status, Part 2 Amendment 2 to second edition, FDAM status, Part 2 COR 1, COR status, Part 3 Version 2, FDAM status, Part 4 COR 1, COR status, Part 5
March 2001 (Singapore)	Call for proposals for new tools for audio coding [N3992] Call for proposals for new tools for video compression technology [N4065] Call for proposals for interpolator compression [N4098] WG11 approval of second edition, Part 3 WG11 approval of second edition, Part 4 Amendment 1 to second edition, PDAM, Part 4 WG11 approval of second edition, Part 5

Table 1.2 MPEG-4 time schedule* (Continued)

Date	Event
March 2001 (Singapore) (cont)	Amendment 1 to second edition, PDAM, Part 5 CD, Part 8
July 2001 (Sydney)	Call for proposals for hardware reference code [N4218] Amendment 2 to second edition, FPDAM status, Part 1 Amendment 3 to second edition, PDAM status, Part 1 COR 1, COR status, Part 1 (second edition) WG11 Approval of Edition, Part 2 Amendment 1 to second edition, FPDAM, Part 5 Amendment 2 to second edition, PDAM, Part 5 PDTR, Part 7 FCD, Part 8
December 2001 (Pattaya)	Amendment 4 to second edition, PDAM status, Part 1 Amendment 5 to second edition, PDAM status, Part 1 COR 2, DCOR status, Part 1 (second edition) Amendment 3 to second edition, PDAM, Part 2 COR 1, DCOR status, Part 3 (second edition) Amendment 1 to second edition, FPDAM, Part 4 COR 1, DCOR status, Part 4 (second edition) Amendment 2 to second edition, FPDAM, Part 5 DTR, Part 7 FDIS, Part 8

*Note: For the first edition of each Part of the standard, Version N corresponds to Amendment N-1.

Looking at Table 1.2, it is possible to conclude that the development of Version 1 of the MPEG-4 standard took about four and a half years between issuing the first call for proposals and the publication of the IS (Version 1). Although MPEG is generally considered as a body adopting challenging workplans, notably by its members, it is worthwhile to reflect about the time it takes to develop a standard, and thus on its chances of success against proprietary solutions in such a quickly moving technical landscape, if at least four years is the time a fast standardization body needs to make a standard available to the industry. This is not to speak about the undefined additional time the companies owning the essential patents will take to set the licensing procedure so that industry can start selling products based on that standard.

MPEG-4 was developed by hundreds of experts from many companies and universities around the world who believe that MPEG-4 technology can power the next generation of multimedia products and services. MPEG-4 Version 1 reached FDIS status at the end of 1998 and thus was technically finished by that time. The following amendments increased the capabilities of the standard in a backward-compatible way and are being developed as the industry needs emerge.

1.6 MPEG-4 INDUSTRY FORUM

Because MPEG itself is not allowed to deal with any issues besides the development of technical specifications (notably patent identification and licensing), the industry players interested in the deployment of products and applications based on the MPEG-4 standard decided in the spring of 2000 to create the MPEG-4 Industry Forum (M4IF) [M4IF]. The M4IF is a not-for-profit organization for industrial players who want to manufacture, deploy, or use MPEG-4 technology. The philosophy is that all involved parties, even though competitors in some respect, profit from the standard being accepted and taking off. The major goal of the forum is "to further the adoption of the MPEG-4 standard, by establishing MPEG-4 as an accepted standard among users, application developers, service providers, content creators and end users" [M4IF].

The forum is open to all parties agreeing with the forum's objectives and includes a broad, worldwide representation from consumer electronics, computers, and telecommunications companies as well as research institutions. Also, some of the members are business users of MPEG-4. Among the participants are many small companies that develop or deploy MPEG-4 technology. Membership goes beyond the MPEG constituency, partly because some of the smaller companies cannot comply with the requirements for participation in MPEG (which vary from country to country) and also because some companies do not need to be involved in the development phase.

The activities of M4IF generally start where MPEG stops. This includes issues that MPEG cannot deal with because of ISO rules, such as clearance of patents. According to the M4IF statutes, the M4IF purposes shall be pursued by the following means:

- ☞ Promoting the standard and serving as a single point of information on MPEG-4 technology, products, and services.
- ☞ Initiating discussions leading to the potential establishment of patent pools outside M4IF, which should grant a license to an unlimited number of applicants throughout the world under reasonable terms and conditions that are demonstrably free of any unfair competition; includes studying licensing models for downloadable software decoders, such as Internet players.
- ☞ Organizing MPEG-4 exhibitions and tutorials.
- ☞ Creating industrial focus around the standard, for example, by identifying which MPEG-4 profiles are needed for which market.

M4IF sees the creation of patent pools[11] as one of the most important issues to enable the wide-scale adoption of the MPEG-4 standard. That is why the forum has facilitated a number of meetings of patent holders and intends to continue doing so until the pools are well under way toward being established. M4IF will not directly deal with patents, patent pools, and patent licensing. This is strictly a matter for patent owners to resolve and falls outside M4IF's scope. This means M4IF will not sell any MPEG-4 licenses or even determine the licensing policies. It merely acts as a catalyst in getting holders of essential patents to sit together and establish a portfolio of essential worldwide patents that are necessary for the implementation of the MPEG-4 standard in order to provide all MPEG-4 users with fair, reasonable, nondiscriminatory access to the technology under a single license. Also, M4IF discusses possible licensing principles, applicable to hardware and software products, to better understand the needs for licensing in emerging MPEG-4 application domains.

The goals are realized through the open international collaboration of all interested parties, on reasonable terms applied uniformly and openly. M4IF will contribute the results of its activities to appropriate formal standards bodies if applicable. The business of M4IF is not conducted for the financial profit of its members but for their mutual benefit. Any corporation and individual firm, partnership, governmental body, or international organization supporting the purpose of M4IF may apply for membership. Members are not bound to implement or use specific technology standards or recommendations by virtue of participation in M4IF.

M4IF anticipates holding three physical meetings per year, with a slightly higher frequency in the start-up phase. About 100 people have attended these meetings from all over the world. The instructions to join M4IF are available at the M4IF Web site at *www.m4if.org/join.html*.

1.7 SUMMARY

MPEG-1 and MPEG-2 are successful standards that have given rise to widely adopted commercial products, such as CD-interactive, digital audio broadcasting, and digital television. However, these standards are limited in terms of the functionalities provided by the data representation models used.

The MPEG-4 standard opens new frontiers in the way users will play with, create, reuse, access, and consume audiovisual content. The MPEG-4

---

11. A patent pool is typically a set of patents, belonging to a group of companies, for which a single licensing point exists, facilitating the licensing task of the companies willing to develop applications and products using those patents. The patents in a pool are all relevant to build integrated solutions to address the needs of certain application domains. For example, there is a patent pool for MPEG-2 Systems and MPEG-2 Video main profile@main level tools.

object-based representation approach—in which a scene is modeled as a composition of objects, both natural and synthetic, with which the user may interact—is at the heart of the MPEG-4 technology. Moreover MPEG-4 behaves rather well also in terms of compression. For example, for frame-based video coding, MPEG-4 brings a competing solution from very low bit rates to very high bit rates; in fact there are already MPEG-4 visual levels from 64 kbit/s for the `simple` profile at Level 0 (adopted by the Third Generation Project Partnership [3GPP] consortium for the third-generation mobile networks applications) up to 1,800 Mbit/s for the `simple studio` profile at Level 4.

The MPEG-4 vision and its associated technology provide the means to launch a great diversity of applications, with varying degrees of interactivity, notably for the emerging third-generation mobile networks, for the Internet, and even for digital radio and cable broadcasting networks. It is now up to the application developers and content authors to transform this great technology in content, products, and applications, making the MPEG-4 standard the audiovisual playground of the future.

Whatever will be the success of the MPEG-4 standard in terms of products and applications (likely determined by industrial, economic, legal, and marketing interests), the new concepts underpinning the MPEG-4 standard point in the right direction in terms of representation technology, as they rely on some basic characteristics of the human–world relationship, brought for the first time to the audiovisual representation arena.

1.8 REFERENCES

[Chia97] Chiariglione, Leonardo. "MPEG and Multimedia Communications." *IEEE Transactions on Circuits and Systems for Video Technology*, 7(1): 5–18, February 1997.

[H263] ITU-T Recommendation H.263++. "Video Coding for Low Bit Rate Communication." International Telecommunications Union—Telecommunications Standardization Sector (Geneva), 3, 2000.

[JTC1] JTC1 Home Page. *www.jtc1.org/*

[KoPC97] Koenen, Rob, Fernando Pereira, and Leonardo Chiariglione. "MPEG-4: Context and Objectives." *Signal Processing: Image Communication*, 9(4): 295–304, May 1997.

[M4IF] MPEG-4 Industry Forum Home Page. *www.m4if.org/index.html*

[M532] MPEG Test. *Report of the Ad Hoc Group on MPEG-4 Video Testing Logistics.* Doc. ISO/MPEG M532, Dallas MPEG Meeting, November 1995.

[Marr82] Marr, David. *Vision.* New York: W.H. Freeman, 1982.

[MPEG] MPEG Home Page. *http://mpeg.telecomitalialab.com/*

[MPEG4-1] ISO/IEC 14496-1:2001. *Coding of Audio-Visual Objects—Part 1: Systems*, 2d Edition, 2001.

[MPEG4-2] ISO/IEC 14496-2:2001. *Coding of Audio-Visual Objects—Part 2: Visual*, 2d Edition, 2001.

[MPEG4-3] ISO/IEC 14496-3:2001. *Coding of Audio-Visual Objects—Part 3: Audio*, 2d Edition, 2001.

[MPEG4-4] ISO/IEC 14496-4:2001. *Coding of Audio-Visual Objects—Part 4: Conformance Testing*, 2d Edition, 2001.

[MPEG4-5] ISO/IEC 14496-5:2001. *Coding of Audio-Visual Objects—Part 5: Reference Software*, 2d Edition, 2001.

[MPEG4-6] ISO/IEC 14496-6:2000. *Coding of Audio-Visual Objects—Part 6: Delivery Multimedia Integration Framework (DMIF)*, 2d Edition, 2000.

[N1172] MPEG Video. *MPEG-4 Video Verification Model 1.0*. Doc. ISO/MPEG N1172, Munich MPEG Meeting, January 1996.

[N1177] MPEG Convener. *MPEG-4 Project Description*. Doc. ISO/MPEG N1177, Munich MPEG Meeting, January 1996.

[N1214] MPEG Audio. *MPEG-4 Audio Verification Model 1.0*. Doc. ISO/MPEG N1214, Florence MPEG Meeting, March 1996.

[N1315] MPEG Convener. *Call for Proposals and PPD on SNHC Tools*. Doc. ISO/MPEG N1315, Tampere MPEG Meeting, July 1996.

[N1364] MPEG SNHC. *MPEG-4 SNHC Verification Model 1.0*. Doc. ISO/MPEG N1364, Chicago MPEG Meeting, September 1996.

[N1375] MPEG Convener. *Verification Model (VM) Development and Core Experiments*. Doc. ISO/MPEG N1375, Chicago MPEG Meeting, September 1996.

[N1397] MPEG Convener. *Call for Proposals on Synthetic Audio*. Doc. ISO/MPEG N1397, Chicago MPEG Meeting, September 1996.

[N1499] MPEG Convener. *MPEG-4 Call for Proposals*. Doc. ISO/MPEG N1499, Maceió MPEG Meeting, November 1996.

[N1714] MPEG Convener. *Call for Proposals on Identification and Protection of Content in MPEG-4*, Doc. ISO/MPEG N1714. Bristol MPEG Meeting, April 1997.

[N1919] MPEG Convener. *Call for Proposals for an MPEG-4 Intermedia Format*. Doc. ISO/MPEG N1919, Fribourg MPEG Meeting, October 1997.

[N271] MPEG Convener. *New Work Item Proposal (NP) for Very-Low Bitrates Audiovisual Coding*. Doc. ISO/MPEG N271, London MPEG Meeting, November 1992.

[N2724] MPEG Requirements. *MPEG-4 Applications*. Doc. ISO/MPEG N2724, Seoul MPEG Meeting, March 1999.

[N3157] MPEG Convener. *Call for Proposals for an MPEG-4 Textual Format*. Doc. ISO/MPEG N3157, Maui MPEG Meeting, December 1999.

[N3341] MPEG Convener. *Call for Proposals for a Generic Animation Framework of Synthetic Objects*, Doc. ISO/MPEG N3341. Noordwijkerhout MPEG Meeting, March 2000.

[N3543] MPEG Convener. *Call for Proposals for IPMP Solutions*. Doc. ISO/MPEG N3543, Beijing MPEG Meeting, July 2000.

[N3574] MPEG Convener. *Call for Proposals on Multi-Users Worlds Technology*. Doc. ISO/MPEG N3574, Beijing MPEG Meeting, July 2000.

[N3671] MPEG Convener. *Call for Proposals for New Tools to Further Improve Video Coding Efficiency*. Doc. ISO/MPEG N3671, La Baule MPEG Meeting, October 2000.

[N3908] MPEG Video. *MPEG-4 Video Verification Model,* Version 18. Doc. ISO/MPEG N3908, Pisa MPEG Meeting, January 2001.

[N3992] MPEG Convener. *Call for Proposals for New Tools for Audio Coding.* Doc. ISO/MPEG N3992, Singapore MPEG Meeting, March 2001.

[N4065] MPEG Convener. *Call for Proposals for New Tools for Video Compression Technology.* Doc. ISO/MPEG N4065, Singapore MPEG Meeting, March 2001.

[N4098] MPEG Convener. *Call for Proposals for Interpolator Compression.* Doc. ISO/MPEG N4098, Singapore MPEG Meeting, March 2001.

[N4218] MPEG Convener. *Call for the Submission of Hardware Reference Code for MPEG-4—Part 9: Reference Hardware Description.* Doc. ISO/MPEG N4218, Sydney MPEG Meeting, July 2001.

[N4319] MPEG Requirements. *MPEG-4 Requirements.* Doc. ISO/MPEG N4319, Sydney MPEG Meeting, July 2001.

[N4333] MPEG. *Multimedia Framework (MPEG-21)—Part 1: Vision, Technologies and Strategy.* Draft Technical Report, Doc. ISO/MPEG N4333, Sydney MPEG Meeting, July 2001.

[N4400] MPEG. *JVT Terms of Reference.* Doc. ISO/MPEG N4400, Pattaya MPEG Meeting, December 2001.

[N4427] MPEG. *Coding of Audio-Visual Objects—Part 8: Carriage of MPEG-4 Contents Over IP Networks.* Final Draft International Standard, Doc. ISO/MPEG N4427, Pattaya MPEG Meeting, December 2001.

[N4454] MPEG Test. *Results of Subjective Assessment of Responses to Video Call for New Tools to Further Improve Coding Efficiency.* Doc. ISO/MPEG N4454, Pattaya MPEG Meeting, December 2001.

[N4466] MPEG Video. *JVT Coding Joint Working Draft.* Doc. ISO/MPEG N4466, Pattaya MPEG Meeting, December 2001.

[N4500] MPEG Convener. *Terms of Reference.* Doc. ISO/MPEG N4500, Pattaya MPEG Meeting, December 2001.

[N4505] MPEG. *MPEG-4 Overview.* Doc. ISO/MPEG N4505, Pattaya MPEG Meeting, December 2001.

[N4508] MPEG Requirements. *Requirements for JVT.* Doc. ISO/MPEG N4508, Pattaya MPEG Meeting, December 2001.

[N4509] MPEG Requirements. *Overview of the MPEG-7 Standard.* Doc. ISO/MPEG N4509, Pattaya MPEG Meeting, December 2001.

[N4511] MPEG. *MPEG-21 Overview.* Doc. ISO/MPEG N4511, Pattaya MPEG Meeting, December 2001.

[N4554] MPEG. *Coding of Audio-Visual Objects—Part 7: Optimized Visual Reference Software.* Draft Technical Report, Doc. ISO/MPEG N4554, Pattaya MPEG Meeting, December 2001.

[N998] MPEG AOE. *Proposal Package Description (PPD): Revision 3.* Doc. ISO/MPEG N998, Tokyo MPEG Meeting, July 1995.

[N999] MPEG Test. *MPEG-4 Testing and Evaluation Procedures Document.* Doc. ISO/MPEG N999, Tokyo MPEG Meeting, July 1995.

[PeKo96] Pereira, Fernando, and Rob Koenen. "Very Low Bitrate Audiovisual Applications." *Signal Processing: Image Communication,* 9(1): 55–77. November 1996.

[PeKo99] Pereira, Fernando, and Rob Koenen. "MPEG-7: Status and Directions," (A. Puri and T. Chen, Eds.). *Multimedia Systems, Standards and Networks*. New York: Marcel Dekker, 1999.

[Pere99] Pereira, Fernando. "MPEG-4: Why, What, How and When?" *Signal Processing: Image Communication, Tutorial Issue on the MPEG-4 Standard*, 15(4–5): 271–279. December 1999.

MPEG-4 Overview

by Olivier Avaro, Rob Koenen, and Fernando Pereira

Keywords: MPEG-4 design goals, multimedia architecture, audiovisual object representation, MPEG-4 tools, multimedia standards, MPEG-4 applications

The MPEG-4 project started in 1992 with the goal of setting a standard for *Very Low Bit Rate Audiovisual Coding* [N271]. The initial participants were companies from the consumer electronics, computer, and telecom industries, as well as academia. There was the promise of great new coding methodologies that would greatly improve (video) compression efficiency over the existing hybrid DCT-based coding schemes. The participants discussed whether MPEG-4 should be centered on *Very Low Bit Rates* or on *Very High Compression Efficiency*. Unfortunately, the "Great Plan" was built on quicksand, which became apparent in July of 1994 at the Grimstadt MPEG meeting. At that point, MPEG altered the goals of MPEG-4 considerably. After reassessing developments in the multimedia arena, MPEG redefined the goals to align them with the anticipated evolution in the production, delivery, and consumption of audiovisual content, as summarized in Table 2.1. The *linear* approach that had dominated the multimedia industry at all stages of the content lifecycle was replaced by a *nonlinear, object-based, interactive* way of dealing with audiovisual data. As a consequence, to reflect the focus of the work more accurately, the name of the MPEG-4 project was changed to *Coding of Audio-Visual Objects* [N998]. The scope of the MPEG-4 standard is defined in more detail in the MPEG Requirements document [N4319]. At the time of writing (February 2002), the MPEG-4 standard was

Table 2.1 Evolution of multimedia content production, delivery, and consumption

	Traditional	**MPEG-4**
Production	• Mostly 2D content • Natural content, produced with cameras and microphones • Composition (mixing) of objects at production stage	• 2D and 3D content • Content with hybrid natural and computer-generated audio and video • Rectangular and preseg-mented video content • Explicit coding of composition of separate objects
Delivery	• Few networks carry audio-visual information (satellite, LAN, ISDN) • Many networks have their own content representation scheme • Communications and delivery over homogeneous networks	• Virtually all networks carry audiovisual information • Common content representa-tion scheme • Communications and delivery over heterogeneous networks (multiple network types)
Consumption	• Mostly passive content • Content precomposed	• Composition of objects at con-sumption stage • Information is read, seen, and heard in an interactive and personalized way • More and more information is audiovisual and digital

largely finalized, although some additional tools for advanced functionality were still under development. A complete overview of the MPEG-4 technology is provided in the MPEG-4 Overview [N4505], which is updated at every MPEG meeting—about once every quarter.

2.1 DESIGN GOALS

The main MPEG-4 design goals, as formulated in 1994, can be summarized as follows:

1. To provide the technological foundations to represent multimedia content for a wide range of multimedia services and networks. To provide a common representation basis encompassing the interactive, broadcast, and conversational paradigms. To allow for seamless content usage across these different types of services.

2. To extend the interactivity found in text and images to content having a temporal dimension. To provide for this type of content client-side inter-activity (i.e., without the need for a return channel), as well as server-side interactivity, exploiting an on-line connection.

3. To extend multimedia content usage from high-bandwidth networks to all types of networks, including those with low bandwidth and high error rates. MPEG calls this *Universal Access*.

4. To integrate different media types into a single framework, giving each its own optimized coded representation.

5. To offer rich interactivity to content authors and content users.

6. To allow for consumption of the audiovisual content that respects the usage rights that are attached to it.

For all its complexity and power, the MPEG-4 standard is based on a very simple concept: an audiovisual scene is composed of audiovisual objects whose properties are defined using a scene description language. This concept allows

☞ Interaction with individual elements in the audiovisual content, named *audiovisual objects*

☞ Adaptation of the coding scheme on a per-object basis

☞ Easy reuse and customization of the audiovisual content

Audiovisual objects can differ in nature. They can be audio-only objects like a multichannel audio content or speech, or video-only objects, like traditional rectangular movies or arbitrarily shaped video objects. Objects can be natural, like audiovisual data captured from a microphone or from a camera, or synthetic, like text and graphics overlays, animated faces, or synthetic music. They can be 2D, like a Web page, or 3D, such as spatialized sound and 3D virtual worlds.

The scene description provides the spatial and temporal relationships between the audiovisual objects. These relationships can be purely 2D or 3D, or a mixture of 2D and 3D (for instance, a 2D "subscene" mapped in a 3D virtual world). The scene description includes commands specifying the behavior and interactive options embodied in the audiovisual objects and scenes. In addition, MPEG-4 specifies protocols to modify and animate the scene over time (for instance, adding or deleting objects, or changing objects' properties and behavior). The scene description is dynamic, which allows incremental build-up of the scene (as opposed to waiting for a download and then playing).

All information representing the audiovisual objects and the audiovisual scene description are provided as compressed binary streams, and all of these streams can be synchronized.

A possible audiovisual scene including natural and synthetic objects is shown in Figure 2.1. This figure shows how various visual and audio streams are composed on the top of a fixed still picture background (an object in its own right), according to a scene description stream. The diagram introduces the concept of *object descriptor* (OD). An object descriptor is a structure that lays the links between the scene description and the individual streams with media content. More details on these object descriptors may be found in Chapter 3.

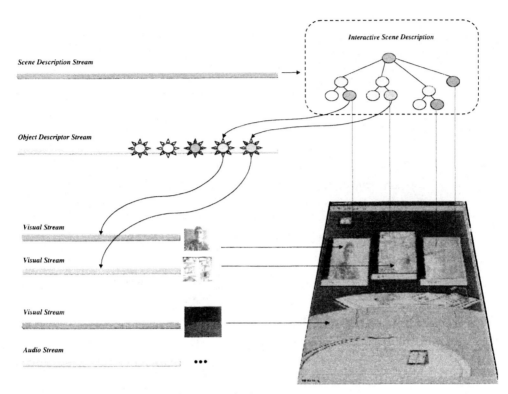

Interactive Scene Description

Scene Description Stream

Object Descriptor Stream

Visual Stream

Visual Stream

Visual Stream

Audio Stream

Fig. 2.1 Interactive audiovisual scene with objects.

2.2 AN END-TO-END WALKTHROUGH

This section explains the high-level workings of the MPEG-4 standard using an end-to-end walkthrough of an MPEG-4 session. Although the MPEG-4 standard is designed to be used in many environments, the way MPEG-4 content is produced, delivered, and consumed often follows the same path. Such a typical path is shown in Figure 2.2, which highlights the main interoperability elements of the MPEG-4 standard as well as their respective position in the chain. While this walkthrough provides a typical scenario, MPEG-4 also can be used in other scenarios that are not covered by the walkthrough in Figure 2.2.

At the start of the walkthrough are content authors, who produce audiovisual content with the authoring tools of their preference. Part of the content creation process may be live; needless to say, though, off-line creation is also important in many MPEG-4-based applications. The content creation process can be separated into two major steps: authoring and publishing. Authoring involves the production of the audiovisual data, including the media, the scene description, and the interaction. Publishing involves adapting the content to the constraints imposed—by networks and terminals, for example.

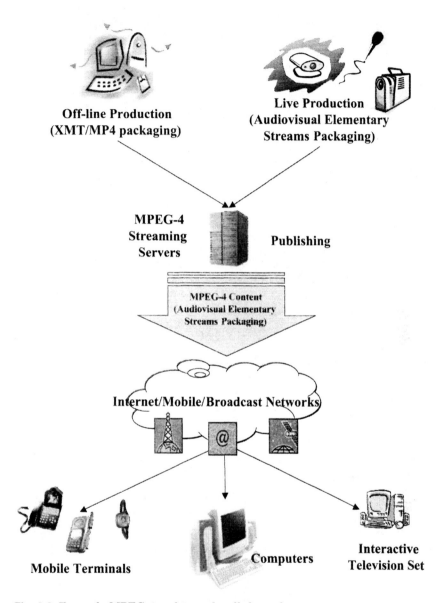

Fig. 2.2 Example MPEG-4 end-to-end walkthrough.

While streaming content is produced using MPEG-4's binary streaming formats, off-line multimedia content can be delivered to servers in an interoperable way using the MPEG-4 eXtensible MPEG-4 Textual format (XMT), as described in Chapter 6, or using the binary MP4 file format, as described in Chapter 7. The choice between these two depends on the freedom that authors want to offer users (or other authors) at the next stage of the delivery chain.

XMT provides a structured and extensible representation of the scene description, allowing much flexibility for further manipulation or adaptation of this data (e.g., for automated processing like customization). It may contain additional information such as metadata, for instance, to express constraints on the adaptation to the delivery network. This makes it an appropriate format for exchanging content that needs to be further processed; XMT is therefore useful for the exchange of content between authors.

MP4 is more rigid than XMT in the sense that the scene description data is binary coded and hence hard to alter afterwards. The binary form is obviously more compact but also more deterministic with regard to what the users will see and hear, which is useful to ensure that an author's creations are presented as intended.

While MP4 and XMT files are the natural interoperability formats between authoring tools and MPEG-4 servers, this does not mean that the content will be stored as MP4 or XMT files on the server. Server implementations may have other ways to represent the content, optimized for specific software and hardware. What matters is that the served streams be compliant with the standard. But MPEG-4 servers can also directly use MP4 or XMT files to serve content on various networks. If XMT is used, the server needs to perform more than streaming; it will also need to encode the scene description data into binary form.

What goes out of the server are streams of data containing MPEG-4 content named *elementary streams*. The content of these elementary streams is discussed later in this chapter. What is important at this stage is that MPEG-4 audiovisual scenes can be split in several elementary streams, that these streams can be carried on different networks and that end terminals receiving these streams from different networks are capable of reconstructing the transmitted data in a synchronized manner. When content is produced live, it can be delivered to servers as interoperable elementary streams.

One of MPEG-4's design goals is to cover a wide range of access conditions, so that content can be created once and played on any network. This is achieved by abstracting the content delivery layer with an interface as specified in the Delivery Multimedia Integration Framework (DMIF) corresponding to MPEG-4 Part 6 [MPEG4-6]. This interface is named the DMIF Application Interface (DAI).

When streaming content, the MPEG-4 interoperability points are the specific formats of (and constraints on) the individual elementary streams, and the compliance to the walkthrough (i.e., the series of steps that has to be performed to access the elementary streams) defined by the DAI. In principle, what happens below the DAI is outside the scope of the MPEG-4 standard. In this way, the multimedia representation can be kept completely independent from the way it is delivered. There are exceptions to this rule: an efficient low-complexity multiplexing tool (FlexMux) and a dedicated file format (MP4) were developed because MPEG-4 needed specific tools for its transport (see Chapter 7).

In order to transport the MPEG-4 content in existing environments, network-specific transport mechanisms have been defined. Transport in MPEG encompasses both wire formats and file formats. Currently, these transport mechanisms are available:

☞ Storage of MPEG-4 content in MP4 files [MPEG4-1]

☞ Carriage of MPEG-4 content on the Internet [MPEG4-8, GENT01]

☞ Carriage of MPEG-4 content in MPEG-2 transport streams [MPEG2-1]

These mechanisms are described in detail in Chapter 7. The spectrum of MPEG-4 consumption devices includes desktop computers, mobile devices, and interactive television sets. This last device nicely illustrates the various ways MPEG-4 content can be consumed.

Assume that the interactive TV set in Figure 2.2 receives through a satellite connection an MPEG-2 transport stream containing several MPEG-2 digital TV programs. Using the MPEG-2 extension that specifies the carriage of MPEG-4 on MPEG-2 transport streams [MPEG2-1], MPEG-4 content related to such a TV program can be sent along with the linear programming. This can provide the user with an enhanced interactive experience, such as a local interaction with a 3D model of a car in an ad or a multimedia electronic program guide. Note that all the interactivity is sent in the broadcast stream and a return path is not required.

Assume now that the TV set is also connected to the Internet. It is then possible to augment the broadcast experience with client-server functionality as well as to send rich media that can be blended with the basic TV program. One can imagine a range of services encompassing program enhancements with video clips streamed from the Internet on-demand up to multiuser games related to the TV programs with votes, 3D chats, and interaction with the scenario of the broadcast content.

2.3 TERMINAL ARCHITECTURE

The overall architecture of an MPEG-4 terminal is depicted in Figure 2.3. At the bottom of the figure is the storage or transmission medium; this refers to the lower layers of the delivery infrastructure. MPEG-4 data can be transported on a variety of delivery systems as described in Section 2.2. This includes MPEG-2 transport streams, RTP/UDP over IP, and MPEG-4 (MP4) files. MPEG-4 data can also be sent over other transport mechanisms, such as Asynchronous Transfer Mode (ATM) or Digital Audio Broadcast (DAB), using one or a combination of the previous stacks of protocols.

Most of the currently available transport layer systems provide native means for multiplexing information. There are, however, a few exceptions, such as RTP Internet connections. In addition, the existing multiplexing mechanisms do not always fit MPEG-4 needs in terms of low delay, or they

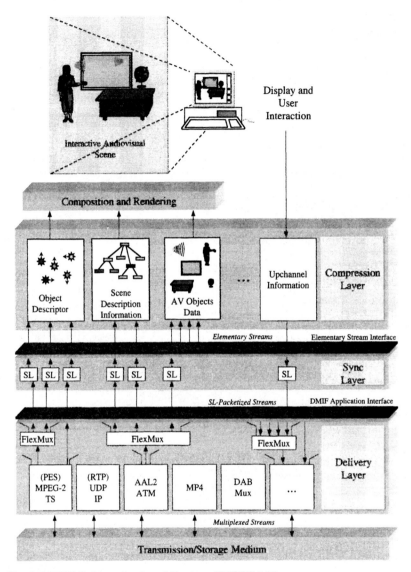

Fig. 2.3 MPEG-4 terminal architecture [MPEG4-1].

may incur prohibitive overhead in handling the expected large number of streams associated with an MPEG-4 session. As a result, MPEG-4 has defined a multiplexing tool, FlexMux, that can be used on top of the existing transport delivery layer (see Chapter 7).

The delivery layer provides a number of elementary streams to the MPEG-4 terminal. These streams can be audiovisual objects data, scene description data, control data in the form of object descriptors, or metadata describing the content or associating intellectual property rights to it.

Note that not all the streams have to be downstream (server to client); upstream channels can also exist, for the purpose of conveying data from the terminal to the transmitter or server. MPEG-4 standardizes the mechanisms that trigger the transmission of such upstream data at the terminal, as well as the formats in which they are transmitted back to the sender.

It is important that there be a common mechanism for conveying timing and framing information that applies to all types of elementary streams. The basic MPEG-4 timing model and the Sync Layer (SL) provide this functionality (see Chapter 3). The SL is a flexible and configurable packetization tool that allows the inclusion of timing, fragmentation, and continuity information in the associated data packets. Such information is attached to data units called *access units*, which comprise complete presentation units (e.g., an entire video frame or an audio frame).

Elementary streams are sent to their respective decoders, which process the data and produce composition units (e.g., a decoded video frame or, in MPEG-4 terminology, a Video Object Plane [VOP]). In order for the receiver to know what type of information is contained in each stream, control information in the form of ODs is used. The ODs describe the properties of elementary streams and are used to associate one or more elementary streams to each of the audio and visual objects in the scene. Through these ODs, the terminal can identify the content being delivered to it. Without an OD, the terminal wouldn't know how to use the elementary stream. ODs may also contain—or point to—Object Content Information (OCI). OCI elements can contain data such as content classification, rating, language, and keywords. (Because it was ready to use first, OCI forms sort of a precursor to MPEG-7.) MPEG-4 scenes can also contain MPEG-7 streams through a dedicated elementary stream type. Detailed descriptions of synchronization in MPEG-4 and of the OD framework are given in Chapter 3.

Advanced synchronization mechanisms (described in Chapter 4) augment the basic MPEG-4 timing model to permit synchronization of multiple streams and objects that can originate from a number of sources. The Flextime toolkit allows the definition of simple temporal relationships among MPEG-4 objects, such as CoStart, CoEnd, and Meet, as well as the specification of constraints for the timing relationship between MPEG-4 objects, as if the objects were on springs. The duration of an object is represented by the length of the spring, and these springs can be expanded or contracted as needed, to adjust the duration of each of the objects (adjusting the playback speed). Flextime is modeled after the W3C's Synchronized Multimedia Integration Language (SMIL) [SMIL01], and one of the reasons to include it in MPEG-4 was to provide compatibility with SMIL content.

In the context of an MPEG-4 audiovisual scene, at least one of the streams must contain the scene description information associated with the content. The scene description information defines the spatial and temporal position of the various objects, their dynamic behavior, and any interactivity features embedded in the content. As mentioned above, the (coded) data that represents an audiovisual object is carried in separate elementary streams. The scene description contains *pointers* to ODs when it refers to a particular audiovisual object. The MPEG-4 scene description language, called *binary format for scenes* (BIFS) [MPEG4-1], is further explained in Chapter 4. It complements the OD framework that defines the objects and their characteristics.

A key feature of the scene description is that, since it is carried in its own elementary stream(s), it can contain full timing information. This implies that the scene can be dynamically updated over time, a feature that provides considerable power for content creators. There is also a special lightweight mechanism to animate parts of the scene: *BIFS Anim*. This animation is accomplished by coding, in a separate stream, only the parameters that need to be updated.

The system's compositor uses the scene description information to aggregate the various natural and synthetic audiovisual objects and render the final scene for presentation to the user. MPEG-4 natural video objects (see Chapter 8) include rectangular and shaped, scalable and error-resilient video as well as still pictures [MPEG4-2]. Synthetic visual objects can be 2D and 3D meshes, as well as animated 3D faces and bodies as described in Chapter 9 and specified in both MPEG-4 Visual [MPEG4-2] and MPEG-4 Systems [MPEG4-1].

The palette of the MPEG-4 Audio tools includes speech coding from 2 to 24 kbit/s (see Chapter 10), general audio coding from 4 to 64 kbit/s/channel and above (see Chapter 11), as well as synthetic audio such as text-to-speech synthesis and synthetic music coding (see Chapter 12). All these tools are specified in MPEG-4 Audio [MPEG4-3], and the interfaces exposed at the scene description level are specified in MPEG-4 Systems [MPEG4-1].

The scene description tools provide mechanisms to capture user or system events. In particular, these mechanisms allow the association of events to user operations on desired objects and they can—in turn—modify the behavior of the object. Event processing is the core mechanism with which application functionality and differentiation/customization can be provided. In order to provide flexibility in this respect, MPEG-4 allows the use of ECMAScript (also known as JavaScript) scripting within the scene description. Use of scripting tools is essential in order to access state information and implement sophisticated interactive applications.

MPEG-4 also defines a set of Java language Application Programming Interfaces (APIs), generally called MPEG-J, through which access to an underlying MPEG-4 player can be provided to Java applets (called *MPEGlets*). These MPEGlets can form the basis for very sophisticated applications, opening up rich possibilities for audiovisual content creators to author attractive content. Chapter 5 provides a complete description of the MPEG-J application engine.

In addition to the new interactive features for end users, the object-based content structure provides great benefits to content creators as well. The use of an object-based structure, where composition happens at the receiver, considerably simplifies the content creation process. Starting from a set of audio-visual objects, it is easy to define a scene description that combines these objects in a meaningful presentation. An essentially similar approach is used in HTML and Web browsers, which makes it easy for nonexpert users to create their own content. The fact that the content's structure survives the process of coding and distribution also allows for the reuse of the objects in the scene. For example, content filtering and searching applications can be easily implemented using ancillary OCI carried in object descriptors. Finally, when allowed by the content author, consumers with the right tools can easily extract and reuse individual objects.

2.4 MPEG-4 TOOLS

This section provides a brief overview of the tools available in the MPEG-4 standard. A more detailed description can be found in the MPEG-4 Overview [N4505]. Each of these tools is explained in detail in one of the remaining chapters of this book.

The tools are organized in one of four categories—Systems, Visual, Audio, and DMIF—corresponding to the four technology specification Parts of the MPEG-4 standard. The tools listed cover the state of the standard as of February 2002. Since that date, more tools may have been standardized in MPEG-4; for up-to-date information, consult the MPEG Web site, *mpeg.telecomitalialab.com* [MPEG].

2.4.1 Systems Tools

The major tools standardized in MPEG-4 Systems [MPEG4-1] are

☞ **Object descriptor framework:** The OD framework defines the relationship between the individual elementary streams and the media objects in the scene (e.g., the audio or video streams associated with an object representing a participant in a videoconference). ODs provide information such as the elementary streams available to represent a given media object, the characteristics of the decoders needed to consume the elementary streams, and the location (possibly a URL) of the elementary streams data. ODs include some *Intellectual Property Management and Protection* (IPMP) information in the form of data sets covering the identification of intellectual property rights associated with the media objects. See Chapter 3 for further detail.

☞ **Systems decoder model:** The systems decoder model provides the basic synchronization and streaming features of the MPEG-4 standard. The model defines the initialization and continuous management of the receiving terminal's buffers. It also defines the timing-identification, synchronization, and recovery mechanisms. For further details see Chapter 3.

☞ **Binary format for scenes:** BIFS describes the spatio-temporal arrangements of the MPEG-4 objects in the scene. The scene description provides a rich set of functionality for 2D and 3D composition as well as text and graphics primitives. It also provides forms of interactivity (e.g., general event handling and routing between objects in the scene) among events triggered by the user or the scene, as well as client- and server-based interaction. See Chapter 4 for details.

☞ **MPEG-J:** MPEG-J allows the use of Java programs to complement the logic and programmatic part of MPEG-4 content. These programs are called MPEGlets when they are transmitted along with the multimedia content. MPEG-J defines interfaces to various aspects of the terminal and networks in the form of Java APIs. MPEG-J also defines a delivery mechanism allowing MPEGlets and other Java classes and objects to be streamed separately. Chapter 5 provides greater detail.

☞ **Extensible MPEG-4 textual format:** XMT is a framework for representing MPEG-4 scene descriptions using a textual syntax. The XMT format allows the content authors to exchange their content with other authors or with tools or service providers, and it facilitates interoperability with both the X3D format being developed by the Web3D consortium [X3D01] and the Synchronized Multimedia Integration Language (SMIL) format from the W3C consortium [SMIL01]. See Chapter 6 for more detail.

☞ **Transport tools:** Although transport of multimedia content is not in the core activity of the standard, MPEG-4 defines two transport tools, MP4 and FlexMux, for which there was no clearly identified standardization body where they could have been standardized. MP4 is a tool for storing MPEG-4 data in a file (the MPEG-4 file format). FlexMux is a tool for the interleaving of multiple streams into a single stream, including timing information. See Chapter 7 for a discussion of these tools.

☞ **IPMP "hooks":** APIs that provide an interface to IPMP systems and a way to identify which IPMP system was used. The IPMP system itself is not specified by MPEG. At the time of writing (February 2002), MPEG-4 was developing an extension to these MPEG-4 hooks that will provide more digital rights management capabilities as well as more flexibility for the IPMP system [N4270]. The goal of this work is to allow end users to consume protected content in a transparent manner and to provide renewability of IPMP systems (see Chapter 3).

2.4.2 Visual Tools

The major tools standardized in MPEG-4 Visual [MPEG4-2] are introduced here.

☞ **Video compression tools:** Algorithms providing the efficient compression of video for bit rates between 5 kbit/s and 1 Gbit/s and for resolutions from sub-QCIF to studio editing resolutions (4k×4k pixels).

Progressive and interlaced video are supported. Compact coding of textures with a quality adjustable from *acceptable* for very high compression ratios up to *near lossless* is supported, notably targeting texture mapping on 2D and 3D meshes. Random access of video provides functionalities such as pause, fast forward, and fast reverse of stored video. Different types of scalability are provided, including complexity scalability at the encoder and decoder, object scalability, spatial scalability, temporal scalability, and quality scalability. See Chapter 8 for more detail.

☞ **Robustness in error-prone environments:** Error resilience is supported to assist the access of images and video over a wide range of storage and transmission media. This includes the useful operation of image and video compression algorithms in error-prone environments at low bit rates (e.g., less than 64 kbit/s). There are coding tools that address both the bandwidth-limited nature and error-resilience aspects of wireless networks. See Chapter 8 for more detail.

☞ **Fine grain scalability (FGS):** FGS allows small quality steps by adding or deleting layers of extra information with fine granularity (quality scalability). It is useful in a number of environments, notably for streaming purposes, but also for dynamic (*statistical*) multiplexing of pre-encoded content in broadcast environments. See Chapter 8 for more detail.

☞ **Shape and alpha channel coding:** Efficient techniques are provided for coding of *binary* and *gray-scale* or *alpha* shape information. A binary alpha map defines whether a pixel belongs to an object; in a binary alpha map, a pixel can be *on* or *off*. A gray-scale map offers the possibility of defining a certain transparency for each pixel. See Chapter 8 for more detail.

☞ **Face and body animation:** The face and body animation tools specify parameters for defining, calibrating, and animating synthetic faces and bodies. The 3D models themselves are not standardized by MPEG-4; only the calibration and animation parameters are. See Chapter 9 for more detail.

☞ **Coding of 2D meshes:** MPEG-4 has tools for mesh-based coding and animated texture transfiguration. These tools comprise 2D Delaunay and regular mesh formalisms, with motion tracking of animated objects, motion prediction, and suspended texture transmission with dynamic meshes. There is also geometry compression for motion vectors (2D mesh compression with implicit structure and decoder reconstruction). See Chapter 9 for more detail.

☞ **Coding of 3D meshes:** MPEG-4 includes tools for coding 3D polygonal meshes. Polygonal meshes are widely used as a general representation of 3D objects. The underlying technologies compress the connectivity, geometry, and properties such as shading normals, colors, and texture coordinates of 3D polygonal meshes. See Chapter 9 for more detail.

Natural video coding tools are presented in further detail in Chapter 8, and synthetic coding tools are described in Chapter 9.

2.4.3 Audio Tools

This section lists MPEG-4's tool set for audio coding. Speech and general audio coding tools are explained in more detail in Chapters 10 and 11, respectively; tools for coding of synthetic audio and speech are described in Chapter 12. The major tools standardized in MPEG-4 Audio [MPEG4-3] are

☞ **Speech coding:** MPEG-4 Audio supports speech coding at bit rates from 2 kbit/s up to 24 kbit/s, using two algorithms: a Code Excited Linear Prediction (CELP) coder and a parametric coder. With variable bit-rate coding, even lower bit rates can be achieved—down to an average of 1.2 kbit/s. Low delay is possible for communications applications. The parametric Harmonic Vector Excitation Coding (HVXC) coder allows the user to modify speed and pitch during playback. If the CELP coder is used, a change of the playback speed can be achieved by using an additional tool for effects processing. See Chapter 10 for more detail.

☞ **General audio coding:** Most of the MPEG-4 general audio coding tools are based on the filterbank coding approach as used in MPEG-2 Advanced Audio Coding (AAC) [MPEG2-7]. Building on this technology, MPEG-4 added new tools with the target of improving the compression performance and supporting new functionalities. For example, the Transform-domain Weighted Interleave Vector Quantization (TwinVQ) coder supports operation at very low bit rates (e.g., 6 kbit/s/channel). MPEG-4 is the first audio coding standard to support scalability modes (e.g., fine grain scalability with a granularity down to 1 kbit/s/ch). A special low delay coding mode is available for communication applications among others. MPEG-4 also specifies a parametric audio coder, Harmonic and Individual Lines plus Noise (HILN), that makes a decomposition of the input signal into distinct sound components, namely a harmonic tone, individual sinusoids, and a noise component. This coder can operate at bit rates down to about 4 kbit/s, provides bit-rate scalability, and can be combined with the MPEG-4 HVXC parametric speech coder. See Chapter 11 for more detail.

☞ **Synthesized audio coding:** Structured audio coding tools provide a powerful framework in which score-based control information can be used to control "musical instruments" described with a special language. If limited resources are available, synthetic sound can be generated using a standardized wavetable format. The popular MIDI format is also supported. See Chapter 12 for more detail.

☞ **Synthesized speech coding:** Text-to-Speech (TTS) tools allow a text, possibly with prosodic parameters (pitch contour, phoneme duration, and so on), to be used to generate intelligible synthetic speech. Scalable TTS coders' bit rates range from 200 bit/s to 1.2 kbit/s. This tool also enables the prosody of the original speech to be used; lip synchronization control with phoneme information; trick mode functionality; international language and dialect support for the text; international symbol support for phonemes; support for specifying age, gender, and speech rate of the speaker; as well as support for conveying Facial Animation Parameter (FAP) bookmarks.

2.4.4 DMIF Tools

The major tools standardized in DMIF, Part 6 of the MPEG-4 standard [MPEG4-6] are

☞ **DMIF-Application Interface (DAI):** The DAI is a transparent application interface allowing access to multimedia content, irrespective of whether the peer is a remote interactive peer, a broadcast, or a local storage entity. The communication model underlying these APIs allows for continuous Quality of Service (QoS) monitoring, specific QoS queries, and notification of QoS violations. It also enables the simultaneous access, presentation, and synchronization of MPEG-4 content carried through different delivery technologies.

☞ **DMIF Signaling Protocol:** The DMIF Signaling Protocol is a generic session-level protocol designed for streaming multimedia data. The protocol is used to configure the delivery protocol stacks at the peer ends. The protocol stack may also include the FlexMux tool specified in MPEG-4 Systems [MPEG4-1]. The protocol was conceived with an eye to the future of networking technologies, and it enables features that are not readily available with current techniques. Such features include QoS and resource management and operation over a variety of networks, including heterogeneous networks.

See Chapter 7 for further details on DMIF tools.

2.4.5 Other MPEG-4 Tools

Strictly speaking, the following tools are not part of the MPEG-4 standard, but they allow the transport of MPEG-4 content on various delivery networks. See Chapter 7 for details.

☞ **Carriage of MPEG-4 on MPEG-2 Systems:** This extension to the Systems Part of the MPEG-2 standard [MPEG2-1] specifies how MPEG-2 Systems streams may carry individual MPEG-4 audio and visual elementary streams as well as audiovisual scenes with their associated streams. A distinction is made between MPEG-4 individual elementary streams (only MPEG-2 Systems tools are used) and an MPEG-4 audiovisual scene with its associated streams (MPEG-4 Systems stream management tools are used along with MPEG-2 Systems tools).

☞ **Carriage of MPEG-4 on IP networks:** This specification of the Internet Engineering Task Force (IETF) describes a payload format for transporting MPEG-4 encoded data using the Real-Time Transfer Protocol (RTTP) [GENT01]. This specification is complemented by Part 8 of the MPEG-4 standard, "Carriage of ISO/IEC 14496 contents over IP networks" [MPEG4-8], which provides a framework for the carriage of MPEG-4 content over IP networks as well as guidelines for designing payload format specifications for the detailed mapping of MPEG-4 content into several IP-based protocols.

2.4.6 Profiles and Levels

MPEG-4 would be too unwieldy for most applications if any compliant decoder had to be able to support all of the tools listed above. Therefore, MPEG has defined profiles and levels for the MPEG-4 standard, much like it did for its predecessors MPEG-1 Audio (which has three *layers* of audio coding) and MPEG-2 Audio and Video.

Profiles define subsets of tools that can be used for a large class of applications and services. Profiles are defined for specific tool dimensions (e.g., visual or audio) and not for combinations of these dimensions. This means that application developers can choose the tools they need by picking, for each tool dimension, the profile that best suits the application. Profiles exist for these dimensions:

☞ Visual

☞ Audio

☞ Graphics

☞ Scene graph

☞ Object descriptors

☞ MPEG-J

Levels further limit the complexity of the tools within a given profile (see Appendixes A–D). Conformance points of the MPEG-4 standards are defined as a given profile at a given level. Conformance to the standard can be tested at these conformance points for the various profiling dimensions. Conformant terminals for a given profile and level will interoperate. MPEG-4 profiles and levels are further described in Chapter 13.

2.5 MPEG-4 AND OTHER MULTIMEDIA STANDARDS

The multimedia standards and solutions in place before the MPEG-4 standard was developed influenced the development of MPEG-4. Among the standards and solutions that MPEG-4 has used and referenced, the following are especially relevant:

1. **ISO/IEC MPEG-1 Video** [MPEG1-2] and **Audio** [MPEG1-3], **ISO/IEC MPEG-2 Video** [MPEG2-2] and **Audio** [MPEG2-3], and **ITU-T H.263** [H263]: MPEG-4 builds on these standards to develop the MPEG-4 data formats for natural audio and video objects. A simple version of MPEG-4 video [MPEG4-2] is compatible with H.263 baseline, the video coding format for H.323 terminals. At the MPEG-4 Systems level, MPEG-4 scenes can include MPEG-1 and MPEG-2 audio and video streams.

2. **ISO/IEC MPEG-1 Systems** [MPEG1-1], **ISO/IEC MPEG-2 Systems** [MPEG2-1] and **ITU-T H.223** [H223]: MPEG-4 borrows heavily from the MPEG-1 and MPEG-2 Systems concepts for synchronization and buffer management, adding fundamentally new tools on top of these robust mechanisms. Finally, the MPEG-2 Systems standard [MPEG2-1] supports the transport of MPEG-4 content using MPEG-2 transport mechanisms. The FlexMux tool borrows concepts embodied in recommendation H.223.

3. **Web3D Virtual Reality Modeling Language (VRML'97)** [VRML97]: MPEG-4 bases its scene description solution on the approach specified by VRML'97. MPEG-4 BIFS provides such additional functionality as the integration of streams in the scene, 2D scene graph capabilities, integration of 2D and 3D scene description constructs, advanced audio features such as environmental spatialization, an integrated timing model between the scene graph and the streams, scene update and animation protocols to modify the multimedia description in time, and, in the best of MPEG tradition, compression efficiency for the scene description.

4. **Apple's QuickTime** [QUICKTIME]: Among the several file formats for storing, streaming, and authoring multimedia content that were proposed after MPEG called for file format proposals in October of 1997, QuickTime was selected as the basis for the development of the MPEG-4 file format, known as *MP4*. (None of the proposed formats fulfilled all the identified requirements.)

During the development of the MPEG-4 standard, other standardization bodies and forums have developed technology that served as background for the MPEG-4 standard. The most important ones are

1. **Sun's Java** [JAVA96a, JAVA96b]: MPEG-4 offers a Java-based programmatic environment, MPEG-J (see Chapter 5). MPEG-J serves to extend the content creator's capability to incorporate complex controls and data processing mechanisms along with the BIFS scene representations and elementary media data. At the presentation end, the MPEG-J environment intends to enhance the end user's ability to interact with the content. MPEG-J is based on Java applications (MPEGlets) delivered along with the multimedia content in dedicated elementary streams.

2. **W3C XML** [EXML01]: MPEG-4 offers an eXtensible Markup Language (XML)-based representation of the scene description, named XMT (eXtensible MPEG-4 Textual format). XMT comes in two flavors: a low-

level representation that exactly mirrors the BIFS representation (XMT-A) and a high-level representation (XMT-Ω) that is closer to the author's intent and can be mapped on the low-level XMT format. XMT-A provides interoperability between VRML/X3D [X3D01] and MPEG-4, and XMT-Ω provides interoperability between SMIL [SMIL01] and MPEG-4.

In parallel to the development of the MPEG-4 standard, other standardization bodies and industry consortia have developed tools and applications that address some of the MPEG-4 objectives. There are also some proprietary formats that do the same. When *open* alternatives exist (technical issues aside) the mere fact of being *closed* is a significant disadvantage of proprietary solutions for the content industry. The MPEG-4 standard enables different companies to separately develop authoring tools, servers, or players, opening up the market to independent product offerings. This competition should spur a fast proliferation of content and tools that can interoperate.

In addition, none of the alternatives to MPEG-4 offers a solution at the level of sophistication provided by MPEG-4: composition capabilities, interactive features, and range of codecs. MPEG-4 is a real-time solution at its very heart. All of the elements, from scene description to individual audio and data streams, can be tightly synchronized—a unique feature of the standard.

There is a strong demand for interoperability: interoperability between products from different providers, interoperability between different application spaces, and interoperability between industries. This demand, which will only grow, is a strong argument in favor of MPEG-4.

Among the several technologies that could be seen as alternatives to parts of MPEG-4, the most relevant are

1. **W3C Synchronized Multimedia Integration Language (SMIL)** [SMIL98, SMIL01]: SMIL is an XML-based language that allows authors to write 2D interactive multimedia presentations. Using SMIL 2.0, an author can describe the temporal behavior of a multimedia presentation, associate hyperlinks with media objects, and describe the layout of a presentation on a screen. The SMIL syntax and semantics can be used with other XML-based languages, in particular those that need to represent timing and synchronization. For example, SMIL 2.0 components are used for integrating timing into the W3C Scalable Vector Graphics format [SVG01].

2. **W3C Scalable Vector Graphics (SVG) format** [SVG01]: SVG is a language for describing 2D graphics in XML. SVG allows for three types of graphic objects: vector graphic shapes (e.g., paths consisting of straight lines and curves), images, and text. Graphical objects can be grouped, styled, transformed, and composited into previously rendered objects. The feature set includes nested transformations, clipping paths, alpha masks, filter effects, and template objects.

3. **ETSI Digital Video Broadcasting–Multimedia Home Platform (DVB-MHP)** [MHP01]: The MHP specification issued by the DVB industry

consortium defines a generic interface between interactive digital applications and the terminals on which those applications are executed. This interface decouples different provider's applications from the specific hardware and software details of different MHP terminal implementations.

As shown in Figure 2.4, the XMT format facilitates interoperability with the X3D, SMIL, and SVG specifications. One of the XMT design goals has been to maximize the overlap with SMIL, SVG, and X3D to enable content authors to compile in MPEG-4 the content they have already produced in these formats, given explicit authoring constraints.

Not all SMIL and SVG tools are supported by XMT because some tools were already defined in MPEG-4. Replicating the tools would have put an extra burden on MPEG-4 terminals. Still, most of the functionality of SMIL and SVG is supported by XMT. In addition, MPEG-4 supports features that are not supported by these formats. For instance, MPEG-4 is built on a true 2D and 3D scene description, including the event model, as extended from VRML. SMIL does not support graphics such as spheres and circles, nor was it designed to support the important MPEG-4 scene updates features ("update the color of the text at time t," for example). These differences are easily explained: SMIL was designed to allow player composition, whereas MPEG-4 is for object composition. Another important point of departure is that in SMIL, temporal layout of elements (e.g., <par> and <seq>) provides the most basic underlying structure of the document, and spatial layout, which is declared up front, can be seen as a separate layer providing a *loose* integration of space into time. In MPEG-4, time and space are more closely integrated, and space is the most basic part of a node definition.

Regarding DVB, MPEG-4 is the solution to link the terminals currently deployed for interactive television and the middle term implementation of the DVB-MHP. MPEG-4 can be used to enhance the functionality currently deployed on the market with graphical animations and the streaming of video, audio, speech, text, and graphics. A profile of the MPEG-4 specification could be used to define a low complexity interactivity engine for the set-top

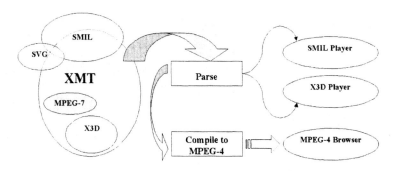

Fig. 2.4 Relationships among XMT, SMIL, and X3D.

box. Terminal manufacturers then could grow in a backward compatible way, according to their market strategies, by including more sophisticated MPEG-4 tools. Arguably the most complex tool in MPEG-4 is the Java-based MPEG-4 application engine for running *MPEGlets*. The implementation complexity of MPEGlets is similar to that of MHP, which also supports Java. The introduction of Java in the terminals would then be the point of convergence between the two standards. High-end set-top boxes can directly include both the MHP and MPEG-4 standards. Indeed, these standards address different problems: MHP can be seen as an operating system, whereas MPEG-4 would be a specific player for multimedia content running on that operating system.

2.6 MPEG-4 APPLICATIONS

The goal of MPEG-4 tools is to enable the deployment of applications for users. Listing *all* the applications enabled by MPEG-4 is impossible. The limits of MPEG-4 applications depend only on the creativity of the application developers. Some of the application areas for which the MPEG-4 standard provides interesting functionality are

☞ Streaming multimedia over the Internet/Intranet

☞ Mobile communications

☞ Digital multimedia broadcast

☞ Content-based storage and retrieval

☞ Interactive media distributed on (optical) storage

☞ Infotainment

☞ Real-time communications

☞ Studio and television postproduction

☞ Surveillance

☞ Virtual meetings

A description of many example MPEG-4 applications can be found in the MPEG-4 Applications document [N2724]. The remainder of this section details the use of MPEG-4 in three specific application scenarios: multimedia portals, interactive broadcast, and multimedia conferencing and communities. The application snapshots below were produced by real applications or prototypes. These examples are intended to illustrate possible uses of MPEG-4 technology and not to prescribe how the tools should be used.

2.6.1 Multimedia Portals

In this scenario, the client is a multimedia terminal connected to the Internet. An example of such a terminal is a personal computer (PC) with multimedia features. The content/application (it is sometimes hard to tell the difference in

MPEG-4) may have been received over the network from a remote server or loaded from a hard disk, CD-ROM, or DVD-ROM. The MPEG-4 player may well be configured as a *plug-in* for a standard Web browser.

A possible application using this scenario is a 3D shopping portal, as shown in Figure 2.5. In this application, the user enters a 3D virtual shop; navigates it; examines products, possibly presented by a video animation (as on the bottom left of the picture) or modeled in 3D (e.g., the 3D model of a video camera).

After collecting information about a product of interest, the user decides to receive more information via a direct link to the vendor and finally starts a real-time communication with the vendor (the person seated near the camera), who can give more details about the product.

MPEG-4 provides the tools for such integration of content with the notions of mixed 2D and 3D scenes. Real-time presentations of streamed content, like a 2D segmented video, can easily be included in the scene, at any given moment in the interaction.

2.6.2 Interactive Broadcasting

In this scenario, the MPEG-4 terminal is, for example, a home set-top box or a high-end home theater, connected to a high-bandwidth broadcast network.

Fig. 2.5 MPEG-4-based next generation of portals (courtesy of Andreas Graffunder, T-Systems Nova, European project IST SoNG).

With the advent of digital broadcasting, the broadcast networks are no longer limited to the conventional satellite or cable networks. The Internet can now be considered a broadcast network too. The key concept of MPEG-4 of *create once, access everywhere* allows content creators and service providers to make their content available across the entire range of delivery systems.

Figure 2.6 shows simple features such as the interactive enhancement of the broadcast streams. Part (a) of the figure shows a news program enhanced with synchronized graphics. The news program is coded using a traditional MPEG-2 system such as a DVB application. The graphics are coded as MPEG-4 graphics and streamed to the receiver using MPEG-4 carriage over MPEG-2 Systems (an extension of MPEG-2 Systems, [MPEG2-1]). The user can interact with the graphics and navigate in the news menu to have access to background information. Part (b) of Figure 2.6 shows a similar application, in which the graphics are used in a sports event to display enhanced data about a soccer match. The user can select a player, a team, or specific roles (e.g., a forward) on a team, which are then tracked and highlighted in real time. Streamed graphics are frame-to-frame synchronized with the MPEG-2 content, which is unique to MPEG-4.

Figure 2.7 shows enhanced features within the same context but in an Internet- and PC-based broadcast environment. The application exploits MPEG-4 shape-coded video along with streamed graphics to show personalized advertisements during a sports event. No other standard supports such functionality.

The enhancement features described above can easily be expanded (with 3D graphics, text-to-speech synthesis, face and body animation, for example)

(a)

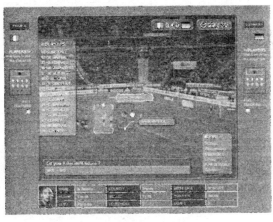

(b)

Fig. 2.6 Interactive broadcasting ([a] courtesy of Gianluca De Petris, Tilab et al., IBC demonstration from the Advanced Interactive Content Initiative, and [b] courtesy of Pierre Pleven, Symah Vision).

Fig. 2.7 Interactive broadcasting with ad manipulation.

to support many applications, such as interactive home shopping, enriched documentary programming, advanced electronic services and program guides, interactive advertisements, interactive entertainment like sports programs or quiz shows, Web-like content viewing, and demographically focused programming.

2.6.3 Multimedia Conferencing and Communities

In this application, illustrated in Figure 2.8, the terminal could be a multimedia PC (equipped with camera, microphone, and speakers) or part of a video-conferencing system. In fact, these two kinds of terminals could very well be communicating in the same virtual environment. The main idea here is that of a shared communication space and the use of MPEG-4 to represent the shared data.

Users connect to the conference site in the same way they would connect to any other Web site (i.e., through a specific address or URL). They first receive the MPEG-4 data representing the shared space. Other participants in this multimedia conference session may already be connected and be represented in the shared space by streamed data. The application supports much more than simply observing the shared space: users can project themselves into the shared space using audiovisual data streams captured from a camera and microphone or a 3D avatar (synthetic representation of a person) animated by speech.

Fig. 2.8 Multimedia conferencing and communities (top left, courtesy of Ananda Allys et al., France Telecom R&D, Oxygen project; top right, courtesy of John K. Arthur et al., Telenor, Eurescom Venus project; bottom, courtesy of Peter Schickel, blaxxun Interactive, European project IST SoNG).

As shown in Figure 2.8, the environment can be as simple as a 2D scene (snapshot on the top left), or it can be a shared 3D environment with multiple users (snapshot on the top right), or it can also be a huge virtual community with many people (snapshot at the bottom).

2.7 SUMMARY

This chapter provided a brief overview of the MPEG-4 standard, starting with the design goals that have driven the development of the standard. Next, it described an end-to-end MPEG-4 creation, delivery, and consumption walk-through using MPEG-4 player anatomy and tools. The chapter positioned the MPEG-4 standard in the marketplace with regard to its predecessors', competitors', and partners' technology. This overview concluded by describing three example applications that can benefit from the MPEG-4 technology.

MPEG-4 is designed to provide an integrated multimedia framework allowing for rich, interactive experiences in many different environments. MPEG-4's core concept is the audiovisual scene: a composition of audiovisual objects according to a scene description.

Contrary to its predecessors, MPEG-4 not only specifies client-side tools (the so-called *MPEG-4 player*). It also specifies tools for interchange between content authors and between content authors and service providers like the MPEG-4 file format, MP4, and the MPEG-4 extensible textual format, XMT. The end-to-end walkthrough also highlights the delivery-independent nature of MPEG-4 content.

MPEG-4 provides a unique framework for the representation of rich multimedia content. Its core design is firmly rooted in proven concepts that underpin such successful solutions as MPEG-2, H.323, VRML, and QuickTime. It integrates state-of-the-art technology such as Java and XML. It provides bridges to similar technology such as SMIL and a smooth evolutionary and integration path in future products based on DVB-MHP.

In Web portals, MPEG-4 provides a rich multimedia experience that goes beyond what can be done with pre-MPEG-4 *plug-ins*. In the area of interactive broadcast, MPEG-4 exploits its close relationship with MPEG-2 transport and allows a frame-by-frame synchronization of rich content enhancements. Finally, the newest tools developed in MPEG-4 allow the shared experience of virtual spaces populated with avatars and agent-animated virtual characters.

2.8 REFERENCES

[EXML01] W3C. Extensible Markup Language (XML) 1.0. W3C Recommendation. October 2000. *www.w3.org/TR/REC-xml/*

[GENT01] Gentric P., et al. *RTP Payload Format for MPEG-4 Streams.* Internet Draft. *draft-gentric-avt-mpeg4-multiSL-01.txt*

[H223] ITU-T Recommendation H.223. *Multiplexing Protocol for Low Bit Rate Multimedia Communication.* March 1998.

[H263] ITU-T Recommendation H.263. "Video Coding for Low Bit Rate Communication," International Telecommunications Union - Telecommunications Standardization Sector, Geneva, 1, 1995; 2 (H.263+), 1998; 3 (H.263++), 2000.

[JAVA96a] Gosling, J., B. Joy, and G. Steele. *The Java Language Specification.* Reading, MA: Addison-Wesley, September 1996.

[JAVA96b] Lindholm, T., and F. Yellin. *The Java Virtual Machine Specification.* Reading, MA: Addison-Wesley, September 1996.

[MHP01] Digital Video Broadcasting. *Multimedia Home Platform (MHP) 1.1*. June 2001. *www.mhp.org/technical_essen/html_index.html*

[MPEG] MPEG Home Page, *mpeg.telecomitalialab.com/*

[MPEG1-1] ISO/IEC 11172-1:1993. *Information Technology—Coding of Moving Pictures and Associated Audio for Digital Storage Media at up to About 1.5 Mbit/s—Part 1: Systems*, 1993.

[MPEG1-2] ISO/IEC 11172-2:1993. *Information Technology—Coding of Moving Pictures and Associated Audio for Digital Storage Media at up to About 1.5 Mbit/s—Part 2: Video*, 1993.

[MPEG1-3] ISO/IEC 11172-3:1993. *Information Technology—Coding of Moving Pictures and Associated Audio for Digital Storage Media at up to About 1.5 Mbit/s—Part 3: Audio*, 1993.

[MPEG2-1] ISO/IEC 13818-1:2000. *Information Technology—Generic Coding of Moving Pictures and Associated Audio Information—Part 1: Systems*, December 2000.

[MPEG2-2] ISO/IEC 13818-2:2000. *Information Technology—Generic Coding of Moving Pictures and Associated Audio Information—Part 2: Video*, December 2000.

[MPEG2-3] ISO/IEC 13818-3:1998. *"Information Technology—Generic Coding of Moving Pictures and Associated Audio Information—Part 3: Audio*, 1998.

[MPEG2-7] ISO/IEC 13818-7:1997. *Information Technology—Generic Coding of Moving Pictures and Associated Audio Information—Part 7: Advanced Audio Coding*, 1997.

[MPEG4-1] ISO/IEC 14496-1:2001. *Coding of Audio-Visual Objects—Part 1: Systems*, 2d Edition, 2001.

[MPEG4-2] ISO/IEC 14496-2:2001. *Coding of Audio-Visual Objects—Part 2: Visual*, 2d Edition, 2001.

[MPEG4-3] ISO/IEC 14496-3:2001. *Coding of Audio-Visual Objects—Part 3: Audio*, 2d Edition, 2001.

[MPEG4-6] ISO/IEC 14496-6:2000. *Coding of Audio-Visual Objects—Part 6: Delivery Multimedia Integration Framework (DMIF)*, 2d Edition, 2000.

[MPEG4-8] ISO/IEC 14496-8:2000. *Coding of Audio-Visual Objects—Part 8: Carriage of MPEG-4 Contents Over IP Networks*. Final Committee Draft, Doc. ISO/MPEG N4282, Sydney MPEG Meeting, July 2001.

[N271] MPEG Convener. *New Work Item Proposal (NP) for Very-low Bitrates Audiovisual Coding*. Doc. ISO/MPEG N271, London MPEG Meeting, November 1992.

[N2724] MPEG Requirements. *MPEG-4 Applications*. Doc. ISO/MPEG N2724, Seoul MPEG Meeting, March 1999.

[N4270] MPEG Systems. *Text of PDAM ISO/IEC 14496-1:2001/AMD3*. Doc. ISO/MPEG N4270, Sydney MPEG Meeting, July 2001.

[N4319] MPEG Requirements. *MPEG-4 Requirements*. Doc. ISO/MPEG N4319, Sydney MPEG Meeting, July 2001.

[N4505] MPEG. *MPEG-4 Overview*. Doc. ISO/MPEG N4505, Pattaya MPEG Meeting, December 2001. See *mpeg.telecomitalialab.com/standards/mpeg-4/mpeg-4.htm* for the most recent version.

[N998] MPEG AOE. *Proposal Package Description (PPD)—Revision 3*. Doc. ISO/MPEG N998, Tokyo MPEG Meeting, July 1995.

[QUICKTIME] Apple Computer. *QuickTime File Format Specification*. May 1996. *www.apple.com / quicktime / resources / qtfileformat.pdf*

[SMIL01] *Synchronized Multimedia Integration Language (SMIL)*, 2.0 Specification, W3C Recommendation. August 7, 2001. *www.w3.org / TR / smil20 /*

[SMIL98] *Synchronized Multimedia Integration Language (SMIL)*, 1.0 Specification, W3C Recommendation. June 15, 1998. *www.w3.org / TR / REC-smil /*

[SVG01] W3C. *Scalable Vector Graphics (SVG)*, 1.0 Specification, W3C Proposed Recommendation. July 2001. *www.w3.org / TR / 2001 / PR-SVG-20010719 / index.html*

[VRML97] ISO/IEC FDIS 14772-1:1997. *Information Technology—Computer Graphics and Image Processing. The Virtual Reality Modeling Language (VRML)—Part 1: Functional Specification and UTF-8 Encoding.* 1997.

[X3D01] Extensible 3D (X3D) Graphics Working Group. *VRML 200x-X3D*. 2001. *www.web3d.org /*

Object Description and Synchronization

by Carsten Herpel

*Keywords: object description, stream description, OCI, IPMP,
content access, synchronization, system decoder model*

An MPEG-4 presentation may consist
of many audiovisual objects that are conveyed through a number of audiovisual streams to a receiver, either in a truly streaming manner or stored in a
file. An important task in structuring an MPEG-4 presentation is the description of all streams, including information such as media codecs, bit rates, and
so on used to encode them as well as the relations between streams. Additional
information about streams may be desired for semantic description or for
implementing digital rights management. Further to the stream description,
means for a synchronized presentation of all streams are required, including a
timing model and a buffering model for the MPEG-4 terminal. The way
MPEG-4 provides all these functionalities is presented in this chapter. These
tools are specified in MPEG-4 Part 1: Systems [MPEG4-1].

3.1 OBJECT DESCRIPTORS: ENTRY POINTS TO MPEG-4 CONTENT

Part 2 [MPEG4-2] and Part 3 [MPEG4-3] of the MPEG-4 standard (corresponding to Chapters 8 to 12 in this book) specify the coding tools to generate compressed audio and visual streams. Each semantic visual or audio object may be
coded into one or more such streams. An MPEG-4 presentation that consists of

multiple such objects (and, hence, many streams) needs mechanisms to describe all the objects in the presentation. One part of this mechanism is embodied in the object descriptor framework [MPEG4-1, Clause 8]. All components of an MPEG-4 presentation are discovered recursively through object descriptors (ODs). The details of this approach are explained in this chapter.

Object description enumerates only the streams in a presentation and specifies how they relate to media objects. In order to assemble those media objects into a specific audiovisual scene, most MPEG-4 presentations also include a scene description. The scene description format, called *binary format for MPEG-4 scenes* (BIFS), is explained in detail in Chapter 4. In order to understand the present chapter, one must understand the general role of scene description and its conceptual difference from object description.

The scene description conveys the spatio-temporal layout of the media objects in the scene—that is, it captures the work of the creative content designers. Object description serves an asset management purpose; it allows locating streams and retrieving information about them. This information may be produced during authoring of the scene description, but—different from the scene description—it also may be subsequently amended or changed, for example, when media streams for a presentation are relocated to a different media server.

The basic concept of an OD is that of a container aggregating all the useful information about the corresponding object. This information is structured in a hierarchical manner through a set of subdescriptors. Each OD describes streams that are referred from a specific location in the scene description. In the case of audio streams, for example, this could be an AudioSource node (see Figure 3.1). The media streams that are related to this single AudioSource node constitute the *object* or *media object* that is described by the OD. More than one media stream can be associated with the object, as will be seen.

3.1.1 Syntactic Description Language

In order to document the syntax of the OD framework, the MPEG-4 Systems standard—as well as this chapter—uses the Syntactic Description Language (SDL). SDL is specified in [MPEG4-1, Clause 14]. This language is a convenient shorthand to express bitstream syntax of inherently object-oriented structures in a C++ kind of way. The strict syntax definition of SDL has the advantage that it enables the fully automatic generation of a code stub for the bitstream parser from the SDL code in [MPEG4-1].[1]

A variable definition in SDL typically includes a directive to parse the value for this variable from a bitstream. Specifically, parsing length bits off

---

1. The other Parts of the MPEG-4 standard continue to use a less rigid pseudo-C notation for bitstream specification. This is mostly due to the fact that SDL has been developed in the same timeframe as the other Parts of the MPEG-4 standard.

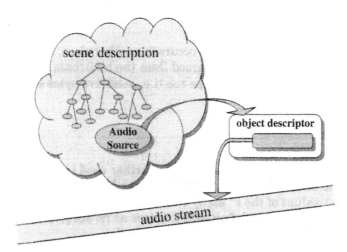

Fig. 3.1 BIFS node (AudioSource) pointing to a(n audio) stream through an OD.

the bitstream into a variable called element_name with a certain type is expressed as follows:

```
[aligned] type[(length)] element_name [= value];
```

where type is any of the data types—bit for arbitrary bit sequences, double for floating values, int for integer or unsigned int for unsigned integer. Square brackets express optional elements. For example, if length is missing, it means that the variable is not parsed from the bitstream but rather is a locally defined variable as in any programming language. The optional aligned attribute causes the reading of length bits to start from the next byte boundary, making the variable byte-aligned. If the variable element_name is assigned a value, this causes the parser to assert that the value read from the bitstream actually corresponds to this *value*. In the example,

```
aligned bit(32) picture_start_code=0x00000100;
```

the parser is asked to parse 32 bits from the next byte boundary and to check that this variable corresponds to the hexadecimal value 100.

Often, the number of bits for a certain syntax element is itself signaled in the bitstream. Therefore, it is permitted to use variables in the length field, as in the second line of the next example:

```
unsigned int(3) precision;
int(precision) DC;
```

A number of variables that constitute the data for an object may be arranged into classes:

```
class Foo {
    unsigned int(5) aVal;
```

```
    bit(3) bVal;
}
```

Each time an instance of class `Foo` occurs in another class, a five-bit value `aVal` and a three-bit value `bVal` are parsed from the bitstream. For example, class `Bar` will parse five elements of type `Foo` (i.e., read five bytes from the bitstream) with the following code:

```
class Bar {
    Foo aFoo[5];
}
```

The array elements read into `aFoo` can be further used in the SDL code of class `Bar` with the familiar syntax `aFoo[i].aVal` and `aFoo[i].bVal`, yielding the `aVal` and `bVal` values of the ith array element.

After those fairly simple definitions, it comes as no surprise that classes may inherit from other classes, as in the next example:

```
class Foobar extends Foo {
    int(8) cVal;
}
```

`Foobar` now actually has the elements `aVal`, `bVal`, and `cVal`. In other words, parsing of an instance of class `Foobar`, as in the next example,

```
class ParseAfoobar {
    Foobar aFoobar;
}
```

results in reading two bytes (5 + 3 + 8 bit) that are accessible as `aFoobar.aVal`, `aFoobar.bVal`, and `aFoobar.cVal`.

Furthermore, classes can be identified with a tag value (0×23 in the following example) that is parsed from the bitstream before reading any element within the class:

```
class ClassWithID : aligned bit(8) classTag=0x23 {
    Foobar aFoobar;
}
```

So, if an instance of `ClassWithID` is expected, as in

```
class ParseAclassWithID {
    int (8) a
    ClassWithID aCwID;
}
```

an eight-bit value other than 0×23 is read from the bitstream after reading the eight bits of `a`, and thus the parser knows that an error has occurred. Although this may be useful to know, the major use of tags is the discrimination of different classes that may occur in the bitstream at the same location. Those classes would have to derive from a common base class:

```
class ClassesWithID : aligned bit(8) classTag=0x23 .. 0x33 {
    Foobar aFoobar;
}
```

in a manner like:

```
class FirstClass extends ClassesWithID : aligned bit(8) classTag=0x23
    {
    int(8) aaValue;
}
class SecondClass extends ClassesWithID : aligned bit(8)
    classTag=0x24 {
    int(8) bbValue;
}
```

giving unique class tags to `FirstClass` and `SecondClass` in the range of tags specified for classes that inherit from `ClassesWithID`.

Now, instances of the derived classes of `ClassesWithID` can be parsed through

```
class ParseClassesWithID {
    int (8) a
    ClassesWithID aCwID;
}
```

If a value of 0x23 is read after the eight bits of `a`, an instance of `FirstClass` will be read; if it is 0x24, it will be an instance of `SecondClass`; in all other cases with tag values between 0x25 and 0x33, an instance of the base class `ClassesWithID` will be read.

Base classes also can be used to organize a set of classes and provide a common name scope for their tag values. For this purpose, similar to C++, abstract classes can be defined that cannot be instantiated (i.e., they cannot be read from the bitstream) and which do not have tag values of their own. The derived classes can then populate the full range of tag values, as given by the type declaration of the tag. So, in the following example,

```
abstract class MyClassesWithID : aligned bit(8) classTag=0 {
    Foobar aFoobar;
}
class FirstClass extends MyClassesWithID : aligned bit(8)
    classTag=0x23 {
    int(8) aaValue;
}
class SecondClass extends MyClassesWithID : aligned bit(8)
    classTag=0x24 {
    int(8) bbValue;
}
class OtherClass : aligned bit(8) classTag=0x24 {
    int(8) ccValue;
}
class ParseClassesWithID {
    int (8) a;
    OtherClass aClass;
    MyClassesWithID aCwID;
}
```

`OtherClass` may reuse the tag value of 0×24, as it is not derived from `Class-esWithID`. `ParseClassesWithID` parses an integer variable `a` and one instance of `OtherClass` before parsing an instance of `MyClassesWithID`. Here, the parser has to encounter either a value of 0×23 or 0×24 to identify a `FirstClass` or `SecondClass` instance. Otherwise, the parser would assume it has encountered an unknown derived class from `ClassesWithID`. Nevertheless, parsing would have to stop here with an error, because parsing of that unknown class is not possible.

Therefore, a special type of classes exists that is both expandable and can be skipped easily, as the overall length of this class is encoded. The length indication is conveyed immediately after the tag (if any) and before any parsable variable.

```
abstract expandable (228-1) class ClassesWithID : aligned bit(8)
    classTag=0 {
    Foobar aFoobar;
}
```

The derived classes of this version of `ClassesWithID` would have a length indication following the class tag, limiting the class size to $2^{28}-1$. When an instance of such a derived class is encountered with a length bigger than what can be computed from the class definition, the parser knows that some trailing data are conveyed in that class which, possibly, it cannot parse but can ignore and continue parsing after skipping that data. When using expandable classes, the parser does not have to fail when parsing an unknown derived class out of a number of derived classes from a base class. The length indication is helpful here to skip such a class instance.

These basic[2] SDL principles outlined above are used in the subsequent sections to introduce the basic syntactic structures of the OD framework.

3.1.2 Object Description

The syntax specification of all descriptors [MPEG4-1, Clause 8.6] is derived from a common base class, called `BaseDescriptor` (see Table 3.1). It specifies that descriptors always are identified by a numeric tag of eight-bit length. Furthermore, it limits the length of descriptors to $2^{28}-1$ bytes and, through the use of an expandable class definition, specifies that each derived descriptor

Table 3.1 Base Descriptor: The template for other descriptors

```
abstract aligned(8) expandable(2^28-1) class BaseDescriptor : bit(8) tag=0 {
    // empty. To be filled by classes extending this class.
}
```

---

2. Features of SDL that are not needed for the code fragments given in this book have been omitted.

defined in the MPEG-4 specification can be extended in the future by syntax elements that are added at the end of the descriptor.

The syntax of the OD [MPEG4-1, Clause 8.6.3] presented in Table 3.2 shows that ObjectDescriptor inherits from ObjectDescriptorBase which in turn inherits from BaseDescriptor (Table 3.1). The latter is another empty abstract base class for different types of ODs.

For labeling, each OD has an ObjectDescriptorID. This ID allows the unique identification of a set of streams in an MPEG-4 presentation. Additionally, it allows the scene description stream, BIFS, to refer to a specific descriptor.

An OD may further contain either only a URL or an actual OD. If it includes only a URL, it means that the actual OD can be found at the location pointed to by the URL. This is useful for managing large assets of media, in which not all media and their descriptions are located on a single media server. Basically, URLs would allow assembling one MPEG-4 presentation from content distributed all over the globe in a manner comparable to today's compound Web pages, which assemble elements from multiple Web servers.

The content of the OD (in the absence of a URL) is a series of subdescriptors. These subdescriptors describe four elements:

☞ Individual elementary streams (ESs)
☞ Semantic information about an object (OCI[3] descriptor)
☞ Hooks for content access management (IPMP[4] descriptor pointer)
☞ A placeholder for future extension descriptors

ES description is detailed first, with the OCI and IPMP frameworks coming later in the section about semantic description and access management.

Table 3.2 OD syntax

```
class ObjectDescriptor extends ObjectDescriptorBase : bit(8)
    tag=ObjectDescrTag {
    bit(10) ObjectDescriptorID;
    bit(1) URL_Flag;
    const bit(5) reserved=0b1111.1;
    if (URL_Flag) {
        bit(8) URLlength;
        bit(8) URLstring[URLlength];
    } else {
        ES_Descriptor esDescr[1 .. 255];
        OCI_Descriptor ociDescr[0 .. 255];
        IPMP_DescriptorPointer ipmpDescrPtr[0 .. 255];
    }
    ExtensionDescriptor extDescr[0 .. 255];
}
```

---

3. Object content information.

4. Intellectual property management and protection.

3.1.3 Stream Description

Each individual stream—or, more precisely, each *elementary stream*—is described by an elementary stream descriptor (ES descriptor or ESD; see Table 3.3) specified in [MPEG4-1, Clause 8.6.5]. An ES is conceived as a flow of (possibly compressed) data that originates from a single source in the sender and terminates in a single sink at the receiver. In a simple case, a media encoder (audio or video) and the corresponding media decoder are communicating through one ES (media encoder/decoder 1 in Figure 3.2). However, a single media decoder also may consume multiple ESs concurrently, specifically in case of a scalable encoding (media encoder/decoder 2 in Figure 3.2). Formally, a set of one or more ESs that are jointly decoded to recover the media object makes up a *media stream* in the MPEG-4 context.

Each ESD provides detailed information about the associated stream, including its location, type, encoding, bit rate, buffer requirements, and decoder specific information. As a comparison, only a subset of this information could be conveyed with multipurpose Internet mail extensions (MIME) [RFC2045] or the session description protocol (SDP) [RFC2327], mechanisms often used in Internet streaming applications for stream description.

Looking at the ESD elements in detail, stream location is provided first. This is done through one of two mechanisms: It can be done solely by means of an elementary stream ID (ES_ID). In that case, it is assumed that the media

Table 3.3 ESD syntax

```
class ES_Descriptor extends BaseDescriptor : bit(8) tag=ES_DescrTag {
    bit(16) ES_ID;
    bit(1) streamDependenceFlag;
    bit(1) URL_Flag;
    bit(1) OCRstreamFlag;
    bit(5) streamPriority;
    if (streamDependenceFlag)
        bit(16) dependsOn_ES_ID;
    if (URL_Flag) {
        bit(8) URLlength;
        bit(8) URLstring[URLlength];
    }
    if (OCRstreamFlag)
        bit(16) OCR_ES_Id;
    DecoderConfigDescriptor decConfigDescr;
    SLConfigDescriptor slConfigDescr;
    IPI_DescrPointer ipiPtr[0 .. 1];
    IP_IdentificationDataSet ipIDS[0 .. 255];
    IPMP_DescriptorPointer ipmpDescrPtr[0 .. 255];
    LanguageDescriptor langDescr[0 .. 255];
    QoS_Descriptor qosDescr[0 .. 1];
    RegistrationDescriptor regDescr[0 .. 1];
    ExtensionDescriptor extDescr[0 .. 255];
}
```

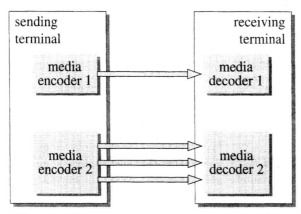

Fig. 3.2 Media encoder and decoder can be connected by one (case 1) or more (case 2) elementary streams.

server storing this stream will be able to connect this ES_ID to some stream resource in an application-specific way. Second, the URL field can be used. This allows locating ESs in a more versatile manner. In both cases, however, the actual description of the stream is done by the ESD at hand.

The flags after ES_ID signal whether the related syntax elements are present in the ES descriptor in question. The streamPriority element signals the relative importance of the streams within a presentation: The higher the value, the more important the stream. This information is mainly of use for a stream server or a gateway that can evaluate the stream priority values for all streams in a presentation and then group them into different transmission channels accordingly. In other words, the semantics of this field must be determined within a specific application scope.

As mentioned before, ODs and ESDs are containers for information about the associated object and ES, respectively. Therefore, the subsequent syntax elements relate to a number of different features that are introduced in the sections to come. First, the stream dependence mechanism defines how streams hierarchically depend on each other by pointing to the ES_ID of the stream upon which this one directly depends. The usage will be further discussed in the stream-dependence section. The OCRstreamFlag and OCR_ES_Id allow tying streams to a specific time base by pointing to the ES_ID of the stream that conveys this time base information. The usage will be discussed in the section on synchronization.

The remainder of the ESD is a series of subdescriptors with more specific and optional information. Details on IPI_DescrPointer, IP_Identification-DataSet, IPMP_DescriptorPointer, and language descriptors are given in the IPMP and OCI sections. SLConfigDescriptor describes synchronization-related properties and is explained in the synchronization section. The remaining subdescriptors are covered in the following sections.

3.1.3.1 Decoder Configuration The `DecoderConfigDescriptor` (Table 3.4) provides setup information required by the decoder to process the stream in question. First, the stream is roughly classified through the `streamType` value as visual, audio, scene description, and so on. Then, the `object-TypeIndication` gives further information on the actual coding employed, such as JPEG, MPEG-1, MPEG-2, and MPEG-4. For legacy stream types (MPEG-1, MPEG-2), profile and level information is directly included, as applicable.

The `upStream` flag is set for streams that provide upstream information from the client terminal back to the server. Attaching such a stream to an OD, in addition to a regular stream flowing from server to client, means that this object has its own back channel.

The remaining values in `DecoderConfigDescriptor` indicate the buffer size required for the decoding buffer as well as the maximum and the average bit rates that should be expected when receiving the stream.

The `DecoderSpecificInfo` descriptor serves as an opaque container from the MPEG-4 Systems point of view and carries information about the decoder configuration needed to decode the stream properly.

Finally, the `profileLevelIndicationIndexDescriptor` allows associating this stream to one or more complexity categories specified for the overall MPEG-4 presentation, as detailed in the complexity section below.

3.1.3.2 Quality of Service Descriptor Comparable to the `streamPriority` element in the ESD, the quality of service (QoS) descriptor is mostly of use for the media server and for gateways that intelligently forward streams or sets of streams. The QoS descriptor consists of either an (application-dependent) index of a predefined QoS scenario or a set of QoS qualifiers (Table 3.5). QoS qualifiers are defined as derived classes from the abstract `QoS_Qualifier` class.

Table 3.4 Decoder Configuration Descriptor syntax

```
class DecoderConfigDescriptor extends BaseDescriptor : bit(8)
     tag=DecoderConfigDescrTag {
   bit(8) objectTypeIndication;
   bit(6) streamType;
   bit(1) upStream;
   const bit(1) reserved=1;
   bit(24) bufferSizeDB;
   bit(32) maxBitrate;
   bit(32) avgBitrate;
   DecoderSpecificInfo decSpecificInfo[0 .. 1];
profileLevelIndicationIndexDescriptor profileLevelIndicationIndexDescr
     [0..255];
}
```

Table 3.5 QoS Descriptor syntax and QoS Qualifier definition

```
class QoS_Descriptor extends BaseDescriptor : bit(8) tag=QoS_DescrTag {
    bit(8) predefined;
    if (predefined==0) {
        QoS_Qualifier qualifiers[];
    }
}

abstract aligned(8) expandable(2^28-1) class QoS_Qualifier : bit(8)
    tag=0x01..0xff {
    // empty. To be filled by classes extending this class.
}
```

A number of QoS qualifiers have been predefined to express the following:

☞ Maximum end-to-end delay for the stream in microseconds
☞ Preferred end-to-end delay for the stream in microseconds
☞ Allowable loss probability of any single AU as a fractional value between 0.0 and 1.0
☞ Maximum allowable number of consecutively lost AUs
☞ Maximum size of an AU in bytes
☞ Average size of an AU in bytes
☞ Maximum arrival rate of AUs in AU/second

This set of qualifiers may need application-specific extension, depending on the selection of a specific QoS framework.

3.1.3.3 Registration Descriptor Finally, a registration descriptor (Table 3.6) is provided to enable MPEG-4 users to unambiguously carry non-MPEG-4 ESs. Although MPEG does not (prefer to) open up the set of allowed ESs, it is acknowledged that applications will still have the need to convey such private streams. The registration descriptor provides a method to unambiguously identify the characteristics of these streams.

As in other such cases, ISO will issue a Call for Registration Authority, a body that will subsequently manage the space of formatIdentifier values and make sure that they are uniquely assigned. The second field in the descriptor may hold any additional information, as requested by the owner of a certain formatIdentifier.

Table 3.6 Registration Descriptor syntax

```
class RegistrationDescriptor extends BaseDescriptor : bit(8)
    tag=RegistrationDescrTag {
    bit(32) formatIdentifier;
    bit(8) additionalIdentificationInfo[sizeOfInstance-4];
}
```

3.1.4 Stream Relationship Description

The mechanisms to describe a single ES for a media object have been introduced. Describing an object with a single ES description consists in creating an ESD and placing it within an OD. However, the OD framework has some more features that are specified in [MPEG4-1, Clause 8.7]. For example, it is possible to describe various alternative representations of the same media object.

In a simple case, a single media object is encoded in different qualities. The corresponding media streams can be generated independently, including for each of them an ESD with the necessary parameters, such as coding format and bit rate. Those ESDs are simply put together in one single OD. The client terminal will then select one of the streams and request it from the server for playback (see Figure 3.3).

In some application contexts, scalable coding is more adequate than providing a set of independent quality streams—for example, in the case of multicast distribution, in which the client cannot request a specific quality from the server. A base quality (base layer) is provided by a first ES that can be decoded by the decoder independently. Depending on the concrete scalability algorithms used by the media codec, this quality can be improved, possibly in multiple steps, corresponding to the multiple ESs (enhancement layers) that are generated. Here it is important to know in which order to pass the set of streams to the media decoder (see Figure 3.4). This is achieved with the stream-dependency mechanism of the ESD. Each stream points to the one it depends on; more precisely, it points to the ES_ID of that stream using the dependsOn_ES_ID syntax element. The client terminal in that case will select one or more of these streams for playback, starting, of course, from the base layer stream.

Both in the case of different quality alternatives for a media object and in the case of scalable coding, the receiving terminal has a choice. The terminal needs to figure out which of the streams to use, depending on its capabilities. The terminal can do this by simply evaluating all the ODs and using custom

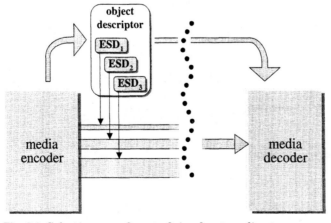

Fig. 3.3 Selecting one of a set of simulcast media streams.

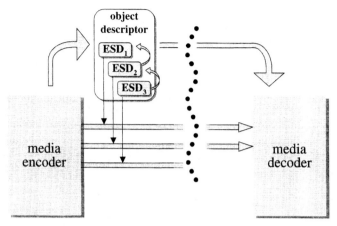

Fig. 3.4 Selecting a set of scalable streams.

rules to decide what to do. More likely, however, profile and level definitions are used for this purpose, as will be seen in the next section.

3.1.5 Content Complexity Description

The question of content complexity is important and has to be answered up front before an MPEG-4 terminal can decide whether it will attempt to decode and present that content at all. For a single media stream, it may be simple to examine the OD for that stream in all its advertised variations, and then to decide which one the terminal is capable of presenting. But now assume that the content consists of a large number of media objects embedded into a large scene. In that case, an examination of all related ODs would be quite arduous. Because of that, another mechanism has been implemented to answer the question of content complexity. It is basically a set of labels indicating profiles and levels for the content. The profile and level concepts will be addressed in detail in Chapter 13.

The profile and level indications are conveyed by means of two types of descriptors: a derivative of the OD called the *initial object descriptor* (IOD, Table 3.7), and the *extension profile level descriptor* (Table 3.8).

The IOD is identical to the regular OD, except that it features the said elements for signaling the different profiling dimensions for scene graph, audio, visual, graphics, and OD (see Chapter 13). Additionally, a flag (includeInlineProfileLevelFlag) signals whether these profiles and levels correspond to the sum of the complexity of the overall scene including all sub-scenes[5] (includeInlineProfileLevelFlag=1), or whether those complexities

---

5. Scene descriptions can be recursively included in each other using the Inline node, which will be introduced in Chapter 4.

Table 3.7 Initial OD syntax

```
class InitialObjectDescriptor extends ObjectDescriptorBase : bit(8)
    tag=InitialObjectDescrTag {
  bit(10) ObjectDescriptorID;
  bit(1) URL_Flag;
  bit(1) includeInlineProfileLevelFlag;
  const bit(4) reserved=0b1111;
  if (URL_Flag) {
      bit(8) URLlength;
      bit(8) URLstring[URLlength];
  } else {
      bit(8) ODProfileLevelIndication;
      bit(8) sceneProfileLevelIndication;
      bit(8) audioProfileLevelIndication;
      bit(8) visualProfileLevelIndication;
      bit(8) graphicsProfileLevelIndication;
      ES_Descriptor esDescr[1 .. 255];
      OCI_Descriptor ociDescr[0 .. 255];
      IMP_DescriptorPointer ipmpDescrPtr[0 .. 255];
  }
      ExtensionDescriptor extDescr[0 .. 255];
}
```

Table 3.8 Extension Profile Level Descriptor syntax

```
class ExtensionProfileLevelDescriptor() extends BaseDescriptor :
    bit(8) ExtensionProfileLevelDescrTag {
    bit(8) profileLevelIndicationIndex;
    bit(8) ODProfileLevelIndication;
    bit(8) sceneProfileLevelIndication;
    bit(8) audioProfileLevelIndication;
    bit(8) visualProfileLevelIndication;
    bit(8) graphicsProfileLevelIndication;
    bit(8) MPEGJProfileLevelIndication;
}
```

are signaled by their own `InitialObjectDescriptors` or `ExtensionProfileLevelDescriptors`.

The `ExtensionProfileLevelDescriptor` is necessary because a single complexity indication for an MPEG-4 presentation may not be sufficient. As explained, ODs allow aggregating multiple representations of the same content with different characteristics. Therefore, a signaling means for the complexity of these different representations in terms of profiles and levels is required. This is achieved by attaching multiple `ExtensionProfileLevelDescriptors` to the initial OD, each describing one of these representations and uniquely labeled by a `profileLevelIndicationIndex`. Additionally, this descriptor serves to signal MPEG-J profiles (see Chapter 5) that did not exist in Version 1 of the MPEG-4 Systems standard.

A stream that belongs to a given representation (or complexity category) may use the `profileLevelIndicationIndexDescriptor` (Table 3.9) from its `DecoderConfigDescriptor` to reference the appropriate `profileLevelIndicationIndex`.

Table 3.9 Profile Level Indication Index Descriptor syntax

```
class ProfileLevelIndicationIndexDescriptor () extends BaseDescriptor
:    bit(8) ProfileLevelIndicationIndexDescrTag {
     bit(8) profileLevelIndicationIndex;
}
```

3.1.6 Streaming ODs

The previous sections explained how media objects and media streams are defined and described through ODs and ESDs. They also detailed how to describe different encoded versions of the same media object and how to indicate the complexity of these objects. The next step is to communicate the ODs with all this information from the sender to the client terminal.

In MPEG-4 the focus is on enabling dynamic multimedia content. This led to a design in which ODs can be sent, updated, or deleted at any time during an MPEG-4 presentation. In fact, ODs are sent within their own OD stream [MPEG4-1, Clause 8.5]; that is, they are generally processed in the same way as other media streams or the scene description streams.

This MPEG-4 approach is quite different from existing mechanisms for signaling stream properties. Both SDP [RFC2327], in conjunction with the real-time streaming protocol (RTSP) [RFC2326] used in some Internet streaming applications, and the ITU-T Recommendation H.245 [H245] used for videotelephone and videoconferencing applications perform the signaling of stream properties only at the beginning of the multimedia presentation or communication.

Within the OD stream, a set of OD commands can be conveyed:

- OD update
- OD remove
- ESD update
- ESD remove
- OD execute

The OD update command conveys a set of ODs or initial ODs. Adding and updating an OD are treated the same way. For each updated OD, the terminal will check whether existing streams were removed from it or new streams have been added. Through an OD update, all the auxiliary information related to OCI and IPMP can be updated, too.

The ESD update makes it possible to convey individual ESDs and, therefore, plug an ES into an OD after that OD is already conveyed to the receiving terminal. It advertises at any stage during a presentation that an additional version of a media object just became available—for example, an enhanced quality version. Removing ESDs or ODs indicates future unavailability of the associated streams.

The OD execute command exists in order to let the terminal know upfront for which streams a decoder needs to be set up immediately. In the absence of such OD execute commands, the terminal may set up decoders only when the scene description actually requests the playback of some media stream. In that case, an extra delay for setting up the decoder must be expected, which will be implementation-dependent. The need for the OD execute command was established very late in the standardization process. Therefore it is found only in the second edition of the MPEG-4 Systems standard [MPEG4-1].

All these commands follow a syntax similar to that of the descriptors encountered before. They are given in Table 3.10 for completeness. All commands derive from the abstract class `BaseCommand`. As mentioned earlier, an abstract class defines a name scope for the class tags. So, `BaseCommand` defines

Table 3.10 Syntax of OD commands

```
abstract aligned(8) expandable(2^28-1) class BaseDescriptor : bit(8) tag=0 {
    // empty. To be filled by classes extending this class.
}

class ObjectDescriptorUpdate extends BaseCommand : bit(8)
      tag=ObjectDescrUpdateTag {
    ObjectDescriptorBase OD[1 .. 255];
}

class ObjectDescriptorRemove extends BaseCommand : bit(8)
      tag=ObjectDescrRemoveTag {
    bit(10) objectDescriptorId[(sizeOfInstance*8)/10];
}

class ES_DescriptorUpdate extends BaseCommand : bit(8)
      tag=ES_DescrUpdateTag {
    bit(10) objectDescriptorId;
    ES_Descriptor esDescr[1 .. 255];
}

class ES_DescriptorRemove extends BaseCommand : bit(8)
      tag=ES_DescrRemoveTag {
    bit(10) objectDescriptorId;
    aligned (8) bit(16) ES_ID[1..255];
}

class ObjectDescriptorExecute extends BaseCommand : bit(8) tag=
      ObjectDescriptorExecuteTag {
        bit(10) objectDescriptorId[(sizeOfInstance*8)/10];
}
```

a new name scope for the command tags that is distinct from the name scope for descriptor tags.

3.1.6.1 Streaming Configuration Information in ODs

In previous standards (such as MPEG-2), decoder configuration information (like coding format and coding options) is repeated regularly as part of the media ES itself in order to allow the random access to a stream, as indicated in Figure 3.5a. In case of MPEG-2, some of this configuration information is additionally repeated in the Systems layer [MPEG2-1]. In MPEG-4, it is suggested to convey the decoder configuration information within the DecoderSpecificInfo element (Table 3.4), avoiding any duplication of this information in the ES itself.

This approach makes it possible to adapt the delivery rules for the decoder configuration information to the delivery medium. In case of unicast delivery without the ability to randomly tune in, decoder configuration information is delivered up front and no repetition is needed. In case of a broadcast medium, the ODs can easily be repeated as often as needed within their OD stream, as indicated in Figure 3.5b. The arrows in this principal drawing indicate that it is possible to maintain a temporal link between the OD stream and the media streams through the use of time stamps.

Unfortunately, this ideal approach cannot always be followed, because the MPEG-4 toolbox approach allows the use of MPEG-4 Visual or Audio coding

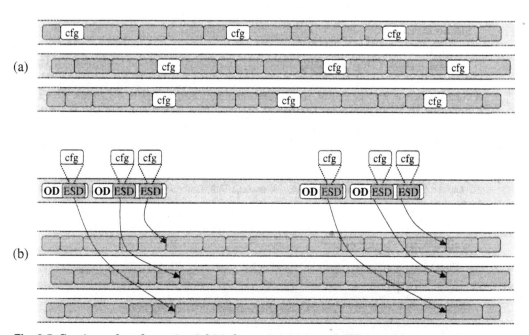

Fig. 3.5 Carriage of configuration (cfg) information (a) in each ES (possibly repeated), as in MPEG-2, and (b) in the decoder-specific info descriptor within a separate OD stream, as in MPEG-4.

tools independently of the Systems tools, which include the ODs. In the absence of ODs, the ES syntax has to revert to the approach shown in Figure 3.5a.

3.1.7 Linking a Scene to Its Media Streams

It was mentioned before (Figure 3.1 and Figure 3.2) that ODs are the link between an individual media object as represented in the scene description and its associated media stream(s). Given that both ODs and the scene description are conveyed in streams, it becomes apparent that further rules are needed to associate an entire scene description stream to an OD stream [MPEG4-1, Clause 8.7.2].

3.1.7.1 Association of OD Streams and Scene Description Streams   OD streams are usually associated with a scene description (BIFS) stream, as that stream conveys the information on how to assemble the various media streams described within the OD stream into an MPEG-4 presentation. The means to associate OD and BIFS streams is another OD. The ESDs for an OD stream and its associated scene description stream are put into one OD or IOD and the scene description stream is set to depend on its OD stream (Figure 3.6).

It is also possible to split the scene description information into more than one stream—basically, to do a scalable encoding of the scene description. The order of relevance of the various BIFS streams is again signaled through the stream dependency mechanism in the OD (Figure 3.7), as done for scalable media streams.

Splitting a single BIFS stream in this way is notably different than partitioning a scene in subscenes through Inline nodes (see Chapter 4). It just means that some of the scene description data (called BIFS commands) that construct a scene are sent in one stream and some in the other(s). The BIFS decoder in the receiving terminal merges the BIFS commands from those streams in the right temporal order prior to decoding. It is up to the authoring tool to use this feature in a meaningful way, for example to support error resilience for the scene description data.

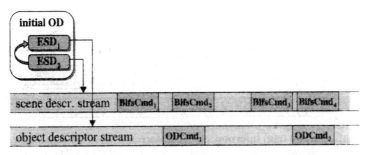

Fig. 3.6 Associating a scene description stream with its OD stream.

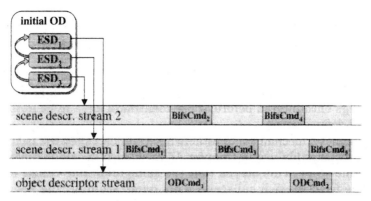

Fig. 3.7 Scalable scene description.

3.1.7.2 Association of MPEG-4 Presentations

Another real-world question is how to author small MPEG-4 presentations and plug them together into bigger presentations later on. As an example, consider a main audiovisual show that is presented together with an overlaid ad banner including ani-mated graphics and its own stamp-sized video stream (see Figure 3.8).

The authors for these two items are likely to work independently, but the overall presentation still needs to be assembled from both parts. Notably, in

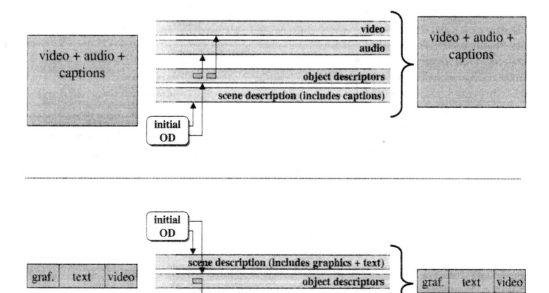

Fig. 3.8 Authoring two independent scenes: a main audiovisual program (top) and a banner-ad-like portion consisting of graphics, text, and video elements (bottom).

MPEG-4 this must be possible without reediting either content, because in fact the assembly will take place at the receiver. Unfortunately, independent editing means that both partial presentations may have used the same numerical identifiers to mean different things.

The proper solution for such problems is the definition of name scopes for the various parts of the content. This means, similar to the name scopes for descriptor tags encountered before, those identifiers have to be unique only within a well-defined scope.

BIFS defines a mechanism to establish subscenes, called the Inline node (as will be further explained in Chapter 4). Each such subscene constitutes a separate name scope. This means that the same identifier values for BIFS nodes can be used in each subscene. This name scoping rule needs to be extended to the references from BIFS to OD and from OD to media streams, which are usually made through OD_ID and ES_ID.

Consequently, name scopes in the OD domain are defined the following way: Each OD that points to BIFS and OD streams opens a new name scope. In fact such an OD will be either the initial OD of a scene or an OD that is pointed to by an Inline node. So this translates back to the BIFS rule that each Inline node opens a new name scope.

For the example in Figure 3.9, this means that OD_IDs and ES_IDs, as

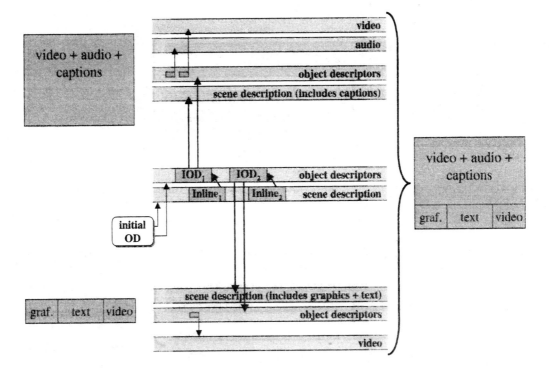

Fig. 3.9 Assembling a scene from two subscenes.

well as nodeIDs, routeIDs, and so on in BIFS, can be assigned independently in both individual scenes. The initial ODs (see Figure 3.8) for both scenes open one unique name scope for each scene. In these conditions, there will be no conflict when the scenes are composed in a combined presentation.

The process to compose the two scenes into one is the following: Another piece of scene description has to be authored that *inlines* the other two scenes as subscenes, defining the right place on the screen for each of them, of course. The initial ODs of the two subscenes will become part of the OD stream for the composite scene, as indicated in Figure 3.9. Starting from the new initial OD, the receiver will be able to reconstruct the composite scene.

3.1.8 MPEG-4 Content Access Procedure

In the first part of this chapter, all the elements necessary to access MPEG-4 content have been introduced. Content access always starts with an initial OD (see the top left of Figure 3.10). The IOD contains pointers to at least the two

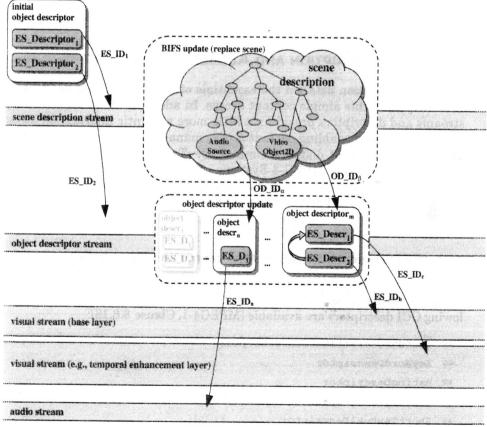

Fig. 3.10 MPEG-4 content access procedure.

essential streams: a scene description (BIFS) stream and an OD stream. The ODs in the OD stream allow the location of the additional media streams used in the scene and have all the information needed to initialize the corresponding media decoders. The streams are then referenced from the BIFS scene through pointers to their respective ODs.

This content access procedure [MPEG4-1, Clause 8.7.3] has been formalized in the MPEG-4 standard as a series of calls to an API, called the DMIF Application Interface (DAI). This API is specified in Part 6 of the MPEG-4 standard, *Delivery Multimedia Integration Framework* [MPEG4-6]. The DAI is an abstraction layer that enables the description of the content access procedure independently of the actual transport and signaling protocols used.

MPEG-4 does not define the transport or delivery of its coded data. Therefore, depending on the concrete MPEG-4 application scenario, different means may exist to communicate the IOD to the receiver. For example, this could be achieved by publishing URLs for IODs on Web pages, by using a suitable signaling protocol such as RTSP [RFC2326] or ITU-T H.245 [H245], or by embedding the IOD in an MPEG-2 transport stream [MPEG2-1]. Some of those mechanisms will be further explored in Chapter 7.

3.2 SEMANTIC DESCRIPTION AND ACCESS MANAGEMENT

The previous section detailed the essentials of MPEG-4 stream description necessary to enable simple content access. In addition, there are auxiliary streams and descriptors providing a little more semantic information about media objects and enabling content access management in order to prevent unauthorized content distribution and use. Two major types of auxiliary streams are defined in MPEG-4 Systems—OCI streams and IPMP streams. The intent of OCI is clear: It provides what is called *metadata*, describing objects semantically. IPMP provides hooks to control the legitimate usage of MPEG-4 content.

3.2.1 Object Content Information: Meta Information About Objects

A set of descriptors and a stream type have been defined to carry information about the media object in general: OCI descriptors and OCI streams. The following OCI descriptors are available [MPEG4-1, Clause 8.6.18]:

- ☞ ContentClassificationDescriptor
- ☞ KeyWordDescriptor
- ☞ RatingDescriptor
- ☞ LanguageDescriptor
- ☞ ShortTextualDescriptor
- ☞ ExpandedTextualDescriptor

☞ ContentCreatorNameDescriptor

☞ ContentCreationDateDescriptor

☞ OCICreatorNameDescriptor

☞ OCICreationDateDescriptor

☞ SmpteCameraPositionDescriptor

☞ SegmentDescriptor

☞ MediaTimeDescriptor

Content is classified, for example, by genre. Classification schemata need to be registered with a registration authority to make sure that all implementers of MPEG-4 systems have access to this information. The same approach holds for the rating descriptor that is supposed to indicate the suitability of the content for different groups of audiences. Descriptors for keywords, short text, and expanded text are all freestyle but presumably in increasing order of complexity. Name of content creators and OCI creators as well as the creation dates for those are self-explanatory. The camera position descriptor inherited from SMPTE [S315M] allows keeping track of the camera position for that media stream. Finally, the segment descriptor and media time descriptors are additions that allow assigning a name to a temporal segment of the media stream based on the definition of a media timeline (see section on synchronization).

Any of those descriptors can be attached to an OD. This means that all the ESs referenced through this OD share the same (static) object content information. The only information that can be attributed to a single stream within one single OD is the language, because it should be possible to convey different language variants of the same content as one logical media object, as presented in Figure 3.11. This is why the LanguageDescriptor features as a subdescriptor of the ESD (see Table 3.3).

OCI also may be dynamically changing over time—for example, the keywords and textual descriptor. In that case, a separate OCI stream may be used to convey that information [MPEG4-1, Clause 8.4]. Each OD can refer to one OCI stream at most. Within an OCI stream, an OCI_Event message is used to wrap each set of descriptors that is to be conveyed jointly. The OCI_Event has the added functionality that the embedded set of OCI descriptors can now be associated with a temporal segment of the media stream, identified through start time and duration (see Figure 3.12). The SmpteCameraPosition-Descriptor is a typical example of a descriptor that mostly makes sense when conveyed in an OCI stream, as it may be assumed that camera position changes frequently over time.

Concluding, it should be mentioned that OCI was established at a time when no other MPEG metadata format was available. Today the standardization effort for the MPEG-7 multimedia description framework is well advanced [N4509]. All concepts expressed in OCI could equally well be expressed in MPEG-7. In fact, a label for MPEG-7 streams has already been reserved. So, instead of an OCI stream, an MPEG-7 stream can be associated

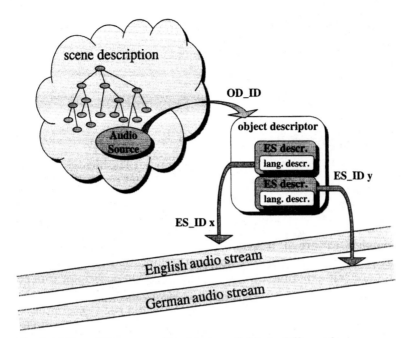

Fig. 3.11 Describing one audio object available in different languages.

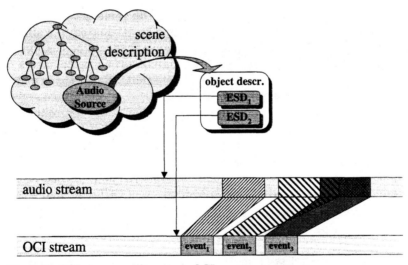

Fig. 3.12 OCI stream with various OCI events.

with an MPEG-4 media stream. It is hard to predict how OCI and MPEG-7 will coexist. It is conceivable that future MPEG-4 applications will rely more on MPEG-7 than on MPEG-4 OCI as a metadata format because of its larger capabilities, but it may also be that applications with few metadata requirements continue to use OCI.

3.2.2 Intellectual Property Management and Protection

A number of methods exist to encrypt, watermark, or otherwise protect digital content. MPEG-4 provides for an IPMP framework that defines sufficient hooks to implement such protection, without being overly prescriptive, by standardizing a specific IPMP tool set.

The IPMP framework consists of a fully standardized *intellectual property identification (IPI) data set* as well as *IPMP descriptors* and *IPMP streams*, which provide a standard shell to be filled with proprietary data used to drive equally proprietary IPMP systems.

The IPI data set descriptors [MPEG4-1, Clause 8.6.9] are an optional part of an ESD (see Table 3.3). The following descriptors are available:

- ☞ **ContentIdentificationDescriptor:** This is a vehicle to convey standardized identifiers for content, such as International Standard Book Number (ISBN), International Standard Music Number (ISMN), or Digital Object Identifier (DOI).

- ☞ **SupplementaryContentIdentificationDescriptor:** This descriptor allows adding freestyle content information in the form of attribute-value pairs. Notably, this information should be related just to the IP situation of this content.

If multiple audiovisual objects within one MPEG-4 session are identified by the same IPI information, the IPI data set may consist solely of a pointer to another ES—that is, to its `ES_ID`—carrying the IPI information.

Beyond the IPI data set, the core IPMP framework [MPEG4-1, Clauses 8.3, 8.8] builds on a modified model of an MPEG-4 terminal, as depicted in Figure 3.13. This model adds a number of control points in the data path. The meaning of *control point* is IPMP system-specific and might translate, for example, to decrypting or enabling of ESs or to reports about an ES.

IPMP descriptors and IPMP streams are used to configure the IPMP system in this terminal model; that is, they provide information that might help to decrypt ESs or that contain authorization or entitlement information. Multiple IPMP systems can coexist in such a terminal, each driven by its own IPMP descriptors and streams. This approach has been followed in acknowledgment that service providers are unlikely to adopt a single IPMP system.

IPMP descriptors are carried in the OD stream in separate `IPMP_DescriptorUpdate` commands—that is, not as elements within ODs or ESDs. Within the ODs and ESDs, IPMP descriptor pointers can be used to associate the OD or ESD with an IPMP descriptor (see Table 3.2 and Table 3.3). If associated with an OD, it means that the IPMP descriptor is relevant for all streams described by that OD. If associated with an ESD, it will only be valid for that stream.

IPMP streams have been designed as a complementary method to convey IPMP information in a streaming fashion. This may be confusing at the first glance, as IPMP descriptors can be updated over time. However, IPMP

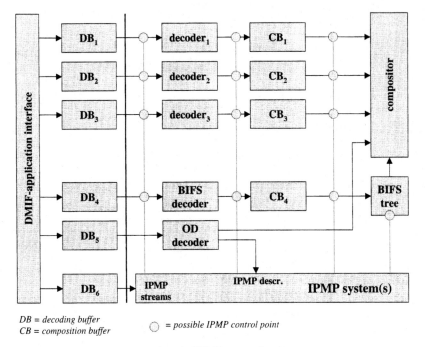

DB = decoding buffer
CB = composition buffer   ◯ = possible IPMP control point

Fig. 3.13 MPEG-4 terminal with IPMP control points.

streams are separate from the OD stream, which has the advantage that
IPMP information can more easily be kept separate from the original MPEG-4
information. Furthermore, the OD parser (which, at least conceptually, is not
interested in IPMP information) is not bothered by data that it just has to
pass through to the IPMP system.

3.3 TIMING MODEL AND SYNCHRONIZATION OF STREAMS

In the previous section we saw how to associate a set of streams with one
another in order to describe a multimedia presentation composed of a large set
of such streams. The presentation of these streams in a coordinated manner is
basically governed by the scene description. Therefore, a clearly defined notion
of time must be established between the scene description and all the media
streams. Then, a mechanism is needed to convey such timing information.
This is achieved by the synchronization layer syntax specified in [MPEG4-1,
Clause 10]. The synchronization layer often is abbreviated as *sync layer* or SL.

3.3.1 Modeling Time

Time appears to be the most natural thing in the world—it passes by even
when we aren't thinking of it. Unfortunately, in the context of multimedia

streaming, a lot of thought has to be dedicated to the proper definition of the notion of time.

Assume an MPEG-4 presentation that is prepared for later streaming. Usually, most timing information will be relative within the presentation. If, either in the scene description or in a media stream, absolute (wall clock) time values are used, it means that such values in most cases need to be modified at the time such a presentation is actually streamed.

On the other side, there is no absolute time dependency if the scene description consists of commands that request to show, for example, the video stream "now," or if it instructs to "show video in 10 seconds" (see Figure 3.14). So, the first important thing to note is that *time in MPEG-4 is always relative.* Now we will explore what is needed to make this approach work.

Making time relative is an easy task if a simple temporal reference point exists. For example, in any playback from a local file or in a unicast streaming presentation, the client processes the presentation from its start. In that case, the start of the presentation makes a great reference point.

However, MPEG was concerned with defining an approach that works both in the cases mentioned above and in the case of broadcast or multicast playback. In that case, a client may not be aware of the start of the presentation. The only known point in time is when the client tunes into the broadcast. Unfortunately, this point is different for each client and unknown to the sender. So it should be concluded that the definition of a time "now" within the scene description should depend neither on any "start of presentation" nor on wall clock time. Instead, the point in time when a portion of the scene description data is received by the terminal is taken as reference (Figure 3.15).

More precisely, the reference point is defined as the point in time when a given BIFS command is due for composition in the terminal. That point in time is known through a *time stamp*. The above definition is appropriate for

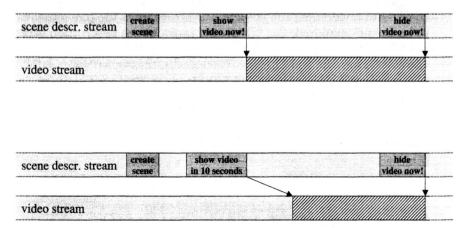

Fig. 3.14 The relative nature of time in BIFS and media streams.

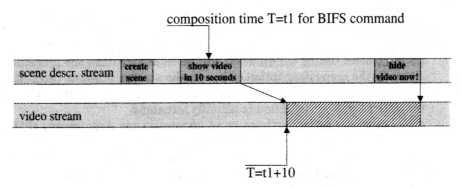

Fig. 3.15 Time relative to composition time of a BIFS command.

most cases, but it does not allow expressing that a certain event is bound to happen at, say, 8:00 p.m. sharp. However, such an access to wall clock time can still be accomplished either on the server side or at the client terminal. As soon as the server knows when a scene is bound to be streamed, the absolute time value can be trivially translated into a relative one by server-side editing of the scene description, making use of the wall clock time at the server. Furthermore, accessing the receiving terminal's wall clock time may be possible through a piece of program code, or a script (see Chapter 4 for details), conveyed as part of the scene description.

3.3.2 Time Stamps and Access Units

Despite all the good definitions of relative time, relations between streams (an importar t concept) are still missing: How is it actually known that two events in two different streams are supposed to happen at the same time?

Time stamps associated with these stream events accomplish this task. In order to understand the concept, the structure of the individual ESs needs to be further explored: For all stream types, scene description streams, OD streams, OCI, IPMP, or media streams, there are discrete portions of data related to a specific point in time. Generically, these portions of data are called *access units* (AUs; see Figure 3.16). So, each ES is actually modeled as a sequence of access units. The size and content of AUs completely depend on the media coder used. In an MPEG-4 video stream, for example, it is the coded representation of a *video object plane* (VOP), the arbitrarily shaped equivalent of a video frame. AUs are the data elements to which time stamps can be attached.

elementary stream	access unit 1	AU 2	AU 3	......	AU n	

Fig. 3.16 Elementary stream as a sequence of AUs.

Two different types of time stamps can be attached to an AU:

☞ **Decoding time:** The decoding time stamp for an AU indicates the point in time at which all its data has to be available in the receiver and ideally be decoded at once.

☞ **Composition time:** The composition time stamp indicates the time at which the decoded AU becomes available for composition and subsequent presentation.

Generally, the decoding time is the same as the composition time under the model assumption of instantaneous decoding. However, in case of a video stream there is a difference between those two time stamps if video frames for algorithmic reasons need to be decoded not in their composition order. This is the case with predicted frames (or predicted VOPs for MPEG-4 video) in case bidirectional prediction is used (see Chapter 8). If there is no algorithmic reason for them to differ, there is no distinction between decoding and composition time.

With time stamps available, one can specify that two AUs from different streams are to be decoded or composed at the same time when their respective time stamps show the same value.

3.3.3 Packetizing Streams: The Sync Layer

ESs must be packetized for transport in most delivery scenarios, no matter whether they are to be transported over IP, embedded into a digital TV signal (i.e., an MPEG-2 transport stream), or transmitted over a wireless link. Therefore, MPEG-4 defines a common packetization layer that contains the time stamps. This layer is called the *sync layer* (SL) and the resulting packetized streams are called *SL packetized streams*. Each SL packet consists of a header and the payload. The payload is, at most, one AU, but an AU may be split across multiple SL packets.

The SL packet header is configurable, so that all unneeded header elements can be configured away to allow for maximum compression efficiency. Configuration is done through the SL configuration descriptor (see Table 3.11), the last element of the ESD we will discuss.

The complete syntax listing of the actual SL packet header is suppressed here. The SL packet header syntax elements are discussed instead by looking at the `SLConfigDescriptor` elements. The SL packet header syntax basically uses the elements defined in the `SLConfigDescriptor` to conditionally enable the like-named syntax elements or to determine the number of bits to be parsed from the bitstream, as shown in the syntax excerpt in Table 3.12.

Due to the large number of configuration options, the `predefined` field allows the selection of a complete set of the subsequent options with one value. The set of options associated with a `predefined` value must be registered with ISO or defined in an application-related standard.

Table 3.11 Sync Layer Configuration Descriptor syntax

```
class SLConfigDescriptor extends BaseDescriptor : bit(8)
      tag=SLConfigDescrTag {
    bit(8) predefined;
    if (predefined==0) {
        bit(1)  useAccessUnitStartFlag;
        bit(1)  useAccessUnitEndFlag;
        bit(1)  useRandomAccessPointFlag;
        bit(1)  hasRandomAccessUnitsOnlyFlag;
        bit(1)  usePaddingFlag;
        bit(1)  useTimeStampsFlag;
        bit(1)  useIdleFlag;
        bit(1)  durationFlag;
        bit(32) timeStampResolution;
        bit(32) OCRResolution;
        bit(8)  timeStampLength;        // must be ≤ 64
        bit(8)  OCRLength;              // must be ≤ 64
        bit(8)  AU_Length;             // must be ≤ 32
        bit(8)  instantBitrateLength;
        bit(4)  degradationPriorityLength;
        bit(5)  AU_seqNumLength;        // must be ≤ 16
        bit(5)  packetSeqNumLength;  // must be ≤ 16
        bit(2)  reserved=0b11;
    }
    if (durationFlag) {
        bit(32) timeScale;
        bit(16) accessUnitDuration;
        bit(16) compositionUnitDuration;
    }
    if (!useTimeStampsFlag) {
        bit(timeStampLength) startDecodingTimeStamp;
        bit(timeStampLength) startCompositionTimeStamp;
    }
}
```

Table 3.12 Excerpt of SL Packet Header syntax

```
aligned(8) class SL_PacketHeader (SLConfigDescriptor SL) {
    if (SL.useAccessUnitStartFlag)
        bit(1) accessUnitStartFlag;
            :
            :
            if (SL.useTimeStampsFlag) {
                bit(1) decodingTimeStampFlag;
                bit(1) compositionTimeStampFlag;
            }
            :
            if (decodingTimeStampFlag)
                bit(SL.timeStampLength) decodingTimeStamp;
        :
        :
}
```

The `accessUnitStartFlag` and `accessUnitEndFlag` are elements that permit splitting an AU into multiple SL packets. If that is not desired, these flags are not needed. Most media coding schemes in the MPEG-4 toolbox specify random AUs that can be flagged using the `randomAccessPointFlag`. If a stream consists of random AUs only, this can be signaled up front in the `SLConfigDescriptor` by `hasRandomAccessUnitsOnlyFlag`. Furthermore, byte padding for payload with noninteger byte length can be signaled with a `paddingFlag` and associated `paddingBits`, in order to allow unpacking and correct forwarding of such a payload to the media decoder.

The `idleFlag` indicates that a stream will temporarily stop sending data. The flag, if set, should disable error concealment mechanisms that could otherwise assume that subsequent packets have been lost because of errors. This is mostly useful for audio streams, as video decoders may use means such as freezing the last frame in such a case. In any case, error concealment is not normative and so use of this flag is not normative either.

A loss of SL packets can be detected through the `packetSequenceNumber`. A separate counter exists for AUs, the `AU_sequenceNumber`. This serves the sole purpose of repeating AUs in a controlled way—for example, for error resilience or improved random access to the stream. This important feature is also nicknamed *BIFS carousel*, as it was invented to repeat the BIFS scene description.

The BIFS carousel algorithm works using both the `randomAccessPointFlag` and the `AU_sequenceNumber`, assuming that the decoder of a stream is in either of two modes: Either it is searching for a random access point to start (or restart after an error) decoding or it is already successfully decoding the stream. In the latter case, the decoder can directly exploit the incremental information added by nonrandom AUs. In fact, it would not be advisable to frequently send a complete update of the scene in a random AU to a decoder that is fully synchronized, given the possible complexity of a BIFS scene.

Therefore, the BIFS carousel logically repeats a nonrandom AU in a special manner to provide random access. The fully synchronized decoder should ignore this repetition, because it has already successfully decoded the AU with a given `AU_sequenceNumber`. The repeated AU additionally sets the `randomAccessPointFlag`. Indeed, it must contain a complete update of the scene including all the incremental updates of the nonrandom AUs since the last random access point. With this mechanism, frequent random access points can be established both in scene description streams and OD streams (see Figure 3.5b), without burdening the decoder with additional CPU-intensive work after it has gained access to the stream.

There is a lot of flexibility configuring the `decodingTimeStamp` and `compositionTimeStamp` elements conveyed in the SL header. These time stamps are basically counters with no relation to any wall clock time. Both the `timeStampLength` (the number of bits spent for each time stamp) and the `timeStampResolution` (the number of clock ticks per second) can be set in a wide range in order to adapt to both very low bit-rate applications and high bit-rate or studio applications. The `OCRLength` and the `OCRResolution` can be configured independently.

In some cases, a stream is homogeneously partitioned in AUs of constant duration; for example, each AU might contain a coded audio frame with 24 ms of audio data. Such an `accessUnitDuration` and `compositionUnitDuration` (in case there is not a one-to-one mapping between both) that is constant throughout a stream can be signaled in the `SLConfigDescriptor`, together with the `timeScale` that defines the fraction of a second that corresponds to `accessUnitDuration = 1`.

Specifically in the case of constant AU duration, it is not necessary to tag every AU with a time stamp while maintaining synchronization. It can be indicated with specific flags (`decodingTimeStampFlag` and `compositionTimeStampFlag`; see Table 3.12) in the SL packet header whether time stamps are present.

Finally, time stamps can be omitted altogether and replaced by a start time for decoding of the stream, `decodingStartTime`, and composition of the first composition unit, `compositionStartTime`, signaled in the `SLConfigDescriptor`.

SL packetized streams are transported using an underlying transport protocol stack. The design goal is to convey the same SL packetized stream on various transport protocols. However (as will be explored in depth in Chapter 7), in reality modifications to the SL packetized stream may be required. Therefore, rules exist on how to retrieve the SL packetized stream from the transport format, if so desired. Still, the SL packetized stream serves as a conformance point for MPEG-4, constrained by OD profiles and levels indicating the allowed ranges for the parameters introduced in this section.

3.3.4 Distributed Content, Time Bases, and OCR Streams

A previous example showed how different portions of content can be authored by independent authors and then merged at a later stage (Figure 3.8 and Figure 3.9). It became clear that name scoping is important here to allow such authoring independence. The same is true in the temporal domain. In a context like this it cannot be assumed that both content authors are aware of a common timeline. This is true both in terms of absolute time values (which MPEG-4 does not support, anyway) and in terms of the speed of time. The speed of time (or *time base*) of the two content authors may be off by just a small amount. This means there is drift between them in the long run, as visualized in Figure 3.17.

As an example, assume that the crystals controlling two different encoders are off by 0.1%, which may be realistic for nonprofessional equipment. This means that after one hour of playback, there will be a drift between the two streams of 3.6 seconds. If the receiver relies on the time stamps for establishing the synchronization between the streams, it will now have a buffer with 3.6 seconds' worth of data for the faster stream that it could not yet play out.

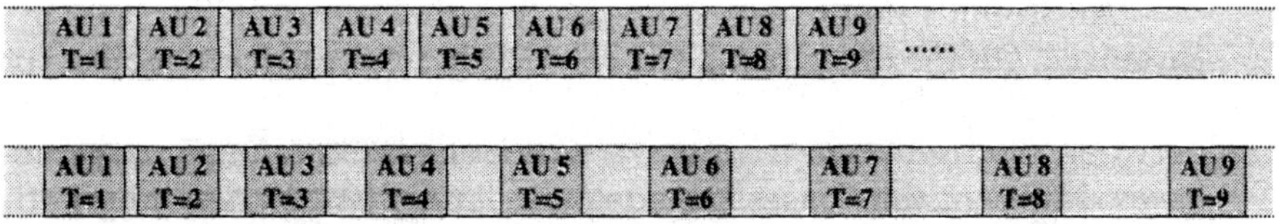

AU 1	AU 2	AU 3	AU 4	AU 5	AU 6	AU 7	AU 8	AU 9	
T=1	T=2	T=3	T=4	T=5	T=6	T=7	T=8	T=9	......

AU 1	AU 2	AU 3	AU 4	AU 5	AU 6	AU 7	AU 8	AU 9
T=1	T=2	T=3	T=4	T=5	T=6	T=7	T=8	T=9

Fig. 3.17 Drift between different time bases.

Drift is not an issue if one of the partial scenes is played back from a local file in sync with the other scene that is broadcast live. Here, the time base of the set of live streams can be taken as the time base for the prerecorded second scene. But if both scenes are fed to the receiver as live broadcasts, the drift pictured in Figure 3.17 (of course, quite exaggerated) will occur.

In order to deal with this problem, another type of time stamp exists: the *object clock reference* (OCR) time stamp. OCR time stamps are sent on a regular basis and contain the current reading of the sender's time base at the time when the OCR time stamp is placed in the stream and sent out. Assuming a delivery with fixed end-to-end delay, this allows the receiver to estimate the speed of the encoder time base. This is basically a clock recovery problem that can be solved by a phase locked loop (PLL) algorithm. The precision of these estimates is largely dependent on the precision (i.e., resolution) of the OCR values.

When the receiving terminal has estimated all the time bases, it can express the time stamps of all streams as readings of one master time base through a simple *affine transform*. Consequently, the temporal relation of all streams is now well established. Taking the example in Figure 3.17, this means that the terminal now knows that it is supposed to play AU9 of the first stream at the same time as AU7 of the second stream, despite the fact that their time stamps differ.

OCR time stamps do not have to be present in each SL packetized stream if multiple streams are slaved to the same time base. Such slaving can be indicated in the ESD (Table 3.3) using the OCR_ES_ID value to indicate the ES_ID of the ES that actually carries the OCR information. It is even possible to dedicate a special stream, a so-called *clock reference stream*, to the sole purpose of conveying these OCR time stamps. In that case, all media streams attached to that time base will point to the ES_ID of the clock reference stream using their OCR_ES_ID fields.

All this signaling is needed because MPEG-4 allows composite content with streams tied to different time bases. In that case, there will be one stream with OCR values for each distinct time base. Formally, these time bases are called *object time bases*, to indicate that they do not have system-wide scope.

Summarizing, it can be said that each object time base is established at the receiver by evaluating a corresponding OCR stream. The decoding and composition time stamps of the AUs within all the media streams are then points on the respective object time lines.

All streams that obey a *single* object time base can always be firmly synchronized. On the other hand, the relation between *different* object time bases is undetermined in the sense that it cannot be expected to have lip sync between an audio and a video stream generated using independent time bases. Different time bases are adequate for content that is inherently loosely synchronized—for example, the audio and video streams from different videoconference participants displayed in one MPEG-4 scene. Of course, the individual pairs of audio and video streams should be tightly synchronized in that case.

Application specifications may require that only content synchronized to one single master time base should be offered to a receiver as a single presentation. That would ease handling of time in a receiver terminal considerably while putting more constraints on the content generation. This is the usual model applied in classical TV broadcast.

3.3.5 Media Time

In order to complete the set of timing definitions in the MPEG-4 specification, we must first admit that it is often important to be able to say, "Play this video stream from T = 20 minutes." So, a mechanism is needed to establish *media time* (or, in MPEG-2 terms, *normal play time*) for a stream.

In case of unicast connections or local playback from file, the implicit assumption that the start of a stream corresponds to media time zero is sufficient. However, in a broadcast application with random tune-in capability, another mechanism is needed. This is the media time descriptor briefly mentioned in the OCI section. Media time descriptors are carried in an OCI stream that accompanies the media stream. The correspondence between the media time line and the object time line is established with the following definition: *The composition time (a value on the object time line) of the AU that conveys the media time descriptor equals the media time value conveyed therein,* as shown in Figure 3.18.

The reason for the inclusion of the media time in the OCI stream is that it becomes easier to edit these media times independently from the media stream. For example, media assets will be used in different productions; hence, the same frame of a video stream may correspond to different media times in those different productions. Furthermore, in broadcast applications, where streams might continue endlessly, associating a media timeline may be inconvenient altogether.

3.3.6 System Decoder Model

An MPEG-4 terminal will, in many respects, have limited resources. The amount of processing power and memory and perhaps the number of concurrent processes are some of those resources. In order to exploit the capabilities of a given terminal optimally without exceeding the available resources, it is

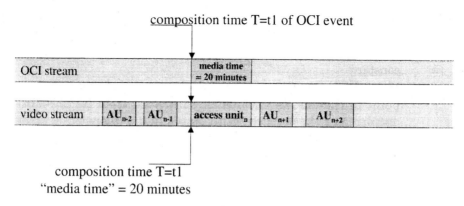

Fig. 3.18 Associating media time to object time via the media time descriptor.

helpful to have a model of this terminal. In fact, the complete model is composed of several elements. The profile and level mechanism mentioned earlier is one of them (see Chapter 13). Another element is the System Decoder Model (SDM), which models the ideal behavior of an MPEG-4 terminal with respect to buffering and timely processing of streamed data. The structure of the decoder model is depicted in Figure 3.19. It assumes that data for each SL packetized ES is delivered to its individual decoder buffer DB_n. Depacketization occurs at the input to this buffer, so that the DB_n in the model only buffers the raw ESs. Each DB_n connects to a decoder. In case of scalable coding of some media, more than one decoding buffer will connect to a single decoder. The decoded output is placed in a composition memory from which the compositor may fetch it.

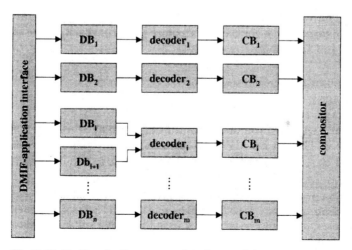

Fig. 3.19 Buffers in the system decoder model.

The important assumptions of the SDM are these:

☞ **Fixed end-to-end delay:** For each individual ES, the time for transmission of data from the stream encoder output to the decoder buffer input is constant.

☞ **Instantaneous decoding:** An AU that is bound to decoding at a given point in time is instantaneously removed from its decoding buffer, decoded, and the decoded result placed in the composition memory. This implies that the complete AU must be delivered to the decoding buffer at the decoding time. The size of each decoding buffer can be set through the bufferSizeDB field in the DecoderConfigDescriptor in order to adapt to the needs of each stream.

The decoded AUs are called composition units and are placed in the composition memory (buffer) CB_n of the corresponding stream. The size of the composition memory is not managed explicitly as in the case of the decoder buffer, where a precise size value can be signaled.[6] The model assumes that each composition unit stays available in the composition memory until the composition time of the subsequent composition unit is reached. During that period of time, the compositor may access the composition unit as often as required.[7] Depending on the composition frame rate, a composition unit may be visible in zero, one, or more composition frames. Apparently, this concept is only relevant for visual information, however. Differences in sampling rate of audio composition units are typically dealt with by proper sample rate conversion.

Notably, the modeling of the MPEG-4 terminal behavior as presented so far is complete from an MPEG-4 Systems perspective, despite the fact that it does not take into account the behavior induced by the delivery layer (i.e., the transport protocol stack). Specifically, the jitter due neither to multiplexing of streams nor to the violation of the *constant end-to-end delay* assumption is accounted for. The terminal model needs to be extended appropriately by a specific *DMIF instance* for a concrete transport protocol stack. Some examples will be detailed in Chapter 7.

3.4 SUMMARY

This chapter has explained the means to manage and synchronize potentially large numbers of ESs in an MPEG-4 presentation. ODs describe sets of streams related to a single media object. The OD is also the link between the

---

6. The composition memory size required for a given presentation depends on the terminal implementation. For a specific implementation, the composition memory size for a given presentation may be estimated from the profile and level indications.

7. Note that visual composition is not part of the MPEG-4 specification, whereas audio composition is specified in MPEG-4 Systems.

streaming resources and the scene description that will be detailed in the next chapter. Object content information adds further detail to the description of each media object, and the IPMP framework allows attaching proprietary systems to identify and protect content based on the intellectual property rights associated to it.

Timing behavior and synchronization of ESs is based on a system decoder model and supported by well-known time stamp mechanisms, embodied in an SL syntax, that are enhanced to allow multiple time bases within a single presentation. Furthermore, the relation of time events within the scene description to the time stamps of the associated media streams is defined. Finally, it is noted that the SL packetized streams represent an intermediate layer that is further encapsulated in a transport protocol stack, as will be explained in Chapter 7.

3.5 REFERENCES

[H245] ITU-T Recommendation H.245. *Control Protocol for Multimedia Communication*. February 2000. *www.itu.int/itudoc/itu-t/rec/h/h245.html*

[MPEG2-1] ISO/IEC 13818-1:2000. *Generic Coding of Moving Pictures and Associated Audio Information—Part 1: Systems*. 2000.

[MPEG4-1] ISO/IEC 14496-1:2001. *Coding of Audio-Visual Objects—Part 1: Systems*, 2d Edition, 2001.

[MPEG4-2] ISO/IEC 14496-2:2001. *Coding of Audio-Visual Objects—Part 2: Visual*, 2d Edition, 2001.

[MPEG4-3] ISO/IEC 14496-3:2001. *Coding of Audio-Visual Objects—Part 3: Audio*, 2d Edition, 2001.

[MPEG4-6] ISO/IEC 14496-6:2000. *Coding of Audio-Visual Objects—Part 6: Delivery Multimedia Integration Framework (DMIF)*, 2d Edition, 2000.

[N4509] MPEG Requirements. *Overview of the MPEG-7 Standard*. Doc. ISO/MPEG N4509, Pattaya MPEG Meeting, December 2001.

[RFC2045] *RFC 2045, 2046 (and others) on Multipurpose Internet Mail Extensions (MIME)*. November 1996. *www.ietf.org/rfc/rfc2045.txt*

[RFC2326] Schulzrinne, H., A. Rao, and R. Lanphier. *Real Time Streaming Protocol (RTSP)*. November 1996. *www.ietf.org/rfc/rfc2326.txt*

[RFC2327] Handley, J. *SDP: Session Description Protocol*. April 1998. *www.ietf.org/rfc/rfc2327.txt*

[S315M] SMPTE 315M-1999. *Television–Camera Positioning Information Conveyed by Ancillary Data Packets*. 1999.

BIFS: Scene Description

by Jean-Claude Dufour

Keywords: hierarchical scene description, VRML, binary encoding, scene examples, node reference

The biggest difference in MPEG-4, with respect to MPEG-2 and MPEG-1, is the notion of object. An MPEG-2 program typically is constituted of two MPEG-4 objects: one full-screen video object, and one audio object. MPEG-4 content, on the other hand, may be built with any number of audiovisual objects (subject to the constraints imposed by the profile@level definitions, see Chapter 13). Each object can be of any of a large number of object types, such as rectangular video, natural audio, video with shape, synthetic face or body, generic 3D objects, speech, music, synthetic audio, text, or graphics. As these objects are from either the visual or aural worlds, there is a need for visual and aural composition of objects. There are many ways to compose objects on the screen[1] and in sound space, including spatial and temporal composition. Spatial composition information includes items such as these: Is this a flat (2D) world or a 3D world? Where does this object go? In front of or behind this other object? With which transparency and with which mask? Does it move? Does it react to user input? Sound composition information includes some of the preceding as well as some specific twists,

---

1. *Screen* refers generically to the part of the terminal that is in charge of conveying visual information to the user: TV screen, computer monitor, head-up display, and so on.

such as room effects. Temporal composition information includes the time (relative to the scene or to another object's time) when an object starts playing. So, there is a need to specify object composition directives. As what the viewer sees and hears is called a *scene*, these directives are gathered into the so-called *scene description*.

This chapter presents the principles of MPEG-4 scene description, called *binary format for scenes*, or *BIFS*. BIFS is the scene description format designed by the MPEG Systems subgroup [MPEG4-1] to complement the OD framework presented in Chapter 3. Whereas the OD framework defines the objects and their characteristics, BIFS defines how the objects are combined together for presentation.

The development of the MPEG-4 scene description format started when the virtual reality modeling language (VRML) [VRML97] was gaining momentum in the 3D community. VRML was brought to the MPEG Systems subgroup as a candidate for the core of the MPEG-4 scene description tool and became the kernel of MPEG-4 BIFS. Being a 3D textual language for download-and-play, VRML did not address many of the MPEG-4 scene description requirements [N4319]; thus there was need for many improvements and additions, which are described hereafter.

This chapter will not deal with the audio part of the scene description, beyond a short presentation. The audio part of the scene description, due to its complexity and close relation with synthetic audio tools, is described in Chapter 12. Moreover, the mesh, face, and body representation tools, although closely related to BIFS nodes, are described in detail in Chapter 9. All BIFS samples are given in XMT-A format, presented in Chapter 6.

4.1 BASICS OF BIFS

After an introduction to the basic concepts of BIFS, the main BIFS features are presented in this section, from the usage point of view.

4.1.1 Scene and Nodes

What the user of the MPEG-4 terminal sees and hears is the *scene*. The scene has a visual and an audio component. The visual component (i.e., what is on the screen) can be a 2D or a 3D world. A 2D scene is one in which objects have the same visual size regardless of their being in front or in the back. This type of representation is sometimes called 2.5D because, although the objects are *flat*, they have a depth order, which is a first step toward the third dimension. 3D fanatics would call a 2D world *the orthographic projection of a 3D world.*

The sound component of a scene can have one source (mono), two sources (stereo), or more (e.g., 5 + 1 channels).

One option for organizing objects into a scene could have been to use a simple display list with a set of composition properties per object. It has been found beneficial to build the scene as a hierarchical structure or *scene tree*, with the visible or audible objects being leaves of the tree or *leaf nodes*, and composition directives being attached to branching points or *grouping nodes*. Indeed, the tree structure allows objects to be grouped semantically: For example, the scene representing a room could contain one group of objects per furniture element; each further element would be further decomposed into further subgroups, and so on, recursively. In such a scene tree, the composition properties of a leaf node are the accumulation of the composition properties of all parent (higher-level) nodes. For example, translation is attached to a grouping node (Transform2D, introduced later), and the translation of a leaf node is the result of the addition of the translations attached to any of its parents up to the root node.

In the simplified scene tree of Figure 4.1, the following nodes can be seen: a Movie node, a Sound node, the Root node (which is a grouping node), and a Transform node (which sets the position of the Text node). The leaf nodes are bounded with a wide line and correspond to the audiovisual objects in the scene.

The scene has a single root node, which has to be a grouping node. The most frequent grouping nodes are these:

☞ **Group:** This is a simple container.

☞ **Transform:** This defines a 3D coordinate system for its children that is relative to the coordinate systems of its parents or ancestors.

☞ **Transform2D:** This defines a 2D coordinate system for its children that is relative to the coordinate systems of its ancestors.

☞ **OrderedGroup:** This defines the rendering order of its children regardless of their order within its children field in 2D, or their distance to the viewpoint in 3D.

Some nodes have names or IDs so they can be designated, and others are anonymous because they will never need to be reused or modified. As a DEF assignment gives a name to a node in VRML, so it gives a numeric ID to a node in BIFS. This ID can then be referred to through a USE clause (see the

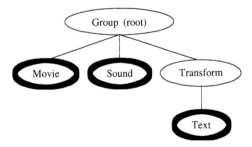

Fig. 4.1 Simplified scene tree.

Magnifying Glass example in Section 4.2.6). In the scene tree, wherever a node is needed, a USE reference to a node of the same type can be inserted. Examples of USE are reusing a font style, an appearance, multiple identical instances of an object such as the columns of a temple, and so on. Because multiple references to the same node are allowed, the scene is not really a tree but a directed a-cyclic graph.[2]

4.1.2 Fields and ROUTES

Nodes are built of primitive components called *fields*. Fields are the attributes and the interface of the nodes. A field has the following characteristics:

☞ A **value,** with possibly a default value.

☞ A **type of the value,** as described in Table 4.1. The prefix SF means *single field* (i.e., a field with a single value, even if the value has two, three, or four components), as opposed to MF for *multiple field* (a field that contains an array of values).

☞ A **type of behavior,** which tells whether the field is constant (field), modifiable (exposedField), only a source of events (eventOut), or only a target of events (eventIn). An exposedField may be both source and target of events; an initial value may be specified for field and exposed-Field, not for eventIn and eventOut;

☞ A **name,** which is a plain string for documentation purposes, as it does not appear in the binary stream.

Table 4.1 Field types

Field type	Description of the value of the field
SFBool	A Boolean
SFColor	An RGB color
MFColor	An array of colors
SFFloat	A float (32 bits)
MFFloat	An array of floats
SFImage	An image specified as width, height, number of pixel components, and array of pixels[*]
SFInt32	An integer (32 bits)
MFInt32	An array of integers

2. A directed graph is a graph whose edges are ordered pairs of vertices. That is, each edge can be followed from one vertex to the next. A directed a-cyclic graph is a directed graph where no path starts and ends at the same vertex.

Table 4.1 Field types (Continued)

Field type	Description of the value of the field
SFNode	A node, or a reference to a node
MFNode	An array of nodes or of references to nodes[†]
SFRotation	A quadruplet of floats, allowing the encoding of rotations in 3D space
MFRotation	An array of quadruplets of floats
SFString	A UTF-8 string[‡]
MFString	An array of UTF-8 strings
SFUrl	Either a numbered reference to an OD, or a plain URL in a string[**]
MFUrl	An array of SFUrls as described above
SFTime	A double value containing a time in seconds[††]
MFTime	An array of times
SFVec2f	A pair of floats, encoding the X and Y coordinates in 2D space
MFVec2f	An array of pairs of floats
SFVec3f	A triplet of floats, encoding the X, Y, and Z coordinates in 3D space
MFVec3f	An array of triplets of floats
SFCommand-Buffer	A set of untimed BIFS-Commands (see Section 4.1.5.1)

[*]Authors should be aware that using the SFImage field of the PixelTexture node means transmitting uncompressed images. SFImage and PixelTexture are only present for backward compatibility with VRML.

[†]A node containing an MFNode field is called a grouping node.

[‡]UTF-8 is a standard for string encoding. It is defined in ISO 10646-1:2000 Annex D and also described in RFC 2279 as well as Section 3.8 of the Unicode 3.0 standard.

[**]In the bitstream, SF/MFUrl has a specific encoding; in the scene tree, SF/MFUrl appears as an SF/MFString, and a reference to OD 3 appears as the string "od:3."

[††]The origin of time is the beginning of the stream. See the discussion on media time in Chapter 3.

Some node fields are active and emit events. For example, the BIFS timer, TimeSensor, emits float events, which are values in the interval [0,1] indicating where the scene is within the current cycle of this timer.

ROUTES are the means to transmit these events to the appropriate target field. ROUTES are composed of one source field and one target field, each field specified by the DEF ID of the node to which it belongs, followed by the rank or index of the field in the node specification. ROUTES connect one field to another of the same type. Whenever the value of the first field changes, the mechanism of the ROUTE mirrors the change in the other field.

4.1.3 Node Types

There are many kinds of nodes in addition to the grouping nodes that create the structure of the scene tree. These are explained shortly.

The Shape node deserves special explanation. Any entity in the scene that should be visible has to be attached to a Shape node. The Shape node abstracts the *geometry* or *form* of the visible entity from its *appearance* characteristics. The geometry field of the Shape node contains a geometry node (2D or 3D), for example, Rectangle, Circle, Box, Sphere, Bitmap. The appearance field of the Shape node contains an Appearance node, which carries a Material node specifying color, transparency, and other aspect attributes, as well as texture-related nodes for optional texture information.

The visual media nodes are texture nodes (that is, nodes that are contained in an Appearance node). The ImageTexture and MovieTexture nodes provide a connection between a media stream and the scene.

The audio equivalent of the Shape node are the Sound and Sound2D nodes. They are containers for an audio subtree, as explained in Chapter 12. The sensor nodes allow the creation of interactive scenes (i.e., scenes whose aspect or behavior varies depending on user actions). All sensor nodes but the TimeSensor node interpret user interaction data coming from a pointing device such as a mouse. The sensors map such data to sets of events on specific eventOut fields. When such fields are routed to the appropriate targets, user interaction can trigger an action—for example, move an object in a plane within bounds or rotate an object as a sphere or a cylinder.

The interpolator nodes allow the mapping of a number between 0 and 1, representing a fraction, to any set of values by piecewise linear interpolation. They are used mostly in conjunction with a TimeSensor node, which emits such fraction events. All numeric values can be interpolated. By interpolating a set of positions, an object can be moved over time.

The use of Script or Conditional nodes further expands the possibilities of interaction. A Script node is a node interface wrapped around an ECMA[3]-Script (or JavaScript) function [ECMA99]. This allows complex transcoding of events. A Conditional node contains a (constant) set of changes to the scene—for example, popping a menu in front of the scene. Execution of the set of changes in a Conditional node is triggered by an event.

A PROTO defines a new node type in terms of already defined (built-in or prototyped) node types. It is a form of BIFS macro.

---

3. European Computer Manufacturers Association (ECMA) is an international industry association founded in 1961, dedicated to the standardization of information and communication systems.

4.1.4 Subscenes and Hyperlinks

Two scenes can be related in different ways. A subscene is a scene that has been designed to fit inside another scene. For example, a subscene may contain the description of a room, within a scene that describes a house. The rationale behind splitting a scene into one main scene and some subscenes can be one of these:

- ☞ The complete scene is too big to be managed easily.
- ☞ Parts of the scene need to be modified more often than others.
- ☞ Different authors design the various subscenes.
- ☞ The subscene needs to be protected from modifications of the main scene, or vice versa.

The way to insert a subscene into a scene is to use an `Inline` node. The `Inline` node is a grouping node that reads its children data from another scene, whose location is defined by the content of the `Inline` URL field. The subscene streams may be in the same package, come from the same server as the main scene, or be independent from the main scene (see Chapter 3 for a description of URLs and `Inlines` from the stream management point of view).

The `Inline` node establishes absolute barriers between the main scene and the subscene. The node ID scopes are totally separate,[4] and therefore there is no way for actions on nodes of one of the scenes to influence the nodes in the other scene.

Another type of relationship between two scenes is hyperlinking. The `Anchor` node is a grouping node whose children act as hotspots and which contains a URL field. A mouse click on `Anchor`'s children geometries has the same result as a mouse click on a link of an HTML page: The target of the URL, which should be another MPEG-4 scene, becomes the active scene in the terminal. Thus a Web site can be constructed with MPEG-4 scenes.

4.1.5 Scene Changes

One of the greatest assets of BIFS is its dynamic quality. Scenes can be designed to change over time. So a scene has a set of states, similar to the successive frames of a video. And in the purest MPEG tradition, BIFS scenes have initial states similar to video I (intra-coded) frames, and predictive states similar to video P (predicted-coded) frames. Scene I frames are hereafter called the *initial scene*, and Scene P frames are called *scene updates*. A BIFS scene description stream is an MPEG-4 elementary stream (see Chapter 3).

---

4. The scope of node IDs is usually restricted to the scene—that is, a node ID is *known* only within the same scene. There is no way to refer to a node otherwise than through its node ID. Thus, for two scenes, if the node ID scopes are separate, no action in one scene can change anything in the other scene.

The initial scene is packaged into an AU, which is a random access point, and scene updates are packaged into subsequent AUs, which are not random access points. The initial scene contains first a list of PROTOS (see Section 4.3.2), then the initial scene tree, and finally a list of ROUTES.[5] All scene updates, which occur at the same time, go into one single AU.

There are two mechanisms for changing a scene:

1. **BIFS-Commands** are *single* changes to the scene, packaged in AUs of the scene description ES. Changing the color of one object or the position of another may be done with a BIFS-Command.

2. **BIFS-Anim** streams are separate streams of structured changes to a scene. Animating a mesh or a face may be done with BIFS-Anim.

BIFS-Anim and BIFS-Commands should not be confused with scene animations, which are created with TimeSensors and interpolators. Scene animations are intrinsic changes (i.e., described inside the scene itself), and thus the changes are already present in the terminal before the animation starts. No external scene change is sent to or read by the terminal in the case of scene animations.

4.1.5.1 BIFS-Commands The basic BIFS-Commands are Insert, Delete, and Replace. These commands can be applied to a node, a ROUTE, a field, or a value in a multiple field. Only a node or a ROUTE with an ID can be modified by BIFS-Commands. To modify a field, the container node needs to have an ID. See Table 4.2 for a complete picture of BIFS-Commands.

In Table 4.2 the *Command* column contains the type of action. The *Target* column contains the object on which the action is to be executed. The *Where* column contains, when the target is a multiple field, the position of the target value. The *What* column indicates what is inserted or by what the target value is replaced. Finally, the *Description* column describes the action to be performed.

BIFS-Commands can be found in two places: BIFS AUs and the buffer of Conditional nodes. The time stamp of the BIFS AU enclosing a BIFS-Command determines the time at which the command should be executed. The reception of the appropriate event by the Conditional node triggers the execution of the timeless BIFS-Commands in its buffer. In other words, the time stamp of the triggering event becomes the composition time stamp of the BIFS-Commands contained in the buffer of the Conditional node.

All fields of type exposedField and eventIn are modifiable. Modifying an exposedField changes the value of this field and all other fields routed from

---

5. Current practice is that there is only one random access point in the scene ES, at the beginning. This means that the random access information (the so-called *carousel*) is provided in another ES. This choice is made because the carousel parameters are dependent on the delivery scenario, but not on the scene itself.

Table 4.2 BIFS-Commands

Command	Target	Where	What	Description
Insert	field children of grouping node <nodeID>	First, last, or at <index>	<node>	Inserts the <node> in the children field of the node <nodeID> at the specified position.
Insert	field <fID> of node <nodeID>	First, last, or at <index>	<value>	Inserts the <value> in the <fID> field of the node <nodeID> at the specified position.
Insert	–	–	ROUTE	Inserts the ROUTE in the scene.
Delete	<nodeID>	–	–	Deletes all instances of the node <nodeID>.
Delete	field <fID> of node <nodeID>	First, last, or at <index>	–	Deletes the value in the <fID> field of the node <nodeID> at the specified position.
Delete	<routeID>	–	–	Deletes the ROUTE <routeID>.
Replace	<nodeID>	–	<node>	Replaces all instances of node <nodeID> by the <node>.
Replace	field <fID> of node <nodeID>	–	<value>	Replaces the (whole) value of the <fID> field of the node <nodeID> by <value>.
Replace	field <fID> of node <nodeID>	First, last, or at <index>	<value>	Replaces one value of the <fID> multiple field of the node <nodeID> by <value> at the specified position.
Replace	<routeID>	–	<ROUTE>	Replaces a ROUTE <routeID> by <ROUTE>.
Replace	Scene	–	–	Defines a new scene.

this field. Modifying an eventIn triggers the same mechanism as when a value is routed to this eventIn.

4.1.5.2 BIFS-Anim The BIFS-Anim framework is composed of three elements: the *animation mask*, the *animation frames*, and the AnimationStream node. The nodes whose fields need to be animated are given a DEF ID. An ordered list of the fields to be animated is built and placed into the animation mask. Lists of values of the corresponding fields are placed into successive animation frames. The animation mask constitutes the DecoderSpecificInfo (see Chapter 3) of the BIFS-Anim decoder. The animation frames constitute the AUs of the BIFS-Anim ES. In order to allow the user to interact with the animation, an AnimationStream node is inserted into the scene: Through sensors routed to the AnimationStream node or through BIFS-Commands, the animation can be started or stopped.

Only numeric field types can be animated: SF/MFInt32, SF/MFFloat, SF/MFRotation, SF/MFColor, SF/MFVec2f, and SF/MFVec3f. An animation frame can send values in *intra* or in *predictive* mode. In intra mode, the field values are quantized (see Section 4.1.8). In predictive mode, the difference between the quantized current value and the last transmitted value is coded. The encoding is performed using an adaptive arithmetic coder.

4.1.6 Scene Rendering

The possibility of efficiently rendering both 2D and 3D scenes within the same renderer has yet to be demonstrated.[6] Optimizations for 2D rendering are orthogonal to, or even incompatible with, optimizations for 3D rendering. Thus, it has been decided to make the rendering context of an MPEG-4 scene explicit: The scene is declared either 2D or 3D.

To allow the use of both 2D and 3D objects within the same scene, specific nodes have been designed. The Layer2D and Layer3D nodes define a local 2D and 3D rendering context, respectively. The CompositeTexture2D and CompositeTexture3D nodes define off-screen rendering contexts which can be used as a texture to be mapped onto other objects.

The appropriate rendering method (or context) for a scene can be determined by looking at the top node of the scene. A Layer2D or OrderedGroup root declares the scene as 2D. A Layer3D or Group root declares the scene as 3D.[7]

Once the default rendering context is established, the author may want to escape to the other rendering context for a part of the scene. Examples of use are these:

☞ Showing a 3D object in a 2D scene, as in an e-shop application, where the proposed item can be viewed from all angles and manipulated

☞ Showing a 2D object in a 3D scene—for example, an HTML-like page in a 3D environment

☞ Using a rendered scene as a texture to be mapped onto a 2D or 3D object

Display units can be meter or pixel. The origin of the coordinate system is at the center of the rendering area; the x-axis is positive to the right, and the y-axis is positive upward. The width of the rendering area represents −1.0 to +1.0 (meter) or −w to +w (pixel, when the rendering area width is 2w) on the x-axis (see Figure 4.2). The extent of the y-axis is determined by the aspect ratio[8] of the rendering area.

---

6. Rendering is the process of creating the image presented to the user, starting from the scene description and the decoded objects.

7. Group can also be used in a 2D context, and OrderedGroup in a 3D context, but not as a root of the scene tree.

8. The aspect ratio of a rectangle is its width divided by its height.

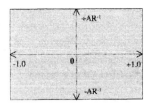

Fig. 4.2 Meter coordinate system (AR stands for aspect ratio).

4.1.7 Binary Encoding

MPEG-4 scenes are encoded in binary form, which is one of the main improvements of BIFS over VRML. As mentioned earlier, the scene description is sent to the receiving terminal as any other ES and, as such, it is split into AUs. The BIFS AUs are *Command Frames*. Command Frames are composed of any number of BIFS-Commands with the same composition time stamp. The initial scene is encapsulated in a Replace Scene command (see Table 4.2), which contains the top node and possibly PROTOs and ROUTES. Following commands can be any of the types described in Table 4.2.

BIFS nodes are essentially encoded as a Node Data Type tag to recognize the kind of node (e.g., TimeSensor), possibly an ID (DEF) for later reference, followed by the fields with nondefault values. The fields may be encoded as field index–field value pairs, or as a Boolean table defining which field values are encoded, followed by the field values. (See Figure 4.3 for a simplified binary structure of the node.) The Node Data Type tag length, value, and meaning are context-dependent. For each SFNode or MFNode field, a Node Data Type is given, defining the list of allowed subnodes. The tag length is defined by the number of possible subnodes.

BIFS fields are encoded by default as 1 bit for SFBool, 32 bits for *integer* and *float* values, and 64 bits for *time* values. SFVec2F, SFVec3F, and SFRotation are pairs, triplets, and quadruplets of floats, respectively. MF types are encoded either as *lists* (branch above in Figure 4.3) or as *vectors* (branch below in Figure 4.3). A list description consists of a list of (ID, value) pairs separated by 1 bit meaning *stop or more*. A vector description contains 1 mask bit per field in the current node: Each mask bit set to 1 is followed by the value of the corresponding field. BIFS field values may be quantized in order to increase the compression ratio.

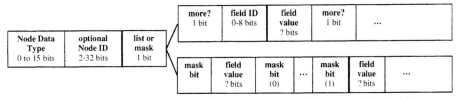

For the nth mask bit: 0 - no value for field n; 1 - value follows for field n

Fig. 4.3 Simplified node binary structure.

Typical encoded node sizes are very small for nodes like Group (apart from the encoding of its children), most of the bits being used by the Node ID. Thus, to avoid wasting bits, it is advisable to use as few Node IDs as possible. For nodes with *float* or *coordinate* fields, most of the bits are used for the encoding of those floats, whereas the node structure encoding is almost negligible.

4.1.8 Quantization

Quantization is the key to achieving good compression rates when using scenes with lots of coordinates (i.e., polygons or interpolators). For each numeric field of each node type, a quantization type is specified, according to the way the value is used.

For each quantization type, and unless stated otherwise, the quantization is linear between minima and maxima. The target number of bits, as well as the minima and maxima per category, may be set through the use of a QuantizationParameter (QP) node.

The QP node can be placed in any grouping node and takes effect either on all further sibling nodes (and their children) or just on the next sibling node (and its children). Where a QP is active, another QP can be defined, which then supersedes locally the effect of the previous QP. The previous QP becomes active again at the end of the scope of the new QP—that is, at the end of the siblings of the new QP.

To give a simple example, a 2D cartoon designed with only polygons has a size roughly proportional to the number of vertices. Because vertices are limited to [0,2000] in both dimensions and can be compressed from 2x32 bits, which is the size of an unquantized SFVec2F, to 2x11 bits (2^{11} = 2048), a compression ratio of roughly 3 to 1 can be achieved. Table 4.3 describes the quantization types available for values within MPEG-4 nodes.

4.2 BASIC BIFS FEATURES BY EXAMPLE

In this section, scenes of increasing complexity are described, in order to introduce the basic BIFS features in a gradual manner. To describe concisely and yet precisely the building of a scene, a textual BIFS representation is needed. The MPEG-4 XMT-A format was the natural choice, although it is only introduced in Chapter 6.

In the first three examples, elements of a reformulation of DVD content as MPEG-4 content are explored.

4.2.1 Trivial Scene

The most trivial and yet useful scene in MPEG-4 content may be the typical *MPEG-2 scene*, or the plain DVD movie chapter. The visual part is made of a

Table 4.3 Quantization types

Quantization type	Typical use of the field
3D positions	`Vec3f` used as translation or vertex
2D positions	`Vec2f` used as translation or vertex
Drawing order	Single pseudo-depth value
`SFColor`	RGB value
Texture Coordinate	2D coordinate for the placement of the texture with respect to the object
Angle	Single value
Scale	Single positive value
Interpolator keys	Single value in the [0, 1] interval
Normals	Normalized `Vec3f`, quantized with a special method based on quaternions [HeBa97]
Rotations	Quadruplet of floats, quantized with a special method based on quaternions
Object Size 3D	`Vec3f` with positive values
Object Size 2D	`Vec2f` with positive values
Linear Scalar Quantization	Integers
`CoordIndex`	Special case for small integers

rectangular full-screen video, and the audio part is a stereo sound. This is a 2D scene with no interactivity.

Figure 4.4 shows an abstract representation of the scene tree. The full XMT representation of the scene tree corresponding to the *Trivial Scene* example is presented following Figure 4.4. The rightmost column presents the size of the binary encoding in bits; the fields that do not need to be encoded because their value is equal to the default value are shown in italics.

Lines 1 and 17 are the traces of the BIFS-Command that creates the initial scene (see Section 4.1.5). The root node is Layer2D (line 2), which defines the rendering context to be 2D. Lines 3 to 6 define the audio part of the scene, a 2D sound node containing an AudioSource pointing to the stream described by ObjectDescriptor 3 (not shown), starting immediately and playing until the end of the stream. Lines 7 to 15 define the visual part of the scene—the video. All visual elements are placed as geometries or appearances in a Shape node. Its appearance field contains an Appearance node specifying the visual attributes (e.g., material and texture) to be applied to the geometry. Its geometry field contains a geometry node.

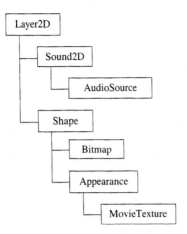

Fig. 4.4 Scene tree abstract for the *Trivial Scene*.

```
 1    <Replace> <Scene>                                    10
 2     <Layer2D> <children>                                12
 3      <Sound2D> <source>                                 11
 4       <AudioSource url="od:3" startTime="0.0"           29
 5                           stopTime="-1.0"/>             65
 6      </source> </Sound2D>                                1
 7      <Shape>                                             8
 8        <geometry> <Bitmap/> </geometry>                 7
 9        <appearance>                                      1
10         <Appearance> <texture>                          6
11          <MovieTexture url="od:4" loop="false"         28
12                  startTime="0.0" stopTime="-1.0"/>      65
13         </texture> </Appearance>                        2
14        </appearance>                                     1
15      </Shape>                                            1
16     </children> </Layer2D>                               1
17    </Scene> </Replace>                                   2
```
_____

                         Total size in bits: 250

The geometry node here is a Bitmap node. Bitmap is an MPEG-specific node allowing efficient 2D rendering by making trivial the mapping of the texture specified in the appearance field. By default, a texture is stretched to fit onto the geometry to which it is applied. On a Bitmap node, the texture is mapped as is, without transformation. As the resizing of a video is a very expensive operation, using a Bitmap with the video texture is much less resource-intensive than using a Rectangle. The video, described in Object-Descriptor 4 (not shown) and referred by the MovieTexture node in line 11, is assumed to have the same size as the screen. Placing the Shape node directly in the Layer2D node without any Transform2D implies that the Shape is rendered at the center of the screen. The video starts immediately, never stops (i.e., is played until its end), and does not loop. This scene description has a total binary size of 32 bytes.

4.2.2 Movie with Subtitles

A slightly more complex example involves the same scene with additional subtitles. The initial scene is almost the same as the preceding. The XMT representation of the first part of the scene tree corresponding to this example is presented in the following:

```
1    <Replace> <Scene>
2     <Layer2D> <children>
3      <Sound2D> <source>
...          ...
6      </source> </Sound2D>
7      <Shape>
...          ...
15     </Shape>
16     <Transform2D translation="0 -100"> <children>
17      <Shape>
18       <geometry>
19        <Text DEF="subtitle" string="">
20         <fontStyle>
21          <FontStyle family="'Serif'" size="20.0"
22              justify="'MIDDLE' 'MIDDLE'"/>
23         </fontStyle>
24        </Text>
25       </geometry>
26       <appearance> <Appearance>
27        <material>
28         <Material2D emissiveColor="1 1 1"/>
29        </material>
30       </Appearance> </appearance>
31      </Shape>
32     </children> </Transform2D>
33    </children> </Layer2D>
34   </Scene> </Replace>
```

The additional parts with respect to the previous example are lines 16 to 32. A Text node (line 19) with ID subtitle has been added and placed as geometry in a Shape node. The Shape node is placed in a Transform2D node (line 16), which positions the Text node in the bottom part of the screen, assumed to be of size 320x240. The ID subtitle will be transformed into a number by the XMT-to-BIFS encoder. The Text node is empty at the beginning (line 19). The font is Serif, with a size of 20 pixels. The content of the justify field (line 22) means that the text will be centered horizontally and vertically around the reference point, which is 100 pixels down from the center of the

screen. The appearance of the Shape node contains, as material (line 28), a
Material2D node, which defines the text color as white.

The following XMT fragment shows the second part of this example, corresponding to the scene updates.

```
35    <par begin="3s">
36     <Insert atNode="subtitle" atField="string"
37            position="BEGIN"
38            value="'Here&apos;s looking at you, Blue Eyes'"/>
39    </par>
40    <par begin="8s">
41     <Delete atNode="subtitle" atField="string"
42            position="END"/>
43    </par>
44    <par begin="9s">
45     <Insert atNode="subtitle" atField="string"
46            position="0"
47            value="'Cheers!'"/>
48    </par>
49    <par begin="13s">
50     <Replace atNode="subtitle" atField="string"
51            Position="0" value="'foo bar'"/>
52    </par>
...    ...
```

This part of the scene description is composed of insertion, deletion,
and replacement of the string in the Text node. The single changes are encapsulated in the timed construct <par> </par>. The first command inserts one
string at the beginning of the string array after 3 seconds. The second command deletes the last (and only) string at 8 seconds. Another string is inserted
at 9 seconds, at a position specified by a number (the index is 0-based). The
string is then replaced at 13 seconds.

The size and position of the text and the size of the screen would allow it
to display two strings, which, with this justify mode (line 22), would be displayed centered, one below the other.

The FontStyle node has many options, notably these:

☞ The **family** field specifies a list of possible font families; the terminal
  uses the first family that corresponds to an installed font.

☞ The **horizontal**, **leftToRight**, and **topToBottom** fields control the direction of writing and of justification, if any.

☞ The **justify** field specifies a combination of BEGIN, FIRST, MIDDLE, and
  END, which define the position of the text with respect to the reference
  point.

☞ The **language** field specifies the context of the language for the text
  string.

☞ The **size** field specifies the size (in meters or pixels) of the maximum
  height of the characters.

☞ The **spacing** field specifies the spacing between adjacent lines, relative to size.

☞ The **style** field specifies the character style; it can take the values PLAIN, BOLD, ITALIC, or BOLDITALIC.

4.2.3 Icons and Buttons

In the previous two examples, an equivalent of the DVD chapters has been defined. In order to pursue the parallel, the mechanisms behind the DVD menu must be explored. In order to create an icon, the construct presented in the following is needed.

```
1    <Shape>
2      <geometry> <Bitmap/> </geometry>
3      <appearance>
4        <Appearance>
5          <texture>
6            <ImageTexture DEF="N8" url="od:2" />
7          </texture>
8        </Appearance>
9      </appearance>
10   </Shape>
```

The snippet is very close to the one used to display the video in the *Trivial Scene* example. The difference comes from the use of the ImageTexture node. The ImageTexture node is basically a MovieTexture node designed for a video with one single frame. So the main field is the url field that points to the stream described by Object Descriptor 2, which contains an image.

In order to create a button, three images are needed: the normal image of the button, the image of the button when the mouse is over it, and the image of the button when the mouse button is pressed (i.e., when the button is pushed; see Figure 4.5). A mechanism to display the right image at the right time also is needed—hence, the construct presented in the following. Because the XMT listing is verbose, the repetitive parts of the listing are given in a kind of shorthand.

```
1    <Group>
2      <children>
3        <TouchSensor DEF="N6" enabled="true"/>
4        <Switch DEF="N7" whichChoice="0">
5          <choice>
6            ... normal state image ...
7            ... over image ...
8            ... down image ...
9          </choice>
10       </Switch>
11     </children>
12   </Group>
13   <Conditional DEF="N10">
14     <buffer>
```

```
15         <Replace atNode="N7" atField="whichChoice" value="0"/>
16       </buffer>
17     </Conditional>
18     ... Conditional N11 with value 1 ...
19     ... Conditional N12 with value 2 ...
20     <ROUTE fromNode="N6" fromField="isOver"
21            ToNode="N11" toField="activate"/>
22     ... ROUTE N6.isOver to N10.reverseActivate ...
23     ... ROUTE N6.isActive to N12.activate ...
24     ... ROUTE N6.isActive to N11.reverseActivate ...
```

Lines 1 to 12 pertain to the visual part of the button. Lines 6, 7, and 8 are the constructs for the three button images (see the images in Figure 4.5). The container of the images, a Switch node, has the function of showing only one of the three images at a time. The single visible child of Switch is the one whose index is the value of the whichChoice field.

A TouchSensor node tracks the location and state of the mouse and detects when the user points at a geometry contained by the TouchSensor's parent group. In the XMT fragment presented above, the TouchSensor in line 3 applies to the visible child of the Switch. The isOver field becomes true when the mouse is over the visible child, and false when the mouse leaves. The isActive field becomes true when the mouse button is pressed over the visible child.

The Conditional nodes N10, N11, and N12 (lines 13–19) capture the following three actions: make visible the normal image, the over image, and the click image, respectively, by an appropriate change of the whichChoice field of the Switch node. The TouchSensor fields are routed (lines 20–24) to the Conditional nodes. When the mouse enters the button zone, isOver becomes true, which is routed to N11.activate and N10.reverseActivate. Because an activate field reacts only to true values, and reverseActivate reacts only to false values, only the buffer of Conditional N11 is executed, thus changing the visible image to the over image (index 1). When the mouse leaves the zone, isOver becomes false and the buffer of Conditional N10 is executed, thus changing the visible image back to the normal image (index 0). In the same way, mouse button down triggers isActive true, which executes the buffer of Conditional N12 and changes the visible image to the click image (index 2). Conversely, mouse button up triggers isActive false, executes the buffer of N10, and changes to the normal image.

Normal state   Mouse Over state   Mouse Click state

Fig. 4.5 Three states of the button.

All of this may look overly complex. However, it is very simple to implement, does not consume many resources, and yet is quite powerful.

Similar BIFS/XMT constructs can be used to create, for example,

☞ A highlightable menu element, by switching between a piece of text and the same text with a different color on top of a rectangle;

☞ A menu, with a set of the above elements;

☞ A popup menu by switching the above menu on and off, and so on.

Apart from the TouchSensor node, other sensor nodes interpret user input and translate it into scene events, notably these:

☞ The **PlaneSensor** node maps pointer motion on a 3D world into 2D translation in a plane, with optional minimum, maximum, and automatic offset.

☞ The **CylinderSensor** node maps pointer motion on a 3D world into a rotation on an invisible cylinder.

☞ The **SphereSensor** node maps pointer motion on a 3D world into a rotation on an invisible sphere.

☞ The **DiscSensor** node maps pointer motion on a 2D world into a rotation around a circle.

☞ The **PlaneSensor2D** allows dragging objects in a 2D world, with optional minimum, maximum, and automatic offset.

4.2.4 Slides and Transitions

Consider a set of slides, linked with simple transitions. A slide is here defined as any rectangular piece of scene, which may contain any object, image, video, text, or a set of BIFS geometries building a drawing. In the example shown in Figure 4.6, a gray rectangle builds the background; on top of it, a bitmap, a text, and an icon build up the currently shown slide.

Available 2D geometry nodes, usable for constructing more complex slides, are these:

☞ The **Rectangle** and **Circle** nodes allow rectangles and circles to be created. To create an ellipse, you need to place a Circle in a Transform2D node with a scale for the eccentricity and an angle for the orientation of the principal axis.

☞ The **IndexedLineSet2D** node is a set of straight lines. The main fields of this node are coord, the set of coordinates for the vertices of the lines, and coordIndex. The (–1)-separated lists of indices in coordIndex specify the list of vertices connected by lines. Other fields allow the separate coloring of each line. In the absence of a value for the coordIndex field, each vertex is connected to its neighbor in the coord field.

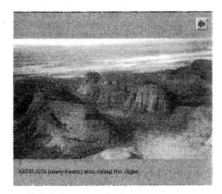

Fig. 4.6 Example of a slide.

☞ The **IndexedFaceSet2D** node is a set of polygons. In the same way as above, the vertices can be reused in many polygons through the use of coordIndex, and by default only one polygon is defined. The polygons are always filled. Other fields allow the separate coloring of each polygon, or each vertex, independently. The texture position also can be specified per polygon.

☞ The **Curve2D** node creates a curve or a set of curves. It is the only node in this list that is not a simple restriction of a VRML node to 2D. Its fields are listed here:

✗ **point:** the list of vertices

✗ **fineness:** an indication of the smoothness of the curve

✗ **type:** an optional specification of how to use the vertices

With no type value, the vertices build an Nth order Bezier spline: This curve passes through the first and last vertices, and all other vertices are control points. If the type field contains a set of integers ranging from 0 to 3, this specifies the creation of a more complex object, consisting of straight lines and cubic splines. The values are fed into an automaton that starts at the first vertex. A value of 0 indicates *move to the next vertex without drawing*. A value of 1 indicates *draw a straight line to the next vertex*. A value of 2 or 3 indicates *draw a cubic spline using 2 or 3 points as control points and target.*

Available appearance features are

☞ The **Material2D** node defines the color, the transparency, the "filledness" of 2D geometries but for IndexedLineSet2D (never filled) and IndexedFaceSet2D (always filled), and the way the lines are drawn.

☞ The **LineProperties** node is a subnode of Material2D defining the line-drawing properties: line width, line color, and line style (for which five dotted styles are qualitatively defined).

☞ The **Material (3D)** node defines ambientIntensity (albedo), diffuse-Color (paint), emissiveColor (glowing object), shininess (presence and size of shiny spots), specularColor (color of the shiny spots), and transparency.

☞ The **Texture** nodes allow any kind of image to be used as texture mapped onto an object. Texture nodes are ImageTexture, MovieTexture, Pixel-Texture (a variant of ImageTexture), CompositeTexture(2D) (a 3D or 2D subscene used as texture), and (very likely in a future amendment) MatteTexture (for complex visual effects).

☞ The **TextureTransform** node controls the placement of the texture on the object. Control variables are translation, scale, rotation, and center of rotation.

Suppose now that simple transitions must be designed for the visualization of a set of slides. A *fly transition* is a simple trajectory assigned to one of the slides. The XMT fragment for a possible slide transition is presented in the following:

```
1   <TimeSensor DEF="N69" cycleInterval="3.0" enabled="true"
2       Loop="false" startTime="0.0" stopTime="3.0"/>
3   <PositionInterpolator2D DEF="N46" key="0.0 1.0"
4       KeyValue="0.0 0.0 800.0 600.0"/>
5   <Layer2D size="800.0 600.0"> <children>
6     <Transform2D DEF="N1" translation="0.0 0.0"> <children>
7        ... the slide content ...
8     </children></Transform2D>
9   </children></Layer2D>
10  <ROUTE fromNode="N69" fromField="fraction_changed"
11         toNode="N46" toField="set_fraction"/>
12  <ROUTE fromNode="N46" fromField="value_changed"
13         toNode="N1" toField="translation"/>
```

The first two lines define the timing of the transition: from 0 to 3 seconds. The next two lines define the specific *fly transition* used here: The trajectory starts at the center (0,0) and finishes outside the frame (800x600 is the assumed size of the screen for this example), so this is a *fly out*. The Layer2D node (line 5) around the Transform2D node, which contains the slide, clips the slide to an 800x600 rectangle. The ROUTEs connect the TimeSensor to the PositionInterpolator2D, and the PositionInterpolator2D to the translation field of the slide's Transform2D. The TimeSensor node is a timer: It generates, as often as necessary, time events (a float value in the [0,1] interval) that inform the PositionInterpolator2D of the progression of the transition. The PositionInterpolator2D node translates the time event to a position by piecewise linear interpolation: The values in the key field correspond to the values in the keyValue field, and intermediate values are linearly interpolated.

A *box in transition* is a transition in which the next slide appears inside a growing rectangle on top of the previous slide. The new slide is not zoomed but clipped by a rectangle of increasing size. The Layer2D node can implement a clipping region. The next XMT fragment shows the implementation of a *box in transition.*

```
1    <TimeSensor DEF="N69" cycleInterval="3.0" enabled="true"
2       loop="false" startTime="0.0" stopTime="3.0"/>
3    <PositionInterpolator2D DEF="N46" key="0.0 1.0"
4       keyValue="0.0 0.0 800.0 600.0"/>
5    <Layer2D DEF="N2" size="0.0 0.0"> <children>
6       <Transform2D DEF="N1" translation="0.0 0.0"> <children>
7          ... the slide content ...
8       </children></Transform2D>
9    </children></Layer2D>
10   <ROUTE fromNode="N69" fromField="fraction_changed"
11          toNode="N46" toField="set_fraction"/>
12   <ROUTE fromNode="N46" fromField="value_changed"
13          toNode="N2" toField="size"/>
```

The only difference with the *fly transition* is that the PositionInterpolator2D is connected to the size field of the Layer2D instead of the translation field of the Transform2D below it, thus increasing the size of the visible part of the slide instead of moving the slide away.

The PositionInterpolator2D node allows the animation of 2D positions or sizes. Other interpolators that can be used to build other types of transitions follow:

☞ The **ScalarInterpolator** node animates a single value—for example, the transparency field of Material or Material2D nodes in a *fade transition.*

☞ The **PositionInterpolator** node animates a 3D coordinate or size—for example, the translation field of a Transform node containing a 3D object to move the object on a trajectory.

☞ The **ColorInterpolator** node animates a color—for example, the emissiveColor field of a Material node corresponding to a Sphere to create a pulsating effect.

☞ The **CoordinateInterpolator(2D)** node animates a 3D (2D) line set, face set, or curve by replacing the set of vertices with interpolated versions—for example, to create complex morphing transitions.

☞ The **NormalInterpolator** node animates normal vector sets—for example, to create variations on the bump effect on a surface.

☞ The **OrientationInterpolator** node animates 4D values representing rotations—for example, the rotation field of a Transform node—in order to rotate the objects contained in the Transform.

4.2.5 Simple 3D Scene

Showing spatial properties of molecules is a simple and useful application of 3D scenes. Figure 4.7 shows a 3D scene with the abstract representation of the water molecule. The corresponding XMT fragment follows:

Fig. 4.7 3D model of a water molecule.

```
1      <Replace><Scene>
2        <Group><children>
3          <Viewpoint fieldOfView="0.5" orientation="-0.3 0.9 0.0 1.5"
                   position="2.7 1.0 -0.2" description="Render Camera"/>
4          <Background groundColor="0.2 0.2 0.4" skyColor="0.2 0.2
                            0.4"/>
5          <SpotLight color="1.0 1.0 1.0" direction="-0.4 -0.7 -0.7"
                   cutOffAngle="1.59" intensity="1" location="0.8 1.2 1.1"/>
6          <Transform translation="-0.509 -0.113 -0.179"><children>
7            <Shape>
8              <geometry><Sphere radius="0.305"></geometry>
9              <appearance><Appearance><material>
10               <Material ambientIntensity="0.9"
                          diffuseColor="1.0 0.038 0.095"/>
11             </material></Appearance></appearance>
12           </Shape>
13         </children></Transform>
14         <Transform ...
15             ... the first blue sphere ...
16         </Transform>
17         <Transform ...
18             ... the second blue sphere ...
19         </Transform>
20         <Transform rotation="1.0 0.0 0.0 2.269"
                      translation="-0.509 -0.443 0.235"><children>
21           <Shape>
22             <geometry>
23               <Cylinder height="0.728" radius="0.018"/>
24             </geometry>
25             <appearance><Appearance><material>
26               <Material diffuseColor="0.0 0.0 0.0"
                          ambientIntensity="0.926" transparency="0.7"/>
```

```
27              </material></Appearance></appearance>
28            </Shape></children>
29          </Transform>
30          <Transform ...
31              ... the second link cylinder ...
32          </Transform>
33        </children></Group>
34      </Scene></Replace>
```

This 3D example shows some new features with respect to the previous 2D examples, notably these:

☞ The top node is a **Group** node, signaling a 3D rendering context (line 2).

☞ A **Viewpoint** node defines the initial viewpoint, from which the user can navigate. If many viewpoints are defined, the user can jump from one viewpoint to another (line 3).

☞ A **SpotLight** defines the lighting conditions of the object (line 5). With this node the light comes from a point within a sector of space. Other types of lights are DirectionalLight (light comes from an infinite distance and applies only to siblings and their children) and PointLight (light comes from a point and applies to all objects within reach).

☞ The nodes used here are without the 2D suffix and have more options: Background (line 4), Transform (line 6), and Material (line 10 and 26). Moreover, 3D geometries are used.

Here is a short introduction to 3D geometry nodes, which can be used to create 3D scenes:

☞ The **Sphere** node (line 8) describes a sphere.

☞ The **Cylinder** node (line 23) describes a cylinder.

☞ The **Box** node describes a box.

☞ The **Cone** node describes a cone.

☞ The **ElevationGrid** node is an array of height values specifying the height of a surface above each point of a rectangular grid; it is used to represent realistic terrain, such as hills, mountains, and valleys.

☞ The **Extrusion** node creates a volume based on a 2D cross section extruded along a 3D spine; the cross section can be scaled and rotated at each spine point. This node is used to represent many types of objects, such as surfaces of revolution; bottles; a ship; the hull of a plane; a corkscrew; and anything bent, twisted, or tapering.

☞ The **IndexedFaceSet** node describes a 3D polygon.

☞ The **IndexedLineSet** node describes a set of lines in 3D.

☞ The **PointSet** node describes a set of points in 3D.

☞ The **Text** node describes a set of strings, as in 2D.

Fig. 4.8 More complex 3D scene.

Figure 4.8 shows a more complex 3D scene, built mainly with 20 IndexedFaceSet nodes. In this scene there are approximately 25,000 vertices. The scene has been automatically generated from a model described in a proprietary language and, as such, is not optimized. The actual form of the content depends on whether the authoring tool is based on meshes, extrusions, constructive solid geometry, or solid geometry design logic [Niel00, Rotg96].

One big difference between a 2D and a 3D MPEG-4 player is the default user interaction processing. When the user sees the pumpkins in Figure 4.8, unless he or she can interact with them, presenting them in 3D does not bring about any benefit compared to a 2D presentation. When showing a 3D scene with no interactivity, a 3D terminal will offer, by default, 3D navigation. With the mouse, the viewpoint can be moved to create the effect of walking in the scene, examining the object by rotating it, and so on. If interactivity is designed into the scene, that interactivity will take precedence.

Constructing 3D scenes in BIFS is not different from constructing such scenes in VRML [VRML97], beyond the differences coming from MPEG-4 extensions to VRML, as described in Section 4.1. The design of 3D objects (geometry and appearance), interactivity using sensors and ROUTEs, and animations with interpolators and lighting are strictly the same in BIFS and in VRML. As a result, the reader may find in [HaWC96] more specific help about the construction of 3D scenes.

Fig. 4.9 Magnifying Glass scene.

4.2.6 Magnifying Glass

Let's return to the 2D domain to analyze a more complex example: An image or a video is shown in the middle of the screen; a round object can be dragged with the mouse across the screen; and, behold, the object behaves as a magnifying glass on top of the image or video (see Figure 4.9).

The next XMT fragment corresponds to the *Magnifying Glass* example. The object to drag is a Circle (line 9) with a wide yellow border (lines 13–15), placed in a Group node (line 4) together with a PlaneSensor2D node (line 5). The PlaneSensor2D has a translation_changed field that can be connected (line 22) to the translation of the Transform2D node on top of the Circle, thus achieving the dragging. In line 5, autoOffset avoids a jump when the user clicks on the Circle, and min/maxPosition defines the maximum extent of the dragging.

```
 1   ...
 2   <Transform2D DEF="N9"> ... the main image ... </Transform2D>
 3   ...
 4   <Group><children>
 5     <PlaneSensor2D DEF="N1" autoOffset="true" enabled="true"
                 maxPosition="300 300" minPosition="-300 -300 "/>
 6     <Transform2D DEF="N4" translation="0.0 0.0"><children>
 7       <Shape>
 8         <geometry>
 9           <Circle radius="130.0"/>
10         </geometry>
11         <appearance><Appearance><material>
12           <Material2D DEF="N2" binaryID="2" filled="false">
13             <lineProps>
```

```
14                    <LineProperties lineColor="1 0.8 0.2" width="16.0"/>
15                  </lineProps>
16                </Material2D>
17              </material></Appearance></appearance>
18            </Shape>
19          </children></Transform2D>
20        </children></Group>
21        ...
22        <ROUTE fromNode="N1" fromField="translation_changed"
               toNode="N4"    toField="translation"/>
23        <Transform2D DEF="N0"><children>
24          <Layer2D DEF="N8" size="200.0 200.0"><children>
25            <Transform2D DEF="N6" scale="3.0 3.0"><children>
26              <Transform2D USE="N9"/>
27            </children></Transform2D>
28          </children></Layer2D>
29        </children></Transform2D>
30        <Valuator DEF="N7" Factor1="-3.0" Factor2="-3.0"/>
31        ...
32        <ROUTE fromNode="N1" fromField="translation_changed"
               toNode="N7"    toField="inSFVec2f"/>
33        <ROUTE fromNode="N7" fromField="outSFVec2f"
               toNode="N6"    toField="translation"/>
34        <ROUTE fromNode="N1" fromField="translation_changed"
               toNode="N0"    toField="translation"/>
```

The magnifying effect is implemented by a `Transform2D` node with a scale factor of 3 (line 25). The `Transform2D` will have as single child a `USE` of the main image (line 26). In order for the magnified image to follow the area below the glass, the `USE` node must be translated by the magnifying glass movement multiplied by −3. This translation is implemented by modifying the `translation` field of the scaling `Transform2D` node. The multiplication by −3 can be achieved with a `Valuator` node (line 30): The `Valuator` node allows type casting and affine operations on events.

4.3 ADVANCED BIFS FEATURES

In this section, some advanced BIFS features are presented.[9] The first feature, *scripting*, is introduced through an example in which the `Script` node is used to achieve complex interaction and create a panorama. Next, data encapsulation and model reuse with `PROTO` are presented. Finally, a way to deal with the unpredictability of text layout by using the `Layout` and `Form` nodes is explained.

---

9. The subjective difference between *basic feature* and *advanced feature* is that all users of BIFS are expected to have to use *basic features*, whereas only a subset of users will need *advanced features*. This should not be confused with the somewhat arbitrary usage of the term "advanced" in the standard.

4.3.1 Scripting

This example features a panorama—a picture with a very wide field of view. In this case, the panorama has a field of view of more than 360° and has been created from a set of pictures taken from a single vantage point, which have been stitched together to create a seamless band. The expected effect is to give the viewer the impression of standing in the middle of the scene and allow him or her to look around. The viewer sees through the viewport (see Figure 4.10) and may move the viewport to the left or the right to simulate looking right or left, respectively. When the viewport comes to one end of the picture, it must be moved to the other end of the picture. To avoid any visible transition, the left and right ends of the panorama show the same (duplicated) piece of view.

The implementation of this effect requires the creation of a viewport that is smaller or equal in width to the duplicated overlap. When the viewport reaches the left edge, it must be *wrapped around* to the right edge, and vice versa. The function that checks both bounds and wraps is encapsulated in a Script node. The source of events will be a set of TimeSensor nodes triggered by the entrance of the mouse in two zones (the *go left* and *go right* zones in Figure 4.11), implemented as a TouchSensor on a transparent Rectangle, or as a ProximitySensor2D node.

The XMT fragment corresponding to the panorama effect is presented in the following:

```
1    <Group><children>
2      <Transform2D DEF="T"/> ...the image... </Transform2D>
3      <ProximitySensor2D DEF="P" center="-0.75 0" size="0.5 2"/>
4      <ProximitySensor2D DEF="Q" center="0.75 0" size="0.5 2"/>
5      <TimeSensor DEF="A" cycleInterval="0.2" startTime="0"
                    stopTime="-1" enabled="false"/>
6      <TimeSensor DEF="B" cycleInterval="0.2" startTime="0"
                    stopTime="-1" enabled="false"/>
7      <Script DEF="S">
8        <field name="left" vrml97hint="eventIn" type="Time"/>
9        <field name="right" vrml97hint="eventIn" type="Time"/>
10       <field name="var" vrml97hint="field"
              type="Vector2" vector2Value="0 0"/>
11       <field name="pos" vrml97hint="eventOut" type="Vector2"/>
12       <url>...the script below goes here...</url>
13     </Script>
14   </children></Transform2D>
15   <ROUTE fromNode="P" fromField="isActive"
            toNode="A"    toField="enabled"/>
16   <ROUTE fromNode="Q" fromField="isActive"
            toNode="B"    toField="enabled"/>
17   <ROUTE fromNode="A" fromField="cycleTime"
            toNode="S"    toField="left"/>
18   <ROUTE fromNode="B" fromField="cycleTime"
            toNode="S"    toField="right"/>
19   <ROUTE fromNode="S" fromField="pos"
            toNode="T"    toField="translation"/>
```

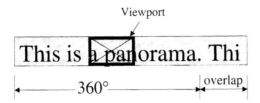

Fig. 4.10 View of the whole panorama.

Fig. 4.11 Navigation zones on the screen.

The script is listed separately for clarity: The programming language used is ECMA-Script [ECMA99], which is very close to JavaScript. Assuming that the image is five times wider than the viewport, the script would be the following ECMA-Script fragment:

```
javascript:
function left(value){
    var -= 0.1;
    if (var <= -2.0) var = 2.0;
    pos = var;
}
function right(value){
    var += 0.1;
    if (var >= 2.0) var = -2.0;
    pos = var;
}
```

When the pointer enters one of the ProximitySensor2D zones (lines 3–4 above), an isActive true event is sent (ROUTEs at lines 15–16) to the corresponding TimeSensor node (lines 5–6), which starts emitting cycleTime events every 0.2 seconds. The corresponding function of the script (line 7) is then activated. The current position is increased or decreased, checked against the bound, and routed (line 19) to the translation field of the Transform2D node containing the image. When the pointer leaves the zone, the ProximitySensor2D emits an isActive false event, which stops the corresponding TimeSensor.

4.3.2 Encapsulation and Reuse

One of the easy ways to influence the binary size of the scene description is to reuse already existing pieces of the scene. Of course, the DEF/USE mechanism already allows you to point at the same object from different parts of the scene tree. Because DEF/USE only creates references, there is only one *copy* of the node. If the node has, for example, interactive behavior, all the places in the scene tree in which this node appears react together. You may want to have in the scene local copies of the same model, so that each copy reacts independently from the others. This is what PROTO does.

A PROTO defines a new node type in terms of already defined (built-in or prototyped) node types. In other words, PROTOS are a kind of macro. PROTOS can be instantiated many times, for a much smaller cost (in bits) than that of resending a copy. PROTOS also provide data protection, as only the interface fields of the PROTO are directly accessible from the rest of the scene, and not the body of the PROTO.

PROTOS have a *name*, an *interface* consisting of a list of interface fields, and a *body*. In the body, any field can be set to the value of one field of the PROTO interface by an IS reference. The IS reference establishes the equivalent of a ROUTE between one field in the body of the PROTO and one field of the PROTO interface. IS references are the only way to give to the rest of the scene indirect access to the body of a PROTO.

Each interface field has a *name*, a *type*, a *behavior* (called vrml97hint), and *mandatory default values*. The example of a PROTO XMT fragment presented next creates a sort of table, with a table top and four legs.

```
 1    <ProtoDeclare name="SimpleTable">
 2      <field name="topColor" type="Color" vrml97hint="exposedField"
             colorValue="1 0 0"/>
 3      <field name="legColor" type="Color" vrml97hint="exposedField"
             colorValue="0 1 0"/>
 4      <field name="legSize" type="Float" vrml97hint="exposedField"
             floatValue="0.1"/>
 5      <Transform translation="0 0.6 0"/><children>
 6        <Shape>
 7          <geometry><Box size="1.4 0.2 1.4"/></geometry>
 8          <appearance><Appearance><material>
 9            <Material DEF="TopCol"> <IS>
              <connect nodeField="diffuseColor"
                     protoField="topColor"/>
              </IS> </Material>
10          </material></Appearance></appearance>
11        </Shape>
12      </children></Transform>
13      <Transform translation="-.5 0 -.5"/><children>
14        <Shape DEF="Leg">
15          <geometry>
16            <Cylinder DEF="L" height="1"> <IS>
                <connect nodeField="radius" protoField="legSize"/>
              </IS> </Cylinder>
17          </geometry>
18          <appearance><Appearance><material>
19            <Material DEF="LegCol"> <IS>
                <connect nodeField="diffuseColor"
```

```
                      protoField="legColor"/>
                  </IS> </Material>
20            </material></Appearance></appearance>
21          </Shape>
22        </children></Transform>
23        <Transform translation=".5 0 -.5"/><children>
24          <Shape USE="Leg"/>
25        </children></Transform>
26        <Transform translation="-.5 0  .5"/><children>
27          <Shape USE="Leg"/>
28        </children></Transform>
29        <Transform translation=".5 0  .5"/><children>
30          <Shape USE="Leg"/>
31        </children></Transform>
32      </ProtoDeclare>
33      <ProtoInstance name="SimpleTable">
34        <fieldValue name="topColor" value="0 0.5 0.5"/>
35        <fieldValue name="legSize" value="0.2"/>
36      </ProtoInstance>
```

Lines 1 through 32 are the PROTO declaration, with three field declarations and then the five objects: table top, first leg, and three copies of the first leg. The IS references (lines 9, 16, and 19) define the connections between the field containing the reference and the PROTO interface field of the name declared in the attribute. For example, in line 9 there is a connection akin to a ROUTE between the diffuseColor field of the TopCol Material node and the topColor interface field of the PROTO. Lines 33 through 36 are the PROTO instance. The topColor is changed to blue-green and the table has thicker legs than the default (0.2 instead of 0.1); legColor is unchanged (see Figure 4.12).

Fig. 4.12 Instance of PROTO defined simple table.

4.3.3 Text Layout

One of the big difficulties with text in MPEG-4 is the lack of predictability. The scene author cannot know which font the user's terminal is going to use. To start with, only three fonts are guaranteed to be present in the MPEG-4 terminal: Serif, Sans, and Typewriter. But what if these three fonts are not enough for the author? When using other font families, the terminal will make a best effort at substitution by an available font, and no size or alignment can be guaranteed across all MPEG-4–compliant terminals.

The nodes Layout and Form have been designed to help the author achieve satisfactory layout results with objects of unpredictable sizes, mostly texts with substituted fonts. The Layout node allows the arrangement of objects in a paragraph: Pieces of text and graphics are typeset in wrapped rows or columns, with the usual options for justification and scrolling. The Form node is targeted at more complex arrangements, such as tables or alignments.

The Layout node is a grouping node with all the fields of the FontStyle node. It arranges its Text children nodes according to their own FontStyles and its non-Text children nodes, according to the value of its own fields. Other layout options are these:

☞ The **size** field specifies the size of the Layout frame—that is, the rectangular space within which the children are laid out.

☞ The **wrap** field specifies whether children are allowed to wrap to the next row (or column) after the edge of the layout frame is reached.

☞ The **smoothScroll** field controls whether scrolling is smooth or line-by-line.

☞ The **loop** field controls whether, when the objects have completely scrolled out of the Layout frame, they should reappear progressively on the other side.

☞ The **scrollVertical** field controls the direction of the scrolling.

☞ The **scrollRate** field controls the rate of scrolling in meter or pixel per second.

The Form node also is a grouping node; it lays out its children nodes according to alignment or distribution constraints. Its fields are these:

☞ The **groups** field defines the groups that will be used in the layout process, by −1-separated lists of children indices.

☞ The **constraints** field defines the constraints that are applied in sequence.

☞ The **groupsIndex** field defines the −1-separated lists of groups on which the constraints apply; the nth constraint applies to the nth list of groups.

The Form children are placed in groups and each constraint applies to a group of objects. The laying-out principle is this: *Objects start in the middle of the* `Form` *local coordinate system (0,0), and the constraints are applied in sequence.* Available constraints are alignment of the center of the objects or their left, right, top, or bottom edge, possibly with additional space, and spread within a group or within the `Form` frame, all of that horizontally or vertically. The index 0 is for the frame of the `Form` node; the indices for the children and groups start at 1.

As an example, imagine that three objects have to be aligned in a frame:

☞ The first object is the largest (an image of unknown size) and needs to be above the two others; it constitutes the first row.

☞ The two other objects are smaller and need to be vertically aligned; they constitute the second line.

☞ The second object should be left-aligned with the first object.

☞ The third object should be right-aligned with the first object.

☞ The two lines need to be evenly spread vertically within the frame.

Figure 4.13 shows the scene regarding this example, which corresponds to the following XMT fragment:

```
1     <Form size="800.0 600.0"
              groups="1 -1 2 -1 3 -1 2 3 -1"
              constraints="'AL' 'AR' 'SVin'"
              groupsIndex="1 2 -1 1 3 -1 1 4 -1">
2       <children>
3       ... object 1, the image ...
4       ... object 2, the legend ...
5       ... object 3, the copyright statement ...
6       </children>
7     </Form>
```

Fig. 4.13 Form example.

The groups field defines the groups G1: image, G2: legend, G3: copyright, and G4: legend+copyright. The constraints and groupsIndex fields define the constraints C1: align left (code AL) G1 and G2, C2: align right (code AR) G1 and G3, and C3: spread vertically (code SVin), in the Form frame, G1 and G4.

4.4 A Peek Ahead on BIFS

The following BIFS tools were being finalized at the time this book was written. They are specified in Amendments 1 and 2 to the second edition of the MPEG-4 Systems standard [MPEG4-1]. Their importance in terms of enabling applications justifies this anticipation.

4.4.1 Media Nodes

It was found quite early in the design of BIFS that the control features of some nodes inherited from VRML were impossible to use in the same way across all delivery scenarios. Rather than making the viewing experience for the same content different for different delivery scenarios, it was decided to restrict the semantics of the fields startTime and stopTime of the MovieTexture and Sound nodes and to design new, more complex nodes that would work the same way in all scenarios. These new nodes are MediaSensor, MediaControl, and MediaBuffer. These nodes are the only ones that deal with media time. The notion of *stream segment* is also linked with the Media nodes. Stream segments are named portions of a stream. They are defined by a Segment-Descriptor (see Chapter 3).

Scene and media time are defined as the object time base of the scene description and media stream, respectively. See Section 3.3.5 for a discussion of media time.

The MediaSensor node monitors the availability and presentation status of a stream or a stream segment. When the media arrival time is uncertain, it allows the synchronization of some scene events to the arrival of the media—for example, popping up the MovieTexture only when "enough" of the video has arrived. The MediaSensor functionality can be used to implement part of the FlexTime nodes (see Section 4.4.3).

The MediaControl node controls the playback and, hence, delivery of a stream object (i.e., an elementary stream [ES] or stream segment; see Chapter 3). It allows the selection of a time interval within a stream object for playback, modification of the playback direction and speed, as well as prerolling and muting of the stream. The MediaControl node is a way to change the relation between a media time and the scene time.

The MediaBuffer node allows local caching of media streams. This allows storage of clips for interactive playback or looping. Typical uses are the channel logo on television or the complex video animation of a fancy button. This is similar but more generic than the AudioBuffer functionality described in Chapter 12.

4.4.2 New Sensors

Looking at the interactivity-related nodes, one would think that the only device allowing interactivity is the pointer—be it the mouse, joystick, or digitizing table—with a single button. There is no way to use keyboards, game pads, haptic devices, data gloves, or motion capture within a BIFS scene. New devices, or new types of devices, are being developed and will be developed in the future. Moreover, different devices will be used on different types of terminals, ranging from handheld terminals with low capabilities to multimedia terminals, like personal computers with high capabilities.

The InputSensor node was designed with this in mind. It is a low-footprint, generic way to define user input and depends on a few elements external to BIFS, notably these:

☞ The **User Interaction Stream:** a new type of ES, purely local to the terminal

☞ The **User Interaction Decoder:** a new decoder with a front end receiving data from the device and a back end similar to that of a BIFS decoder—either producing BIFS composition units or directly modifying the scene tree

The fields of the InputSensor node are these:

☞ **enabled:** a Boolean value controlling the activity of the node

☞ **url, SFUrl:** referring to a user interaction stream

☞ **interactionBuffer:** has the same type as a conditional buffer (SFCommandBuffer; i.e., contains a set of untimed BIFS commands)

Upon insertion of InputSensor, the user interaction stream decoding pipeline is set up, as shown in Figure 4.14.[10]

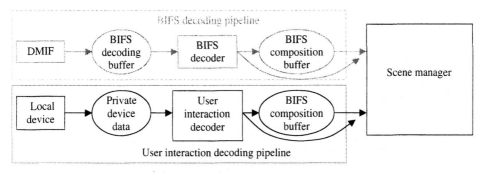

Fig. 4.14 User interaction stream decoding pipeline.

---

10. For more information about the decoding pipeline, see Chapter 3.

The arrow between the BIFS and user interaction decoders and the scene manager in Figure 4.14, bypassing the BIFS composition buffer, means that in some implementations the decoder directly calls the scene manager API instead of building an actual BIFS composition buffer. The private device data takes the form of a list of primitive data types: for example, SFVec2f (i.e., two floats) for a position and SFBool for the state of a button. The interaction-Buffer should hold two Replace commands, the first for a SFVec2f field and the second for a SFBool field. The user interaction decoder is responsible for reading the content of interactionBuffer, matching it with the incoming data, and then enacting the scene changes.

Imagine you want to have a StringSensor node so you can use, inside the scene, the name of keys pressed or released by the viewer. This functionality can be achieved with the following virtual node definition:

```
StringSensor {
   exposedField SFBool     enabled            TRUE
   eventOut     SFString    keyName
   eventOut     SFTime      eventTime
}
```

The following XMT fragment describes the equivalent InputSensor implementation.

```
1    <InputSensor url="od10">
2      <buffer>
3        <Replace atNode="Text10" atField="string" value="'''">
4      </buffer>
5    </InputSensor>
```

In the virtual StringSensor definition, the enabled and isActive fields can be mapped directly onto the corresponding field of the InputSensor node. The keyName field represents one string. The keyboard device plug-in is responsible for getting the operating system events and sending to the user interaction decoder one string encoding the keyboard status—that is, PRESS_SHIFT_A.

In the InputSensor implementation, whenever the user presses a key, the field string of the node Text10 will display the value of the last key pressed or released.

Obviously, the StringSensor defined above is not general. MPEG needs to standardize more precisely the mapping of useful devices onto the Input-Sensor node, and the first of these mappings will be for the keyboard; however, with InputSensor, neither changes in the bitstream syntax nor new profiles will be necessary to add support for new input devices.

4.4.3 FlexTime

The FlexTime model provides the ability to synchronize, within certain limits, objects with different time bases. For example, the FlexTime model allows one to:

☞ Display together two pieces of media coming from two different servers or from the Web with no predictability of the response time and quality of service: the first object to arrive is buffered until the other object arrives or a time limit is reached; and

☞ Synchronize a change in the scene with the actual reception of a media or its end: a MovieTexture node can be inserted or activated when the beginning of the video stream has arrived and been decoded. The MovieTexture node can also be deleted at the end of the video. The actual time of start and end may depend on the network characteristics, and the video may come from another source, which the scene author has no control over.

FlexTime consists of two nodes—namely, the TemporalTransform, which defines the flexible objects, and TemporalGroup, which applies constraints on them. The TemporalTransform is a grouping node that assigns temporal properties and applies temporal transformations, to children nodes or one ES referred by its url field. A startTime and optimalDuration can be defined, along with minimum and maximum stretching values, as well as preferred modes of stretching and shrinking. A speed field also can be used to control the time base of the TemporalGroup, or to know its evolution under the ministrations of a TemporalGroup. TemporalTransform can thus be used alone to interactively control a complex BIFS animation with many TimeSensors (possibly interlocked), which may be very hard to do directly.

TemporalGroup also is a grouping node, but it can only contain Temporal-Transforms or TemporalGroups. Its costart, coend, and meet fields determine the operation performed on the flexible children. Thus, a functionality similar to SMIL's <par> and <seq> elements is achieved [SMIL98].

4.4.4 ServerCommand

The ServerCommand node allows the author to design scenes, which may send messages to anywhere, typically to the server. The message is a string, the recipient is a URL, and the trigger can be any Boolean *true* event. The URL defines an upstream channel, on which the message is sent. The application potential is big: Useful messages to send back to the server are the name of the requested video, the customer key or password, information about the user's actions to store in the user's profile, and so on.

4.5 PROFILES

As all MPEG-4 applications may not require all of the BIFS tools, profiles have been defined to segment the BIFS functionality into meaningful subsets. Two profile dimensions related to BIFS have been created: *graphics* and *scene graph*.

The graphics profiles specify the allowed graphics tools—that is, the set of all intrinsically visible nodes—plus Shape and all the appearance-related nodes. The scene graph profiles specify the allowed scene graph (structural)

tools. Nodes that have no intrinsic visible component—together with ROUTES, BIFS-Commands, and BIFS-Anim—are the BIFS elements with which scene graph profiles are created.

An additional difficulty with respect to graphics profiles comes from the presence of BIFS nodes, made necessary by the use of some visual or audio profiles. For example, if the Simple Face visual profile is specified, the Face node may be used in the scene regardless of its presence in the selected graphics profile.

For the details on graphics and scene graph profiles and levels, see Chapter 13 and Appendixes C and D.

4.6 ALL BIFS NODES

This section includes a set of tables with reference to all of the BIFS nodes, notably with a short explanation for the many nodes that did not find their way into one of the examples presented in this chapter. The nodes are organized using the following classes:

- ☞ Table 4.4 General-use nodes
- ☞ Table 4.5 Visual nodes
- ☞ Table 4.6 Animation-related nodes
- ☞ Table 4.7 Interactivity-related nodes
- ☞ Table 4.8 Texture-related nodes
- ☞ Table 4.9 Geometry nodes
- ☞ Table 4.10 Audio nodes (see Chapter 12)
- ☞ Table 4.11 SNHC visual-related nodes (see Chapter 9)

Some node descriptions include a 2D or 3D label. Nodes without tags can be used in both 2D and 3D scenes. Nodes with a 2D or 3D tag belong only to 2D or 3D subscenes, respectively.

Table 4.4 General-use nodes

Name	Description
Anchor	Hyperlink: This grouping node directs the browser to another scene when one child is clicked on.
Group	Simplest grouping node: See Section 4.2.5.
Inline	Grouping node: Reads its children from another scene.
LOD	3D: Specifies different versions of an object with increasing levels of detail or complexity, and provides hints allowing browsers to automatically choose the appropriate version based on the distance from the user—like a Switch whose whichChoice field is managed automatically by the browser.

Table 4.4 General-use nodes (Continued)

Name	Description
NavigationInfo	3D: Contains information describing the physical characteristics of the viewer's avatar and viewing model.
QuantizationParameter	Controls the quantization of the data fields; see Section 4.1.8.
Switch	Grouping node: Displays one of its children only; see Section 4.2.3.
TermCap	Queries the resources of the terminal, in order to achieve simple adaptive content by ROUTEing the result to a Switch node.
WorldInfo	Contains documentation-only information about the scene.

Table 4.5 Visual nodes

Name	Description
Appearance	Organizes the appearance of a Shape; see Section 4.2.1.
ApplicationWindow	Allows an external application, such as a Web browser, to exist within the MPEG-4 scene graph.
Background	3D: Used to specify a color backdrop that simulates ground and sky, as well as a background texture or panorama that is placed behind all geometry in the scene and in front of the ground and sky.
Background2D	2D equivalent of Background.
Billboard	3D: Grouping node that modifies its coordinate system so that it always faces the viewer.
Collision	3D: Grouping node that specifies the collision detection properties for its children.
Color	Set of RGB colors to be used in the fields of another node.
Coordinate	Set of 3D coordinates to be used in the coord field of vertex-based 3D geometry nodes.
Coordinate2D	Set of 2D coordinates to be used in the coord field of vertex-based 2D geometry nodes.
DirectionalLight	3D: Light coming from an infinitely far source; see Section 4.2.5.
Fog	3D: Provides a way to simulate atmospheric effects by blending objects with a color based on the distances of the various objects from the viewer.
FontStyle	Defines the visual characteristics of text; see Section 4.2.2.
Form	2D: Controls the layout of objects; see Section 4.3.3.
Layer2D	Transparent rendering rectangular region on the screen where a 2D scene is drawn; clips to its size, if specified; see Section 4.2.6.

Table 4.5 Visual nodes (Continued)

Name	Description
Layer3D	Transparent rendering rectangular region on the screen where a 3D scene is drawn.
Layout	2D: Controls the layout of objects as words in a paragraph; see Section 4.3.3.
LineProperties	2D: Controls the drawing of lines; see Section 4.2.4.
Material	Controls the drawing of 3D objects; see Section 4.2.4.
Material2D	Controls the drawing of 2D objects; see Section 4.2.4.
MaterialKey	When used in conjunction with Bitmap, makes transparent all pixels of the Bitmap whose color is similar to a given color.
Normal	Set of 3D surface normal vectors to be used in the vector field of some geometry nodes.
OrderedGroup	Grouping node allowing the management of the drawing order; as a top node, specifies a 2D rendering context.
PointLight	3D: Light coming from a point; see Section 4.2.5.
Shape	Organizes the visual characteristics of an object; see Section 4.2.1.
SpotLight	3D: Restriction of a PointLight to a sector of space; see Section 4.2.5.
Transform	3D: Grouping node that defines a 3D coordinate system for its children relative to the coordinate systems of its ancestors; see Section 4.2.5.
Transform2D	2D: Grouping node that defines a 2D coordinate system for its children relative to the coordinate systems of its ancestors; see Section 4.2.2.
Viewpoint	3D: Defines a specific location from which the user may view the scene.

Table 4.6 Animation-related nodes

Name	Description
AnimationStream	Control node for an animation stream; see Section 4.1.5.2.
ColorInterpolator	Interpolates colors; see Section 4.2.4.
CoordinateInterpolator	3D: Interpolates sets of 3D vertices; see Section 4.2.4.
CoordinateInterpolator2D	2D: Interpolates sets of 2D vertices; see Section 4.2.4.
NormalInterpolator	3D: Interpolates sets of normal vectors; see Section 4.2.4.
OrientationInterpolator	3D: Interpolates rotations; see Section 4.2.4.

Table 4.6 Animation-related nodes (Continued)

Name	Description
PositionInterpolator	3D: Interpolates 3D positions or sizes; see Section 4.2.4.
PositionInterpolator2D	2D: Interpolates 2D positions or sizes; see Section 4.2.4.
ScalarInterpolator	Interpolates scalar values; see Section 4.2.4.
TimeSensor	Generates time events for animation purposes; see Section 4.2.4.

Table 4.7 Interactivity-related nodes

Name	Description
Conditional	Interprets a buffered bit string of BIFS-Commands when it is activated; see Section 4.2.3.
CylinderSensor	3D: Maps pointer motion onto a cylinder; see Section 4.2.3.
DiscSensor	2D: Maps pointer motion to a rotation; see Section 4.2.3.
PlaneSensor	3D: Maps the pointer motion onto a plane in 3D; see Section 4.2.3.
PlaneSensor2D	2D: Allows the dragging of an object in 2D; see Section 4.2.3.
ProximitySensor	3D: Generates events when the viewer enters, exits, or moves within a region in space defined by a box.
ProximitySensor2D	2D: Generates events when the pointer enters, exits, or moves within a rectangle—like a TouchSensor that would not need a sibling Rectangle; see Section 4.3.1.
Script	Animation/interactivity: Used to program behavior in a scene; see Section 4.3.1.
SphereSensor	3D: Maps pointer motion onto a sphere; see Section 4.2.3.
TouchSensor	Maps pointer motion onto scene related events; see Section 4.2.3.
Valuator	Simple type-casting mechanism for events; see Section 4.2.6.
VisibilitySensor	3D: Detects visibility changes of a rectangular box as the user navigates the world; typically used to stop costly animations when invisible.

Table 4.8 Texture-related nodes

Name	Description
CompositeTexture2D	Makes a 2D scene usable as a texture on an object; see Section 4.2.4.
CompositeTexture3D	Makes a 3D scene usable as a texture on an object; see Section 4.2.4.

Table 4.8 Texture-related nodes (Continued)

Name	Description
ImageTexture	Allows the use of an image as a texture; see Section 4.2.4.
MovieTexture	Allows the use of a movie as a texture; see Section 4.2.4.
PixelTexture	Allows the use of a picture, transmitted within the scene, as a texture; see Section 4.2.4.
TextureCoordinate	Specifies a set of 2D texture coordinates used by vertex-based geometry nodes to map textures to vertices.
TextureTransform	Defines a 2D transformation that is applied to texture coordinates.

Table 4.9 Geometry nodes

Name	Description
Bitmap	MPEG-specific geometry node allowing efficient 2D rendering by making trivial the mapping of the texture specified in the appearance field; see Section 4.2.1.
Box	3D: Defines a box; see Section 4.2.5.
Circle	2D: Defines a circle or disk; see Section 4.2.4.
Cone	3D: Defines a cone; see Section 4.2.5.
Curve2D	2D: Defines a Bezier curve; see Section 4.2.4.
Cylinder	3D: Defines a cylinder; see Section 4.2.5.
ElevationGrid	3D: Array of height values specifying the height of a surface above each point of a rectangular grid; see Section 4.2.5.
Extrusion	3D: Defines a volume based on a two-dimensional cross section extruded along a 3D spine; see Section 4.2.5.
IndexedFaceSet	3D: Defines a polygon; see Section 4.2.5.
IndexedFaceSet2D	2D: Defines a polygon; see Section 4.2.4.
IndexedLineSet	3D: Defines a wire frame; see Section 4.2.5.
IndexedLineSet2D	2D: Defines a wire frame; see Section 4.2.4.
PointSet	3D: Defines a set of points; see Section 4.2.5.
PointSet2D	2D: Defines a set of points; see Section 4.2.4.
Rectangle	2D: Defines a rectangle; see Section 4.2.4.
Sphere	3D: Defines a sphere; see Section 4.2.5.
Text	Defines a set of text strings; see Section 4.2.2.

Table 4.10 Audio nodes

Name	Description
AcousticMaterial	Attaches acoustic and visual properties to surfaces (planar polygons) defined by an IndexedFaceSet node that is either a sibling or exists in a subgraph of a sibling of an AcousticScene node.
AcousticScene	Defines a rendering region for DirectiveSound nodes and for building acoustic rooms.
AudioBuffer	Takes audio clips from incoming audio streams for interactive use.
AudioClip	Retrieves audio clips.
AudioDelay	Delays the playing of sound a defined amount of time (mainly for synchronization purposes).
AudioFX	Allows arbitrary signal-processing functions defined using structured audio tools to be applied to the audio defined by the children of this node.
AudioMix	Mixes together several audio signals in a simple, multiplicative way.
AudioSource	Adds streaming sound to a BIFS scene.
AudioSwitch	Selects a subset of audio channels from its children.
DirectiveSound	3D: Obtains sound source directivity and advanced room acoustics modeling features in a 3D scene.
ListeningPoint	3D: Specifies the reference position and orientation of a listener for spatial audio presentation—defaults to the active Viewpoint, if any.
PerceptualParameters	Contains information about the perceptual properties of room acoustics that are applied to DirectiveSound.
Sound	3D: Specifies a simple spatial presentation of a sound within a 3D world.
Sound2D	2D: Specifies the spatial presentation of a sound within a 2D scene.

Table 4.11 SNHC visual-related nodes

Name	Description
BAP	Defines the current look of a body by means of body animation parameters.
BDP	Customizes the proprietary body model of the decoder to a particular body, or decodes a body model along with the information on how to animate it.
Body	Organizes the definition and the animation of a body.

Table 4.11 SNHC visual-related nodes (Continued)

Name	Description
BodyDefTable	Defines the behavior of body animation parameters on a down-loaded body by specifying displacement vectors of vertices inside IndexedFaceSet objects as a function of a combination of BAPs.
BodySegment-ConnectionHint	Defines the connection information of segments as a hint for maintaining connected surfaces.
Expression	Defines the expression of the face as a combination of two expressions from the standard set of expressions.
Face	Defines and animates a face in the scene.
FaceDefMesh	Allows for the deformation of an IndexedFaceSet as a function of the amplitude of a FAP as specified in the related FaceDefTables node.
FaceDefTables	Defines the behavior of a facial animation parameter on a down-loaded face by specifying the displacement vectors for moved vertices.
FaceDefTransform	Defines which field (rotation, scale, or translation) of a Transform node is updated by a facial animation parameter, and how the field is updated.
FAP	Defines the current look of the face by means of expressions and FAPs, and gives a hint to TTS controlled systems on which viseme to use.
FDP	Defines the face model to be used at the terminal, either by calibration of the terminal internal model or by decoding a new face model.
FIT	Allows a smaller set of FAPs to be sent, through the use of a rational polynomial mapping between parameters.
Hierarchical3DMesh	Represents multiresolution polygonal models with multiple levels of detail, smooth transition between consecutive levels, and hierarchical transmission through an independent ES.
Viseme	Defines a blend of two visemes from a standard set of 14 visemes.

4.7 SUMMARY

In this chapter, the scene description format specified in MPEG-4 Systems [MPEG4-1] has been presented. After introducing the basic concepts, a set of scenes of increasing complexity served as the basis for presenting the different kinds of nodes and their features. Important new BIFS features, currently being finalized, were previewed. Finally, a reference section gathers a short but exhaustive description of all BIFS nodes. For another view on BIFS, see [Sign99] or Chapter 14 of [PuCh00].

4.8 REFERENCES

[ECMA99] ECMA-262 Standard Specification, ECMAScript Language Specification, 3d Edition, December 1999.

[HaWC96] Hartman, Jed, Josie Wernecke, and Rikk Carey. *The VRML 2.0 Handbook.* Reading, MA: Addison-Wesley, 1996.

[HeBa97] Hearn, Donald, and Pauline Baker. *Computer Graphics*, pp. 419–420, 617–618, Upper Saddle River, NJ: Prentice Hall, 1997.

[N4319] MPEG Requirements. *MPEG-4 Requirements.* Doc. ISO/MPEG N4319, Sydney MPEG Meeting, July 2001.

[Niel00] Nielson, G. "Volume Modelling," *Volume Graphics* (M. Chen, A. Kaufman, and R. Yagel, Eds.), pp. 29–48. London: Springer, 2000.

[MPEG4-1] ISO/IEC 14496-1:2001. *Coding of Audio-Visual Objects—Part 1: Systems,* 2d Edition, 2001.

[PuCh00] Puri, Atul, and Tsuhan Chen (Eds.). *Multimedia Systems, Standards and Networks.* New York: Marcel Dekker, 2000.

[Rotg96] Rotge, J.-F. *Principles of Solid Geometry Design Logic.* Proceedings of the CSG 96 conference, Winchester, UK, 1996.

[Sign99] Signes, J. *Binary Format for Scenes (BIFS): Combining MPEG-4 Media to Build Rich Multimedia Services.* Proceedings Visual Communications and Image Processing '99, San Jose, California, 1999.

[SMIL98] *Synchronized Multimedia Integration Language (SMIL),* 1.0 Specification, W3C Recommendation. June 15, 1998. *www.w3.org/TR/1998/REC-smil-19980615*

[VRML97] ISO/IEC 14772-1:1997. *The Virtual Reality Modeling Language, www.vrml.org/Specifications/VRML97*

CHAPTER 5

MPEG-J: MPEG-4 and Java

by Viswanathan Swaminathan, Alex MacAulay,
Gianluca De Petris

Keywords: application engine, adaptive session, MPEG-J, Java in
MPEG-4, resource management, algorithmic control, and
algorithmic interactivity

*M*PEG-4 terminals may vary from high-quality entertainment/TV set-top boxes to wireless and handheld devices. The computational capabilities and the network (bandwidth) resources available at these terminals differ considerably. Without MPEG-J, these wide variations in resources would make it very hard to achieve the dream of *create once, run anywhere.*

Recent technological advancements made it hard to perceive the boundaries between a terminal that decodes audiovisual streams and a terminal that runs a computer program. On one hand, this makes the resources available at an MPEG-4 terminal vary wildly but, on the other hand, this provides an opportunity to control the audiovisual presentation at the client terminal through (computer) programs. This is the vision behind MPEG-J.

These programs can make complex decisions to modify the audiovisual presentation based on the conditions at the client. The modifications are aimed at adapting the presentation to the client's capabilities. Since these are computer programs, complex logic can be used to detect static and (time-varying) dynamic conditions at the terminal. For example, the program can discern the capabilities of a PDA as opposed to a desktop computer and can detect extreme conditions such as network bottleneck problems or declines in the available computing cycles of a processor.

The modifications to the presentation also can be the result of a complex (algorithmic) response to user interactivity (for example, the response to a user move in a tic-tac-toe game). The parametric BIFS scene description mechanism (see Chapter 4) provides the scene creator with a wide range of choices, suitable for most multimedia presentations. However, one may want to associate a behavior with a scene based on a complex algorithm. This can be achieved by associating a part of the scene to a routine written in a procedural language.

MPEG-4 specifies coding tools for audio and visual (not just video) objects. The Systems part of the MPEG-4 standard [MPEG4-1] specifies how these objects that are encoded in the elementary bitstreams are positioned and synchronized with other objects. This is called the *presentation engine* in MPEG-4. This engine decodes and builds a presentation from the scene description (BIFS) and object description (OD) streams using any associated decoded audiovisual streams. For simplicity, this can be viewed as an interpreter of a parametric description. The objective of MPEG-J is to add to the presentation engine an application engine that can associate complex programmatic behavior with the presentation based on user input and terminal conditions. The main requirements for such an application engine are

1. Ability to execute computer programs or applications from downloadable compiled code
2. Ability to run the same code on different hardware and software platforms available at the terminal
3. A well-defined security model
4. Ability to run multiple applications synchronously without conflicts
5. Ability of mechanism to notify and handle, if necessary, exceptional or drastic conditions
6. Availability of interfaces to detect static and dynamic information about the terminal environment, the computational capabilities, and the network bandwidth capabilities
7. Availability of interfaces to get information about the presentation: objects in the scene, elementary streams, decoders, and so forth
8. Availability of interfaces that can be used by the program to access, modify, or even create parts of the presentation
9. Ease of application writing

Java[1] satisfies these requirements, and in addition is widespread and object oriented; it also provides exception handling, event mechanism, polymorphism, reflection, and so on, making application creation easy [GoJS96, LiYe96]. These were the main reasons that led to the choice of Java in MPEG as the programming language for the applications bundled with the encoded

---

1. Refers to both the programming language and the virtual machine.

media and of the Java virtual machine (JVM) as the platform to execute the embedded Java applications that would exist in all compliant MPEG-J terminals. In order to run the applications written in Java, an implementation of a JVM must exist in the terminal. Compiled Java byte code is interpreted by the JVM, enabling the application to run without necessitating any platform-specific changes or recompiling.

Java is a simple, object-oriented programming language, which makes developing applications easy. The object-oriented nature of Java lets programmers think in terms of reusable objects of data and interfaces to the objects instead of programming constructs. For further reading on object-oriented programming, please refer to [Booc94].

When a Java program is compiled, the compiler generates platform-independent object code that can be executed on any processor given the presence of a Java run-time system, or JVM. The JVM interprets the architecture-neutral compiled Java code on the fly. Java is designed to be easy to interpret and to translate efficiently to machine code by the JVM.

Java is designed to be robust and secure. Security here does not refer to the content (as in MPEG-4 IPMP) but to the security of the terminal executing Java code. It is important to secure the terminal (the JVM implementation), which executes the Java program that has been compiled on a different machine. The security model in Java ensures that malignantly compiled code cannot cause any damage to the terminal.

Java is multithreaded, which makes it possible for the program to perform more than one operation simultaneously. Java is also dynamic, which means that new libraries can be added without stopping the execution of the program. This feature is very useful because applications can be kept running and new features can be added progressively as the corresponding software becomes available.

Minimal background knowledge of the Java programming language or some other programming language is required in order to understand some MPEG-J concepts and code examples in this chapter. Nevertheless, this requirement is kept to a bare minimum, so that even a reader with no object-oriented programming basics can catch the philosophy and the main concepts of MPEG-J.

The objective of MPEG-J, specified in MPEG-4 Systems, Version 2 [AMD1-1], is to define a Java *application engine* (complementing the presentation engine) that specifies how the above-mentioned applications are contained in a bitstream and executed at the client terminal. Toward this objective, MPEG-J defines a set of APIs that enable the application to interact with the terminal and its environment. MPEG-J also defines the delivery of the application to the client terminal and the scope, life cycle, and security restrictions. It should be noted that a Java application is composed of a set of classes that are compiled to an intermediate language called *byte code*, and that each class can have one or more instances of it called *objects*. A delivery mechanism is defined in MPEG-J to carry the Java byte code (*classes*) and Java objects (these objects are different from the audiovisual objects in the scene graph) that make an *MPEG-J application*. The life cycle and scope of an MPEG-J application is defined in the standard along with a strict security model, which protects the MPEG-4 terminal from malicious applications, such as viruses.

Because a routine written in a procedural language is not very useful if it cannot sense the external environment and react, a way to communicate with the terminal had to be defined. This way, every MPEG-J application creator can count on the presence on any terminal of the same communication interfaces (interoperability requirement). For this reason, MPEG-J also defines a set of Java APIs to access and control the underlying MPEG-4 terminal that plays the *MPEG-4 audiovisual session*. Here an interface is a function or a procedure call that can be made by the program. An MPEG-J application that uses these APIs can be embedded in the content as an *MPEG-J elementary stream* similar to audio and visual streams. Such an application carried with the content is called an *MPEGlet*.

Almost all the MPEG-J applications that were considered ask for the capability to sense network and performance conditions to act on the presentation engine in different ways. This is the rationale behind the MPEG-J APIs for programmatic access to the scene, network, and decoder resources. After sensing the environment and with knowledge about the current presentation, the application can gracefully adapt the presentation to varying static or dynamic terminal conditions. Currently, most clients adapt to extreme resource conditions using some ad hoc policy. Using MPEG-J, an optimal policy can be enforced by the content creator by embedding this as an MPEGlet along with the content. MPEG-J also offers enhanced interactivity with the users, thanks to the possibility of associating complex algorithms with user interaction. Content creators can include MPEGlets to respond to user triggers. In addition, a subset of the original scene based on client capabilities, user preferences, or user interactions can be displayed.

Figure 5.1 shows the building blocks in an MPEG-J-enabled MPEG-4 terminal. BIFS defines the binary format for the parametric description of the scene. This is used to build the scene and present it because BIFS describes the spatio-temporal relationships of the audiovisual objects in the scene (see Chapter 4). The part of the terminal that interprets BIFS and its related streams and composites and presents them is called the *presentation engine*. The *execution engine* decompresses the coded audiovisual streams. MPEG-J defines an *application engine*, which offers programmatic access and control to the presentation and execution engines. Specifically, MPEG-J enables an application to access and modify the MPEG-4 scene and, to a limited extent, the network and decoder resources of the terminal. The MPEG-J APIs (the platform-supported Java APIs), together with the JVM, form the application engine that facilitates the execution of the delivered application [SwFe00, M2566, FSPS00, N3545]. This application engine allows the content creators to include programs in their content that intelligently control the MPEG-4 terminal depending on client conditions [M2566].

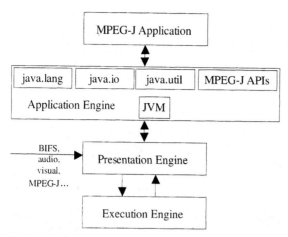

Fig. 5.1 MPEG-J application engine.

Because the MPEG-J programs (MPEGlets) are usually carried separately, they can be reused with other audiovisual scenes. It is also possible to have these programs resident on the terminal, provided that they are written in a way that is independent of the scene and the elementary streams associated with it.

Some examples where the MPEG-4 scene can adapt to varying available resources at the terminal using the programmatic control of MPEG-J are

1. Instead of degrading an entire scene when resources are low, a lower priority video object (e.g., picture-in-a-picture in a newscast) in the scene can be replaced with a still texture object.

2. Depending on whether a wireless phone, a PDA, a poorly powered out-of-date PC, or a top-of-the-line PC is used, the video to be played adapts itself to the environment. For example, the same content uses high-quality five-channel surround sound audio or low-quality mono audio depending on the receiving device. This content adaptation can also be based on user preferences.

3. An educational video is streamed to the user's desktop or handheld device. After the user completes each chapter, a quiz is administered. The difficulty level varies based on user responses.

The next section introduces the MPEG-J *architecture*, wherein all the components that make the Java application engine are defined. The execution rules, delivery mechanism, and security limitations of the MPEG-J applications are also discussed. Section 5.2 introduces the MPEG-J APIs defined in the MPEG-J specification [AMD1-1]; the functionalities provided by these APIs are discussed in detail with example Java code to illustrate how to use them. Section 5.3 discusses some possible application scenarios for MPEG-J.

Section 5.4 introduces the design of the MPEG-J Reference Software, and finally Section 5.5 summarizes this chapter. To get readers started quickly on using the MPEG-J APIs, Appendix E provides Java code·samples that contain detailed examples of MPEGlets using the MPEG-J APIs.

5.1 MPEG-J ARCHITECTURE

This section introduces the MPEG-J architecture and gives a brief overview on how MPEG-J applications are delivered and executed at the terminal. This section will also explain the delivery and execution mechanisms, scope, life-cycle and security of MPEG-J applications.

Figure 5.2 illustrates a simplified architecture of an MPEG-4 player as specified in MPEG-4 Systems, Version 1 [MPEG4-1]. BIFS data in the BIFS elementary streams together with object descriptors (ODs) in the OD streams is used to build the scene graph and associate the nodes in the scene graph with the corresponding audiovisual (media) elementary streams. The BIFS decoder is used to decode the scene description information. The decoded media data is buffered in the composition buffers that feed the *compositor* and *renderer.* Composition is the act of spatially reconstructing the scene from the objects using the scene description information. Rendering is the act of drawing/presenting a composed scene.

The architecture of an MPEG-J enabled MPEG-4 player is shown in Figure 5.3 [AMD1-1]. The parametric presentation engine of MPEG-4 Systems, Version 1, forms the lower half of Figure 5.3. The *MPEG-J application engine* (upper half of Figure 5.3) is used to control the presentation engine through well-defined application execution and delivery mechanisms [N3545]. The MPEG-J application uses the MPEG-J APIs for this control.

Applications can be local to the terminal or carried in the MPEG-J stream and executed in the terminal. While the term *MPEGlet* is used to refer to a streamed MPEG-J program, a local program is generically called *MPEG-J application* [AMD1-1]. MPEGlets are required to implement the (Java applet-

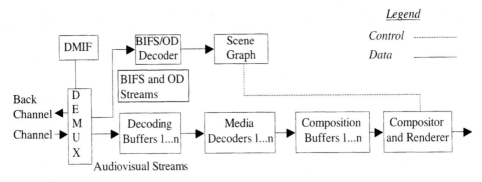

Fig. 5.2 Simplified MPEG-4 player architecture (non-MPEG-J enabled).

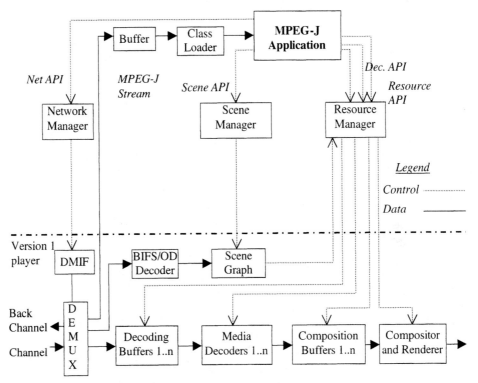

Fig. 5.3 MPEG-J-enabled MPEG-4 player architecture.

like) MPEGlet[2] interface[3] defined in the MPEG-J specification [AMD1-1]. The MPEGlets and all other associated MPEG-J classes and objects[4] are carried in an MPEG-J elementary stream. The received MPEGlet is loaded by the *class loader* (shown in Figure 5.3) and started at time instants prescribed in the time stamps of the MPEG-J stream. The MPEG-J application uses the scene, network, and resource managers to control the scene, network, decoder, and terminal resources. In Figure 5.3 the data flow is shown using solid lines while the dotted lines show the control flow.

An MPEG-J session is initiated when the MPEG-4 player receives an MPEG-J OD. The *MPEG-J decoder* (composed of the buffer, class loader, and API implementations) handles the arriving Java classes and objects. There can be more than one MPEGlet in an MPEG-J stream. Each time a class is

---

2. This font will be used for code, names of Java classes, interfaces, or packages.

3. Java interfaces are a collection of methods (functions), all of which will be implemented by any Java class that implements the interface.

4. Java objects are instances of the Java classes. A Java object contains all the data specific to that instance of the class. Java has a mechanism to serialize Java objects for transport (e.g., if Integer is a class, the variable v=3 is an instance).

received, it is loaded at the time instants specified by its time stamps. Any class that implements the MPEGlet interface is an MPEGlet. When an MPEGlet is received by the decoder, the decoder performs the prescribed initiations and starts the execution as a separate thread in the terminal. This will be further explained in the next section.

5.1.1 MPEGlets

Any MPEG-J application that implements the MPEGlet interface defined in the MPEG-J specification [AMD1-1] is an MPEGlet.[5] Particularly, the term *MPEGlet* is used for applications that are streamed to the terminal. However, most of the concepts defined in this section are applicable both to local and streamed MPEG-J applications.

As MPEG-J allows computer programs to be streamed and executed at the MPEG-4 terminal, it becomes necessary to define a set of rules for executing the programs: MPEGlets, in this case. This section describes how the MPEGlets are identified, bootstrapped, and executed at the terminal. This section also describes how multiple MPEGlets can be executed at the same time without conflicts.

5.1.1.1 Life Cycle   After an MPEGlet is received, the terminal needs to know exactly how it has to execute it. For this purpose, the MPEG-J specification defines the life cycle of an MPEGlet. The MPEGlet interface defines four methods (functions), namely init(), run(), stop(), and destroy(). The life cycle and the states of MPEGlets are implicitly defined by these methods in the interface. The rules as to when these methods have to be called by the terminal are defined to enable a terminal to start, run, and stop an MPEGlet. Those readers who are familiar with Java applets would recognize that the MPEGlet is very similar to a Java applet.

As in the case of applets, the MPEGlets implement the init(), run(), stop(), and destroy() methods specified in the MPEGlet interface. The author of an MPEGlet would write the appropriate code for all the methods defined in the MPEGlet interface. All the initialization of resources form the init() method, while reclaiming the same resources forms the destroy() method. The run() method will be executed as a new thread and hence will contain all the algorithmic control mechanisms to programmatically control the MPEG-4 terminal. Moreover, the author would put all the code to halt the MPEGlet in stop(). After the coding is completed, the author compiles it to generate a class file (Java class) out of the MPEGlet. This class file would then be delivered to the terminal. At the terminal, when the class is received, it is loaded according to its time stamps. Additionally, the terminal

---

5. In this chapter, the term *MPEGlet* is used both for the Java interface and the Java class (or an application) that implements that interface.

can determine if the class implements the MPEGlet interface through Java primitive function calls. This determines whether a given class is an MPEG-J application or a supporting class.

The deadline by which each class has to be loaded is given by its time stamp. If it is an MPEGlet, it is also instantiated and the init() method is called exactly at that deadline. Once initiated, a separate thread is created by the MPEG-J decoder and the run() method is called.

5.1.1.2 Name Scope To ensure there are no naming conflicts in MPEG-J applications, the name scope of MPEGlets has to be defined. For example, two classes using the same name or two MPEGlets with the same name would cause loading and hence execution problems. This section describes the name scope limits identified by the MPEG-J specification to avoid such problems [AMD1-1].

The OD defines the name scope of audiovisual objects in an MPEG-4 scene. This also applies to the MPEG-J classes and objects. All the identifiers (e.g., classID) are interpreted within the name scope of that MPEG-J elementary stream and its OD. The OD that the MPEG-J elementary stream is associated with sets the name scope of the MPEGlet. This is consistent with the scope definition of Inline scenes in MPEG-4 Systems (see Chapter 4).

The managers shown in Figure 5.3 set the name scope of the MPEGlet by identifying its object descriptor. In practice, this is done using the reference of the MPEGlet passed in the constructor[6] of the MpegjTerminal, a class specifically defined among the MPEG-J APIs.

The following Java code snippet[7] shows an example of an MPEGlet passing its own reference while creating the MpegjTerminal object:

```
public class SceneExample implements MPEGlet
{
    MpegjTerminal mpegjTerminal;
    public void init()
    {
        mpegjTerminal = new MpegjTerminal(this);
[...]
```

Although the MPEGlet must pass a reference to itself in the constructor, a local application may use the zero-argument constructor of the MPEG-J terminal to imply the root of the name scope, as in the following example:

```
public class AtlanticDemo extends WindowAdapter {
    MpegjTerminal MpegjTerminal = new MpegjTerminal();
[...]
```

---

6. Constructor is a special function (method) defined in all Java classes which is used to tell the executing system to generate or construct a new instance of that class with specified attributes. A class can have multiple constructors.

7. All the code samples in this chapter are written in Java.

5.1.2 Delivery

Typically the mechanisms by which the computer programs are delivered to terminals that execute them is very different from how the audiovisual data is delivered. Inherently, audiovisual data is very time aware and, in most cases, tolerant of losses, whereas compiled computer programs are intolerant of losses and are not time aware. The MPEG-J specification defines the mechanisms for carrying MPEG-J data (MPEGlets and all associated Java classes and objects) using the delivery mechanisms (access units and elementary streams with time stamps) defined for audiovisual data in MPEG-4 (see Chapter 3). Interpretation of time stamps for MPEG-J data, the dependencies among MPEG-J data, and how this data is carried in an MPEG-J elementary stream are explained in this section.

5.1.2.1 Time Stamp Semantics   MPEG-4 audiovisual data carry two types of time stamps. One is the time at which a decoder should start decoding an access unit called the *Decoding Time Stamp* (DTS). Similarly, the *Composition Time Stamp* (CTS) is the time instant at which the decompressed data has to be composited into the scene for the presentation. For MPEG-J, this time stamp pair does not have a direct meaning. Thus the MPEG-J specification defined new semantics for DTS and CTS.

Two types of time stamps are used in MPEG-J streams. These two together provide a time window in which each class has to be made available and loaded. The first time stamp, *Start Loading Time Stamp*, signals the time after which the class *may be* loaded by the class loader into the MPEG-J decoder. This avoids name scope and resource conflicts. All the classes that are used by a given class should also be loaded by this time stamp of the given class. This is carried in the SL header as the DTS. The second time stamp, *Load By Time Stamp*, signals the deadline by which the class *has to* be loaded. This is carried in the SL header as the CTS. It is recommended that the temporal window specified by these time stamps be made large enough to handle problems caused by different loading times on different terminals.

5.1.2.2 Class Dependency   MPEGlets can use API calls (functions) that are defined in a Java class either in the platform (`java.io`, `java.util`, `java.lang`) or in the MPEG-J specification [AMD1-1]. MPEGlets can also use function calls that are defined in an application-specific class. An application-specific class can, in turn, use API calls defined in other application-specific classes. Here the MPEGlet depends on the first application-specific class which, in turn, depends on another application-specific class. In Java, when a class is loaded, the JVM can figure out that it needs a dependent class and can load it automatically if it is available. However, if the classes are carried in an MPEG-J elementary stream to an MPEG-J terminal with no back channel, such late recursive loading can result in unpredictable loading times for applications or simply even break the loading due to the unavailability of classes at the terminal. The MPEG-J specification sets some simple rules to avoid such eventualities.

A terminal should start loading a class only after loading all the classes it depends on. This should be reflected correspondingly in the time stamps of the classes. The MPEG-J header (introduced in Section 5.1.2.3) signals the dependencies by listing the identifiers of all the dependent classes. Alternatively, all the dependent data can be packaged using the Java archiving (JAR) format.

5.1.2.3 MPEG-J Elementary Stream

MPEG-J data comprises Java classes and Java objects (instances of classes carrying a state). MPEG-J data is delivered to the MPEG-J terminal as part of an MPEG-J elementary stream. Elementary streams carry access units (see Chapter 3). Each Java class or object can form an MPEG-J access unit. Alternatively, all the class files that depend on each other to constitute an MPEG-J application along with the associated object data can be packaged together as a single access unit. The packing format used for this is the *zip* format or the *JAR* packaging format supported by Java.

Further, the data (Java classes and objects) that make the MPEG-J access unit can be compressed using the *ZLIB* [ZLIB] compression format (supported by the JAR and zip tools). Although the same tool (zip or JAR) usually supports both the compression and the packaging, either compression or packaging can be used without the other. For example, a single class file can be compressed, while a packaged access unit containing a number of class files may not need to be compressed.

MPEG-J adds a lightweight header to the Java classes, Java objects, or the JAR package that compose a single access unit. This header is used to specify the identification, type (class or otherwise), compression, packaging, and class dependency information. All of this information is used by the MPEG-J decoder to correctly decode the class and run it at the right time. Although most of the information is already present in the class or object files and can be programmatically obtained using Java, this forms valuable information for all the layers of the player that may not be Java aware.

5.1.3 Security

This section illustrates the salient features of the security model in MPEG-J. The MPEG-J specification defines a security model for the MPEGlets [AMD1-1] that is similar to the security model for Java applets.

Given the flexibility of computer programs, it is possible to write one to be malignant to a terminal that is executing it, especially a terminal executing a precompiled computer program. The MPEG-J security model protects the MPEG-4 terminal from misbehaving MPEGlets. A security manager has to be implemented at the MPEG-J terminal to enforce the restrictions.

The MPEG-J security model imposes the following security restrictions:

☞ MPEGlets are forbidden to load libraries or define native methods.

☞ MPEGlets cannot read or write files on the terminal.

☞ Starting a program and reading certain system properties (not allowed by the Capability API) is prohibited.

☞ MPEGlets can only use their own Java code, MPEG-J APIs, the Java APIs provided by the execution environment, and the classes that have been loaded from the MPEG-J stream.

It is completely up to the security manager to relax any restrictions that are imposed by the security model. Any MPEG-J application that violates the security model is noncompliant and cannot be expected to run on compliant MPEG-J terminals.

This MPEG-J security model is unrelated to the Intellectual Property Management and Protection (IPMP) tools included in MPEG-4 Systems standard [MPEG4-1] because IPMP is aimed at defining the hooks to ensure that the rights associated with media content can be protected. In a nutshell, IPMP is about protecting the content from *bad* terminals and users, while the MPEG-J security model is about protecting the terminal from *bad* content. Both are complementary for a secure global solution.

5.2 MPEG-J APIs

So far the way MPEGlets and all the associated data are received and executed at the MPEG-J terminal has been analyzed. It has been seen that MPEGlets can make API calls that are defined in the Java packages supported by the platform. For those readers familiar with Java packages,[8] `java.io`, `java.lang`, and `java.util` are three Java packages that have to be supported by the Java platform present in the terminal. In other words, all MPEG-J–enabled terminals must normatively support these Java packages in their implementation. The MPEGlets can also make function calls that are defined in application-specific classes. In the last section, a mechanism for delivering all the dependent classes at the terminal was presented. The third set of interface calls that the MPEGlets or any supporting application-specific class can make are the APIs defined by the MPEG-J specification [AMD1-1]. These are the *MPEG-J APIs*. This section describes the MPEG-J APIs.

While the Java platform packages provide some generic features, the MPEG-J APIs support the features that are specific to MPEG-J applications. The MPEG-J APIs complement the Java platform APIs to meet the requirements of MPEG-J applications. The algorithms embedded in the MPEGlets

---

8. Java packages are a collection of classes and interfaces for logical organization. Each Java package has its own name scope, and it is possible to limit access to some private interfaces within a Java package. Java packages can also contain other subpackages.

are implemented using the interfaces defined by the MPEG-J specification. Using the MPEG-J APIs, the applications have programmatic access to the scene, terminal, and network resources. The MPEGlets use the MPEG-J APIs to sense the environment at the terminal and modify the audiovisual scene.

MPEG-J defines a variety of categories of API. Each category provides a specific functionality and is organized as a Java package. The following packages are defined in the MPEG-J specification [AMD1-1]:

☞ org.iso.mpeg.mpegj

☞ org.iso.mpeg.mpegj.scene

☞ org.iso.mpeg.mpegj.resource

☞ org.iso.mpeg.mpegj.decoder

☞ org.iso.mpeg.mpegj.net

In addition, MPEG-J refers to some APIs already defined in DAVIC (Digital Audiovisual Council) specifications, namely Service Information and Section Filtering [DAVI98].

The org.iso.mpeg.mpegj package is a collection of all the classes and interfaces that are necessary to bootstrap a session by an MPEGlet. The interfaces and the classes defined in the org.iso.mpeg.mpegj package are called the *Terminal APIs*. The Terminal APIs provide an MPEGlet with a first point of access to the terminal. The two main interfaces defined in this package are the MPEGlet and the MpegjTerminal. An application session bootstraps by constructing an MpegjTerminal and gaining access to the appropriate managers (see Figure 5.3). The Terminal APIs are discussed in detail in Section 5.2.1.

The org.iso.mpeg.mpegj.scene package is a collection of all the Java classes and interfaces defined in the MPEG-J specification with which the audiovisual objects in the MPEG-4 scene can be accessed and modified. This package contains low-level interfaces to access and modify the nodes of the scene graph corresponding to an MPEG-4 presentation. These interfaces constitute the *Scene APIs*. The Scene APIs allow the MPEG-J application to monitor the scene for certain events and modify the scene programmatically based either on events or user triggers. Events can be sent and received from specific nodes in the scene. The scene can be monitored for changes. Nodes can be created or (certain previously identified nodes) can be modified or deleted using the Scene APIs. Using these APIs, the scene can be altered based on an algorithm that can react in a complex way to the environment of the terminal or user triggers. The Scene APIs are very powerful and will be examined in Section 5.2.2.

The package org.iso.mpeg.mpegj.resource contains all the classes and interfaces necessary to monitor the static and dynamic resources at the terminal. These are referred to as *Resource APIs*. The Resource APIs provide hooks for resource management at the MPEG-4 terminal. The Resource APIs provide access to the capability manager to obtain static and dynamic capabilities. An event mechanism is provided by the Resource APIs to monitor for certain eventualities in the terminal. The mechanism to handle those events can be embedded in the bitstream as *event handler classes*. It also provides access to a decoder associated with a particular node in the scene and hooks to replace a particular decoder with another available and compatible decoder. However, it should be noted that MPEG-J does not allow embedding a new decoder in the MPEG-J bitstream. More details about these features are given in Section 5.2.3.

The `org.iso.mpeg.mpegj.decoder` consists of the classes and interfaces defined by the MPEG-J specification necessary to control a decoder resource in the terminal. It has already been seen that an MPEGlet could obtain access to a decoder associated with a node in the scene using the Resource APIs. The *Decoder APIs* define the decoder interface and provide the means to control the decoder. Apart from simple start/stop control, the Decoder APIs also support obtaining information about decoders, attaching and detaching elementary streams, and so forth. Details of the Decoder APIs will be discussed in Section 5.2.4.

The `org.iso.mpeg.mpegj.net` contains a set of high-level interfaces for simple control and query of the network channels associated with the MPEG-J terminal. These are called the *Net* or *Network APIs*. Using the Net APIs, an application may query the terminal about sessions, channels, quality of service, and so forth. These APIs interact with the DMIF layer (see Chapter 7) to obtain information about network sessions and channels. The Net APIs support enabling and disabling of network channels. Details of the Network APIs are discussed in Section 5.2.5.

MPEG-J also normatively refers to Service Information, Section Filtering, Resource Notification, and MPEG Component APIs specified by DAVIC [DAVI98]. These APIs are relevant only when the MPEG-4 data is transported over MPEG-2 Systems [MPEG2-1] because they are used for accessing information in the MPEG-2 Systems transport streams. MPEG-2 Systems transport streams may contain the Service Information tables defined by the Digital Video Broadcasting (DVB) consortium [DVB], including information useful for the terminal to display electronic program guides. The Service Information tables can be accessed using these DAVIC APIs. Application-specific private data can be transported in MPEG-2 Systems transport streams as private sections. The Section Filtering APIs can be used to access these private sections. These APIs are discussed in detail in Section 5.2.6.

The rest of this section describes the different categories of APIs in detail. The features supported by these APIs are illustrated with short Java code snippets. Universal Modeling Language (UML) representations of the various packages will be presented. UML helps to concisely describe the design of the different categories of APIs. Along with the more elaborate code samples in Appendix E, the UML diagrams can help you quickly start using the MPEG-J APIs. The following sections and Appendix E require a little more Java knowledge than the rest of this chapter. For a fuller understanding of this material, refer to the initial chapters in [ArGH00].

5.2.1 Terminal APIs

All the classes and interfaces defined in the org.iso.mpeg.mpegj package are referred to as *Terminal APIs*. They provide a bootstrapping mechanism for the MPEG-J applications. The Terminal APIs provide abstractions of the MPEG-J application, the MPEG-J terminal, and the communication between them. Here details about the main components of the Terminal APIs, the functionalities supported by them, and examples of their use are offered.

Figure 5.4 shows the UML representation of all the interfaces and classes in the org.iso.mpeg.mpegj package that make the Terminal APIs. This includes the MPEGlet interface, the MpegjTerminal class, and the descriptor interfaces (ObjectDescriptor, ESDescriptor, and DecoderConfigDescriptor). The ResourceManager, SceneManager, and NetworkManager shown in Figure 5.4 are defined in the Resource, Scene, and Network APIs sections, respectively.

The MPEGlet interface was introduced and described in Section 5.1.1. The MPEGlet interface is based on the Java applet. The MPEGlet extends (or is derived from) the Runnable interface defined in the platform Java package java.lang. This implies that each MPEGlet will be executed as a separate

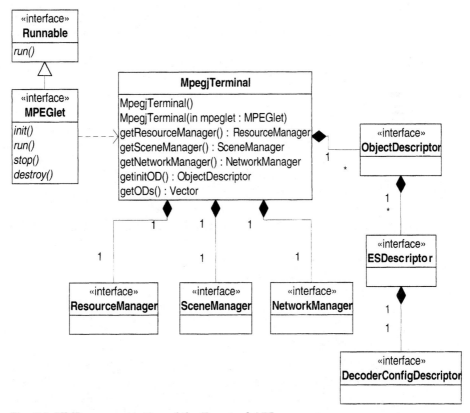

Fig. 5.4 UML representation of the Terminal APIs.

execution thread on the terminal facilitating simultaneous operation of multiple MPEGlets at the terminal. The semantics of the `init()`, `run()`, `stop()`, and `destroy()` methods were described in Section 5.1.1.

The `DecoderConfigDescriptor`, `ObjectDescriptor`, and `ESDescriptor` interfaces form the abstractions of the corresponding MPEG-4 Systems descriptors [MPEG4-1], which provide vital information about the MPEG-4 session. MPEGlets obtain information about the current MPEG-4 session in the terminal through these interfaces.

The Java class `MpegjTerminal` forms the key abstraction of the control elements present in the MPEG-J Terminal. This class provides the first point of contact for the MPEGlet. The `MpegjTerminal` provides information about the elementary streams, object descriptors, and decoder configurations and sets the context (name scope) of the MPEGlets and local MPEG-J applications. Access to the different managers is provided to the MPEGlets through `MpegjTerminal`.

5.2.1.1 Setting the Namescope and Context `MpegjTerminal` is instantiated by each MPEG-J application. This should not be interpreted as creating a new terminal for each MPEGlet but just as a handle to the environment the terminal offers. As has already been seen, an `MpegjTerminal` implementation gives the appropriate managers to the MPEGlet. The terminal, along with the managers, controls the environment (e.g., the name scope) of the MPEGlet. Section 5.1.1.2 presented the name scope of MPEGlets and MPEG-J applications and gave examples of how an MPEGlet and a local MPEG-J application name scope are set.

5.2.1.2 Obtaining the Descriptors `MpegjTerminal` provides the MPEGlet with information about the MPEG-4 session. This information is passed in the form of `ObjectDescriptor`, `ESDescriptor`, and `DecoderConfigDescriptor` interfaces using the API calls of `MpegjTerminal`. Information about elementary streams, their types, and decoder information can be obtained using these APIs.
Example:

```
ObjectDescriptor iod = terminal.getinitOD();
```

5.2.1.3 Gaining Access to the Managers The `MpegjTerminal` class provides functions to gain access to all the managers, namely, `SceneManager`, `ResourceManager`, and `NetworkManager`.
Example:

```
SceneManager mgr = terminal.getSceneManager();
```

5.2.2 Scene APIs

The Scene APIs provide programmatic access to the scene graph that is being rendered by the MPEG-4 terminal. These APIs can be used to examine the current state of the scene graph, to modify elements of the scene graph, and to notify when an element of the scene graph changes.

The Scene APIs have been designed to give efficient and fine-grained control of the scene graph. Higher-level libraries may be built on top of these APIs to provide application-specific functionality, such as 3D gaming libraries or multiuser engines. These libraries can be written purely in Java and therefore may take advantage of the flexible nature of Java byte-code to be downloaded on demand as a plug-in or included as part of the content in the MPEG-J stream.

Figure 5.5 shows the UML representation of the main components of the Scene APIs. The Scene, SceneManager, and SceneListener are the most important interfaces dealing with the scene graph associated with the MPEG-4 presentation. All the communication between the application and the nodes of the scene graph is made through the Node interface. The fields of the nodes are accessed and modified through the FieldValue interfaces. Different types of the FieldValue interfaces are shown in Figure 5.5 although the list is not exhaustive. Please refer to the MPEG-J API specification for the other types [AMD1-1]. NodeValue is an important FieldValue, which is used along with the Node and the NewNode interfaces to create and add a new node to the scene graph. The other main functionalities include copying a new node, deleting an existing node, reading the value of a field, notifying a field value change, sending an event to field, and modifying a ROUTE.[9] In the following sections, a detailed top-down description of the Scene APIs is provided, showing how an application can use the Scene APIs to access the scene graph; the main components are detailed, and some code snippets to achieve the previously mentioned functionalities are given. More detailed examples are provided in Appendix E.

5.2.2.1 SceneManager Interface   The SceneManager interface is the top-level entry point of the Scene APIs. An instance of the SceneManager can be obtained from the MpegjTerminal class by calling the getSceneManager() method. The application can use the SceneManager to get a reference to the current MPEG-4 scene as a Scene object.[10] This will be used by the application to access and modify the audiovisual presentation.

5.2.2.2 SceneListener Interface   In Java, an interface for the listener is defined ahead of time. A mechanism is usually defined to register all the classes that implement that listener interface to an event generator. When the desired event occurs, the component that generates the events (event generator) notifies all the registered listener classes.

---

9. Refer to Chapter 4 on BIFS for an explanation of ROUTES in the scene.
10. The Scene object is a Java object and is not to be confused with audiovisual objects in the scene. A Java object contains all the data associated with a single instance of a Java class. In this case, all the data (i.e., values associated with the variables) in the class that implements the Scene interface is obtained by the application.

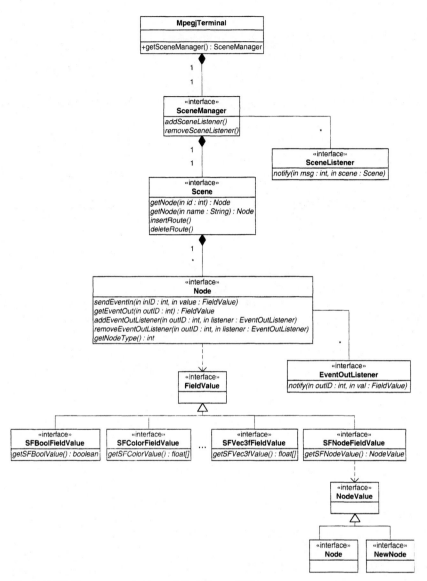

Fig. 5.5 UML representation of the Scene APIs.

The MPEG-J application needs to implement the `SceneListener` interface in order to get a reference to the scene. The application registers itself as a listener with the `SceneManager` by calling the `addSceneListener` method. When the scene is ready for access by the application, the listener is notified with a SCENE_READY message, giving the application a reference to the `Scene` object. Each time the scene is replaced (typically by a BIFS `Scene Replace` command), the listener is notified with a SCENE_REPLACED message. Finally,

when the scene is no longer available, the listener is called with the
SCENE_REMOVED message.

The following code shows how an MPEGlet obtains a reference to the
Scene by implementing the SceneListener interface:

```
public class MyMpeglet implements MPEGlet, SceneListener {
    private Scene m_scene = null;
    public void run() {
        m_terminal.getSceneManager().addSceneListener(this);
    }
    public void notify(int message, Scene scene) {
        switch (message) {
        case SceneListener.Message.SCENE_READY:
            System.out.println("Scene is ready");
            m_scene = scene;
            break;
        case SceneListener.Message.SCENE_REPLACED:
            System.out.println("Scene has been replaced");
            break;
        case SceneListener.Message.SCENE_REMOVED:
            System.out.println("Scene has been removed");
            break;
        }
    }
    ...
}
```

5.2.2.3 Scene Interface   The application can access the nodes and modify the
ROUTEs in the scene via the Scene interface. Each BIFS node in the scene may
have an associated number (called the node ID) that uniquely identifies that
node within the scene. The getNode method in the Scene interface can be
called with a node ID to obtain a reference to the corresponding node. Note
that only nodes with an associated node ID are accessible to the application.
The design decision to limit access only to nodes with a node ID gives the terminal an opportunity to optimize the scene composition, knowing that some of
the nodes without an ID will never change.

BIFS nodes may also have an associated node name, but only if the
BIFS USENAME feature is enabled for that scene. Such nodes can be accessed
by calling getNode with either the node ID (an integer) or the node name (a
string).

Modifying ROUTEs   The application can add or remove ROUTEs from the scene
by calling the insertRoute and deleteRoute methods of the Scene interface.[11]
The following example inserts a new ROUTE between the string fields of two
Text nodes:

---

11. The insertRoute and deleteRoute methods as well as the ability to get a node by
    name were added to the standard as corrigenda (see ISO/IEC 14496-1:2000 COR1).

```
Node firstNode = m_scene.getNode(1);
Node secondNode = m_scene.getNode(2);
m_scene.insertRoute(
    firstNode, EventOut.Text.string,
    secondNode, EventIn.Text.string
);
```

5.2.2.4 Node Interface As shown in Figure 5.5, the Node interface provides access to a node in the scene and is where most of the action occurs in the Scene APIs. The Node interface has methods for reading and writing the values of fields (getEventOut and sendEventIn) and for registering a listener on a field that is notified when the field changes (addEventOutListener and removeEventOutListener). The details of the Node interface are described in the following sections.

5.2.2.5 Node Types If the application does not know the type of the node (for example, whether the node is an OrderedGroup or Transform2D), it can call the getNodeType method in the Node. The value returned is an integer value equal to the SFWorldNode node data type identifier of the node as given in the node coding tables of Annex H in the MPEG-4 Systems specification [AMD1-1]. The values are enumerated as integer constants in the NodeType interface to aid code readability. For example, to check if a node is a Text node:

```
if (firstNode.getNodeType() == NodeType.Text) {
    System.out.println("The node is a Text node");
}
```

5.2.2.6 Field IDs Integer constants are used to indicate which fields of a node are being referenced. The integer constants are based on the node coding tables of Annex H of the MPEG-4 Systems specification [AMD1-1]. When sending an event to a field with the sendEventIn method, the "in ID" of the field is used. When reading the value of a field with the getEventOut method or when a listener is notified of an event out, the "out ID" of the field is used. When creating new nodes, the "all ID" of the field is used to identify the field. These constants are defined in the EventIn, EventOut, and Field interfaces, so it is a simple matter of using one of the constants in these interfaces rather than looking up the node coding tables. For example, EventIn.Anchor.children is equal to 2, EventOut.Anchor.children is equal to 0, and Field.Anchor.children is equal to 2.

5.2.2.7 Field Values The FieldValue interface represents the value of a field at an instant of time. This is a useful abstraction for getting or setting the state of a field. An instance of this interface is returned by the getEventOut method and passed to the sendEventIn method, both of the Node interface. An instance of this interface is also given to the listeners of a field when the field changes. FieldValue is in fact a superinterface for a collection of interfaces that correspond to every possible type of field, ranging from

SFInt32FieldValue representing the value of a 32-bit integer to MFNode-
FieldValue representing a vector of nodes. Each interface has only a single
method that returns the value of the field. The name of this method always
takes the form get*Type*Value in which *Type* is the name of the field type. For
example, SFInt32FieldValue has a method called getSFInt32Value that
returns an integer. For multiple valued types, the method returns an array of
instances of the corresponding single valued interface. So, for example,
MFInt32FieldValue has a method called getMFInt32Value that returns an
array of SFInt32FieldValues.

Reading the Value of a Field   The getEventOut method of the Node interface reads the current value of an eventOut or exposedField. [12] For example,
to read the current value of the scale field of a Transform2D node:

```
FieldValue scaleValue =
    myNode.getEventOut(EventOut.Transform2D.scale);
float[] scale =
    ((SFVec2fFieldValue) scaleValue).getSFVec2fValue();
System.out.println(
    "Scale is " + scale[0] + ", " + scale[1]);
```

Notification of a Field Change   Rather than continuously polling a field to
get its latest value, the application can add a listener on a field and be notified
when the field changes. When the field is no longer of interest to the application, it can remove the listener. It is important to note that the listener may be
called inside the main rendering loop of the terminal and therefore the notification method should avoid doing any significant amount of processing.

The following example shows a listener being added and removed from
the touchTime eventOut of a TouchSensor node:

```
class MyMpeglet implements MPEGlet, EventOutListener {
    private Node m_touch;
    public void init() {
        ...
        m_touch.addEventOutListener(
            EventOut.TouchSensor.touchTime, this);
        ...
    }

    public void destroy() {
        ...
        m_touch.removeEventOutListener(
            EventOut.TouchSensor.touchTime, this);
        ...
    }

    public void notify(int outID, FieldValue newValue) {
```

12. See Chapter 4 for more information on eventOut, eventIn, and exposedField.

```
                         System.out.println("Touch time was " +
                             ((SFTimeFieldValue) newValue).getSFTimeValue());
                 }
                 ...
             }
```

Sending an Event to a Field   The only way to modify the scene graph via the Scene APIs is by using the `sendEventIn` method of the `Node` interface. It is important to note that the `sendEventIn` method does not return until the requested modification has been committed to the scene. In this way, the application can be sure that the change has been made when control returns to the application.

Because each field type is represented by an interface and it is not possible to instantiate an object from an interface without a class, this question is often raised: "How is it possible to pass a `FieldValue` (such as `SFInt32-FieldValue`) to the `sendEventIn` method if it is not possible to instantiate it?" The answer is, of course, "by implementing the interface!" One way to do this is to use an anonymous class that returns the desired value. For example, the following code sets the `whichChoice` field of a `Switch` node to 3:

```
             myNode.sendEventIn(
                 .EventIn.Switch.whichChoice,
                 new SFInt32FieldValue() {
                     public int getSFInt32Value() { return 3; }
                 }
             );
```

5.2.2.8 Node Values   The Scene APIs provide the powerful ability to work with fields that contain nodes, enabling the application to

1. Duplicate a reference to a node from one part of the scene to another
2. Delete a node
3. Create a new node (using the `NewNode` interface)

The basic form of a node reference is the `NodeValue` interface. This is the type returned by the `getSFNodeValue` method of `SFNodeFieldValue`. In the rest of this section, these features will be illustrated with examples.

Copying Nodes   A node reference can be duplicated from one part of the scene to another (equivalent to the USE mechanism in BIFS). This is achieved by getting the node reference using `getEventOut` and then sending it to the destination field using `sendEventIn`. The following example reads the value of the `children` field of a `Group` node and adds all its children to the `children` field of another `Group` node.

```
             FieldValue kids =
                 oneGroup.getEventOut(EventOut.Group.children);
             anotherGroup.sendEventIn(EventIn.Group.addChildren, kids);
```

Deleting Nodes   A node can be deleted by simply sending an event with an `SFNodeFieldValue` that returns null. For example, to delete the `geometry` of a `Shape` node:

```
shapeNode.sendEventIn(
    EventIn.Shape.geometry,
    new SFNodeFieldValue() {
        public NodeValue getSFNodeValue() { return null; }
    }
);
```

All the children of a `Group` node can be deleted by sending an `MFNode-FieldValue` that returns an empty array:

```
groupNode.sendEventIn(
    EventIn.Group.children,
    new MFNodeFieldValue() {
        public SFNodeFieldValue[] getMFNodeValue() {
            return new SFNodeFieldValue[0];
        }
    }
);
```

Creating Nodes   A newly created node (of an existing node type) can be added to the scene graph by sending an event with an `SFNodeFieldValue` that returns a `NewNode`. The `NewNode` interface has three methods that allow the application to describe the new node to the terminal:

1. **`getNodeType`**, which must return the type of the new node
2. **`getNodeID`**, which must return the ID of the node (or zero if it does not have an ID)
3. **`getField`**, which is given a field ID and must return the initial value of the corresponding field in the new node (or null to indicate that the default value of the field is to be used)

It is up to the application to implement the `NewNode` interface—the terminal calls back the interface to extract the information about the node that it must create. The `NewNode` provides a static description of the new node that is used only during the creation of the node.

This example sets the geometry of an existing `Shape` node to a new `Rectangle` node with a size of 200x100 using a custom class that implements the `NewNode` interface:

```
class Rectangle implements NewNode {
    private float m_width, m_height;
    Rectangle(float width, float height) {
        m_width = width;
        m_height = height;
    }
    public getNodeType() {
        return NodeType.Rectangle;
```

```
            }
            public getNodeID() {
                return 0;
            }
            public getField(int id) {
                if (id == Field.Rectangle.size) {
                    return new SFVec2f(m_width, m_height);
                } else {
                    return null;
                }
            }
        }
    shapeNode.sendEventIn(
        EventIn.Shape.geometry,
        new SFNodeFieldValue() {
            public NodeValue getSFNodeValue() {
                return new Rectangle(200, 100);
            }
        }
    );
```

5.2.2.9 Scene APIs and Other MPEG-4 Scene Access Mechanisms   The
MPEG-J Scene APIs have certain aspects in common with other mechanisms
available in MPEG-4 to access the scene graph, but for each case there is a distinction and therefore a focus on a different application area. In the following,
these distinctions are briefly illustrated.

A BIFS update stream can make the same kind of modifications to the
scene graph as can be made through the Scene APIs, except that the modifications are *server-based* and therefore cannot be derived (at least not instantaneously) from the local interaction of a user with the scene graph or a local
condition (event or exception). BIFS update streams are typically intended for
time-based updates that are independent of the particular state of the terminal. Also, the Scene APIs are intended to change the scene based on embedded
intelligence in the application.

The Conditional BIFS node can make the same kind of modifications as
the Scene APIs and can be triggered by local interaction with the user. However, the Conditional node can provide only a finite set of possible modifications to the scene because each Conditional node is preprogrammed with a
single action; this does not happen with the Scene APIs. Thus the Conditional node can be used only for simple graphical interfaces and limited
interactivity.

The BIFS Script node provides a scene graph interface to scripts using
the ECMA Script language (see Chapter 4). This interface is roughly comparable in features to the MPEG-J Scene APIs. The Script node can support
reasonably complex interfaces and user interaction. Due to the inherent
nature of lightweight scripting languages, however, it typically suffers from
poorer code maintainability, flexibility, and scalability in comparison to the
MPEG-J Scene APIs.

5.2.3 Resource APIs

The `org.iso.mpeg.mpegj.resource` package (also called Resource APIs) helps the MPEG-4 session adapt itself to varying terminal resources. The Resource APIs are used for regulation of performance by providing a centralized facility for managing resources. The Resource APIs in the `org.iso.mpeg.mpegj.` resource package can be used to monitor the system resources, to listen to exceptional conditions through the event mechanism, and to handle such eventualities. The main components of the resource package are the `ResourceManager` class, the `CapabilityManager` class (which implements the `StaticCapability`, `DynamicCapability`, `DynamicCapability-Observer`, and `TerminalProfileManager` interfaces), and the interfaces supporting `EventMechanism` for `Renderer`, `RendererFrame`, and decoder events. In this section, the main functionalities offered by each of these components are described and illustrated by code snippets. More detailed examples of the Resource APIs can be found in Appendix E.

5.2.3.1 `ResourceManager`   The `ResourceManager` is the entry point to the Resource APIs, providing access to the `CapabilityManager`, the renderer, and the decoders. Figure 5.6 shows the UML class diagram for the `ResourceManager` and its related classes and interfaces. The `ResourceManager` allows the application to gain access to the decoder associated with a node in the scene graph. Once the decoder is accessed, the `ResourceManager` also provides the `getDecPriority` and `setDecPriority` functions to get and set decoder priorities. The following example shows how to get a decoder priority:

```
decoder = resourceManager.getDecoder(m_node);
decoder.getDecPriority();
```

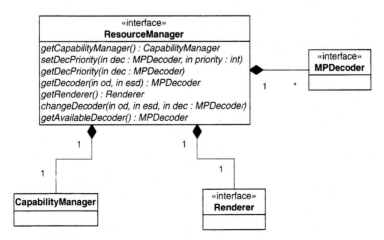

Fig. 5.6 UML representation of the Resource APIs (`ResourceManager` and related interfaces).

There are a number of other functionalities that are offered once the
`MPDecoder` is obtained. These will be further discussed in the section on
Decoder APIs.

The application can use the `ResourceManager` to access and register
event handlers with the `Renderer` for renderer events:

```
// getting the renderer from the Resource Manager
try {
    m_renderer = resourceManager.getRenderer();
} catch( RendererNotFoundException rnfe) { }

// to add event listener to the Renderer
if( m_renderer != null )
    m_renderer.addMPRendererMediaListener(renderer_EH);
```

The `ResourceManager` also provides access to the `CapabilityManager`.
The `CapabilityManager` is used to access, monitor, and modify the static and
dynamic capabilities of the terminal:

```
CapabilityManager terminalCM; // declaration
...
try {
    terminalCM = resourceManager.getCapabilityManager();
} catch(CapabilityManagerNotFoundException ex){
}
```

5.2.3.2 `CapabilityManager` The `CapabilityManager` is responsible for providing access both to dynamic and to static terminal capabilities. The distinction between static and dynamic terminal capabilities is reflected in the API
through the `StaticCapability` and the `DynamicCapability` interfaces. The
`DynamicCapability` and the `StaticCapability` interfaces provide the application access to the dynamic and static capabilities, respectively. The `Terminal-`
`ProfileManager` gives information about the profile of a terminal. The
`DynamicCapabilityObserver` facilitates monitoring the dynamic capabilities
and notification of drastic changes. Figure 5.7 depicts the `CapabilityManager`
and all the interfaces it implements. In this section, these interfaces are presented and examples are provided to illustrate how to use the APIs.

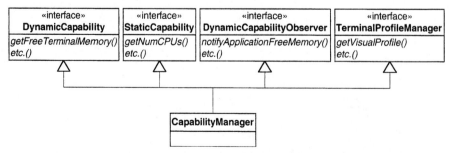

Fig. 5.7 UML representation of the Resource APIs (`CapabilityManager`).

Static capabilities are defined as characteristics of the terminal that do not vary between one MPEG-J session and another. The MPEGlet reads them only once because they will not change during the session. Here's an example of retrieving a static capability:

```
jlist = new JList(terminalCM.getAudioDrivers());
```

On the other hand, capabilities that are dynamic might vary during the MPEGlet's lifetime, according to many concurrent events happening in the terminal. These include changes of the decoded elementary streams, changes in the scene due to BIFS updates, actions performed by the MPEGlet itself, and so on. Here's an example of retrieving a dynamic capability:

```
maxVal= (int)(terminalCM.getTotalApplicationMemory()/1024);
```

DynamicCapabilityObserver is an interface defined to facilitate notification of changes to the dynamic capabilities. This interface can be used to register and deregister for dynamic changing events. For example:

```
// call this.update() when CPU load goes above 80%
terminalCM.notifyTerminalLoad(80, this);

// call myClass.update() when network load goes over 90%
terminalCM.notifyTerminalNetworkLoad(90, myClass);
```

The TerminalProfileManager provides the means by which the application can obtain the profile information (see Chapter 13). The TerminalProfileManager provides a facility that allows applications to find out the profiles supported by the terminal where the application runs. Once an application knows them, it can decide how to behave and what capabilities can operate in the terminal environment. For example, to get the scene graph profile:

```
switch(terminalCM.getSceneDescriptionProfile()) {
    case ...
```

5.2.3.3 Events In the previous section, a mechanism of notifying the application on certain changes in the dynamic terminal capabilities was presented using the DynamicCapabilityObserver. The Resource APIs also present another event mechanism for renderer and decoder events. In this case, the events are already identified (for example, MISSED_FRAMES). If an application is interested in a particular event, it has to implement a listener and register that listener with the corresponding EventGenerator. For example, if an application is interested in MPRendererMediaEvent, it implements an MPRendererMediaListener and registers it with the implementation of the MPRendererEventGenerator. When the event occurs, the event handler, written as part of the registered Listener class, will be called. The application can handle the event intelligently in the event handler. The event handler can be part of the application that is delivered to the terminal in an MPEG-J elementary stream. Three types of events are supported: MPDecoderMediaEvents, MPRendererMediaEvents, and MPRendererFrameEvents.

For each decoder, the `ResourceManager` would have an instantiation of a class that implements `MPDecoder`. These decoder instantiations generate the different defined events for various conditions in the terminal (`MPDecoder-MediaEvents` such as `STREAM_START`, `STREAM_OVERFLOW`, etc.). This is illustrated in the next section along with the Decoder APIs.

The `Renderer` generates two types of events (see Figure 5.8) to which the application can listen: *media events* (which indicate a decoder underflow or a missed frame) and *frame events* (which indicate that a frame has been rendered). The `Renderer` provides notification of exceptional conditions (during rendering) and can provide notification of frame completion when an application registers with it for this kind of event.

5.2.4 Decoder APIs

The Decoder APIs enable an application to control a decoder associated with a particular node in the scene graph. These APIs were introduced in Section 5.2. Here further details on the Decoder APIs illustrate the functionalities provided. Complete detailed Java code samples of how to use the Decoder APIs are available in Appendix E.

Figure 5.9 shows a UML representation of the Decoder APIs. The `MPDecoder` interface is the only major component. Each decoder in the terminal is accessed via the `MPDecoder` interface, which provides the ability to access, monitor, and control (start/stop) the decoder. An application can register with the `MPDecoder` to be notified when certain events such as decoding buffer overflow or underflow occur. The supported events are called the `MPDecoderEvents`. To register and notify when the events occur, `MPDecoderEventGenerator` and `MPDecoderEventListener` are defined in the Decoder APIs.

The `MPdecoder` object associated with a specific node can be queried through the `getDecoder()` method of the `ResourceManager` class as shown in the following example:

```
try {
    decoder = resourceManager.getDecoder(m_node);
} catch( DecoderNotFoundException  dnfe) { }
  catch (BadNodeException bne) { }
```

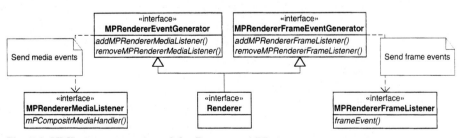

Fig. 5.8 UML representation of the Resource APIs (`Renderer` events).

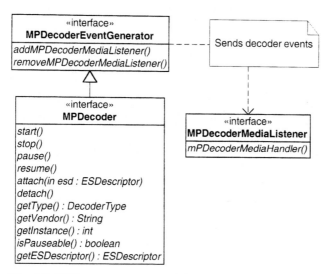

Fig. 5.9 UML representation of the Decoder APIs.

The obtained MPDecoder interface can be used to control the decoder. The MPDecoder interface allows starting, stopping, pausing, and resuming a decoder. It also provides a method (isPauseable) to query if the decoder can be paused. Examples are

```
// to stop a decoder
decoder.stop();
// restarting again
decoder.start();
//Pause the decoder if it can be paused
if (decoder.isPauseable())
    decoder.pause();
```

The MPDecoder interface facilitates obtaining the ESDescriptor of the currently attached stream. An elementary stream associated with an ESDescriptor can be attached to the decoder using the attach() method. The currently attached elementary stream can be detached using the detach() method.

The decoder attached to a specific node and elementary stream (defined by an ESDescriptor) can be changed to another decoder of the same type. This is done using the changeDecoder() function call of the ResourceManager. A list of available decoders of a specific type can be obtained and one of them can be chosen to replace a current decoder. Typically, this is useful when one of the decoders is known to perform better than the other. The following code snippet shows how this can be accomplished:

```
try {
    resourceManager.changeDecoder( node, decoder)
} catch ( DecoderNotFoundException dnfe ) { }
  catch( BadNodeException bne) { }
```

5.2.5 Network APIs

Through the Network APIs, MPEG-J applications can interact with the network entities of the MPEG-4 player such as the associated network channels and services. These APIs allow control of the network component of the player. Due to the level of abstraction provided by the MPEG-J Network APIs, the applications can access a network service without being aware of the details of the network connections being used (LAN, WAN, broadcast, local disks, etc.). To avoid architectural inconsistencies and duplication of tools, MPEG-J Network APIs forbid full arbitrary usage of the DMIF (see Chapter 7) and Sync Layers (see Chapter 3).

Figure 5.10 shows the UML representation of the Network APIs. The main components of the Network APIs are the NetworkManager, DMIFMonitor, and ChannelController interfaces. The NetworkManager has a DMIFMonitor, which provides basic information about the currently opened network sessions and channels, and a ChannelController, which can be used to enable and disable individual channels. The ServiceSessionDescriptor and the ChannelDescriptor abstract the information about a network service or a channel.

The NetworkManager is the access point for the application to the network resource. The NetworkManager can be retrieved from the MpegjTerminal class as shown in the following example:

```
try {
    netManager = MpegjTerminal.getNetworkManager();
} catch(NetworkManagerNotFoundException ex) {
}
```

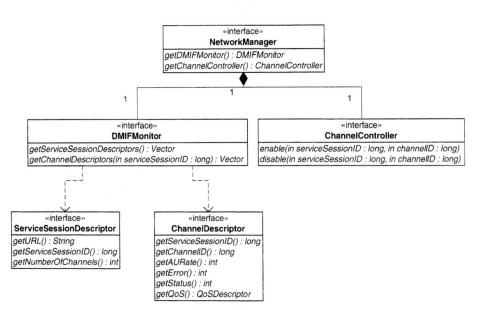

Fig. 5.10 UML representation of the Network APIs.

The functionality provided by the current Network APIs can be split into two major groups: *network query* and *channel control*. The rest of this section describes these functionalities in more detail with code snippets illustrating how to use the Network APIs to achieve these functionalities. For more detailed examples, please refer to Appendix E.

5.2.5.1 Network Query Network query corresponds to the ability to perform requests to the network module in order to get statistical information about the DMIF resources used by the MPEG-4 player. This is important for adapting the audiovisual session to different network conditions. One can retrieve DMIF information through the DMIFMonitor, which can be obtained from the NetworkManager as shown in the following example:

```
dmifMonitor = netManager.getDMIFMonitor();
// get the info from the dmif monitor
Vector sInfo = dmifMonitor.getServiceSessionDescriptors ();
Vector cInfo = dmifMonitor.getChannelDescriptors (0);
// for session 0
```

Channel information can be retrieved using a type cast to the object that describes the channels:

```
for ( int k = 0; k < cInfo.size(); k++ ) {
   long channelId =
     ((ChannelDescriptor)cInfo.elementAt(k)).getChannelId();
   long ESid = channelId;
   ....
}
```

The quality of service associated with that DMIF channel can be retrieved using suitable methods of the ChannelDescriptor interface:

```
for ( int k = 0; k < cInfo.size(); k++ ) {
QoSDescriptor qos = ((ChannelDescriptor)cInfo.elementAt(k)).getQoS();
   Vector qualifiers = qos. getQualifiers() ;
   // use getQualifierCount() to get the number
   // qualifiers is a vector of Qualifier objects, pairs of
   // QualifierTag and values
   ....
}
```

5.2.5.2 Channel Control A simple channel control mechanism is also provided. Using this feature, an MPEG-J application can temporarily disable or enable existing elementary stream channels without any negative influence on the rest of the player. This functionality fits one of the MPEG-J general requirements: the capability to allow graceful degradation under limited or time-varying resources. Here's an example of disabling a channel:

```
ChannelController chanController = netManager.getChannelController();
...
chanController.disable(0,channelId);
// disable ES_ID channelID of session 0
```

5.2.6 Service Information and Section Filtering APIs

The Service Information (SI) and Section Filtering APIs are normatively referred from the DAVIC 1.4.1 specification, Part 9 [DAVI98]. A Main profile–compliant MPEG-J terminal should also implement the normative APIs referred in this section; see Section 5.2.7 on MPEG-J profiles.

5.2.6.1 Service Information API
The SI API (contained in the org.davic. net.dvb.si package) allows interoperable applications to access service information data from MPEG-2 Systems streams. Service information is metadata describing the content of the stream, such as the scheduling of TV programs, genre, and so on. One example of an enabled application is an electronic program guide. This API is relatively high level, allowing applications to access information from the SI tables in a clean and efficient way [DVB].

5.2.6.2 MPEG-2 Section Filter API
The objective of the MPEG-2 Section Filter API (contained in the org.davic.mpeg.sections package) is to provide a general mechanism allowing the access to data held in MPEG-2 Systems private sections. This allows interoperable access to data that is too specialized to be supported by the high-level Service Information API or that is not actually related to service information. The complete definition of the MPEG-2 Section Filter API is available in Annex E of the DAVIC 1.4 specification, Part 9 [DAVI98].

5.2.6.3 Resource Notification API
The Section Filter API uses a Resource Notification API from the org.davic.resources package. This API provides a standard mechanism for applications to register interest in scarce resources and to be notified of changes in those resources or removal of those resources by the terminal. The complete definition of this API can be found in Annex F of the DAVIC 1.4 specification, Part 9 [DAVI98].

5.2.6.4 MPEG Component API
Various MPEG-related APIs use an MPEG Component API in the org.davic.mpeg.sections package. This API provides a way of referring to standard features related to MPEG-2 streams. The definition of the MPEG Component API is available in Annex G of the DAVIC 1.4 specification, Part 9 [DAVI98].

5.2.7 MPEG-J Profiles

Two MPEG-J profiles (see Chapter 13) are defined in the MPEG-J specification [AMD1-1]: the Personal and Main profiles. These profiles differ in terms of the set of MPEG-J APIs that has to be implemented at an MPEG-J–compliant terminal. The MPEGlets can use only those APIs that are supported by the intended profile.

The `Personal` profile addresses a range of constrained devices from mobile and portable devices up to personal computers. Examples of such devices include cell phones, PDAs, and multimedia computers. The Terminal, Scene, Resource, Decoder, and Network APIs are supported by the `Personal` profile.

The `Main` profile is a superset of the `Personal` profile; it addresses the broadcast-oriented entertainment consoles. Examples of such devices include set-top boxes and digital TVs. This profile includes all the MPEG-J APIs. Notice that the support of Service Information and Section Filtering APIs is the only difference between the `Main` and the `Personal` MPEG-J profiles. No levels are defined for either MPEG-J profile.

5.3 APPLICATION SCENARIOS

Because MPEG-J uses Java, a rich programming language, the applications possible are limited only by the ingenuity of the content creator.[13] The real challenge ahead is to provide intelligent authoring tools for creating MPEG-J content. Such authoring tools would provide a set of high-level constructs that could generate at least part of the Java byte code automatically to embed along with the content. Some MPEG-J application scenarios are presented in the following sections.

5.3.1 Adaptive Rich Media Content for Wireless Devices

In this context, adaptation refers to varying network and workload conditions. Although adaptation is a generic functionality that refers to terminals and networks of any power and speed, it is particularly useful for terminals with low capabilities, including PDAs connected with mobile phones and integrated devices (mobile and PDA). In a scenario where different power devices and wireless networks (e.g., GSM, GPRS, UMTS) coexist, content providers may want to decide which parts of the scenes they want to sacrifice and which are essential for the message they want to convey. Without an MPEG-J programmatic control, this decision would be left to the terminal or to a gateway between the server and the terminal.

5.3.2 Enhanced Interactive Electronic Program Guide (EPG)

BIFS functionalities can be used to embed EPGs in broadcast TV programs as MPEG-4 scenes providing rich multimedia content and good interaction. MPEG-J APIs provide the means to get this information and interact with the BIFS scene and the user in order to format the retrieved content, integrate it

---

13. Within the bounds of the MPEG-J architecture.

with other MPEG-4 multimedia information, and display it on request, according to user preferences. While, traditionally, this could be done in a fixed way by the terminal vendor, MPEG-J provides each user the means to view his or her own interactive EPG. Different kinds of EPGs can be sent out by the different service providers, retrieved in the public domain or created by skilled users. Users can choose the EPG they prefer the same way they choose their favorite computer program for a given purpose. Moreover, user preferences can be stored and dynamically updated by an MPEGlet that sorts TV programs according to the specific user's preferences.

5.3.3 Enriched Interactive Digital Television

Most TV programs wishing to carry interactive multimedia information can benefit from MPEG-J. For example, content creators can give home users the possibility of answering questions asked in a show through multiple choices. At the end of the show, the answers can be evaluated and the users with the best scores rewarded. A trusted environment with authentication (using a back channel) may be necessary to make it secure.

5.3.4 Content Personalization

One of the most promising application areas in the immediate future is content personalization, in which knowledge of user tastes and preferences is exploited to modify the presentation. As an example, if so instructed, a user's agent could automatically choose the language of a multilanguage program or the most interesting subjects in a (stored) news program along with a favorite display order or even the most appropriate advertisements. These content-personalization MPEGlets can act as the user's agents to access the scene graph and modify it in order to adapt content to that user's taste.

5.4 REFERENCE SOFTWARE

Like all the other parts of the MPEG-4 standard, MPEG-J has an associated reference software implementation consisting of a body of source code that implements the MPEG-J specification [AMD1-1]. Its primary purpose is to demonstrate and test the capabilities of the specification, but it can also serve as a reference or starting point for developers wishing to implement the specification. This reference software is included in MPEG-4 Part 5, Version 2 [AMD1-5].

The MPEG-J reference software is written in a mixture of C++, C, and Java. It takes advantage of the MPEG-4 reference software [MPEG4-5], building on top of the base MPEG-4 player (see Chapter 14). However, this is done in a way that is independent of the implementation details of the base player by creating a strict player-independent interface boundary between the base player and the MPEG-J software. This approach provides significant benefits since it buffers the MPEG-J implementation from the evolution of APIs in the base player. In addition, it also creates the opportunity for the implementation to be used with other MPEG-4 players, which would only need to write some glue code to take advantage of the MPEG-J implementation.

The basic steps in the execution of the MPEG-J reference software are

1. The player detects an MPEG-J stream, opens the stream, and creates an MPEG-J decoder for it.
2. A JVM is started.
3. For each module in the MPEG-J software (each module typically corresponds to one of the Java packages that make up the MPEG-J APIs):

   ✗ The player loads the dynamic library for that module.
   ✗ The player creates a table of functions (*jump table*) with pointers to functions that it implements and will be called by the MPEG-J software (such as a function to get a reference to a BIFS node given its node ID).
   ✗ The player calls the entry point of the module's dynamic library, giving it a pointer to the jump table.
   ✗ Inside the entry point, the module fills in pointers in the jump table for functions that are implemented by the module and are called by the player (e.g., a notification of an eventOut from a field in the BIFS scene).

4. When the MPEG-J decoder receives an access unit, it loads the class (or classes) in the access unit using a custom Java class loader.
5. If the access unit contains an MPEGlet class to be loaded, the decoder creates an instance of that class, calls its init method, creates a thread, and runs the MPEGlet in the new thread.
6. When the MPEGlet calls one of the methods in the MPEG-J APIs—as most of the methods in the MPEG-J software are implemented as native methods—this normally calls a native function in the corresponding module's dynamic library. Normally, the native function will need to call a function in the jump table to make a request to the player.

Figure 5.11 shows the basic structure of the MPEG-J reference software and the relation between the MPEG-J software, the base player, and the Java virtual machine.

5.5 SUMMARY

This chapter introduced the use of Java to control some specific features of an MPEG-4 player. Although a wide range of applications is possible using the declarative language expressed by BIFS, all of those applications that require

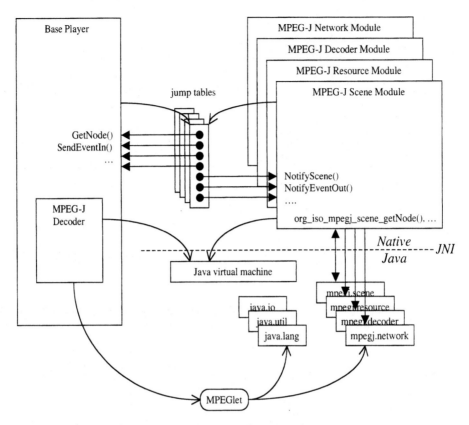

Fig. 5.11 Basic structure of the MPEG-J reference software.

a complex algorithm to be associated with a specific part of the scene or a user action (e.g., a simple game, a quiz answer, an educational test) need to be in the form of programs embedded along with the content.

The architecture of the MPEG-J was presented, emphasizing the role of the application engine. The application engine includes the delivery of an MPEG-J application (compiled Java code) to the terminal and execution of the application based on certain rules. In order to communicate with the MPEG-4 player, new APIs specific to MPEG-4 were defined: Terminal, Scene, Resource, Decoder, and Network. These APIs (classes and interfaces grouped into packages) and their usage were described in detail and illustrated with examples. Service Information and Section Filtering APIs, normatively included from the DAVIC specification, were introduced.

MPEG-J is a major step toward delivering intelligent and interactive content. MPEG-J puts in place all the tools necessary for content creators to start producing even more interesting MPEG-4 content.

5.6 REFERENCES

[AMD1-1] ISO/IEC 14496-1:2001, Amendment 1. *Coding of Audio-Visual Objects—Part 1: Systems*, 2001.

[AMD1-5] ISO/IEC 14496-5:2001, Amendment 1. *Coding of Audio-Visual Objects—Part 5: Reference Software*, 2001.

[ArGH00] Arnold, Ken, James Gosling, and David Holmes. *The Java Programming Language*, 3d Edition. Reading, MA: Addison-Wesley, 2000.

[Booc94] Booch, Grady. *Object-Oriented Analysis and Design with Applications*, 2d Edition. Redwood City, CA: Benjamin/Cummings, 1994.

[DAVI98] DAVIC 1.4.1:1998. *Part 9: Information Representation.*

[DVB] ETSI. *Specification for Service Information (SI) in DVB Systems.* Doc. EN 300 468 V1.3.1.

[FSPS00] Fernando, Gerard, Viswanathan Swaminathan, Atul Puri, Robert L. Schmidt, Gianluca De Petris, and Jean Gelissen. "Java in MPEG-4 (MPEG-J)," *Advances in Multimedia: Standards, Systems, and Networks* (A. Puri and T. Chen, Eds.). New York: Marcel Dekker, 2000.

[GoJS96] Gosling, James, Bill Joy, and Guy Steele. *The Java Language Specification.* Reading, MA: Addison-Wesley, September 1996.

[LiYe96] Lindholm, Tim, and Frank Yellin. *The Java Virtual Machine Specification.* Reading, MA: Addison-Wesley, September 1996.

[M2566] Courtney, J., P. Shah, J. Webb, G. Fernando, and V. Swaminathan. *Adaptive Audiovisual Session Format.* Doc. ISO/MPEG M2566, Stockholm MPEG Meeting, July 1997.

[MPEG2-1] ISO/IEC 13818-1:2000. *Generic Coding of Moving Pictures and Associated Audio Information—Part 1: Systems*, 2000.

[MPEG4-1] ISO/IEC 14496-1:2001. *Coding of Audio-Visual Objects—Part 1: Systems*, 2d Edition, 2001.

[MPEG4-5] ISO/IEC 14496-5:2001. *Coding of Audio-Visual Objects—Part 5: Reference Software*, 2d Edition, 2001.

[N3545] Fernando, Gerard, Atul Puri, Viswanathan Swaminathan, Robert L Schmidt, Pallavi Shah, and Keith Deutsch, *Architecture and API for MPEG-J.* Doc. ISO/MPEG N3545, Dublin MPEG Meeting, July 1998.

[SwFe00] Swaminathan, Viswanathan, and Gerard Fernando. *MPEG-J: Java Application Engine in MPEG-4.* ISCAS 2000, Geneva, Switzerland, May 2000.

[ZLIB] Home Page. *www.cdrom.com / pub / infozip / zlib*

Extensible MPEG-4 Textual Format

by Michelle Kim and Steve Wood

Keywords: VRML, X3D, XML, SMIL, XMT-Ω, XMT-A, MPEG-7, interoperability, deterministic mapping, exchangeable authoring format, MPEG-4 conformance, authoring tool, scalable content, automated authoring

*T*he eXtensible MPEG-4 Textual (XMT) format is a framework for representing MPEG-4 Systems content [MPEG4-1] and associating audiovisual media streams with that content using a textual syntax. XMT is designed to facilitate the creation and maintenance of MPEG-4 multimedia content, whether by human authors or by automated machine programs. The textual representation of MPEG-4 content has high-level abstractions that allow authors to exchange their content easily with other authors or authoring tools, at the same time preserving semantic intent. XMT also has low-level textual representations covering the full scope and function of MPEG-4 Systems [MPEG4-1]. The XMT language uses the W3C XML standard [EXML00] as the basis for the textual representation. XMT is also designed to facilitate interoperability with the extensible 3D (X3D) specification [X3D01] developed by the Web3D consortium as the next generation of Virtual Reality Modeling Language (VRML) [VRML97], and with the Synchronized Multimedia Integration Language (SMIL) 2.0, a recommendation from the W3C consortium [SMIL01]. For brevity, SMIL 2.0 will be referred to in this chapter simply as SMIL; if the reference is specifically to SMIL 1.0, it will be written as such. Also, at the time of this writing, the MPEG standardization process is not yet finished for XMT, although it is nearing completion (FPDAM status in July 2001) [AMD2]. So the XMT syntax and samples as provided here are based on the latest version of the specification that has been approved by the MPEG committee; but it is subject to change.

6.1 OBJECTIVES

MPEG-4 is an ISO/IEC standard developed by MPEG for communicating interactive audiovisual scenes. The standard defines a set of tools that provide the binary coded representation of individual audiovisual objects, text, graphics, and synthetic objects. The interactive behaviors of these objects and the way they are composed in space and time to form an MPEG-4 scene are dependent on the scene description, which is coded in a binary format known as binary format for scenes (BIFS—presented in Chapter 4). The audiovisual streams are defined as elementary streams (ESs) and managed according to the object descriptor (OD) framework (see Chapter 3, which describes the relationships both between the scene description and the streams and among the streams themselves). In addition, the OD framework defines additional streams for object content information (OCI), MPEG-J, and intellectual property management and protection (IPMP).

XMT is a textual representation of MPEG-4 Part 1: Systems [MPEG4-1], using the extensible markup language (XML) [EXML00]. MPEG-4 Systems structures are represented as XML elements, with attributes of the element containing its values. For example, the BIFS Material node is represented by the element <Material>. The Material node's emissiveColor field becomes an attribute of the <Material> element—that is, <Material emissiveColor="1.0 0.0 0.0"/>.

The XMT language has grammars that are specified using the W3C XML Schema language [SCHM01]. The grammars contain rules for element placement, attribute values, and so on. The rules, for example, allow XMT to state that the <LineProperties> element can be contained only in a <Material2D> element. These rules for XMT, defined using the XML Schema language, follow the MPEG-4 Systems binary coding rules and help ensure that the textual representation can be coded into correct binary according to the MPEG-4 Systems coding rules [MPEG4-1]. The MPEG-4 Conformance Testing specification [MPEG4-4] provides conformance points to ensure that a conformant receiver can decode and render the content.

Although the MPEG-4 specification includes Parts for Systems, Visual, and Audio tools, XMT is a textual format for the Systems Part only. All constructs in the Systems specification have parallels in the XMT textual format. For the Visual and Audio Parts [MPEG4-2, MPEG4-3], XMT provides a means to reference external media streams of either pre-encoded or raw audiovisual binary content. Although XMT does not contain a textual format for audiovisual media, it does contain hints in a textual format that allow an XMT tool to encode and embed the audiovisual media into a complete MPEG-4 presentation.

MPEG-4 allows the representation of a wide variety of content from simple audiovisual formats to highly interactive, complex, animated presentations.

The binary coded representation of such content can be complex and often cannot easily be *reverse-engineered* to represent the content author's original intentions. Although the nodes and routes can be understood as such, in looking at a particular piece of content one might ask, what exactly did the original author have in mind (intend) with that particular combination of nodes and routes? It may become obvious if one knows that this was intended to be a *SnakeWipe*[1] *transition* effect between an image and a video. In XMT, it can be directly defined as such. But when mapped to MPEG-4 Systems, there might be many nodes, routes, animation streams, and so on that can be used to achieve this visual effect; and reverse-engineering the author's original high-level intention to create such a SnakeWipe may not be easy, especially as there is often more than one way to achieve the same net effect with MPEG-4 tools.

XMT is thus designed to provide an exchangeable format between content authors while preserving an author's intentions in a high-level textual format. In addition to providing a suitable, author-friendly abstraction of the underlying MPEG-4 technologies, another important consideration for XMT design has been to respect the existing practices of content authors familiar with other standards, such as the Web3D X3D [X3D01], VRML [VRML97], W3C SMIL [SMIL01], and HTML [HTML99]. This will allow authors familiar with creating multimedia content in other environments to gain rapid entry to MPEG-4 writing in XMT by reusing their skills and practices.

XMT is suitable for many uses, including manually authored content as well as machine-generated content using multimedia database material and templates. XMT can be encoded and stored in an exchangeable binary format such as the MPEG-4 file format [MPEG4-1] (see Chapter 7), commonly known as the mp4 format. XMT also can be encoded directly into the MPEG-4 Systems streams it represents and immediately transmitted to clients for playback. In such a live encoding scenario, there are XMT delivery hints to assist this process.

XMT comprises two levels of representation, known as *XMT-Ω* and *XMT-A*. XMT-Ω (omega), also known as XMT-O, is a high-level format based on SMIL 2.0 [SMIL01]. XMT-A (alpha) is a direct XML representation of the MPEG-4 Systems binary format [MPEG4-1]. As MPEG-4 Systems incorporated much of VRML, the XMT-A textual format has much in common with X3D, the next generation of VRML defined in XML. Interoperability with other standards was a key design goal for XMT.

6.2 CROSS-STANDARD INTEROPERABILITY

The XMT format can facilitate content interchange between SMIL players [SMIL01], VRML [VRML97] or X3D [X3D01] players, and MPEG-4 players.

---

1. A SnakeWipe transition conceals one media (e.g., image) and reveals a new media in a pattern that winds from side to side, creating a visual snakelike effect.

The XMT-Ω format can be preprocessed and played directly by a SMIL player; preprocessed to the corresponding X3D nodes and played back by a VRML/X3D player; or compiled to an MPEG-4 representation, such as an mp4 file (the exchangeable binary file format defined by MPEG-4), and played by an MPEG-4 player. Figure 6.1 presents a graphical description of the XMT interoperability capabilities it facilitates. MPEG-7, also shown in Figure 6.1, is an emerging standard for describing multimedia content [N4509]. The integration of MPEG-7 with XMT will enable the metadata markup of MPEG-4 content in the textual format and facilitate applications such as the content-based retrieval of MPEG-4 objects.

As shown in Figure 6.1, XMT has XML textual representations in common with SMIL and X3D, as depicted by the overlapping shapes, and can contain MPEG-7 descriptions of the content. A SMIL player is designed to directly play the SMIL textual format, and hence can simply parse and play the compatible subset of XMT with minimal changes to the textual format. The same is true for a VRML/X3D player for its compatible subset. For an MPEG-4 player, the textual format is compiled into the binary representation for playback.

6.3 XMT TWO-TIER ARCHITECTURE

As mentioned, XMT consists of two levels of textual syntax and semantics: the XMT-Ω format (XMT-O) and the XMT-A format.

XMT-Ω is a high-level abstraction of MPEG-4 tools that was designed based on W3C SMIL [SMIL01], an XML-based language that allows authors to create dynamic, interactive, multimedia presentations. For example a video can be represented simply by <video src="MyVideo.mpeg4"/>, which says that this is a video media object in which the source (src=) for the video is from a location (file) named MyVideo.mpeg4. This is in contrast to MPEG-4 Systems where first an InitialObjectDescriptor is required and then a scene containing several BIFS nodes (including a MovieTexture node) would

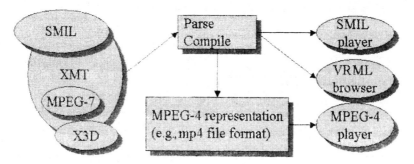

Fig. 6.1 XMT interoperability with other relevant standards.

be created, not forgetting the `ObjectDescriptors` necessary to reference the video stream.

Using SMIL, authors can describe the temporal behavior and spatial layout of multimedia presentations, as well as associate hyperlinks with the media objects in that presentation. The XMT-Ω language, being based on SMIL, allows MPEG-4 content to be expressed this way. SMIL contains several functional areas—such as timing, synchronization, transitions, and media—that are divided into one or more XML modules. The aim of such modularity is to allow the incorporation of these functions, at that granularity, into other XML-based host languages. A host language is defined as an XML language that integrates (or *hosts*) SMIL modules. The SMIL modules are not totally independent, as some have dependencies on other modules. However, SMIL identifies these dependencies as well as provides rules and guidelines for the integration of the modules into other host languages. XMT-Ω integrates many of the SMIL modules into its language.

When multimedia content is represented in XMT-Ω, one or possibly more mappings may exist that can represent the same content when coded in binary to MPEG-4 Systems. Such mappings can be represented in XMT-A by various combinations and sequences of XMT-A elements that will all represent the same content as that expressed in the higher-level XMT-Ω terms. From this it can be concluded that there is no deterministic mapping between XMT-Ω and XMT-A. That is, given a construction at one level there may be no single construct at the other level that uniquely represents the same content through a mapping. This is due to the fact that high-level author intentions can be expanded to more than one sequence of low-level constructs to produce the same outcome. There are many tools in MPEG-4 Systems, and these can be used to best suit the player and its capabilities as determined by the MPEG-4 profile and level to which it is compliant for each profiling dimension. However, with such nondeterminism comes flexibility. XMT-Ω content can be converted and targeted to MPEG-4 players of such varying capability. For example, animation in a scene for a simple player may be mapped using BIFS-Commands; for a more capable player, the animation might be mapped more efficiently to a BIFS-Anim (animation) stream (see Chapter 4). Flexibility can also be taken advantage of by authoring tool vendors to distinguish themselves among competitors—for example, by providing superior optimized mappings, reducing bandwidth requirements, and so on—while still preserving the author's intent for the content.

Moreover, for those authors who want to control the implementation of certain portions of their presentations, XMT provides an escape mechanism from XMT-Ω to XMT-A. The escape mechanism enables content authors to mix and match the two formats, XMT-Ω and XMT-A. So, rather than express the content in XMT-Ω, an author can choose XMT-A and effectively bypass whatever mapping the particular XMT implementation provides.

XMT-A is an XML-based textual representation of the full set of tools in MPEG-4 Systems [MPEG4-1] and closely mirrors its binary representation.

The goals of the XMT-A format were to provide a textual representation of MPEG-4 Systems binary coding; to support a deterministic, one-to-one mapping to the binary representations for conformance; and to be interoperable with the X3D specification [X3D01] being developed for VRML 200x (X3D) by the Web3D consortium.

XMT-A provides a textual representation of BIFS, both the commands for scene updates and the nodes themselves. The XMT textual representation of the nodes is fully compatible with the XML representation of the large subset of the X3D (VRML) that is supported by MPEG-4 Systems. MPEG-4 Systems scene description originally used VRML 97 as a base and added further nodes for synthetic representation, audio, and 2D support. VRML 97 is an ASCII textual representation, and its follow-on (X3D) is based on XML. Currently, the X3D specification has additional functionalities not present in MPEG-4; hence, it is a large subset of overlap rather than complete coverage. The XML for the additional nodes in MPEG-4 Systems has been based on the same representation philosophy defined by X3D, which originally guided their XML conversion from VRML plain text. Such commonality will facilitate any future harmonization efforts between MPEG-4 and X3D.

XMT-A also has textual representations of features unique to MPEG-4 Systems not found in X3D, such as the OD framework, including object descriptors (ODs) and commands, OCI descriptors and events, and IPMP descriptors and messages (see Chapter 3). The ODs are used to associate scene description components (nodes) to the ESs that contain the corresponding coded data. These include visual streams, audio streams, and animation streams that update elements of the scene more efficiently than BIFS-Commands for complex animations. XMT-A also includes a textual representation to allow MPEG-J streams to be created either from Java classes or from zipped files.

Within XMT-A and XMT-Ω, common elements that can be used at either level—such as MPEG-7 descriptions, authoring elements, encoding hints, delivery hints, and publication hints—have been identified and collected into a common section used within both formats that is designated *XMT-C*. Authoring elements and the various hints are features that have been designed as part of the XMT language to facilitate authoring.

Figure 6.2 provides a schematic description of the XMT two-tier architecture. XMT-Ω consists of a set of SMIL-compatible modules (such as timing, animation, and transitions), including the xMedia module that provides MPEG-4 specific extensions to SMIL's Media Object module [SMIL01]. XMT-Ω can be converted into XMT-A, or compiled directly into the MPEG-4 Systems binary format. XMT-A contains the textual representation of MPEG-4 nodes, ODs, and so on, with references to media data. The common elements in XMT-C include metadata tools, encoding hints, and authoring elements. XMT-A is then shown encoded into the MPEG-4 Systems binary representation and stored in the mp4 exchangeable file format.

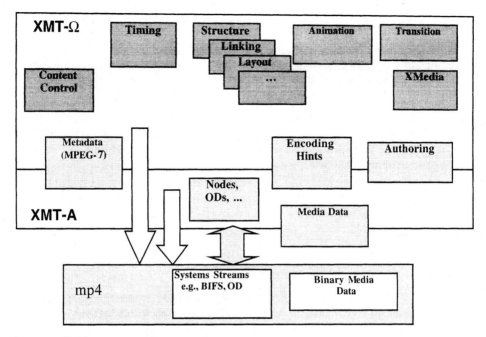

Fig. 6.2 XMT two-tier architecture.

6.4 XMT-Ω FORMAT

The goals of the XMT-Ω format are to provide ease of use for content creation, to facilitate content exchange between authors and authoring tools, and to provide a content representation that is compatible and interoperable with SMIL 2.0 [SMIL01].

In MPEG-4, objects are represented as nodes in a scene graph, with some nodes connected to audio or visual ESs, with their interactive behaviors described using a routing mechanism that connects two fields together from an event output to an event input (see Chapter 4). The XMT-Ω format describes audiovisual objects and their relationships at a higher level, wherein content requirements are expressed in terms of the author's intent, using media, timing, and animation abstractions, rather than by coding explicit node and route connections. These higher-level constructs facilitate, among other aspects, content exchange between authors and authoring tools and also content repurposing. An authoring tool would compile XMT-Ω into MPEG-4 content by mapping these constructs into BIFS, ODs, media streams, and so on, along with any appropriate audio or visual media compression or conversions that may be required. Media sources can be of a variety of formats native to the machine on which the authoring tool is executing, and it is the responsibility of the tool, during the compilation phase, to convert media to suitable target formats for MPEG-4, using appropriate bit rates and so on.

In converting the XMT-Ω format to MPEG-4, there is not necessarily only one mapping possible. MPEG-4 nodes and routes are powerful tools and often there is more than one way to represent XMT-Ω constructions. Also, as MPEG-4 nodes can be *wired* together with routes in many combinations, it often is difficult to reverse-engineer an author's intent from a collection of nodes and routes. When confronted by such content, potentially containing many nodes and routes, the reauthoring and maintenance can be quite challenging, if the high-level view of that presentation must be inferred. XMT-Ω, however, provides a high-level view with high-level authoring constructs and thus facilitates exchange of content between tools or authors, rapid content repurposing, reauthoring, and ongoing maintenance of content.

Recognizing that some authors may want to access low-level nodes and routes, however, XMT-Ω allows the embedding of XMT-A node and route definitions within an identified low-level escape section, to create custom media constructs.

6.4.1 Reusing SMIL in XMT-Ω

To create the XMT-Ω format, as a high-level abstraction and representation of content authors' intentions, SMIL is used as a basis [SMIL01]. SMIL is an XML-based language that allows authors to write interactive multimedia presentations. The main strengths of SMIL are that its constructs are self-describing; it is based on XML, which provides an excellent format for interchange of data among different applications; it is relatively easy to author; and it is a language familiar to many content creators. The language also is extensible, so that new objects or metadata can be inserted easily into the representation.

The basis of the XMT-Ω format is a subset of the modules defined by SMIL, which have been reused and thus where the semantics are compatible. In addition, a new set of elements has been designed for XMT-Ω that express a high-level view of MPEG-4 content. Unlike SMIL, the XMT-Ω format is not specifically designed as a playback format; rather, it is intended that the constructs are mapped and compiled to a binary MPEG-4 representation. However, XMT-Ω also can be processed for exchange with SMIL or translated to X3D within the limits of compatible capabilities.

To enable the reuse of SMIL-defined functions, the SMIL language has been composed into a number of functional areas that have been broken down into a finer granularity of modules. A module contains XML elements and attributes necessary to express the particular function for that module. These modules—comprising XML elements, attributes, and attribute values—can then be combined and brought together in other host languages, such as XMT-Ω. XMT-Ω is referred to as a host language because it integrates the modules within a larger set of XML representation tools, including the specific XMT-Ω XML constructs.

SMIL provides requirements and guidelines for integration of these modules into other host languages. The guidelines are designed to provide a consistent usage of the modules. XMT-Ω follows these rules, as much as possible, to preserve compatibility with SMIL, and both adhere to the semantics of the modules as well as their syntax. A key difference to note here is that SMIL is a language with syntax and semantics that have been designed first, and then implementations for it followed. Also, whereas MPEG-4 is an existing binary specification and implementation, XMT-Ω is being developed as a high-level language to represent MPEG-4 in the form of SMIL. With an existing binary format and an existing high-level language compatibility target, this required careful design decisions, as unique challenges were faced during the creation of the XMT-Ω language.

As XMT-Ω must be mapped (compiled) into MPEG-4, maintaining the semantics of certain behaviors specified by SMIL can be difficult, or overly complicated. However, it is the authoring tools' responsibility to maintain correct semantics during the mapping. To achieve satisfactory mappings, an authoring tool can use all the power of the MPEG-4 representation, including scripting and MPEG-J [MPEG4-1]. A major constraint for the mapping, however, is the requirement to limit the use of MPEG-4 tools to only those included in the MPEG-4 profile(s) and level(s) for which the presentation is being created.

The mappings of the semantics can be either static or dynamic in nature. Static mappings capture the semantics for deterministic behavior that can be fully evaluated at authoring time. Dynamic mappings require runtime support of MPEG-4 player mechanisms and would be utilized to support non-deterministic behavior such as unpredictable user interactions whose timing cannot be fully evaluated a priori into a fixed static timeline.

Table 6.1 presents a brief overview of the functional areas of SMIL and their reuse in XMT-Ω. Modules were included in XMT-Ω when it was considered that the SMIL module semantics and syntax could be appropriately reused in MPEG-4. When this was not the case, new modules were designed, but in the spirit of SMIL. For example, the SMIL Layout modules contain many options that are not conducive to managing the visual content in MPEG-4 dynamically. As such, a Layout module more suitable for MPEG-4 was created in XMT-Ω.

From the large set of modules outlined previously, the remainder of this section on XMT-Ω will focus on the Media, and Timing and Synchronization modules, as these provide the basic core of the XMT-Ω format. Media and timing also are the foundation of SMIL 1.0 functionality, and hence of SMIL 2.0 functionality as well.

6.4.2 Extensible Media (xMedia) Objects

SMIL provides a useful abstraction for multimedia. However, SMIL concerns itself with the representation of multimedia player composition rather than

Table 6.1 Relation between SMIL and XMT-Ω modules

SMIL module	XMT-Ω module
Animation	XMT-Ω incorporates the SMIL Animation modules. These modules support dynamic updating (animation) of attributes. The SMIL language defines semantics based on having a base value as well as a presentation value of an attribute available. The fields of MPEG-4 nodes in a scene, however, have only one value. Events propagated into fields by routing, and so on, overwrite the existing values, making preserving SMIL-compatible semantics the responsibility of authoring tools.
Content Control	XMT-Ω incorporates SMIL Content Control modules that allow content choice expressions and selection based on test attributes. These constructs may be mapped to MPEG-4 binary features such as alternate streams, which, for example, can be selected based on language. Content control can be used to select alternate content at compile time, for example, by author-predefined attribute settings or by interactive dialogs to prompt for attribute values.
Layout	XMT-Ω does not incorporate SMIL Layout modules. However, XMT-Ω defines its own layout module that is consistent with the hierarchical, tree-structured, spatial layout and groupings intrinsic to MPEG-4.
Linking	XMT-Ω does not incorporate SMIL Linking modules. However, XMT-Ω defines support for linking that can be mapped to the BIFS `Anchor` node.
Media	XMT-Ω incorporates most of SMIL Media modules. XMT-Ω defines new media elements that extend the set of SMIL media types to include MPEG-4-specific media, both 2D and 3D. XMT-Ω also defines media augmentation elements (for effects such as illumination and fog), a media group element, and a custom media element integrating the XMT-A constructs.
Meta-Information	XMT-Ω incorporates SMIL Meta-Information modules and also supports embedding of MPEG-7 metadata representations.
Structure	XMT-Ω does not incorporate the SMIL Structure module but defines a compatible structure module and containment rules that facilitate content interchange with SMIL and minimize any processing necessary for such content exchange.
Timing and Synchronization	SMIL contains an extensive, comprehensive set of Timing and Synchronization modules. XMT-Ω incorporates most of these modules.
Time Manipulations	XMT-Ω incorporates the SMIL Time Manipulations module, which permits time transformations. XMT-Ω also defines a new Flexible Time Manipulations module to represent MPEG-4 FlexTime tools.
Transitions	SMIL supports high-level transitions with effects defined by SMPTE [SMPTE93] as well as effects defined by SMIL. XMT-Ω incorporates some of these Transitions modules.

basic multimedia object composition. As such, SMIL coordinates the temporal and spatial layout of media players but does not yet standardize many facilities to manipulate the internal characteristics of the media being played by the media players. SMIL does provide the `mediaRepeat` attribute as one standard facility; another is the `<param>` element. Although `<param>` is a standard facility, in terms of syntax, it uses name and value pairs to provide parameters to a media player and as such its specification is media-player-dependent.

On the other hand, MPEG-4 is focused on both scene composition as a combination of multimedia objects and the manipulation of the fundamental properties of these objects to create rich, interactive, dynamic presentations.

For example, SMIL would handle a text media object by passing the media data to a text player (renderer) capable of handling the associated mime-type.[2] It would let the text player concern itself over any fonts, styles, kerning, colors, and so on, and the representation of the text and any attributes would be part of the internal media data structure for that mime-type and opaque to SMIL. However, an MPEG-4 text object includes font, style, alignment, colors, and so on, and it specifies the representation and data streams, so that an MPEG-4 player is intimately aware of the detail and fundamental properties of media objects.

MPEG-4 contains audio, image, visual, and text media like SMIL. It also includes 2D media elements similar to those that are described in scalable vector graphics (SVG) [SVG01], such as the `<rectangle>` and `<circle>` elements. SVG also uses modules defined by SMIL—for example the Content Selection and Animation modules. The Animation module, in fact, was a joint development between the W3C SYMM and SVG working groups.

Recognizing both the similarities and differences between SMIL and MPEG-4, the XMT-Ω media elements are based on the SMIL media elements and have been extended to include additional child elements[3] to represent the fundamental properties of the media. Thus, XMT-Ω defines a set of extensible media (xMedia) elements as basic building blocks, representing multimedia objects, that can be combined in complex spatial and temporal layouts and whose fundamental properties can be animated to create the rich, dynamic, interactive content that MPEG-4 is all about.

An xMedia object is defined by an element, such as `<img>` or `<rectangle>`, which abstracts geometry containing media-specific geometric property attributes as well as timing attributes defined by SMIL for temporal layout. Spatial properties of an xMedia object can be defined further by a set of common child elements, such as `<transformation>`, `<material>`, `<chromakey>`, `<texture>`, `<light>`, and `<hotspots>`. These child elements of an xMedia

---

2. A mime-type is a Multipurpose Internet Mail Extensions standard-defined media type, for example, text/plain and image/jpeg.

3. A child element is an element contained inside another element (its parent) in the document tree.

object can be used to define either 2D or 3D properties, as the elements provide a combined set of attributes for this purpose. The `<transformation>` element allows the object to be positioned and scaled, and so on, and maps to the BIFS `Transform` or `Transform2D` node. The `<material>` element defines part of the xMedia object's appearance and maps to the BIFS `Material` or `Material2D` node. `<material>` can contain an `<outline>` element, but this is only used for 2D xMedia objects such as `<rectangle>`. The `<texture>` element defines the other part of the xMedia object's appearance and maps to the BIFS nodes that can be placed in the `texture` and `textureTransform` fields of the `Appearance` node. `<light>` is for 3D and provides illumination of the xMedia object.

An xMedia element abstracts MPEG-4 Systems and audiovisual streams and, as such, is a high-level abstraction for the BIFS, OD framework, media streams, and so on to which XMT-Ω is mapped.

The following are XMT-Ω examples of xMedia objects, wherein the `<img>` media object is fully compatible with the one defined in the SMIL Media modules. The `<img>` media object simply represents an image in which the `src` attribute references the media data itself.

```
<img src="landscape.jpg"/>
<rectangle size="160 120"/>
<circle radius="80"/>
<curve points="0 0; 100 100; 200 200; 60 50; 10 10"/>
```

xMedia objects in XMT-Ω cover the complete range of basic 2D and 3D visual objects from MPEG-4 Visual [MPEG4-2] as well as MPEG-4 Audio [MPEG4-3].

To the xMedia objects has been added the `<hotspots>` element, which allows further xMedia elements (such as `<circle>` and `<rectangle>`) to be used as hotspot links. The `<hotspots>` element permits any xMedia element to be added onto any other media elements and be designated as a hotspot. In MPEG-4, any visual shape can have a `TouchSensor` attached and so can be used as a hotspot. Both 2D and 3D visual objects can be used, so a wider range of shapes is available for use in comparison to the functionally similar to the `<area>` element in SMIL. The objects used as hotspots can be timed and animated to provide dynamic, changing hotspots over time.

6.4.3 Timing and Synchronization

XMT-Ω elements are temporally arranged and synchronized using the SMIL Timing and Synchronization modules. The terms SMIL timing and XMT-Ω timing (or simply XMT timing) are used interchangeably in this section, but they are distinguished when necessary. The syntax and semantics of timing elements and attributes are according to the SMIL specification [SMIL01].

The Timing modules define both elements and attributes that coordinate and synchronize the presentation of media objects, animations, and transitions across time. Three synchronization elements, called *time containers* and presented in Table 6.2, support timing by grouping their child elements into coordinated timelines.

The Timing modules also specify attributes to individually control an element's timing behavior. The essential timing attributes are presented in Table 6.3.

The following shows an example of the use of the `<par>` and `<seq>` time containers. Note that the `begin` attribute defaults to 0 seconds when not specified. The `<par>` will begin at t = 5 seconds, and at that point the `<rectangle>` xMedia object and the `<polygons>` xMedia object in the `<seq>` will start playing. At t = 1 seconds, the `<img>` will begin as well, so that three objects are playing in parallel. At t = 10 seconds, the `<rectangle>` will end, and at t = 12 seconds the `<img>` will end, leaving only the `<polygons>` object playing. At t = 15 seconds, the `<polygons>` object will end and the next element in the sequence, the `<circle>`, will begin to play for its 2-second duration. At t = 17 seconds, the sequence ends with the `<circle>`, and the `<par>` ends too.

```
<par begin="5s">
  <rectangle size="10 10" dur="5s"/>
  <img src="my.jpg" begin="1s" dur=6s"/>
  <seq>
    <polygons coord="10 10; 34 45; 23 12" dur="10s"/>
    <circle radius="50" dur="2s"/>
  </seq>
</par>
```

Table 6.2 Synchronization elements

Element	Description
`<par>`	Plays one or more child elements allowing *parallel* playback.
`<seq>`	Plays the child elements one at a time, in sequence.
`<excl>`	Plays one child at a time, but does not impose any specific order.

Table 6.3 Essential timing attributes

Attribute	Description
`dur`	Specifies the duration of an element.
`begin`	Specifies the beginning time of an element in a variety of ways, ranging from simple clock times to event-based occurrences (for example, a mouse-click).
`end`	Specifies the end time of an element in a variety of ways, like the begin time.
`min`	Specifies the minimum active duration of an element.
`max`	Specifies the maximum active duration of an element.

XMT-Ω also supports the additional Timing attributes presented in Table 6.4.

SMIL Timing and Synchronization support is broken down into a number of modules that form semantically related elements, attributes, and attribute values. In reusing the SMIL Timing and Synchronization modules, XMT-Ω supports most of the SMIL Timing constructs, including time containers, sync-arcs, event-based timing, stream synchronization dependencies, and so on. XMT-Ω also fully incorporates the SMIL Time Manipulations module.

The XMT-Ω elements that support timing attributes include xMedia elements, animation elements, time container elements, and the group element.

6.4.3.1 Event Timing
XMT-Ω, like SMIL, has event-based timing wherein the beginning or end of an object can be triggered by events. Events are essential to supporting nondeterministic behavior to provide engaging interactive content, whether it is for 2D or 3D content.

As an example, the following XMT-Ω fragment shows a yellow circle that begins playing (appears) when an object called myButton is clicked.

```
<circle radius="200" begin="myButton.click">
  <material color="yellow"/>
</circle>
```

Table 6.4 Additional timing attributes

Attribute	Description
repeatDur, repeatCount	Specifies that an element playback will be repeated by specifying either a total active duration for the repeat or a repeat count.
fill	Specifies how an element that has ended behaves if the container it is within is still active—that is, how any remaining time is to be filled by that element. This can be used to fill gaps in a presentation, for example, by freezing the final state: fill = "freeze".
endsync	Controls the active duration of the par and excl containers with respect to the active duration(s) of child element(s). It forces the time container to end when the selected child element(s) ends.
restart, restartDefault	Controls the restart behavior of an element, that is, the ability of an element to begin multiple times within its time container.
syncBehavior, syncBehaviorDefault	Defines the runtime synchronization behavior for an element, allowing independent or synchronized (locked) timing.
syncMaster	If set to true, this forces other elements in the time container to synchronize their playback to this element.

6.4.3.2 XMT-Ω Events   In addition to providing basic events, such as mouse click, mouse up, mouse down, mouse over, and mouse out, XMT-Ω also supports more advanced events concerned with the interaction of MPEG-4 objects in both 2D and 3D spaces. Table 6.5 lists XMT-Ω behaviors and their corresponding event symbols (partly XMT-Ω native and partly from SMIL).

XMT has defined unique event symbols to support events not defined in SMIL, but it also supports some of the event symbols defined by the SMIL language, as listed in Table 6.5. In some cases, there are two symbols for the same event—for example, `click` and `activateEvent`. An authoring tool should support both and treat them as synonyms. `click` is potentially more readable, but `activateEvent` is more interoperable with the SMIL language and better supports the concept of accessibility. Dependent on how the content will be used, an author can choose to use either symbol.

Both 2D and 3D xMedia objects can generate events such as `click`, `mouseup`, `mousedown`, `mouseout`, and `mouseover`; `viewable`, `near`, and `collide` are primarily for 3D content. For example, to detect the collision of 3D media objects with other 3D media objects, the objects are placed as child elements within a `<group>` that has the `collide` attribute set to true. An XMT-Ω

Table 6.5 XMT-Ω behaviors and event symbols

Behavior/event	XMT-defined event symbols	SMIL-defined event symbols
Mouse click	`click`	`activateEvent`
Mouse up	`mouseup`	
Mouse down	`mousedown`	
Mouse out	`mouseout`	`outOfBoundsEvent`
Mouse over	`mouseover`	`inBoundsEvent`
Mouse drag	`mousedrag`	
Detecting visibility of an object	`viewable`	
Detecting proximity of an object	`near`	
Detecting collision of objects	`collide`	
Observed beginning of object		`beginEvent`
Observed end of object		`endEvent`
Object has begun a repeat cycle		`repeatEvent`

`<group>` allows two or more xMedia objects to be grouped together and spatially related so that they can be acted upon simultaneously. A `<group>` also allows the detection of collisions among the 3D xMedia objects it contains. The following sample XMT fragment has an audio sound that starts playing when the two objects, the sphere and the box, collide.

```
<head>
  <layout metrics="pixel">
    <topLayout width="640" height="480"/>
  </layout>
</head>
<body>
  <group id="ballBox" collide="true">
    <sphere id="ball" radius="8" dur="20s"/>
      <transformation translation="-100 50 30">
        <animateMotion to="0 0 0" dur="5s"/>
      </transformation>
    </sphere>
    <box id="crate" size="20 20 20" dur="20s"/>
  </group>
  <audio src="crash.wav" begin="ballBox.collide"/>
</body>
```

6.4.4 Time Manipulations

XMT-Ω incorporates the SMIL Time Manipulations module, providing `speed`, `accelerate`, `decelerate`, and `autoReverse` attributes to allow the manipulation of the flow of time.

The `speed` attribute allows the rate of playback to be altered such that, in the following example, the content of the sequence will play twice as fast relative to the parent time. Hence each video will play twice as fast as normal, and the sequence will last for only 14 seconds.

```
<seq speed="2.0">
  <video dur="10s" .../>
  <video dur="18s" .../>
</seq>
```

To define a simple acceleration or deceleration of time for the element, the attributes `accelerate` and `decelerate` are used. Their value is expressed as a fraction of the simple duration in the range 0.0 to 1.0, where the acceleration occurs at the start of the animation and the deceleration at the end. So, in the following example, the image will be moved over its path and accelerate from standstill over the first 3 seconds (0.25 of the 12-second duration) to its maximum velocity. For 6 seconds it will run at constant rate, and then finally it will decelerate smoothly to a stop during the last 3 seconds. Such motion is often termed *ease-in ease-out* and makes the animation look more realistic.

```
<img ...>
  <animateMotion dur="12s"
                 accelerate=".25" decelerate=".25" ... />
</img>
```

Finally, the `autoReverse` attribute allows oscillations, wherein time flows back and forth, allowing swinging (pendulums), pulsing, and bouncing to be easily expressed—which, when coupled with `accelerate` and `decelerate`, can provide very convincing animations.

FlexTime Support   The MPEG-4 FlexTime model [AMD1] is fully supported by XMT-Ω. The SMIL time containers, `<par>` and `<seq>`, are used to specify temporal relationships (such as co-occur and meet) between xMedia elements. FlexTime can provide simple lower and upper bounds on the playback duration of an element using the SMIL `min`/`max` attributes such that those constraints can be satisfied during playback where network delays are not predictable. Beyond this simple bounding (or clipping) of the duration, FlexTime also can provide a more sophisticated mechanism to compensate for nondeterministic delays by allowing the stretching or shrinking of the playback duration of an element, using application-specified flexibility constraints as given by the content author. The `flexBehavior` attribute specifies an ordered list of preferred modes by which the playback active duration can be lengthened (stretch modes) or shortened (shrink modes). For a continuous media, such as video, slowing down or speeding up the element time accomplishes the stretching or shrinking of playback duration, respectively. For audio, more complex techniques to preserve pitch, or to eliminate or add distributed silence, can be used in addition to simple repetition or truncation as the preferred modes allow. For discrete media, such as still images, circles, and so on, the stretching or shrinking of active duration is trivially achieved merely by adjusting duration itself.

The `flexBehavior` attribute works in conjunction with the `min` and `max` attributes. The `min` and `max` attribute values in a *flexed* element provide upper and lower limits to which the active duration can be adjusted (flexed) to meet the synchronization requirements. The `flexBehaviorDefault` attribute allows the default flex behavior to be set such that when the `flexBehavior` is not explicitly specified, the default value assumed can be controlled.

In summary, the FlexTime support in XMT allows an element to adjust its playback speed, according to the preferred modes of stretching or shrinking, by allowing the length of the effective duration to be altered to anywhere between the `min` and `max` limits. Such adjustments are made to preserve temporal alignment (synchronization) using flexible, application-level (author-defined) rules to recover from nondeterministic delays.

6.4.5 Animation

XMT-Ω incorporates the SMIL Animation modules to allow attributes of xMedia objects to be animated to create dynamically changing content. Animation features can be mapped to MPEG-4 in various ways. The simpler `<set>` element, which sets a value into an attribute, can be mapped to the MPEG-4 `Replace Field` BIFS update command. The more complex `<animate>`, `<animateColor>`, and `<animateMotion>` can be mapped to MPEG-4 `TimeSensors`, `Interpolators`, and `ROUTEs`. `BIFS-Anim` streams also can be used where appropriate.

Animation elements are timed like xMedia elements, using the same set of timing attributes. This allows a wide variety of timed dynamic behaviors, including event-based timing for the animation.

An animation element can explicitly target the element whose attribute is to be animated, or it can be left implicit. The implicit target element is the parent element containing the animation element in the document instance.

The SMIL Animation module defines attributes to have a base value, the value exposed by the underlying object model, and a presentation value, the value the attribute takes as it is dynamically updated during the presentation. Unlike SMIL players, MPEG-4 players do not maintain both a base value and a dynamic presentation value for attributes (fields). MPEG-4 supports only field replacement, whether done by `ROUTEs`, by `BIFS-Commands`, or by `BIFS-Anim` streams. To preserve SMIL Animation semantics, the XMT authoring tool will be required to manage and track the attributes state such that the SMIL semantics can be honored at runtime. In straightforward use cases, this simply means resetting the attribute value back to its initial state after the animation is complete.

The following example shows a green rectangle whose color is changed to blue using the `<set>` element. The rectangle plays for 12 seconds in total. At t = 3 seconds, the color is changed to blue, which lasts for 6 seconds; then the change in attribute value ends, and the color reverts to green. This can be mapped to BIFS with a `Replace Field` command at t = 3 seconds to set the color to blue, and another at t = 9 seconds to set it back to green.

```
<rectangle size="100 150" dur="12s">
  <material color="green" filled="true">
    <set attributeName="color" to="blue"
        begin="3s" dur="6s"/>
  </material>
</rectangle>
```

The following example shows a red circle whose color is linearly interpolated between the colors red to yellow and then to green. The circle plays for a total of 15 seconds. At t = 5 seconds, the color will begin to change from red to yellow such that at t = 9 seconds (5 seconds + 0.4 x 10 seconds) the color has reached yellow and begins to change to green, which it reaches at t = 15 seconds. The color animation can be mapped to a BIFS `ColorInterpolator` used with a `TimeSensor` having a `cycleInterval` of 10 seconds.

```
<circle radius="160" dur="15s">
  <material color="red" filled="true">
    <animateColor attributeName="color"
                  values="red; yellow; green"
                  keyTimes="0; .4; 1" calcMode="linear"
                  begin="5s" dur="10s"/>
  </material>
</circle>
```

Finally, the next example shows a circle whose radius is linearly interpolated from 160 to 320 and back to 160. The circle plays for a total of 15 seconds. At t = 0 seconds, the radius will begin to change from 160 to 320 such that at t = 7.5 seconds (0.5 x 15 seconds), the radius is 320. Then the radius will start to change back from 320 such that it reaches the final value of 160 at t = 15 seconds. The animation of the circle's radius can be mapped to a BIFS `Scalar-Interpolator` used with a `TimeSensor` having a `cycleInterval` of 15 seconds.

```
<circle radius="160" dur="15s">
  <animate attributeName="radius"
           values="160; 320; 160"
           keyTimes="0; .5; 1" calcMode="linear"
           dur="15s"/>
  <material color="red" filled="true" />
</circle>
```

6.4.6 Spatial Layout

In XMT-Ω, the spatial properties for visual media content can be defined for a single xMedia object with the `<transformation>` element, or by using a `<group>` element to apply such properties to more than one object.

XMT-Ω also defines its own Layout module, using the SMIL Layout elements `<layout>`, `<topLayout>`, and `<region>` as a basis. xMedia objects can then be arranged in a hierarchical layout similar to SMIL. Note that whereas XMT-Ω does not incorporate SMIL Layout modules, the XMT-Ω-defined Layout, although unique to XMT, is syntactically and semantically compatible with SMIL and approximates to SMIL Hierarchical Layout with center registration and alignment.

6.4.7 XMT-Ω Examples

This section contains some examples of the XMT-Ω high-level format and mappings into MPEG-4 Systems.

Example 1: Rectangle with color animation after mouse click The first example shows a rectangle whose color changes over a 6-second duration where the color animation begins when the rectangle is clicked on.

Using the XMT-Ω format, a rectangle is defined of size 50x50 (a square), whose midpoint is positioned at coordinate x = 40 y = 75, using a child `<transformation>` element. A child `<material>` element provides the rectangle with a color, specifying that it is to be drawn as a filled shape rather than an outline. The `<material>` then has an `<animateColor>` child element that describes a linear interpolation, over a 6-second duration, using three color values, and beginning when the mouse is clicked.

```
<rectangle id="mySquare" size="50 50">
  <transformation visibility="true"translation="40 75"/>
  <material color="#ee0000"filled="true">
```

```
        <animateColor attributeName="color"
           dur="6s" begin="mySquare.click"
           values="#ee0000; #ffcc45; #ffffff"
           keyTimes="0; 0.3; 1"calcMode="linear"/>
      </material>
   </rectangle>
```

This XMT-Ω format example is now reworked using XMT-A syntax to show how it could potentially be mapped to MPEG-4 nodes and routes. The OD framework elements, such as the `InitialObjectDescriptor`, which would be necessary to represent the complete MPEG-4 content, have been omitted for clarity.

To create the scene, a `BIFS-Command` to `Replace Scene` is needed. To provide the root node for the scene, a top-level node is needed. Here an `Ordered-Group` node is used to create the necessary 2D context for the `Rectangle`. Then there is an MPEG-4 `Switch` node so the entire xMedia object can be hidden or shown (`visibility = "true"` attribute above).

The XMT-Ω `<rectangle>` is the basic pattern of a `Shape` containing `appearance` and `geometry`, where `geometry` is a `Rectangle` and `appearance` contains a `Material2D` describing the color. The `Rectangle` (`Shape`) is set under a `Transform2D` to position it. This is opposite to XMT-Ω, in which the `<transformation>` is a child element of `<rectangle>`.

To sense mouse activity, a `TouchSensor` is needed; to define the duration of the color change, a `TimeSensor` is needed. Then there is a `ColorInterpolator` to animate the color. To make the behavior work, the `TouchSensor` `touchTime` is routed to the `TimeSensor` `startTime` to start the `TimeSensor` when the mouse is pressed. Then the `fraction_changed` output of the `Time-Sensor` is routed to the `set_fraction` input of the `ColorInterpolator`. Finally, the `value_changed` output of the `ColorInterpolator` is routed to the `emissiveColor` field of the rectangle's `Material2D` (not forgetting, of course, to `DEF` the required nodes so that they can be identified for the `ROUTE`s).

```
        <Replace>
          <Scene>
            <OrderedGroup>
              <children>
                <Switch whichchoice="0">
                  <choice>
                    <Transform2D translation="40 75">
                      <children>
                        <Shape>
                          <appearance>
                            <Appearance>
                              <material>
                                <Material2D
                                  DEF="SquareMat"
                                  emissiveColor="0.93 0.0 0.0"
                                  filled="true" />
                              </material>
                            </Appearance>
                          </appearance>
```

```
          </appearance>
          <geometry>
            <Rectangle size="50 50"/>
          </geometry>
        </Shape>
        <TouchSensor DEF="Touch" />
        <TimeSensor  DEF="Timer"
                     cycleInterval="6" />
        <ColorInterpolator
                     DEF="Coloring"
                     key="0.0 0.3 1.0"
                     keyValue="0.93 0.0  0.0,
                               1.0  0.93 0.27,
                               1.0  1.0  1.0"/>
      </children>
     </Transform2D>
    </choice>
   </Switch>
  </children>
 </OrderedGroup>
 <ROUTE fromNode="Touch" fromField="touchTime"
        toNode="Timer"   toField="startTime"/>
 <ROUTE fromNode="Timer" fromField="fraction_changed"
        toNode="Coloring"toField="set_fraction />
 <ROUTE fromNode="Coloring"fromField="value_Changed"
        toNode="SquareMat" toField="emissiveColor"/>
 </Scene>
</Replace>
```

Example 2: BIFS, OD, and media stream mapping   An example showing BIFS, OD, and media stream mapping is presented in the following:

```
<seq begin="20s">
  <img id="Image1" dur="5s" />
  <img id="Image2" dur="2s" />
  <img id="Image3" dur="3s" />
</seq>
```

This example shows a segment of a presentation that begins at t = 20 seconds and has three images that are to be shown one after the other in sequence. This is a typical image slideshow example. The first image starts at t = 20 seconds and lasts for the specified duration of 5 seconds. At t = 25 seconds, the second image starts and lasts for 2 seconds. The third and final image starts at t = 27 seconds and has a duration of 3 seconds. This makes a total overall duration for the <seq> time container of 10 seconds.

This example is mapped to the following MPEG-4 BIFS, OD, and media streams. The mapping shows the sequence of commands, but for clarity, the detail of the nodes contained in the BIFS stream and the detail of OD content have been omitted in order to focus on the sequence of commands.

    t = 20s   OD Update for Image1 media stream

    t = 20s   Media data for Image1

t = 20s   BIFS-Command to insert nodes for Image1

t = 25s   OD Delete for Image1 media stream and OD Update for Image2 media stream

t = 25s   Media data for Image2

t = 25s   BIFS-Command to delete nodes for Image1 and insert nodes for Image2

t = 27s   OD Delete for Image2 media stream and OD Update for Image3 media stream

t = 27s   Media data for Image3

t = 27s   BIFS-Command to delete nodes for Image2 and insert nodes for Image3

t = 30s   BIFS-Command to delete nodes for Image3

t = 30s   OD Delete for Image3 media stream

Note that the timing assumes instantaneous delivery and would require modification by an MPEG-4 server to assure that media data transmission times are compensated for.

Example 3: Yellow rectangle positioned in a rectangular display area   The next example is a complete simple example showing a yellow rectangle of size 100×100, positioned at coordinates –5,10 from the center of a 640x480 display area. The metrics attribute on layout allows either meter or pixel metrics to be used for the scene and corresponds to the pixelMetrics field of the BIFS DecoderSpecificInfo.

```
<XMT-O>
  <head>
    <layout metrics="pixel">
       <topLayout width="640" height="480"/>
    </layout>
  </head>
  <body>
    <par>
      <rectangle dur="5s" id="Rectangle" size="100 100">
        <transformation translation="-5 10"/>
        <material color="yellow"/>
      </rectangle>
    </par>
  </body>
</XMT-O>
```

The following conversion has the MPEG-4 Systems Initial Object Descriptor (IOD) containing an ESD for the BIFS stream. The BIFS stream contains a Replace Scene command that creates the Rectangle followed 5 seconds later by a Delete command to remove it, providing a mapping for the duration of 5 seconds. The Switch is to allow the media object to be hidden and shown by animating the `visibility` attribute of the `<transformation>` element, although in this particular example the `visibility` attribute is not animated. The mapping, however, is general enough to handle that.

```
<XMT-A>
  <Header>
    <InitialObjectDescriptor objectDescriptorID="IOD">
      <Profiles ODProfileLevelIndication="None"
          sceneProfileLevelIndication="Unspecified"
             graphicsProfileLevelIndication="Unspecified" />
      <Descr>
        <esDescr>
          <ES_Descriptor ES_ID="IODBifs">
            <decConfigDescr>
              <DecoderConfigDescriptor bufferSizeDB="auto"
                            objectTypeIndication="MPEG4Systems1"
                            streamType="3">
                <decSpecificInfo>
                  <BIFSConfig nodeIDbits="auto"
                                  routeIDbits="auto"
                                  pixelWidth="640"
                                  pixelHeight="480"
                                  pixelMetric="true"/>
                </decSpecificInfo>
              </DecoderConfigDescriptor>
            </decConfigDescr>
            <slConfigDescr>
              <SLConfigDescriptor timeStampLength="auto"
                          timeStampResolution="auto"
                          useAccessUnitStartFlag="true" />
            </slConfigDescr>
          </ES_Descriptor>
        </esDescr>
      </Descr>
    </InitialObjectDescriptor>
  </Header>
  <Body>
    <par begin="0.0">
      <Replace>
        <Scene>
          <OrderedGroup>
            <children>
              <Switch DEF="TheRectangle" whichChoice="0">
                <choice>
                  <Transform2D translation="-5.0 10.0">
                    <children>
                      <Shape>
                        <geometry>
                          <Rectangle size="100 100"/>
                        </geometry>
                        <appearance>
                          <Appearance>
                            <material>
```

```
                            <Material2D
                                emissiveColor="1.0 1.0 0.0"
                                filled="true" />
                        </material>
                    </Appearance>
                  </appearance>
                </Shape>
              </children>
            </Transform2D>
          </choice>
        </Switch>
      </children>
    </OrderedGroup>
  </Scene>
 </Replace>
</par>
<par begin="5s">
 <Delete atNode="TheRectangle" />
</par>
</Body>
</XMT-A>
```

6.5 XMT-A FORMAT

The XMT-A format provides a direct textual representation of MPEG-4 Systems [MPEG4-1]. At the core of XMT-A is a textual representation of the scene graph of nodes that MPEG-4 has in common with X3D. Not only do MPEG-4 and X3D have most nodes in common, XMT-A and X3D also share compatible textual formats for these nodes. Such compatibility facilitates content interchange and interoperability.

XMT-A contains a direct textual representation of both the MPEG-4 Systems OD framework and the BIFS (see Chapters 3 and 4). The scene description declares the spatial-temporal relationship of audio, visual, and graphics objects. The OD framework specifies the ES resources that provide the time-varying data for the scene and consists of a set of descriptors allowing the identification, description, and association of ESs both to each other and to objects in the scene description.

When MPEG-4 content is represented as an XMT-A document, the XML document instance, as it is known, will be structured according to defined rules for element placement. These rules are defined using the XML Schema language specification developed by the W3C [SCHM01] and are in accordance with the MPEG-4 Systems binary coding specification [MPEG4-1].

6.5.1 Document Structure

An XMT-A document instance has a single optional <Header> element followed by a single <Body> element. The <Header> element may contain <meta>

description elements and the MPEG-4-specific element for the `<Initial-ObjectDescriptor>`. The `<InitialObjectDescriptor>` is the initial access point for MPEG-4 content that also signals profile and level information so that the receiving terminal can determine if it can process the content. The document structure is very similar to the X3D document structure to facilitate interoperability of XMT-A and X3D.

6.5.2 Timing

Rather than defining a new timing construct, XMT-A uses one of the SMIL time containers, the `<par>` element, to express timing. The `<par>` element can be used to group multiple commands, events, or messages that must occur at the same time. A `<par>` element is timed using only a `begin` attribute. Whereas the SMIL language (and XMT-Ω) allows the `begin` attribute to be present on many other elements, in particular media elements, XMT-A allows only a `begin` attribute on the `<par>` element to specify the execution (`begin`) time of `BIFS-Commands`, OCI messages, and so on. Unlike SMIL and XMT-Ω, XMT-A does not allow `begin` time negative values. The following fragment shows the `<par>` syntax.

```
<par begin= "">
  <!--One or more commands/messages/events
      and/or <par> elements -->
</par>
```

However, like SMIL, the `<par>` elements also can contain other `<par>` elements. The time given by the `begin` attribute or a nested `<par>` element is relative to the time of its immediate parent `<par>` time container. So, in the following example, the outer `<par>` starts at a time of t = 5 seconds; and the nested `<par>`, with a `begin` attribute time of 4 seconds, starts 4 seconds after its parent—that is, at t = 9 seconds. This is useful to preserve relative time so that a whole block of related elements can be moved in time as a group. If in the following example a command in the outer `<par>` was to insert a rectangle into the scene and in the nested `<par>`, 4 seconds later, was to delete it, then the rectangle would be visible for 4 seconds starting at t = 5 seconds. By changing the time on the outer `<par>` to 15 seconds, the whole block is shifted in time but is still visible for 4 seconds.

```
<Body>
  ...
  <par begin="5s">
    ...
    <par begin="4s"> <!-- begins at 9 seconds -->
      ...
    </par>
      ...
  </par>
    ...
</Body>
```

An XMT-A document instance has an implied top level `<par begin="0s">`, immediately inside the `<Body>` as an overall outer `<par>` (another way of saying this is that the body functions are a `<par>` with `begin = 0 seconds`). The `<par>` elements need not appear ordered in time in the document instance; indeed, nesting of `<par>` elements will often preclude this.

The `<par>` element may contain the following:

- `<par>`
- `BIFS-Commands`
- OD Commands
- IPMP Messages
- OCI Events
- MPEG-J Stream Headers

For any given stream, the commands, messages, or events that are to be executed at the same time will be coded into the stream in the order in which they appear in the document. For example, for a given BIFS stream, all `BIFS-Commands` for the same time will be coded into a single `CommandFrame`, a single AU, in document order. OD commands, IPMP messages, and OCI events for the same streams follow this rule as well and will be coded into single AUs.

6.5.3 Scene Description

XMT-A contains a textual representation of MPEG-4 BIFS (see Chapter 4). This includes the nodes for scene description, routes that connect fields between nodes, and `BIFS-Commands` to deliver and update the scene.

6.5.3.1 Nodes   The XMT-A representation of MPEG-4 BIFS nodes in XML follows the same rules as X3D used to create its XML representation, and hence is compatible with X3D. To this, MPEG-4 adds attributes for deterministic binary encoding, as well as elements and attributes for use by authors and authoring tools to augment the authoring process. These additional attributes and elements are extensions such that the base representation remains compatible.

To convert MPEG-4 BIFS nodes and their fields to XMT-A XML elements and attributes, the following algorithm is used:

1. Each node is converted to an XMT-A element with the same name.
2. For each field in the node:
   a. If the type of the field itself is a node, then the field is again converted to an XMT-A element having the same name as the field. This element will be a child element that can contain node(s) of the same type as the field.

b. If the type is not a node but a basic type, such as SFInt32,

   i. If the type is a Field or an exposedField, the field is made into an attribute of the element, preserving its name. If the field has no default value in MPEG-4, the attribute use is optional. If it has a default value, the attribute use is default with the same value.

   ii. If the type is eventIn or eventOut, the field is omitted from the representation, as these types of fields are not encoded and hence need not be represented as attributes of the element. These fields act as inputs and outputs for the nodes only when they are active in the scene.

The Conditional node is an exception to the preceding rule, wherein the buffer field, although a non-node field, is converted to an element so that it can contain one or more BIFS-Command elements in this XML representation.

The following example depicts the conversion of the BIFS Material2D node to illustrate this rule. First the Material2D node, as it can be found in the MPEG-4 Systems specification [MPEG4-1], is presented. The node has four fields: Three are basic fields and the fourth, lineProps, is of type node.

```
Material2D {
    exposedField SFColor emissiveColor 0.8, 0.8, 0.8
    exposedField SFBool  filled         FALSE
    exposedField SFNode  lineProps      NULL
    exposedField SFFloat transparency   0.0
}
```

The Material2D node is thus converted into the XML Schema fragment shown next. The Material2D node becomes an element with the name Material2D. The lineProps field, being a node, becomes a child element named lineProps that can contain a LineProperties node type. The emissiveColor, filled, and transparency fields become optional attributes with default values. The additional elements and attributes are for authoring and to represent the BIFS DEF and USE node mechanisms (see Chapter 4).

```
<element name="Material2D">
  <complexType>
    <sequence>
      <element name="lineProps" minOccurs="0"
               form="qualified">
        <complexType>
          <sequence>
            <element ref="xmta:LineProperties"/>
          </sequence>
        </complexType>
      </element>
    </sequence>
    <attribute name="emissiveColor" type="xmta:SFColor"
               use="optional" default="0.8 0.8 0.8"/>
    <attribute name="filled"        type="xmta:SFBool"
               use="optional" default="false"/>
```

```
    <attribute name="transparency"  type="xmta:SFFloat"
              use="optional" default="0"/>
    <attributeGroup ref="xmta:DefUseGroup"/>
  </complexType>
</element>
```

Some examples of the <Material2D> element are given next. The <Material2D> element would normally be contained in the <material> element child of an <Appearance> node. In the following examples, the <Material2D> element is shown alone to emphasize the examples. The first example shows the <Material2D> with emissiveColor and filled attributes being explicitly stated, whereas the transparency attribute will default to 0, as it is not present. The second example shows a child LineProperties node in the lineProps field. Finally, the third and fourth examples show how a node can be created and used elsewhere in the scene using the MPEG-4 DEF and USE mechanisms. In these examples, a Material2D is created with an identifier, specified by the DEF attribute; the Material2D node thus defined can later be used again in the scene by referencing the original node through the USE attribute. This allows a reuse of nodes so that many instances can be created from a single definition.

```
<Material2D emissiveColor="1.0 0.1 0.78" filled="true"/>

<Material2D emissiveColor="1.0 0.1 0.78">
  <lineProps>
    <LineProperties lineColor="1.0 1.0 0.0"
                         width="2.0"/>
  </lineProps
</Material2D>

<Material2D DEF="BrightRed" emissiveColor="1.0 0.0 0.0"/>

<Material2D USE="BrightRed"/>
```

The following is a more complete example showing the entire <Body> of an XMT-A document instance that has a single rectangle of size 100×100 in the scene at a location of –5,10.

```
<Body>
  <par begin="0.0">
    <Replace>
      <Scene>
        <OrderedGroup>
          <children>
            <Transform2D translation="-5.0 10.0">
              <children>
                <Shape>
                  <geometry>
                    <Rectangle size="100 100"/>
                  </geometry>
                  <appearance>
                    <Appearance>
```

```
                    <material>
                        <Material2D
                            emissiveColor="0.0 0.3 0.8"
                            filled="true"/>
                    </material>
                </Appearance>
              </appearance>
            </Shape>
          </children>
        </Transform2D>
      </children>
    </OrderedGroup>
  </Scene>
</Replace>
</par>
</Body>
```

6.5.3.2 ROUTES The MPEG-4 ROUTE mechanism, used to connect fields together, is represented by the <ROUTE> element in XMT-A. An optional DEF ID attribute names the ROUTE and allows the ROUTE to be deleted or replaced by a different ROUTE at a later time.

The following example shows two ROUTEs. The first connects the emissiveColor of a Material2D node belonging to one Shape node to that of another. If MyRectangle's color is changed, MyCircle's color will change to be the same via the route. The second route is similar but has an ID, specified by the DEF attribute, so that it can be deleted (or replaced) at a future point in time.

```
<ROUTE fromNode ="MyRectangleMaterial"
       fromField="emissiveColor"
       toNode   ="MyCircleMaterial"
       toField  ="emissiveColor"/>

<ROUTE DEF="NamedRoute"
       fromNode ="MySquareMaterial"
       fromField="emissiveColor"
       toNode   ="MyIFSMaterial
       toField  ="emissiveColor"/>
```

6.5.3.3 BIFS-Commands In MPEG-4 Systems, there are 11 BIFS-Commands [MPEG4-1]. The Insert, Delete, and Replace commands can be used for nodes, indexed values, and ROUTEs, making nine commands. In addition, Replace can be used on fields; finally, there is a ReplaceScene command.

In XMT-A, these commands are represented by three basic commands: <Insert>, <Delete>, and <Replace>. The MPEG-4 Replace Scene command is, in XMT-A, <Replace> with a child <Scene> element, the <Scene> element being compatible with X3D. As in BIFS, the XMT-A <Insert>, <Delete>, and <Replace> commands can be used on nodes, values in multiple value fields (indexed values), or ROUTEs. In addition, <Replace> can act on a whole multiple value field.

Insert   The `<Insert>` command represents node, indexed value, and ROUTE insertion. For `<Insert>`, the `atField` attribute defaults to a value of "children" and the `position` attribute defaults to a value of "END" to facilitate inserting (adding) a node to a group.

```
<Insert atNode="" atField="children"
        position="BEGIN | END | n" value="" >
  <!-- Insert may contain nodes (including subtrees)
       and/or ROUTEs -->
</Insert>
```

Delete   The `<Delete>` command represents node, indexed value, and ROUTE deletion. For `<Delete>`, the `atField` attribute defaults to a value of "children" but the `position` attribute has no default.

```
<Delete atNode="" atRoute="" atField="children"
        position="BEGIN | END | n">
  <!-- Delete may contain ROUTEs -->
</Delete>
```

Replace   The `<Replace>` command represents node, field, indexed value, ROUTE, and scene replacement. For `<Replace>`, neither the `atField` nor the `position` attributes have default values.

```
<Replace atNode="" atRoute="" atField="children"
         position="BEGIN | END | n" value="">
  <!-- Replace may contain nodes (including subtrees)
       and/or ROUTEs and/or Scene -->
</Replace>
```

Examples   The following examples show how the commands can be used.

The first example has two objects (shapes) inside a new OrderedGroup, defined as OrderedGroupB, which are inserted at the beginning of the children field of an OrderedGroup already in the scene, called OrderedGroupA. The objects are a Circle and a Rectangle, respectively; both have a default appearance, as no appearance has been given.

```
<Insert atNode="OrderedGroupA" position="BEGIN">
  <OrderedGroup DEF="OrderedGroupB">
    <children>
      <Shape>
        <geometry>
          <Circle radius="40"/>
        </geometry>
      </Shape>
      <Shape>
        <geometry>
          <Rectangle size="20 10"/>
        </geometry>
      </Shape>
    </children>
  </OrderedGroup>
</Insert>
```

The next example inserts two shapes into OrderedGroupX and also inserts a ROUTE. The Rectangle will be inserted at the beginning of the OrderedGroupX children and the Circle at the end. The ROUTE is inserted to connect the filled field of two Material2D nodes together. Unlike the previous example, which is encoded to a single BIFS-Command, the following <Insert> example is encoded into three BIFS commands in a single Command-Frame. The <Circle> has a DEF ID so that subsequent BIFS-Commands can refer to it; the <Shape> containing the <Circle> has a DEF ID as well.

```
<Insert atNode="OrderedGroupX" position="BEGIN END">
  <Shape>
    <geometry>
      <Rectangle size="120 80"/>
    </geometry>
  </Shape>
  <Shape DEF="MyCircle">
    <geometry>
      <Circle radius="25" DEF="Circle25"/>
    </geometry>
  </Shape>
  <ROUTE fromNode="Mat2D-A" fromField="filled"
         toNode="Mat2D-B"   toField="filled" />
</Insert>
```

The following example inserts a single value into a multivalue array field; this is known in MPEG-4 Systems as an IndexedValue insertion. The value 3 is inserted at position (index) 1 in the multivalue array field colorIndex. After the insert there will be one more value in that array, with the previous array values from position 1 and above being shifted to position 2 and above.

```
<Insert atNode="TheIndexedFaceSet2D" atField="colorIndex"
        value="3" position="1" />
```

The next example uses XMT-A to express more than one Insert in a single command. It inserts colorIndex values 3, 6, 2, 8, and 11 at positions 1, 4, BEGIN, END, and 6. Note that values are inserted in the order listed, and the position specified is the position in the multivalue array field after the insertion of all the preceding values in the list. For example, position 4 is the new position after the first value has been inserted at position 1, and so on. The example will be encoded into five BIFS-Commands for IndexedValue field insertion.

```
<Insert atNode="TheIndexedFaceSet2D" atField="colorIndex"
        position="1 4 BEGIN END 6"
        value="3 6 2 8 11" />
```

The following examples would delete the nodes that were inserted in the previous examples, and delete the single inserted indexed value.

```
<Delete atNode="OrderedGroupB" />
<Delete atNode="MyCircle" />
<Delete atNode="TheIndexedFaceSet2D"atField="colorIndex"
        position="1" />
```

The `Replace` command, like `Insert` and `Delete`, can act on nodes, `ROUTE`s, and indexed values. `Replace` can also replace entire individual fields as well as the complete scene. Examples of the latter two cases follow.

The first example replaces a simple field, the radius of a circle geometry. The second replaces the entire `colorIndex` multivalue field of an `IndexedFaceSet2D` node. The third example replaces the `geometry` field, which is of type node, of a `Shape` node called `MyCircle` by a new geometry, in this case a `Rectangle`.

```
<Replace atNode="Circle25" atField="radius" value= "40" />

<Replace atNode="TheIndexedFaceSet2D" atField="colorIndex"
         value= "1 2 3 3 2 2 1" />

<Replace atNode="MyCircle" atField="geometry">
  <Rectangle size="25 25" />
</Replace>
```

The final example shows a `Replace Scene` command defining a complete new scene. One `Circle` shape is defined and used again, but under a `Transform2D` at another location. A `TimeSensor` and `PositionInterpolator2D` are connected by `ROUTE`s to the `Transform2D translation` field so that the second `Circle` shape instance will move back and forth between its starting point at location 50,50 and 100,100 over a period defined by the `cycleInterval` attribute of the `TimeSensor`.

```
<Replace>
  <Scene>
    <OrderedGroup>
      <children>
        <Shape DEF="Circle50">
          <geometry>
            <Circle radius="50"/>
          </geometry>
          <appearance>
            <Material2D DEF="MatCircle50" filled="true"/>
          </appearance>
        </Shape>
        <Transform2D DEF="T2DCircle50"
                     translation="50 50"/>
          <children>
            <Shape USE="Circle50"/>
          </children>
        </Transform2D>
        <TimeSensor DEF="Timer" cycleInterval="6.0"
                    loop="true"/>
        <PositionInterpolator2D DEF="Mover"
           key="0.0 0.5 1.0"
           keyValue="50 50, 100 100, 50 50"/>
      </children>
    </OrderedGroup>
  </Scene>
</Replace>
```

```
        <ROUTE fromNode="Timer"  fromField="fraction_changed"
               toNode="Mover"    toField="set_fraction" />
        <ROUTE fromNode="Mover"     fromField="value_changed"
               toNode="T2DCircle50" toField="translation" />
    </Scene>
</Replace>
```

6.5.4 Object Descriptor Framework

XMT-A contains a textual representation of the MPEG-4 OD framework (see Chapter 3). This includes the ODs, commands to update the descriptors, and also the representation for ES data.

6.5.4.1 Descriptors ODs identify and describe ESs and can be associated with an audiovisual scene description. ODs allow access to MPEG-4 content.

An OD contains one or more ESDs that provide type, configuration, and other information for the streams. ODs are themselves conveyed in ESs. Each OD is assigned an identifier, an object descriptor ID (OD_ID). This identifier is used to associate audiovisual objects in the scene to a particular OD, and hence to the ESs contained therein.

ES descriptors (ESDs) also have a unique identifier, an elementary stream ID (ES_ID), and contain information about the stream type, configuration information for the decoding process, and dependencies between streams for scalable, layered codecs. Alternate stream representations also may be present, and language descriptors describe language-based alternatives—for example, audio streams in different languages. ESDs also contain information for transmission in the form of synchronization layer configuration, as well as QoS requirements.

OCI and IPMP descriptors as well as other descriptors defined in MPEG-4 Systems are represented in XMT-A.

To convert the MPEG-4 Systems descriptor syntax to an XML representation, the following guidelines were followed for XMT-A:

1. Each descriptor is converted to an element having the same name. Then simple fields—e.g., bit(5) and bit(8)—are converted to attributes with the name of the field, unless the field is part of an if construct. In the latter case, the field is converted to an attribute of an element, wherein the element has the name (or a name very close to that) of the if condition flag.

2. Where a descriptor contains an array of descriptors, the array is converted to an element with the same name as the array that can contain the descriptors of that given type. Arrays of simple fields are converted to a single attribute that can contain a list of values.

The following example shows the InitialObjectDescriptor binary syntax as presented in MPEG-4:

```
class InitialObjectDescriptor extends ObjectDescriptorBase
{
  bit(10) ObjectDescriptorID;
  bit(1) URL_Flag;
  bit(1) includeInlineProfileLevelFlag;
  const bit(4) reserved=0b1111;
  if (URL_Flag)
  {
    bit(8) URLlength;
    bit(8) URLstring[URLlength];
  }
  else
  {
    bit(8) ODProfileLevelIndication;
    bit(8) sceneProfileLevelIndication;
    bit(8) audioProfileLevelIndication;
    bit(8) visualProfileLevelIndication;
    bit(8) graphicsProfileLevelIndication;
    ES_Descriptor          esDescr[1..255];
    OCI_Descriptor         ociDescr[0..255];
    IPMP_DescriptorPointer ipmpDescrPtr[0..255];
  }
  ExtensionDescriptor extDescr[0..255];
}
```

Following is an illustration of the XMT-A representation using an XML fragment to show the key points of the conversion rules of fields into attributes and elements.

```
<InitialObjectDescriptor
  objectDescriptorID=""
  includeInlineProfileLevelFlag="" >
  <URL URLstring=""/>
  <Profiles
    ODProfileLevelIndication=""
    sceneProfileLevelIndication=""
    audioProfileLevelIndication=""
    visualProfileLevelIndication=""
    graphicsProfileLevelIndication=""/>
  <Descr>
    <esDescr>
      <!-- 1 to 255 ES_Descriptors          -->
    </esDescr>
    <ociDescr>
      <!-- 0 to 255 OCI_Descriptors          -->
    </ociDescr>
    <ipmpDescrPtr>
      <!-- 0 to 255 IPMP_DescriptorPointers -->
    </ipmpDescrPtr>
  </Descr>
  <extDescr>...</extDescr>
</InitialObjectDescriptor>
```

6.5.4.2 Commands XMT-A includes descriptor commands for `ObjectDe-scriptor`, `ES_Descriptor`, and `IPMP_Descriptor` update and remove. These commands are timed using the `<par>` timing element in the same way that `BIFS-Commands` are timed. Mapping MPEG-4 Systems syntax to XMT-A representation follows the descriptor guidelines given previously.

The following example shows the XMT-A representation for `<ObjectDe-scriptorUpdate>` and `<ObjectDescriptorRemove>`. For `<ObjectDescriptor-Remove>`, the `objectDescriptorId` attribute is a list of IDs.

```
<ObjectDescriptorUpdate>
  <OD>
    <!-- 1 to 255 ObjectDescriptors and/or
         InitialObjectDescriptors  -->
  </OD>
</ObjectDescriptorUpdate>

<ObjectDescriptorRemove objectDescriptorId="" />
```

6.5.4.3 Elementary Streams In XMT-A, ES data, including audiovisual media data, are referenced and associated with ESDs using the `<Stream-Source>` element. In XMT-A, some ESs, such as BIFS and OD, have textual representations wherein the streams are created when the content is encoded to the MPEG-4 binary format. Other ESs, such as video and audio, have no textual representation of the media itself, and external media sources are referenced to provide the stream data itself.

When using externally referenced sources for video, audio, and so on, these sources do not have to be in the final compressed media format for the stream. An XMT-A authoring tool may support media transcoding or encoding from a source to a target format; basic support, however, requires only that a tool can handle media in the correct target format. To encode or transcode a media source, an `<EncodingHints>` element contains information to assist or provide hints on how the media should be coded. XMT-A authoring tools are not limited by the XMT-A representation, but may have other platform dependencies that limit the range of media types supported in practice.

ESs that are represented textually within the XMT-A document are BIFS and OD streams as well as OCI, IPMP, and MPEG-J. Although with MPEG-J the stream is partly internal—that is, there is a textual representation of the MPEG-J stream headers—the MPEG-J stream contents themselves are composed of Java classes or zipped files that are referenced as external data. For these streams a `<StreamSource>` element may still be used within the `<ES_Descriptor>` to supply `<EncodingHints>`; an example is a BIFS stream, but here the `url` attribute would be omitted as no external stream reference is required (XMT-A embeds the textual representation of that stream directly).

To create MPEG-4 OCI, IPMP, and MPEG-J streams, there are textual representations of `OCI_Event`, `IPMP_Message`, and `JavaStreamHeader`. Using

these elements, complete ESs can be created with timing again provided by
`<par>` elements. The textual representation of an `<OCI_Event>` follows.

```
<OCI_Event eventID=""
  absoluteTimeFlag="true|false"
  startingTime="" duration="" >
  <OCI_Descr>
    <!-- 1 to 255 OCI_Descriptors -->
  </OCI_Descr>
</OCI_Event>
```

6.5.5 Deterministic Mapping of XMT-A

To ensure interoperability, MPEG defines conformance points that allow
MPEG functionalities, in a given implementation, to be tested for conformity
to the specification(s) [MPEG4-4]. For XMT-A the conformance includes having the implementation under test creating an mp4 file from the XMT textual
representation and then comparing it to the one created by the XMT reference
software [MPEG4-5]. Such comparison is limited to the binary encoded XMT-A
representations—for example, BIFS, OD, and OCI. It does not include comparison of the video or audio ES data; but it does include all the OD framework
descriptors that describe it.

When converting (coding) XMT-A into the MPEG-4 Systems binary format (BIFS, ODs, OCI, etc.), the use of alternate MPEG-4 coding schemes—for
example, list versus vector—is possible, which will produce alternate binary
representations that are all legally valid. To ensure that content is coded consistently and also to support deterministic coding for conformance, the following rules are used:

1. Timed elements for any single stream—that is, BIFS-Commands, OD commands, and so on that, according to the `<par>` timing, are to be executed
   at the same time—must be encoded in the order in which they appear in
   the XMT document. All other elements are to be encoded in the order in
   which they appear in the XMT document, unless the encoding requires a
   particular sequence. For example, XMT-A often does not mandate a coding order of child elements within a construct, but MPEG-4 binary representation does; for example, ObjectDescriptor, where the sets of
   descriptors esDescr, ociDescr, and ipmpDescrPtr can be in any order in
   XMT-A but are in a defined order in MPEG-4 binary coding. However,
   within each set, the descriptor order of the array will be preserved. Also,
   any ordering must be preserved if more than one command (and so on) is
   coded into a single AU.

2. Where there are binary coding alternatives (for example, in BIFS) there
   are encoding hints (`<BIFSEncodingHints>` for the BIFS case) to control
   which alternative is selected. For example, in BIFS there are coding
   options to allow binary list or vector coding, so there are encoding hints

in XMT-A to control the final encoding from this textual representation into binary.

3. Elements are identified in XMT by an ID (a name), and this name can be a readable, meaningful name. Such IDs often must be coded into binary IDs—for example, `NodeIDs` or `ObjectDescriptorIDs`. Normally the XMT authoring tool can create binary for all the names and ensure their uniqueness. However, to ensure a given conversion to a specific binary value, the `binaryID` attribute has been added to elements whose names must be converted to binary, and any authoring tool must respect the binary value and use it if so provided. Such explicit conversion is used for conformance but may also be useful where known IDs are required—for example, MPEG-J connection to the scene can be done by known IDs for the nodes.

6.5.6 Interoperability with X3D

This section compares Web3D X3D documents with XMT-A documents to demonstrate the level of interoperability that can be achieved between X3D and MPEG-4 XMT for content that is supported by both standards and hence is exchangeable.

Both X3D and XMT-A formats have a single optional `<Header>` element followed by a single `<Body>` element. The `<Header>` element contains 0 or more `<meta>` elements, and unique to XMT-A is the MPEG-4-specific element for the `<InitialObjectDescriptor>`.

Inside the `<Header>`, an X3D document would go directly into a `<Scene>` element. MPEG-4 XMT-A, however, has the `<Scene>` inside a `<Replace>` BIFS-Command.

Table 6.6 compares the XMT-A and X3D representations to illustrate the high degree of compatibility and the small number of changes necessary to go from X3D to MPEG-4 XMT-A, and vice versa.

To completely convert a document instance from X3D to XMT-A (or vice versa), the outer `<X3D>` or `<XMT-A>` element, with schema namespace references,[4] must be changed accordingly.

Like X3D, `<ROUTE>` elements can only be placed inside the `<Scene>` element before the closing `</Scene>` and cannot be included inside other elements within the `<Scene>`. This is unlike VRML, which does not have such a restriction on ROUTE placement within its ASCII text representation.

Note that an X3D `<Scene>` does not need a `<Group>` at the top level, whereas MPEG-4 XMT-A requires a top-level node such as `<Group>`, `<OrderedGroup>`, `<Layer2D>`, or `<Layer3D>` as the root of the scene graph. If the X3D scene does not have a single `<Group>` at the root, one must add this when converting to XMT-A.

---

4. A schema namespace reference is the pointer (uniform resource identifier) to locate and identify the schema for XMT-A and X3D, respectively.

Table 6.6 XMT-A versus X3D

XMT-A	X3D
\<Header\>	\<Header\>
\<meta\>	\<meta\>
\</meta\>	\</meta\>
\<InitialObjectDescriptor .../\>	
\</Header\>	\</Header\>
\<Body\>	
\<Replace\>	
\<Scene\>	\<Scene\>
\<!-- The scene contents --\>	\<!-- The scene contents --\>
\</Scene\>	\</Scene\>
\</Replace\>	
\</Body\>	

Note also that in X3D, visual and audio sources are referred directly by URLs. Although MPEG-4 can express the URLs in an identical manner, it is more likely that a conversion would create ODs for these media types and replace the source URL references by OD_IDs.

6.6 SUMMARY

This chapter described the eXtensible MPEG-4 Textual Format (XMT) framework. The XMT framework consists of two levels of textual syntax and semantics: the XMT-A format, providing a one-to-one deterministic mapping to MPEG-4 Systems binary representation, and the XMT-Ω format, providing a high-level abstraction of XMT-A to content authors so they can exchange content with other authors while preserving their original semantic intent. XMT-A provides interoperability between VRML/X3D and MPEG-4, and XMT-Ω provides interoperability between SMIL and MPEG-4.

6.7 REFERENCES

[AMD1] ISO/IEC 14496-1:2001. *Amendment 1. Coding of Audio-Visual Objects—Part 1: Systems*, 2001.

[AMD2] ISO/IEC 14496-1:2001. *Amendment 2. Coding of Audio-Visual Objects—Part 1: Systems*. FPDAM, Doc. ISO/MPEG N4268, Sydney MPEG Meeting, July 2001.

[EXML00] W3C. *Extensible Markup Language (XML) 1.0*. W3C Recommendation. October 2000. *www.w3.org/TR/REC-xml/*

[HTML99] W3C. *HTML 4.01 Specification*. W3C Recommendation. December 1999. *www.w3.org/TR/html4*

[MPEG4-1] ISO/IEC 14496-1:2001. *Coding of Audio-Visual Objects—Part 1: Systems*, 2d Edition, 2001.

[MPEG4-2] ISO/IEC 14496-2:2001. *Coding of Audio-Visual Objects—Part 2: Visual*, 2d Edition, 2001.

[MPEG4-3] ISO/IEC 14496-3:2001. *Coding of Audio-Visual Objects—Part 3: Audio*, 2d Edition, 2001.

[MPEG4-4] ISO/IEC 14496-4:2001. *Coding of Audio-Visual Objects—Part 4: Conformance Testing*, 2d Edition, 2001.

[MPEG4-5] ISO/IEC 14496-5:2001. *Coding of Audio-Visual Objects—Part 5: Reference Software*, 2d Edition, 2001.

[N4509] MPEG Requirements. *Overview of the MPEG-7 Standard*. Doc. ISO/MPEG N4509, Pattaya MPEG Meeting, December 2001.

[SCHM01] W3C. *XML Schema Part 0: Primer*. W3C Recommendation. May 2001. *www.w3.org/TR/xmlschema-0/*

[SMIL01] W3C. *Synchronized Multimedia Integration Language (SMIL 2.0)*. W3C Recommendation. August 2001. *www.w3.org/TR/smil20/*

[SMPTE93] Society of Motion Picture and Television Engineers. *Transfer of Edit Decision Lists*. SMPTE 258M. 1993.

[SVG01] W3C. *Scalable Vector Graphics (SVG) 1.0 Specification*. W3C Proposed Recommendation. July 2001. *www.w3.org/TR/2001/PR-SVG-20010719/index.html*

[VRML97] ISO/IEC FDIS 14772-1:1997. *Information Technology—Computer Graphics and Image Processing. The Virtual Reality Modeling Language (VRML)—Part 1: Functional Specification and UTF-8 Encoding*. 1997.

[X3D01] Extensible 3D (X3D) Graphics Working Group. VRML 200x-X3D. 2001. *www.web3d.org/*

Transporting and Storing MPEG-4 Content

by Carsten Herpel, Guido Franceschini, and David Singer

Keywords: delivery, transport, storage, DMIF, file format, hinting, RTP, payload format, multiplex, MPEG-2 Systems, FlexMux, signaling, session management, channel management, video over IP

*M*PEG-4, like its predecessors MPEG-1 and MPEG-2, defines a format for audiovisual content representation. The way this coded content is conveyed from A to B is, to some extent, independent of the format itself. Still, the transport of the coded content is important, as the MPEG-4 format is intended as a basis for multimedia communication and entertainment applications.

MPEG-2, which is at the core of all digital TV standards around the world, defines both a content representation format for audio and video data [MPEG2-3, MPEG2-2] and a way to multiplex the coded data with signaling information into one serial bitstream [MPEG2-1]. This same multiplexed bitstream is then differently modulated onto physical carriers for terrestrial, satellite, or cable transmission.

During the design process of the MPEG-4 standard, it became clear that this approach should not be followed. At the time MPEG-2 was conceived, almost no digital infrastructure existed, but today there are a number of different transport options for digital multimedia data. Most prominently, such transport options are provided by the MPEG-4 predecessor MPEG-2 itself [MPEG2-1] and by the protocols that form the Internet (IP [RFC0791] and others). Other options in the telecom domain include Asynchronous Transfer Mode (ATM) [ATMF] and ITU-T's H.223 videoconferencing multiplex [H223].

So, the MPEG committee decided to provide a clean interface between the coded content representation and the transport layer, instead of defining yet another transport multiplex protocol for a number of concurrent data streams. This approach requires, however, the definition of some adaptations of the MPEG-4 streams to the existing transport protocols in order to make sure that a synchronized real-time delivery of data streams is possible. Such *glue definitions* are made in collaboration with the standardization body responsible for the specification of the respective transport protocol. Specifically, the MPEG committee itself has amended the MPEG-2 Systems specification to cover the transport of MPEG-4 over MPEG-2. MPEG also is working with the Internet Engineering Task Force (IETF) to define the transport of MPEG-4 over IP.

Content storage is another important topic. Storage can occur in various container formats, the same way that content is transported or streamed over various protocols. However, a singular storage format is important when it comes to content exchange. Therefore, MPEG has developed an *MPEG-4 file format,* largely based on Apple's QuickTime [QUICKTIME] multimedia storage format. QuickTime had been identified as a suitable starting point for this development by the usual MPEG process of setting up requirements [N4319], issuing a call for proposals, and evaluating the technical submissions based on these requirements. The MPEG-4 file format serves a variety of purposes, depending on the ways in which it is used. These purposes include storage of captured content, preparation and editing of content, storage on a server (either for download or streaming), and storage of content for interchange (exchange). Because the same format serves all these uses, it is a life-cycle format: Files may be taken from a server, for example, and easily reedited.

A content creator who wants to show a video clip in a window of the current multimedia presentation does not care much whether the video stream is played back from a local file or is streamed over the network. In terms of functionality, opening and playing a stream is similar in both cases. This finding has been captured into an abstraction of the content delivery functionality for MPEG-4, called the *Delivery Multimedia Integration Framework (DMIF)* [MPEG4-6]. Thus, in MPEG-4, *delivery* is a term that encompasses both storage (file format) and transmission protocols. The main virtue of DMIF is an interface that provides homogeneous access to storage or transport functionalities, no matter whether a stream is hosted in a file or delivered over the network. This interface is called the *DMIF-Application Interface,* even though it sounds much more pleasant to call it *Delivery Application Interface (DAI)* which captures its role quite well.

Another content delivery tool defined by MPEG-4 is called *FlexMux.* The FlexMux is a simple syntax to interleave data from various streams. Because all relevant delivery stacks support multiplex, use of this tool is envisaged only in cases where the native multiplex of that delivery mechanism is not flexible enough.

This chapter will detail the delivery abstractions defined by DMIF, and then explain the FlexMux tool, followed by in-depth coverage of the MPEG-4 file format as well as the specification of MPEG-4 over MPEG-2 Systems transport; finally, a summary of the ongoing work to define the transport of MPEG-4 over IP is provided.

7.1 DELIVERY FRAMEWORK

Part 6 of the MPEG-4 standard (DMIF) [MPEG4-6] is dedicated to the specification of the MPEG-4 delivery framework. This framework is a model that hides to its upper layers the details of the technology being used to access the multimedia content, and it supports virtually any known communication scenario (e.g., stored files, remotely retrieved files, interactive retrieval from a real-time streaming server, multicast, broadcast, interpersonal communication).

The delivery framework provides, in ISO/OSI terms, a session layer service (layer 5). This is further referred to as *DMIF layer*, and the modules making use of it are referred to as *DMIF users*. The DMIF layer manages *sessions* (e.g., associated to overall MPEG-4 presentations) and *channels* (e.g., associated to individual MPEG-4 elementary streams), and allows for the transmission of both user data and commands. The data transmission portion is often referred to in the literature as *user plane*, while the management side is referred to as *control plane*. The term *DMIF instance* is used to indicate an implementation of the delivery layer for a specific delivery technology.

In the DMIF context, the different protocol stack options are generally named transport multiplexers—or, in short, *TransMuxes*; specific instances of a TransMux, such as a User Datagram Protocol (UDP) [RFC0768] socket or an MPEG-2 PES (described in Section 7.4.1.1), are named *TransMux channels*. Within a TransMux channel several streams can be further multiplexed, and MPEG-4 specifies a suitable multiplexing tool, the *FlexMux*. The need for an additional multiplexing stage (the FlexMux) derives from the wide variety of potential MPEG-4 applications, in which even huge numbers of MPEG-4 elementary streams (ESs) could be used at once. This is a somewhat new requirement specific to MPEG-4; in the IP world, for example, the Real-Time Transport Protocol (RTP) [RFC1889] that is often used for streaming applications normally carries one stream per socket. However, in order to more effectively support the whole spectrum of potential MPEG-4 applications, the usage of the FlexMux in combination with RTP is being considered jointly between IETF and MPEG (see Section 7.5.4).

Figure 7.1 shows some of the possible stacks that could be used within the delivery framework to provide access to MPEG-4 content. Reading the figure as it applies for the transmitter side: ESs are first packetized and packets are equipped with information necessary for synchronization (timing)—SL packets. Within the context of MPEG-4 Systems, the Sync Layer (SL) syntax is used for this purpose (see Chapter 3). Then, the packets are passed through the DAI; they possibly get multiplexed by the MPEG-4 FlexMux tool, and

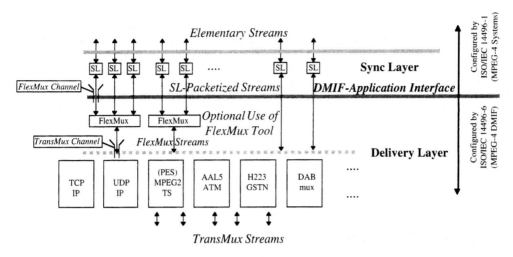

Fig. 7.1 User plane in an MPEG-4 terminal [MPEG4-6].

finally they enter one of the various possible TransMuxes—for example TCP/ IP sockets or MPEG-2 Systems.

To some extent an analogy between DMIF and the File Transfer Protocol (FTP) [RFC0959] exists, although DMIF is a framework (providing a uniform API on top of a variety of transport mechanisms) whereas FTP is a specific protocol. In both cases, a session is opened first. Then, in the FTP case, a selection of files is made by the application on top of the FTP service, and the download of those files is started by means of FTP. In the DMIF instead, a selection of ESs is made by the application on top of the DMIF layer, and channels for carrying them are created by means of DMIF. This operation might be trivial (as in the case of locally stored files) or involve network signaling. Because ESs are media with an associated timing, not just raw data, DMIF cannot handle them just as FTP does with files: That is why channels are created, which subsequently carry the actual content of the ESs.

In order to control the flow of the ESs, commands such as PLAY, PAUSE, RESUME, and related parameters need to be conveyed as well: Such commands are considered by DMIF as user commands, associated to channels. Such commands are opaquely managed (i.e., not interpreted by DMIF and just evaluated at the peer entity): This allows the stream control protocol(s) to evolve independently from DMIF. When the Real-Time Streaming Protocol (RTSP) [RFC2326] is used as the actual control protocol, the separation between user commands and signaling messages vanishes as RTSP deals with both channel setup and stream control. This separation is also void, for example, when directly accessing a file.

The delivery framework also is prepared for QoS management. Each request for creating a new channel might have associated certain QoS parameters, and a simple but generic model for monitoring QoS performance has been introduced as well. The infrastructure for QoS handling does not include, however, generic support for QoS negotiation or modification: Such features are too advanced for an area in which not even the basics appear to be widely agreed upon and understood.

Of course, not all the features modeled in the delivery framework are meaningful for all scenarios: For example, it makes little sense to consider QoS when reading content from local files. Still, an application making use of the DMIF service as a whole need not be further concerned with the details of the actually involved scenario.

The approach of making the application running on top of DMIF totally unaware of the delivery stack details works well with MPEG-4: Multimedia presentations can be repurposed with minimal intervention. Repurposing means here that a certain multimedia content can be generated in different forms to suit specific scenarios: for example, a set of files to be locally consumed, or broadcast/multicast, or even interactively served from a remote, real-time streaming application. Combinations of these scenarios are also enabled within a single presentation. The only delivery-sensitive bits in the entire MPEG-4 content are those identifying the location of the content itself: URLs embedded in the object descriptor (OD; see Chapter 3) portion of the content are used by DMIF to determine the correct mechanism to access that content and then retrieve it. Of course, no magic is applied here, and not all MPEG-4 presentations can be repurposed—as, for example, a broadcast event does not allow for remote interactivity.

7.1.1 DMIF-Application Interface

The DAI represents the boundary between the session layer service offered by the delivery framework and the application making use of it, thus defining the functions offered by DMIF. In ISO/OSI terms, it corresponds to a Session Service Access Point.

The entity that uses the service provided by DMIF is termed DMIF user and is hidden from the details of the technology used to access the multimedia content.

The DAI comprises a simple set of primitives that are defined in the standard in their semantics only.[1] Actual implementations of the DAI need to assign a precise syntax to each function and related parameters, as well as to extend the set of primitives to include initialization, reset, statistics monitoring, and any other housekeeping function. Because such features have no relevance in the modeling of the delivery framework, they have been left out of the scope of the specification.

---

1. A DAI syntax for the C programming language was later added as an informative annex to the DMIF specification [MPEG4-6], thus providing a syntax for the defined primitives and parameters in that specific language.

DAI primitives can be categorized into five families, analyzed in the following:

☞ **Service primitives:** Create or destroy a service.

☞ **Channel primitives:** Create or destroy channels.

☞ **QoS monitoring primitives:** Set up and control QoS monitoring functions.

☞ **User command primitives:** Carry user commands.

☞ **Data primitives:** Carry the actual media content.

In general, all the primitives being presented have two different although similar signatures (i.e., variations with different sets of parameters): one for communication from the DMIF user to the DMIF layer and another for the communication in the reverse direction; the second is distinguished by the `Callback` suffix and, in a retrieval application, applies only to the remote peer. Moreover, each primitive presents both IN and OUT parameters, meaning that IN parameters are provided when the primitive is called, whereas OUT parameters are made available when the primitive returns. Of course, a specific implementation may choose to use nonblocking calls and to return the OUT parameters through an asynchronous callback: This is the case for the implementation provided in the MPEG-4 reference software [MPEG4-5].

Some primitives also use a `loop()` construct within the parameter list: This indicates that multiple tuples of those parameters can be exposed at once—for example, in an array.

7.1.1.1 Service Primitives Service primitives are used to create a new service based on a URL, or to destroy a previously created service. The two primitives—`DA_ServiceAttach()` and `DA_ServiceDetach()`—defined and their (four) signatures follow:

☞ DA_ServiceAttach (IN: parentServiceSessionId, URL, uuDataInBuffer, uuDataInLen; OUT: response, serviceSessionId, uuDataOutBuffer, uuDataOutLen)

☞ DA_ServiceAttachCallback (IN: serviceSessionId, serviceName, uuDataInBuffer, uuDataInLen; OUT: response, uuDataOutBuffer, uuDataOutLen)

☞ DA_ServiceDetach (IN: serviceSessionId, reason; OUT: response)

☞ DA_ServiceDetachCallback (IN: serviceSessionId, reason; OUT: response)

The `DA_ServiceAttach()` primitive allows, by means of the input URL parameter, to determine the correct DMIF instance to use. Only the URL scheme (the part before the `://`) is parsed for this task. Because URLs may be relative, the identifier of the parent service (`parentServiceSessionId`) is provided as well, to enable the reconstruction of the absolute URL before initiating the parsing (if necessary).

At the peer side (for a peer-to-peer scenario), the fields in the corresponding Callback primitive are slightly different, as part of the (absolute) URL has already been parsed and consumed along the way: serviceName is the remaining URL portion that is meaningful at this point.

Given, for example, the URL dtcp://<hostname>/<service path and params>, the scheme dtcp allows the originating side to determine the correct DMIF instance to use (in this case, the one for the default DMIF signaling protocol), the <hostname> portion is used at the originating side to determine the peer address, and the <service path and params> portion becomes the serviceName parameter.

User-to-user data might be carried as well, in both flow directions. This data is, in this case as well as in all other occurrences, opaquely managed by DMIF. Opaquely means that this data is not interpreted, but merely transported. However, control protocols being used over the wire might, in principle, constrain this data to be limited in size. For the case of MPEG-4 applications, the uuDataInBuffer is empty, but the uuDataOutBuffer contains the initial object descriptor (IOD).

Once a service has been created, and is therefore addressable within DMIF, either side gets a locally meaningful identifier to access it (serviceSessionId). Such identifier is created within the local DMIF layer, and a DMIF user is guaranteed no ambiguity when using it. The serviceSessionId (not the URL or serviceName) is subsequently used every time any operation on that service is requested. As an example, see the signature of DA_ServiceDetach() when the service has to be destroyed. In terms of software engineering, the serviceSessionId is a unique handle.

The mapping between the serviceSessionId and the actual resources corresponding to the URL or ServiceName is internally managed by DMIF: This way the variety of the mapped resources, highly dependent on the delivery technology being used (e.g., files, MPEG-2 transport streams, remote server) is fully confined within the delivery framework.

7.1.1.2 Channel Primitives Channel primitives are used to create new channels based on application-defined identifiers, or to destroy a previously created channel. The two primitives defined—DA_ChannelAdd() and DA_Channel-Delete()—and their (four) signatures follow:

☞ DA_ChannelAdd (IN: serviceSessionId, loop(channelDescriptor, direction, uuDataInBuffer, uuDataInLen); OUT: loop(response, channelHandle, uuDataOutBuffer, uuDataOutLen))

☞ DA_ChannelAddCallback (IN: serviceSessionId, loop(channelHandle, channelDescriptor, direction, uuDataInBuffer, uuDataInLen); OUT: loop(response, uuDataOutBuffer, uuDataOutLen))

☞ DA_ChannelDelete (IN: loop(channelHandle, reason); OUT:
loop(response))

☞ DA_ChannelDeleteCallback (IN: loop(channelHandle, reason);
OUT: loop(response))

Each primitive can create or destroy multiple channels at once: Particularly in the case of networked scenarios, this allows reducing the signaling overhead without adding complexity to the mapping between DAI primitives and the actual selected protocol.

For DA_ChannelAdd(), each channel is identified by means of user-to-user data, which is opaquely managed by DMIF. For the case of MPEG-4 applications, the uuDataInBuffer will contain the ES identifier (ES_ID; see Chapter 3). For broadcast, multicast, or file retrieval scenarios, this ES_ID is used by the (MPEG-4 aware) DMIF instance to identify the location of the corresponding ES; for peer-to-peer scenarios, no MPEG-4 awareness is present in the DMIF instance, and the ES_ID will be evaluated only at the peer application.

A channel request also is characterized by a direction parameter, to indicate whether the channel is supposed to work downstream, upstream, or both, and by a channelDescriptor, which in practice represents a door for exposing any further information. More specifically, channelDescriptors carry QoS and application-specific descriptors.

Application-specific descriptors are, in the case of MPEG-4, the SYNC_GROUP and the SL_CONFIG_HEADER descriptors [MPEG4-1]. The first one intends to provide information on the synchronization relationship among the media being carried in the requested channels, to favor the mapping to protocols making assumptions on this aspect (namely the RTSP, see Section 7.5.1). The second descriptor intends to favor the usage of protocol stacks in the user plane, which provide features similar to those provided by the MPEG-4 SL, but with a different syntax: By means of this descriptor, which actually carries the MPEG-4 Systems-defined SLConfigDescriptor, the DMIF instance is able to extract the SL information from the protocol headers and convert it into a uniform representation in DA_Data() (see Section 7.1.1.5): This is the case, for example, for RTP.

QoS descriptors are meant to provide the set of QoS requisites for the channel in question. A few QoS descriptors have been tentatively defined, but a robust and proven QoS model is still missing in the literature: The DAI thus provides an infrastructure to host current and future QoS descriptors.

Once a channel has been created and is therefore addressable within DMIF, either side gets a locally meaningful identifier to access it (channelHandle). Such an identifier is created within the local DMIF layer, and a DMIF user is guaranteed no ambiguity when using it. The channelHandle (not the uuDataInBuffer, such as ES_ID) is subsequently used every time any operation on that channel is requested. As an example, see the signature of DA_ChannelDelete() when the channel has to be destroyed. In terms of software engineering, the channelHandle is a unique handle.

The mapping between the `channelHandle` and the actual resources corresponding to it is internally managed by DMIF: This way the variety of the mapped resources, highly dependent on the delivery technology being used (tracks in MP4 files; see Section 7.3 on the MPEG-4 file format, MPEG-2 packet identifiers [PIDs], and possibly FlexMux channel numbers in MPEG-2 transport streams, socket addresses, and selected protocol stack for IP communication), is fully confined within the delivery framework.

7.1.1.3 QoS Monitoring Primitives   QoS monitoring primitives are used to set up and control QoS monitoring functions. The two primitives defined and their signatures[2] are these:

☞ DA_ChannelMonitor (IN: channelHandle, qosMode; OUT: response)
☞ DA_ChannelEvent (IN: channelHandle, mode, qosReport)

The QoS monitoring model enabled by these simple primitives is generic and applicable to a wide range of requirements. By issuing a `DA_ChannelMonitor()` primitive, the DMIF user requests the DMIF layer to provide QoS reports in three different ways:

☞ Almost synchronously, through a single call to the callback primitive `DA_ChannelEvent()`;
☞ Whenever the QoS requirements expressed as part of the `channelDescriptor` in `DA_ChannelAdd()` have been violated; and
☞ Regularly, at intervals as specified in a field of `qosMode`.

Figure 7.2 shows the various ways to operate with the QoS monitoring primitives.

The `qosReport` field provided through the `DA_ChannelEvent()` callback includes a list of QoS descriptors, each providing the value of that particular QoS parameter.

Both primitives—`DA_ChannelMonitor()` and `DA_ChannelEvent()`—refer to a single channel: It is up to the DMIF instance to determine the QoS statistics valid for that specific channel—that is, in case several channels are multiplexed together in a single TransMux channel.

The exact mapping between the `DA_ChannelMonitor()` and `DA_ChannelEvent()` primitives and the actual mechanisms available to monitor the network performance (e.g., the Real-Time Control Protocol [RTCP] associated with RTP/UDP/IP sockets) is not defined within DMIF, but it is believed that enough information is available to make this feasible in a concrete implementation. Also, by means of QoS descriptors, the semantic information flowing through the DAI can be easily extended to accommodate further needs.

---

2. These two primitives are one-way each; therefore, no callback signatures exist.

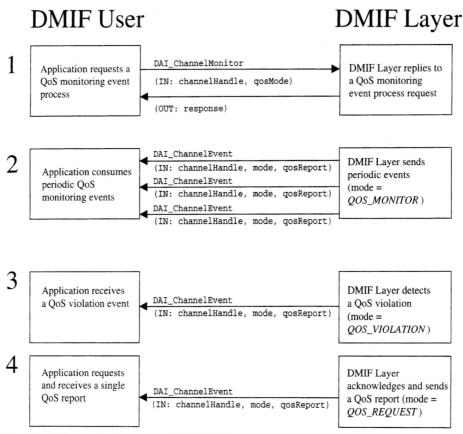

Fig. 7.2 QoS monitoring events [MPEG4-6].

7.1.1.4 User Command Primitives

User command primitives are used to carry application-defined commands associated with existing channels. The two primitives defined—DA_UserCommand() and DA_UserCommandAck()—and their (four) signatures follow:

☞ DA_UserCommand (IN: uuDataInBuffer, uuDataInLen, loop(channelHandle))

☞ DA_UserCommandCallback (IN: uuDataInBuffer, uuDataInLen, loop(channelHandle))

☞ DA_UserCommandAck (IN: uuDataInBuffer, uuDataInLen, loop(channelHandle) ; OUT : response, uuDataOutBuffer, uuDataOutLen))

☞ DA_UserCommandAckCallback (IN: uuDataInBuffer, uuDataInLen, loop(channelHandle) ; OUT : response, uuDataOutBuffer, uuDataOutLen))

Whenever a command is issued by a DMIF user, it is carried opaquely by DMIF toward the peer entity; depending on the application needs, a response might be expected. Without an explicit support from the underlying layers, this would force additional complexity in the application protocol in order to correlate commands and responses. This justifies the existence of two variations of DA_UserCommand(), which differ in that the first one—DA_UserCommand()—does not provide support for carrying back an acknowledge to the command, whereas the second form—DA_UserCommandAck()—does.

In both cases, the uuDataInBuffer parameter carries the command, whereas the channelHandle(s) indicates the channel(s) to which that command applies: As specified previously, media content is not anymore referred through ES_IDs or analogous identifiers, but only by means of the corresponding channelHandle(s). Also, channelHandles are locally assigned by DMIF and are unique within the space of an application: Thus, there is no further (strict) need to scope them through the serviceSessionId, although some implementations might prefer to do so.

In the case of MPEG-4, uuDataInBuffer for the DA_UserCommandAck() primitive will contain commands such as PLAY or PAUSE, which are typically triggered by media nodes such as MovieTexture, or by the MediaControl node. The DA_UserCommand() primitive will instead carry any kind of data and be triggered by the ServerCommand node.

7.1.1.5 Data Primitives Data primitives are used to carry the actual media content. There is a single primitive defined but, different from the other cases, this primitive has multiple signatures:

- ☞ DA_Data (IN: channelHandle, streamDataBuffer, streamDataLen)
- ☞ DA_Data (IN: channelHandle, streamDataBuffer, streamDataLen, appDataBuffer, appDataLen)
- ☞ DA_DataCallback (IN: channelHandle, streamDataBuffer, streamDataLen, errorFlag)
- ☞ DA_DataCallback (IN: channelHandle, streamDataBuffer, streamDataLen, appDataBuffer, appDataLen, errorFlag)

The first signature is of generic use: DMIF just delivers whatever it retrieves—for example, from a UDP/IP socket—without further inspecting the higher layer headers.

However, in the case of MPEG-4, this form presents a problem: MPEG-4 Systems specifies an SL and syntax to represent its semantics. When considering the delivery of such information over a network, there are situations in which the syntax should be transformed to better integrate with existing and vastly used protocols. This is the case of RTP, but not only that: MPEG-2 packetized ESs (PESs) are another example, and even the MP4 file format, in a wider sense, provides a different syntax to represent the same semantic information specified by the MPEG-4 SL.

In order to keep the details of the syntax representing the SL information within the borders of the delivery framework, it was decided to define the
second form of the `DA_Data()` primitive. In this variation, new `app-
DataBuffer/appDataLen` fields are inserted, to contain an application defined
structure. The usage of this second form of `DA_Data()` is triggered by the presence of application-specific `channelDescriptors` in the `DA_ChannelAdd()`
primitive that created the channel.

In particular, whenever the `channelDescriptor` `SL_CONFIG_HEADER` is
provided in the `DA_ChannelAdd()`, a structure describing all the SL information is used as `appDataBuffer` parameter in `DA_Data()`. This way, the DMIF
instance is appointed the task of extracting the relevant information from the
appropriate headers, which are protocol-stack-dependent, and of converting it
to such uniform internal representation. The DAI does not specify the exact
syntax of this structure, so that it can evolve independently from DMIF.

The `streamDataBuffer` exposed in the two forms of `DA_Data()` slightly
differs: When `appDataBuffer` is not provided, `streamDataBuffer` includes the
protocol headers of those layers in the stack that are on top of the pure transport function—for example, the RTP or PES headers; when `appDataBuffer` is
provided instead, `streamDataBuffer` carries only the pure media payload as
extracted—for example, from the RTP or PES packet.

7.1.2 DMIF Network Interface

The MPEG-4 delivery framework is intended to support a variety of communication scenarios while presenting a single, uniform interface to the DMIF user.
It is then up to the specific DMIF instance to map the DAI primitives into
appropriate actions to access the requested content. In general, each DMIF
instance will deal with very specific protocols and technologies, such as the
MP4 file format, the broadcast of MPEG-2 transport streams, or the communication with a remote peer. In the latter case, however, a significant number of
options in the selection of control plane protocols exists: This variety justifies
the attempt to define a further level of commonality among the various
options, making the final mapping to the actual bits on the wire a little bit
more focused (see Figure 7.3).

This is the idea behind the DMIF Network Interface (DNI), an additional
API internal to the delivery framework specified along the same lines as the
DAI—that is, only semantics are defined.

The DNI captures a few generic concepts that are potentially common to
peer-to-peer control protocols: the usage of a reduced number of network
resources (such as sockets) into which several channels would be multiplexed,
and the ability to discriminate between a peer-to-peer relation (network session)
and different services possibly activated within that single session (services).

The separation between network sessions and services is related to the
fact that several URLs, as provided by a DMIF user, may actually refer to a single
remote host. Depending on the overall application architecture, even separate

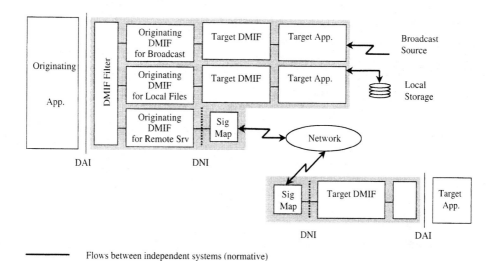

Flows between independent systems (normative)

Flows internal to a single system (either informative or out of DMIF scope)

Fig. 7.3 DAI and DNI in the DMIF architecture [MPEG4-6].

processes (i.e., separate DMIF users within a terminal) might require services from the same remote host. Although not universally true, it could be convenient to maintain a single network session (e.g., for billing purposes) encompassing the several services being run. It is worth recalling, anyway, that the DNI follows a model that helps determine the correct information to be delivered between peers but by no means defines the bits on the wire: If the concepts of sharing a TransMux channel among several streams or of the separation between network session and services are meaningless in some context, that is fine, and does not contradict the DMIF model as a whole.

The mapping between DAI and DNI primitives has been specified in [MPEG4-6]; as a consequence, the actual mapping between the DAI and a concrete protocol can be determined as the concatenation of the mappings between the DAI and DNI and between the DNI and the selected protocol. The first mapping determines how to split the service creation process into two elementary steps, and how multiple channels managed at the DAI level can be multiplexed into one TransMux channel (by means of the FlexMux tool). The second protocol-specific mapping is usually straightforward and consists of placing the semantic information exposed at the DNI in concrete bits in the messages being sent on the wire.

In general, the DNI captures the information elements that need to be exchanged between peers, regardless of the actual control protocol being used.

In the terminology used in the MPEG-4 specification [MPEG4-6], and adopted here as well, the *originating DMIF* represents the entity initiating the message exchange and the *target DMIF* represents the corresponding remote peer.

DNI primitives can be categorized into five families, analyzed in the following:

☞ **Session primitives:** Create or destroy a session.

☞ **Service primitives:** Create or destroy a service.

☞ **TransMux primitives:** Create or destroy a TransMux channel carrying one or more streams.

☞ **Channel primitives:** Create or destroy a (Flexmux) channel carrying a single stream.

☞ **User command primitives:** Carry user commands.

In general, all the primitives being presented have two different but similar signatures, one for each communication direction: The `Callback` suffix indicates primitives that are issued by the lower layer. Different from the DAI, for the DNI the signatures of both the normal and the associated `Callback` primitives are identical. As for the DAI, each primitive presents both IN and OUT parameters, meaning that IN parameters are provided when the primitive is called, whereas OUT parameters are made available when the primitive returns. As for the DAI, the actual implementation may choose to use nonblocking calls, and to return the OUT parameters through an asynchronous callback. Also, some primitives use a `loop()` construct within the parameter list: This indicates that multiple tuples of those parameters can be exposed at once (e.g., in an array).

7.1.2.1 Session Primitives   Session primitives are used to create a new session based on the addresses of the calling and called peer (the last one is extracted from the absolute URL provided at the DAI), or to destroy a previously created session. The two primitives available are these:

☞ `DN_SessionSetup[Callback] (IN: networkSessionId,`
   `calledAddress, callingAddress, compatibilityDescriptorIn; OUT:`
   `response, compatibilityDescriptorOut)`

☞ `DN_SessionRelease[Callback] (IN: networkSessionId, reason;`
   `OUT: response)`

The very first action performed in establishing a relation between two peers is the invocation of `DN_SessionSetup`.

The `calledAddress` is extracted from the URL provided by the DMIF user in the `DA_ServiceAttach()`. The `callingAddress`, if required, can be automatically computed by the originating DMIF entity and does not require further information exchange with the DMIF user through the DAI. The `networkSessionId` is a network-wide unique identifier assigned by the originating DMIF entity: This identifier should be large enough to guarantee uniqueness within a network. Its syntax will depend on the network type: As an example, over an IP network a `networkSessionId` could be defined as (or simply associated to) the tuple `<peer1IpAddress, peer1IpPort, peer2IpAddress, peer2IpPort,`

`IpProtocol>` identifying the socket providing the DMIF signaling channel. The `networkSessionId` is subsequently used every time any operation on that session is requested—for example, when the session has to be destroyed, as shown in the signature of `DN_SessionRelease()`.

This primitive considers as well the exchange of terminal capabilities. Some protocols, such as H.245 [H245], require terminal capability exchange at the beginning of a session, and this is the appropriate place to perform it.

7.1.2.2 Service Primitives
Service primitives are used to create a new service based on the portion of the absolute URL that follows the peer address, or to destroy a previously created service. The two primitives available are these:

☞ `DN_ServiceAttach[Callback]` (IN: `networkSessionId, serviceId, serviceName, ddDataIn()`; OUT: `response, ddDataOut()`)

☞ `DN_ServiceDetach[Callback]` (IN: `networkSessionId, serviceId, reason`; OUT: `response`)

Whenever a `DA_ServiceAttach()` has been invoked, the `DN_ServiceAttach()` primitive is invoked as well, provided that the appropriate network session with the remote peer has been previously established.

The `serviceId` is an identifier assigned by the originating DMIF entity to identify this specific service within the network session. As a matter of fact, the `serviceSessionId` used at the DAI is mapped one to one within the DMIF instance into the tuple `<networkSessionId, serviceId>`.

The `serviceName` is extracted from the URL originally provided by the DMIF user and identifies the required service within the remote host. The `ddDataIn()` contains the `uuDataInBuffer` provided by the DMIF user in the `DA_ServiceAttach()`. The `ddDataOut()` contains the `uuDataOutBuffer` provided in the reply by the DMIF user to the target peer (i.e., the IOD).

Figure 7.4 shows the overall service setup procedure for the most generic case using DAI and DNI concepts. At the originating side, the DMIF user (Application in the figure) invokes `DA_ServiceAttach()`, indicating the URL (step 1): The URL itself was acquired from a previous action (e.g., Web browsing, from an OD). Based on the URL scheme, the appropriate DMIF instance is activated within the DMIF layer. The DMIF instance determines with its own criteria whether a useful network session already exists with the remote peer indicated in the URL: If not, a new network session is generated (steps 2 and 3). At this point, a `networkSessionId` (nsId in the figure) representing that session exists and is known by both peers. The format of this identifier depends on the actual signaling protocol used: For example, the tuple `<IP source address & port, IP destination address &port, IP protocol>` might serve this purpose.

A specific service is then requested (step 4) within that network session, and marked with an end-to-end significant `serviceId`: The `serviceName` is the remaining portion (i.e., excluding scheme and address sections) of the original URL.

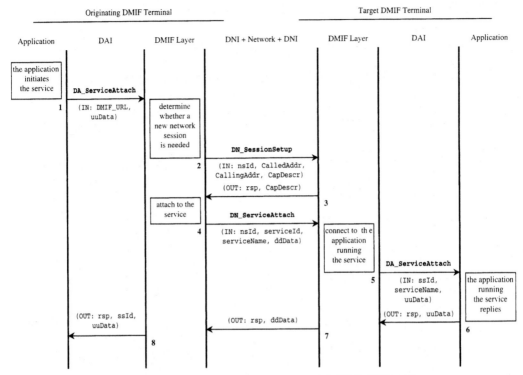

Fig. 7.4 Initialization of a service in a remote interactive DMIF [MPEG4-6].

At the peer side, a locally unique ssId is associated to the requested service, and the request is forwarded to the DMIF user (Application) in charge of serving it (step 5); its reply (step 6) allows completing the procedure (steps 7 and 8).

As a final result, a locally unique ssId exists at both peers, for further communications across the DAI, and end-to-end significant nsId and serviceId exist for further communications across the network. Note that the walkthrough gets simplified for scenarios in which a network session can only include one single service.

7.1.2.3 TransMux Primitives TransMux primitives are used to create a new TransMux channel, or to destroy a previously created TransMux channel. The three primitives available follow:

☞ DN_TransMuxSetup[Callback] (IN: networkSessionId, loop(TAT, qosDescriptor; resources()); OUT: loop(response, resources()))

☞ DN_TransMuxRelease[Callback] (IN: networkSessionId, loop(TAT); OUT: loop(response))

☞ DN_TransMuxConfig[Callback] (IN: networkSessionId, loop(TAT, ddDataIn()); OUT: loop(response))

The TransMux Association Tag (TAT) is an identifier assigned by the originating DMIF entity to identify a specific TransMux channel within the network session. Each peer will maintain a mapping between the TAT identifier and the actual network resources associated to it, so that each further reference to the TransMux channel will use just the TAT—for example, when the TransMux channel has to be destroyed, as shown in the signature of `DN_TransMuxRelease()`. The TAT syntax might differ, depending on the control protocol being used.

Whenever a new TransMux channel is being created, a `qosDescriptor` describing the QoS requisites can be specified. As for the DAI, the `qosDescriptor` provides a generic container for current and future description of QoS requisites. A TransMux channel is associated as well to a set of `resources()`. The `resources()` specify the details of the TransMux channel—for example, the IP address and number as well as the protocol stack to be used: By defining a generic `resources()` structure, and not an IP-based structure, the DNI is applicable to network environments other than IP (e.g., ATM).

In addition to the primitives to create and destroy a TransMux channel, a third primitive allows for its reconfiguration. This is intended to support the reconfiguration of multiplexing tools used within the TransMux channel: The configuration of such tools is normally incrementally generated in the process of adding multiplexed (e.g., FlexMux) channels to a TransMux channel (see `DN_ChannelAdd()` below), but from time to time it might require updates without affecting the presence of multiplexed channels.

7.1.2.4 Channel Primitives Channel primitives are used to create new (e.g., FlexMux) channels within a TransMux channel, or to destroy a previously created channel. The three primitives available are these:

☞ `DN_ChannelAdd[Callback]` (IN: networkSessionId, serviceId, loop(CAT, channelDescriptor, direction, ddDataIn()); OUT: loop(response, TAT, ddDataOut()))

☞ `DN_ChannelAdded[Callback]` (IN: networkSessionId, serviceId, loop(CAT, channelDescriptor, direction, TAT, ddDataIn()); OUT: loop(response, ddDataOut()))

☞ `DN_ChannelDelete[Callback]` (IN: networkSessionId, loop(CAT, reason); OUT: loop(response))

Each primitive may create or destroy multiple channels at once: This allows reducing the signaling overhead without adding complexity to the mapping between DNI primitives and the actual selected protocol.

The Channel Association Tag (CAT) is an identifier assigned by the originating DMIF entity to identify a specific channel within the network session. Each peer will maintain a mapping between the CAT identifier and the actual resources associated to it, including the TAT, so that each further reference to the channel will use just the CAT—for example, when the channel has to be destroyed, as shown in the signature of `DN_ChannelRelease()`. The CAT syntax might differ depending on the control protocol being used.

As a matter of fact, the `channelHandle` used at the DAI is mapped one to one within the DMIF instance into the tuple `<networkSessionId, CAT>`: networkSessionId and `CAT` are themselves further mapped to the resources they represent.

Whenever a new channel is being created, a `channelDescriptor` describing application-specific as well as QoS requisites can be specified. As for the DAI, the `channelDescriptor` provides a generic container for current and future description of application-specific and QoS requisites: It actually replicates the analogous information exposed at the DAI. Also, the `direction` parameter replicates the corresponding DAI parameter, and `ddDataIn()` actually conveys the `uuData` exposed at the DAI. The `ddDataIn()` parameter also includes further information to stack and configure additional multiplexing stages (e.g., the MPEG-4 defined FlexMux tool). Once again, this parameter provides a generic infrastructure for describing the protocol stack, and the description of new multiplexing tools can be easily integrated.

Not all signaling protocols allow establishing upstream or downstream flows indifferently. Thus, in some cases it is necessary for the originating peer to ask the target peer to set up the required network resources (the TransMux channel). This is why two different primitives for adding channels have been defined. With `DN_ChannelAdd()`, the originating peer asks the target peer to create the necessary TransMux channels, and for each requested channel gets in reply the associated TAT. With `DN_ChannelAdded()`, the originating peer provides instead the target peer with the TATs as well, as it has been already able to create the necessary TransMux channels.

Figure 7.5 and Figure 7.6 show the overall channel setup procedure for the most generic case using DAI and DNI concepts. In Figure 7.5, the DMIF user (Application) at the originating side invokes `DA_ChannelAdd()`, indicating the channels required and providing for each such channel information concerning QoS parameters, the ES to be carried, and its direction (step 1). The DMIF instance previously activated, and identified by means of the `ssId` parameter, executes the call. This DMIF instance assigns a CAT for each requested channel, and forwards the request to the peer (step 2).

At the peer side, the DMIF instance associates a locally unique `channelHandle` (`chId`) to each requested channel, and then forwards the request to the DMIF user (Application) associated to the service (step 3) and gets its reply (step 4). At this point, the DMIF instance determines with its own criteria whether existing TransMuxes (e.g., sockets) can be used to carry the requested channels. If not, new TransMuxes are created (steps 5 and 6), each associated with a TAT. The procedure continues with the reply sent by the DMIF instance at the target side conveying in particular, for each requested channel, its CAT and the TAT of the TransMux carrying it (step 7). For each successfully established channel, the DMIF instance at the originating side associates a locally unique `channelHandle` (`chId`) and finally replies to the DMIF user.

As a final result, for each established channel a locally unique `chId` exists at both peers, for further communications across the DAI, and end-to-

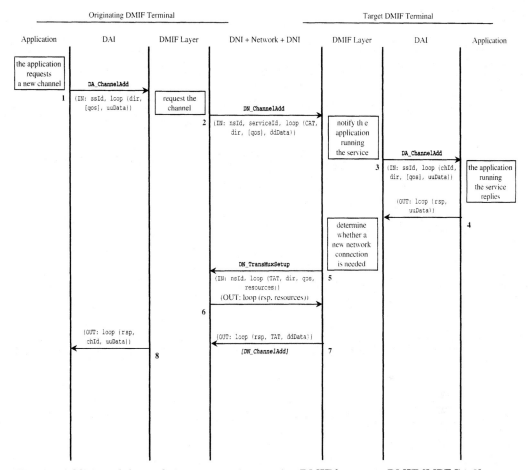

Fig. 7.5 Addition of channels in a remote interactive DMIF by target DMIF [MPEG4-6].

end significant CAT and TAT exist for further communications across the network.

Figure 7.6 differs from Figure 7.5 in that the decision to add new Trans-Muxes is made at the originating side, but the final result holds valid. Note that the walkthroughs get simplified for scenarios in which a TransMux can only include one single channel.

7.1.2.5 User Command Primitives   User command primitives are used to carry application-defined commands associated to existing channels. The two primitives available follow:

☞ DN_UserCommand[Callback] (IN: networkSessionId, ddDataIn(), loop(CAT))

☞ DN_UserCommandAck[Callback] (IN: networkSessionId, ddDataIn(), loop(CAT) ; OUT : response, ddDataOut())

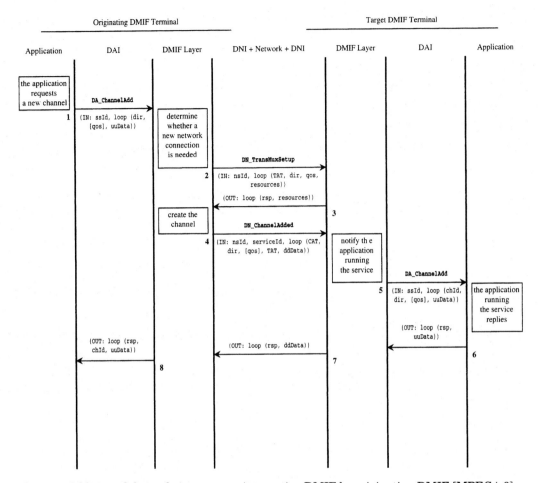

Fig. 7.6 Addition of channels in a remote interactive DMIF by originating DMIF [MPEG4-6].

The mapping between these primitives and the corresponding DAI primitives is trivial: The networkSessionId and CATs identify, respectively, the network session and channels the command applies to; the ddDataIn() and ddDataOut() parameters convey the corresponding uuData parameters as exposed at the DAI.

Both forms of user command primitives are defined, as at the DAI, to support both one-way commands and commands with acknowledgment.

7.1.3 Existent Signaling Protocols

Having defined the DAI and the DNI, the mapping to existing signaling protocols becomes quite straightforward. This exercise has been actually completed for a couple such protocols, namely, ATM Q.2931 [ATMF] and ITU-T H.245 [H245]. In the first case, the mapping was simple and did not require any interaction with the standardization body defining it; on the contrary, some extensions to the H.245 protocol were deemed necessary, and a relation with ITU-T was established to include such additions in H.245 Version 6 [H245]. Also, for the second case, some constraints were imposed to avoid inserting excessive complexity for the support of unlikely scenarios. Thus, for example, the FlexMux tool is not supported, as it has been considered unlikely that a huge number of ESs would be used in an interpersonal communication application that is the major H.245 application domain. Also, the URL field in DA_ServiceAttach() will carry just the remote host address, without specifying a serviceName within the remote host naming space.

In addition to those protocol mappings, the access to MPEG-4 content through the RTSP protocol is of great interest. In this case, the usage scenario focuses on retrieval services over IP: More information on this topic is given in Section 7.5.

RTSP is missing the support for back channels. Back channels are upstream channels that are supposed to deliver feedback information to the source of an ES flowing downstream. This feedback might, for example, provide further information on the hidden parts of a 3D object (based on local interaction), thus allowing the source to better exploit CPU and network resources by omitting the detailed coding of those parts; or it might request the retransmission of missing objects or of portions of a video stream that are essential for further decoding.

In order to step aside from all the difficulties of integrating existing technologies, and to ensure proper support to any MPEG-4 application, DMIF has defined its own *default* signaling protocol, which derives directly from the DNI semantics and is therefore not further described here; this protocol has been detailed for IP and ATM networks in [MPEG4-6]. In particular, the mechanism to set up sessions and TransMux channels has been made explicit.

By means of the DMIF Default Signaling Protocol, multimedia applications (namely MPEG-4 applications) are enabled with both retrieval and interpersonal communication models, FlexMuxes, back channels, and the like.

The DMIF Default Signaling Protocol has been designed consistently with the MPEG-2 DSMCC-UN protocol [MPEG2-6], which was specified with the goal of enabling video retrieval services across heterogeneous networks (i.e., a network path involving sections with different transport technologies, such as MPEG-2 TS and IP). This was achieved by means of a sophisticated resource management scheme, which has been simplified in the DMIF Default Signaling Protocol. As a consequence, the DMIF Default Signaling Protocol does not currently support heterogeneous networks, although it is designed to possibly evolve in that direction [MPEG4-6].

To sum up, there are mappings to currently deployed protocols such as H.245 and RTSP that, unfortunately, have limitations that would make some MPEG-4 tools unusable in these contexts. In order to overcome these limitations, DMIF has defined its own mechanism to support the full range of possible MPEG-4 content also across networks. Because the functional coverage of

H.245 and RTSP is wide enough to be appealing for the industry and the introduction of a novel signaling protocol is not easy, it may be expected that first MPEG-4 products will be based on such existing signaling protocols.

7.2 FLEXMUX TOOL

The FlexMux tool provides a simple way to interleave data from different ESs into one serialized bitstream. Of course, this is what every multiplex does. However, some of the existing multiplex tools do not have enough flexibility to accommodate the MPEG-4 needs to adapt to a wide range of compression tools and, hence, bit rates, as well as the potentially large number of streams. MPEG-4 requires, for example:

- ☞ Low multiplexing overhead in order not to spend the bit rate saved through high compression in the multiplex stage,
- ☞ Low multiplexing delay in order to facilitate low end-to-end delay applications, and
- ☞ Variable packet size to cope with the variable access unit (AU) size of the compression tools.

Therefore, the FlexMux has been defined as an optional tool that can be used when needed. Usually that would be as a multiplexing stage prior to the multiplex provided by the selected DMIF instance, as indicated in Figure 7.1.

Despite the fact that the FlexMux is a delivery layer tool, the FlexMux specification is part of the MPEG-4 Systems standard [MPEG4-1], as that part covers all the user plane functionality—that is, the syntax how to convey or store (in case of the MPEG-4 file format) the actual data.

The FlexMux packet syntax written down in Table 7.1 using the Syntactic Description Language (SDL; see Chapter 3) might look rather complex; however, it is mainly a packet format consisting of a packet identifier, called index, a length indication, and a payload as depicted in Figure 7.7 (left). The syntactic complexity derives from the fact that the index value is used to indicate the semantics of the remainder of the FlexMux packet. As the FlexMux is intended as a complementary multiplex for low-bit-rate streams, the length indication is only 8 bits long, allowing FlexMux packets with a length up to 255 bytes.

Index values between 0 and 237 directly translate to FlexMux Channel (FMC) numbers, which label the individual SL-packetized streams. The FlexMux payload in this case consists of an SL packet (as defined in Chapter 3). This mode, called *simple mode*, is most suitable for streams that produce SL packets of variable size, as their length is simply signaled in the FlexMux packet header. In the simple mode, the multiplex overhead per packet amounts to 2 bytes.

Index values between 240 and 255 indicate another mode, dubbed *Mux-Code mode*, allowing the concatenation of SL packets originating from multi-

Table 7.1 FlexMux packet syntax

```
class FlexMuxPacket (MuxCodeTableEntry mct[], FlexMuxTimingDescriptor FM) {
    unsigned int(8) index;
    bit(8) length;
    if (index<238) {
        SL_Packet sPayload;
    } else if (index == 238) {
        bit(FM.FCR_Length) fmxClockReference;
        bit(FM.fmxRateLength) fmxRate;
    } else if (index == 239) {
        bit(8) stuffing[length];
    } else {
        bit(4) version;
        const bit(4) reserved=0b1111;
        multiple_SL_Packet mPayload(mct[index-240]);
    }
}
```

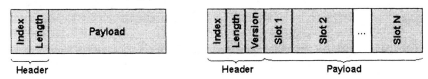

Fig. 7.7 FlexMux packet in simple mode (left) and MuxCode mode (right).

ple streams in one single FlexMux packet with a total overhead of only 3 bytes (Figure 7.7, right). This mode is useful especially for low-bit-rate speech coding, in which the payload size itself can be just a few bytes, so that even 2 bytes overhead may be too much. The 16 index values in the said range of 240 to 255 are not interpreted as a label for a stream but as a pointer to a so-called MuxCodeTableEntry (MuxCode = index − 240). The length value in the FlexMux packet header for a MuxCode packet signals the overall packet length. The MuxCodeTableEntry signals the length of the individual SL packets concatenated in one single FlexMux packet.

A MuxCodeTable with up to 16 entries serves as memory for up to 16 layouts (or templates) of MuxCode FlexMux packets. The structure of a MuxCodeTableEntry is given in Table 7.2. The MuxCode identifies the table entry. The version number allows dynamic changes of a MuxCodeTableEntry while making sure that the correct version of the table entry is referenced in the FlexMux packet header, where the version field is present as well.

The remainder of the MuxCodeTableEntry defines the pattern of the *payload slots* within one FlexMux packet with that MuxCode. Each slot is defined by the FlexMux channel number and its length in bytes. It is visible from the design of the MuxCodeTableEntry syntax that it was done by compression

Table 7.2 `MuxCodeTableEntry` syntax

```
aligned(8) class MuxCodeTableEntry {
    int     i, k;
    bit(8) length;
    bit(4) MuxCode;
    bit(4) version;
    bit(8) substructureCount;
    for (i=0; i<substructureCount; i++) {
        bit(5) slotCount;
        bit(3) repetitionCount;
        for (k=0; k<slotCount; k++){
            bit(8) flexMuxChannel[[i]][[k]];
            bit(8) numberOfBytes[[i]][[k]];
        }
    }
}
```

experts, as it applies some type of compression itself; the table entry could have been a simple list of tuples (`flexMuxChannel[i]`, `numberOfBytes[i]`). Foreseeing that the packet layout may follow a regular pattern (like "A,B, A,B, A,B, C,D,E, C,D,E") its representation can be optimized by tagging each set of slots that repeats itself, called a *substructure*, with a repetition count. In the preceding example, "A,B" and "C,D,E" would be these substructures, with two and three slots, respectively, repeated three and two times, respectively.

Apparently, the `MuxCodeTableEntries` have to be conveyed from the sender to the receiver in some way, for example, using a special message or descriptor. The details of this mechanism are left to the specification of the MPEG-4 adaptation for individual DMIF instances. An example will be given in the section about MPEG-4 transport over MPEG-2 Systems. Despite the fact that `MuxCodeTableEntries` can be changed during the transmission of an MPEG-4 presentation, the most obvious use of the MuxCode mode is to optimize the overall compression for sets of streams with fixed SL packet size and fixed bit rate each—that is, not using such dynamic updates. It is quite challenging to devise a decision process for such dynamic updates, given that packet size statistics of several streams need to be taken into account as well as the cost for sending the changed `MuxCodeTableEntries`.

Both the simple and MuxCode modes of FlexMux may be combined in a single FlexMux stream carrying multiple SL-packetized streams. Figure 7.8 (top) shows an example with streams 1, 2, and 3 using MuxCode mode and streams 4 and 5 using simple mode. Presumably, streams 1–3 have fixed bit rate and packet sizes whereas streams 4 and 5 are more variable. In fact, FlexMux would even allow streams 1–3 to use both MuxCode and simple modes, for example, if these streams needed to include occasional overhead information that is sent in addition to the regular packet schedule in extra packets. In other words, the *regular* packets for streams 1–3 could use Mux-Code packets, and *extra* packets could use simple mode packets, as indicated

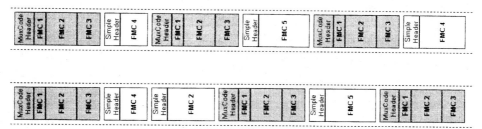

Fig. 7.8 Examples for combining both FlexMux modes in one FlexMux stream (top) and different modes for different ESs (bottom); ES in FMC = 2 uses both modes.

for FMC 2 in the bottom FlexMux stream in Figure 7.8. Note that the streams are exclusively identified by the FlexMux channels (FMCs) that are either obvious from the `index` value in simple mode (FMC = index) or derived from the template information in the `MuxCodeTableEntry`.

7.2.1 Timing of a FlexMux Stream

The complete syntax of a FlexMux packet (see Table 7.1) includes two more special channels, identified by `index` values 238 and 239. `Index = 239` indicates a FlexMux channel that is used to insert stuffing bytes where necessary. Stuffing may be needed to achieve a constant bit rate for a FlexMux stream, making delivery timing easier. However, a statistical multiplexing approach would push stuffing down the protocol stack as far as possible, so that bandwidth that is not used by one FlexMux stream could be reused by a peer stream that is multiplexed into the same delivery multiplex, say an MPEG-2 transport stream.

FlexMux channel 238 is dedicated to timing information. Timing information is required if the delivery timing of a FlexMux stream will be controllable. Two elements are carried in FMC 238 packets: the FlexMux Clock Reference (FCR) and the instant bit rate of the FlexMux stream. The FCR has nearly the same semantics as the Object Clock Reference (OCR) in a single SL-packetized stream. That is, it is a sample of an object time base at the point in time this FlexMux packet is being created. Regularly inserted FCR values, together with the current bit rate (`fmxRate` in Table 7.1), allow determining the delivery timing of the FlexMux stream. The `fmxRate` field gives the piecewise constant bit rate for the FlexMux stream until the next `fmxRate` occurs. The use of these timing fields is optional, as the underlying delivery mechanism might already provide delivery-timing signaling.

A `FlexMuxTimingDescriptor` (see Table 7.3) is provided to configure the timing-related elements used in FMC 238. The `FCR_ES_ID` assigns an `ES_ID` to this stream so that it can be referenced as a clock reference stream (see Chapter 3). `FCRResolution` and `FCRLength` provide the resolution (in ticks per second) and length of the `fmxClockReference` elements. `FmxRateLength` provides the length of the `FmxRate` elements.

Table 7.3 `FlexMuxTimingDescriptor` syntax

```
aligned(8) class FlexMuxTimingDescriptor {
    Cbit(16)  FCR_ES_ID;
    Cbit(32)  FCRResolution;
    Cbit(8)   FCRLength;
    Cbit(8)   FmxRateLength;
}
```

Control of the delivery timing is used mainly in application scenarios where a low end-to-end delay or precise buffer fullness control is desired. In fact, these are the two sides of the same medal: low end-to-end delay does require both small receiver-side buffers and an unjittered, precisely timed delivery.

The System Decoder Model (SDM; introduced in Chapter 3) assumes such unjittered delivery for each ES. However, it disregards the jitter that is necessarily introduced by multiplexing a set of streams. So, in order to model the system more realistically, SDM must be extended to cover the delivery layer on which MPEG-4 streams are transported. Logically, this part of the model resides in the DMIF instance for the given delivery layer.

SDM extension is done for the FlexMux in the following way: Each FlexMux channel, n, is attributed a FlexMux Buffer, FB_n, of a known size, as indicated in Figure 7.9. This association can be signaled using the `FlexMux-BufferDescriptor` (Table 7.4). The rate, Rbx, at which an FB_n is filled is deter-

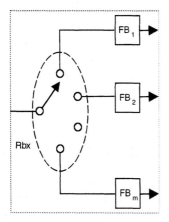

Fig. 7.9 FlexMux buffer model [MPEG4-1].

Table 7.4 `FlexMuxBufferDescriptor` syntax

```
aligned(8) class FlexMuxBufferDescriptor {
    Cbit(8)   flexMuxChannel;
    Cbit(24)  FB_BufferSize;
}
```

mined by the maximum bit rate of the associated ES, as indicated in the ES descriptor. As with the basic SDM compliance, it is the sender's task to ensure that the FlexMux buffers never overflow. Or, phrased positively, the sender knows how to interleave data coming from different streams, being sure they will arrive in time, relative to all the other streams.

In Section 7.4.3.5, it will be seen how this model is further extended to cover the whole MPEG-4 over MPEG-2 delivery stack.

Both the `FlexMuxBufferDescriptor` and the `FlexMuxTimingDescriptor` need to be conveyed *out-of-band*—that is, outside the actual FlexMux stream, and, as for the `MuxCodeTableEntries`, this will be specified in the MPEG-4 adaptation of each DMIF instance. A proposal for in-band transport of these descriptors has been made, as that would give the additional benefit of establishing the FlexMux stream as a self-supported exchange format for multiplexed MPEG-4 streams, including delivery timing, independent of any delivery layer.

A potential use for in-band descriptor delivery would be applications that need to convey the same preproduced presentations both on MPEG-2 and on IP, or even those that need to forward content arriving on one delivery stack to another one. That would be largely eased, as a necessary gateway could simply act as a media pump, retrieving data from one delivery instance and pumping it out again through the second delivery instance, driven by the time stamps embedded in the FlexMux stream and, of course, taking into account the information in the embedded descriptors.

So, the rationale for in-band transport of FlexMux descriptors is the ease of streaming MPEG-4 presentations across heterogeneous networks. A generic solution to ease the exchange of MPEG-4 presentations, not necessarily in a streaming manner, already exists and is discussed in Section 7.3.

7.3 MPEG-4 FILE FORMAT

Historically, prior to the MPEG-4 standard, MPEG has not had an explicit file format. MPEG-1 and MPEG-2 content typically is exchanged as files that represent, in a sense, a stream ready to be delivered. These files usually have embedded absolute time stamps, and the data often has been fragmented with a particular transport in mind (e.g., MPEG-2 Systems transport stream). These characteristics can make random access difficult and editing or reuse of the streams hard without decoding, demultiplexing, and then rebuilding the stream after editing. In MPEG-4, no single transport protocol is *preferred*, so fragmenting the data to suit the packet size of a specific protocol was not acceptable.

The MPEG committee sought a *life-cycle* file format—one in which the files could be used when capturing media, editing it, and combining it; when serving the media as a file download or as a stream; and when exchanging partial or complete presentations. This need for a life-cycle format is not met in many simple file format designs. For example, as noted previously, the design approach of MPEG-2, in which a stream is simply recorded to a file, makes editing hard. A more flexible design and approach were needed.

The design approach and principles embodied in the QuickTime file format from Apple [QUICKTIME] were closest to meeting these needs. After a requirements definition phase and a call for proposals, both design principles and details from QuickTime were adopted and refined to build the *MP4 file format*.

Perhaps the most unusual aspect of the MP4 design is the separation of the media data itself—the video data, for example—from data about that media data, often called *metadata*. Metadata might include timing information, the number of bytes occupied by a video frame, the location of the file in which it is stored, its position within that file, and so on. This timing and structural data is kept separate from the media data itself, in compact tables. So, for example, neither the length in bytes of a video frame nor its time stamps are stored adjacent to that frame, but in tables in the metadata part of the file. This is what is meant when the MP4 format is described as *nonframing*.

To further facilitate editing, the timing data is relative, not absolute. This means that time stamps do not have to be rewritten when, for example, a frame is inserted or deleted. Also, the interrelationships between streams are compactly represented in the metadata. In addition, media data is stored in its *natural* or base state, not preferring any one transport protocol or system: This means, for example, video frames are stored whole, not fragmented to any fixed size.

When the data must be streamed, special instructions for the protocol, optionally stored in the file, instruct streaming servers in the process of fragmenting and time stamping the data. These instructions are known as *hints*. As a result, the MP4 file format is, on first inspection, more complex than simple single-function file formats. However, as a format, MP4 is also more powerful than single-function formats and can be used for a wide variety of purposes.

With these principles in mind, this section looks first at how time is managed, then at how storage and space are managed in the context of the MP4 file format. Then the way MPEG-4 Systems concepts and structures are handled in the file format is explained. The special support for streaming is then detailed, and the MP4 section concludes with a printout of a small example file. This printout provides an example that can be reviewed as the format is described.

7.3.1 Temporal Structure: Tracks and Streams, Time and Durations

An MPEG-4 presentation is composed of a number of streams. In MP4, each stream is stored as a separate track in a file. A *track* in a file represents a timed series of media—successive video frames, for example. Each track has a *track identifier*, which maps to the ES_ID. A track is treated as a sequence of *samples,* which are AUs, in MPEG-4 terms—the smallest entities to which timing can be attributed (see Chapter 3). So, a sample in a video track is a coded frame or VOP; in an audio track, a compressed frame of audio (not a single uncompressed audio sample); in a BIFS track, a set of instructions for one-time instant; and so on.

For each sample, the file format records the time *difference* between its time stamp and the time stamp of the next sample—its *duration.* This relative timing makes the process of editing much easier, as sections of the timeline may be moved around without disturbing their internal structure. Indeed, the file format has a facility for expressing precisely this operation. A set of structures known as an *edit list* can provide an explicit *map* of how the timeline of a single track should be mapped into the timeline of the presentation. Edit lists allow for the insertion of *empty time* and the use (and reuse) of sections of the timeline. So, a repeated piece of media can be defined once and used repeatedly by reference.

MPEG-4 video follows earlier MPEG standards and uses the concept of bidirectionally predicted frames or VOPs, called B-frames or B-VOPs. To decode a B-frame, the decoder needs three frames: the closest non-B-frame that will be displayed before it (called here F[p]), the closest non-B-frame that will be displayed after it (called here F[s]), and the B frame being decoded itself (called here F(b)). The decoder must be given these frames not in the order they are displayed but in the order that they must be decoded: Frame F(s) is given to the decoder before the B-frame F(b). To support this, and to manage decoding buffers in some terminals, MPEG separates the concepts of decoding time and composition (display) time. In the MP4 file format, frames are stored always in decoding order, and by default the decoding and composition time stamps are the same. However, an optional table can provide a positive offset from the decoding time to the composition time, when they differ. The reordering for display purposes is managed by this composition time offset.

Each track has a time scale, which defines how fast its clock ticks. Using an appropriate time scale enables precise definition of timing. For example, usually the time scale of an audio track will be chosen to match its sampling rate, to enable sample-accurate timing (e.g., a time scale of 22,050 ticks per second for 22.05 kHz sampled audio). In video, a time scale of 30,000 ticks per second, with media sample duration of 1,001 ticks, exactly defines NTSC video (often, but incorrectly, referred to as 29.97 frames per second).

In many sequences of media, some of the samples (frames) can be decoded without knowing any other samples: In video, these are known as I-frames or I-VOPs. In general, such samples provide useful random access or synchronization points. The indices of these samples are stored in a table, for each track, to enable random access and editing.

Sometimes, when seeking, it can be useful to find extra random access points. The file format provides a facility called *shadow sync* samples. These samples provide an alternative random access point (I-frame or sync sample) for a sample that is not a random access point.

Streams stored in the same file are normally assumed to be synchronized together. However, the file format provides explicit facilities to indicate the synchronization relationships (clock reference relationships) between streams, if that is needed. This is done by means of *track references*, which are described in more detail below.

When a file is delivered as an MPEG-4 stream, the timing is conveyed by the MPEG4 SL, or by the transport layer being chosen (such as RTP on TCP/IP networks). Because MP4 files do not record the actual time stamps (as noted previously), the time stamps must be computed. This computation proceeds in three steps:

1. If the track has an edit list, this list is used to build an overall time map of the stream times from the implicit timeline of the sample times.

2. The decoding time stamp of each AU is then computed by adding the *durations* of the preceding samples in that track; this time is then mapped according to the edit list.

3. If the track has explicit composition time information, it is represented by the offset from the decoding time computed in step 2. The composition offset must be added to the decoding time, to obtain the composition time, and then again mapped through the edit list information (if any).

This process is shown in Figure 7.10. In Figure 7.10, a simple timeline of 10 samples is shown with an edit list. The edit list contains four edits: an empty edit, then two edits each containing three samples, and finally an edit containing two samples (marked empty, 3 sample edit, and 2 sample edit, respectively). The computation of the time stamps for the second sample involves (a) the duration of the empty edit; (b) the duration of the preceding sample in the edit, which gives the decoding time for this sample; and (c) the composition offset for this sample, which, added to the decoding time, gives the composition time.

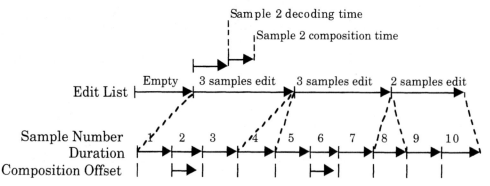

Fig. 7.10 Calculating time stamps for a sample.

7.3.2 Physical Structure: Atoms and Containers, Offsets and Pointers

This section introduces the concepts that underpin the physical structures in the MP4 file. All data stored in MP4 files is organized in atoms. An *atom* is a container that has an explicit *type* and *length*. So, for example, the synchronization points discussed previously can be found in a table stored in each track, in an atom of type stss; this atom contains a table listing, in ascending order, the numbers of the samples that are random access, or sync, points.

Atoms can contain other atoms. Typically, an atom either serves as a container or has fields that define particular concepts, although this is not a general rule. Any media data (samples) are stored in a media data atom (of type mdat).

All the metadata in a presentation is stored in the *movie* structure, the atom of type moov. Unused space in a file is designated with an atom of type free. The three atoms mdat, moov, and free are generally the only three atoms that occur at the top-level of a file. Of these, only the moov atom has internal structure. Inside the mdat atom, the media data itself is stored. Clearly the content of a free atom is irrelevant.

The moov atom contains a header atom and then a set of tracks. The movie header (mvhd) gives basic information about the presentation—the date it was last edited, for example, and its overall duration. The metadata for each track is stored in an atom of type trak.

Like the movie header, the trak atom has a track header, which also has creation and edit dates, a time scale for this track, and the overall duration of the track. This trak atom in turn contains a series of atoms, which define the media and how it may be handled.

There are four important structures in the trak atom. They are the edit list, the data reference information, the handler information, and the sample table.

The *edit list* is related to timing and was discussed previously. The *data reference information* defines whether the media data used in this track is stored in the same file or in other files (referenced by URL). This ability to refer to media data in many files is important when composing a project: Media from a library of files may be used by reference, and only copied when a self-contained file is needed. The *handler information* provides the first level of differentiation of the tracks; the handler type identifies the general kind of information in a track, such as visual, audio, BIFS, and so on. The *sample table* provides the detailed information about each sample—AU—within the track. It is structured as a set of atoms that are formatted as tables. These atoms define both the physical location of each sample (frame of video, for example) and its timing. To keep these tables compact, a variety of techniques is used. One that is used to compact the location and size information relies on the observation that several frames from the same track are often stored contiguously, even when data from various tracks is interleaved. This run of contiguous samples is called a *chunk*. The sample-to-chunk table provides the mapping from sample number to chunk. The position of each chunk is recorded as a chunk offset (using 32 or 64 bits), which is measured from the beginning of the file in which the chunk resides. The length, in bytes, of each sample also is recorded, in the sample size table. Therefore, by using the data reference from the track, the chunk offset, and the sizes of the preceding samples in the same chunk, it is possible to find the data file containing the sample, which may be a file referenced by URL from the MP4 file itself, the chunk within that file, the offset of the sample within the chunk, and the size of the sample itself.

Figure 7.11 shows an example of a file that starts with the media data. In this media data, the frames of video and frames of audio are stored. There are three video chunks, containing three, two, and two samples (frames or VOPs) each. There are two audio chunks, containing two samples each. In this example, there also is some unused data. This unused data might be a video frame that has been edited and now stored elsewhere in the file, or any other kind of data that is no longer needed.

Note that because of the way this works, there is no need for the samples to use all the data in a file: There may be unreferenced data between the chunks. This can be used to skip unwanted data or the headers of a foreign file format, for example. Nor do the chunks have to occur in any specific order; data can be collected from disparate places in the files as it is needed. This means that much editing can be done by reference, simply changing tables and pointers.

The final structural concept that needs to be introduced is that of a *sample description*. A terminal wishing to display an MPEG-4 presentation needs to know how to decode each track: not only whether the data is visual or audio, but also how to set up each decoder to *understand* the coded data (e.g., whether the audio is AAC or HVXC, for example). This means knowing its corresponding audio or visual object type and decoder specific information. As with all the metadata encountered so far, this information is kept in a table. There is a set of sample descriptions in each track, and each sample is implicitly tied to exactly one of them. The name of the sample description *names* the

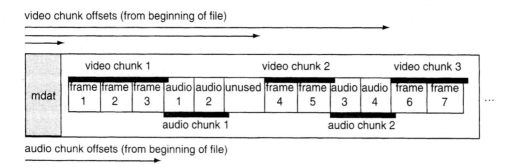

Fig. 7.11 Chunk offsets.

media format, and the parameters to that sample description parameterize the decoder as needed.

7.3.3 MPEG-4 Systems Concepts in MP4

So, how do the general concepts introduced previously map into the concepts introduced in MPEG-4 Systems? In order to facilitate editing, the descriptors for each stream are stored in the track that contains the stream. This makes it easy to move a stream from one file to another, for example. However, when a presentation is streamed, these descriptors are delivered to the terminal in an OD stream, or in the IOD (see Chapter 3). To avoid duplicating the descriptors, a special technique is used inside MP4 files: Neither the IOD nor the OD stream stored in the file contains ES descriptors (ESDs). Instead, a reference to the appropriate tracks is placed in the ODs.

A new MP4 structural concept is used here. Each track has a table of typed references to other tracks. The OD track has track references to those tracks for which it will carry setup information. In that OD track, at the time when an ESD is needed, a structure is stored instead which refers to the track from which the ESD is needed. In the process of streaming, the reference is replaced by the actual descriptor from the referenced track.

The initial *hook* into an MPEG-4 streamed presentation is the IOD. This descriptor, too, would normally contain the ESDs for the streams that must be initially decoded. As with the OD stream, these are not stored in the IOD; instead, again, the track ID of the initial tracks is stored in the file, and these structures must be replaced by the appropriate ESDs when the file is streamed.

The other MPEG-4 streams (and, indeed, streams that are integrated with MPEG-4, such as MPEG-7) follow these structures without needing any specific support from the file format. For example, BIFS streams—both animation and update—use these structures, including the concept of random access, or sync, samples. A random access point in BIFS performs a `Replace Scene` command.

7.3.4 MPEG-4 Track Types and Storage

In the MP4 file format, determining how to decode a track involves two levels of decision. First, what basic kind of track is it: audio, visual, and so on? And then, precisely how are the samples coded: what codec—MPEG-4 object type—and with what parameters and settings (decoder-specific configuration)?

The first decision is made by examining the *handler type;* along with the handler type, there is a type-specific header atom within the track structures for each kind of track. The code points shown in Table 7.5 are used for the handler type and the type-specific header atom.

The last line in Table 7.5 is included for completeness; clock reference streams are not usually stored within the file. Instead, if a particular stream-

Table 7.5 MP4 handler and header atom types

Stream type	Handler type	Header atom type
VisualStream	vide	vmhd
AudioStream	soun	smhd
ObjectDescriptorStream	odsm	nmhd
SceneDescriptionStream	sdsm	nmhd
MPEG7Stream	m7sm	nmhd
ObjectContentInfoStream	ocsm	nmhd
IPMP Stream	ipsm	nmhd
MPEG-J Stream	mjsm	nmhd
Hint Track	hint	hmhd
ClockReferenceStream	crsm	nmhd

ing transport requires a clock reference stream, it is constructed when a file is prepared for streaming, or by the streaming server itself.

7.3.5 Hinting

The penultimate line in Table 7.5 declares hint tracks. These are not MPEG-4 media tracks at all; instead, they are the key to understanding how efficient streaming may be done from a file in which the media data has been stored without regard to the needs of a particular streaming protocol.

Streaming media data involves sending media to a client system *just in time*, using small amounts of buffering. If packets get lost in transit, the client usually has to carry on; even if a retransmission is requested, there is no assurance it will arrive in time. The way in which the decoder can cope with this loss is critically dependent on the way the media data was originally placed into packets. For example, if a media sample (such as a video frame) is split arbitrarily into packets, any one of those packets being lost will usually make the rest of the frame, or perhaps the entire frame, unusable, and packets that did arrive will have to be discarded.

However, this problem may be alleviated by packetizing the media data carefully. If appropriate *boundary* points are found in the media, then the client often can recover all the data from the packets that successfully arrived. This can make a major difference to the user experience. For example, if a stream divides each video frame into six packets, and there is 10% loss, almost half the frames will lose at least one packet $(1 - 0.9^6)$. If all those frames are discarded, the user experience is of very degraded video (approximately 50%

missing). If, instead, the parts of those frames that arrive are displayed, the user experience is of the full frame rate, with 10% degradation.

The problem in finding these boundary points is twofold: Scanning the media data to find these boundaries is time consuming, and the way to do it requires the knowledge of the coding system being used—format parsing. If a media server must execute these rules as it streams, it will slow down; and if new rules are introduced, then every server must be updated.

Hint tracks provide a bridge between the media data and the packet data. In a media track, a sample is an audio frame—a media sample. In contrast, a hint sample tells the server how to make a packet or set of packets. Hint samples reference the media tracks they are packetizing and contain instructions, such as "insert these 4 header bytes into the packet," "copy 200 bytes from offset 43 in sample 267 of my media track," and so on. By following these instructions, the server can rapidly build valid packets without knowing either the coding system or the packetization rules for that coding system. Special authoring tools called *hinters* generate these hint tracks, after the editing process is complete. Hinters are specific both to the format of the media data and to the requirements of a specific transport protocol. The format-specific knowledge required has been moved from the server to the hinter. This means that the hinter is at liberty to compute the boundaries carefully and perform other optimizations, transformations, or corrections that are needed. So it would normally be the hinter that constructs any required clock reference stream, for example. And likewise the hinter would transform the OD track into an OD stream, replacing track references by the required ESDs from those tracks.

Of course, servers always have the freedom to know the parsing rules and packetize on the fly; this is why they are called hints—hints that the server is free to ignore.

Hint tracks are structured like media tracks. Just as media tracks must cope with differing coding systems, so hint tracks must manage multiple streaming protocols. Each server must find the hint tracks for the protocol it uses, and ignore the rest.

As a result, the same file contains both the original media—unpacketized, in its native format—and support for potentially multiple streaming formats. Indeed, by careful use of the *data references*, various MP4 files can be built that use the same media database in different ways. A hinted file can always be taken from a streaming server, the hint tracks deleted, and then edited, rehinted, and replaced on the server. This is part of the life-cycle management that MP4 offers.

The format of the hint samples is, of course, protocol-specific. Because this parallels the way that the format of media samples is codec-specific, the same structure is used. The sample description both names the protocol (and hence defines the format of the hints) and provides any necessary parameters to the server.

At the time of writing, the format of hint tracks for the RTP protocol is being standardized. There is also a proposed format of hint tracks for MPEG-2 Systems transport streams.

7.3.6 Atoms

At this point, all the basic MP4 concepts are in place: how time and space are handled, how MPEG-4 structures are mapped into the MP4 file format, and the basic concept of hinting. This section presents the atoms that make up an MP4 file.

All atoms start with a 4-byte `size`, and a 4-byte `atom type`. The type is usually four printable characters (this makes debugging and inspection of files much easier). The `size` is the length of the entire atom in bytes, including this 8-byte header. Two special values of the `size` field are relevant: 0 means that this atom extends to the end of the file, and 1 means that a 64-bit `size` field immediately follows the `type` field. This allows for an occasional very large atom.

Many atoms are defined to have a `version` and `flags` field, of 1 and 3 bytes respectively, following the `type`. These are called *full atoms* in the specification. A few atoms have more than one version, and use the `version` field to indicate which variant is present. The `flags` field is used by some atoms for purposes specific to that atom.

7.3.6.1 Top-Level Atoms   Three kinds of atoms occur at the top-level of MP4 files (not contained in any other atom):

1. **moov:** The movie atom `moov` is the container for all the metadata in the file. A movie atom is mandatory in an MP4 file; it can be anywhere in the top-level sequence of atoms. The `moov` atom is a pure container atom; it has no fields of its own, and it contains other atoms.

2. **mdat:** If there is media data in this file, it is stored in an `mdat` (media data) atom.

3. **free** or **skip:** If there is spare space, that is in atoms of type `free` or `skip`.

7.3.6.2 Inside the Movie Atom   A movie atom `moov` contains these atoms:

☞ A movie header atom `mvhd`

☞ Optionally, an IOD atom `iods`

☞ A track atom `trak` for each track in the presentation

The movie header has bookkeeping fields for the creation and modification time of the movie itself. It also records the duration of the presentation (the duration of its longest track, considering edits), and defines the time scale (in ticks per second) for the movie and track duration values. Finally, it has a `helper` field that suggests the next free track identifier. The track atom is a pure container atom; there may be any number of them, one for each track in the movie.

7.3.6.3 Inside a Track Atom   A track atom `trak` contains these atoms:

☞ A track header atom `tkhd`

☞ Optionally, an edit-list atom `edts`

☞ Optionally, a track reference atom `tref`

☞ A media atom `mdia`

The media atom in turn contains:

☞ A handler atom `hdlr`

☞ A media information atom `minf`, which contains:

✗ A header atom specific to the kind of media in the track

✗ A data reference atom `dref`, with a set of data references

✗ The sample table `stbl` for the track, with the sample timing and location information

Like the movie atom, the track atom also has a header, the track header `tkhd`. This header contains the track identifier, the track's duration, and creation and modification stamps for this track.

If the track has links to other tracks—for example, from hint tracks to media, or from OD tracks to the ES tracks they manage—then a track reference atom `tref` is in the track. This track reference atom contains a set of atoms, whose atom type identifies the kind of reference and whose content is a simple list of track IDs. For example, a synchronization dependency is indicated by a reference of type `sync`, an OD dependency by a reference of type `mpod`.

Likewise, any edit list for the track is in the track atom, after the header, in an atom of type `edts`. Each edit maps a section of the media timeline into the track timeline. Each edit in the list contains a media start time and duration; the edits, placed end to end, form the track timeline. A media time of −1 indicates the insertion of *empty time* for the indicated duration.

Finally, the track atom holds a container for the media declarations, the media atom `mdia`. Inside the media atom, the type of track is declared by means of a handler declaration in a `hdlr` atom, and then the media information container atom, `minf`, provides data on the media itself.

The media information atom, of type `minf`, has three main constituent atoms. The first is a header specific to the kind of media: a visual media header, for example. (The complete list of media headers is included in Table 7.5.) The second is the set of data references (URL file locations) in a `dref` atom for the actual media samples used in this track. And the third is the sample table atom `stbl` containing all the timing data and structural data for the media.

7.3.6.4 Inside the Sample Table A sample table atom `stbl` contains these atoms:

- ☞ A sample description atom `stsd`
- ☞ A sample size atom `stsz`
- ☞ A sample-to-chunk atom `stsc`
- ☞ A chunk offset atom `stco` (with 32-bit offsets) or `co64` (with 64-bit offsets)
- ☞ A time-to-sample atom, `stts`, providing the sample durations
- ☞ An optional composition offset atom, `ctts`, providing composition time offsets
- ☞ A sync sample atom `stss`
- ☞ Optionally, a shadow sync sample atom `stsh`

The sample table contains several atoms. They fall into two groups: those that define layout (size and position) and those that define time and other characteristics.

The sample description table `stsd` has been referred to before; for media tracks, it contains the MPEG-4 ESDs corresponding to that track, and for hint tracks, the name and parameters of the protocol.

The size of each sample is declared in the sample size atom `stsz`. This atom has a count of the number of samples in the track. If the samples are of constant size in bytes, then a single field provides that size. If the size varies, this field is not used, and the atom contains a vector with the sizes, one per sample.

As noted previously, the position information is compacted by using the notion of chunks. Within an MP4 file, all the samples within one chunk are stored contiguously, and they use the same sample description. The sample-to-chunk table `stsc` compactly codes the map from sample number to chunk number and sample description number. For each chunk, the number of samples in that chunk (and the sample description number used) is recorded. The declarations for consecutive chunks using the same values are compacted together.

Then, a chunk offset table `stco` or `co64` records the offset from the beginning of the containing file of each chunk. Given this data, and the size of the preceding samples in the same chunk (which are contiguously stored), a sample may be located. Chunk offsets are in bytes, and may be 32 or 64 bits, depending on the atom type.

Samples are stored in the track in decoding order. The decoding time stamp of a sample is determined by adding the duration of all preceding samples (in the applicable edits). The time-to-sample table `stts` gives these durations. This table is run-length coded; each entry specifies a duration and a count of the consecutive samples having that duration.

If the samples have different decoding and composition (display) times, then the composition time table `ctts` gives the positive offsets from decoding to composition time. This table is run-length coded, just like the decoding time table.

Finally, the sync table defines which samples are sync points (I-frames in video), by listing their sample numbers in order. The shadow sync table stsh provides alternative samples (AUs) which are random access points, for samples that are not (random access points).

7.3.7 Random Access

Seeking is accomplished primarily by using the atoms contained in the sample table atom. If an edit list is present, it must also be consulted. To seek a given track to a time T, where T is in the time scale of the movie header atom, the following operations have to be performed:

1. First, locate the sample nearest that time. If the track contains an edit list, determine which edit contains the time T by iterating over the edits. The start time of the edit in the movie time scale must then be subtracted from the time T to generate T', the duration into the edit in the movie time scale. T' is next converted to the time scale of the track's media to generate T''. Finally, the time in the media scale to use is calculated by adding the media start time of the edit to T''. The time-to-sample atom for a track indicates what times are associated with which sample for that track. This atom is used to find the first sample prior to the given time.

2. Then, locate the nearest preceding random access point (I-frame or sync sample). This step requires consulting two atoms. The sync sample table indicates which samples are, in fact, random access points. Locate the sync sample that is at or prior to the sample number obtained in step 1. The absence of the sync sample table indicates that all samples are synchronization points, and makes this problem easy. The shadow sync atom gives the opportunity for a content author to provide samples that are not delivered in the normal course of delivery, but which can be inserted to provide additional random access points. This improves random access without affecting bit rate during normal delivery. This atom maps samples that are not random access points to alternative samples that are. This table must also be consulted, if present, to find the first shadow sync sample prior to the sample in question. Having consulted the sync sample table and the shadow sync table, it is necessary to seek whichever resultant sync sample is at or prior to the sample found in step 1—the sample that should be displayed, that is at or prior to the seek time.

3. At this point, the sample that will be used for random access is known; the sample-to-chunk table can thus be used to determine in which chunk this sample is located.

4. Knowing which chunk contains the sample in question, the chunk offset atom has to be used to figure out where that chunk begins. Starting from

this offset, the information contained in the sample-to-chunk atom and the sample size atom has to be used to figure out where within this chunk the sample in question is located. This is finally the desired information.

7.3.8 An MP4 Example

This section prints out a tiny example file (with comments), which can be compared and cross-checked with the explanations given in previous sections. This file contains only two tracks (video and audio). Some fields have not been printed, to shorten and simplify the example.

```
moov          the movie atom, containing all the meta-data
  mvhd        the movie header introduces the presentation
    CREATION-TIME        Jan 9th 2001 5:23pm
    MODIFICATION-TIME    Jan 9th 2001 5:23pm
    TIME-SCALE           600          in ticks per second
    DURATION             2425         the units are the time-scale
    NEXT-TRACK-ID        3            a suggested unused track identifier
  iods
    ...   an initial object descriptor
  trak
    tkhd
      CREATION-TIME        Jan 9th 2001 5:23pm
      MOD-TIME             Jan 9th 2001 5:23pm
      TRACK-ID             1
      DURATION             2406    the units are the movie time-scale
    edts
    elst
      ENTRY-COUNT    1
      EDITS        duration 2406, offset 0
          the whole timeline, simply mapped
    mdia
    mdhd
      CREATION-TIME        Jan 9th 2001 5:23pm
      MOD-TIME             Jan 9th 2001 5:23pm
      TIME-SCALE           600          each track has its own scale
      DURATION             2425
    hdlr
      TYPE                 vide
      NAME                 Video Media Handler
    minf
      vmhd
      ...          the type-specific media header
      dinf         data references tell us where the media data is
      dref
          ENTRY-COUNT    1
          REFS           a flag indicates the data is in this file
      stbl           the sample table gives timing and location info
      stsd           sample descriptions contain ES Descriptors
          ENTRY-COUNT    1
          DESCRIPTIONS   1 video ES descriptor is stored here
```

```
    stts                the decoding times are stored here as durations
        ENTRY-COUNT     1
        TIMETOSAMPLE    97 samples each 25 ticks in duration
    stss                the sample number of I-frames are stored here
        ENTRY-COUNT     19
        SYNCSAMPLES     1, 4, 17, 32, ... are key frames
    stsc                the mapping from sample number to chunk number
        ENTRY-COUNT     2
        SAMPLETOCHUNK   chunk 1 contains 1 sample, sample description 1
                all other chunks contain 2 samples, sample description 1
    stsz
        DEFSAMPLESIZE   0    the default sample size is unused here
        ENTRY-COUNT     97
        SAMPLESIZES     3276, 3280, 3280, 48...
    stco                the file offset of each chunk
        ENTRY-COUNT     17
        CHUNKOFFSETS    6944, 23332, 27240, 30816, 35226, 40922...
trak            another track, containing audio
  tkhd
  CREATION-TIME      Jan 9th 2001 5:23pm
  MOD-TIME           Jan 9th 2001 5:23pm
  TRACK-ID           2
  DURATION           2194
  edts
  elst
      ENTRY-COUNT    1
      EDITS          duration 2194, offset 0
                     the whole timeline simply mapped

  mdia
  mdhd
      CREATION-TIME  Jan 9th 2001 5:23pm
      MOD-TIME       Jan 9th 2001 5:23pm
      TIME-SCALE     44100    note the change of time-scale
      DURATION       161280
  hdlr
      TYPE           soun
      NAME           Sound Media Handler
  minf
      smhd
      ...
      dinf
      dref
          ENTRY-COUNT    1
          REFS           a flag indicates the data is in this file
      stbl
      stsd
          ENTRY-COUNT    1
          DESCRIPTIONS   1 sound ES descriptor
      stts
          ENTRY-COUNT    1
          TIMETOSAMPLE   140 samples of duration 1152
      stsc
          ENTRY-COUNT    2
          SAMPLETOCHUNK  chunk 1 and later contains 25 samples
```

```
                             final chunk has 23 samples
        stsz
            DEFSAMPLESIZE   104          constant size samples
            ENTRY-COUNT     140
        stco
            ENTRY-COUNT     16
            CHUNKOFFSETS    2888, 4864, 4968, 6840, 23672, 25703...
mdat
    DATA...         in here are the audio and video frames
```

7.3.9 Summary of MP4

The MP4 format meets the needs defined by the MPEG committee through a series of techniques that are not common in file format design. These techniques include the following:

1. Nonframing metadata principle: Metadata is not physically adjacent to the media data it describes.
2. Handling timing information by means of relative numbers (durations and composition offsets) rather than absolute numbers.
3. Being able to store media data spread over several files.
4. Locating media data by means of data offsets and length information; the offsets are not necessarily in time order.
5. Handling streaming protocols through optional hint tracks.

As a result, the MP4 format is a life-cycle format. The same format—even the same file—can be used for capture, editing, local presentation, download, and streaming, and indeed may be recycled around from the streaming server back to editing. As one of the proponents of an alternative technology put it, "It is possible to design a format which is better than MP4 for any one specific purpose; but unlike such single-function formats, MP4 is acceptably good at every function." This aspect of MP4, together with its flexibility and extensibility, has made it the basis for the Motion JPEG 2000 file format as well [MJP2], and (at the time of writing) it is being explored for further MPEG standards as well.

7.4 TRANSPORTING MPEG-4 OVER MPEG-2

The MPEG-2 Systems standard [MPEG2-1] defines two methods of multiplexing data from a set of ESs into one serial bitstream. The resulting streams are called *transport stream (TS)* and *program stream (PS)*. The TS was designed mainly for broadcast-type applications and is at the heart of most digital TV standards specified worldwide, including Europe's DVB [DVB], America's ATSC [ATSC], and Japan's ARIB [ARIB] standards. The program stream is used for storage-type applications, for example on DVDs.

Due to the widespread availability of the MPEG-2 infrastructure, notably in broadcasting environments, it was important to make sure that MPEG-4 data can be easily embedded there. Indeed, the MPEG-2 design team was farsighted enough to leave room for bitstream syntax extensions that could be exploited for an *MPEG-4 over MPEG-2* amendment to MPEG-2 Systems, which has subsequently become part of the second edition of the MPEG-2 Systems standard [MPEG2-1].

Transport of MPEG-4 data over MPEG-2 Systems comes in two flavors:

☞ **Stream-based:** MPEG-4 audio or MPEG-4 video streams are simply added into an MPEG-2 multiplex. This makes it possible to design services that use MPEG-4 compression instead of MPEG-2 compression for these streams.

☞ **Scene-based:** Complete MPEG-4 presentations including scene description and ODs are carried over MPEG-2 Systems. This option is most interesting to design services that offer rich extensions to the underlying basic MPEG-2 service.

Both options are discussed in detail in the following sections, after a short introduction to the MPEG-2 Systems standard [MPEG2-1].

7.4.1 Brief Introduction to MPEG-2 Systems

In order to more easily understand the section about MPEG-4 transport over MPEG-2, a certain familiarity with MPEG-2 Systems is necessary. Thus, a few basics on MPEG-2 Systems are introduced here, without making an attempt to replace an MPEG-2 Systems tutorial text.

7.4.1.1 Packetized Elementary Stream   Each ES in MPEG-2 Systems is first packetized and wrapped in a structure called a Packetized Elementary Stream (PES). The resulting PESs are then interleaved into the TS or PS. The PES is an intermediate layer to facilitate the conversion of content between TSs and PSs in order to use the content both on broadcast networks and on the DVD or other storage media.

Elementary streams are sliced into *PES packets,*[3] each one starting with a PES packet header. The PES packet header is a lengthy structure with many options. In this context, the most important features of the PES packet header are that it

☞ Conveys the length of the PES packet,

☞ Identifies the content with a `stream_id`, and

---

3. MPEG-2 Systems [MPEG2-1] does not give many rules how to do this; thus it is largely application dependent.

☞ Conveys presentation and decoding time stamps (PTS and DTS) for the content.

7.4.1.2 Program Stream   A PS consists of a sequence of PES packets, which are grouped into packs by the interspersed pack headers (see Figure 7.12). In other words, within a PS, the PES packets are the primary multiplexing entity for the different streams. The stream_id serves both as a stream number and as an identifier for the type of content found in a specific PES packet. Therefore, within the range of available stream_ids, for example, the number of video streams is limited to 16 and the number of audio streams to a maximum of 32.

Furthermore, a number of special streams can be identified with stream_id. Most importantly in this context is the Program Stream Map (PSM). Looking at the PSM syntax in Table 7.6, it can be noted, without going into full detail, that there are first a number of elements related to the entire program, including a set of custom descriptors. Then, in the second loop, the PSM contains further information about the streams tied together in this PS. For each stream, an association of the stream_id with a stream_type is made. The stream_type identifies precisely the coding format of that stream; in the case of audio and video streams, for example, it is possible to distinguish between MPEG-1 and MPEG-2 coding. Again, custom descriptors for each stream can be added; this fact has been exploited for defining the carriage of MPEG-4 over MPEG-2.

7.4.1.3 Transport Stream   The encapsulation of PES packets in an MPEG-2 TS is conceptually quite different from the PS, as can be easily seen in Figure 7.13. A TS consists of fixed-length 188-byte TS packets. Because the TS packet header can have variable length, the payload takes what is left over from the total 188 bytes. So, PES packets are not left intact but are chopped in pieces of maximum 184 bytes, as 4 bytes is the shortest TS packet header.

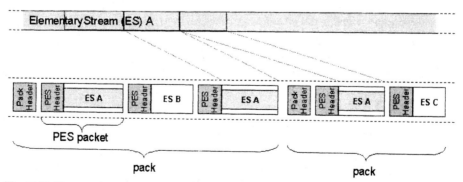

Fig. 7.12 Encapsulating PES in a program stream.

Table 7.6 *Program Stream Map* syntax

Syntax	Number of Bits
program_stream_map() {	
packet_start_code_prefix	24
map_stream_id	8
program_stream_map_length	16
current_next_indicator	1
Reserved	2
program_stream_map_version	5
Reserved	7
marker_bit	1
program_stream_info_length	16
for (i = 0; i < N; i++) {	
descriptor()	
}	
elementary_stream_map_length	16
for (i = 0; i < N1; i++) {	
stream_type	8
elementary_stream_id	8
elementary_stream_info_length	16
for (i = 0; i < N2; i++) {	
descriptor()	
}	
}	
CRC_32	32
}	

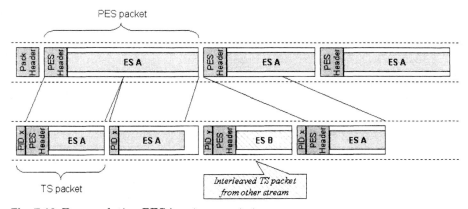

Fig. 7.13 Encapsulating PES in a transport stream.

Multiplexing of payloads from different streams is governed by the Packet IDentifier (PID) in the TS packet header. All PES packets originating from a single ES will end up in TS packets with one specific PID value.

With 13-bit PID values, the TS has the capability to convey a potentially large number of independent programs, each consisting of a number of ESs. In order to understand the content within a MPEG-2 transport stream, a hierarchical structure of Program Specific Information (PSI) exists. The Program Association Table (PAT) conveys the list of programs and points to a Program Map Table (PMT) for each single program, which conveys information about the ESs constituting that program. For the purpose of this chapter, it is important to note that the PMT is conceptually the same as the PSM within the PS and, so, it includes some global information about the program followed by a list of associations of PIDs to stream_type elements for each stream with the same functionality as introduced previously.

The PMT is split for transport in a series of one or more program map sections, as shown in Table 7.7. The same, so-called *MPEG-2 section syntax* is used to convey PAT and other PSI tables in chunks that are well suited for regular insertion into the TS. Sections allow building enumerated lists of elements (using section_number) that can be updated with the version indicated

Table 7.7 *Program Map Table* syntax

Syntax	Number of bits
TS_program_map_section() {	
table_id	8
section_syntax_indicator	1
'0'	1
Reserved	2
section_length	12
program_number	16
Reserved	2
version_number	5
current_next_indicator	1
section_number	8
last_section_number	8
Reserved	3
PCR_PID	13
Reserved	4
program_info_length	12
for (i = 0; i < N; i++) {	
descriptor()	
}	
for (i = 0; i < N1; i++) {	
stream_type	8
Reserved	3
elementary_PID	13
Reserved	4
ES_info_length	12
for (i = 0; i < N2; i++) {	
descriptor()	
}	
}	
CRC_32	32
}	

in version_number. Furthermore, it can be indicated whether a certain version of a section is current or will only become valid soon, using the current_next_indicator.

7.4.1.4 Timing Model

To some extent, MPEG-4 has inherited its *timing model* from MPEG-2 Systems. The major difference is that MPEG-4 does not define a transport multiplex at all and therefore the MPEG-4 System Decoder Model (see Chapter 3) is by design only a subset of the System Target Decoder (STD) defined in MPEG-2 Systems. The transport stream STD (T-STD) defines the delivery schedule, $t(i)$, for each byte, i, of the transport stream and its buffering after demultiplexing into the different streams. The three-stage buffering (TB_n, MB_n, EB_n), as indicated in Figure 7.14, reflects

- ☞ **TB_n:** Slow-down buffering in order to adapt the high TS bit rate to the lower ES bit rate;
- ☞ **MB_n:** Multiplex buffering in order to cope with the multiplex jitter; and
- ☞ **EB_n:** ES buffering according to the needs of the audio and video compression tools.

For audio and systems streams that typically have much lower bit rates than the video streams, no distinction is made between a multiplex and an ES buffer. Both functions are incorporated in B_n. Without explaining Figure 7.14 in detail, it can be noted that the rates (RX, Rbx values) at which data is transferred between buffers as well as the sizes of the buffers TB_n, MB_n, and EB_n are precisely defined so that it can be guaranteed that AUs ($A_n[j]$) can be

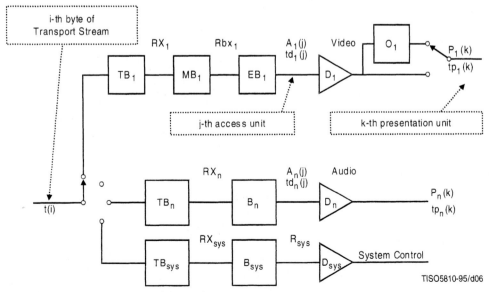

Fig. 7.14 MPEG-2 T-STD [MPEG2-1].

decoded at the proper time $td_n(j)$. Note that the distinction of three buffers in the model does not mean that an implementation actually has to have these three buffers.

The MPEG-4 System Decoder Model introduced in Chapter 3 comprises only the decoding buffers for the ESs—that is, the functionality of the EB_n of the T-STB—as [MPEG4-1] does not define any multiplexing.

MPEG-4 allows presentations composed of streams that do not share a single time base, whereas MPEG-2 requires the streams within each program to be slaved to a single time base. Therefore, the MPEG-4 concept of an object time base conveyed through OCR only roughly maps to MPEG-2 Program Clock References (PCR), in the case of TSs, or System Clock References (SCR), in the case of PSs [MPEG2-1]. PCR or SCR are used to determine the delivery timing of a multiplexed stream, either one program out of a TS or of a PS. Knowing the nominal frequency of the system clock, the delivery schedule for each byte in the multiplex can be determined through evaluation of the SCR/PCR stamps and additional information about the instantaneous bit rate of the stream. The client terminal may implement a Phase-Locked Loop (PLL) circuit fed with the PCR/SCR stamps, getting a precise recovery of the clock speed.

In MPEG-4, the OCR could be used to determine the delivery timing for a single SL-packetized stream carrying those OCRs. Although this is not particularly useful for a single stream, OCRs in context of a FlexMux stream (then called FCRs) do indeed serve that purpose, as explained earlier in this chapter. The main use of OCRs in the context of individual SL-packetized streams is initializing the object timeline and tracking that timeline for drift versus the terminals' system clock through long-time observation.

Equipped with this overview of the MPEG-2 Systems functionality, it is now easy to understand how MPEG-4 content is carried in both MPEG-2 TS and PS.

7.4.2 Transport of MPEG-4 Elementary Streams over MPEG-2 Systems

The transport over MPEG-2 Systems of isolated MPEG-4 ESs is limited to streams that do not rely on MPEG-4 Systems functionality—this means audio and video streams. With this mode, it is possible to design applications that use MPEG-4 compression instead of MPEG-2 compression, for example, because of the higher coding efficiency or the new features, like arbitrary shape coding. Moreover, even some limited scene composition functionality can be realized. For example, the MPEG-4 video syntax [MPEG4-2] specifies how a number of video objects are overlaid without need for a BIFS scene description. However, interactivity (by using buttons or objects that can be moved around on the screen) cannot be realized in this mode; or, more precisely, the standardized MPEG-4 Systems tools for this purpose are not available. Of course, an application might choose some non-MPEG-4 tools to perform the composition of video objects conveyed in both MPEG-4 streams and MPEG-2 streams.

In order to enable the transport of MPEG-4 media streams over MPEG-2 Systems, the `stream_id` values for video and audio streams within the PES header have been redefined to allow MPEG-1, MPEG-2, and MPEG-4 compressed audio and video streams. The fact that media streams are actually MPEG-4 coded is expressed by newly defined `stream_type` values carried in the program stream map (within a PS) or PMT (within a TS), respectively. Furthermore, an `MPEG-4_video_descriptor` or `MPEG-4_audio_descriptor` in the second descriptor loop of PSM/PMT conveys profile and level information about the stream—that is, about the precise coding tools used. In line with standard MPEG-2 practice, but contrary to the usual MPEG-4 practice, there are no data alignment rules for chopping an MPEG-4 ES into PES packets. Remember that in MPEG-4 there is always an alignment between AUs (e.g., audio frames or VOPs in video) and the SL packets that carry them. MPEG-4 audio ESs are wrapped in low-overhead MPEG-4 Audio Transport Multiplex (LATM), defined within MPEG-4 Audio [MPEG4-3], before being packetized in PES. LATM is necessary to interleave stream header information (sampling frequency, bit rate, etc.) with the actual audio ESs. When MPEG-4 Systems is used, such information is carried as part of the ODs, so LATM is not needed.

Concerning synchronization, the MPEG-4 stream is treated similarly to any MPEG-2 stream in the same program—that is, encoders are supposed to directly express the decoding and presentation time for AUs with respect to the MPEG-2 Systems timeline. Those times are then coded into DTS and PTS values in the PES header for the MPEG-4 stream.

The STD constraints are relaxed for MPEG-4 streams. Usually in MPEG-2, transmission of one coded video frame cannot take longer than 1 second. MPEG-4 object-based coding, on the other hand, might be used, for example, to create scenes with high-resolution background objects that only need to be refreshed occasionally. Therefore, the maximum transmission duration for a coded frame is increased to 10 seconds.

7.4.3 Transport of MPEG-4 Scenes over MPEG-2 Systems

The previously described way for adding MPEG-4 streams to an MPEG-2 program does not encompass the standardized MPEG-4 scene composition capabilities (see Chapter 4), which are especially attractive to create highly interactive content that transforms an otherwise noninteractive broadcast into a completely new experience. A number of proprietary solutions for interactive applications exist, for example, based on the OpenTV [OPENTV] or CanalPlus (MediaHighway) [MEDIAH] platforms, but MPEG-4 proposes a standardized way to enable them, adding features that have not been available in any of the proprietary solutions. The changes made to the MPEG-2 Systems standard to pave the way for such applications are detailed in the following. First, the options for stream transport are discussed, followed by the additional descriptive elements and the changes to the STD.

7.4.3.1 Conveying SL-Packetized Streams in PES As soon as complete MPEG-4 presentations, including a scene description, are to be conveyed, the default formats to embed into the delivery layer are SL-packetized streams, in order to maintain the timing relations established between the streams. SL-packetized streams are mapped to PES such that one SL packet constitutes the payload of one PES packet. A new `stream_id` value in the PES header has been assigned to identify a PES stream carrying an SL-packetized stream. Of course, all SL packets coming from a single stream are to be mapped to a single PES stream, as identified by the PID in case of transport stream. As said earlier, in the PS `stream_id` serves both as a stream type indication and as a stream number. Therefore a PS can only convey one single SL-packetized stream, as only one `stream_id` value has been set aside.[4] How to overcome this restriction is discussed below.

A number of constraints exist for the timing information in SL-packetized streams that are to be embedded in PES. OCR time stamps must have a resolution of 90 kHz/k, with k being an integer bigger than or equal to 1. Other time stamps (CTS, DTS) must have the same resolution or that resolution divided by an integer number. Because all time stamp fields (CTS, DTS, and OCR) are modulo counters—that is, they wrap at some point in time—it is furthermore mandated that all time stamps must have a length such that they wrap after the same time, hence removing potential time stamp ambiguity.

The timing of the SL-packetized streams is slaved to the MPEG-2 TS/PS timing in the following way: All SL-packetized streams must contain OCR time stamps. Whenever an SL packet header has an OCR, its enclosing PES packet must have a presentation time stamp. The value of that PTS expresses the same time on the MPEG-2 timeline as the OCR value on the object timeline. Hence, both timelines are tightly coupled, despite the fact that their respective time stamps may have different resolutions and offsets.

7.4.3.2 Conveying FlexMux Streams in PES Especially for the case of PSs, which do not allow carrying more than one SL-packetized stream, as well as for the case of a set of low-bit-rate streams, it is necessary to introduce a second multiplex layer into the PES so that a number of MPEG-4 ESs can share a single PES. The FlexMux introduced earlier in this chapter is used for that purpose. A PES packet simply contains an integer number of FlexMux packets. A new `stream_id` value indicates that this PES conveys a FlexMux stream.

The timing constraints mentioned in the previous section apply here as well, with the difference that the single SL-packetized streams do not carry OCRs. Instead the FlexMux stream conveys FCR stamps, as described in the FlexMux section. The link between the MPEG-4 timeline, expressed by FCR,

---

4. This is because there are only a small number of unassigned values of the `stream_id` syntax element left over.

and the MPEG-2 timeline is made again by requiring that a PES packet conveying an FCR in its payload must also carry a PTS. Both values express the same point in time in the two different time lines.

7.4.3.3 Conveying FlexMux Streams in Sections

The MPEG-2 section syntax defined mostly for PSI descriptors (see Table 7.7) is suitable as well for the carriage of certain MPEG-4 streams. Both scene description and OD streams may be rather static in simple MPEG-4 presentations—that is, initially a scene description and the associated ODs are sent, with no further update. In order to enable random access into such a presentation, the scene description and ODs have to be regularly repeated. A section can be repeated (`version_number` and `section_number` remain unchanged) so that this syntax can be used as a vehicle for such a feature. Therefore, a FlexMux stream carrying only scene description and OD stream(s) may be encapsulated in a TS using the section syntax rather than the PES syntax. As before, a section must carry an integer number of FlexMux packets.

It should be noted, however, that this vehicle for the repetition of information necessary for random access is somewhat redundant with a repetition that is directly done at the SL. Repetition of BIFS or OD AUs may be performed with SL tools as well, using the AU sequence number syntax element as an indicator for repetition (see Chapter 3).

7.4.3.4 Describing a Program with MPEG-4 Components

In addition to the rules defining how to embed SL-packetized streams in a TS or PS, some modifications to the MPEG-2 descriptive framework are needed. The MPEG-4 streams need to logically become part of an MPEG-2 program. As mentioned before, MPEG-2 programs and their elements are described through the PSM or PMT in PS and TS, respectively. Here, new `stream_types` have been defined for MPEG-4 content that simply state that the stream is an "SL-packetized stream or FlexMux stream carried in PES packets" or an "SL-packetized stream or FlexMux stream carried in ISO/IEC14496_sections." All other information about the content can be retrieved using the new descriptors specified in [MPEG2-1] and introduced below and in the MPEG-4 ODs.

Assigning `ES_IDs` to Streams In Chapter 3, it was mentioned that it is a task for the application to resolve an MPEG-4 stream identifier, the `ES_ID`, to an actual stream location. For a PES stream that conveys an SL-packetized stream, the `SL_descriptor` (see Table 7.8) achieves this task by associating the stream's `ES_ID` to the delivery layer identifiers—that is, `stream_id` (PS) or PID (TS), respectively. For this purpose, the `SL_descriptor` is included in the descriptor loop for that stream—that is, the second descriptor loop in Table 7.6 (PS) and Table 7.7 (TS), respectively.

For a FlexMux stream encapsulated in PES or in sections, a whole set of `ES_IDs` needs to be associated with that stream. In that case, a `FMC_descriptor` (see Table 7.9) is included in the descriptor loop of PSM/PMT

Table 7.8 `SL_descriptor` syntax

Syntax	Number of bits
SL_descriptor () {	
descriptor_tag	8
descriptor_length	8
ES_ID	16
}	

Table 7.9 `FMC_descriptor` syntax

Syntax	Number of bits
FMC_descriptor () {	
descriptor_tag	8
descriptor_length	8
for (i=0; i<descriptor_length; i += 3) {	
ES_ID	16
FlexMuxChannel	8
}	
}	

instead of the `SL_descriptor`. The `FMC_descriptor` achieves a two-stage association, associating both each stream (i.e., `ES_ID`) within the FlexMux to a FlexMux channel and the whole FlexMux stream to a PID or `stream_id`, respectively.

The MuxCode mode of FlexMux (as introduced earlier in this chapter) allows even better compression of the multiplex overhead for a FlexMux stream. Even though this may not be relevant for MPEG-4 transport over MPEG-2, a mechanism to carry the MuxCode definitions, the `MuxCode-TableEntries`, has been added for completeness. If a FlexMux stream uses MuxCode, the corresponding PSM/PMT entry needs to carry both an `FMC_descriptor` and a `MuxCode_descriptor` (see Table 7.10).

It is conceivable that not only MPEG-4 streams but also some MPEG-2 audiovisual streams that are part of the program are referenced in the BIFS scene description, as they may have to be composed together. In that case they,

Table 7.10 `MuxCode_descriptor` syntax

Syntax	Number of bits
Muxcode_descriptor () {	
descriptor_tag	8
descriptor_length	8
for (i = 0; i < N; i++) {	
MuxCodeTableEntry ()	
}	
}	

too, need an ES_ID. For that purpose, an External_ES_ID_descriptor (see Table 7.11) exists that is put in the descriptor loop in the PSM/PMT of that MPEG-2 program element.

Accessing the Content With the preceding descriptors, each MPEG-4 stream and even the MPEG-2 streams, if desired, are tagged with an ES_ID value. Now a way to access the MPEG-4 content has to be defined. As described in Chapter 3, the IOD is key to this. So, the means to convey the IOD and its semantics in the context of MPEG-2 Systems are needed. Hence, the IOD is encapsulated in an IOD_descriptor (see Table 7.12), which is conveyed in the first descriptor loop of the PSM/PMT that carries information about the program as a whole.

Multiple MPEG-4 presentations can be associated with an MPEG-2 program. Therefore, each IOD_descriptor obtains a label that is unique either within a given program or within the TS or PS at large.

7.4.3.5 MPEG-2 STD Extensions

It has already been mentioned that MPEG-4 and MPEG-2 time stamps are strictly coupled in this specification [MPEG2-1]. Therefore, MPEG-4 streams also benefit from the strict delivery timing defined in MPEG-2. To reflect this, the STD diagram presented in Figure 7.14 is extended by two branches, for SL-packetized streams and FlexMux streams, respectively.

Figure 7.15 shows just the additional branches for FlexMux streams and SL-packetized streams in case of the T-STD, omitting the MPEG-2 branches. Additional descriptors exist to dimension the FlexMux buffers FB_{nm} as well as the multiplex buffers MB_n. For each FlexMux or SL-packetized stream, a

Table 7.11 External_ES_ID_descriptor syntax

Syntax	Number of bits
External_ES_ID_descriptor () {	
descriptor_tag	8
descriptor_length	8
External_ES_ID	16
}	

Table 7.12 IOD_descriptor syntax

Syntax	Number of bits
IOD_descriptor () {	
descriptor_tag	8
descriptor_length	8
Scope_of_IOD_label	8
IOD_label	8
InitialObjectDescriptor ()	
}	

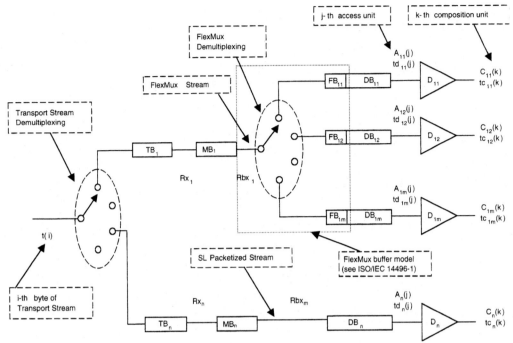

Fig. 7.15 MPEG-2 STD extensions for MPEG-4 transport [MPEG2-1].

branch 'n' consisting of a TB_m and an MB_m is inserted in the T-STD and a multiplex buffer descriptor is needed, defining the size of the associated MB_n and the rate Rx_n at which data leaks out of the TB_n buffer. For FlexMux streams, the FlexMux buffer descriptor defines just the size of the associated FB_{nm}. The leak rate Rbx_m out of MB_n varies corresponding to the maximum bit rate for each elementary stream within the FlexMux stream as signaled in its ESD descriptor. In that sense, FB_{nm} are not separate buffers but simply extend the decoding buffers DB_{nm} in size, as needed to perform proper demultiplexing.

7.5 TRANSPORTING MPEG-4 OVER IP

A number of proprietary formats exist to convey streaming, real-time data over IP. Each *owner* of such a format may make use of MPEG-4 stream types in their context. Of course, such an approach cannot be considered as the basis for an open standardization process. On the other hand, unlike the transport of MPEG-4 content over MPEG-2, a specification for the transport of MPEG-4 over IP could not be accomplished within the MPEG community alone, because specifications in the realm of IP are typically dealt with by the IETF, even though their requests for comments (RFCs) do not have the formal status of an International Standard. By 1996, the IETF had created RTP [RFC1889] in an attempt to facilitate the transport of real-time data over IP. Transport of MPEG-4 streams makes use of RTP as well. A short overview of RTP and the protocols associated with it to control a streaming media application is given first in the following, before detailing how MPEG-4 data is mapped to this framework.

7.5.1 Brief Introduction to Streaming over IP

Conveying real-time data over IP requires a data packetization protocol that allows synchronization, RTP, and protocols for discovering and controlling the real-time presentations. The Session Description Protocol (SDP) [RFC2327] allows describing the content of the streaming presentation, much like ODs (MPEG-4) or program-specific information (MPEG-2). The RTSP [RFC2326] provides presentation control, like START, PAUSE, STOP.

Starting with RTP, in brief, it associates packets of data with time stamps and allows establishing the notion of a continuous, timed data stream from a source to a receiver. RTP can be carried on top of the User Datagram Protocol (UDP) [RFC0768] and IP [RFC0791], even though RTP/TCP/IP and RTP/HTTP/TCP/IP variants are also used when the guaranteed delivery of the Transmission Control Protocol (TCP) [RFC0793] is desired or when the only way through a firewall is by using the Web's HTTP [RFC2068].

The layers underneath RTP determine the delivery constraints imposed on the data, as RTP, despite its name, does not at all guarantee the real-time delivery of any data. Use of TCP implies that usually all data packets sent will also be received. However, the end-to-end delay will vary vastly, as lost packets are resent until they have been successfully received. UDP, on the other hand, does not have any such built-in mechanism. A lost packet is lost, and it is up to the application to deal with this. This behavior allows continuous sending of packets at a regular pace, which is essential for real-time data.

RTP itself consists of two major building blocks:

☞ A packet syntax for the data transport, and
☞ The RTCP.

RTCP is used to monitor the delivery performance and, hence, the quality of service (QoS) of a connection. For example, receiver reports allow feeding back to the sender the number of lost packets in a certain time frame. The sending application can then attempt to adjust its transmission parameters accordingly. The abstracted DMIF interface to communicate this functionality between delivery layer and application was discussed at the beginning of this chapter.

The RTP packet has a header (see Figure 7.16) with the following major items:

☞ **Marker bit** (M), indicating a special condition in the data stream, such as the last packet of a video frame;
☞ **Payload type** (PT), indicating the codec used for this stream;

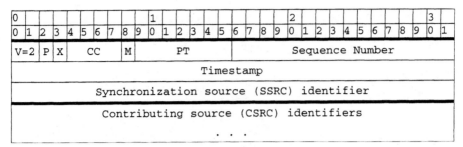

Fig. 7.16 RTP packet header [RFC1889].

☞ **Sequence number**, to reorder packets delivered out of order;

☞ **Time stamp**, indicating the sampling time of the first data in the packet; and

☞ **Unique identifier**, of the source of this data stream (SSRC).

The remaining header items are not important in the context of this brief overview.

Usually, the mapping of a certain type of content like MPEG-4 to RTP is defined in a so-called *RTP payload format*, an accompanying RFC that specifies how that content is to be packetized in RTP packets. Payload formats exist for a large number of non-MPEG-4 formats, ranging from H.263 [RFC2429] to MPEG-2 video streams [RFC2250].

The SDP is a textual protocol that helps to discover both which streams form part of a streaming media session and what payload formats are applied to them. A short example description is given in Table 7.13 to explain the essence of the SDP syntax.

SDP may contain textual information about the session, including its title (s= line). Moreover, it identifies the transport on which the session is being streamed (c= line); in the case of Table 7.13, the transport is on the Internet, using IP Version 4 and address 224.2.17.12. Packets are supposed to be discarded after a time-to-live of 127 router hops. Starting and end time of the presentation (t= line) can be given as Network Time Protocol (NTP) [RFC1305] values. Then the media streams are identified (m= lines) as being

Table 7.13 Excerpt of an SDP description for an audiovisual session

```
s=A sample session
c=IN IP4 224.2.17.12/127
t=2873397496 2873404696
m=audio 49230 RTP/AVP 96 97
m=video 51372 RTP/AVP 31
m=application 32416 udp wb
a=rtpmap:96 L8/8000
a=rtpmap:97 L16/8000
```

audio, video, and a white board (a graphical scratch pad) application. The port numbers on the originating device are given as the second parameter in the m= lines, followed by the protocol used for that stream on top of the transport. This protocol is RTP, with a specific audiovisual profile (AVP) [RFC1890] in the case of audio and video and plain UDP in the case of the white board application. The numbers at the end of the m= lines correspond to the payload type (PT) values found in the RTP packets for those streams. According to the AVP, the value 31 indicates the ITU-T H.261 video codec. The audio stream allows explaining two more features. First, all PT values 96 and larger require the dynamic association to a concrete payload type. For the audio stream in the example of Table 7.13 there are two numbers, indicating that both these formats can be used (and switched) within the stream. The mapping of PT values to concrete audio codecs is then done through the a=rtpmap: lines. In this case, it is trivial 8-bit and 16-bit PCM with a sampling rate of 8 kHz.

SDP descriptions of a session can be communicated in multiple ways to a potential customer for that session. For example, the SDP description can be sent by email or can form part of a Web page. Due to a specific MIME [RFC2045] type, an application can automatically detect and process the SDP data. The most common use of SDP, however, is as part of the RTSP.

RTSP is another textual protocol that is, in spirit, similar to the HTTP. Instead of controlling the delivery of Web pages, RTSP controls the delivery of real-time streams. In today's Web browsers, a URL of format rtsp://<source>/<file> invokes an RTSP handler that is part of a media player (e.g., Real or QuickTime). The handler will communicate with a peer entity on the server through a number of commands, most importantly DESCRIBE, SETUP, PLAY, PAUSE, and TEARDOWN.

When a client encounters an RTSP URL, it sends a DESCRIBE to the server to retrieve the SDP description of the session. In on-demand scenarios, the client then requests the SETUP of the transmission, indicating its reception port numbers. PLAY and PAUSE are used to control the synchronous playback of all streams belonging to the session identified by the RTSP URL. Optionally, a starting time different from the beginning of the presentation can be requested. Finally, TEARDOWN terminates the session.

In the case of a multicast session, individual recipients usually cannot apply stream control. In that case, RTSP can still be used to retrieve the description of a presentation, but specific protocols have to be followed in order to join the multicast presentation—for example, the Internet Group Management Protocol (IGMP) [RFC2236].

7.5.2 Transport of Elementary Streams over IP

After this brief overview of IETF specifications for streaming over IP, it may have become clear that the first task to perform within the MPEG-4 context in view of its transport over IP was to define payload formats for MPEG-4 content. In principle, anyone can approach IETF to propose an RTP payload format for any of the different coding tools that are housed under the umbrella of the MPEG-4 specification. Therefore, a framework specification [MPEG4-8] that does not define any RTP payload formats has been developed by MPEG and submitted to IETF for consideration as a best practice guide. MPEG-4 Part 8 [MPEG4-8] proposes some rules that payload formats for individual MPEG-4 ES types (video, audio/AAC, etc.) should follow to achieve uniformity with a generic payload format for the transport of SL-packetized streams (see below), in order to maximize content reusability.

As an example of RTP payload formats for individual stream types, RFC 3016 [RFC3016] specifies the payload format for both MPEG-4 video and audio ESs. RFC 3016 has been proposed by parties that consider using just those stream types in the context of third-generation wireless applications. Unlike PES encapsulation of ESs, RFC 3016 requires a specific alignment of headers (e.g., VOP headers and video-packet headers) with RTP packet boundaries. These constraints are necessary to exploit the built-in error-resilience features in MPEG-4—for example, the video packet syntax.

The MPEG-4 video syntax defines video packets that are independently decodable portions of a VOP (see Chapter 8). Such a video packet or an integer number of them should be mapped to a single RTP packet. This confines the consequences of the loss of an RTP packet to that single video packet(s) (see Figure 7.17, top), whereas a single RTP packet loss with ill-sized and ill-aligned video packets could render both the two adjacent video packets useless (see Figure 7.17, bottom).

Note that the MPEG-4 video payload format does not require any addi-

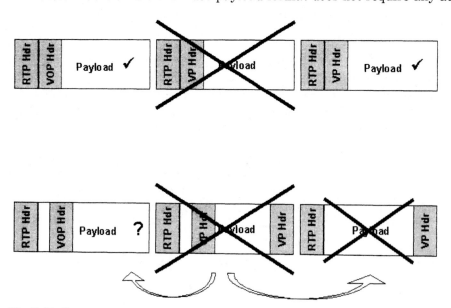

Fig. 7.17 Consequences of a lost RTP packet, with good (top) and inadequate (bottom) video packet alignment.

tional header information in the RTP packet to exhibit good error resilience. Payload formats for other video codecs, such as ITU-T H.261 [RFC2032] and H.263 [RFC2429], have compensated for the weaker built-in error resilience by additional side information in the RTP header. However, this approach comes at the price that video streams have to be understood by the entity doing the RTP encapsulation, whereas in the MPEG-4 case no knowledge of the coding format is necessary at that stage.

Similar considerations for intelligent packetization, often termed application-level fragmentation (ALF; first introduced by [CLTE90]), apply for the MPEG-4 audio payloads, where the unit to be preserved is called audioMux-Element. As for the case of MPEG-2 carriage, the LATM format is used to interleave audio header information and the actual stream.

7.5.3 Transport of SL-Packetized Streams over IP

Under the assumption that an intelligent, ALF-type packetization can be done indiscriminately for all MPEG-4 audiovisual codecs without the need for delivery layer modifications, it appears beneficial to define one single payload format that is applicable for all MPEG-4 stream types. In that case, the SL-packetized stream is being used as the entity to map to RTP. SL packets are deemed to be ALF-type packets whose payload need not be further inspected prior to RTP packetization.

All information necessary to control RTP packetization options is conceptually derived from the SL packet headers. *Conceptually* means here that a sender implementation does not have to create an SL-packetized stream first, before performing the mapping to RTP. As results from the RTP description, there is a lot of overlap between RTP and SL: Both supply time stamps and sequence numbers as well as means to indicate AU (frame) boundaries. Therefore, the RTP payload format for SL-packetized streams proposes to remove some of that redundancy by defining a mapping of some SL packet header fields to their RTP header equivalents.

Such a generic payload format, developed within MPEG, has been proposed to IETF [GENT01]. With the caveat that it was work not yet fully finalized as of October 2001, it is presented here. The generic payload format has two basic modes of operation:

- ☞ **Single SL packet mode:** One SL packet is mapped to one RTP packet.
- ☞ **Multiple SL packet mode:** An integer number of SL packets are concatenated into one RTP packet.

Conceptually, the single SL packet mode would be sufficient, as the AVT [AVT] group in IETF that is maintaining RTP and its payload formats encourages payload format specifications wherein each RTP packet contains exactly one independently decodable ALF packet of data. However, concatenation of RTP packets is needed for the case of small payloads, as for speech codecs, which are in the order of 20 bytes. Otherwise, the packetization overhead induced by IP, UDP, and RTP headers (roughly 40 bytes) becomes overwhelming. According to the AVT group, concatenation of multiple such RTP packets into one underlying UDP packet would then be done by other mechanisms (as discussed, for example, in [TCRTP]), which are independent of the actual RTP payload format. Because this work on a generic RTP multiplexing specification had not concluded by the time [GENT01] was proposed to IETF, the single SL packet mode was complemented with the multiple SL mode.

In order to efficiently represent multiple concatenated SL packets, each SL packet is disassembled into three sections within the RTP packet (see Figure 7.18):

☞ Mapped SL Packet Headers (MSLH section)

☞ Remaining SL Packet Headers (RSLH section)

☞ Concatenated SL Packet Payloads (SLPP section)

Within these three sections, data for the SL packets is conveyed in the original order of the SL packets.

The composition time stamp (CTS) of the first or single SL packet header becomes the RTP time stamp and is removed from the SL header. The marker bit is set to 1 if all SL packets in the RTP packet convey a complete (or the end of an) AU.

The following, partially optional elements constitute the mapped SL packet header (MSLH) for each SL packet:

☞ **PayloadSize,** indicating the size of the SL payload in bytes

☞ **Index/IndexDelta,** indicating the (differential) packet sequence number for this SL packet

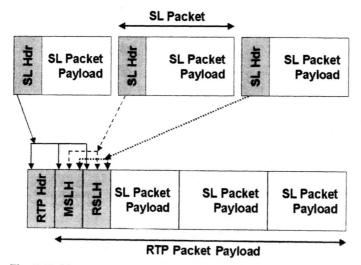

Fig. 7.18 Mapping of multiple SL packets into an RTP packet.

☞ **CTSFlag,** indicating whether this SL packet has a CTS

☞ **CTSDelta,** which is included only if CTSFlag=1, indicating the CTS offset w.r.t. the previous SL packet

☞ **DTSFlag,** indicating whether this SL packet has a DTS

☞ **DTSDelta,** which is included only if DTSFlag=1, indicating the DTS offset w.r.t. the previous SL packet

For the single SL packet mode, only the sequence number and the time stamp difference to obtain the decoding time stamp (DTS; if needed) are coded in the MSLH. For the multiple SL mode, first the overall length of the MSLH section is conveyed (see Figure 7.19). Then, for each SL packet, an MSLH consisting of the previously listed values follows. For the first SL packet, its packet sequence number is coded as Index, and the differential IndexDelta to the previous SL packet sequence number is coded for subsequent SL packets. Furthermore, the DTSs and CTSs (except for the first SL packet) are coded differentially to the RTP time stamp—which, in turn, corresponds to the first CTS.

Note that SL packet sequence numbers could have been mapped to their RTP header counterparts; however, by not doing so, they can be applied for interleaving of the SL packets across multiple RTP packets. In order to avoid redundancy in an IP-only application and when using the single SL packet mode, the SL header can be configured to omit the sequence numbers right away in favor of RTP sequence numbers, without loosing any functionality.

As explained in Chapter 3, all elements of the SL header, including time stamps, are configurable in size in order to adapt to a wide range of applications. These elements of the MSLH are, of course, not defined through the SLConfigDescriptor; however, to optimize compression here as well, the size of these elements can be signaled using SDP. This approach has been chosen because the information in the MSLH is deemed to be useful no matter whether an application is MPEG-4 Systems aware or not.

All other elements of the SL packet header(s), if present, form the remaining SL header, RSLH. If an application is being tailored toward RTP delivery only, it is conceivable that SL packet headers will be configured so that the RSLH remain void, as those optional SL packet header elements like degradationPriority or idleFlag are not strictly needed for many applications.

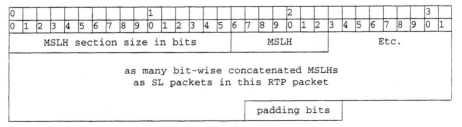

Fig. 7.19 MSLH section structure for the multiple SL mode.

The SL packet disassembly approach requires an extra effort from the applications that need to (re)generate SL-packetized streams rather than directly generating RTP packets from ESs, and vice versa. However, it makes it easy for MPEG-4 Systems–unaware receivers to ignore the SL packet headers and directly retrieve the ESs. For example, a compatibility mode is defined with RFC 3016 in which MSLH and RSLH are empty and the RTP packets for both payload formats are the same.

7.5.4 FlexMux Streams over IP and Timing Models

As for the transport of MPEG-4 over MPEG-2, it is conceivable to map an entire FlexMux stream onto an RTP session in order to save resources, including communication sockets between server and client devices as well as bandwidth due to less packetization overhead. Furthermore, it would become possible to signal a delivery timing for the FlexMux stream, much like the way it is defined in the RTP payload format for MPEG-2 TSs [RFC2038].

This approach, although having been considered for a long time, has not yet attained support from the IETF, as it would violate the timing model of RTP. That timing model assumes that only the presentation (or sampling) time of media is really relevant for a synchronized presentation. Given the jittery behavior of today's Internet, control of packet delivery through specific DTS is deemed meaningless. However, this implicitly assumes that the receiver takes care of this jitter by buffering incoming streams for a relatively long time, in the order of a few seconds, before starting a presentation. This way, sudden variations of delivery bit rate can be compensated for. Of course, intelligent implementations should be able to estimate the delivery jitter in order to still keep the prebuffering to a minimum.

The heuristic delivery-timing model of RTP is in strong contrast to the typically deterministic timing models favored by MPEG. The MPEG-2 transport stream is an example for which, at least theoretically, the delivery time for every single byte can be computed. Trying to preserve this feature on current, best-effort Internet is certainly meaningless in itself. However, in some application scenarios, such streams might be delivered over IP for immediate or later playback on a (strictly timed) MPEG-2 device. Then, the timing needs to be maintained over IP.

This same argument can be applied to MPEG-4—and the FlexMux stream could be the vehicle to have one single, strictly timed MPEG-4 stream for transport over IP. A first draft specification [M7066] that technically is quite similar to the FlexMux over MPEG-2 specification discussed earlier in this chapter has been presented to IETF. [M7066] does not resolve the jitter issues on today's IP networks; therefore, an already available alternative to such a specification may be the near real-time transmission of an MPEG-4 file that has all the appropriate hinting information to allow subsequent strictly timed real-time delivery on another delivery layer, like MPEG-2. If AVT concludes its work on Tunneling Multiplexed Compressed RTP [TCRTP], the need for a FlexMux over IP specification because of communication channel resource considerations might also go away.

Concluding, it can be observed that there are multiple options for the transport of MPEG-4 over IP that are being pursued with varying intensity, depending on the business relevance of the underlying application scenarios. Such scenarios exist for individual MPEG-4 media codecs—for example, in the domain of next-generation wireless services or DSL services—and, therefore, the specification of the transport of such media streams over RTP is more advanced than the generic MPEG-4 over IP specification or even the FlexMux over IP specification.

7.6 SUMMARY

This chapter has explored in detail the general approach and some specific mechanisms for the delivery of MPEG-4 presentations. First, the delivery layer abstraction of the DMIF was introduced, which specifies the interface functionalities that are required from any transport protocol for the delivery of MPEG-4 presentations. Two practical applications of the DMIF abstraction, or DMIF instances, were subsequently introduced: the specification of the transport of MPEG-4 presentations on MPEG-2 Systems and on IP. The characteristics of each of these delivery layer adaptations were highlighted. Furthermore, the delivery-related functionalities specified within the MPEG-4 Systems standard (namely, the FlexMux tool and the MPEG-4 file format) were introduced. The file format is another DMIF instance, yet a very special one, as it constitutes the generic exchange format for MPEG-4 presentations. MPEG-4 presentations can be adapted for delivery over many transport protocols as needed for future MPEG-4 applications.

7.7 REFERENCES

[ARIB] Japanese Association of Radio Industries and Businesses Home Page, *www.arib.or.jp*

[ATMF] ATM Forum Specifications, *www.atmforum.com*

[ATSC] Advanced Television Systems Committee Specifications, *www.atsc.org*

[AVT] Audio-Video Transport Group of IETF Home Page, *www.ietf.org/html.charters/ avt-charter.html*

[CLTE90] Clark, D., and D. L. Tennenhouse. "Architectural Considerations for a New Generation of Protocols." *SIGCOMM Symposium on Communications Architectures and Protocols,* pp. 200–208. Philadelphia: IEEE. September 1990.

[DVB] Digital Video Broadcast Specifications, *www.dvb.org*

[GENT01] Gentric, P., O. Avaro, A. Basso, S. Casner, R. Civanlar, C. Herpel, Z. Lifshitz, Y. K. Lim, C. Perkins, and J. van der Meer. *RTP Payload Format for MPEG-4*

Streams (work in progress). July 2001. *www.ietf.org / internet-drafts / draft-ietf-avt-mpeg4-multiSL-01.txt*

[H223] ITU-T Recommendation H.223. *Multiplexing Protocol for Low Bit Rate Multimedia Communication.* March 1998.

[H245] ITU-T Recommendation H.245. *Control Protocol for Multimedia Communication.* July 2001.

[M7066] Curet, D., et al. *RTP Payload Format for FlexMux Streams.* Doc. ISO/MPEG M7066, Singapore MPEG Meeting, March 2001.

[MEDIAH] MediaHighway. Information about platform, *www.canalplus-technologies.com /*

[MJP2] ISO/IEC 15444-3. *Coding of Still Pictures: Motion JPEG 2000 Specification.* 2001.

[MPEG2-1] ISO/IEC 13818-1:2000. *Information Technology: Generic Coding of Moving Pictures and Associated Audio Information—Part 1: Systems,* 2000.

[MPEG2-2] ISO/IEC 13818-2:2000. *Information Technology: Generic Coding of Moving Pictures and Associated Audio Information—Part 2: Video,* 2000.

[MPEG2-3] ISO/IEC 13818-3:1998. *Information Technology: Generic Coding of Moving Pictures and Associated Audio Information—Part 3: Audio,* 1998.

[MPEG2-6] ISO/IEC 13818-6:1998. *Information Technology: Generic Coding of Moving Pictures and Associated Audio Information—Part 6: Extensions for Digital Storage Media Command and Control,* 1998.

[MPEG4-1] ISO/IEC 14496-1:2001. *Information Technology: Coding of Audio-Visual Objects—Part 1: Systems,* 2001.

[MPEG4-2] ISO/IEC 14496-2:2001. *Information Technology: Coding of Audio-Visual Objects—Part 2: Visual,* 2001.

[MPEG4-3] ISO/IEC 14496-3:2001. *Information Technology: Coding of Audio-Visual Objects—Part 3: Audio,* 2001.

[MPEG4-5] ISO/IEC 14496-5:2001. *Information Technology: Coding of Audio-Visual Objects—Part 5: Reference Software,* 2001.

[MPEG4-6] ISO/IEC 14496-6:2000. *Information Technology: Coding of Audio-Visual Objects—Part 6: Delivery Multimedia Integration Framework (DMIF),* 2000.

[MPEG4-8] FCD of ISO/IEC 14496-6. *Information Technology: Coding of Audio-Visual Objects—Part 8: Carriage of ISO / IEC 14496 Contents over IP Networks.* Doc. ISO/MPEG N4282, July 2001.

[N4319] MPEG Requirements. *MPEG-4 Requirements.* Doc. ISO/MPEG N4319, Sydney MPEG Meeting, July 2001.

[OPENTV] OpenTV Home Page, *www.opentv.com /*

[QUICKTIME] Apple Computer. *QuickTime File Format Specification.* May 1996. *www.apple.com / quicktime / resources / qtfileformat.pdf*

[RFC0768] Postel, J. *User Datagram Protocol.* August 1980. *www.ietf.org / rfc / rfc0768.txt*

[RFC0791] Postel, J. *Internet Protocol.* September 1981. *www.ietf.org / rfc / rfc0791.txt*

[RFC0793] Postel, J. *Transmission Control Protocol.* September 1981. *www.ietf.org / rfc / rfc0793.txt*

[RFC0959] Postel, J., and J. Reynolds. *File Transfer Protocol*. October 1985. *www.ietf.org / rfc / rfc0959.txt*

[RFC1305] Mills, D. *Network Time Protocol (Version 3) Specification and Implementation*. March 1992. *www.ietf.org / rfc / rfc1305.txt*

[RFC1889] Schulzrinne, H., S. Casner, R. Frederick, and V. Jacobson. *RTP: A Transport Protocol for Real Time Applications*. *www.ietf.org / rfc / rfc1889.txt*

[RFC1890] Schulzrinne, H. *RTP Profile for Audio and Video Conferences with Minimal Control*. January 1996. *www.ietf.org / rfc / rfc1890.txt*

[RFC2032] Turletti, T., and C. Huitema. *RTP Payload Format for H.261 Video Streams*. October 1996. *www.ietf.org / rfc / rfc2032.txt*

[RFC2038] Hoffmann, D., G. Fernando, and V. Goyal. *RTP Payload Format for MPEG1 / MPEG2 Video*. October 1996. *www.ietf.org / rfc / rfc2038.txt*

[RFC2045] RFC 2045, 2046 (and others). *Multipurpose Internet Mail Extensions (MIME)*. *www.ietf.org / rfc / rfc2045.txt*

[RFC2068] Fielding, R., J. Gettys, J. Mogul, H. Nielsen, and T. Berners-Lee. *Hypertext Transfer Protocol—HTTP / 1.1*. January 1997. *www.ietf.org / rfc / rfc2068.txt*

[RFC2236] Fenner, W. *Internet Group Management Protocol, Version 2*. November 1997. *www.ietf.org / rfc / rfc2236.txt*

[RFC2250] Hoffman, D., G. Fernando, V. Goyal, and M. Civanlar. *RTP Payload Format for MPEG-1 / MPEG-2 Video*. January 1998. *www.ietf.org / rfc / rfc2250.txt*

[RFC2326] Schulzrinne, H., A. Rao, and R. Lanphier. *Real Time Streaming Protocol (RTSP)*. April 1998. *www.ietf.org / rfc / rfc2326.txt*

[RFC2327] Handley, M., and V. Jacobson. SDP: Session Description Protocol. April 1998. *www.ietf.org / rfc / rfc2327.txt*

[RFC2429] Bormann, C., L. Cline, G. Deisher, T. Gardos, C. Maciocco, D. Newell, J. Ott, G. Sullivan, S. Wenger, and C. Zhu. *RTP Payload Format for the 1998 Version of ITU-T Rec. H.263 Video (H.263+)*. October 1998. *www.ietf.org / rfc / rfc2429.txt*

[RFC3016] Y. Kikuchi, T. Nomura, S. Fukunaga, Y. Matsui, and H. Kimata. *RTP Payload Format for MPEG-4 Audio / Visual Streams*. November 2000. *www.ietf.org / rfc / rfc3016.txt*

[TCRTP] Thompson, B., T. Koren, & D. Wing. *Tunneling Multiplexed Compressed RTP (TCRTP)*. July 2001. *www.ietf.org / internet-drafts / draft-ietf-avt-tcrtp-04.txt*

CHAPTER 8

Natural Video Coding

by Michael Wollborn, Iole Moccagatta, and Ulrich Benzler

Keywords: object-based video coding, motion compensated hybrid video coding, shape coding, error resilience, scalability, texture coding

*T*he coding of natural video content, including still images, has been under investigation in research and industry for a long time. But what does it mean, to *code* video content? In early times, the only goal was to represent a video scene, be it moving or not, in a format allowing transmission or storage. And, because storage capacity and bandwidth were and still are restricted resources, the main and ultimate objective of video coding for a long time was *compression*. In this sense, the coding of video content aimed mainly at representing it using as few data as possible, at a certain visual quality, or vice versa, to maximize the visual quality at a given bit rate.

This coding paradigm is reflected in the early video coding standards, such as H.261 [H261] and MPEG-1 Video [MPEG1-2]. Of course, they also included application-driven functionalities such as random access, scalability, and others; however, the idea of improving compression was always the driving force behind all the activities. At the same time, several research activities [e.g., Kunt85, MuHO89, SaTM95] investigated and proposed a different approach than the well-established *coding of rectangular frames*—namely, the so-called *object-based* or *region-based* video coding. Although one objective of these proposals was higher compression, the principal ideas opened a much broader scope of possible functionalities and applications. Influenced by both the well-established video-compression standards and the new ideas of coding (not the rectangular window to the visual world, but its content), the MPEG-4 natural video coding activities began in 1995 with the aim of combining the benefits of both approaches: high compression over a broad range of bit rates and innovative content-based functionalities.

This chapter gives an overview of the MPEG-4 natural video coding tools [MPEG4-2]. However, it does not describe the complete standard in all details—that would require a book of its own. Therefore, the emphasis is on the general concepts and on those tools that are newer, compared to already existing technology and standards. For more details on specific topics, the reader is referred to the references.

8.1 GENERAL OVERVIEW

In this section, an overview of the MPEG-4 natural video coding tools is given. The functionalities and respective application scenarios are described briefly, followed by a short description of the principal MPEG-4 video coding approach.

8.1.1 Functionalities and Application Scenarios

The MPEG-4 Visual standard [MPEG4-2] comprises tools for coding natural visual sources—that is, sequences of images (video)—as well as still images (visual texture). Whereas the video coding is based on the well-known motion-compensated hybrid DCT coding scheme, the visual texture coding (VTC) is based on wavelet transform and zero-tree coding.

In addition to previous standards that addressed only coding efficiency, the MPEG-4 video coding tools provide several additional functionalities (see also Chapter 1): coding of arbitrarily shaped objects; efficient compression of video sequences and still images over a wide range of bit rates; spatial, temporal, and quality scalability; and robust transmission in error-prone environments.

These advanced features greatly aid content creators in generating rich video and thus multimedia content. In particular, the ability to *code objects of arbitrary shape and size* goes beyond the scope of all previous video and image coding standards. It allows content creators to overcome the limits of frame-based content coding and opens the doors for a multitude of innovative applications in the future.

The ability to efficiently compress still textures and videos over a wide range of bit rates using the MPEG-4 tool set greatly simplifies the content creator's job by providing one-stop shopping opportunity for nearly lossless to lossy compression while delivering the best quality for the bit rate required by the application.

Scalable video coding is of interest for various applications. For mobile communication, a hierarchical transmission system can be used, in which the base layer signal is especially well protected against errors in the transmission channel by applying robust forward-error correction to the base layer bitstream. For video server applications, scalable coding allows users to access the video at different bit rates, without the need for multiple coding and storage of separate video bitstreams in different resolutions or qualities.

To enable the transmission over wireless and wired networks with severe error conditions, the MPEG-4 Visual standard has adopted *error-resilience coding* tools. Depending on the transmission channel, these tools can replace or complement channel coding to improve quality of service (QoS) [M4575].

Finally, the MPEG-4 VTC provides an efficient tool for the compression of still images, based on the wavelet transform and zero-tree coding. It can be combined with the MPEG-4 shape coding tools and thus provides, with this respect, the same features as the video coding tools. The combination of VTC with the natural video coding tools allows the coding of scenes comprising both moving and still images by means of a single standard.

8.1.2 Basic Principles

One important principle in ISO/MPEG standardization activities is that the encoder is not standardized—only the bitstream it produces and the procedure for decoding it (i.e., the decoder). The reason is that for interoperability, it is not necessary to standardize the encoder. As a positive consequence, there are several degrees of freedom and optimization left for building the encoder. This gives one the possibility of improving the performance of the coding algorithm by introducing new (non-normative) techniques while the standard is already fixed, and it also leaves room for competition in building better encoders. This principle holds true also for MPEG-4 natural video coding. Therefore, in the remainder of this chapter, mostly decoding procedures will be described. Only when it is essential for the understanding of the technical issues will encoding algorithms be explained.

As mentioned in the introduction, MPEG-4 natural video coding tools comprise a whole tool-set for different applications and use cases. Moreover, these tools can be combined to code different sources of natural video. Not all of the tools are required for every application, and in some cases, it may be preferable if a decoder not support certain tools (e.g., for reasons of complexity or computational load). On the other hand, for reasons of interoperability, a decoder should at least support a clearly defined set of tools. Thus, for the MPEG standards, *profiles* are defined. For each profile, the tools to be supported are specified, and each decoder that claims to be compliant to a certain profile must support all of these tools. Further, for each profile different *levels* are specified. A level sets certain complexity bounds—for example, for the required memory, number of objects, or the bit rate that can be handled by respective decoders. In order to define a conformance point for an MPEG decoder, in addition to the profile the level must be specified. This means that for compliant decoders a profile and a level must be defined. More details about MPEG-4 conformance and profiling can be found in Chapter 13.

In Figure 8.1, the complete set of decoders described in the MPEG-4 Visual standard [MPEG4-2] is shown. The decoders printed in gray (face and body animation decoding and [2D and 3D] mesh decoding) are not described in this chapter, but in Chapter 9. Also the scene composition is not explained here, but in Chapter 4. All of the other elements shown in Figure 8.1 (except *for the DeMUX, which is a simple demultiplexer) are explained in detail in the* following sections.

Analyzing Figure 8.1, several MPEG-4 natural video coding features can be highlighted. First, it is possible to code both still images (referred to as visual textures) and video sequences with the standard's tool-set. Moreover, taking only the two blocks Motion Decoding and Compensation and Texture Decoding results in the structure of the well-known (frame-based) motion-compensated hybrid coder—actually, it is even a block-based one. Adding Shape Decoding to the tool-set converts the entire MPEG-4 Visual standard into an object-based video coding solution. This means that the coded content is no longer restricted to be rectangular, but can be of arbitrary shape, which is coded by means of the MPEG-4 shape coding tools. In this mode, the shape is used also for decoding of motion and texture and for motion compensation.

It may seem that the usage of shape, and thus the possibility for content-based functionalities, is the only innovation of MPEG-4. However, this is not the case; the *rectangular* or *frame-based* parts of the video coding algorithms also have been greatly improved—for example, with increased compression

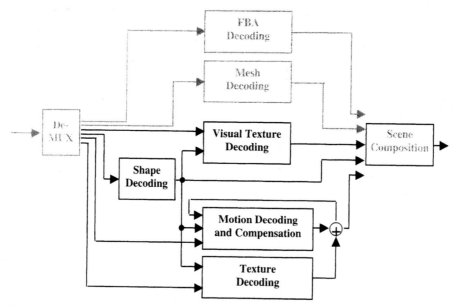

Fig. 8.1 Overview of MPEG-4 natural video decoding tools.

efficiency and error resilience. This will be described in detail in the remaining sections. Nevertheless, the object-based approach is a major issue in MPEG-4, and therefore the basic ideas and principles are described in the following section.

First, it has to be noted that even in the object-based mode, the coding of the images of a video sequence in MPEG-4 is done blockwise. This means that each image is subdivided into square blocks, whose content is then coded. In order to understand this principle, some terminology for MPEG-4 video coding is introduced and explained, using the simple exemplary scene shown in Figure 8.2.

In MPEG-4, each scene to be coded can be composed of one or several *visual objects* (VOs). In case of video sequences, the VO is of type `video` and is defined as a sequence of *video object planes* (VOPs). In the following, a VO of type `video` is also referred to as *video object*. Each VOP is an instance of the video object at a certain time instance, t. This means that a video object is composed of a sequence of VOPs, much as an image sequence is composed of a sequence of images. Using that analogy, the VOP corresponds to a single image, and the video object corresponds to an image sequence. The main difference is that a VOP can be either rectangular or of arbitrary shape. In the latter case, it is defined not only by its Y, U, and V components but also by its shape.[1] The representation and coding of the shape will be described in detail later.

The complete hierarchy of an MPEG-4 video bitstream is shown in Figure 8.3. The top level is a visual sequence (VS) with several VOs. Each VO of type `video` comprises one or (in case of scalable video coding) more *video object layers* (VOLs). Each layer corresponds to a certain spatial resolution or image quality, as will be explained in Section 8.4. Further, a set of successive VOPs

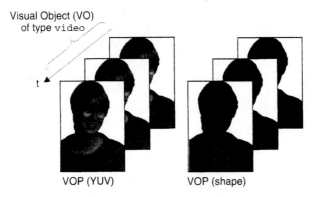

Visual Object (VO)
of type `video`

t

VOP (YUV)   VOP (shape)

Fig. 8.2 Visual object (VO) of type `video`, represented by a sequence of video object planes (VOPs), which are composed of Y, U, and V texture matrixes and a shape matrix in case of an arbitrarily shaped VO, as shown. Only the nontransparent pels, marked black in the shape matrix, belong to the VO.

1. In other contexts, the VO shape matrix is also referred as the *alpha plane*.

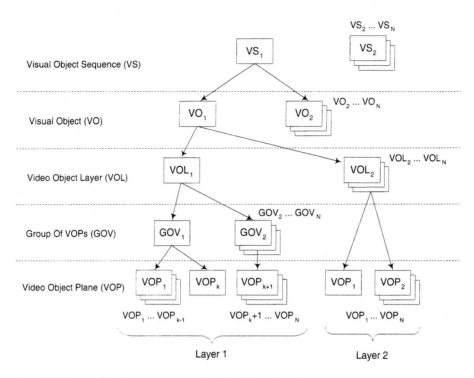

Fig. 8.3 Hierarchical structure of the MPEG-4 video bitstream.

can be clustered in order to form a *group of VOPs* (GOV). The GOV carries header information, which is useful for random access and resynchronization (e.g., in case of channel errors). The lowest level of the hierarchy is the VOP, as shown in Figure 8.3.

Just as previous coding standards do not specify how a video sequence is captured and even encoded, MPEG-4 does not standardize how a video object (and in particular the arbitrary shape information) is generated. The applied segmentation algorithm may be fully automatic (e.g., as described in [SalP94, NeCR97, MecW98]) or existing blue screen technology may be used. Also, user-assisted systems (as described in [AOWM98, MCMM99, KJKL99]) can be applied to generate arbitrarily shaped video content of high visual quality.

For coding purposes, each VOP is processed blockwise. Therefore, the smallest surrounding rectangle of the VOP is built, the *VOP bounding box*. This bounding box is extracted from the scene for each time instance and used to represent the corresponding VOP by means of its texture and shape information (see Figure 8.4).

For coding, the bounding box is divided into square blocks of size 16×16 luminance pels (and corresponding chrominances), the *macroblocks* (MBs). Therefore, the VOP bounding box is extended to match horizontal and vertical

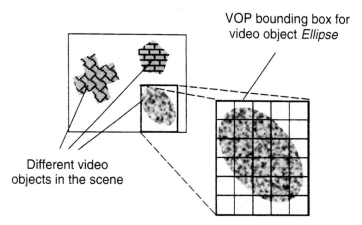

Fig. 8.4 Exemplary scene with three different video objects (*Cross*, *Pentagon*, and *Ellipse*); enlarged is the VOP bounding box for the video object Ellipse, divided into macroblocks of size 16×16 pels.

size of multiples of 16 pels (as the smallest surrounding rectangle to the object does not necessarily have multiples of 16 pels dimensions). Three different types of MBs are distinguished in the VOP bounding box, as shown in Figure 8.5.

1. **Transparent:** MBs that are completely outside the VOP. For these MBs, there is no YUV data to be coded; due to the shape information, the receiver knows that these MBs are *transparent* (i.e., not visible in the decoded scene).

2. **Opaque:** MBs that are completely inside the VOP. These MBs are processed as in the well-known block-based hybrid coders—that is, they are either *intra* coded using their YUV values or *inter* coded applying motion-compensated prediction and coding of the prediction error. The processing of these blocks is described in Section 8.2.

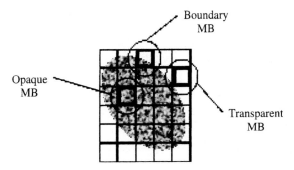

Fig. 8.5 MB types within the VOP boundary box.

3. Boundary: MBs that are at the boundary of the VOP. For processing these boundary MBs, specific tools for coding of arbitrarily shaped objects have been developed. These tools are presented in Section 8.3.

In addition to the two sections mentioned previously, which describe the MPEG-4 coding of rectangular video objects and of arbitrarily shaped video objects, there are sections describing specific MPEG-4 video coding tools with more detail. In Section 8.4, scalable video coding is explained. In particular, those parts of the scalable coding that are new compared to the state-of-the-art scalable video coding known, for example, from MPEG-2 Video [MPEG2-2], are emphasized. In Section 8.5, specific tools for interlaced video coding, error-resilient coding, dynamic resolution coding, sprite coding, and high-quality texture coding are described. Finally, Section 8.6 describes the MPEG-4 visual texture coding tool, which is an efficient wavelet-based tool for coding of rectangular as well as arbitrarily shaped still images.

8.2 CODING OF RECTANGULAR VIDEO OBJECTS

In MPEG-4 video coding, two main coding modes are distinguished: the coding of rectangular video objects and the coding of arbitrarily shaped video objects. In this section, the coding of rectangular video objects is described.

In principle, a rectangular video object is identical to a *frame* that is used in well-known video coding standards such as MPEG-1 and MPEG-2 [MPEG1-2, MPEG2-2, HaPN96], H.261 and H.263 [H261, H263, Rijk96]. Also, the coding principle of rectangular MPEG-4 video coding is the same as in these standards—namely, motion-compensated block-based hybrid coding. However, some significant improvements have been introduced by applying new tools or improving known algorithms. Therefore, this section describes the root coding algorithm only briefly, and concentrates on the new tools and algorithms introduced by MPEG-4 Visual [MPEG4-2].

8.2.1 Overview

For coding of rectangular video objects, only motion and texture information is transmitted to the decoder. Shape information is not required, and therefore in this case, no shape coding and decoding are necessary. In this respect, the MPEG-4 video coding solution is a block-based motion-compensated hybrid coder, as shown in the block diagram of Figure 8.6.

Figure 8.6 shows a block diagram of the decoding process for MPEG-4 video. Because no shape information is required for rectangular video objects, the corresponding modules in the block diagram are depicted in gray. The remaining blocks are those for motion decoding and compensation and for texture decoding. Compared with known block-based hybrid coding solutions, all

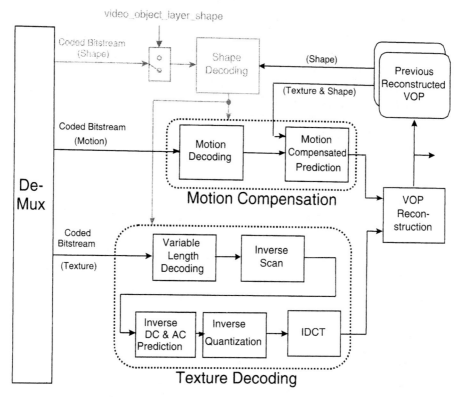

Fig. 8.6 MPEG-4 video decoding block diagram (shape decoding is in gray, as it is not required for rectangular video objects).

of these blocks have been significantly improved by introducing new or enhanced tools and algorithms.

Because the block-based hybrid coding principle is well known and copiously described in the literature (see, e.g., [CFHM71, Giro87, MuPG85, NetH88]), only a short description of the overall coding technique is given here. Each image of a video sequence is considered as one (rectangular) VOP. The coding of each VOP is performed blockwise—that is, each VOP is divided into macroblocks of 16×16 luminance pels size (with the corresponding chrominances, which depend on the colour subsampling format). Two major coding modes are distinguished:

1. *Intra* **mode:** The luminance (Y) and chrominance (U,V) values for each MB are coded independently of previous or future VOPs.

2. *Inter* **mode:** The difference of the MB luminance and chrominance values with respect to prediction values obtained from previous or future VOPs is coded.

The advantage of the inter mode is that, in general, the images in a video sequence do not change their content completely between two images (or VOPs, speaking in MPEG-4 terminology). Therefore, not the complete image but only the differences to the previous or future images (or combinations of them), referred to as the *reference image*, need to be transmitted to the decoder. Because the decoder knows the reference image(s), it can reconstruct the current image by adding the received difference signal. This principle is also referred to as *predictive coding*—that is, the encoder first calculates prediction values and then transmits only the *prediction error.*

In motion-compensated block-based hybrid coding, this prediction takes into account the motion within the captured scene and is therefore called *motion-compensated prediction.* The motion information is calculated and coded as side information, so that the decoder can carry out the motion-compensated prediction. The representation of the motion information depends on the specific coding algorithm. In MPEG-4, two motion modes are defined at the MB level:

1. *1MV mode:* One motion vector (MV) is transmitted for the complete macroblock (i.e., for each 16×16 luminance pels block).

2. *4MV mode:* The macroblock is subdivided into four blocks of 8×8 luminance pels, and one MV is transmitted for each of these blocks (i.e., in total, four MVs for the complete MB).

A VOP can be coded using one of two prediction modes: For *predicted VOPs* (P-VOPs), the reference image is the previously decoded intra coded VOP (I-VOP) or P-VOP, whereas for *bidirectionally predicted VOPs* (B-VOPs) both the previous and the subsequent I- or P-VOP can be used for prediction.

The computation of the motion information is referred to as *motion estimation.* Because it is an encoder issue, the standard does not specify how to carry out this step, as however it is done, it does not impact interoperability (providing that the syntax and semantics are respected); however, exemplary algorithms can be found in an informative annex of MPEG-4 Visual [MPEG4-2] and in the informative part of the MPEG-4 Reference Software [MPEG4-5]. Some specifically optimized algorithms can also be found in the MPEG-4 Optimised Reference Software [N4554].

For the coding of the texture (i.e., the luminance and chrominance information), the same algorithms are used for the intra and inter modes. The difference is that in the intra mode actual texture values are coded, whereas in the inter mode the coded values represent the prediction error (i.e., difference values). The texture coding is based on a 2D 8×8 *discrete cosine transform* (DCT; see, e.g., [NetH88, MPEG4-2]), with subsequent quantization and entropy coding of the transform coefficients. In addition, predictive coding is used for some of the transform coefficients, referred to as AC/DC prediction.

In the following sections, the motion-compensation tools and algorithms introduced by the MPEG-4 video standard are described. Then, the improvements regarding the texture coding are explained in detail.

8.2.2 New Motion-Compensation Tools

In this section, new and improved tools and algorithms for motion compensation in block-based hybrid coding, introduced by the MPEG-4 Visual standard [MPEG4-2], are described. Therefore, first the temporal prediction structure is described, as it is essential for the understanding of the following sections. Then, the following tools and algorithms are explained:

☞ **Quarter-pel motion compensation:** Algorithm for motion compensation using MVs with an increased resolution of one-quarter pel instead of only half- or full-pel resolution used in previous standards such as H.261, H.263, MPEG-1 Video, and MPEG-2 Video. This allows a significantly improved prediction and thus decreases the prediction error.

☞ **Global motion compensation:** A single set of motion parameters is coded for the complete VOP, representing the VOP global motion. These parameters can be used alternatively to the (local) MVs of an MB for motion compensation. This tool is particularly important for sequences with a large portion of global motion (e.g., caused by a moving camera) and for nontranslational motion like zoom or rotation.

☞ **Direct mode in bidirectional prediction:** This is an improvement of the bidirectional motion-compensated prediction, which uses the MVs of neighboring P-VOPs in order to decrease the bit rate required for coding the B-VOP MVs. It is a generalization of the "PB frames" introduced by H.263 [H263].

Those readers who are more familiar with the MPEG-4 Visual standard may miss one tool in this list—namely, the *overlapped block motion compensation* (OBMC). Because at the time of writing OBMC is not included in any of the MPEG-4 Visual object types and thus cannot be used by any MPEG-4–conformant application up to now, this tool is not described in this book. The interested reader is instead referred to the literature [MPEG4-2, OrcS94].

8.2.2.1 Temporal Prediction Structure
In order to understand the algorithms and terminology of motion-compensated prediction, it is important to clarify the temporal prediction structure. For the MPEG-4 video case, this structure is depicted in Figure 8.7.

As shown in Figure 8.7, an I-VOP does not refer to any other VOP (has no reference VOP), as it is completely encoded using the intra coding mode. P-VOPs refer to the so-called *forward reference*, which is defined as the most recently decoded I- or P-VOP in the past for which information has been trans-

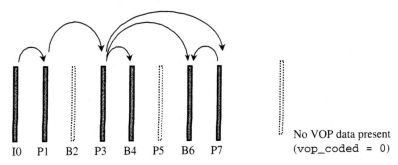

I0   P1   B2   P3   B4   P5   B6   P7   No VOP data present
                                                  (vop_coded = 0)

Fig. 8.7 Temporal prediction structure of MPEG-4 motion-compensated hybrid coding. Intra VOPs are marked as Ix, predictive VOPs as Px, and bidirectionally predicted VOPs as Bx. The arrows describe the direction of the prediction. The number x denotes the display order of the VOPs, not their order of transmission [MPEG4-2].

mitted (signaled by the condition[2] vop_coded=1). Finally, B-VOPs may refer to the forward reference (as just defined) as well as to the *backward reference*, which is defined as the most recently decoded I- or P-VOP in the future. In the case that no information has been transmitted that can be used as backward reference (i.e., vop_coded=0 for the backward reference), only the forward reference is used for the prediction of the B-VOP.

In Figure 8.7, the *display order* of the VOPs is shown. It should be noted that this order is not identical to the *transmission order*. As the B-VOPs may make reference to an I- or P-VOP that is to be displayed *in the future* with respect to their own display time, these backward-reference VOPs must be transmitted before the corresponding B-VOPs. For the example shown in Figure 8.7, the transmission order would be I0 P1 P3 B2 P5 B4 P7 B6. Because in that example no information is transmitted for VOPs B2 and P5 (i.e., for these VOPs vop_coded=0), VOPs B6 and P7 make reference to VOP P3. The omission of VOP B2 has no influence on the referencing structure, as B-VOPs are not used as reference for motion-compensated prediction.

8.2.2.2 Quarter-Pel Motion Compensation The MVs that are used for the motion-compensated prediction must be transmitted to the decoder. Therefore, the resolution of the MVs must be limited to a finite number. In the early motion-compensation algorithms (e.g., H.261 [H261]), only full-pel MVs were used—that is, the components of an MV were restricted to an integer number. Because the real motion between two successive images of a sequence is, in general, not exactly described by steps of one pel, this restricted resolution leads to errors in the motion-compensated prediction and thus to an increased prediction error.

---

2. The parameter vop_coded is a syntax element that is transmitted at VOP level; if it is set to zero, this signals that no more information is transmitted for this VOP— that is, the VOP is "empty" (no VOP data present).

The next generations of motion-compensation algorithms used an increased resolution of half-pel—that is, the components of the MVs could take full- and half-pel values. This was the case for MPEG-1 Video, MPEG-2 Video, and H.263; this tool increased the coding efficiency in terms of compression significantly [Giro93]. One major component of the half-pel motion compensation is the sample interpolation filter. It is required, as in the case of half-pel MVs, that these refer to a position in the reference image (or VOP) between two image samples (in fact, *virtual samples*). Therefore, the prediction value cannot be taken directly from the reference image, but has to be calculated from the neighboring samples. The half-pel sample interpolation structure used in MPEG-4 [MPEG4-2] is shown in Figure 8.8.

As shown in Figure 8.8, for MVs with half-pel resolution there are different positions in the reference image where they may point. First, in case of position a, both MV components are actually full-pel, so no sample interpolation is necessary. Second, in the case of the positions b and c, one of the MV components is actually half-pel; therefore, the sample value for it is calculated from the neighboring samples by bilinear interpolation—that is, $b = (A + B + 1)/2$[3] and $c = (A + C + 1)/2$ (all values are luminance or chrominance values). In case of position d, both MV components are half-pel, and the sample values are again calculated by bilinear interpolation from all four neighboring samples—that is, $d = (A + B + C + D + 2)/4$.

After half-pel motion compensation was established, many attempts were made to further increase the resolution of the MVs. However, in most cases the prediction error could not be reduced significantly by using quarter-pel MVs, whereas the bit rate for motion information was increased. In MPEG-4, it was shown that an increased MV resolution beyond half-pel increases the coding efficiency. This is due to an improved sample interpolation process that takes into account aliasing components in the images [BenW96]; this process is described below.

The process of quarter-sample interpolation can be subdivided into two steps. In the first step, half-pel samples are calculated by an improved interpo-

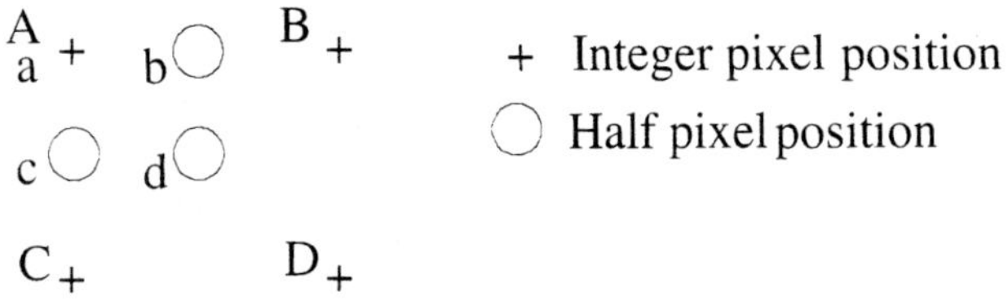

Fig. 8.8 Sample interpolation for half-pel MVs.

---

3. In the context of this and the following equations, the sign / means an integer division with truncation of the result toward zero. For example, 7/4 and –7/–4 are truncated to 1 and –7/4 and 7/–4 are truncated to –1. The exact sample interpolation process additionally includes a rounding control; details can be found in [MPEG4-2].

lation. Here, not only the two or four neighboring samples are used for (bilinear) interpolation. Instead, eight neighboring samples in each direction (i.e., horizontal and vertical) are taken into account. For the interpolation, a finite impulse response (FIR) filter is used. The half-pel sample interpolation structure is depicted in Figure 8.9 for the horizontal direction.

The sample values for the positions a and c are now calculated by applying an 8-tap FIR filter with the following filter values that are specified in the MPEG-4 Visual standard [MPEG4-2]:

$$[-8/256, 24/256, -48/256, 160/256, 160/256, -48/256, 24/256, -8/256]$$

Using these filter coefficients, the value for the position a results:

$$a = (-8 \times A4 + 24 \times A3 - 48 \times A2 + 160 \times A1 +$$
$$160 \times B1 - 48 \times B2 + 24 \times B3 - 8 \times B4) / 256$$

For the vertical case, the process is similar using neighboring sample values from top and bottom positions. In the case of the central position (i.e., when both MV components are half-pel), the filtering is applied first in the horizontal and then in the vertical direction.

In the second step, the quarter-pel sample values are calculated by a subsequent interpolation using the half-pel sample values calculated in the first step. Therefore, bilinear interpolation of the quarter-pel sample values is used as described in the following, based on the sample arrangement as shown in Figure 8.10.

In this case, the positions A, B, C and D in Figure 8.10 mark the half-pel sample positions and values as they result from the previously described FIR

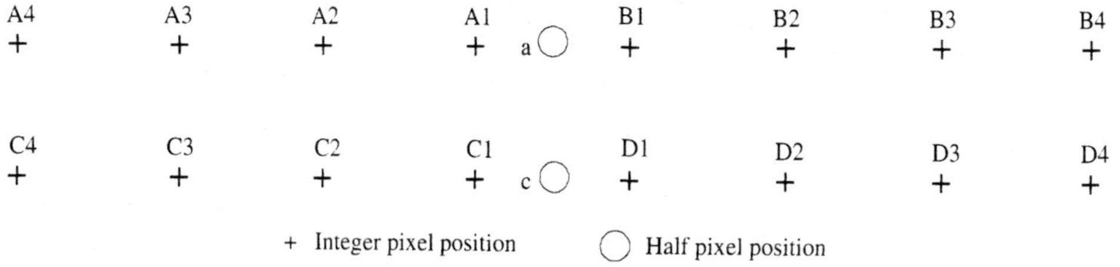

Fig. 8.9 Improved half-pel sample interpolation as first step for quarter-pel sample interpolation (only horizontal direction).

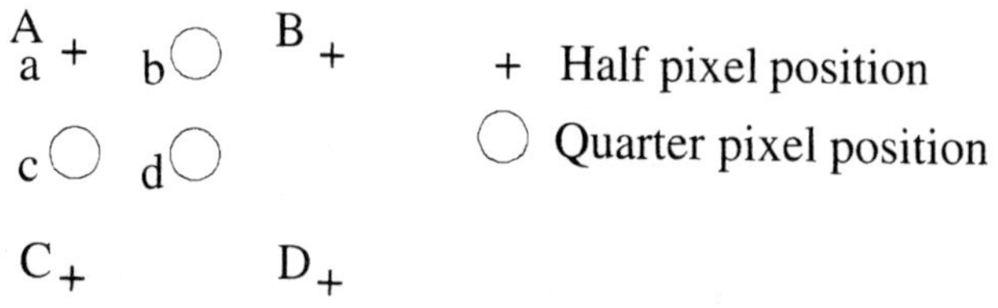

Fig. 8.10 Sample arrangement used for quarter-pel sample value interpolation using the half-pel sample values from the improved half-pel interpolation step.

interpolation procedure. Now, the quarter-pel sample values are calculated by bilinear interpolation from these half-pel sample values: $b = (A + B + 1)/2$, $c = (A + C + 1)/2$, and $d = (A + B + C + D + 2)/2$.

For motion-compensated prediction in half-pel mode, normally a block of $(8 + 1) \times (8 + 1)$ pels is read from the reference VOP. The position of this *reference block* is defined by the decoded MV. The reference block comprises one more row and one more column than the 8×8 block to be predicted, in order to allow subpel bilinear sample interpolation at the border of the block. For quarter-pel motion compensation, because of the application of an 8-tap FIR filter, three additional pels outside the boundary of the reference block are required for the interpolation. Therefore, the reference block is symmetrically extended at the block boundaries by three samples using the so-called *block boundary mirroring,* as shown in Figure 8.11.

The block boundary mirroring was introduced in order to reduce the required memory bandwidth of the motion-compensated prediction process. Without the mirroring, the additional three values outside the reference block—and thus a total of $(8 + 7) \times (8 + 7)^4 = 225$ samples—would have to be read from the reconstructed VOP memory. The block boundary mirroring decreases the number of samples to $(8 + 1) \times (8 + 1) = 81$, resulting in 64% reduction of the memory bandwidth.

8.2.2.3 Global Motion Compensation When global motion (e.g., due to camera motion) is present in a video scene, it would be reasonable to use only one set of motion parameters for the motion-compensated prediction of a complete VOP. However, the global motion parameters may not always be applicable for

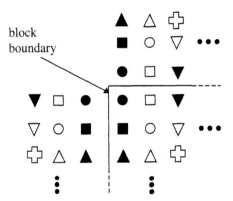

Fig. 8.11 Block boundary mirroring: The reference block used for motion-compensated prediction is symmetrically extended at the boundary.

---

4. Inside the block eight samples, and outside the block four samples in the direction of the sub-pel sample position and three at the other side of the block, both in horizontal and vertical direction.

the complete VOP; in general, they lead to quite good prediction results for certain areas of the VOP, whereas for other areas, the prediction may lead to significant prediction errors. Therefore, in MPEG-4, *global motion compensation* (GMC) is applied alternatively to the (local) block-based motion compensation [M615].

When GMC is used, the global motion parameters are transmitted in the VOP header. Then, for each macroblock, the encoder decides if the GMC parameters or the MB estimated MV(s) are used, depending on the resulting prediction error for the macroblock. The information—whether for an MB, the GMC, or the *local* motion compensation—is applied and is transmitted as side information for each macroblock.

The global motion parameters in MPEG-4 are transmitted as a set of up to four MVs for one VOP. In addition, for each of these MVs the corresponding reference position in the current VOP is specified. For example, the GMC MVs could be located at the four corners of the VOP, as shown in Figure 8.12.

Taking the example given in Figure 8.12, the global motion compensation of an MB can be described as follows: For each pel in the MB, an MV is calculated from the GMC MVs and their reference position. This can be interpreted as a kind of *interpolation* of the GMC MVs for each position in the MB. The resolution of these interpolated MVs is specified by the encoder, in order to guarantee identical results at encoder and decoder. For each pel of the MB, a motion-compensated prediction is then carried out using the corresponding interpolated MV. This process is also referred to as a *warping* process. It is used not only for the global motion compensation but also for the coding of sprites, as described in Section 8.5.4.

In case of sub-pel resolution of the warping MVs, the sample values are calculated by bilinear interpolation, as shown in Figure 8.13.

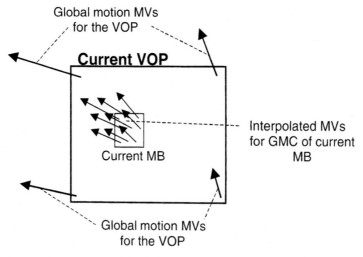

Fig. 8.12 Global motion compensation of a macroblock in MPEG-4.

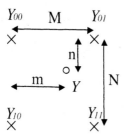

Fig. 8.13 Sample arrangement for bilinear sample interpolation as used in global motion compensation of an MB.

Here, Y_{00}, Y_{01}, Y_{10}, and Y_{11} denote the sample values from the reference VOP, and Y denotes the prediction value required located at a position, as shown in Figure 8.13, for which the interpolation is carried out as follows:

$$Y = \frac{N-n}{N} \cdot \frac{M-m}{M} \cdot Y_{00} + \frac{N-n}{N} \cdot \frac{m}{M} \cdot Y_{01} + \frac{n}{N} \cdot \frac{M-m}{M} \cdot Y_{10} + \frac{n}{N} \cdot \frac{m}{M} \cdot Y_{11}$$

8.2.2.4 Direct Mode in Bidirectional Motion Compensation In bidirectional motion compensation, two VOPs can be used for the prediction: the previously decoded I- or P-VOP (the forward reference) and the subsequently decoded I- or P-VOP (the backward reference). Four different prediction or compensation modes are distinguished:

☞ **Forward mode:** Only the forward reference is used for prediction, and thus only one MV is transmitted.

☞ **Backward mode:** Only the backward reference is used for prediction, and thus only one MV is transmitted.

☞ **Interpolative mode:** Both forward and backward references are used for prediction, and thus two MVs are transmitted; the resulting prediction value is interpolated from both reference values.

☞ **Direct mode:** Both forward and backward references are used for prediction; however, the required MVs are derived from the MV of the colocated macroblock in the backward-reference VOP, and only a correction term called *delta vector* is transmitted.

Whereas the first three modes were used in the MPEG-2 Video standard [MPEG2-2], the direct mode was introduced by MPEG-4.[5] Therefore, it is described in more detail in the following; for the other modes, the reader is referred to the literature [MPEG2-2].

---

5. The direct mode in MPEG-4 Visual is a generalization of the PB frames used in the H.263 standard [H263].

The basic idea of the direct mode is to exploit the knowledge of the MVs between the forward- and the backward-reference VOPs, as depicted in Figure 8.14.

The current VOP in Figure 8.14 is the B-VOP B1, and the position of the current macroblock is shown in the figure. Further, the decoder already knows the backward-reference VOP P3 and its associated MVs, as it was decoded before the current VOP. In the direct mode, the knowledge of these MVs is used to derive the forward and backward MV for the prediction of the current macroblock.

Therefore, the MV of the *colocated macroblock* in the backward-reference VOP P3 is used. The colocated macroblock is defined as the macroblock that has the same horizontal and vertical index as the current macroblock in the B-VOP. To get the MVs of the current macroblock, the MV is scaled, depending on the *position in time* of the B-VOP in question between the two reference VOPs. Further, a correction term called *delta MV* is added to the result of the scaling. The delta MV is the only motion information for a B-VOP macroblock that is transmitted to the decoder in the direct mode.

The scaling factors are calculated from the time differences between the two reference VOPs and the current B-VOP. They are defined as follows:

$$TRB = display\_time(currentVOP) - display\_time(forward\_ref)$$
$$TRD = display\_time(backward\_ref) - display\_time(forward\_ref)$$

The scaling is performed under the assumption of linear motion between the two reference VOPs by linear interpolation. In the same step, the delta MV (MV_D) is added.[6] Following this process, the B-VOP MVs are defined as follows:

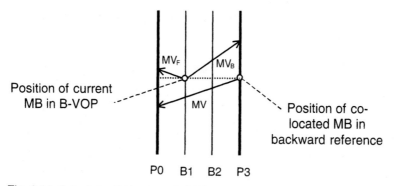

Fig. 8.14 Principle of direct mode bidirectional motion compensation. The forward and backward motion vectors MV_F and MV_B are derived from the displacement vector MV of the colocated macroblock in the backward-reference VOP.

---

6. The MV scaling and adding of the delta MV are done independently for the x and y component of the MV.

$$MV_F = \frac{TRB}{TRD} \cdot MV + MV_D \; ; \quad MV_B = \frac{TRB - TRD}{TRD} \cdot MV + MV_D$$

For the example in Figure 8.14, the forward MV would be $1/3MV+MV_D$ and the backward MV would be $-2/3MV+MV_D$.

For the motion compensation in B-VOPs, only the 1MV mode may be used—that is, only one MV may be transmitted for one macroblock for each prediction direction. The only exception is the direct mode: In the case that the colocated MB in the backward-reference VOP is coded with the 4MV mode (i.e., four MVs are available), for the current MB in direct mode four MVs are used for the corresponding four 8×8 blocks of the current MB in the B-VOP. Due to this rule, the direct mode is the only possibility to use the 4MV mode with bidirectional motion compensation in MPEG-4.

If the current MB is coded in *skipped mode* (the modb flag at the beginning of the MB header is set, which means that no other information is coded for this MB), it is also motion compensated using the direct mode, with zero delta MV and no encoded prediction error.

An additional rule that derives from the H.263 PB frames [H263] is the *implicit skipped mode*: If the colocated MB of the backward-reference VOP is *skipped* (i.e., it uses zero vector motion compensation and no prediction error coding, signaled by the not_coded flag in the MB header), no information for the current B-MB is coded, not even the modb flag; the MB is reconstructed by forward MC with zero MV.

This rule can lead to unwanted artifacts, if the encoder makes the decision to skip a P-MB without taking into account that the colocated B-MBs in all directly preceding B-VOPs also will be skipped. Especially in video scenes containing large motion, this problem may occur, as the position of the colocated MBs does *not* take motion in the VOPs into account. A possible way to avoid this is not to use the skipped mode in P-MBs, but to explicitly code the lack of prediction error and the zero MV. This needs six more bits, but it avoids this type of artifact.

8.2.3 New Texture Coding Tools

Texture coding in MPEG-4 video means either the coding of luminance and chrominance values in case of the intra mode, or the coding of prediction error values in case of the inter mode (after the motion-compensated prediction). In extension to the 8-bit amplitude resolution used in MPEG-2 Video [MPEG2-2] (and 9 bits for prediction error values), MPEG-4 Visual supports the use of 4 to 12 bits per pixel for luminance and chrominance values, the so-called *N-bit* tool. This is signaled to the decoder by the not_8_bit flag and the bits_per_pixel field in the VOL header. The coding of *gray-scale alpha values* (see Section 8.3) is always done with 8 bits, regardless of the (luminance and chrominance) bits_per_pixel field.

For the coding of the luminance and chrominance values (for intra and inter frame mode), the process shown in Figure 8.15 is applied.

In Figure 8.15, f[y][x] denotes either the luminance and chrominance values or the prediction error values of an 8×8 pels block. First, an 8×8 DCT is applied, resulting in the transform coefficients F[v][u], which are then quantized to QF[v][u]. Then, prediction of some of the transform coefficients is performed for intra coded macroblocks. Finally, the two-dimensional matrix of coefficients and prediction differences PQF[v][u] is converted into a one-dimensional vector QFS[n]. This vector is then entropy coded using a variable-length Huffman code.

In the following sections, the quantization, transform coefficient prediction (also referred to as *AC/DC prediction*), and 2D-to-1D conversion process by means of a scanning procedure are described in more detail.

8.2.3.1 Quantization of DCT Transform Coefficients   In order to reduce the required bit rate for texture coding, the DCT transformation coefficients are quantized. Two possible quantization procedures can be applied: The first is derived from the MPEG-2 Video standard [MPEG2-2], and the second one was used in recommendation ITU-T H.263 [H263]. It is decided at the encoder side which of the two methods is used for coding a video object; the quantization method selection is sent as side information. Furthermore, the DC coefficient of an 8×8 block coded in intra mode, which represents its mean luminance or chrominance value, is quantized using a fixed quantizer step size.

The quantization step size is controlled by a specific parameter: the quantiser_scale. It can take values from 1 to 31 (or [$2^{quant\_precision} - 1$] for the not_8_bit case),[7] and is coded once per VOP. In addition, it is possible in spe-

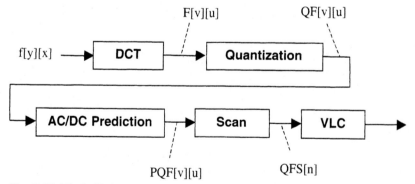

Fig. 8.15 Block diagram of MPEG-4 texture coding for the intra and inter coding modes. The AC/DC Prediction is applied only for MBs in intra coding mode.

---

7. The parameter quant_precision specifies the number of bits used to represent quantizer parameters. Values between 3 and 9 are allowed. When the parameter not_8_bit is 0, and therefore quant_precision is not transmitted, it takes a default value of 5.

cial MB modes to modify the `quantiser_scale` incrementally using additional side information. In the following, both the quantization formula and the inverse quantization formula are provided.

Intra DC Coefficient Quantization   The DC coefficients of intra coded macroblocks are quantized using an optimized, nonlinear quantization method. Here, the value of the quantization step size `dc_scaler` depends on the value of the `quantiser_scale`, as shown in Table 8.1. Here, as a shortcut, the term Qp is used instead of `quantiser_scale`.

The quantization and inverse quantization are then carried out as follows:

$$Quantization: \quad QF[0][0] = F[0][0] // dc\_scaler$$

$$Inverse\ quantization: \quad F[0][0] = QF[0][0] \cdot dc\_scaler$$

Here, the operator `//` denotes an integer division with rounding to the nearest integer, in which half-integer values are rounded away from zero unless otherwise specified. For example, `3//2` is rounded to `2`, and `-3//2` is rounded to `-2`.

First Quantization Method: MPEG Quantization   The first quantization method is derived from the coefficient quantization used in the MPEG-2 Video standard [MPEG2-2]. Therefore, it was nicknamed *MPEG quantization*. Because this nickname does not appear in the MPEG-4 specification, the reader should be aware of this correspondence.

The specific property of this quantization method is that the encoder can take into account the properties of the human visual system. This is possible because the quantization method allows adapting the quantization step size individually for each transform coefficient by means of *weighting matrices*.

In MPEG-4 video, different quantization matrices `W[v][u]` are used for intra and for inter coded macroblocks. Furthermore, either default matrices or new matrices, which are transmitted at the beginning in order to configure the decoder, can be applied. In Figure 8.16, the default matrices for intra and inter coded macroblocks are shown.

Using the quantization matrices, for example, the transform coefficients representing higher spatial frequencies can be quantized more coarsely, without significantly decreasing the visual quality of the decoded image, if properly selected quantization steps are applied. With the coarser quantization of

Table 8.1 Quantization step size `dc_scaler`

`quantiser_scale` (Qp)	1–4	5–8	9–24	25–31
`dc_scaler` (luminance)	8	2Qp	Qp + 8	2Qp - 16
`dc_scaler` (chrominance)	8	(Qp + 13) / 2	(Qp + 13) / 2	Qp - 6

8	17	18	19	21	23	25	27
17	18	19	21	23	25	27	28
20	21	22	23	24	26	28	30
21	22	23	24	26	28	30	32
22	23	24	26	28	30	32	35
23	24	26	28	30	32	35	38
25	26	28	30	32	35	38	41
27	28	30	32	35	38	41	45

16	17	18	19	20	21	22	23
17	18	19	20	21	22	23	24
18	19	20	21	22	23	24	25
19	20	21	22	23	24	26	27
20	21	22	23	25	26	27	28
21	22	23	24	26	27	28	30
22	23	24	26	27	28	30	31
23	24	25	27	28	30	31	33

Default weighting matrix for *intra coded* MBs   Default weighting matrix for *inter coded* MBs

Fig. 8.16 Default weighting matrices for the first quantization method.

some coefficients, the bit rate for their coding can be reduced, so in total the compression efficiency is increased.

For *quantization*, the weighting matrices W[v][u] are incorporated in the process, as shown in the following equation:[8]

$$QF[v][u] = \left(F[v][u] \cdot 16 // W[v][u] - k \cdot quantiser\_scale\right) // \left(2 \cdot quantiser\_scale\right)$$

$$\text{where } k = \begin{cases} 0 & \text{for intra coded blocks} \\ sign(QF[v][u]) & \text{for inter coded blocks} \end{cases}$$

As can be seen in the equation, the quantization is controlled by the value of the weighting matrix W[v][u]. The higher the value, the coarser the quantization. Thus, a different quantization step size can be used for each transform coefficient, if appropriate.

The inverse quantization is carried out using the same matrices, according to the following equation:

$$F''[v][u] = \begin{cases} 0, \text{ if } QF[v][u] = 0 \\ ((2 \times QF[v][u] + k) \times W[w][v][u] \times quantiser\_scale) \,/\, 16, \text{ if } QF[v][u] \neq 0 \end{cases}$$

where :

$$k = \begin{cases} 0 & \text{for } intra \text{ coded blocks} \\ sign(QF[v][u]) & \text{for } inter\ coded \text{ blocks} \end{cases}$$

---

8. It should be noted that, in the case of intra coded MBs, the DC coefficient is coded using the procedure described at the beginning of the "Intra DC Coefficient Quantization" section. In that case, the weighting matrix approach is used only for the AC coefficients of the intra MB.

Here, the operator / denotes an integer division with truncation of the result toward zero. For example, 7/4 and −7/−4 are truncated to 1, and −7/4 and 7/−4 are truncated to −1.

Second Quantization Method: H.263 Quantization   The second quantization method is derived from the coefficient quantization that is used in Recommendation ITU-T H.263 [H263]. Thus, it was nicknamed *H.263 quantization*. Again this nickname does not appear in the MPEG-4 Visual specification.

Unlike the first quantization method, the second method does not apply the weighting matrix technique. Therefore, it is less complex and easier to implement, but it does not allow optimizing the coder with respect to the application of coefficient adaptive quantization.

The quantization process is performed as shown in the following equation:

$$|QF[v][u]| = \begin{cases} |F[v][u]|/(2 \cdot quantiser\_scale) & for\ intra\ coded\ blocks \\ (|F[v][u]| - quantiser\_scale/2)/(2 \cdot quantiser\_scale) & for\ inter\ coded\ blocks \end{cases}$$

The sign of F[v][u] is then incorporated to obtain QF[v][u] as:

$$QF[v][u] = Sign(F[v][u]) \times |QF[v][u]|$$

The inverse quantization is then carried out, applying the following equation:

$$|F''[v][u]| = \begin{cases} 0, & if\ QF[v][u] = 0, \\ (2 \times |QF[v][u]| + 1) \times quantiser\_scale, & if\ QF[v][u] \neq 0,\ quantiser\_scale\ is\ odd, \\ (2 \times |QF[v][u]| + 1) \times quantiser\_scale - 1, & if\ QF[v][u] \neq 0,\ quantiser\_scale\ is\ even. \end{cases}$$

The sign of QF[v][u] is then incorporated to obtain F''[v][u] as:

$$F''[v][u] = Sign(QF[v][u]) \times |F''[v][u]|$$

8.2.3.2 AC/DC Prediction for Intra Macroblocks   For some of the AC and DC coefficients of neighboring blocks, there exist statistical dependencies—that is, the value of one block can be predicted from the corresponding value of one of the neighboring blocks. This is exploited in MPEG-4 video coding by the so-called *AC/DC prediction*. It should be noted that this prediction is applied in only the case of intra coded macroblocks. The idea behind the AC/DC prediction tool is presented in Figure 8.17.

Figure 8.17 shows the current macroblock and two 8×8 luminance blocks x and y to be encoded. The black square dot in the top left corner of each block represents the DC coefficient, the vertical gray rectangle represents the first column of AC coefficients, and the horizontal gray rectangle represents the first row of AC coefficients. AC/DC prediction is carried out only for the

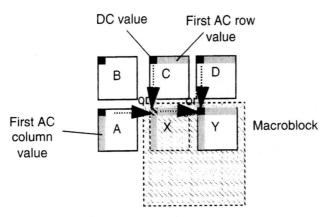

Fig. 8.17 AC/DC prediction process for intra coded macroblocks.

marked coefficients—that is, for the DC coefficient and the first row and first column of AC coefficients.

The prediction is carried out either from the left or from the top neighboring block in an adaptive fashion, but without the coding of further side information. For that, the horizontal and vertical DC gradients around the block to be coded are computed. In the example in Figure 8.17, for selecting the prediction direction (horizontal or vertical) for block X, the DC gradient from B to A and the DC gradient from B to C are compared. Then, the direction with the lower gradient is selected for the AC/DC prediction. Because the already quantized DC coefficients of the neighboring blocks are used for this decision, it can be done at the decoder without coding of any side information (of course, for error-free conditions).

If any of the blocks A, B, or C are outside the VOP boundary or the video packet[9] boundary, or if they do not belong to an intra coded macroblock, their DC coefficient values are assumed to take a value of $2^{(\text{bits\_per\_pixel}+2)}$ (as that corresponds to the DCT transformed mean value of a sample value representation with bits_per_pixel bits), and are used to determine the prediction direction and thus the prediction values.

8.2.3.3 Alternative Scan Modes   After the AC/DC prediction, the DCT coefficients are entropy coded, using a variable-length code table [MPEG4-2]. In order to build the symbols that are entropy coded, the two-dimensional matrix PQF[v][u] of coefficients is transformed into a one-dimensional vector QFS[n]. This vector is then fed into the entropy coder.

For the 2D-to-1D conversion, a *scanning process* is used. This technique also is used in existing standards; however, until now, mostly the well-known

---

9. Video packets are used for error-resilience coding; this concept is presented in Section 8.5.2.1.

zigzag scan was used. In MPEG-4 video, two additional scanning modes are included, which can be used alternatively. The three possible MPEG-4 scanning modes are shown in Figure 8.18.

The principal function of the process is as follows: The transform coefficients (or transform residuals in case of AC/DC prediction) are read from the 2D matrix using the order of the numbers given in the patterns shown in Figure 8.18. The ordered list of the coefficients represents the 1D vector QFS[n], where n runs from 0 to 63. This process also is described by the following small C–program code:

```
for (v=0; v<8; v++)
   for (u=0; u<8; u++)
      QFS[scan_pattern[v][u]] = PQF[v][u];
```

The idea behind these scans is that, in general, with increasing spatial frequency (which goes along with increasing indices u and v), the transform coefficients are very small or even 0, depending on the quantizer scale. The results of the scan should be that the largest coefficients are at the beginning of the vector, and the small and 0 ones are at its end. This is taken into account by the subsequent entropy coding.

In many cases, the zigzag scan will already provide this behavior, and thus is sufficient. However, for certain image structures, there will be a particular preference of vertical or horizontal frequencies, which can be accommodated by applying one of the alternate scans shown in Figure 8.18. Which of the scans is to be used is decided by the following method: For intra coded blocks, if AC/DC prediction is not used, the zigzag scan is selected for all blocks in a macroblock. Otherwise, the DC prediction direction is used to select a scan on a block basis. For instance, if the DC prediction refers to the horizontally adjacent block, alternate-vertical scan is selected for the current block. Otherwise (i.e., the DC prediction refers to the vertically adjacent

0	1	2	3	10	11	12	13
4	5	8	9	17	16	15	14
6	7	19	18	26	27	28	29
20	21	24	25	30	31	32	33
22	23	34	35	42	43	44	45
36	37	40	41	46	47	48	49
38	39	50	51	56	57	58	59
52	53	54	55	60	61	62	63

Alternate horizontal scan

0	4	6	20	22	36	38	52
1	5	7	21	23	37	39	53
2	8	19	24	34	40	50	54
3	9	18	25	35	41	51	55
10	17	26	30	42	46	56	60
11	16	27	31	43	47	57	61
12	15	28	32	44	48	58	62
13	14	29	33	45	49	59	63

Alternate vertical scan

0	1	5	6	14	15	27	28
2	4	7	13	16	26	29	42
3	8	12	17	25	30	41	43
9	11	18	24	31	40	44	53
10	19	23	32	39	45	52	54
20	22	33	38	46	51	55	60
21	34	37	47	50	56	59	61
35	36	48	49	57	58	62	63

Zigzag scan

Fig. 8.18 Alternative MPEG-4 scanning modes for converting the 2D coefficients matrix into a 1D vector of DCT coefficients.

block), the alternate-horizontal scan is used for the current block. For all other block types (in particular, in case of inter coded macroblocks), the 8×8 blocks are scanned using the zigzag scanning pattern.

8.3 CODING OF ARBITRARILY SHAPED VIDEO OBJECTS

As described in Section 8.1.2, each VOP of an arbitrarily shaped video object is represented by its YUV components and, in case of nonrectangular VOPs, by a shape component. In MPEG-4, the shape information is represented by means of *alpha masks.*

An alpha mask defines the level of transparency of a VOP. In a gray-level alpha mask, transparency is represented by an 8-bit integer, resulting in a range of possible values between 0 and 255. The 0 corresponds to *completely transparent pixels* and the 255 to *completely opaque pixels*; in-between values correspond to pixels that are more or less transparent but neither completely transparent nor opaque. In a *binary alpha mask*, only two values are allowed: *completely transparent* (0) and *completely opaque* (255). A video object is only defined for pels whose alpha value (A) is larger than 0—that is, for those pels of the VOP that are not completely transparent.

Both kinds of alpha masks can be used in MPEG-4, depending on the application. Therefore, two shape coding tools for binary and gray-level alpha masks are available; these are described in the following sections. Furthermore, specific tools are used for motion-compensated prediction and texture coding of boundary macroblocks. These also are described in the following section.

8.3.1 Binary Shape Coding

The efficient coding of binary alpha masks is one of the major algorithmic innovations of MPEG-4 video coding. The binary shape coding is performed blockwise—that is, for each MB the alpha values are coded separately. In the following, MBs with binary alpha values also are referred to as *binary alpha blocks* (BABs). Furthermore, although the correct representations of the binary alpha values are 0 and 255, throughout this chapter the values 0 and 1 are used instead, for reasons of simplicity.

The basic idea behind the binary shape coding algorithm is the *context-based arithmetic encoding* (CAE) [BraB00]. It comprises arithmetic encoding of the alpha values depending on the context that has to be computed for all the values to be coded. The alpha values of a BAB are coded pel-wise, using an *arithmetic encoder.* This encoder assigns codewords to the binary value 0 or 1, depending on statistical probability. This probability depends on the *context* of the shape element (alpha pel) to be coded. For each possible context, statistical probabilities have been measured beforehand, using a set of typical binary alpha mask sequences. These probability values, which are part of the

MPEG-4 Visual specification [MPEG4-2], are used as parameters in the arithmetic encoder.

As for YUV values, alpha coding has an intra coding mode and an inter coding mode, which uses motion compensation; unlike in texture coding, the same mode (i.e., intra or inter) must be used for all MBs of a VOP (so no MB-based shape intra refreshment is possible). The shape motion compensation uses MVs, which are independent from the texture MVs. However, shape motion compensation is performed similar to texture motion compensation, with the following additional restrictions:

1. The resolution of the shape MVs is restricted to one pel (i.e., half- or quarter-pel motion compensation is not performed).

2. Motion compensation is always carried out on a complete MB, using only one MV for the complete MB (i.e., the 4MV mode is not used).

3. In case of a B-VOP, either the previous or the subsequent I- or P-VOP is used as reference, depending on which one is nearer to the current VOP in the display order (i.e., no bidirectional shape motion compensation is performed).

For the inter and intra coding modes, different context computation methods are applied. After computation of the context, the arithmetic encoding is the same for both modes.

8.3.1.1 Context Computation for Binary Alpha Blocks For intra coded BABs, a 10-bit context is built for each alpha pel, as illustrated in Figure 8.19a, where $c_k = 0$ for transparent pels and $c_k = 1$ for opaque pels.

For inter coded BABs, temporal redundancy is exploited by using pels from the corresponding motion-compensated BAB in the reference frame (depicted in Figure 8.19b). Here, a 9-bit context is built for each pel to be coded. When building the contexts, some specific rules apply:

☞ Any pels of the context that are outside the VOP bounding box or outside the current video packet (see Section 8.5.2.1) are assumed to be 0 (i.e., transparent).

☞ The context template may cover pels from BABs on the right-hand side of the current BAB, which have not yet been coded and thus are unknown; the values of these unknown pels are defined by repetition of the corresponding border values.

The context value is thus described by a bit pattern of 10 and 9 bits for the intra and inter modes, respectively, as defined by:

$$C = \sum_k c_k \cdot 2^k$$

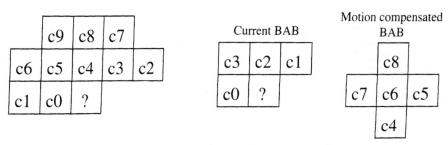

(a) Intra context template   (b) Inter context template

Fig. 8.19 Context templates for binary shape coding in intra and inter coding modes; the ? marks the alpha pel being coded. For the inter context template, the pels ? and $c6$ are located at the same position in the current and the reference pictures.

where c_k are the binary alpha values (0 or 1) in the corresponding position of the context template. Because the values for c_k are binary, the number of possible contexts is $2^{10} = 1024$ for the intra mode and $2^9 = 512$ for the inter mode. Using the context C, the probability of the current alpha pel to be 0 or 1 is estimated by table look-up, using the tables provided in the MPEG-4 Visual standard. This probability is then used for the arithmetic coding of the current alpha value.

In the following, an example of context computation for the intra mode is given using the three example contexts shown in Figure 8.20.

For the example in Figure 8.20, the three contexts[10] and the corresponding probabilities for the pel x_a to be 0 are calculated, using the probability table B-32 in Annex B of the MPEG-4 Visual specification [MPEG4-2].

$$C_{(a)} = (0000000000)_2 \Rightarrow P(x_a = 0 \,|\, C_{(a)}) = 0{,}996$$
$$C_{(b)} = (1111111111)_2 \Rightarrow P(x_a = 0 \,|\, C_{(b)}) = 0{,}004$$
$$C_{(c)} = (0000000011)_2 \Rightarrow P(x_a = 0 \,|\, C_{(c)}) = 0{,}273$$

The counter probabilities for x_a to be 1 can easily be calculated by 1 - P (x_a = 0 | C), as the alpha value can only be 0 or 1. From the example it can be understood how the context works. In case (a), it seems likely that x_a is 0, because the complete context is 0. And because the probability table has been measured from real, typical alpha mask sequences, the probability for this case is very high. This does not mean that x_a is always 0 for that context, but it often will be. This fact, which is reflected by the probability, is taken into account by the arithmetic coding. For case (b), it is exactly the opposite: The

---

10. The contexts in the equations are in binary representation (this is the meaning of the 2 as subscript), where the 0s and 1s are the binary alpha values of the corresponding context pels.

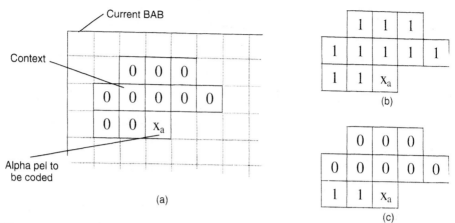

Fig. 8.20 Example of context computation for the intra mode. The alpha pel to be coded is marked by x_a, and three different contexts (a), (b), and (c) are given.

complete context is 1, and thus the probability of x_a to be 0 is very low. Finally, for case (c) things are not that clear; however, one could imagine that the bottom line of the context is the border of an object, whereas the remaining context belongs to the background. In that case, it would be rather likely that x_a is 1, and this is actually reflected by the probability. Of course, there are other possible alpha masks that can cause such a context; therefore, the probability is not as clear as in cases (a) and (b).

When the context and the probabilities have been calculated, the alpha pel is encoded by the arithmetic encoder, as described in the next section.

8.3.1.2 Arithmetic Encoding of Binary Alpha Blocks The literature about arithmetic encoding is manifold, and it does not seem appropriate to offer yet another description in this chapter. Therefore, only the basic principles and the relevant technical details necessary for understanding the MPEG-4 binary shape coding are described. For further details, the interested reader is referred to overview papers [RisL79, Lang84, WiNC87] and, of course, to the MPEG-4 Visual standard itself [MPEG4-2].

The principal idea of arithmetic encoding combines two important approaches for statistical redundancy reduction. First, as in the well-known *Huffman coding*, the probability of the symbols to be coded is taken into account; for events with a high probability, which will occur frequently, a short codeword is assigned. The lower the probability of an event, the longer the codeword used for its representation. Second, the arithmetic encoding allows one to assign, on average, less than one bit to a symbol. One could argue that this is also possible by using Huffman coding, as the Huffman codewords are, in general, assigned to a block of N symbols (using the *N-order extensions* of the source). So, if the codeword length is lower (in bits) than the number N of symbols in the corresponding block, each symbol needs less than one bit. However, in Huffman coding, these blocks of symbols and their joint probability have to be specified beforehand in order to design the codeword tables. This is not necessary in arithmetic encoding, in which the single symbols are encoded, taking into account their corresponding probabilities.

In MPEG-4 binary shape coding, the symbols to be coded are the alpha values of a BAB. Using the context previously described for each alpha value, a probability for being 0 or 1 is calculated. The values and the probabilities are input to the arithmetic encoder, which generates the output bits. Each BAB is encoded separately—that is, after the last pel of a BAB is encoded, the arithmetic code is terminated by a code with 2 or 3 bits, depending on the state of the encoder. This termination is necessary in order to make the arithmetic code *decodeable*; it can be interpreted as a kind of *end of block* symbol, although its algorithmic meaning is somewhat different.

One problem of arithmetic encoders in general is the *start code emulation*. Start codes are used in the video bitstream to signal the beginning of a VOP, a resynchronization point, and so on. They usually consist of a number of consecutive 0s and a final 1. Start codes must be unique throughout the complete bitstream, because otherwise a decoder could not exactly identify the beginning of a VOP (e.g., for random access) or a resynchronization point (e.g., in case of a resynchronization after a transmission error). Usually, all variable- and fixed-length codes in an algorithm are designed in such a way that the emulation of start codes is avoided.

However, because an arithmetic encoder does not have a fixed set of codewords, it can (in theory) generate every possible bit pattern, including start codes. Therefore, in MPEG-4 shape coding a different approach is used to avoid start code emulation due to arithmetic encoding, as shown in Figure 8.21.

The algorithm applied for avoiding start code emulation is very simple. Both encoder and decoder know all start codes that can possibly occur in the bitstream. Therefore, the number of consecutive 0s at the output of the arithmetic encoder is counted. Each time this number exceeds the threshold[11] given by the known start codes, a 1 bit is included into the bitstream; this inclusion of 1 bits is also referred to as *marker bit stuffing*. Because the start codes are also known at the decoder, the stuffed marker bits can be removed from the bitstream at the decoder before arithmetic decoding.

8.3.1.3 Binary Only Shape Coding
For some applications (e.g., clickable maps), the use of binary shape information without associated texture information often is needed, forming a special video object. This is supported by the `binary_only` mode, which can be signaled by the `video_object_layer_shape` parameter in the VOL header. In this mode, only binary shape

---

11. This threshold is chosen such that the maximum number of consecutive 0s that can occur after the bit stuffing is lower than the minimum number of consecutive 0s of all used start codes.

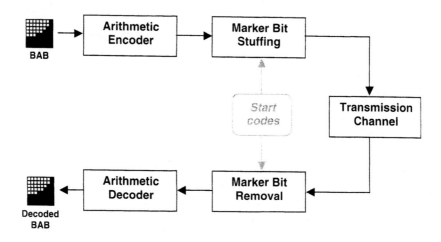

Fig. 8.21 Binary shape coding block diagram, including the stuffing of marker bits to avoid start code emulation.

information is present in the bitstream, coded in the same way as described in the previous sections.

8.3.2 Gray-Level Shape Coding

For coding gray-level or gray-scale alpha planes in MPEG-4, a binary *support mask* m_s is generated, in the first place. In this support mask, a value is set to 0 if the value of the corresponding alpha pel in the gray-level alpha mask a_g is 0; all other values in the support mask are set to 255, meaning that the corresponding alpha pels are not fully transparent. This procedure is described by the following equation:

$$m_s\left(n_x,n_y\right)=\begin{cases}0 & if\ a_g\left(n_x,n_y\right)=0 \\ 255 & otherwise\end{cases}$$

For coding the gray-scale alpha plane, both the support mask and the gray-scale alpha plane values are coded. For the coding of the support mask, the binary shape coding algorithm (as described in the previous section) is used. The gray-scale alpha values are coded as texture data with arbitrary shape, using almost the same coding method as used for the texture luminance channel. This principle is shown in Figure 8.22.

At the decoder, the gray-scale alpha plane and the binary support mask are reconstructed. Then, each pel in the decoded gray-scale alpha plane is set to 0 if the corresponding value in the support mask is 0; the remaining gray-scale plane values are not modified.

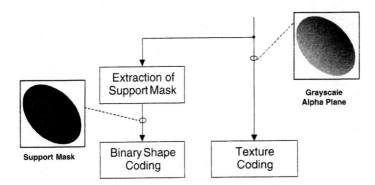

Fig. 8.22 Principal block diagram of gray-scale alpha plane coding: To transmit arbitrarily shaped alpha planes, a binary support mask is coded in addition to the gray-scale alpha values coded as luminance values.

8.3.3 Coding of Boundary Macroblocks

As described in previous sections, in MPEG-4 even arbitrarily shaped VOPs are processed and coded blockwise, using the three different types of macroblocks already presented: *transparent, opaque,* and *boundary*. For transparent MBs, only shape information is transmitted. Opaque MBs are coded as in the rectangular VOP case described in Section 8.2. For the coding of boundary MBs, some special tools are applied, which are described in this section.

8.3.3.1 Motion Compensation for Boundary MBs   In the case of arbitrarily shaped VOs, the reference VOP used for motion-compensated prediction (MCP) is also of arbitrary shape. Thus, it may happen that the MV refers to transparent pels in the reference VOP, which would not lead to efficient prediction results. In order to avoid this, the reference VOP is generated from the previously decoded VOP by a *padding process*. During this process, all transparent pels of the reference VOP are attributed a texture value that is derived from the nontransparent pels in the same VOP. For boundary MBs and for transparent MBs, different padding algorithms are applied. These padding algorithms are normative to guarantee that each decoder generates identical reference VOPs [MPEG4-2].

Boundary MBs are padded by a concatenation of *horizontal padding* and *vertical padding*, in this order. The principle is shown in Figure 8.23 for an example 8×8 block. In this figure, the VOP border alpha pels are denoted as A, B, and so on; VOP inside alpha pels are marked by an x (as their value is not used for the padding process); and the alpha pels outside the VOP are transparent and thus not marked in Figure 8.23.

The process is carried out as follows: First, transparent pels are filled by replicating the corresponding border value on the right- or left-hand side. If there is a VOP border value on each side of the pel, the two values are aver-

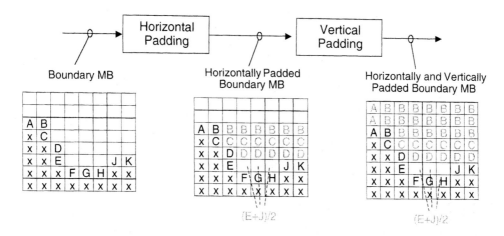

Fig. 8.23 Padding process for boundary MBs: First, transparent alpha pels are *filled* by horizontal replication and then by vertical replication of the border values. In case there are two corresponding border values for a transparent alpha pel, their mean value is used.

aged, as shown in Figure 8.23 for the third row from the bottom (with a resulting value of [E+J/2]). Then, the same process is performed for the vertical direction. Those pels that have been filled during the horizontal padding are now treated as if they would belong already to the VOP; that is, in case they are at the VOP boundary, they are replicated accordingly. Also during this step, if there is a top and bottom border pel, the resulting value is averaged. After the two steps, all pels in the boundary block are padded, and no transparent pels are left.

Transparent MBs, which are placed completely outside the VOP, are padded differently. Those transparent MBs with one or more neighboring boundary MBs are filled by replicating the samples at the border of those boundary macroblocks. If there is more than one neighboring boundary MB, then only one of the boundary MBs is chosen for padding, according to the priority rules shown in Figure 8.24.

In case more than one neighboring boundary MB exists, the transparent MB is padded by replicating upward, downward, leftward, or rightward the row or column of samples from the horizontal or vertical border of the boundary MB with the largest priority number.

Those transparent MBs that have no neighboring boundary MB at all are padded by filling them with the value $2^{bits\_per\_pixel-1}$. For 8-bit luminance and associated chrominances, this implies filling these MBs with the value 128.

8.3.3.2 Predictive Motion Vector Coding For motion-compensated prediction, the MVs that are used must be made available to the decoder. Because, in general, the MVs of neighboring blocks are similar, predictive coding is applied to reduce the bit rate for the motion information. This predictive MV

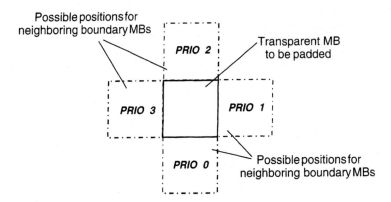

Fig. 8.24 Padding priorities for transparent MBs with one or more neighboring boundary MBs. Boundary MBs have already been completely padded before the padding of transparent MBs.

coding also is applied in Recommendation ITU-T H.263 [H263]. In MPEG-4 video some specific rules are added for the case of arbitrary shape VOPs. The general algorithm and the MPEG-4-specific rules are described in the following section.

For the prediction, up to three neighboring MVs are used, as shown in Figure 8.25.

The selection of the candidates depends on the MV mode used to code the macroblock in question. In case the 1MV mode is used, only one MV is present.

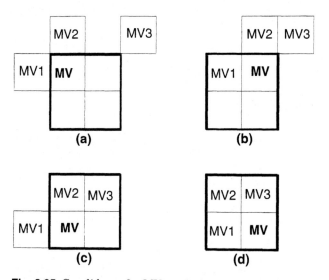

Fig. 8.25 Candidates for MV prediction: (a) candidates for 1MV mode and for the top left 8×8 block if 4MV mode is used, (b) to (d) candidates for the remaining 8×8 blocks if 4MV mode is used.

Here, the candidates shown in Figure 8.25a are used. In the figure, the bold square denotes the current macroblock and the smaller light squares denote the 8×8 blocks of the current macroblock and several 8×8 blocks of neighboring macroblocks. As shown, MVs of neighboring 8×8 blocks are always used as predictor candidates; in the case of a neighboring MB in 1MV mode, the MVs of the corresponding 8×8 blocks are set the same as the MV of the macroblock. In case the 4MV mode is used, the predictor candidates shown in Figure 8.25a–d are used, each one for that 8×8 block of the MB that is marked with **MV**.

The prediction is carried out for each component of the MV separately. For this purpose, the prediction vector value for each component is defined as the median value[12] of the three prediction candidates:

$$P_x = median\ (MV1_x, MV2_x, MV3_x); \quad P_y = median\ (MV1_y, MV2_y, MV3_y)$$

While applying the prediction candidate selection rules shown in Figure 8.25, some special cases may occur near the boundary of a VOP or if *video packets* (see Section 8.5.2.1) are used. For these cases, the following rules additionally apply:

☞ If a candidate predictor is outside the current VOP or outside the current video packet, it is classified as not valid; otherwise, it is set to the corresponding block vector.

☞ If one and only one candidate predictor is not valid, it is set to 0.

☞ If two and only two candidate predictors are not valid, they are set to the third candidate predictor.

☞ If all three candidate predictors are not valid, they are all set to 0.

After calculation of the prediction values, the difference between the actual MV components and the prediction value components is encoded using a variable-length code (Huffman) defined in the standard [MPEG4-2].

In case of B-VOPs, the basic process of predictive MV coding is exactly the same as defined for P-VOPs. The vector predictor for the delta vector is always set to 0, while the forward and backward vectors have their own vector predictors.

8.3.3.3 Texture Coding for Boundary MBs

For the coding of the luminance and chrominance values—that is, the texture of a boundary MB—two algorithms can be applied.[13] In the first algorithm, the transparent pels in the

---

12. The median() function rearranges its arguments in increasing order of the argument values. The result of the function is the argument in the middle position of the ordered list.

13. It should be noted that not all visual object types support both algorithms. In case both algorithms are supported, the encoder can choose which one to use for each video object.

boundary MBs are padded—that is, filled with values according to some rules. When padding is used, the number of luminance and chrominance values to be coded for a boundary MB is the same as for the opaque MB case (i.e., 16×16 luminance pels). The second algorithm, called *shape-adaptive DCT* (SA-DCT) has been defined in MPEG-4 Visual Version 2 (the first amendment to the MPEG-4 Visual standard; see Chapter 1). It is based on the well-known DCT principle, but it takes into account the binary alpha values for the boundary MB and therefore only codes as many values as there are nontransparent alpha pels in the MB. Both ways to code the texture of boundary MBs are described in the following section. Although the padding solution leads to a smaller computational complexity and thus is used for less complex visual object types (see Chapter 13), the SA-DCT solution leads to a higher coding efficiency and thus is used for visual object types targeting higher quality.

Padding of Transparent Pels   In principle, transparent pels of boundary blocks can be filled with arbitrary values between 0 and $2^{\text{bits\_per\_pixel-1}}$. This is acceptable because, after the decoding of the binary shape and the *filled* texture, the transparent pels (filled with some values) will be set to transparent again using the binary shape data. Therefore, from an encoding point of view these pels can be treated as *don't care*, and any value can be filled in there, up to the encoder strategy. For example, the padding techniques used for motion compensation (described in the previous section) could be applied here.

However, many other strategies exist for this kind of padding [TaKM01, ShZM01, Kaup99, MiNK98, OJJC97]. Some of them try to find the optimal padding values under the constraint that the bit rate that is required to code the texture values is minimal. This is certainly the optimal way to do the padding from an efficiency point of view, but the algorithms to *find* (i.e., to estimate) those padding values may be very complex and time-consuming. On the other hand, a simple way is to fill the transparent pels with a fixed value. Here, the *zero-padding* is often used, in particular for coding of the prediction error. The reason is that the prediction error signal is 0 or near to 0 for many pels, so it is quite efficient to fill transparent positions with 0—and this algorithm is of very low complexity.

In summary, the padding of transparent pels for boundary MBs (in relation with texture coding) is not normative, as it does not impact interoperability, and thus each encoder may choose the best strategy for its own purposes. However, in all cases the number of luminance values to code is 16×16 pels, regardless of how many pels are opaque and thus would need to be coded. In certain cases, this may decrease the coding efficiency for boundary MBs, and therefore a second algorithm is described in the following that takes into account the binary alpha values for the boundary MB.

Shape-Adaptive DCT   The principal idea of SA-DCT [KauS98, OJJC97, Kaup99] is to code only the opaque pels within boundary MBs. Therefore, the binary alpha values of the boundary MBs, which are known to the encoder and the decoder through the shape coded data, are taken into account. The algorithm comprises four steps: vertical shifting to the top, vertical one-dimensional (1D) DCT, horizontal shifting, and horizontal 1D DCT of the shifted transform coefficients. The procedure is depicted in Figure 8.26.

In the first step, the luminance or chrominance values x in the boundary MB are shifted vertically to the uppermost position. Then, for each column of this shifted MB a 1D DCT is applied, taking into account the number of opaque pels in each column. This means that for each possible number of pels, a corresponding DCT kernel is used [MPEG4-2]. In the third step, the (intermediate) transform coefficients x resulting from the vertical 1D DCT are shifted to the leftmost position. Finally, for each row of this shifted MB, a horizontal 1D DCT is applied, resulting in the final transform coefficients y.

The resulting transform coefficients are then zigzag scanned, as described in Section 8.2.3.3. However, the zigzag scan process is modified to take into account the transparent pels according to the final shift positions. If the scan covers such a transparent pel, it is omitted in the coefficient vector. Finally, the coefficients are entropy coded, as in the rectangular object case.

In addition to the normal SA-DCT, an enhanced algorithm called *ΔDC-SA-DCT* can be used (see also [KauS98]). This algorithm adds some pre- and postprocessing to the SA-DCT process previously described. In principle, the mean value of the boundary MB opaque pels is calculated and subtracted from all pel values (luminance or chrominance); this results in a 0-mean signal inside the boundary MB. Then, the SA-DCT is carried out on this mean

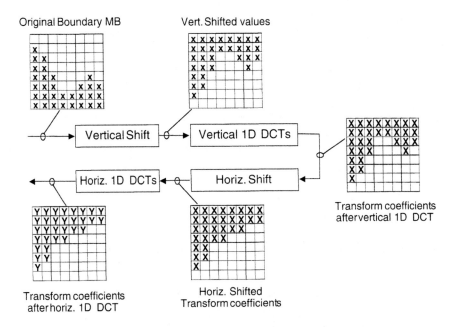

Fig. 8.26 Shape-adaptive DCT processing. The luminance and chrominance values of a boundary MB are transformed into a similar number of DCT coefficients.

free MB. After that step, the calculated mean value of the original MB is transmitted as DC value of the SA-DCT—that is, the DC value resulting from the SA-DCT is overwritten. At the decoder, these steps are inverted accordingly.

8.4 SCALABLE VIDEO CODING

In general, the entire encoded video data is needed for reconstructing the video signal. The encoded data of the *scalable coding schemes* consists of a base and one or more enhancement layer bitstreams. A lower resolution/quality signal is obtained if only the base layer bitstream is decoded. In this case, either the spatial, temporal, or amplitude (SNR) resolution is reduced for the reconstructed video signal. Because MPEG-4 enables the composition of a video scene from several independent video objects, there is also the possibility of decoding only a limited number of video objects (and not the full set of objects), providing *object-based scalability*.

The base layer decoder does not need to know anything about scalability and the upper layers; the base layer bitstream is decoded by using only the nonscalable tools described in the previous sections. The enhancement layer bitstreams contain the complementary information required to reconstruct the higher resolution/quality signals, or the remaining number of video objects in the object-based scalability case.

In this section, only the case of a single enhancement layer is described for simplicity. The extension to the multiple enhancement layers case is straightforward (for the number of enhancement layers allowed in a specific profile@level combination, see Appendix A). A generalized decoder structure for scalable coding is shown in Figure 8.27.

Besides object-based scalability, which comes automatically with the object-based composition model adopted by MPEG-4, the Visual standard [MPEG4-2] offers three tools for scalable natural video coding: (a) spatial scalability, (b) temporal scalability, and (c) SNR fine granularity scalability (FGS). Although temporal scalability can be combined with both spatial scalability and FGS, the combination of spatial scalability and FGS is not (yet) possible.

Both spatial and temporal scalability can be combined with arbitrary shape coding. In this case, the enhancement either of the complete base layer object(s) or of a partial region is possible; this choice is signaled by the enhancement_type flag. In Figure 8.28, an example for these two enhancement types is shown. For enhancement_type = 0 (*full enhancement*), the full base layer object is enhanced, which can be either a rectangular frame or an arbitrarily shaped object. For enhancement_type = 1 (*partial enhancement*), only a part of a base layer object is enhanced in the enhancement layer.[14]

---

14. These two modes are controlled at the video object level, and thus different modes can be used for different objects within the same scene.

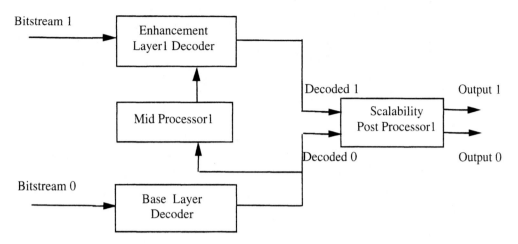

Fig. 8.27 High-level decoder structure for generalized scalability [MPEG4-2].

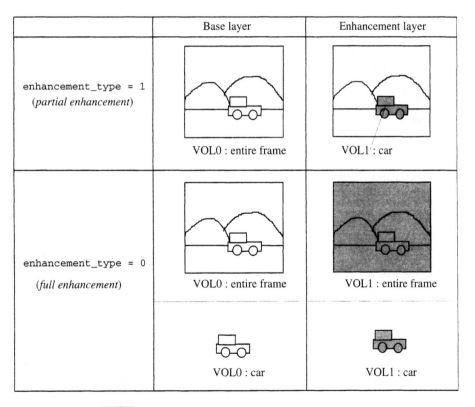

	Base layer	Enhancement layer
enhancement_type = 1 (*partial enhancement*)	VOL0 : entire frame	VOL1ʹ: car
enhancement_type = 0 (*full enhancement*)	VOL0 : entire frame	VOL1 : entire frame
	VOL0 : car	VOL1 : car

: region to be enhanced by an enhancement layer

Fig. 8.28 Video enhancement types. [N3093].

The temporal and spatial scalability tools bear a close resemblance to the corresponding MPEG-2 Video tools [MPEG2-2], but the MPEG-4 SNR FGS follows a substantially different coding scheme compared with MPEG-2 SNR scalability [MPEG4-2, MPEG2-2].

In general, the "advantage of scalability is its ability to provide resilience to transmission errors as the more important data of the lower (*base*) layer can be sent over a channel with better error performance, while the less critical enhancement layer data can be sent over a channel with poor error performance" [MPEG2-2].

Although MPEG-2 Video scalable coding tools allow for other coding standards (e.g., MPEG-1 or H.261) to perform the encoding of the base layer data, this is not possible in MPEG-4 video because of a closer coupling of the base and enhancement layer coding.

8.4.1 Spatial Scalability

In Figure 8.29 the principle behind MPEG-4 spatial scalability with motion compensation is shown. The base layer decoder reconstructs low-resolution VOPs (using only nonscalable coding tools). These are bilinearly upsampled and used to predict the higher-resolution enhancement layer VOPs (*spatial prediction* from the base layer), in combination with the enhancement layer's motion-compensated prediction (*temporal prediction* from the enhancement layer). For the basic functionality of spatial scalability corresponding to MPEG-2 Video, the reader is referred to [HaPN96, WelT94, MPEG2-2]. Here, only the differences regarding the MPEG-4 spatial scalable coding tool will be presented.

Although MPEG-2 Video allows spatial scalability of interlaced video (including a deinterlacing process in the upsampling from the base to the enhancement layer sampling grid), MPEG-4 spatial scalability is only defined for noninterlaced, progressively scanned video. Because MPEG-4 allows the coding of arbitrarily shaped video objects, spatial scalability is also supported for arbitrarily shaped objects (this functionality was added in MPEG-4 Visual, Version 2). In this case, the binary shape information for the enhancement layer is generated according to the *spatial scalability for binary shape* tool (see Section 8.4.1.1) if enhancement_type=0 (*full enhancement*), or has to be encoded in the enhancement layer without scalability if enhancement_type=1 (*partial enhancement*; see the two cases in Figure 8.28). For the spatial prediction, the base layer texture must be padded (see Section 8.3.3.1) before the bilinear texture upsampling.

As in MPEG-2 Video, the MPEG-4 spatial scalability enhancement layer bitstream consists of I-, P-, and B-VOPs (*frames* in MPEG-2 Video), but the possible enhancement layer prediction modes are different (see Figure 8.30):

☞ **I-VOPs:** In contrast to MPEG-2 Video, no spatial prediction is done for the enhancement layer I-VOPs in MPEG-4 Visual. They are decoded just like the base layer I-VOPs, without reference to any other VOP.

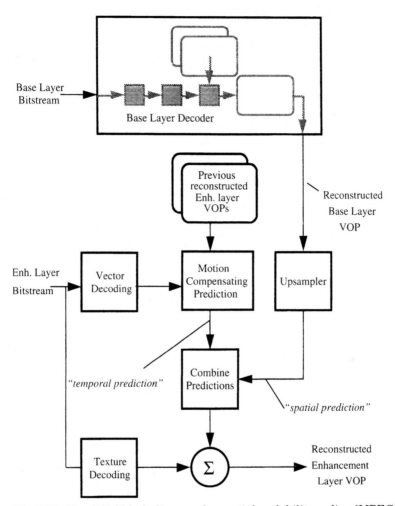

Fig. 8.29 Simplified block diagram for spatial scalability coding [MPEG4-2].

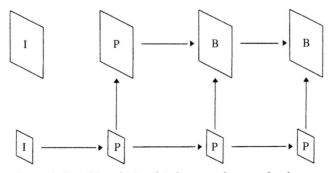

Fig. 8.30 Possible relationship between base and enhancement layers in spatial scalability [MPEG4-2].

☞ **P-VOPs:** They are predicted from the upsampled base layer only (*spatial prediction*), without the use of MVs. Thus they appear more like the I-frames of MPEG-2 spatial scalability.

☞ **B-VOPs:** Backward prediction always points to the upsampled, temporally coincident base layer VOP, and no MV information is transmitted for this direction. Thus, no *real* purely temporal enhancement layer B-VOPs are possible; they appear more like the P-frames of MPEG-2 spatial scalability. The forward prediction uses MV information. It points to the most recently decoded VOP in the enhancement layer, even if this is a B-VOP (this is not allowed in MPEG-2). The modb field shall be present for all macroblocks—that is, the rule that no information for the B-VOP macroblock is encoded if the colocated macroblock in the reference VOP is skipped does not apply for spatial scalable B-VOPs.

8.4.1.1 Spatial Scalability for Binary Shape   For the spatial scalability of binary shape information, the same enhancement layer prediction directions are used as for the texture spatial scalability (see Figure 8.30). For macroblocks in P-VOPs, the prediction is always done from the base layer; this is called the *intra mode*. For macroblocks in B-VOPs, either the intra mode or the *inter mode* is chosen by the enh_bab_type field in the MB header. The inter mode is the same as described in Section 8.2.1, with the exception that the MVs from the base layer are used after scaling them according to the horizontal and vertical shape scaling factors.

For the intra mode, the base layer shape has to be upsampled according to the horizontal and vertical shape sampling factors to match the enhancement layer sampling grid (see Figure 8.31 with the upsampling for horizontal

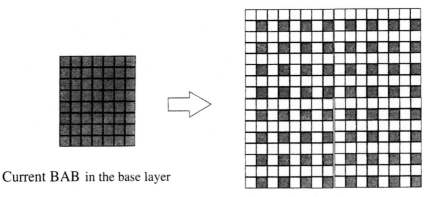

Current BAB in the base layer

Upsampled BAB for spatial prediction

Fig. 8.31 Associated pixel locations and spatial prediction of the current BAB. Gray squares represent the samples simply taken from the base layer; white squares represent the samples to be encoded in the enhancement layer (in this example both horizontal and vertical scaling factors are 2) [MPEG4-2].

and vertical scaling factors equal to 2). This upsampling process is made in a few steps (for details, see [MPEG4-2]).

The order by which the enhancement layer samples (white squares in Figure 8.31) are coded is given for an example in Figure 8.32. The different indices denote the different steps that are needed in the upsampling process if the scaling factors are not powers of 2 (B denotes the positions of the original base layer samples, V_r and H_r the samples repeated in the first upsampling step, and V_{px} and H_{px} the samples inserted by vertical and horizontal upsampling in step x, respectively). For a detailed explanation of these indices, see [MPEG4-2].

For binary shape spatial scalability, the *scan interleaving* (SI) method is used. It consists of several passes in the vertical and horizontal directions

Hp2	Hp2	Hp2	Hp2	Hp2	Hp2	Hp2	Hp2	Hp2	Hp2	Hp2	Hp2	Hp2	Hp2	Hp2	Hp2
Hp1	Hp1	Hp1	Hp1	Hp1	Hp1	Hp1	Hp1	Hp1	Hp1	Hp1	Hp1	Hp1	Hp1	Hp1	Hp1
Hp2	Hp2	Hp2	Hp2	Hp2	Hp2	Hp2	Hp2	Hp2	Hp2	Hp2	Hp2	Hp2	Hp2	Hp2	Hp2
Hr	Hr	Hr	Hr	Hr	Hr	Hr	Hr	Hr	Hr	Hr	Hr	Hr	Hr	Hr	Hr
Hp2	Hp2	Hp2	Hp2	Hp2	Hp2	Hp2	Hp2	Hp2	Hp2	Hp2	Hp2	Hp2	Hp2	Hp2	Hp2
Hp1	Hp1	Hp1	Hp1	Hp1	Hp1	Hp1	Hp1	Hp1	Hp1	Hp1	Hp1	Hp1	Hp1	Hp1	Hp1
Hp2	Hp2	Hp2	Hp2	Hp2	Hp2	Hp2	Hp2	Hp2	Hp2	Hp2	Hp2	Hp2	Hp2	Hp2	Hp2
Vp1	Vr	Vp1	B	Vp1	B	Vp1	B	Vp1	Vr	Vp1	B	Vp1	B	Vp1	B
Hp2	Hp2	Hp2	Hp2	Hp2	Hp2	Hp2	Hp2	Hp2	Hp2	Hp2	Hp2	Hp2	Hp2	Hp2	Hp2
Hp1	Hp1	Hp1	Hp1	Hp1	Hp1	Hp1	Hp1	Hp1	Hp1	Hp1	Hp1	Hp1	Hp1	Hp1	Hp1
Hp2	Hp2	Hp2	Hp2	Hp2	Hp2	Hp2	Hp2	Hp2	Hp2	Hp2	Hp2	Hp2	Hp2	Hp2	Hp2
Vp1	Vr	Vp1	B	Vp1	B	Vp1	B	Vp1	Vr	Vp1	B	Vp1	B	Vp1	B
Hp2	Hp2	Hp2	Hp2	Hp2	Hp2	Hp2	Hp2	Hp2	Hp2	Hp2	Hp2	Hp2	Hp2	Hp2	Hp2
Hp1	Hp1	Hp1	Hp1	Hp1	Hp1	Hp1	Hp1	Hp1	Hp1	Hp1	Hp1	Hp1	Hp1	Hp1	Hp1
Hp2	Hp2	Hp2	Hp2	Hp2	Hp2	Hp2	Hp2	Hp2	Hp2	Hp2	Hp2	Hp2	Hp2	Hp2	Hp2
Vp1	Vr	Vp1	B	Vp1	B	Vp1	B	Vp1	Vr	Vp1	B	Vp1	B	Vp1	B

Fig. 8.32 Example of the decoding order for a horizontal scaling factor of 8/3 and a vertical scaling factor of 5/1 (decoding order, left to right and top to bottom: B from the base layer $\rightarrow V_r \rightarrow V_{p1} \rightarrow H_r \rightarrow H_{p1} \rightarrow H_{p2}$) [MPEG4-2].

(according to the decoding order shown in Figure 8.32) until all enhancement alpha pels are decoded. For each scan pass, the alpha pels are classified according to their position regarding the already decoded reference samples in the reference scan lines (see Figure 8.33). There are three possible cases:

☞ **Predictable case:** Both reference samples (as shown in Figure 8.33) have the same value.

☞ **Level transitional case:** The reference samples have *different values.*

☞ **Exceptional case:** Both reference samples have the same value, but the value to be coded is different.

If the entire BAB consists only of predictable and transitional cases, the BAB type is set to *Transitional BAB,* and only the transitional samples are coded. If one or more exceptional cases exist, the BAB type is set to *Exceptional BAB* and all samples have to be coded.

For the context computation of the samples to code, the context templates presented in Figure 8.34 are used. This information is used to code the sample values in the way described in Section 8.3.1.

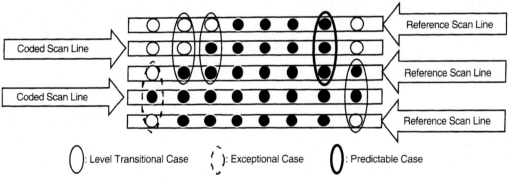

Fig. 8.33 Scan interleaving for scalable coding with a scaling factor of 2 (only the horizontal scan is shown) [MPEG4-2].

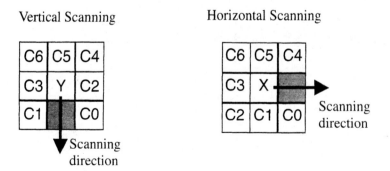

Fig. 8.34 Context templates for the horizontal and vertical scan interleaving directions [MPEG4-2].

8.4.2 Temporal Scalability

In Figure 8.35, the basic principle of temporal scalability is shown. The base layer decoder reconstructs VOPs with low temporal resolution, which are used together with the previously reconstructed enhancement layer VOPs in the enhancement layer's motion-compensated prediction. For the basic functionality of temporal scalability, the reader is referred to [HaPN96, WelT94, MPEG2-2]. Here, only the differences regarding the MPEG-4 temporal scalable coding tool will be presented.

The prediction of the enhancement layer I-, P-, and B-VOPs in MPEG-4 temporal scalability is made in the same way as for the I-, P-, and B-frames in MPEG-2 Video [MPEG2-2], with the exception that no reference to an interlaced base layer is possible; this means the functionality of interlaced-to-progressive temporal scalability is not supported by MPEG-4 Visual. The possible base-to-enhancement layer prediction directions are signaled by the `ref_select_code` transmitted in the VOP header (see Figure 8.35).

For P-VOPs, the `ref_select_code` is either 00, 01, or 10:

☞ **`ref_select_code` = 00:** The prediction reference is the most recently decoded VOP belonging to the same layer.

☞ **`ref_select_code` = 01:** The prediction reference is the previous VOP in display order belonging to the reference layer.

☞ **`ref_select_code` = 10:** The prediction reference is the next VOP in display order belonging to the reference layer.

For B-VOPs, the `ref_select_code` is either 01, 10, or 11:

☞ **`ref_select_code` = 01:** The forward-prediction reference is the most recently decoded VOP belonging to the same layer, and the backward-

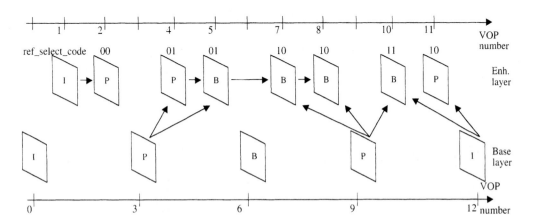

Fig. 8.35 Possible base-to-enhancement layer prediction directions depending on `ref_select_code`.

prediction reference is the previous VOP in display order belonging to the reference layer.

☞ **ref_select_code = 10:** The forward-prediction reference is the most recently decoded VOP belonging to the same layer, and the backward-prediction reference is the next VOP in display order belonging to the reference layer.

☞ **ref_select_code = 11:** The forward-prediction reference is the previous VOP in display order belonging to the reference layer, and the backward-prediction reference is the next VOP in display order belonging to the reference layer.

There are some minor differences for the enhancement layer B-VOPs compared with the base layer B-VOPs in MPEG-4: If the forward reference points to the most recently decoded VOP of the same (i.e., the enhancement) layer, the direct prediction mode shall not be used. As for spatial scalability, the modb field shall be present for all macroblocks—that is, the rule that no information for the B-VOP macroblock is encoded if the colocated macroblock in the reference VOP is skipped does not apply for temporal scalable B-VOPs. And, just like in MPEG-2 Video, the enhancement layer B-VOPs may be used as reference for future temporal prediction.

A special treatment, signaled by the background_composition flag in the VOP header, can be used in the case of arbitrarily shaped temporal scalable video objects. This is relevant for the case of *partial enhancement* (enhancement_type=1; see Figure 8.36 and Figure 8.37), in which only a part (or region) of the VO is temporally enhanced.

If the temporally enhanced region of the video object is moving or changing its shape, a *hole* in the background can develop for which no texture information is available. In this case, a *backward* and a *forward shape* are coded

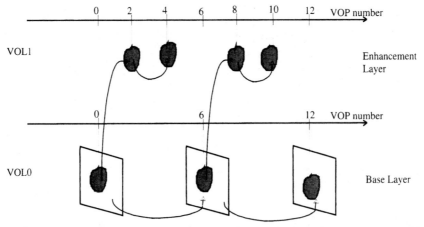

Fig. 8.36 Partial temporal enhancement of a video object using P-VOPs [MPEG4-2].

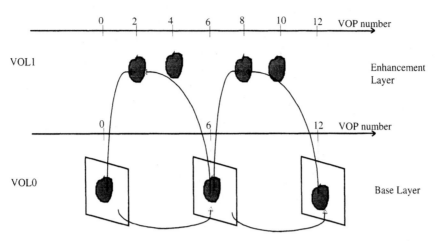

Fig. 8.37 Partial temporal enhancement of a video object using B-VOPs [MPEG4-2].

together with the object and used to select the texture information for this hole by the following padding process (see Figure 8.38):

☞ The part of the background that is covered by neither the backward nor forward shapes is filled with the pixel values taken from the temporally closer VOP.

☞ For the part that is covered by the object in the previous VOP (described by the forward shape) but belongs to the background in the next VOP, the background pixel values are taken from the next VOP.

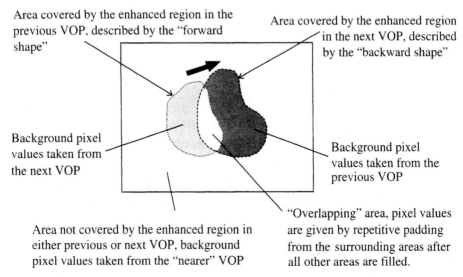

Fig. 8.38 Background composition for partial enhancement temporal scalability [MPEG4-2].

☞ For the part that is covered by the object in the next VOP (described by the backward shape) but belongs to the background in the previous VOP, the background pixel values are taken from the previous VOP.

☞ For the overlapping part of the two shapes, repetitive padding (see Section 8.3.3) is applied after all the other regions are filled.

The decoding process for the backward and forward shapes is the same as the decoding process for the shape of I-VOPs with `binary_only` mode (as described in Section 8.3.1.3).

The background composition is not needed for the *full enhancement* case (`enhancement_type=0`; see Figure 8.39), as here the texture information for the separate background object (VO0) is fully available.

8.4.3 SNR Fine Granularity Scalability

The MPEG-4 FGS coding tool "allows the coverage of a wide range of bit rates for the distribution of video on Internet with the flexibility of using multiple layers, where there is a wide range of bandwidth variation" [N3904]. It is built to allow a separation between the encoding and the distribution process: The encoder produces only one (rather large) enhancement layer bitstream, while the FGS server application truncates this bitstream according to the characteristics of the transmission channel (e.g., bit rate) or the decoder (e.g., computational power).

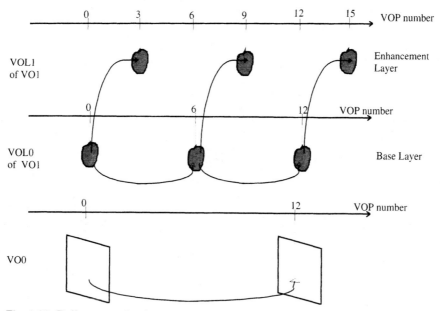

Fig. 8.39 Full temporal enhancement of video object 1 (`enhancement_type=0`, no background composition necessary) [MPEG4-2].

The concept of amplitude scalability (also referred to as quality or SNR scalability) is fundamentally different between MPEG-2 Video and MPEG-4 Visual: In MPEG-2, the requantized reconstruction error of the base layer is coded in the enhancement layer bitstream, using the same quantization mechanisms and VLC tables as the base layer. Both base and enhancement layer data are used to reconstruct the frames from which further temporal prediction is carried out. Thus, a decoder that has access only to the base layer data cannot generate the same reference frames as the encoder, and a so-called *drift* of the prediction signals occurs.

In MPEG-4 SNR FGS, the reconstruction error of the base layer is encoded in the enhancement layer using a *bit plane* representation of the DCT coefficients (see Figure 8.40). The absolute values of the DCT coefficients (Figure 8.40a) are arranged according to their binary representation (Figure 8.40b). Then, the coefficient information is grouped bit plane by bit plane, starting from the most significant bit plane (MSB), until the least significant bit plane (LSB). This is followed by run-length coding of the bits for each bit plane (Figure 8.40c).

First, the MSBs of the enhancement layer signal are encoded into the bitstream for all macroblocks, followed by the second most significant bit planes,

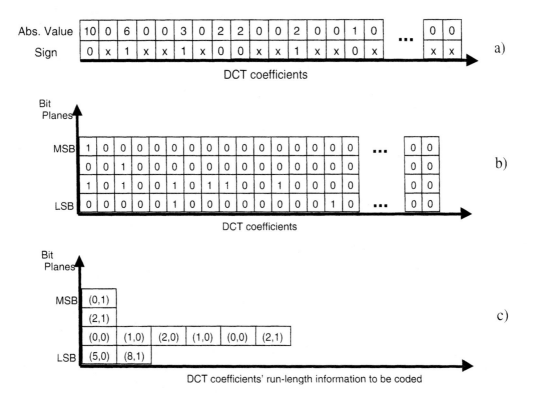

Fig. 8.40 FGS bit plane encoding.

and so on. Thus, it is possible to stop the transmission of the enhancement layer data at any point, while being able to make use of all transmitted data up to that point. In contrast to MPEG-2, only the reconstructed base layer data is used for further temporal prediction, so that no drift of the prediction signals between the encoder and a base layer decoder may occur (see Figure 8.41).

MPEG-4 FGS is defined only for rectangular, nonarbitrarily shaped video objects. To provide functionality similar to the adaptive quantization matrix and to the macroblock-adaptive quantization step size used in the encoding of the base layer, MPEG-4 FGS applies frequency weighting and selective enhancement shifting factors (block *Bit-plane Shift* in Figure 8.41) before encoding the bit planes. This means that for certain MBs in the VOP (regions of interest for which a selectively enhanced representation by the encoder may improve the quality impact) or for the most relevant DCT coefficients (weighting of the DCT coefficients' spatial frequencies) a finer quantization is applied in the FGS enhancement layer. This is done by simply shifting up the corresponding coefficients by one or more bit planes according to a key or mask that is also sent to the decoder. As a result, the most relevant MBs or DCT coefficients in the enhancement layer are coded first, and thus have a higher probability of being used by the decoder if the bitstream is truncated at a certain stage for any reason.

It is possible to combine the FGS bit plane coding with temporal scalability. Then, the prediction for the FGS temporal scalable (FGST) VOPs can only be made from the base layer. Each coded FGST VOP has two separate parts in the bitstream: The first part contains the MV data, and the second part contains the DCT texture data. The syntax for the first part is similar to that of MPEG-4 temporal scalability. The DCT texture data in the second

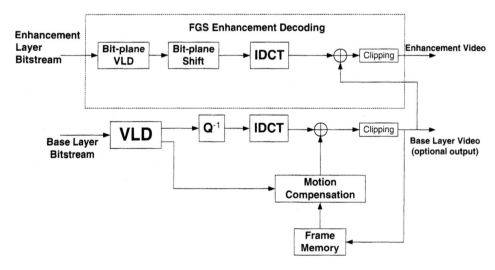

Fig. 8.41 Basic FGS decoder structure [MPEG4-2].

part is encoded using bit plane coding in the same way as for nontemporal scalable FGS.

There are two possibilities for combining FGS and FGST VOPs: (a) by using two separate enhancement layer bitstreams for FGS and FGST (see Figure 8.42), or (b) by using one combined FGS-FGST enhancement layer bitstream (see Figure 8.43). To distinguish between FGS and FGST data in the latter case, the `fgs_vop_coding_type` parameter in the VOP header is used; it is I in the case of an FGS VOP, and P or B in the case of an FGST VOP.

8.5 SPECIAL VIDEO CODING TOOLS

This section describes the MPEG-4 special video coding tools. They are the interlaced, error-resilience, reduced resolution, sprite coding, and high-quality texture coding tools. In addition, the MPEG-4 short video header operational mode is presented in this section.

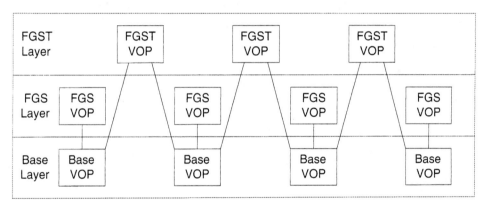

Fig. 8.42 Combining FGS and FGST using two separate enhancement layers [MPEG4-2].

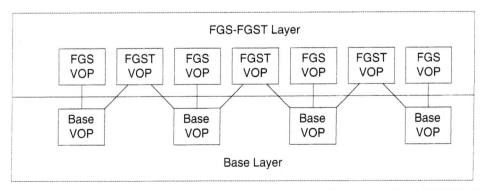

Fig. 8.43 Combining FGS and FGST using one combined enhancement layer [MPEG4-2].

8.5.1 Interlaced Coding

Interlaced video is widely used in TV broadcasting. It still represents a good compromise between frame rate and spatial resolution and there is substantial legacy content. Acknowledging this fact, MPEG-4 has adopted a set of tools to support the compression of arbitrarily shaped interlaced VOPs. Some of these tools are extensions of the MPEG-2 interlaced video coding tools [MPEG2-2], such as frame/field DCT, frame/field motion compensation, and interlaced direct mode. In addition, some MPEG-4 video coding tools have also been extended to support interlaced VOPs, such as field padding and AC/DC prediction. Some of these tools will be briefly described in this section. For more details about all of the interlaced coding tools and their performance, please refer to [MPEG4-2, PXEL00].

8.5.1.1 Frame/Field DCT   In the case of interlaced VOPs with significant motion activity, the pixels in an MB may have vertical correlation smaller than the horizontal one as adjacent lines come from different fields. In this case, better decorrelation is achieved by using field DCT (i.e., performing 8×8 DCT on pixels that belong to the same field) rather than using frame DCT (i.e., performing 8×8 DCT on pixels that belong to different fields). To perform field DCT, the encoder reorders the luminance lines within an MB so that the first eight lines come from the top field and the last eight lines come from the bottom field. This MB is labeled as *field type* MB, whereas the normal MB (i.e., without lines reordering) is labeled *frame type* MB. MPEG-4 field/frame DCT is identical to MPEG-2 field/frame DCT [MPEG2-2].

8.5.1.2 Frame/Field Motion Compensation and Interlaced Direct Mode
Similarly to the DCT case, the motion compensation of interlaced VOPs with significant motion can be improved if performed between fields of the same parity. This concept was exploited in MPEG-2 by introducing field motion compensation, which has been adopted in MPEG-4, as well. When field motion compensation is used, field MVs are calculated for both the top field and the bottom field macroblocks, which are of size 16×8 pels. In B-VOP, the direct mode is extended to better support interlaced VOPs. Interlaced direct mode can be used only for those MBs whose colocated MB in the future-reference VOP uses field motion compensation. For more details on interlaced direct mode, refer to [MPEG4-2, PXEL00].

8.5.1.3 Field Padding   All of these interlaced coding tools can be applied to arbitrarily shaped interlaced VOPs. This is done by using field padding [M2378]. Field padding performs the same repetitive and extended padding used for progressive VOPs (see Section 8.3.3.3), but the vertical padding of the luminance component is conduced separately for each field. This assures that the separation between fields is kept, thus extending to the padded region the correlation existing between pixels of each field.

8.5.2 Error-Resilient Coding

Most of the communication channels used by current applications, and targeted by future ones, to transmit compressed image and video data are error prone. Error-robust source coding techniques have proved helpful in effectively addressing the problem of minimizing the impact of transmission errors. An error-robust source coding technique has the capability to increase the portion of a corrupted bitstream that can be correctly decoded. The larger the correctly decoded portion of data, the more robust the coding is. A robust decoding technique should be able to detect the presence of errors, localize the errors with the best possible precision, and continue the decoding process for the remaining uncorrupted data [DuPe99]. Because of its capability to return to the normal decoding process, error-robust source coding is also referred to as error-resilient coding. The techniques that contribute to build error-resilient coding are referred to as error-resilience tools.

Because of the wide range of applications and communication channels they need to consider, error-resilience techniques at the source coding level face many challenges. They need to deal with a broad range of error conditions: random, burst, or packet loss type. Even very few errors can render the compressed data totally unusable, because many coding techniques are not error robust by nature. For example, predictive coding and variable length coding are able to provide high compression efficiency but are not resilient to such errors. To make things even worse, an error occurring at a certain point in the bitstream can propagate through the bitstream, resulting in the corruption or inability to decode the rest of the entire coded data. Although a solution may be to retransmit the corrupted data, real-time applications often impose a low delay constraint, which renders this solution powerless.

To face these challenges, MPEG-4 has adopted error-resilience tools that provide basic error robustness while satisfying a number of constraints. The error-resilience tools should be easily integrated with the core encoding algorithms and have minimal impact on the coding efficiency. Also, the complexity introduced by these tools should be minimal. Moreover, they should provide a flexible and scalable syntax that can adapt to the characteristics of the transmission channel and allow a trade-off between robustness and efficiency.

As a consequence, the adopted MPEG-4 error-resilience tools are meant to be used together with, not in replacement of, traditional technologies (such as channel coding), which are outside the scope of MPEG-4 Visual. In this scenario, the combined use of traditional channel coding techniques and error-resilience source coding tools reduces the error-resilience overhead while maintaining the same error-resilience efficacy. Such overhead, introduced by the heavy use of channel coding techniques, is reduced by allowing these techniques to leave residual errors in the data stream. The error-resilience efficacy is maintained by using the error-resilience source coding tools to take care of such residual errors. Finally, in coherency with its "specify the minimum" principle (see Chapter 1), MPEG has chosen to minimize the standardization of the error-resilience tools decoding procedure (e.g., error detection, localization, and concealment are free), as its specification is not required to guarantee interoperability. This approach leaves implementers free to choose the best way to use the error-resilience tools, thus improving product differentiation.

Five error-resilience tools are specified in MPEG-4 Visual [MPEG4-2]:

1. Packet-based periodic resynchronization,
2. Data partitioning (DP),
3. Reversible variable-length codes (RVLCs),
4. Header extension code (HEC), and
5. New prediction (NEWPRED).

These tools enable better resynchronization, error localization, data recovery, and error-concealment capabilities and have been adopted in different versions of the MPEG-4 Visual standard. In particular, packet-based periodic resynchronization and DP enable better resynchronization and better error localization. RVLCs and HEC help data recovery, whereas NEWPRED reduces error propagation. The resynchronization and HEC tools may be applied to both video and visual texture coding, whereas the remaining ones may be applied to video only. Resynchronization and HEC for visual texture coding (VTC) are presented in Section 8.6.5. Results of the verification tests accomplished to assess the MPEG-4 video error-resilience tools performance are presented in Chapter 15.

MPEG-4 does not incorporate explicitly an error-resilience mode. Instead, it mandates a list of error-resilience tools that must be supported by all visual object types (see Chapter 13 for more details). Also, the decoding procedure in the presence of transmission errors is outside the scope of the standard. In other words, the standard does not specify which actions the decoder should take when an error is detected (decoder behavior is defined in error-free conditions). The reader may refer to the informative Annex E in the MPEG-4 Visual standard [MPEG4-2] for suggestions on how to detect errors and take advantage of the MPEG-4 error-resilience tools to improve resynchronization, error localization, and data recovery, and to reduce error propagation. Similarly, error concealment is outside the scope of the MPEG-4 Visual standard. Simple informative error-concealment techniques are suggested in Annex E of the MPEG-4 Visual standard [MPEG4-2]. Several state-of-the-art error-concealment techniques based on temporal, spatial, and frequency domain prediction of the lost data also are discussed in [WanZ98].

In addition to using the MPEG-4 error-resilience tools, there are a number of non-normative but standard-compatible encoder-only techniques that can improve the error-resilience performance. An example of such techniques, called *adaptive intra refresh*, is presented in Annex E of the MPEG-4 Visual standard [MPEG4-2] and in [ImuM99].

8.5.2.1 Packet-Based Resynchronization Due to the use of VLCs, compressed bitstreams are particularly sensitive to channel errors. In VLC, the codeword length is implicit, and transmission errors typically lead to an incorrect number of bits being used when decoding. This causes the decoder to lose synchronization with the encoder (see Figure 8.44). If remedial measures are not taken, the lost synchronization causes some of the bits following the corrupted ones to be erroneously decoded, thus degrading the quality of the decoded video data. Eventually, the decoder will flag an error, having encountered an invalid VLD value or an illegal bitstream parameter. The remaining part of the bitstream will become totally unusable.

Resynchronization tools, as the name implies, attempt to reestablish the synchronization between the decoder and the bitstream after an error has been detected. This is achieved by means of resynchronization markers. These markers must be unique codes (i.e., a sequence of bits that cannot be emulated by any code, or combination of codes, used by the encoder). At the encoder side, these markers are inserted into the bitstream prior to transmission. When an error is detected at the decoder side, the decoder searches for the next resynchronization marker. Once it is found, the synchronization is reestablished, thus allowing the correct decoding of the remaining bits. This process is illustrated in Figure 8.44. The data between the resynchronization marker prior to the detected error and the newly found resynchronization marker are generally discarded, but some of this data may also be used. An effective resynchronization tool makes error recovery and concealment easier.

Assuming a block-based coder, one approach to resynchronization is to insert the *resync markers* after a fixed number of encoded MBs. This approach usually is referred to as *periodic spatial resynchronization*. However, in a variable bit-rate coding system, the resynchronization markers are likely to be unevenly spaced in the bitstream.

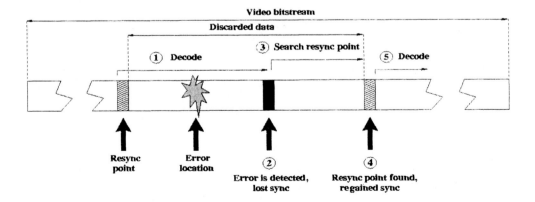

Fig. 8.44 Decoder resynchronization following error detection.

Previous video coding standards (such as H.261 and H.263) implement spatial resynchronization by logically partitioning each of the images to be encoded into units called group of blocks (GOBs) [H261, H263]. GOBs correspond to one or more rows of macroblocks,[15] and may be preceded by a GOB header that contains a resynchronization marker. The number of GOBs in a frame is determined by the frame size. Therefore, if the GOB header is present, it must be inserted at the beginning of a predetermined set of macroblock rows. The only flexibility left to the encoder is to decide, on a GOB-by-GOB basis, whether or not a GOB is preceded by a GOB header (only for H.263, as this choice is not possible in H.261).

An improvement to the periodic spatial resynchronization method is to allow periodic resynchronization at approximately constant bitstream intervals. In this approach, the marker location is not decided by the number of macroblocks but by the number of bits instead: A new marker is inserted after the encoder has generated a predetermined number of bits. To support this approach, MPEG-4 has adopted a packet-based resynchronization solution for its video bitstream syntax.

The MPEG-4 encoder is not restricted to inserting the resynchronization markers only at the beginning of each row of macroblocks, but it has the option of dividing the frame into video packets. Each video packet consists of an integer number of consecutive coded macroblocks preceded by a video packet header. The video packet header contains a resynchronization marker. The video packet can span several rows of macroblocks in the VOP and can even include partial rows of macroblocks. However, each macroblock belongs to only one packet. One suggested mode of operation for an MPEG-4 video encoder is to insert a resynchronization marker periodically, every K bits. However, the length of the video packets is not rigidly fixed; each video packet can have a different length. This implies that the entire frame may consist of a single video packet. Therefore, a VOP may contain one or more video packets. Finally, because the first video packet of a VOP does not have a video packet header (the first video packet header is replaced by the VOP start code, which can also be used to resynchronize the decoder), a VOP may contain none, one, or more video packet headers—that is, none, one, or more resynchronization markers (besides the VOP start code).

Figure 8.45 and Figure 8.46 illustrate the difference between the H.263 spatial periodic resynchronization and the MPEG-4 packet-based resynchronization.

To understand the advantages of the MPEG-4 packet-based approach over the spatial periodic resynchronization approach, consider the case in which there is a significant activity in one area of the VOP. The macroblocks corresponding to this active area generate more bits than the other parts of

---

15. Note that in H.261 the GOB for a common intermediate format (CIF) picture (i.e., 352 pels per line, 288 lines for the luminance, 4:2:0 color format) consists of a rectangular zone with 11×3 MBs.

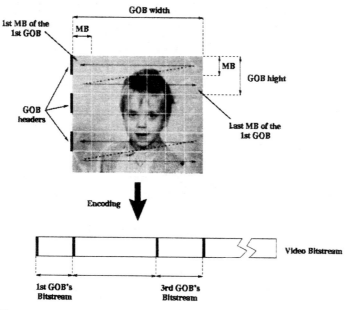

Fig. 8.45 H.263 periodic spatial resynchronization approach: Resynchronization markers are introduced at fixed places.

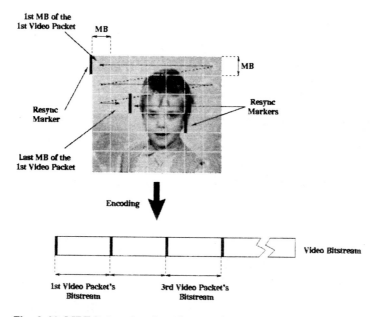

Fig. 8.46 MPEG-4 packet-based resynchronization approach: Resynchronization markers are equally spaced in the bitstream.

the VOP. If the MPEG-4 video encoder inserts the resynchronization markers at uniformly spaced bit intervals, the resynchronization markers are much closer in the high-activity areas and farther apart in the low-activity areas. Thus, in the presence of a short burst of errors, the decoder can quickly localize the error within a few macroblocks in the important high-activity areas of the frame and preserve their visual quality. In the case of H.263, in which the resynchronization markers are restricted to the beginning of the GOBs, the decoder can only isolate the errors to a row of macroblocks, and the error localization is independent of the image content. Version 2 of H.263 has adopted a resynchronization scheme similar to the one used by MPEG-4 (see [H.263], Annex on slice structure mode).

Note that, in addition to inserting the resynchronization markers at the beginning of each video packet, the MPEG-4 encoder removes all dependencies that exist between data belonging to two consecutive video packets. This assures that, even if one of the video packets is corrupted by transmission errors, all of the remaining packets can be decoded. To remove these dependencies, the encoder adds two additional fields to the video packet header: (a) the absolute macroblock number of the first macroblock in the video packet, which indicates the spatial position of the macroblock in the current VOP; and (b) the quantization step for the same first macroblock in the packet. These additional fields are referred to as video packet header (see Figure 8.47). The encoder also modifies the in-picture predictive encoding tools (i.e., AC/DC prediction and MV predictive coding) by limiting the predictions within the video packet boundaries. All other tools remain unchanged.

8.5.2.2 Data Partitioning   MPEG-4 also has adopted a data partitioning mode to enable better resynchronization and error localization. This mode divides or rearranges the bitstream elements (DC and AC coefficients, MVs, etc.) into groups according to their sensitivity to errors. The same error in two different groups can have very different impacts. This observation has been the starting point for early research aiming at partitioning the bitstream into higher- and lower-priority components in the context of asynchronous transfer mode (ATM) networks or other packetized networks [Ghan89]. However, it

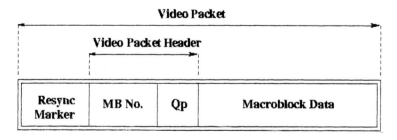

Fig. 8.47 MPEG-4 video packet syntax.

may be difficult, if not impossible, to prioritize the data being transmitted over transmission channels such as existing analog phone lines or wireless networks.

The same observation is at the basis of the MPEG-4 data partitioning mode, which can be used both for video and for visual texture coding. Data partitioning for visual texture coding is described in Section 8.6.5.

After detecting an error in the bitstream and resynchronizing to the next resynchronization marker, most decoders typically discard all data between the two resynchronization markers as being in error. There are two reasons for doing so. The first reason is the uncertainty about the exact location where the error occurred. The second reason is the dependency between different components of the bitstream. As a consequence, even if the corrupted bits can be separated from the uncorrupted ones, their decoding process requires information that is conveyed by the corrupted bits.

Data partitioning reduces the amount of discarded data by providing better error localization and modifying the macroblock syntax so that the correct bits can be decoded. It does so by separating the macroblock data into high-priority and low-priority components, the former being the one that, if corrupted, will most affect the visual quality of the reconstructed macroblock. The modified macroblock syntax is henceforth referred to as *data partitioning syntax* [MPEG4-2]. It should also be noted that MPEG-4 supports a data partitioning syntax for I- and P-VOPs only. In addition, data partitioning has been extended to binary shape information (see [YWJK00] for more details). Finally, data partitioning also has been adopted in 3D mesh coding (see Chapter 9).

For I-VOPs, and for intra macroblocks in P-VOPs, the data partitioning mode organizes the macroblock data within a packet, as shown in Figure 8.48. The resync marker and the video packet header are followed by macroblock coding mode information and six DC coefficients (four for the luminance blocks and two for the chrominance blocks) for each macroblock in the video packet. A DC marker (DCM) with 19 bits signals the end of the first data part. The second part contains the AC DCT coefficients.

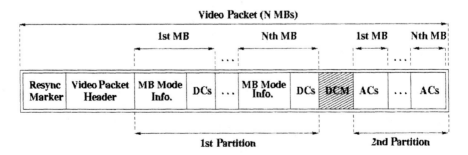

Fig. 8.48 Bitstream organization for I-VOP video packets with data partitioning.

For P-VOPs, the data partitioning mode organizes the macroblock data into a motion part and a texture part, separated by a motion marker (MM) with 17 bits, both preceded by the resynchronization marker and the video packet header, as shown in Figure 8.49. The motion part contains the macroblock syntax elements that are required to decode the MVs—that is, COD, MCBPC, and one or four MVs, this for each macroblock in the video packet. The texture part contains all the remaining macroblock syntax elements that are required to decode the DCT coefficients (i.e., CBPY, DQUANT, and the 64 or less DCT coefficients) for each macroblock in the video packet.

The DCM/MM markers have been computed from the DCT and MV's VLC tables using a search program that assures that each marker has a Hamming distance of 1 from any possible valid combination of the DCT and MV's VLC tables [TMNC99]. Therefore, each marker is uniquely decodable from the DCT and motion VLCs and gives the decoder knowledge about where to stop decoding DC coefficients or MVs before beginning to decode AC coefficients (I-VOPs) or texture information (P-VOPs).

8.5.2.3 Reversible Variable-Length Coding

After synchronization is reestablished, data recovery tools attempt to recover as much as possible of the data between the previous and the current resynchronization marker. How much data recovery can be achieved depends on the underlying coding scheme. In the case of variable length codes (VLCs), all the data between the two resynchronization markers usually is discarded. Reversible variable-length coding (RVLC) alleviates this problem by enabling the decoder to better isolate the errors, thus improving data recovery in the presence of errors [JiaV99, KaiB00].

RVLCs are special VLCs that can be uniquely decoded in both the forward and reverse directions [JiaV98] (see [BHWR00] for details on how the MPEG-4 RVLCs are built). The advantage of these codewords is that when the decoder detects an error (while decoding the bitstream in the forward direction), it can search for the next resynchronization marker and from there decode the bitstream in the backward direction until it encounters a new error

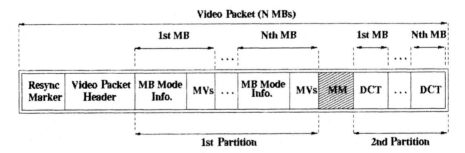

Fig. 8.49 Bitstream organization for P-VOP video packets with data partitioning.

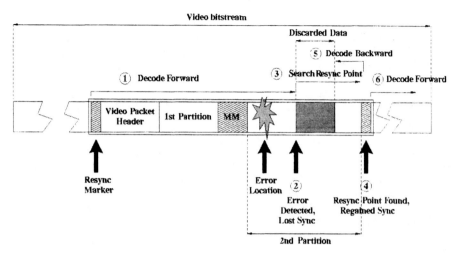

Fig. 8.50 Data recovery with RVLC.

(see Figure 8.50). Based on the location of the two errors, the decoder can recover some of the data that otherwise would have been discarded.

By proper use of training sequences, the RVLCs can be built to match the probability characteristics of the data to be coded, thus maintaining the ability to compactly pack the bitstream while retaining the error-resilience properties. MPEG-4 supports the use of an efficient and robust RVLC table for encoding the DCT coefficients only in conjunction with DP syntax. When RVLC is used, RVLCs are used to encode the AC coefficients (i.e., the second partition) of intra coded macroblocks, and the DCT coefficients (i.e., the texture partition) of inter coded macroblocks. Note that MPEG-4 does not standardize how to decode the RVLCs in the presence of transmission errors. However, some suggested strategies are described in (informative) Annex E of the MPEG-4 Visual standard [MPEG4-2].

8.5.2.4 Header Extension Code A simple approach to improve data recovery is to duplicate part of the data at random locations in the bitstream. Although it can become very inefficient from the compression point of view if used on a large scale, this approach has been proved successful in preserving vital information. An example of such vital information is the VOP header. This header includes information about the VOP spatial dimension, the time stamps associated with its decoding and presentation, and its coding mode (inter or intra). If some of this information is corrupted by channel errors, the decoder has no alternative but to discard the entire VOP.

To reduce the sensitivity of this data, MPEG-4 uses HEC. In each video packet, a 1-bit field called HEC is introduced (see Figure 8.51 for the HEC position in the video packet). When this bit is set to 1, it is followed by the duplicated VOP header information. Because of the trade-off between the overhead

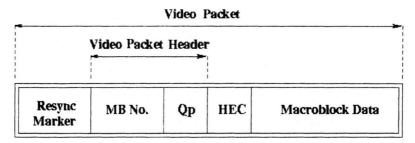

Fig. 8.51 MPEG-4 video packet syntax with HEC.

and error-recovery capability involved with the use of HEC, the MPEG-4 standard does not set any limit on how frequently this tool can be used, and thus it is an encoder choice. Moreover, as for all other MPEG-4 error-resilience tools, the standard does not specify how to use HEC-duplicated information in the presence of errors.

8.5.2.5 New Prediction   Because of the temporal prediction scheme that is at the basis of the motion compensation (MC) hybrid coding used by MPEG-4 video, errors introduced during transmission will affect the visual quality not only of the corrupted VOP but of all the VOPs that use the corrupted one as reference for their temporal predictions. Then, each VOP corrupted will propagate the errors when used as reference for the following VOPs. The error propagation will continue until some intra refreshment occurs.

Cyclic intra refresh or adaptive intra refresh can reduce the error propagation by periodically or selectively intra refreshing the video coding chain [MPEG4-2]. All of these approaches suffer from a number of drawbacks: (a) They cannot adapt to the always-changing conditions of some transmission channel (wireless, etc.), (b) they significantly degrade the coding efficiency, and (c) they provide a slow error recovery.

To improve error recovery, MPEG-4 has adopted a newly developed technique: NEWPRED. NEWPRED provides a fast error recovery for temporal error propagation in real-time coding applications. It uses an upstream channel from the decoder to the encoder, usually referred to as back channel. Based on the decoder information transmitted over the back channel, the encoder can change the reference used by inter VOP coding according to the transmission error effects sent back by the decoder.

If NEWPRED is used, the encoder can change the temporal reference used by inter coded VOPs, avoiding areas it knows have been corrupted by transmission errors. Moreover, the temporal reference consists of a NEWPRED segment (NP segment), a segment being one coded VOP or one video packet. For each decoded NP segment, the decoder transmits a message through the back channel to inform the encoder whether the segment has been correctly decoded or not. Based on this feedback, the encoder can then use only correctly decoded segments for prediction, choosing between the most recent NP segments and spatially colocated but older NP segments. In this case, older means that the NP segment belongs to a frame older then the one usually used as temporal reference. This requires additional memory and implies a loss in coding efficiency, as using another, maybe older NP segment requires additional prediction error information (to code a higher prediction error for the new NP segment) and longer MVs (to point to the older NP segment).

8.5.3 Reduced Resolution Coding

Reduced resolution (RR) coding is one of the visual coding techniques that have been adopted in the first amendment to MPEG-4 Visual (MPEG-4 Visual Version 2). RR brings to MPEG-4 Visual the capability to adapt the resolution of the encoded VOP to its content and to real-time transmission constraints. RR is also referred to as dynamic resolution conversion (DRC).

When encoding a sudden highly active scene, the poor performance of coding tools may result in a sudden and large amount of bits, despite the rate-control effort. To control such a bit-rate spike, the rate-control mechanism may resolve to lower the encoding frame rate by skipping one or more frames. This can result in a serious degradation of the video quality. A possible solution to this problem is to allow the encoder to adaptively control the trade-off between spatial and temporal resolutions. Taking into account that the end user cannot perceive detailed texture on a rapidly moving object, moving areas can be encoded at a reduced spatial resolution without causing significant perceptual degradation.

The MPEG-4 RR tool [NMIM01], similar in concept and functionality to the H.263 *reduced resolution update* mode [H.263], supports the trade-off between spatial and temporal resolutions on a VOP basis. The use of this tool is signaled in the bitstream with a 1-bit flag, transmitted for each VOP, which indicates if the corresponding VOP has been encoded with spatially reduced resolution or not. VOPs encoded with spatially reduced resolution are referred to as *reduced resolution VOPs*.

At the encoder side, the motion estimation for a reduced resolution VOP is carried out using 32×32 pixels luminance macroblocks, each consisting of four 16×16 pixels macroblocks. The 16×16 prediction error blocks are then down-sampled to half resolution. The following encoding process is identical to that of a normal resolution VOP. This technique allows the reduced resolution VOP bitstream syntax to be identical to the syntax of a normal resolution VOP.

At the decoder side, the reduced resolution VOPs are decoded by means of special texture decoding and motion-compensation processes. These special processes are necessary because, mirroring the encoder process, the motion compensation is performed with a macroblock size of 32×32 pixels or block size of 16×16 pixels, instead of a macroblock size of 16×16 pixels and a block size of 8×8 pixels, used for the normal resolution VOPs. Only I- and P-VOPs can be coded as reduced resolution VOPs.

To conclude, this tool helps the encoder stabilize the transmission buffering delay by minimizing the jitter of the amount of output bits per VOP, and to prevent prolonged frame skips. For more information on this tool, including performance in an error-prone environment, the reader is referred to [NMIM01, N2825].

8.5.4 Sprite Coding

Sprite coding (SC) is a well-known and efficient technique for object-based representation and compression of video sequences. It is based on the use of a *sprite*, a special long-term memory associated with one specific object in the scene that contains all pixel information of such object that is visible along the sequence. An obvious example of sprite is a background sprite, also referred to as background mosaic [DufM96], which consists of all the pixels belonging to the background of a scene during a camera-panning recording. Portions of the background may not be visible in certain moments due to the occlusion of the foreground objects or because of the camera motion. However, because the sprite is built using the entire sequence, it contains all parts of the background that were visible at least once. Therefore, the background sprite can be used for direct reconstruction or predictive coding of the background. By recomposing the background with the separately encoded foreground object, each scene can be reconstructed. This process is depicted in Figure 8.52.

Sprite-based coding is supported in MPEG-4 mainly because it provides high coding efficiency in cases such as the one just described [LCLG97]. In MPEG-4 terminology, sprites are classified into offline-built *static* sprites and *dynamic* (or online-built) sprites [M653]. The sprites adopted in MPEG-4 are

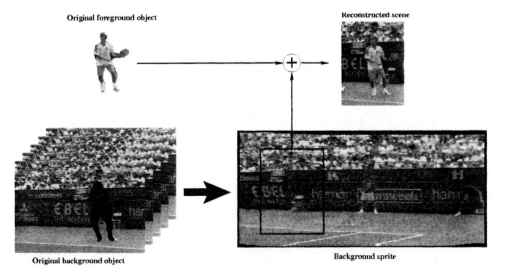

Fig. 8.52 Reconstruction of a scene using its background sprite.

static sprites. Using a static background sprite, the background instantiation at a given instant of time (i.e., the background VOP) can be reconstructed simply by warping or cropping such sprite.

A simplified form of dynamic sprites, called *global motion compensation (GMC)*, has also been adopted in MPEG-4 Visual, Version 2 (refer to Section 8.2.2.3 for more details on GMC). The remaining part of this section will concentrate on offline-built static sprites, referred to from now on simply as *sprites*.

As one may expect, sprite-based coding is suitable for synthetically composed scenes, as in such cases the static sprite is known *a priori*. It also can be used for natural scenes that undergo rigid motion. For such cases, sprite coding has proven to significantly increase coding efficiency.

In MPEG-4 Visual, SC is supported at the VOL layer, and the use of this tool is signaled in the bitstream by a flag in the VOL header. The same flag is used to differentiate between SC and GMC. When using SC, a new VOP coding type, named *Sprite VOP (S-VOP)*, is defined to represent a VOP that is coded using the SC tool.

At the encoder side, the sprite is built offline (i.e., before the encoding process begins) using the original VOPs (i.e., the instantiation in time of the VO that will be coded using such sprite). Using this approach, the entire VO is assumed to be available at the time the sprite is built. The sprite, which is built generally using a global motion estimation algorithm, consists of a large arbitrarily shaped (depending on the motion along the sequence) still texture. Similarly to other arbitrarily shaped VOPs coded with MPEG-4, the texture information of the sprite is represented by one luminance and two chrominance components. The three components are separately encoded, using 4:2:0 subsampling, as for video.

In addition to building and encoding the sprite itself, the MPEG-4 SC technique uses a set of parameters to describe the motion of each S-VOP in the sequence, relative to the sprite. Such parameters represent an estimate of the 2D motion of some S-VOP points (called *reference points*) and are derived using one among a set of transformation models. The transformation models supported by MPEG-4 are the stationary, translation, affine, and perspective models [MPEG4-2]. Each transformation can be defined as either a set of coefficients or the motion trajectories of the reference points. Although the former representation is convenient for performing the transformation, the latter is necessary for encoding the transformation itself. These trajectories (referred to as *sprite motion trajectories* to distinguish them from the P- and B-VOP MVs) are differentially encoded and transmitted at the S-VOP level.

Once the sprite is coded, it may be transmitted to the decoder before the actual video bitstream corresponding to the video sequence. Because the sprite consists of the information needed to code multiple VOPs, it is typically much larger than a single VOP; thus, its transmission can generate a significant delay (i.e., latency). To reduce this drawback, the MPEG-4 SC syntax supports two types of sprites:

☞ **Basic sprites:** These are transmitted in full at the VOL layer, just after the VOL header and before the actual GOV and VOP bitstream layers.

☞ **Low-latency sprites:** These are transmitted piecewise along the video transmission, each piece representing either a portion of the sprite or the information to increase the quality of the sprite or of a portion of the sprite already received.

To support such flexibility, MPEG-4 SC uses two sprite transmission modes:

☞ **Piece mode:** This mode reduces latency by transmitting first the portion of the sprite needed to reconstruct the first few S-VOPs, and transmitting later the remaining pieces, as dictated by the decoding requirements and bandwidth availability.

☞ **Quality update mode:** This mode reduces latency by transmitting first a low-resolution or highly quantized version of the sprite to begin the reconstruction of the S-VOPs, and transmitting later residual sprites to improve the initial sprite quality as bandwidth becomes available.

These two modes may be used separately or in combination. In the piece mode syntax, only the initial pieces of the sprite, together with the size and shape information for the entire sprite, are transmitted at the VOL layer. The remaining portions of the sprite are parceled, together with the corresponding trajectory points, into small pieces. Such pieces are gradually transmitted at the S-VOP layer. It is the duty of the encoder to ensure timely delivery of pieces in such a way that regions of the sprite are always present at the decoder before they are needed. The quality update mode can be of help for those cases in which, due to timing and bandwidth restrictions, some of these sprite pieces may have been delivered at a lower quality than desired. To improve the quality, residuals of these pieces are calculated and transmitted as quality update pieces in a way similar to the transmission of sprite pieces.

When SC is enabled in a VOL, the first VOP of the VOL must be of type intra (i.e., I-VOP), and all other VOPs must be of type S-VOP. The I-VOP carries the entire sprite (basic sprite case), or a portion of it (low-latency case), so its content is not displayed but stored in a sprite buffer and used to reconstruct all the remaining S-VOPs of the same VOL. For basic sprites, the sprite buffer is not changed during the entire VOL decoding process. For low-latency sprites, information carried by the S-VOPs can be used to modify the sprite buffer. In addition, for low-latency sprites, the MB syntax of the sprite pieces depends on the transmission mode. The syntax for object pieces is a subset of the I-VOP syntax, whereas the update pieces use a subset of the P-VOP syntax.

At the decoder side, the S-VOP is reconstructed by directly warping the quantized sprite using the transmitted sprite motion trajectories. Residual error between the original VOP and the warped sprite is not added (and not coded) to the warped sprite. The sprite decoding process is summarized in Figure 8.53. Because it does not impact interoperability, MPEG-4 does not standardize the

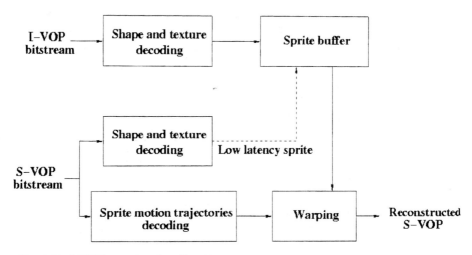

Fig. 8.53 MPEG-4 sprite decoding process.

procedure to generate the sprites. The reader interested in sprite-generation techniques is referred to [LCLG97, SmSO99, GrBS00, JNTV00].

8.5.5 Short Video Header Mode

In the initial stage of the MPEG-4 standardization effort, H.263 [H263] was used as the starting point for the development of the MPEG-4 Visual specification. Therefore, it seemed logical to work in order that an MPEG-4 video decoder should be able to decode an H.263 bitstream in some conditions. This functionality is supported by the short video header mode. An MPEG-4 video decoder operates in this mode when it detects an H.263 baseline bitstream (i.e., a bitstream generated by an H.263 encoder that uses none of the H.263 annexes [H263]). This means that the MPEG-4 Visual standard, when operating in the short video header mode, is backward-compatible with H.263 baseline.[16] And because all compliant MPEG-4 video decoders have to support the short video header mode, MPEG-4 is also forward compatible with H.263 baseline.[17]

8.5.6 Texture Coding for High-Quality Applications

For very high-quality applications, such as video in the studio, special coding tools were added to the MPEG-4 Visual specification [MPEG4-2]; these tools are used in the `Simple` and `Core studio` object types (see Chapter 13):

---

16. Here backward compatibility means that an H.263 decoder can decode the bitstream generated by an MPEG-4 encoder operating in short video header mode.

17. Here forward compatibility means that an MPEG-4 decoder can decode the bitstream generated by an H.263 baseline encoder.

☞ Higher DCT precision (3 additional bits), allowing lossless coding

☞ Uncompressed (PCM) coding mode to avoid data expansion (the coded bitstream data should be equal or less than the input data to be coded)

☞ 4:2:2 and 4:4:4 chrominance sampling (including the corresponding adaptation of the chrominance padding for arbitrarily shaped objects)

☞ Independent chrominance quantizer weighting matrices for the 4:2:2 and 4:4:4 chrominance sampling modes

☞ "Simple transmission" (PCM) mode for binary shape coding (i.e., no context-based arithmetic coding)

☞ Extension of the gray-scale alpha/multiple auxiliary component sample resolution to 4–12 bits

☞ SC extension by `defocusing_control` and `lens_distortion_parameter`

☞ `frame_center_offset` for pan-scan applications

☞ Inclusion of the MPEG-2 Video `4:2:2` profile header extensions (sequence display, quantizer matrix, copyright, picture display, user data)

☞ MPEG-2 compatible MV coding/motion compensation

The precision of the DCT coefficients' coding is enhanced with 3 bits per coefficient, using the 3 most significant bits of the fractional part in the floating-point coefficient representation. The encoding of the coefficients itself is different from both MPEG-2 Video and what is done for the rest of MPEG-4 Visual: 12 different VLC tables are used, from which the one used for a specific symbol to be decoded is chosen depending on the previously decoded symbol. This extends the capabilities of MPEG-4 Visual up to lossless coding. In addition, an uncompressed data mode is introduced (for the cases where the DCT would lead to an expansion of the input data in very high-quality applications), and texture resolutions of 8 or 10 bits per pel are supported.

The chrominance sampling is extended from the 4:2:0 mode, used for all the remaining MPEG-4 Visual object types (see Chapter 13), to the 4:2:2 and the 4:4:4 modes. The chrominance padding (see Section 8.3.3), which is necessary for arbitrarily shaped objects, is extended to support these chrominance sampling modes. Independent chrominance quantizer weighting matrices for both intra and inter MBs allow for a finer distinction between luminance and chrominance quantization when 4:2:2 or 4:4:4 chrominance sampling is used.

The sample resolution for the gray-scale alpha and the multiple auxiliary component (i.e., disparity or depth) information is extended from 8 bits to the range of 4–12 bits, just as the chrominance and luminance resolution for the `not_8_bit` case (see Section 8.2.3). Two new values, `minimum/ maximum_alpha_level`, are introduced to be used for the gray-scale alpha values of the pixels that are fully transparent, according to the binary shape information (`minimum_alpha_level`), or fully opaque (`maximum_alpha_level`). For the remaining MPEG-4 Visual object types, these values are fixed to 0 and 255, respectively.

New sprite parameters (`defocusing_control`, `lens_distortion`, and `lens_center`) can be used to add camera properties to the sprite decoding mechanism. They define filter characteristics that reproduce the physical characteristics of the video capture systems.

The additional `frame_center_offset` parameter set is not used in the decoding process, but for display purposes. It can be used for pan-scan applications, for example, to define which part of a video sequence recorded with an aspect ratio of 16:9 should be shown on a display with an aspect ratio of 4:3.

To reduce the computational complexity of the decoder by omitting the context-based arithmetic decoder, the binary shape is coded using a *simple transmission* (PCM) mode. Furthermore, no B-VOPs are allowed in the studio object types to reduce the necessary VOP reference memory (only one reference VOP needs to be stored) and to facilitate random access and bitstream editing.

The VOL header of the studio object types is extended with the `Sequence_Display`, `Quantizer_Matrix`, `Copyright`, `Picture_Display`, `Time_Code_SMPTE12M`, and `User_Data_MPEG-2` fields, making the VOL header look more like an MPEG-2 `picture_header` (for the meaning of these fields, the reader is referred to [MPEG2-2]). These fields do not affect the decoding process; they are merely informative.

Furthermore, the MV prediction and the motion compensation themselves are similar to the mechanisms used in MPEG-2 Video, making the MPEG-4 Visual studio object types' motion compensation more similar to the MPEG-2 Video `4:2:2` profile than to the other MPEG-4 Visual profiles (i.e., only MPEG-2 motion modes, half-pel motion compensation, and use of the MPEG-2 Video MV prediction scheme instead of the MPEG-4 Visual median predictive MV solution). These changes allow for an easy transcoding between the MPEG-2 Video `4:2:2` profile and the MPEG-4 Visual studio object types. For studio applications, this is more important than being backward compatible to the other MPEG-4 Visual object types. An informative annex to the MPEG-4 Visual standard [MPEG4-2] describes the correspondence between the MPEG-2 Video `4:2:2` profile and the MPEG-4 Visual studio object types' syntaxes.

8.6 VISUAL TEXTURE CODING

In recent years, the telecommunication, computer animation, and multimedia industries have seen an increasing demand for interactive multimedia services over broadband networks. The efficiency of the coding schemes used for compressing the multimedia content, the ability to support alpha-blending, and the flexibility to encode multiple levels of details of a scene in one stream are essential for the success of these emerging services. To address this need, the MPEG-4 Visual standard includes a new VTC technique to support applications that demand high-quality, efficiently coded, and scalable textures.

MPEG-4 VTC is based on wavelet transform and zero-tree coding. Unlike previous standard texture coding techniques that only address coding efficiency, MPEG-4 VTC provides the following additional functionalities:

☞ Efficient compression over a wide range of qualities

☞ Coding of arbitrarily shaped texture objects

☞ Spatial and quality scalability of both rectangular and arbitrarily shaped texture objects

☞ Robust transmission in error-prone environments

☞ Random access

☞ Complexity scalability levels

These advanced features greatly facilitate the content creators' task to generate rich multimedia content. The ability to *efficiently compress textures over a wide range of qualities* using a single coding tool simplifies the content creator's job by providing a one-stop shopping opportunity for lossless and lossy compression[18] while delivering the quality required by his or her application.

The spatial and quality scalability functionality helps to create photo-realistic 2D/3D rendering, in which multiresolution texture can be applied like wallpaper to polygonal meshes to support a wide variation in viewing perspectives. Because the object onto which the texture is mapped can have an arbitrary shape, support for coding *arbitrarily shaped texture* is provided. MPEG-4 VTC also supports the coding of alpha planes because alpha blending is often a required operation in most applications envisioned. In MPEG-4 Visual, Version 2, the support for VTC alpha planes is limited to binary shapes. Meanwhile, this limitation has been removed; thus, MPEG-4 VTC now supports both binary and gray-scale alpha planes. To enable the transmission over wireless and wired networks under severe error conditions, MPEG-4 VTC has adopted error-resilience coding tools.

MPEG-4 VTC also provides efficient means to allow *random access* to some or all objects in a scene. Examples of the random access functionality are the ability to access a set of predefined points in a (scalable or nonscalable) bitstream, and any layer of a scalable bitstream.

Finally, MPEG-4 VTC allows various coding modes with different *levels of complexity*. Low-complexity encoding and decoding is supported. With such complexity modes, MPEG-4 allows a user to strike a trade-off between image quality and computational complexity. The support of various complexity modes is the key to the successful deployment of MPEG-4 on platforms that have different resource requirements. The reader is referred to [N4319] for a complete description of the MPEG-4 requirements, and to [N2724] for a list of potential application scenarios.

---

18. Lossless compression should be read as mathematically lossless compression; lossy compression can be either visually lossless or lossy.

8.6.1 VTC Tools

To meet the requirements described in the previous section, MPEG-4 VTC adopts the following core tools:

- ☞ Wavelet and zero-tree-based compression algorithm to achieve efficient compression
- ☞ Shape-adaptive wavelet coding for compressing arbitrarily shaped still texture objects
- ☞ Three quantization schemes and two wavelet coefficient scanning modes to provide different granularity of spatial and quality scalability for both rectangular and arbitrarily shaped still texture objects
- ☞ Bitstream packetization approach to support error robustness in error-prone environments
- ☞ Tiling scheme to reduce encoding/decoding memory requirements and provide random access

These tools have been adopted in different versions of the MPEG-4 Visual standard. The compression tool is part of MPEG-4 Visual, Version 1, whereas error-resilience, tiling, and the shape-adaptive tools have been adopted in MPEG-4 Visual, Version 2, about a year later.

A block diagram with the MPEG-4 VTC encoder is presented in Figure 8.54. All of the MPEG-4 VTC tools but tiling are presented in this figure. The block diagram depicted in Figure 8.54 assumes that the input data consists of three arbitrarily shaped still texture objects (two children and a ball). The objects' shape and texture components are independently compressed. The scalable shape coding block processes the binary shapes, and the corresponding arbitrarily shaped textures are processed by the shape-adaptive wavelet coding and bitstream packetization block. If the input data consists of one or more rectangular still textures with no shape information, the textures are still processed by the shape-adaptive wavelet coding block, which is also able

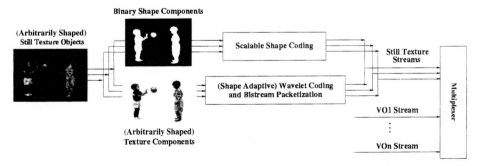

Fig. 8.54 Simplified VTC encoder block diagram.

to code rectangular textures as a special case. For simplicity, this special case is referred to as wavelet coding.

The output of the scalable shape coding and of the shape-adaptive wavelet coding blocks is composed of three elementary streams (ESs), each corresponding to one of the input objects. The ESs are then multiplexed with other visual object streams (VO1 to VOn in Figure 8.54) by the selected multiplexer (MUX).

Each MPEG-4 VTC tool is described in the following. Reflecting the MPEG-4 VTC development, first wavelet coding will be introduced, which consists of the wavelet transform, the explicit/implicit quantization, the prediction (only for the coefficients in the lowest band), the wavelet coefficients scanning, the zero-tree coding, and the adaptive arithmetic coder.

8.6.2 Wavelet Coding

MPEG-4 VTC wavelet coding is a zero-tree compression algorithm [LewK92, Shap93]. A generic zero-tree image compression algorithm consists of three modules: (a) wavelet transform, (b) quantization and zero-tree coding, and (c) entropy coding (i.e., arithmetic coding) of the zero-tree symbols and quantized coefficients, as shown in Figure 8.55.

Wavelet transforms have very good energy compaction properties, which lead to the efficient use of scalar quantizers [DavN99]. However, wavelet-transformed data also show a significant amount of structure, especially in the higher subbands. Because wherever there is structure there is room for compression, advanced wavelet compression algorithms try to exploit this structure to improve compression efficiency. One of the most successful approaches to this problem is based on exploiting the relationship of the wavelet coefficients across subbands. In particular, the relations between coefficients that have little or no energy, together with the notion of coding zeros (null wavelet coefficients) jointly (previously seen in JPEG [PenM93]), are at the base of the zero-tree coders [LewK92].

The zero-tree structure exploits the correlation between a coarse scale coefficient (parent) and its descendants (children) at the finer scales (see Figure 8.56). Because of the self-similarity inherent in the wavelet coefficients, if a wavelet coefficient at the coarse scale is insignificant (zero value), the coefficients of the same orientation at the same spatial location at the finer scales (i.e., all its descendants) are likely insignificant as well. In [LewK92], the likelihood is stretched into certainty, assuming that small or zero coefficients

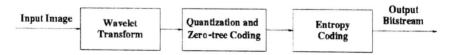

Fig. 8.55 Basic coding steps of a wavelet and zero-tree-based coding algorithm.

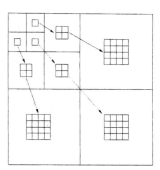

Fig. 8.56 Parent–child relationship of wavelet coefficients.

always have small or zero descendants. Although this approach does not need to transmit the zeros' location (once a zero-tree root location is transmitted, all its descendants are zeros), there is a big hit in coding efficiency when this assumption fails—that is, on the rare occasions where a nonzero coefficient belongs to a zero-tree.

The loss in coding efficiency can be greatly reduced by explicitly transmitting the information that is implicitly contained in the zero-trees. This is achieved by using a *significance map*. A significance map replaces the zero-tree used in [LewK92] and is a binary function whose values determine if a coefficient is significant (i.e., different from zero) or not (i.e., equal to zero). The significance map was first introduced in the *embedded zero-tree wavelet* (EZW) algorithm [Shap93]. For each wavelet coefficient, the EZW coder transmits the significance map values (zero-tree root, isolated zero, and positive/negative significant value) and/or the values of the significant coefficients. The correlation across subbands is now exploited to reduce the significance map bit rate. The map values at each node of the tree are conditionally entropy encoded upon the corresponding value at the parent node.

In addition, the EZW coder uses a successive approximation approach (or bit plane coding) to quantize the wavelet coefficients. This results in an embedded code, which means that a lower bit-rate code (or bitstream) is a prefix of higher bit-rate codes (or bitstreams). Embedded codes carry the progressive transmission or successive refinement property.

Another approach to exploit the correlation between wavelet coefficients at different scales, known as *set partitioning in hierarchical threes* (SPHIT), is presented in [SaiP96]. Instead of building a significance map by scanning the zero-tree subband by subband and labeling its nodes, SPHIT uses the concept of set partitioning to code the location of nonzero coefficients. It also has been shown that both EZW and SPHIT are members of a larger family of tree-structured significance map schemes [DavC97].

The zero-tree algorithm adopted by MPEG-4 combines three variations of the EZW algorithm: the *zero-tree entropy coder* (ZTE) [MSCZ97], its most general case (i.e., the *multiscale zero-tree wavelet entropy coder,* MZTE) [SLHZ99], and the *predictive embedded zero-tree wavelet coder* (PEZW) [Lian97]. MPEG-4 VTC also uses an efficient approach to compress the lowest-resolution (DC) wavelet subband.

In MPEG-4 VTC, a discrete wavelet transform (DWT) [BuGG98] is first applied to the input image to decorrelate the data and to obtain a multiresolution representation of the input image. The DWT decomposes the input image into subbands of varying resolutions. During the coding process, the lowest-resolution subband (DC) is coded separately from the rest of the subbands (AC). The DC coefficients are first uniformly scalar quantized, then DPCM-encoded. This approach was first introduced in PEZW. Each DC quantized coefficient is predicted from its left or top neighbor based on a gradient criterion. The prediction error is finally coded with an adaptive arithmetic coder. The architecture of the shape-adaptive wavelet coder is depicted in Figure 8.57, where the wavelet coding blocks are shown in gray, and the bitstream packetization in white.

The AC bands are coded using a combination of the three zero-tree algorithms mentioned previously. All of them consist of an implicit/explicit quantizer, followed by coefficients scanning and (lossless) entropy coding of the zero-tree symbols and nonzero quantized coefficients.

Three different quantization modes are supported by MPEG-4 VTC: (a) *single quantizer mode,* (b) *multiple quantizer mode,* and (c) *bilevel quantizer mode.* In the single and multiple quantizer modes, the quantization is explicit, whereas in the bilevel quantizer mode the quantization is implicit. The single quantization and multiple quantization modes are based on ZTE and MZTE, respectively. In these two codecs, the EZW bit plane encoding of the wavelet coefficients is replaced with explicit quantization. The single quantization uses only one quantization step to quantize all of the wavelet coefficients. In this mode, when using the traditional zero-tree scanning mode, the bitstream has no embedded property. The multiple quantizer mode uses a variety of quantization steps, ranging from Q_0 to Q_n, in a multistage fashion, as shown in Figure 8.58.

The wavelet coefficients are first quantized with Q_0 and passed on to the next steps: coefficient scanning (CS), zero-tree coding (ZTC), and adaptive

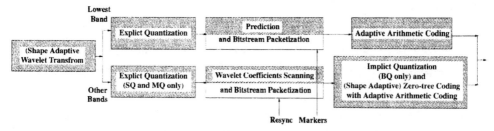

Fig. 8.57 Shape-adaptive wavelet coding block diagram.

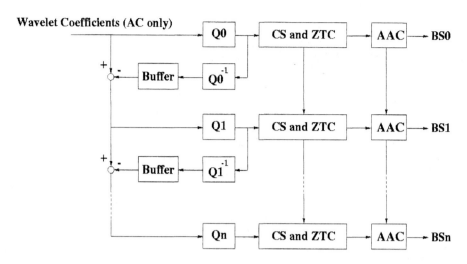

Fig. 8.58 Block diagram for the multiple quantizer mode.

arithmetic coding (AAC). The quantized coefficients also are dequantized and subtracted from the original wavelet coefficients. The Q_0 quantization error is then fed into the next quantization stage, and the process is repeated for all remaining quantizers (Q_1 to Q_n). If the traditional zero-tree scanning mode is used, the bitstream generated at each stage provides one layer of quality scalability. Therefore, the final bitstream, which consists of the combination of the bitstreams generated by each stage (BS_0 to BS_n in Figure 8.58), provides n layers of quality scalability.

The bilevel quantizer mode is based on PEZW and uses a successive approximation approach to implicitly quantize the wavelet coefficients. The result is a fully embedded code with the finest quality scalability granularity.

MPEG-4 VTC also supports either *tree-depth* or *subband-by-subband* scanning order of the wavelet coefficients to go with any of the three quantization modes. In the tree-depth scanning order, which corresponds to the traditional zero-tree scanning, all coefficients of each tree are encoded before starting encoding the next tree. In the subband-by-subband scanning order, all coefficients of the lowest subband are encoded before starting encoding the coefficients of the next subband. The introduction of this scanning order brings spatial scalability in addition to the quality scalability provided by each of the three quantization modes. When using subband-by-subband scanning order, the single quantizer mode provides spatial scalability, the multiple quantizer mode provides spatial scalability and a user-defined granularity of quality scalability, and the bilevel quantizer mode provides spatial scalability and the finest quality scalability granularity.

The three quantizer outputs are fed into a common zero-tree coder. The set of symbols used by MPEG-4 VTC zero-tree coding to code the significant map values are zero-tree root (a node whose coefficient has zero amplitude and is a root of a zero-tree), value zero-tree root (a node whose coefficient has non-zero amplitude and all four children have zero amplitude), isolated zero (a node whose coefficient has zero amplitude), and value (a node whose coefficient has nonzero amplitude). This new set of symbols results in an improved coding efficiency of MPEG-4 VTC with respect to EZW (see [Lian97] for further details). In addition, a differential type of coding is used by the zero-tree coding to handle the outputs of the multistage quantizer used by the multiple quantizer mode. After coding the zero-tree symbols and the values of the first scalability level (i.e., the output of the Q_0 quantizer), the zero-tree symbols and the values of the following scalability levels are encoded as a refinement to the previous level (see [SLHZ99] for more details).

The last stage of wavelet coding consists of the (lossless) entropy coding of the zero-tree symbols, the nonzero quantized coefficient values (magnitude and sign), and the residual values (multiple quantizer mode only) using an adaptive binary arithmetic coder. Context modeling is used to better estimate the probability distribution of the symbols to be encoded. A detailed description of the probability models, their contexts, and their update strategy can be found in the MPEG-4 Visual standard [MPEG4-2].

MPEG-4 VTC has adopted the (9,3) filter, integer version, and the (9,3) filter, floating-point version, as its default integer and floating-point wavelet filters. In addition, users can choose their own custom integers or floating-point wavelet filters, which are transmitted in the bitstream. When using an integer filter, MPEG-4 VTC supports lossless compression.

8.6.3 Shape-Adaptive Wavelet Coding

Shape-adaptive wavelet (SA-wavelet) coding is used for coding arbitrarily shaped objects [MPEG4-2, M4576, KIAK97]. SA-wavelet coding is different from the wavelet coding described in the previous section mainly in its treatment of the shape data. SA-wavelet coding ensures that the number of wavelet coefficients to be coded is exactly the same as the number of pixels in the arbitrarily shaped region. Therefore, the coding efficiency for arbitrarily shaped objects is the same as that for rectangular objects. SA-wavelet coding includes rectangular shapes as a special case, becoming the regular wavelet coding.

As for the regular wavelet coding, in SA-wavelet coding, the wavelet transform is applied to each row/column of the object to be encoded. For arbitrarily shaped objects, symmetric extension and subsampling adjustments are made to take into account the length and the even/odd starting position of the row/column to be transformed. SA-wavelet coefficients are coded using a modified version of the zero-tree-based technique used for rectangular shaped objects. The modification is needed to handle the wavelet trees that have wavelet coefficients corresponding to the pixels outside the shape boundary. For more details about SA-wavelet and its performance, refer to [LiLi00].

8.6.4 Spatial and Quality Scalability

Spatial and quality scalability for the rectangular shaped textures are provided by different combinations of the quantization techniques and scanning methods presented in Section 8.6.2. The wavelet coefficients are scanned in a tree-depth (TD) fashion to obtain a quality scalable bitstream, whereas a resolution scalable bitstream can be achieved using the subband-by-subband scanning. The granularity of these scalabilities can be selected from a wide range of possible levels, as shown in Table 8.2. MAX_wavelet is the number of wavelet decomposition levels, and MAX_SNR is the maximum number of bit planes, which is fixed by the quantization step size.

Figure 8.59 to Figure 8.61 show three different examples of embedded spatial and quality scalabilities. Figure 8.59 shows how M-embedded spatial layers (SP[0] to SP[M-1] in the figure) can be decoded from a single bitstream.

Table 8.2 VTC encoder choices for spatial and quality scalability levels

Quantization modes	Scanning mode	Spatial scalability levels	Quality scalability levels
Single quantizer	Tree-depth	1	1
-	Subband-by-subband	MAX_wavelet	1
Multiple quantizer	Tree-depth	[1, MAX_wavelet]	[1-31]
-	Subband-by-subband	MAX_wavelet	[1-31]
Bilevel quantizer	Tree-depth	1	MAX_SNR
-	Subband-by-subband	MAX_wavelet	[1, MAX_SNR]

Decoded Frame in M Different Spatial Layers

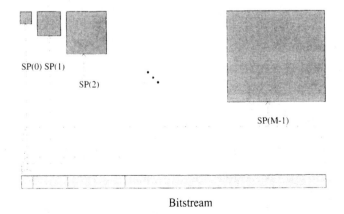

SP(0) SP(1)

SP(2)

SP(M-1)

Bitstream

Fig. 8.59 M layers of spatial scalability [MPEG4-2].

Decoded Frame in N Different Quality Layers

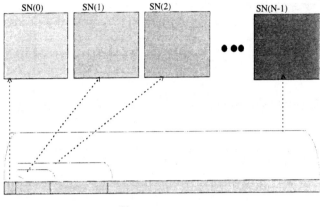

Bitstream

Fig. 8.60 N layers of quality scalability [MPEG4-2].

Decoded Frame in Hybrid Spatial/Quality Layers

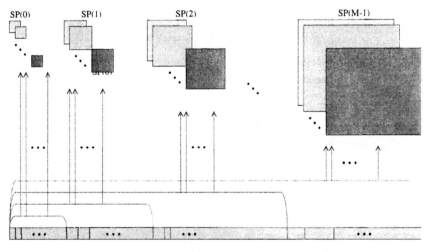

Bitstream

Fig. 8.61 M times N layers of spatial and quality scalability [MPEG4-2].

The smallest resolution layer, SP(0), is reconstructed from the initial portion of the bitstream. Decoding additional bits generates higher spatial resolutions. Similarly, Figure 8.60 shows how N embedded quality layers (SN[0] to SN[N-1] in the figure) can be decoded from a single bitstream. Finally, Figure 8.61 combines the concepts shown in the previous two figures, showing how M times N embedded layers of spatial/quality scalability can be decoded from a single bitstream.

To support spatial and quality scalability for arbitrarily shaped objects, both the texture and shape components need to be scalable. SA-wavelet coding inherits the spatial and quality scalability features of wavelet coding for the texture component, whereas the spatial and quality scalability for the shape component are achieved by using a technique similar to the scalable shape coding technique used for MPEG-4 video data. However, some minor changes are made to improve the coding efficiency of the scalable shape coding. These changes regard the probability tables of the binary arithmetic coding. As in the video case, nonscalable shape coding is used for the base layer, and a binary arithmetic coder based on the scan interleaving method is used to code the enhancement layer [M4040].

Using SA-wavelet coding for the texture and the scalable shape coding for the shape, a decoder can decode an arbitrarily shaped texture at any desired resolution. This enables applications to employ object-based, spatial, and quality scalabilities at the same time. Examples of spatial scalability for rectangular and arbitrarily shaped textures are shown in Figure 8.62 to Figure 8.64.

8.6.5 Bitstream Packetization

To support applications that transmit data in error-prone environments, MPEG-4 VTC needs to be error robust. When a compressed bitstream is trans-

Fig. 8.62 Example of spatial scalability provided by MPEG-4 VTC for the rectangular object *Woman*. These four resolutions have been decoded from the same bitstream.

Fig. 8.63 Arbitrarily shaped texture (on the left, after padding), and corresponding binary shape (on the right).

Fig. 8.64 Example of spatial scalability provided by MPEG-4 VTC for the arbitrarily shaped object in Figure 8.63. These three resolutions have been decoded from the same bitstream.

mitted over a noisy channel, errors are almost inevitable. For a highly compressed bitstream with very limited redundancy, even a single bit error can cause the decoding process to fail. Even with channel coding protection, it is better to have some degree of error resilience at the source coding level, as there are residual errors and different data in the bitstream react to errors differently. By exploiting the knowledge about data-error sensitivity and its impact on the decoded texture, a more flexible solution can be devised than by using channel coding alone.

MPEG-4 VTC has adopted a bitstream packetization approach to make the VTC bitstream robust to channel degradation [MRAC98, MSLC00]. This approach does not affect the VTC spatial and quality scalability features. It requires minimum overhead and thus minimizes the impact on coding efficiency.

The VTC data stream is organized in packets. Each packet consists of a number of coding units. The structure of these units depends on the scanning methods adopted by the encoder. A packet header is inserted into the data stream at the beginning of each packet. The header consists of a resynchronization marker, followed by a 1-bit HEC and by U.First and U.Last. U.First and U.Last denote the unit number of the first and last coding units transmitted in the current packet, respectively. Information vital to the success of the decoding process can be (optionally) repeated after the HEC. The unit number associates the data contained in the unit to its corresponding position in the image. The decoder can use the elements contained in the packet header to (a) detect errors, (b) resynchronize with the encoder, (c) discard corrupted data, and (d) associate the following uncorrupted data to the correct position in the image.

To further increase the error localization and data recovery capabilities, a technique similar to the DP used in video coding has been extended to VTC. (For more information on DP for VTC, refer to [ChSW00, MSLC00].) An example of the VTC packetization performance is presented in Table 8.3 and Table 8.4. The PSNR values shown in these tables are obtained by averaging the mean squared errors over 50 runs of an error-generation program (i.e., with 50 different random seeds). No error concealment is used. Both tables report the results with and without error-resilience technique (i.e., with and without bitstream packetization). The objective quality achieved for the noncorrupted bitstream is 34.49 dB for *Nature*, 36.35 dB for *Woman*, and 32.8 dB for *Café*.[19] As presented in the tables, the objective quality of the corrupted images when no error resilience is used drops dramatically, and the visual content of the image is completely destroyed. The use of error resilience preserves up to 70% of the objective quality (i.e., PSNR) and most of the visual content of the image. The difference in performance reported in the two tables is due to the different nature and intensity of the error conditions: harsh random bit

Table 8.3 Luminance PSNR for subband-by-subband (BB) and tree-depth (TD) scanning for bitstream corruption with random bit errors (BER 10^{-3})

Image	Scanning mode	Average PSNRY (dB) (no error resilience)	Average PSNRY (dB) (with error resilience)
Nature	TD	15.97	18.97
	BB	8.87	20.90
Woman	TD	16.04	21.24
	BB	9.09	24.47
Café	TD	10.10	14.16
	BB	7.95	14.49

---

19. Note that these images are in 4:2:0 color format. The PSNR values reported in this section are calculated on their luminance component only. However, the bit-per-pixel values reported in Figure 8.65 are calculated by compressing both their luminance and chrominance components.

errors (BER 10^{-3}) in Table 8.3, mild burst errors (BER 10^{-3} with 10 ms length and 50% burst corruption) in Table 8.4. To appreciate the difference in visual content, an example of the visual quality obtained under burst errors with and without bitstream packetization (test image *Café*, TD scanning mode case, PSNR values reported in Table 8.4) is shown in Figure 8.65. See [MSLC00] for more details about MPEG-4 VTC packetization performance.

8.6.6 Tiling

MPEG-4 VTC has adopted tiling to support the manipulation of very large textures. Tiling allows reducing the amount of memory required during the encoding and decoding processes. Memory-constrained applications, such as those running on mobile terminals, can benefit from this technology. Moreover,

Table 8.4 Luminance PSNR for subband-by-subband (BB) and tree-depth (TD) scanning for bitstream corruption with burst errors (BER 10^{-3} with 10 ms length and 50% burst corruption)

Image	Scanning mode	Average PSNRY (dB) (no error resilience)	Average PSNRY (dB) (with error resilience)
Nature	TD	18.92	31.54
	BB	13.68	31.25
Woman	TD	19.01	34.65
	BB	9.19	34.17
Café	TD	12.94	31.19
	BB	6.62	24.39

Fig. 8.65 *Café* image* VTC compressed and transmitted with no errors (left); VTC compressed with no error resilience (0.56 bit per pixel), and corrupted with burst errors (center); and VTC compressed with error resilience (0.63 bit per pixel), and corrupted with burst errors (right). The error condition and the objective quality of these images are reported in Table 8.4 (TD scanning mode case).
*This image is in 4:2:0 color format. Only the luminance component is depicted here.

tiling enables the random access capability. Random access means the ability to access and decode an arbitrarily chosen portion of a large compressed image without decoding the entire bitstream. The tiling approach adopted by MPEG-4 VTC allows random access while minimizing the amount of decoded stream along with the random access overhead and time. To support real-time or quasi-real-time applications, the random access time should be proportional to the size of the decoded image.

MPEG-4 VTC uses an independent tiling scheme [M3734]. With this approach, the image is first divided into several tile images, named *subimages*. The subimages are then coded independently using the VTC tool. The subimages' sizes and additional control information are added to the bitstream. The decoder uses this information to decode only the subimages requested by the end user. A block diagram for the tiling encoding and decoding procedure is presented in Figure 8.66.

8.7 SUMMARY

In this chapter, a general overview of the basic MPEG-4 natural video coding approach and its respective tools was presented. First, the wide scope of the MPEG-4 object-based video coding approach was highlighted. Video is no longer restricted to rectangular images; now it is possible to code images and image sequences of arbitrary shape. This provides a number of new functionalities, going beyond pure video compression. The remainder of the chapter was subdivided into two parts: The first dealt with the coding of video sequences, and the second concentrated on the coding of still images, also referred to as *visual texture coding*.

The first part started with a description of the MPEG-4 video tools for the coding of rectangular video sequences. As it is based on the well-known motion-compensated hybrid coding, in particular, the new tools that were

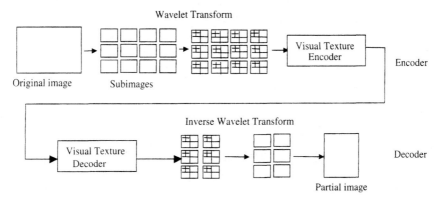

Fig. 8.66 VTC tiling encoding and decoding procedure [MPEG4-2].

introduced by MPEG-4 Visual to improve coding efficiency have been described. These are (among others) quarter-pel motion compensation, global motion compensation, bidirectional motion compensation with direct mode, intra AC/DC prediction, and alternative DCT coefficient scans. The combination of these tools—as used, for example, in the Advanced Coding Efficiency (ACE) profile or the Advanced Simple profile (see Chapter 13)— provides a significantly higher coding efficiency compared to pre-MPEG-4 video coding technology, as has been shown in formal verification tests [N2824]. Based on this first part, the additional object-based coding tools were explained: binary and gray-level shape coding, object-based motion compensation, and object-based texture coding tools such as padding and shape adaptive DCT. Although not aimed at coding efficiency, these tools provide many content-based functionalities, including different quality for different video objects, flexible recomposition of scenes at the receiver side, and access to and manipulation of single video objects in a scene. The MPEG-4 video temporal, spatial, and quality scalability features were described. Based on the MPEG-2 Video scalable coding approach, they can be combined with the shape coding tools and thus applied to arbitrarily shaped objects. Finally, a section addressed special coding tools such as interlaced, error resilience, reduced resolution, and sprite coding, as well as the MPEG-4 short video header operational mode and studio-related tools.

In the second part, the wavelet approach for visual texture coding was described. Because the wavelet-based texture coding and the MPEG-4 shape coding tools can be combined, an efficient coding scheme for arbitrarily shaped visual textures was developed. The wavelet approach is inherently scalable, so the scalable modes for coding of visual textures were presented. Finally, the bitstream packetization (which provides the means for transmissions in error-prone environments) and tiling (which supports manipulation of very large textures) were presented. In summary, the visual texture coding provides a coherent set of tools for efficient, scalable, and error-robust compression of rectangular and arbitrarily shaped textures. Together with the video coding tools, scenes comprising both moving video and still texture objects can be coded with one single standard: MPEG-4 Visual.

8.8 REFERENCES

[AOWM98] Alatan, A., L. Onural, M. Wollborn, R. Mech, E. Tuncel, and T. Sikora. "Image Sequence Analysis for Emerging Interactive Multimedia Services: The European COST 211 Framework." *IEEE Transactions on Circuits and Systems for Video Technology*, 8(7): 802–813. November 1998.

[BenW96] Benzler, U., and O. Werner. *Improving Multi-Resolution Motion Compensating Hybrid Coding by Drift Reduction.* Proceedings of 1996 Picture Coding Symposium (PCS), Melbourne, Australia. March 1996.

[BHWR00] Budagavi, M., W. R. Heinzelman, J. Webb, and R. Talluri. "Wireless MPEG-4 Video Communication on DSP Chips." *IEEE Signal Processing Magazine*, 17(1): 36–53. January 2000.

[BraB00] Brady, N., and F. Bossen. "Shape Compression of Moving Objects Using Context-Based Arithmetic Encoding." *Signal Processing: Image Communication*, 15(7–8): 601–617. May 2000.

[BuGG98] Burrus, C. S., R. A. Gopinath, and H. Guo. *Introduction to Wavelets and Wavelet Transforms: A Primer.* Upper Saddle River, NJ: Prentice Hall. 1998.

[CFHM71] Candy, J. C., M. A. Franke, B. J. Haskell, and F. W. Mounts. "Transmitting Television as Clusters of Frame-to-Frame Differences." *Bell System Technical Journal*, 50(6): 1889–1917. July/August 1971.

[ChSW00] Chai, B.-B., I. Sodagar, and J. Wus. "A New Error Resilience Technique for Image Compression Using Arithmetic Coding." *Proceedings of the 2000 IEEE Conference on Acoustics, Speech, and Signal Processing (ICASSP)*, 4: 2127–2130. 2000.

[DavC97] Davis, G. M., and S. Chawla. "Image Coding Using Optimized Significance Tree Quantization." *Proceedings of the Data Compression Conference*, 387–396. March 1997.

[DavN99] Davis, G. M., and A. Nosratinia. "Wavelet-based Image Coding: An Overview," *Applied and Computational Control, Signals, and Circuits* (B. N. Datta, Ed.). Boston: Kluwer Academic Publishers. 1999.

[DufM96] Dufaux, F., and F. Moscheni. "Background Mosaicking for Low Bitrate Video Coding." *Proceedings of the International Conference on Image Processing (ICIP)*, 1: 673–676. 1996.

[DuPe99] Ducla Soares, L., and F. Pereira. "Error Resilience and Concealment Performance for MPEG-4 Frame-Based Video Coding." *Signal Processing: Image Communication*, 14(6–8): 447–472. May 1999.

[Ghan89] Ghanbari, M. "Two-Layer Coding of Video Signals for VBR Networks." *IEEE Journal on Selected Areas in Communications*, 7: 771–781. June 1989.

[Giro87] Girod, B. "The Efficiency of Motion Compensating Prediction for Hybrid Coding of Video Sequences." *IEEE Journal on Selected Areas in Communication*, 5(7): 1140–1154. August 1987.

[Giro93] Girod, B. "Motion-Compensating Prediction with Fractional-pel Accuracy." *IEEE Transactions on Communications*, 41(4): 604–612. April 1993.

[GrBS00] Grammalidis, N., D. Beletsiotis, and M. G. Strintzis. "Sprite Generation and Coding in Multiview Image Sequences." *IEEE Transactions on Circuits and Systems for Video Technology*, 10(2): 302–311. March 2000.

[H261] Recommendation ITU-T H.261. *Video Codec for Audiovisual Services at p x 64 kbit/s*. CCITT SG XV, COM XV-R37-E. August 1990.

[H263] ITU-T Recommendation H.263. *Video Coding for Low Bit Rate Communication*. International Telecommunications Union–Telecommunications Standardization Sector, Geneva. Ver. 1, 1995; Ver. 2 (H.263+), 1998; Ver. 3 (H.263++), 2000.

[HaPN96] Haskell, B. G., A. Puri, and A. N. Netravali. *Digital Video: An Introduction to MPEG-2*. New York: Chapman and Hall. November 1996.

[ImuM99] Imura, K., and Y. Machida. "Error Resilient Video Coding Schemes for Real-Time and Low-Bitrate Mobile Communications." *Signal Processing: Image Communication*, 14: 519–530. May 1999.

[JiaV98] Jiangtao, W., and J. Villasenor, "Reversible Variable Length Codes for Efficient and Robust Image and Video Coding." *Proceedings of 1998 Data Compression Conference (DCC)*, pp. 471–480. Snowbird, UT. March 1998.

[JiaV99] Jiangtao, W., and J. Villasenor. "Utilizing Soft Information in Decoding of Variable Length Codes." *Proceedings of 1999 Data Compression Conference (DCC)*. Snowbird, UT. March 1999.

[JNTV00] Jasinchi, R. S., T. Naveen, A. J. Tabatabai, and P. Babic-Vovk. "Apparent 3-D Camera Velocity: Extraction and Applications." *IEEE Transactions on Circuits and Systems for Video Technology*, 10(7): 1185–1191. October 2000.

[KaiB00] Kaiser, S., and M. Bystrom. "Soft Decoding of Variable-Length Codes." *Proceedings of 2000 IEEE International Conference on Communications (ICC)*, 3: 1203–1207. 2000.

[Kaup99] Kaup, A., "Object-Based Texture Coding of Moving Video in MPEG-4." *IEEE Transactions on Circuits and Systems for Video Technology*, 9(1): 5–15. February 1999.

[KauS98] Kauff, P., and K. Schuur. "Shape-Adaptive DCT with Block-Based DC Separation and Delta DC Correction." *IEEE Transactions on Circuits and Systems for Video Technology*, 8(3): 237–242. June 1998.

[KIAK97] Katata, H., N. Ito, T. Aono, and H. Kusao. "Object Wavelet Transform for Coding of Arbitrarily-Shaped Image Segments." *IEEE Transactions on Circuits and Systems for Video Technology*, 7(1): 234–237. February 1997.

[KJKL99] Kwak, J., J. G. Jeon, M. Kim, M. Ho Lee, and C. Ahn. *A Semi-Automatic Video Segmentation Method for Object Oriented Multimedia Applications.* 1999 International Technical Conference on Circuits/Systems, Computers and Communications (ITC-CSCC'99), Niigata, Japan. July 13–15, 1999.

[Kunt85] Kunt, M. "Second-Generation Image-Coding Techniques." *Proceedings of the IEEE*, 73(4): 549–574. April 1985.

[Lang84] Langdon, G. G. "An Introduction to Arithmetic Coding." *IBM Journal of Research and Development*, 28: 135–149. 1984.

[LCLG97] Lee, M.-C., W.-G. Chen, C.-L. B. Lin, C. Gu, T. Markoc, S. I. Zabinsky, and R. Szeliski. "A Layered Video Object Coding System Using Sprite and Affine Motion Model." *IEEE Transactions on Circuits and Systems for Video Technology*, 7(1): 130–145. February 1997.

[LewK92] Lewis, A. S., and G. Knowles. "Image Compression Using the 2-D Wavelet Transform." *IEEE Transactions on Image Processing*, 1(2): 244–250. April 1992.

[Lian97] Liang, J. "Highly Scalable Image Coding for Multimedia Applications." *Proceedings of the ACM Multimedia Communication Conference*, Seattle. October 1997.

[LiLi00] Li, S., and W. Li. "Shape-Adaptive Discrete Wavelet Transforms for Arbitrarily Shaped Visual Object Coding." *IEEE Transactions on Circuits and Systems for Video Technology*, 10(5): 725–743. August 2000.

[M2378] Moccagatta, I. *Interlaced Tools for Arbitrarily Shaped Video Objects (P14).* Doc. ISO/MPEG M2378, Stockholm MPEG Meeting. March 1997.

[M3734] Ito, N., S.-Y. Hasegawa, and H. Katata. *Result of the Core Experiment of Tiling for Still Texture Object*. Doc. ISO/MPEG M3734, Dublin MPEG Meeting. July 1998.

[M4040] Son, S. H., D.-S. Cho, J. S. Shin, and J. W. Chung. *Description of Mini Core Experiment on Scalable Shape Coding for Visual Texture Coding Using Version 2 WD Tools*. Doc. ISO/MPEG M4040, Atlantic City MPEG Meeting, October 1998.

[M4575] Moccagatta, I., and H. Chen. *Error Resilience for MPEG-4 Still Texture: Application Areas*. Doc. ISO/MPEG M4575, Seoul MPEG Meeting, March 1999.

[M4576] Fukunaga, S., I. Moccagatta, Y. Nakaya, and S.-H. Son (Eds.). *MPEG-4 Video Verification Model 12.2*. Doc. ISO/MPEG M4576, Seoul MPEG Meeting, March 1999.

[M615] Jozawa, H., K. Kamikura, and A. Sagata. *Technical Description of Video Proposal for MPEG-4 Algorithm Evaluation*. Doc. ISO/MPEG M615, Munich MPEG Meeting, January 1996.

[M653] Dufaux, F. *Background Mosaicking*. Doc. ISO/MPEG M653, Munich MPEG Meeting, January 1996.

[MCMM99] Marcotegui, B., P. Correia, F. Marques, R. Mech, R. Rosa, M. Wollborn, and F. Zanoguera. *A Video Object Generation Tool Allowing Friendly User Interaction*. International Conference on Image Processing 1999 (ICIP '99), Kobe, Japan. October 1999.

[MecW98] Mech, R., and M. Wollborn. "A Noise Robust Method for 2D Shape Estimation of Moving Objects in Video Sequences Considering a Moving Camera." *Signal Processing*, 66(2): 203–217. April 1998.

[MiNK98] Misaka, S., Y. Nakaya, and T. Kinoshita. "Symmetric Padding for Content-based 2D-DCT Coding." *Proceedings of the SPIE—Int. Society for Optical Engineering—Annual Meeting*, San Jose, CA, 3309: 26–35. 1997.

[MPEG1-2] ISO/IEC 11172-2:1993. *Coding of Moving Pictures and Associated Audio for Digital Storage Media at up to About 1,5 Mbit/s—Part 2: Video*. 1993.

[MPEG2-2] ISO/IEC 13818-2:2000. *Generic Coding of Moving Pictures and Associated Audio Information—Part 2: Video*. 2000.

[MPEG4-1] ISO/IEC 14496-1:2001. *Coding of Audio-Visual Objects—Part 1: Systems*, 2d Edition, 2001.

[MPEG4-2] ISO/IEC 14496-2:2001. *Coding of Audio-Visual Objects—Part 2: Visual*, 2d Edition, 2001.

[MPEG4-4] ISO/IEC 14496-4:2001. *Coding of Audio-Visual Objects—Part 4: Conformance Testing*, 2d Edition, 2001.

[MPEG4-5] ISO/IEC 14496-5:2001. *Coding of Audio-Visual Objects—Part 5: Reference Software*, 2d Edition, 2001.

[MRAC98] Moccagatta, I., S. L. Ragunathan, O. Al-Shaykh, and H. Chen. "Robust Image Compression with Packetization: The MPEG-4 Still Texture Case." *Proceedings of the 1998 IEEE 2nd Workshop on Multimedia Signal Processing*, pp. 462–467. Redondo Beach, CA. December 1998.

[MSCZ97] Martucci, S. A., I. Sodagar, T. Chiang, and Y.-Q. Zhang. "A Zerotree Wavelet Video Coder." *IEEE Transactions on Circuits and Systems for Video Technology*, 7(1): 109–118. February 1997.

[MSLC00] Moccagatta, I., S. Soudagar, J. Liang, and H. Chen. "Error Resilient Coding in JPEG-2000 and MPEG-4." *IEEE Journal on Selected Areas in Communications*, 18(6): 1–16. June 2000.

[MuHO89] Musmann, H. G., M. Hötter, and J. Ostermann. "Object-Oriented Analysis-Synthesis Coding of Moving Images." *Signal Processing: Image Communication*, 1(2): 117–138. October 1989.

[MuPG85] Musmann, H. G., P. Pirsch, and H. J. Grallert. "Advances in Picture Coding." *Proceedings of the IEEE*, 73(4): 523–548. April 1985.

[N2604] MPEG Test. *Report of the Formal Verification Tests on MPEG-4 Video Error Resilience*. Doc. ISO/MPEG N2604, Rome MPEG Meeting, December 1998.

[N2724] MPEG Requirements. *MPEG-4 Applications*. Doc. ISO/MPEG N2724, Seoul MPEG Meeting, March 1999.

[N2824] MPEG Test. *Report of the Formal Verification Tests on Advanced Coding Efficiency ACE (Former Main Plus) Profile in Version 2*. Doc. ISO/MPEG N2824 Vancouver MPEG Meeting, July 1999.

[N2825] MPEG Test. *Report of Formal Verification Tests on MPEG-4 Advanced Real Time Simple Profile (Error Robustness, Temporal Resolution Scalability)*. Doc. ISO/MPEG N2825, Vancouver MPEG Meeting, July 1999.

[N3093] MPEG-4 Visual VM. *MPEG-4 Video Verification Model Version 15.0*. Doc. ISO/MPEG N3093, Maui MPEG Meeting, December 1999.

[N3904] ISO/IEC 14496-2:1999/FDAM4. *Information Technology—Coding of Audio-Visual Objects—Part 2: Visual—Amendment 4: Streaming Video Profiles*. Doc. ISO/MPEG N3904, Pisa MPEG Meeting, January 2001.

[N4319] MPEG Requirements. *MPEG-4 Requirements*. Doc. ISO/MPEG N4319, Sydney MPEG Meeting, July 2001.

[N4554] MPEG. *Coding of Audio-Visual Objects—Part 7: Optimized Visual Reference Software*. Draft Technical Report, Doc. ISO/MPEG N4554, Pattaya MPEG Meeting, December 2001.

[NeCR97] Neri, A., S. Colonnese, and G. Russo. *Automatic Moving Objects and Background Segmentation by Means of Higher Order Statistics*. IS&T Electronic Imaging '97 Conference: Visual Communication and Image Processing, San Jose, February 8–14, 1997.

[NetH88] Netravali, A. N., and B. G. Haskell. *Digital Pictures: Representation and Compression*. New York: Plenum Press. 1988.

[NMIM01] Nakagawa, A., E. Morimatsu, T. Itoh, and K. Matsuda. "Dynamic Resolution Conversion Method for Low Bitrate Video Transmission." *IEICE Transactions on Communication*, E84–B(4): 930–940. April 2001.

[OJJC97] Ostermann, J., E. S. Jang, S. S. Jae, and T. Chen. "Coding of Arbitrarily Shaped Video Objects in MPEG-4." *Proceedings International Conference on Image Processing*, 1: 496–499. 1997.

[OrcS94] Orchard, M. T., and G. J. Sullivan. "Overlapped Block Motion Compensation: An Estimation-Theoretic Approach." *IEEE Transactions on Image Processing*, 3(5): 693–699. September 1994.

[PenM93] Pennebaker, W. B., and J. L. Mitchell. *JPEG Still Image Data Compression Standard*. New York: Van Nostrand Reinhold. 1993.

[PXEL00] Panusopone, K., C. Xuemin, R. Eifrig, and A. Luthra. "Coding Tools in MPEG-4 for Interlaced Video." *IEEE Transactions on Circuits and Systems for Video Technology*, 10(5): 755–766. August 2000.

[Rijk96] Rijkse, K. "H.263: Video Coding for Low-Bitrate Communication." *IEEE Communications Magazine*, 34(12): 42–45. December 1996.

[RisL79] Rissanen, J. J., and G. G. Langdon. "Arithmetic Coding." *IBM Journal of Research and Development*, 23: 149–162. 1979.

[SaiP96] Said, A., and W. A. Pearlman. "A New, Fast, and Efficient Image Codec Based on Set Partitioning in Hierarchical Trees." *IEEE Transactions on Circuits and Systems for Video Technology*, 6(3): 243–250. June 1996.

[SalP94] Salembier, P., and M. Pardàs. "Hierarchical Morphological Segmentation for Image Sequence Coding." *IEEE Transactions on Image Processing*, 3(5): 639–651. September 1994.

[SaTM95] Salembier, P., L. Torres, and F. Meyer. "Region-Based Video Coding Using Mathematical Morphology." *Proceedings of the IEEE*, 83(6): 843–857. June 1995.

[Shap93] Shapiro, J. M. "Embedded Image Coding Using Zerotrees of Wavelet Coefficients." *IEEE Transactions on Image Processing*, 41(12): 3445–3462. December 1993.

[ShZM01] Shen, G., B. Zeng, and L. L. Ming. "Arbitrarily Shaped Transform Coding Based on a New Padding Technique." *IEEE Transactions on Circuits and Systems for Video Technology*, 11(1): 67–79. January 2001.

[SLHZ99] Sodagar, I., H.-J. Lee, P. Hatrack, and Y.-Q. Zhang. "Scalable Wavelet Coding for Synthetic/Natural Hybrid Images." *IEEE Transactions on Circuits and Systems Video Technology*, 9(2): 244–254. March 1999.

[SmSO99] Smolic, A., T. Sikora, and J.-R. Ohm. "Long-Term Global Motion Estimation and Its Application for Sprite Coding, Content Description, and Segmentation." *IEEE Transactions on Circuits and Systems for Video Technology*, 9(8): 1227–1242. December 1999.

[TaKM01] Takagi, K., A. Koike, and S. Matsumoto. "Padding Method for Arbitrarily-Shaped Region Coding Based on Rate-Distortion Properties." *Transactions of the Institute of Electronics, Information and Communication Engineers D-II*, J84D-II(2): 238–247. February 2001.

[TMNC99] Talluri, R., I. Moccagatta, Y. Nag, and G. Cheung. "Error Concealment by Data Partitioning." *Signal Processing: Image Communication*, 14: 505–518. May 1999.

[WanZ98] Wang, Y., and Q.-F. Zhu. "Error Control and Concealment for Video Communications: A Review." *Proceedings of the IEEE*, 86(5): 974–997. May 1998.

[WelT94] Wells, N. D., and P. N. Tudor. "Standardization of Scalable Coding Schemes." *Proceedings of the IEEE ISCAS (94) Tutorials*, pp. 121–130. May 1994.

[WiNC87] Witten, I. H., R. M. Neal, and J. G. Cleary. "Arithmetic Coding for Data Compression." *Communications of the ACM*, 30: 520–540. 1987.

[YWJK00] Yao, W., S. Wenger, W. Jiantao, and A. K. Katsaggelos. "Error Resilient Video Coding Techniques." *IEEE Signal Processing Magazine*, 17(4): 61–82. 2000.

CHAPTER 9

Visual SNHC Tools

by Euee S. Jang, Tolga Capin, and Jörn Ostermann

Keywords: graphics, synthetic and natural hybrid coding, face animation, body animation, 2D mesh coding, 3D mesh coding, view-dependent scalability, error resilience

At the end of the last century, 8 out of 10 Hollywood movies that won the Oscar for best visual effects (e.g., *The Matrix, Jurassic Park, Terminator 2*) used graphics elements for photo realistic visual effects [Perr01]. Today, graphics authoring tools like *RenderMan* are widely used by movie studios. Marking its 50 years of technological innovations, graphics, along with natural audiovisual data, turned into a new medium.

In the first half of the 1990s, the MPEG committee successfully launched its MPEG-1 and MPEG-2 standards as efficient tool sets for both audio and video coding. The notion of synthetic elements in MPEG was not ripe until MPEG-4 started in 1994. Until then, the compression of natural audio and video was the major theme in MPEG.

When MPEG-4 started, it was about the time that the Internet became recognized as a new important communication channel. As the Internet allows for *interactivity* between user and content, the interactive Web-based paradigm was about to take over the TV paradigm. Hence, the notion of composing several audiovisual objects in a scene seemed natural to be supported in MPEG-4. At this stage, developing a multimedia coding standard addressing both narrow and broadband communications, allowing the object-based manipulation of multimedia content, both natural and synthetic became the major goal of MPEG-4. This is how synthetic and natural hybrid coding (SNHC) came into the MPEG-4 picture.

While SNHC was on its way toward solving the puzzle of representing and compressing synthetic (or graphics) data, an independent work sprang up to design a virtual reality representation for the Web: the Virtual Reality Modeling Language (VRML). Just like the Hyper Text Markup Language (HTML), VRML provides a way to design a 3D virtual world with interactivity. VRML 2.0 (or VRML97) [VRML97, Web3D] has been officially promoted to ISO/IEC standard (ISO/IEC 14772); it is a well-known technology, useful for many Web-based applications.

VRML 2.0 was a good stimulus for MPEG-4 development and many VRML tools were brought to MPEG-4 (see Chapter 4). The VRML scene graph structure was taken as the starting point for the MPEG-4 binary format for scenes (BIFS) [MPEG4-1] developed by MPEG-4 Systems, which added functionalities such as compression and streaming capabilities. Transmission and storage of synthetic data in VRML remained problematic due to the huge size of such data. At the beginning of MPEG-4, the work scope of SNHC was rather independent from VRML. Because VRML already defined many aspects of SNHC, MPEG-4 decided to incorporate the VRML standard into MPEG-4 Systems [MPEG4-1]. MPEG-4 added streaming of graphics and animation, face and body animation (FBA), and 2D and 3D mesh compression tools [MPEG4-1, MPEG4-2].

The SNHC visual tools presented in this chapter are stand-alone tools for specific applications (e.g., virtual meetings and 3D home shopping). Combining individual tools with more advanced tools into a common framework has recently begun in the context of the so-called MPEG-4 Animation Framework eXtension (AFX) [N4319]. This effort, which should lead to an international standard by the fall of 2002, promises to integrate all synthetic and natural audiovisual tools for rich multimedia applications such as games and commercials. MPEG-4 AFX is to provide advanced graphics features such as new modalities (curved surfaces and subdivision surfaces), high-level animation, and advanced rendering (image-based rendering and multitexturing) [SNHC].

In Section 9.1, an overview of SNHC and its relationship with VRML and X3D is provided. FBA is discussed in Section 9.2. MPEG-4 enables the specification, coding, and animation of 2D meshes as presented in Section 9.3. In Section 9.4, the coding of 3D meshes is presented. Finally, view-dependent scalability is discussed in Section 9.5.

9.1 SNHC OVERVIEW

The work scope of SNHC includes both audio and visual information. Synthetic audio coding tools include text-to-speech (TTS) and structured audio, notably a structured audio orchestra language (SAOL; see Chapter 12). Synthetic audio tools are useful when they are combined with synthetic visual tools, but also as stand-alone. For example, TTS can be used with facial animation to generate the so-called *talking heads* for electronic commerce and entertainment applications. Synthetic visual tools are FBA, 2D and 3D mesh coding, and view-dependent scalability.

Visual texture coding (VTC) was developed by the SNHC group because its need was related to the mapping of textures on top of 2D and 3D mesh models. However, because the visual texture compression tool effectively codes natural images, its presentation is included in Chapter 8, which addresses natural video coding. Moreover, VTC is a stand-alone tool, which can be used without synthetic visual tools.

In conclusion, SNHC provides within MPEG-4 the representation and compression capabilities for synthetic data elements. Because MPEG-4 does not want to discriminate between natural and synthetic content, there is not a Part of the standard specifying SNHC tools and another Part specifying natural data coding tools. Whereas some synthetic audio and visual SNHC tools are specified together with natural audio and natural visual coding tools in the Audio [MPEG4-3] and Visual [MPEG4-2] Parts of the standard, respectively, other SNHC tools related to the synthetic and natural integration are included in the Systems Part of the standard [MPEG4-1].

9.1.1 VRML, X3D: Why Is SNHC Needed?

Conventionally, graphics and synthetic visual elements were considered as image or video data from the beginning of image processing and graphics [ACM]. One good reason for this is that the user only sees the content through 2D (or limited 3D or stereo) displays. In other words, the final product from natural and synthetic visual elements is similar. For example, Pixar's movie *A Bug's Life* was composed of 138 000 frames and stored in three terabytes. In terms of production, it is 100% computer animation. Yet, it is no more than a conventional 2D movie in that it has lost all the 3D features in recording and storage and thus it is consumed like any other movie.

Computer graphics often are compared with their counterpart, computer vision [Leng99]. Whereas the latter tries to find semantically meaningful relations and objects or models in a scene, the former starts with elements (or models) in a 2D or 3D scene. In graphics, the author defines the relationship between objects. These objects and their relationships in a scene are readily available to be extracted from the content. Obviously, this makes content reuse simple and decreases the cost of content production. Natural video, on the other hand, allows for the segmentation of a scene such that parts of the scene can be reused. Clearly, this is technically more complex and reuse is less straightforward.

Conventional (natural) audiovisual objects are authored with their own unique mechanisms and methods (e.g., microphone, video camera). Although created synthetically, synthetic visual elements were processed as natural visual elements for storage and transmission purposes. This is because the advent of high-quality graphics came later than image and video processing. In other words, shooting high-quality natural video was relatively easier than modeling natural-looking high-quality 3D models. Major breakthroughs in rendering technologies in the 1970s opened new ways to apply graphics for art and commercials. Breathtaking special effects powered by graphics are now abundant in everyday movies, but they may still require the use of hundreds of clustered workstations for hours to render seconds-long shots. By the end of the twentieth century, powerful 3D graphics were a luxury affordable only to content providers. And there was no way to handle graphics elements but to process them as natural audiovisual elements. This trend is now changing. Thanks to the continuous improvement of PC performance, what once was possible only with super computers now is possible with a PC. Moreover, the recent boom of 3D games raised the level of standard PC performance. A graphics accelerator is, now, a must-have in a PC.

For natural audiovisual information, there are JPEG, MPEG, and ITU coding standards. For synthetic data, VRML [VRML97] was the first international representation standard. Before VRML, quite a few proprietary file formats were available for different applications (e.g., Computer Aided Design [CAD], animation). As this chapter is being written, VRML is a popular format for virtual reality. Yet, VRML has some critical limitations inhibiting its adoption as the universal file format for 3D graphics [Jang00]:

☞ **Unacceptable file download time:** VRML was not designed to produce compact or compressed files. 3D models, represented by `IndexedFaceSet`, often yield data in the order of MBytes. This is because VRML is a text format, which is good for editing, but bad in terms of compression.

☞ **Lack of streaming support:** A renderer needs to load an entire VRML scene—including graphics, texture maps, animation control, and at least the first couple of seconds of any video present in the scene—before it can start visualizing the scene. This is due to the lack of explicit temporal structure in VRML. Many events are triggered as a result of object motion or collision, and it is not obvious when this will happen. For applications such as interactive games, in which the motion of the objects is impromptu rather than predefined, the streaming capability is a key functionality to minimize the latency time in transmission.

☞ **Slow rendering of complex scenes:** Rendering is a painfully complex and time-consuming process in which most CPU time is spent. This was a common problem for any graphics player when VRML was designed. To overcome or bypass the rendering process, many new technologies (such as image-based rendering and incremental rendering) are proposed today. Advanced rendering techniques (such as freeform shapes with adaptive sampling or motion blur) were considered unimportant at the time VRML was designed. The VRML specification does not support these fast and advanced rendering technologies, but they have become a norm in industry.

☞ **Slow authoring of complex shapes and behaviors:** VRML is a file format for describing graphics content. As such, it also supports a simple script language to define animation and behavior. Because VRML was designed by engineers without support from the major vendors of computer graphics authoring tools, there was no convenient authoring tool available for artists. Therefore, many VRML scenes were authored by writing text in the language of this VRML file format. This added to the perception that VRML is like a programming language and unsuitable for artists. Only recently have major authoring tool vendors supported the import and export of VRML scenes. However, these tools support only the basic features of VRML, such as static geometry import and export.

☞ **Limited support of graphics elements:** In VRML, only a limited number of tools to represent graphics elements are available. Thus, if one wants to use popular, high-level tools like free-form shapes or subdivision surfaces to produce graphics content, one must translate the content to be interpretable to the VRML language. The support of illumination and surface properties is very limited.

☞ **Lack of profiles and guaranteed quality:** Any VRML compliant renderer needs to implement the entire functionality of VRML, including script and behavior support. This leads to a large software packet that is expensive to develop. Furthermore, this delays or limits market penetration. VRML does not set a minimum rendering speed for a compliant renderer. This makes it impossible to create content with guaranteed playback quality. Therefore, the author cannot predict the quality of the user experience.

To overcome the limitations of VRML, in 1997 the Web3D Consortium, responsible for VRML, began a new work item, called *Extensible 3D (X3D)* [WaBS99]. Although X3D was launched as the next generation of VRML97, the major goals of X3D are set as follows [X3DFAQ]:

☞ **Backward compatibility with VRML97:** This will enable reuse of existing content and ease the integration with existing authoring tools.

☞ **Integration with XML:** This will enable authors to simply integrate content from different providers into one scene.

☞ **Componentization:** X3D recognizes the importance of profiles and levels to create content suitable for different platforms. MPEG-4 BIFS will be one of the supported components.

☞ **Extensibility:** X3D wants to be flexible in order to easily integrate new rendering tools, technologies, and graphics elements as they become available in the market.

The motto of X3D is "3D Anywhere" [X3DFAQ]. Considering that the integration with XML is a major target, the motto seems to make sense. The aforementioned limitations of VRML, however, are better solved by the MPEG-4 framework. Streaming and compression have been a stronghold of the MPEG committee. With its support of the basic VRML framework, MPEG-4 is the ideal platform to support 3D graphics along with natural audiovisual data.

Still compact representation and compression of advanced graphics modeling and animation remain a work item to support rich and high-quality animation movies and games. In the context of SNHC, jointly with the Web3D Consortium, MPEG is now developing a new work item called MPEG-4 AFX [SNHC]. AFX is expected to be supported in X3D in the same way that BIFS is supported in X3D—as a component.

According to the MPEG-4 Requirements document [N4319], the major AFX goals are these:

- ☞ **Enhanced texture mapping:** Advanced texture mapping is helpful in bypassing the complex rendering process such as ray tracing for precise illumination modeling. For instance, instead of calculating the illumination from a light source on the surface, one can blend two or more textures representing the original texture and a light map together for rendering purposes. This fast rendering process enables interactive and highly realistic virtual environments that are much less costly to render than the conventional rendering process using ray tracing.

- ☞ **Animation support:** In order to provide high-quality animation of objects, advanced animation tools such as curve interpolators, skeleton-based animation, and free form deformation are being considered on top of the existing key-frame based methods (i.e., interpolators).

- ☞ **High-level shape representation:** Commercially well-known and widespread shape representations (such as parametric surfaces, subdivision surfaces, and solids) are being considered in the framework.

- ☞ **Reusability of scene graph nodes and animation streams:** Because AFX is extending what has been accomplished in MPEG-4, the reusability is an important feature.

- ☞ **Persistence:** Just like any Web page, saving or restoring the current state is an important feature in the framework.

- ☞ **Compression of animated objects:** Compression of animation paths and animated objects is a compulsory feature because of the large amount of animated content.

MPEG-4 AFX is similar to X3D in the sense that it supports advanced features such as multitexturing and NURBS surfaces. MPEG-4 AFX, however, is covering many more advanced graphics features as previously explained. Streaming and compression also are being considered in its goals.

9.1.2 SNHC Visual Tools

In this chapter, the following SNHC visual tools will be addressed: FBA, 2D mesh coding, 3D mesh coding, and view-dependent scalability. FBA allows creating interactive human computer interfaces, e-commerce and kiosk applications, and—provided the video analysis problem could be solved—virtual meeting environments with 2D and 3D virtual heads and associated natural or synthetic speech.

The major applications of 2D mesh coding are object-based video manipulation, indexing, and compression, mainly coinciding with the applications for video coding in MPEG-4 [BTZC99]. 2D mesh coding, however, is a dedicated solution when a synthetic object must be combined with a visual object in a scene. The major importance of 2D mesh coding is that it provides an efficient coding scheme for 2D mesh objects, which conventional video coding cannot provide.

3D mesh coding is the first MPEG-4 tool allowing one to represent and compress 3D models using triangular meshes. In the upcoming AFX work, new and more efficient tools to represent 3D models such as NURBS and subdivision surfaces are under development.

Synthetic visual environments usually contain multiple 2D or 3D objects with complex geometry. In virtual reality applications in which users have individual viewpoints and fields of view, downloading the complete scene graph and its entire elementary streams (ESs) would take an extensive amount of time and complexity. View-dependent scalability provides compression efficiency while lessening the computational burden on the decoder terminal at the cost of increased server load—making this tool unsuitable for large multiuser environments.

9.2 FACE AND BODY ANIMATION

Personal communications are an important type of applications for the MPEG-4 standard. To improve the personal communication capabilities of the MPEG-4 standard, a special visual object was identified targeting the efficient representation of real persons and synthetic characters: the *FBA* object. The FBA object is useful for a wide range of potential applications, enabling direct or indirect personal communications, in contexts such as e-commerce, games, virtual teleconferencing, virtual kiosks, and call centers.

FBA provides tools for model-based coding of video sequences containing human faces and bodies. Instead of representing the faces and bodies as coded 2D raster images, the FBA object uses a 3D synthetic model that can be defined and animated by specific FBA parameters. These parameters may be extracted from real video or synthetically generated, and afterward coded and transmitted. Besides representing real human faces and bodies, the MPEG-4 FBA tool also can represent completely synthetic faces and bodies that do not necessarily resemble humans.

9.2.1 Overview

MPEG-4 specifies a face model in its neutral state, a number of feature points on this neutral face as reference points, and a set of facial animation parameters (FAPs), each corresponding to a particular facial action deforming a face model in its neutral state [AbPe99, BPO00, TO00, WOZ01]. Deforming a neutral face model according to some specified FAP values at each time instant generates a facial animation sequence. The FAP value for a particular FAP indicates the magnitude of the corresponding action (e.g., a big versus a small smile or deformation of a mouth corner). For an MPEG-4 terminal to interpret the FAP values using its face model, it must have predefined model specific animation rules to produce the facial action corresponding to each FAP. The terminal can either use its own animation rules or download a face model and the associated face animation tables (FAT) to have customized animation behavior. Because the FAPs are required to animate faces of different sizes and proportions, the FAP values are defined in FAP units (FAPUs). The FAPUs are computed from spatial distances between major facial features, like mouth width (MW0) on the model in its neutral state (see Figure 9.1).

The simplest representation of a face is a mesh with as many vertices as there are feature points. The vertices are shifted in space as a function of the FAP amplitudes. Usually, face models consist of considerably more vertices than there are MPEG-4 feature points. In this case, each FAP will move several vertices of the face model.

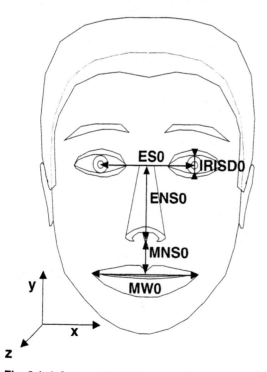

Fig. 9.1 A face model in its neutral state and the feature points used to define FAP units (FAPU). Fractions of distances between the marked key features are used to define FAPUs [MPEG4-2].

Similarly, MPEG-4 specifies a body in its neutral posture standing upright with arms down [CPO00a, TT01]. Feature points define the location of joints of the body—that is, the feature points define the skeleton of the body. Body animation parameters (BAPs) define the rotation of the joints thus enabling the animation of the body. The amplitude of a BAP controls the amount of joint rotation. Consequently, bodies do not require scaling parameters like FAPUs. An MPEG-4 terminal provides a default body model for animation. Alternatively, the content provider can download body definition parameters (BDPs) to customize the body model of the terminal or a new body model. The face of a body is described as an MPEG-4 face model.

The simplest body model is a stick figure with the endpoints of the sticks defined by the position of the feature points. This stick figure is sometimes referred to as the skeleton of the body. Typically, the body models have surfaces defined around the skeleton. These surfaces can be described using one or more meshes.

There is no constraint on the complexity of the 3D models that can be used in MPEG-4 FBA: they can be rather realistic representations of real persons with thousands of vertices, as well as very simple cartoonlike models with few vertices. To achieve the FBA goals, the MPEG-4 Visual standard defines profiles and levels with parameter limits to support the realistic animation of more or less complex FBA models in real time.

MPEG-4 specifies two sets of FBA parameters to define the geometry and animation of 3D face and body models [MPEG4-2]:

☞ **Definition parameters:** These parameters allow the specification of the geometrical shape of the FBA model by means of face definition parameters (FDPs) and body definition parameters (BDPs); moreover, texture data may also be sent to the decoder. These parameters allow the decoder to create an FBA 3D model with specified shape and texture (see Figure 9.2). Typically, texture data is coded using a wavelet coder as defined by the MPEG-4 visual texture coding tool (see Chapter 8).

☞ **Animation parameters:** These parameters allow the definition of face and body animation by means of FAPs and BAPs. FDPs and BDPs typically are transmitted only once, whereas FAPs and BAPs are transmitted once for each frame. MPEG-4 defines 68 FAPs and 186 BAPs to specify the state of the FBA object in one frame during animation. Each FAP/BAP has predefined semantics and defines one degree of freedom. Additionally, 110 parameters are provided in the form of extension animation parameters.

The MPEG-4 FBA specification defines the syntax of the bitstreams and the behavior of the decoders. The way of generating the 3D models (e.g., 3D interactive modeling programs, vision-based modeling, or template-based modeling) is not specified by the standard; the content providers choose their method and model depending on the application. Furthermore, the content

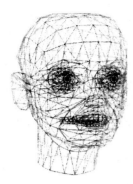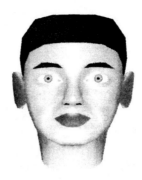

Fig. 9.2 IST face model in its neutral state: (a) polygons only and (b) shaded surfaces. Reprinted with permission from Abrantes, G., and F. Pereira, "MPEG-4 Facial Animation Technology: Survey, Implementation and Results." *IEEE Transactions on Circuits and Systems for Video Technology*, 9 (2), pp. 290–305, March 1999. © 1999 IEEE.

provider also can choose not to send a model to the MPEG-4 terminal, relying on the unknown and proprietary model available at the decoder (if a close control of the model appearance is not essential for the application in question) [AbPe99, BPO00]. MPEG-4 defines a scalable scheme for coding FBA objects; the encoder can choose the adequate coding parameters to achieve a selected bit rate and animated model quality.

The specification of the FBA object is distributed into two separate but related parts of the MPEG-4 standard: Part 1, Systems [MPEG4-1] and Part 2, Visual [MPEG4-2]. MPEG-4 Systems specifies the representation and coding of the model geometry and the methods to adapt the surface of face and body (i.e., FDP and BDP parameters). It also defines the integration of FBA BIFS nodes with other audiovisual objects in the same scene. MPEG-4 Visual specifies the coding of the animation for the models (i.e., FAP and BAP parameters and corresponding coding). MPEG-4 Audio specifies a TTS tool with an interface to the FBA object in order to drive a talking face with a speech synthesizer [MPEG4-3].

In Section 9.2.2, the face model in its neutral state and the body model in its default posture are described. Section 9.2.3 explains the nodes of the FBA object and their integration into a scene graph. The customization of face and body models using definition parameters is presented in Section 9.2.4. In Section 9.2.5, the FAPs and BAPs as well as their coding are presented in detail.

9.2.2 Default Facial Expression and Body Posture

As the first step, MPEG-4 defines a default neutral expression for faces and posture for body models. Every MPEG-4 compliant decoder has to follow similar posture and facial expressions in its default state. The default facial expression has the following properties (see Figure 9.2):

☞ Gaze is in the z-axis direction.

☞ All face muscles are relaxed.

☞ Eyelids are tangent to the iris.

☞ The pupil is one-third of the diameter of the iris.

☞ Lips are in contact; the line of the lips is horizontal and at the same height of lip corners.

☞ The mouth is closed and the upper teeth touch the lower ones.

☞ The tongue is flat, horizontal with the tip of the tongue touching the boundary between upper and lower teeth.

Similarly, the neutral body posture has the following properties (see Figure 9.3):

☞ Standing posture, gaze is in the z-axis direction.

☞ The feet point to the front direction.

☞ The two arms are placed on the side of the body with the palm of the hands facing inward.

☞ The hands point in the y-axis direction, except the thumb, which has 45 degrees' inclination.

9.2.3 FBA Object

FDP and BDP sets are used to define a new FBA model, using the scene graph built through the MPEG-4 BIFS stream (Figure 9.4). BIFS is the binary format for scene description in MPEG-4 (see Chapter 4); it defines semantics and coding techniques for each node and its fields. FBA defines several BIFS nodes: `Body`, `BDP`, `BAP`, `BodyDefTable`, `Face`, `FDP`, `FAP`, `FIT`, `FaceDefTable`, `Face-DefMesh`, and `FaceDefTransform`. The FBA scene graph, which is part of the overall scene graph, and thus is attached to the scene graph hierarchy, has a

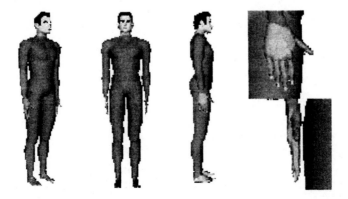

Fig. 9.3 Default body posture [MPEG4-2].

Body node as its root (see Figure 9.4). The following list provides more detail for each node:

☞ The **Body** node encapsulates all the nodes of the virtual character, and contains three children: BDP, BAP, and renderedBody.

☞ The **BDP** node encapsulates the downloadable geometry of the body, together with the joint positions, surface geometry, and deformation tables. The BDP child of the Body node is optional. If it is empty, the FBA decoder must apply the animation to its default 3D model. This requires each FBA decoder to have its own default model; this default model has no constraints but the default state presented in Section 9.2.2. The BDP node has two children: bodySceneGraph and bodyDefTable.

☞ The **bodySceneGraph** node contains a hierarchy of body joints in default posture, together with the attached surfaces. The bodySceneGraph speci-

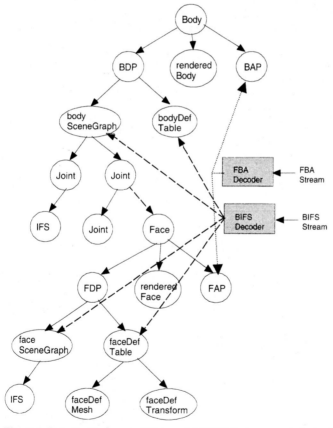

Fig. 9.4 Relation between FBA and BIFS bitstreams and the BIFS scene graph structure. The main purpose of the BIFS stream is to define the FDP and BDP nodes and their children. The FAP and BAP nodes are children of the Face and Body nodes, respectively.

fication is based on the VRML 97 H-Anim 1.1 standard representation of body geometry [Web3D99]. As a minimum, this scene graph contains joint nodes that define parameters such as position, degrees of freedom, and limits on rotation for a joint of the body model. The scene graph also can contain surface meshes. Some meshes may be represented in compressed format using an `Indexed Face Set` (`IFS`) node (Section 9.4). Because the `IFS` node reorders vertices, it requires special attention when using this node together with the `bodyDefTable`.

☞ The **Joint** nodes describe the skeleton hierarchy of the body. This node is based on a `Transform` node, which is used to define the relationship of each body segment to its immediate parent. In addition, the `Joint` node can contain hints for inverse-kinematics systems that want to control the body. These hints include the upper and lower joint limits, the orientation of the joint limits, and a stiffness/resistance value. One field of the `Joint` node points to a `segment` node that describes the hull of the body using `IndexedFaceSet` nodes.

☞ The **Face** node is a descendent of the `bodySceneGraph` node, and is typically attached to the skull base, which is another joint of the body model. It encapsulates the downloaded geometry and texture of the face, the FAPs, and the nodes for deformation of the face. It also contains the children `FDP`, `FAP`, and `renderedFace`.

☞ The **FDP** node allows for customizing or replacing the local model of the FBA decoder with the downloaded model. Similar to the `BDP` node, it is optional. If it is omitted, the default face model of the FBA decoder is used for rendering.

☞ The **faceDefTable** node contains facial deformation tables as its children, and serves to encapsulate all these tables. Deformation tables can be in the form of either `faceDefMesh` nodes or `faceDefTransform` nodes.

☞ The **faceDefMesh** node contains the face model's mesh deformation as a function of FAPs. For example, deformation of the vertices around the character's mouth can be represented as a function of FAPs belonging to the mouth.

☞ The **faceDefTransform** node contains transformation information like rotation scale and translation as a function of FAPs. For example, the rotation of the model's eyeballs can be represented as a function of the FAPs belonging to the eyes.

☞ The **renderedFace** node is the root of the face model that is passed to the compositor of the MPEG-4 terminal for rendering. The children of `renderedFace` are rendered, whereas `faceSceneGraph` stores the downloaded model in its neutral state and is not rendered.

☞ The **FAP** node contains all the FAP amplitudes in its fields; the FBA decoder writes the FAP values in these fields for each frame.

☞ The **FAP interpolation table** (**FIT**) node allows the definition of interpolation rules for the FAPs that have to be interpolated at the decoder.

The 3D model is then animated using the FAPs sent and the FAPs interpolated according to the FIT.

☞ The **bodyDefTable** node defines the behavior of the body surface model as a function of the BAP values. This allows the encoder to customize the body behavior for the whole or specific parts of the 3D model.

☞ The **BAP** node contains the 296 BAPs as its fields, and is used to connect the FBA decoder to the BIFS scene graph. The FBA decoder stores BAP values in these fields for each frame, and the Body node reads these fields to update and render the model.

☞ The **renderedBody** node is the root of the FBA model that is passed to the compositor for rendering. The children of renderedBody are rendered, whereas the bodySceneGraph stores the downloaded model in default posture and thus is not rendered. Because the Face node is a child of bodySceneGraph, the node renderedFace is automatically placed at the right position joining the renderedBody.

The next sections will discuss each of these features in more detail.

9.2.4 FDP and BDP Listing and Coding

FDPs and BDPs allow defining or adapting the 3D face and body models to be animated at the decoder. In the following, the parameters selected by MPEG-4 are listed and their coding is described.

9.2.4.1 Facial Definition Parameters   FDPs are designed to enable the customization of a proprietary 3D facial model at the receiver or the downloading of a new model together with the information about how to animate it [MPEG4-2]. Whereas the first case allows a limited control by the encoder on what the animated face will look like, the second allows for full control of the animation results. Because FDPs are used to configure the facial model, they typically are sent only once per session. Some applications may transmit FDPs multiple times during a session, particularly in broadcast scenarios or for character morphing. FDPs are encoded in BIFS, using the nodes specified in the MPEG-4 Systems standard [MPEG4-1]. To minimize differences in animation results due to differences in decoders' local models and behavior, the encoder should send a sufficient amount of FDP configuration data. FDP data can be used in different ways with different purposes. Four relevant cases regarding the use of FDP data in an MPEG-4 facial animation system can be identified [AbPe99]:

☞ **No FDP data (only FAP data):** Because no FDP data is sent, the 3D model resident at the decoder is animated with the received FAP data without performing any model adaptation.

☞ **Feature points:** A set of feature points, represented by their 3D coordinates, is sent to the receiver with the purpose of calibrating the resident model (see Figure 9.5). These feature points must correspond to a neutral face. Because of potential difficulties in obtaining the coordinates of some feature points, it is not necessary to send all feature points, and thus subsets are allowed. The decoder must adapt the 3D model in a way that the vertices on its resident model corresponding to feature points must coincide with the feature points received. MPEG-4 does not define how this calibration is achieved. Initial work shows that a satisfactory model cali-

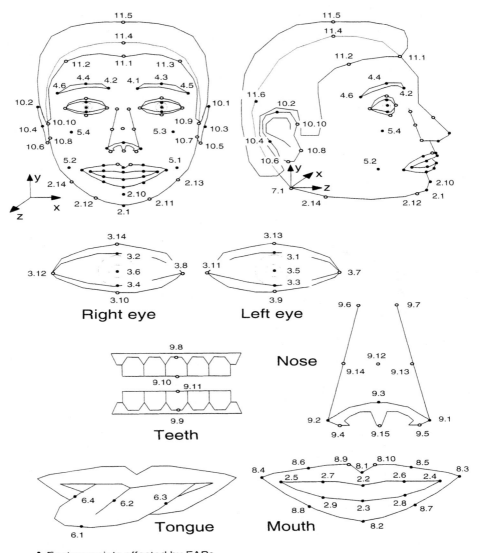

Fig. 9.5 Facial feature points and their grouping [MPEG4-2].

bration is difficult to achieve [ET97, LP99]; we discourage the use of feature points to calibrate a face model.

☞ **Feature points and texture:** It is well known that texture mapping significantly improves the realistic appearance of an animated 3D model. This case considers sending 3D feature points to adapt the model geometry (as explained above) together with texture data and the corresponding calibration information (2D feature points) to allow the decoder to further adapt the model to the texture to be mapped. With this target, one texture and 2D texture coordinates corresponding to some facial feature points are sent. To improve the texture mapping result, the encoder may indicate which type of texture is being sent, either a cylindrical projection (e.g., cyberware texture) or an orthographic projection (e.g., frontal texture). Again, MPEG-4 does not specify how this calibration and mapping are achieved. There are some concerns regarding the quality achieved by mapping texture to the head using only texture coordinates for the feature points.

☞ **faceDefTables (FAT) and new 3D model:** For the encoder to have full control of the animation results, this is the method to be used. To this end, a 3D facial model is downloaded to the receiver, together with features points (to allow the extraction of FAPUs) and FATs that define the FAP behavior for the new model [OH97]. The FAP behavior on the newly received model is specified by indicating which and how the new model vertices should be moved for each FAP. The downloaded model can be composed of multiple meshes, each one with an associated texture.

Deformation Rules using faceDefTables   A faceDefTable defines how a model is deformed as a function of the amplitude of the FAPs. It specifies for each FAP which Transform nodes and which vertices of an IndexedFaceSet node are animated by it and how. A Transform node defines rotation, translation, and scaling for parts of a scene graph. The IndexedFaceSet node defines a set of 3D polygons. The faceDefTables are considered part of the face model. They need to be transmitted to the MPEG-4 terminal whenever the content provider wants to use his own face model. One FAP can animate several Transform and IndexedFaceSet nodes.

Animation Definition for a TransformNode   If a FAP causes simple transformations such as rotation, translation, or scale, a Transform node can describe this animation. For example, the rotation of an eye can be implemented using a Transform node that is part of the scene graph of a face. The faceDefTransform node defines which field (rotation, scale, or translation) of a Transform node is updated by a facial animation parameter, and how the field is updated. For example, the rotation axis of the Transform node for the eye motion would be set to define a rotation around the x-axis or y-axis. The actual rotation is then proportional to the FAP amplitude. For translation and scale, the actual amplitude is proportional to the FAP amplitude and a scale factor defined in the faceDefTransform node.

Animation Definition for an IndexedFaceSet Node If a FAP causes deformations of the face model, the animation results in updating the vertex positions of the affected `IndexedFaceSet` nodes. The affected vertices move along piecewise linear trajectories that approximate flexible deformations of a face. A vertex moves along its trajectory as the amplitude of the FAP varies. By means of a `faceDefMesh` node, the `faceDefTable` defines for each affected vertex its own piecewise linear trajectory by specifying intervals of the FAP amplitude and 3D displacements for each interval, as shown in Figure 9.6.

FAP Interpolation Table The FIT allows a smaller set of FAPs to be sent during a facial animation session, because it permits the encoder to specify interpolation rules for some or all of the FAPs the encoder does not transmit and set for interpolation at the receiver [TCWH99]. The small set of FAPs transmitted then can be used to determine the values of other FAPs, using a rational polynomial mapping between parameters. For example, the top inner lip FAPs can be sent and then used to determine the top outer lip FAPs. To this end, the encoder specifies a graph giving relationships between FAPs for interpolation, and a set of rational polynomial functions that specify which and how sets of received FAPs are used to determine the remaining FAPs. The use of the `FIT` node can decrease the bit rate for a FAP stream by more than 50%.

9.2.4.2 Body Definition Parameters BDPs are used for the customization of the body by the encoder as it becomes able to replace or adapt the model resident at the decoder. BDP parameters define the set of parameters to transform

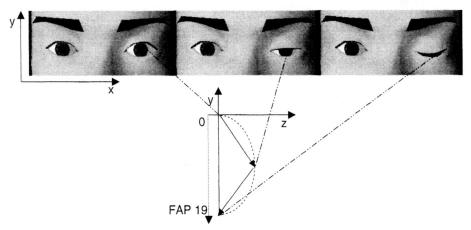

Fig. 9.6 Neutral state of the left eye (left) and two deformed animation phases for the eye blinking (FAP 19). The FAP defines the motion of the eyelid in the negative y-direction (positive FAP values move the vertices downwards); the `faceDefTable` defines the motion of the vertices of the eyelid in the x, y and z directions. Reprinted with permission from J. Ostermann, "Animation of Synthetic Faces in MPEG-4," *Computer Animation '98*, Philadelphia, pp. 49–55, June 1998. © 1998 IEEE.

the default body to a customized body; they include body geometry, calibration of body parts, degrees of freedom, and (optionally) deformation information. Figure 9.7 shows the body feature points [MPEG4-1]. If BDPs are transmitted, the available decoder's body is transformed into a particular body with specific shape and appearance. Normally, the BDPs are transmitted once per session, followed by a stream of compressed BAPs. However, if the decoder does not receive the BDPs, model-independence of BAPs ensures that it can still interpret the BAP stream. *This ensures minimal interoperability in broadcast and teleconferencing applications.*

The encoder can send three types of BDP information. In its simplest version, the decoder receives the location of feature points only. In that case, the decoder scales the individual parts of its body model to match the lengths defined by the feature points. MPEG-4 does not specify an algorithm for doing this. Although the implementation of this feature is straightforward for stick figures as in Figure 9.7, there is no solution available that will adapt any body model to new feature point coordinates and still provide a pleasant body model. In addition to the feature points, the terminal can receive a texture map with texture coordinates for the feature points. Again, it is unclear how to map the texture precisely with such a low number of texture coordinates. Finally, the content provider may decide to download a body model and `bodyDefTables` that define how the model surface is deforming as a function of different BAPs.

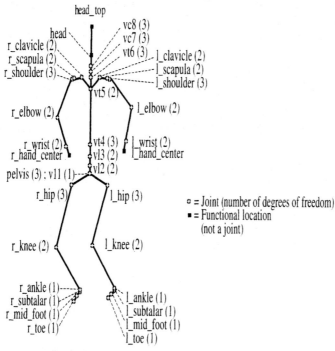

Fig. 9.7 Body feature points.

These tables enable the animation of bodies with seamless surfaces. Downloading a body model into the terminal gives the content provider the most control over the look of the presentation.

Animation Rules using `bodyDefTables` The `bodyDefTable` node is an extension of the `faceDefTable` node. Note that each `faceDefTable` is indexed by a single FAP. This means that either each vertex is controlled by only one FAP (i.e., each vertex appears in only one `faceDefTable`) or each vertex can be controlled by more than one FAP (i.e., a vertex may appear in more than one `faceDefTable`). In the latter case, the final displacement of the vertex is calculated by adding the displacements produced by different tables, with equal weights. In body animation, however, multiple BAPs typically affect the same vertex without equal coefficients (for example, the upper arm vertices may be deformed, based on elbow and shoulder BAPs). For these cases it is not sufficient to add the displacements from different tables. Therefore, the `bodyDefTables` use multiple BAPs to control vertices. A `bodyDefTable` contains *key deformations*, each one consisting of a combination of selected BAPs. A `bodyDefTable` contains the data shown in Table 9.1.

Each row in the `bodyDefTable` corresponds to a combination of BAPs. If the current body posture is not one of these BAP combinations, the vertex deformations have to be interpolated based on these BAP combinations. A linear interpolation technique solves this problem with a low computational overhead. This technique works by representing the current BAP combination as a point in an N-dimensional space. A weighted average between this point and key BAP combinations around it is computed, weights being calculated as distances in the N-dimensional space.

9.2.4.3 FDP and BDP Coding Schemes

The BDP and FDP sets are transmitted as part of the BIFS scene graph to the decoder. The FBA nodes, defined in Section 9.2.3, are transmitted as BIFS nodes and coded using BIFS tools. The fields of the FBA nodes are quantized using a `QuantizationParameter` node, which specifies the quantization category and coding parameters such as minimum and maximum values. BIFS defines different quantization categories for different types of fields: `Position3D`, `Position2D`, `Color`, `TextureCoordinate`,

Table 9.1 Example of a `bodyDefTable`

BAPs				Vertices			
BAP_1	BAP_2	...	BAP_k	$Vertex_1$	$Vertex_2$	...	$Vertex_N$
0	0	.	0	0	0	.	0
0	0	.	100	D_{11}	D_{21}	.	D_{N1}
0	100	.	0	D_{21}	D_{22}	.	D_{N2}
...							

Angle, Scale, InterpolatorKeys, Normals, Rotations, ObjectSize3D, ObjectSize2D, LinearQuantization, and CoordinateQuantization.

As the bodySceneGraph node contains the geometry of the body, it is coded using BIFS tools for coding geometry with generic geometric shapes. However, the other FBA nodes are special nodes for the FBA object, and thus are coded based on a specific syntax.

The bodyDefTable node syntax and the coding scheme for its fields are presented in Figure 9.8. The name of the node is given in bodySceneGraph-NodeName. The name enables other nodes and scripts to refer to this node. The field bapIDs contains the list of BAPs that are used in this table (second row, left cells in Table 9.1). The field vertexIds holds the list of vertices of the body model that will be moved with this table (second row, right cells in Table 9.1). The number of BAP combinations listed in the table is given in the field numInterpolateKeys. The actual list of BAP combinations is given in the field bapCombinations (left columns, rows 3 and higher in Table 9.1). The corresponding vertex displacements are listed in the field displacements (right columns, rows 3 and higher in Table 9.1). A similar coding scheme is used for the faceDefTable node. After quantizing the node field displacements, arithmetic coding is applied to the node.

BAP and FAP nodes are handled in a different way than the other FBA nodes, as they are used for streaming the FBA animation in the form of FAPs and BAPs, rather than for defining any geometry. When FBA tools are used in a scene graph profile (see Chapter 13) together with other nonFBA BIFS nodes, the FBA ES is encapsulated within a BIFS-Anim stream (see Chapter 4). BIFS-Anim is a tool allowing a single stream to update fields in multiple nodes in the transmitted BIFS scene graph. A BIFS-Anim frame with an FBA frame is similar to an FBA frame, with a slight difference: The BIFS-Anim frame consists of a BIFS-Anim frame header and BIFS-Anim frame data. The data field has the same content as the FBA frame data: face animation data followed by body animation data, if any. The BIFS-Anim frame header is also similar to an FBA ES frame header; however, it contains the BIFS-Anim mask to represent the target object, instead of the FBA object mask. This is an N-length Boolean array, in which N is the number of nodes affected by the FBA stream.

Experiments with faceDefTable and bodyDefTable coding have shown that it is possible to create high-quality deformable face and body models. The

```
BodyDefTable [
exposedField   SFString  bodySceneGraphNodeName // String
exposedField   MFInt32   bapIDs                 // Integer, range [1,296]
exposedField   MFInt32   vertexIds              // No quantization, range [0,+I]
exposedField   MFInt32   bapCombinations        // No quantization, range [0,+I]
exposedField   MFVec3f   displacements          // MFVec3f
exposedField   SFInt32   numInterpolateKeys     // No quantization, range [2,+I]
]
```

Fig. 9.8 bodyDefTable node syntax.

model shown in Figure 9.9 requires only 24 kBytes for each node to code the deformation tables in addition to the static body geometry [CPO00a]. Thus, it is possible to effectively code body deformations with low overhead.

9.2.5 FAP and BAP Listing and Coding

FAPs and BAPs enable the animation of 3D face and body models available at the decoder. In the following, these parameters and their coding are described [TO00, CPO00b].

9.2.5.1 Facial Animation Parameters As mentioned before, FAPs were designed to allow the animation of 3D faces, reproducing movements, expressions, emotions, and speech pronunciation. FAPs are based on the study of minimal facial actions and are closely related to muscle actions. The set of FAPs represents a complete set of basic facial movements, allowing terminals to represent most natural facial actions as well as exaggerated, nonhuman-like actions (e.g., useful for cartoonlike animations).

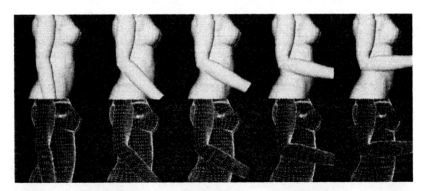
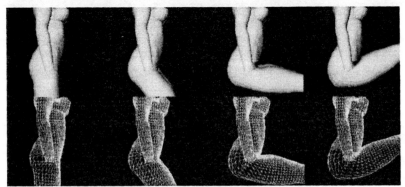

Fig. 9.9 Result of using body deformation tables. Reprinted with permission from Capin, T. K., E. Petajan, and J. Ostermann. "Efficient Modeling of Virtual Humans in MPEG-4," *ICME 2000*, New York: IEEE, pp. TPS9.1. 2000. © 2000 IEEE.

The MPEG-4 FAP set includes 68 FAPs, 66 low-level parameters associated with lips, jaw, eyes, mouth, cheek, nose, and so on, and 2 high-level parameters (FAPs 1 and 2) associated with expressions and visemes. Although low-level FAPs are associated with movements of key facial zones—typically referenced by a feature point (see Figure 9.5), as well as with the rotation of the head and eyeballs—expressions and visemes represent more complex actions, typically associated with a set of FAPs. Low-level FAPs are grouped as defined in Table 9.2. These groups are used to increase the coding efficiency of the FAP coder. The encoder knows the reference feature point for each low-level FAP. Assuming a proprietary model at the decoder, the encoder does not know precisely how the decoder will move the model vertices around that feature point. The encoder knows this only in case it downloads a model to the decoder.

All low-level FAPs involving translational movement are expressed in terms of FAPUs. These units are defined to allow the interpretation of the FAPs on any facial model (normally unknown to the encoder) in a consistent way, producing reasonable animation results. FAPUs correspond to fractions of distances between some key features (e.g., mouth-nose separation, eye separation). The fractional units used for the various FAPs are chosen to allow enough precision for the corresponding FAP. Table 9.3 shows an excerpt from the FAP list, including their numbers and names, short descriptions, the specification of the associated FAPU (e.g., MNS is mouth-nose separation and MW is mouth width; Figure 9.1), whether the FAP is unidirectional or bidirectional, the definition of the movement direction for positive values, the group (see Table 9.2) and subgroup (feature point number within the group) numbers (e.g., according to Table 9.3, FAP 5 animates feature point 3 within group 2 in Figure 9.5), and the default quantization step size (*N/A* means *not appli-*

Table 9.2 FAP groups [MPEG4-2]

Group	Number of FAPs
1: Visemes and expressions	2
2: Jaw, chin, inner lowerlip, cornerlips, midlip	16
3: Eyeballs, pupils, eyelids	12
4: Eyebrow	8
5: Cheeks	4
6: Tongue	5
7: Head rotation	3
8: Outer lip positions	10
9: Nose	4
10: Ears	4

Table 9.3 Excerpt of the MPEG-4 FAP specification table [MPEG4-2]

#	FAP name	FAP description	FAP Units	Uni- or bidirectional	Motion direction	Group number	Subgroup number	Default quantiz. step
1	Viseme	Set of values determining the mixture of two visemes (e.g., pbm, fv, th)	N/A	N/A	N/A	1	N/A	1
2	Expression	Set of values determining the mixture of two facial expressions	N/A	N/A	N/A	1	N/A	1
3	Open_jaw	Vertical jaw displacement (does not affect mouth opening)	MNS	U	down	2	1	4
4	Lower_t_midlip	Vertical top middle inner lip displacement	MNS	B	down	2	2	2
5	Raise_b_midlip	Vertical bottom middle inner lip displacement	MNS	B	up	2	3	2

cable). Assuming that the decoder receives the quantization interval q for a given FAP, the actual amplitude applied to the model is q × {quantizerstep-size} × {FAPU}.

FAPs 1 and 2 are referred to as high-level FAPs. Using the expression FAP, it is possible to select among six different expressions: joy, sadness, anger, fear, disgust, and surprise. The amplitude of two expressions can be animated at the same time. MPEG-4 does not define the mapping of expression amplitude into vertex motion. Therefore, it enables artists to define face models with their own personalities. When animating a group of different faces with low-level FAPs, all faces will show identical implementations of facial expressions, whereas the use of high-level FAPs will allow each face to show its own implementation of the expressions. The visemes are the visual analog to phonemes and allow the efficient rendering of visemes for better speech pronunciation, as an alternative to having them represented using a set of low-level FAPs. However, MPEG-4 allows specifying only two visemes for any given frame to be rendered. This restriction makes the implementation of a coarticulation model for the mouth movements challenging. Coarticulation usually requires knowledge of the sounds made up to 500 ms in the past and sounds to be made up to 500 ms in the future. For such a time period, knowledge of at least three visemes is required [CM93]. MPEG-4 allows indicating, in the FAP stream, the desired gender of the facial model to be used by the decoder. This information does not supersede the FDP information (if available) and is provided only as a *hint*, thus without any normative constraints for the decoder. This can be used in scenarios in which there is no BIFS stream and a 3D face is used together with a text-to-speech system.

9.2.5.2 Body Animation Parameters   BAPs manipulate independent degrees of freedom in the skeleton model of the body to produce animation of body parts. Similar to the face, the remote manipulation of a body model in a terminal with BAPs can accomplish lifelike visual scenes of the body in real time without sending pictorial and video details of the body every frame.

If correctly interpreted, BAPs will produce reasonably similar high-level results in terms of body posture and animation on different body models, also without the need to initialize or calibrate the model. There are 186 predefined BAPs in the BAP set, with an additional set of 110 user-defined extension BAPs. Each predefined BAP corresponds to a degree of freedom in a joint connecting two body parts. These joints include toe, ankle, knee, hip, spine (C1-C7, T1-T12, L1-L5), shoulder, clavicle, elbow, wrist, and the hand fingers (see Figure 9.7). Extension BAPs are provided to animate additional features than the standard ones in connection with body deformation tables (e.g., for cloth animation).

BAPs are categorized into groups with respect to their effect on the body posture (see Table 9.4). Using this grouping scheme has a number of advantages. First, it allows adjusting the complexity of the animation by choosing a subset of the BAPs. For example, the total number of BAPs in the spine is 72, but significantly simpler models can be used by choosing only, for example, the Spine1 group. Second, assuming that not all the animations need all the BAPs, only the active BAPs are transmitted to decrease the required bit rate significantly. This is accomplished by using a mask transmitted every frame with the active BAP groups, as discussed in the next section.

9.2.5.3 FAP and BAP Coding Schemes   Similarly to other MPEG standards, the MPEG-4 standard does not specify the FBA encoding methods, but only the FBA bitstream syntax and semantics, together with the corresponding decoding rules. The goal of using FBA compression is to reduce the bit rate necessary to represent a certain amount of animation data with a certain predefined quality or to achieve the best quality for that data with the available amount of resources (bit rate). FAPs and BAPs are coded at a certain frame rate, indicated to the receiver, which can be changed during the session. Moreover, one or more time instants can be skipped when encoding at a certain frame rate. Because not all the FAPs and BAPs are used for all of the frames, an FBA masking scheme is used to select the relevant FAPs and BAPs for each frame. FAP masking is done using a two-level mask hierarchy. The first level indicates, for each FAP and BAP group (see Table 9.2 and Table 9.4), one of the following four options (2 bits):

1. No FAPs or BAPs are coded for the corresponding group.

2. A second mask is given indicating which FAPs or BAPs in the corresponding group are coded; FAPs or BAPs not selected by the group mask retain their previous value, if any value has been previously set (no interpolation is allowed).

3. A second mask is given indicating which FAPs in the corresponding group are coded; the decoder should interpolate FAPs not selected by the group mask (this mask option is not used for BAPs).

4. All FAPs or BAPs in the group are coded.

Table 9.4 BAP groups [MPEG4-2]

Group	Number of BAPs
1. Pelvis	3
2. Left leg1	4
3. Right leg1	4
4. Left leg2	6
5. Right leg2	6
6. Left arm1	5
7. Right arm1	5
8. Left arm2	7
9. Right arm2	7
10. Spine1	12
11. Spine2	15
12. Spine3	18
13. Spine4	18
14. Spine5	12
15. Left hand1	16
16. Right hand1	16
17. Left hand2	13
18. Right hand2	13
19. Global positioning	6
20. Extension BAPs1	22
21. Extension BAPs2	22
22. Extension BAPs3	22
23. Extension BAPs4	22
24. Extension BAPs5	22

For each FAP/BAP group in the second case, a second mask is transmitted, indicating which FAPs or BAPs in that group are included in the bitstream; in this mask, a 1 indicates that the corresponding FAP/BAP is present in the bitstream and vice versa.

For the high-level FAPs, visemes and expressions, not only a FAP intensity value is sent, as happens for the low-level FAPs. The viseme/expression FAPs allow two visemes/expressions from a predefined set to be mixed together based on the values of two parameters. These parameters are the viseme/expression-selection parameter and the corresponding intensities. To avoid ambiguities, the viseme/expression FAPs can have impact only on low-level FAPs that are allowed to be interpolated at the time the viseme/expression FAP is applied.

There are two modes of coding FAPs and BAPs: as a sequence of FBA object planes, each one corresponding to a certain time instant, or as a sequence of FBA object plane groups, each one comprising a sequence of 16 FBA object planes, also called *segments*. Depending on the chosen mode, FBA object planes or FBA object plane groups, FAPs and BAPs will be coded using a frame-based coding mode or a Discrete Cosine Transform (DCT)-based coding mode. Compared to the frame-based mode, in some conditions the DCT-based mode can give higher coding efficiency at the cost of a higher latency [AbPe99].

Frame-Based Coding Mode   In the frame-based coding mode, FAPs and BAPs are differentially encoded and quantized using an adequate quantization step. Whenever desired, they can be coded without prediction using the *intra coding mode*. The default quantization steps of all animation parameters (see Table 9.3) can be scaled by means of a *quantization scaling factor*, ranging from 1 to 31. Setting the quantization scaling factor equal to 1 minimizes the quantization error and enables lossless coding for those FAPs like expression and shift_jaw with a default quantization step equal to 1. The resulting symbols are then arithmetically encoded. At the decoder, FAP and BAP values are set according to one of three cases: by a value in the bitstream, by a retained value previously set by the bitstream, or by means of decoder interpolation. FAP values set by the bitstream and subsequently masked are allowed to be interpolated only if an explicit indication is sent. For the high-level FAPs, the intensities are encoded as for the other FAPs, but the visemes/expression selection values are differentially encoded without arithmetic encoding (and no quantization).

DCT-Based Coding Mode   In the DCT-based coding mode, 16 FBA object planes are buffered and DCT encoded, using intra or inter coding (see Figure 9.10). The DC coefficient is then differentially coded and quantized using a quantization step, which is one-third of the quantization step for the AC coefficients. It is possible to adjust the quantization steps to be used by means of a *scaling index*, varying from 0 to 31, indexing a table with scaling factors for the default quantization step values. The coefficients are then coded using Huffman tables specified in the standard. High-level FAP intensities and selection values are only differentially encoded without entropy coding.

Figure 9.11 compares the coding performance of the frame-based and DCT-based FAP coders. The Peak Signal to Noise ratio (PSNR) is measured by comparing the amplitude of the original and coded FAP averaging over all FAPs. This PSNR does not relate to picture quality but to the smoothness of

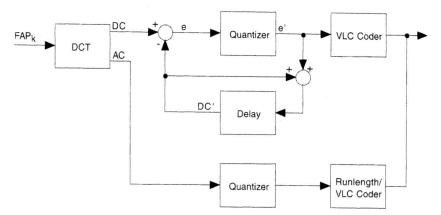

Fig. 9.10 Block diagram of the FAP encoder using DCT. DC coefficients are predictively coded. AC coefficients are directly coded. Reprinted with permission from J. Ostermann, "Animation of Synthetic Faces in MPEG-4," *Computer Animation '98*, Philadelphia, pp. 49–55, June 1998. © 1998 IEEE.

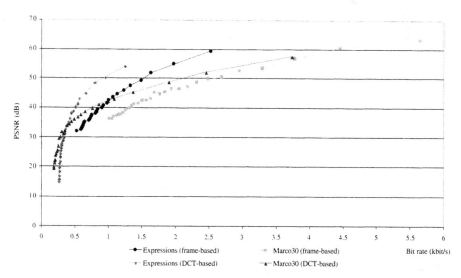

Fig. 9.11 Rate distortion performance of the frame-based and DCT-based coding modes of FAPs for the sequences *Marco30* (30 Hz) and *Expressions* (25 Hz). Reprinted with permission from Abrantes, G., and F. Pereira, "MPEG-4 Facial Animation Technology: Survey, Implementation and Results." *IEEE Transactions on Circuits and Systems for Video Technology*, 9 (2), pp. 290–305, March 1999. © 1999 IEEE.

temporal animation. In contrast to the frame-based coder, the DCT-based coder is not able to code FAPs with near lossless quality. At low data rates, the DCT coder requires up to 50% less data rate than the frame-based coder at the price of an increased coding delay. This advantage in coding efficiency disappears with increasing fidelity of the coded parameters.

Independently of the FBA coding mode used, all MPEG-4 FBA decoders are required to generate at the output an FBA model, including all of the feature points defined in the MPEG-4 Visual specification [MPEG4-2], even if some of the feature points will not be affected by any information received from the encoder.

9.2.6 FBA and Text-to-Speech Interface

MPEG-4 acknowledges the importance of TTS synthesis for multimedia applications providing an interface to a proprietary TTS synthesizer (see Chapter 12). A TTS stream contains text in ASCII and optional prosody in binary form. The decoder decodes the text and prosody information according to the interface defined for the TTS synthesizer. The synthesizer creates speech samples that are handed to the compositor. The compositor presents audio and, if required, video to the user. In the MPEG-4 Visual standard, the encoder is expected to send a FAP stream containing FAP numbers and amplitudes for every frame, to enable the decoder to produce the desired facial actions. Because the TTS synthesizer can behave like an asynchronous source, synchronization of speech parameters with facial expressions of the FAP stream usually is not given, unless the encoder transmits prosody with timing information for the synthesizer.

Another output TTS interface sends the phonemes of the synthesized speech as well as start time and duration information for each phoneme to a phoneme/bookmark-to-FAP converter. The converter translates the phonemes and timing information into face animation parameters that the face renderer uses to animate the face model. In addition to the phonemes, the synthesizer identifies bookmarks in the text that convey nonspeech-related FAPs to the face renderer. The timing information of the bookmarks is derived from their position in the synthesized speech. Because the facial animation now is driven completely from the text input to the TTS synthesizer, transmitting a FAP stream to the decoder is optional. Furthermore, synchronization is achieved as the talking head is driven by the speed of the asynchronous proprietary TTS synthesizer.

To allow for simple bookmarks, each bookmark must describe for one FAP at a time the transition from the current FAP amplitude to a target FAP amplitude. Simply applying a FAP of constant amplitude and resetting it after a certain amount of time does not allow for realistic face motion. Therefore, the bookmark-to-FAP converter creates the appropriate transitions between current and target amplitudes [TO00].

9.3 2D MESH CODING

MPEG-4 is an object-based coding standard. In MPEG-4 Visual [MPEG4-2], the object-based video coding of natural video data is performed by means of a block-based texture and motion coding scheme together with a shape information coding scheme. A different approach to object-based video can be taken through 2D meshes or 2D planar graphs with triangles. Combined with natural video, a 2D mesh object can provide a unique way to support 2D mesh-based video coding.

The functionalities provided by the MPEG-4 2D mesh coding tool are these:

☞ Efficient compression of 2D meshes
☞ Texture mapping on top of 2D meshes
☞ Text overlays
☞ Image and graphics overlays (augmented reality)

2D mesh coding is designed mostly for video manipulation; thus, 2D meshes often are used together with natural images and videos, mapped on the 2D meshes. So a 2D mesh-based application can have the basic architecture shown in Figure 9.12. The video/image to be mapped on the 2D mesh can be MPEG-4 coded or coded according to any other (MPEG-4) accepted still image type, such as JPEG [MPEG4-1]. The choice of the proper image format

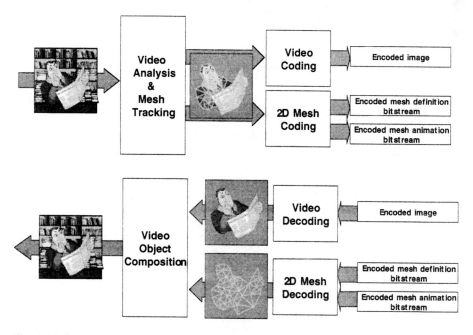

Fig. 9.12 Basic architecture of 2D mesh-based coding (upper) and decoding (lower).

may depend on the application. 2D mesh coding can be useful for the following applications [WL94, N4319, TBTG98, BTZC99]:

☞ Video object compression

☞ Video object tracking

☞ Content-based video indexing and retrieval (e.g., motion-based queries)

☞ 2D animation

☞ Augmented reality

9.3.1 2D Mesh Object

2D mesh coding is a tool to encode a 2D mesh object. A 2D mesh object is a 2D deformable shape (2D mesh), which will be used to create a synthetic visual object (video or still texture) based on the 2D mesh. A 2D mesh object also is called a 2D dynamic mesh, because it can support video coding by moving the vertices of the mesh. The 2D mesh is located on the image plane of the texture it animates. Therefore, the position of the vertices can be defined in pels and in fractions thereof. Polygons other than triangles are not allowed in MPEG-4 2D mesh coding. The topology of a 2D mesh object does not change over time, which means that the connectivity of vertices remains the same. Hence, a 2D mesh object (see Figure 9.13) consists of the following data:

1. **Connectivity:** The edge information on how vertices are connected to form triangles. In a 2D mesh object, the connectivity is implicitly determined, as only two triangulation methods are allowed: (a) uniform or (b) Delaunay. The uniform triangulation (UT) is used to handle rectangular video objects, whereas the Delaunay triangulation (DT) is used for arbitrarily shaped video objects.

Fig. 9.13 2D mesh overlaid on top of a video object (*Mother&Daughter*).

2. **Geometry:** The 2D coordinates of the mesh vertices. Three vertices form a triangle in a 2D mesh object. The set of vertices forms a finite mesh—in other words, a mesh with a boundary.

3. **Motion:** The temporal difference of the vertices' positions between the current and the reference frames. The motion data will determine the deformation of a given 2D triangular mesh.

It is worth noting that the 2D mesh object is supported only in MPEG-4, not in VRML [VRML97]. This is due to the fact that VRML supports only 3D and not 2D data. Although 2D can be realized in 3D, processing 2D with one less coordinate than 3D is a more efficient and compact way to process the data. (More details on the 2D support in MPEG-4 can be found in Chapter 4.)

A 2D mesh object is a subset of the `IndexedFaceSet2D` node in MPEG-4, which is basically a set of polygons (see Chapter 4). `IndexedFaceSet2D` is a 2D subset of its 3D counterpart: the `IndexedFaceSet` node. As shown in Table 9.5, `IndexedFaceSet2D` does not support 3D-related properties such as `normals`, `creaseAngle`, and `solid` as these properties are meaningful only in 3D. For instance, the field `solid` indicates if the given mesh object is a solid (or volumetric) object, in which 2D objects do not have the notion of volume. And an `IndexedFaceSet2D` node needs only two components in its coordinate system: x and y coordinates. More details on the `IndexedFaceSet` node can be found in Section 9.4 on 3D mesh coding.

Although `IndexedFaceSet2D` is designed to represent any type of 2D mesh with polygonal faces, the 2D mesh object supports only triangular meshes (see Table 9.6). In a 2D mesh object, there is no need to transmit the connectivity information, except some header information, to indicate if the given mesh is uniform or Delaunay triangulated. `IndexedFaceSet2D` does not have a dedicated compression scheme; rather, the general BIFS quantization-based compression is used, whereas the 2D mesh object uses a specific differential compression tool, which is more efficient. The 2D mesh object includes the information on the mesh geometry as well as the mesh animation (motion). `IndexedFaceSet2D` is rather the representation for a static 2D model, and it needs `Interpolators` for its animation. In fact, when decoded, the bitstream of a 2D mesh object will be used to generate an `IndexedFaceSet2D` object and its corresponding animation stream.

9.3.2 Coding Scheme

The objective of the 2D mesh coding tool is to encode the mesh geometry as well as the mesh deformation over time (connectivity is implicitly determined). More specifically, the 2D mesh decoder can be depicted as shown in Figure 9.14, explained in the following. As MPEG-4 video coding supports intra and inter coding of the VOPs corresponding to a video object, 2D mesh coding encodes the initial mesh geometry as an intramesh object plane (I-MOP) and the mesh deformation over time as an intermesh object plane (P-MOP).

Table 9.5 Relationship between the `IndexedFaceSet2D` and `IndexedFaceSet` nodes

Field name	IndexedFaceSet2D	IndexedFaceSet
set_colorIndex	Supported	Supported
set_coordIndex	Supported	Supported
set_normalIndex	Not supported	Supported
set_texCoordIndex	Supported	Supported
color	Supported	Supported
coord	Supported	Supported
normal	Not supported	Supported
texCoord	Supported	Supported
ccw*	Not supported	Supported
colorIndex	Supported	Supported
colorPerVertex	Supported	Supported
convex	Supported	Supported
coordIndex	Supported	Supported
creaseAngle	Not supported	Supported
normalIndex	Not supported	Supported
normalPerVertex	Not supported	Supported
solid	Not supported	Supported
texCoordIndex	Supported	Supported

*The ccw field specifies if the points which define a face are ordered counterclockwise (TRUE) or clockwise (FALSE).

Table 9.6 Differences between the `IndexedFaceSet2D` node and a 2D mesh object

	IndexedFaceSet2D	2D Mesh Object
Mesh type	Polygon	Triangles only
Connectivity	Generic	Uniform or Delaunay
Compression	General BIFS compression	Dedicated differential compression
Applications	General 2D	Video object manipulation, search, and indexing
Animation	Through Interpolator node	Dedicated differential coding

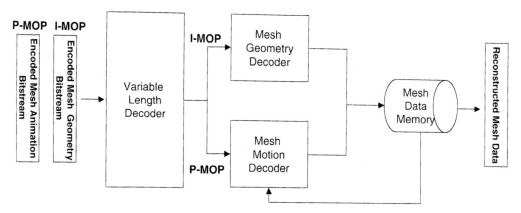

Fig. 9.14 2D mesh decoding architecture.

9.3.2.1 Mesh Connectivity In a 2D mesh object, the connectivity information describing how vertices form triangles is implicitly determined without the need to encode any connectivity information. Two triangulation methods can be used in MPEG-4 2D mesh coding: uniform and Delaunay triangulation [MPEG4-2].

Uniform Triangulation Uniform triangulation is better suited for rectangular video objects, although it can also be used for arbitrarily shaped video objects. In UT, the vertices are located in x and y grids, such that two neighboring triangles forming a rectangle exist. For each axis, a 10-bit unsigned integer is assigned to describe the number of vertices, horizontally and vertically. This allows the creation of meshes with up to 1 million vertices and triangles, which is considered enough for the applications envisioned. An 8-bit unsigned integer is used to specify the length of the horizontal and vertical intervals between neighboring vertices with a half-pixel accuracy, which means the largest triangle can be 128 pixels wide, horizontally and vertically.

With UT, the vertices are by definition spaced regularly, such that all the triangles in the mesh have the same size. Because the vertices are located at evenly spaced horizontal and vertical grids, the triangulation process resumes to a kind of splitting rectangles process. There are four ways to triangulate the rectangles, as shown in Figure 9.15: (1) backslashes (\ \ \); (2) slashes (/ / /); (3) alternated backslash and slash (\ / \ /); and (4) alternated slash and backslash (/ \ / \). A 2-bit code is used to signal the rectangles' splitting method to the decoder.

Delaunay Triangulation Although UT can be applied to any type of video object, an arbitrarily shaped video object can be better coded with a mesh in which there are *boundary vertices*. Boundary vertices are those located at the boundary of the 2D mesh. For example, the 2D mesh shown in Figure 9.15 has 12 boundary vertices and 4 inside vertices. Between two boundary vertices,

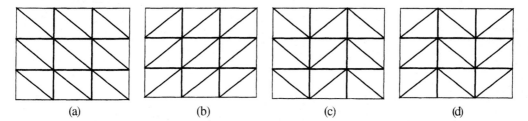

Fig. 9.15 Four types of uniform triangulation.

there exists an edge that is incident to only one triangle (normally an edge between two vertices is incident to two triangles).

For all of the given vertices, there are many ways to form a triangular mesh by triangulation. One popular method is the Delaunay triangulation [Fari96]. Together with the Dirichlet tessellation [Fari96], the Delaunay triangulation guarantees producing the largest minimal angle, which will make triangles close to being equilateral. Close-to-equilateral triangles guarantee better texture mapping performance and better deformation capabilities, without causing the flip of triangles. As shown in Figure 9.16a, a basic form of DT can be thought of as a splitting process of a given quadrilateral, such that the resulted two triangles are close to equilateral triangles. Figure 9.16c shows the result when DT is applied to a star-shaped mesh object.

DT guarantees a unique triangulation for a given set of vertex points, except for the case shown in Figure 9.17. When four vertices form a square, there is no unique DT; this case is called *neutral sets*. In 2D mesh coding, only the triangulation connecting the second and fourth point of a square is permitted as shown in Figure 9.17. There are many ways to perform the Delaunay triangulation [BKOS97]. One of the methods is the following:

1. With the given (inside or boundary) vertices, perform a triangulation, randomly. When forming a triangle out of three vertices, the triangle must reside inside the mesh boundary.

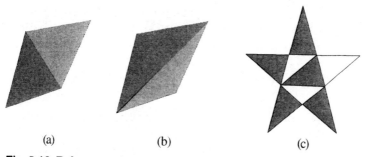

Fig. 9.16 Delaunay triangulation: (a) DT for a quadrilateral; (b) Non-DT for a quadrilateral; and (c) DT for a star-shaped mesh.

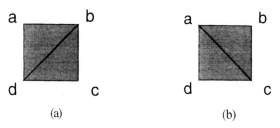

Fig. 9.17 Neutral sets: (a) allowed triangulation and (b) prohibited triangulation.

2. Inspect each and every edge of the triangulation if it is locally Delaunay. An interior edge would have two triangles and associated four vertices. In other words, the edge would be one of the two diagonals of the given quadrilateral (see Figure 9.16). If locally Delaunay, the edge will be shorter in length than the other diagonal in the quadrilateral. If not locally Delaunay, replace the edge with the other diagonal.

3. Repeat step 2 until all the edges are locally Delaunay.

9.3.2.2 Mesh Geometry Once the type of connectivity is decided, the geometry information for the given triangular mesh needs to be coded. For a given session, the geometry information regarding the given mesh needs to be transmitted only once. This information is transmitted as an I-MOP. The deformation of the given mesh over time can be described as temporal differences of the geometry, or *geometry motion*. This information is transmitted as a P-MOP.

It is worthwhile to note that there is no quantization involved in the process of coding mesh geometry and mesh motion. The current MPEG-4 specification allows the mesh geometry precision to have half-pixel accuracy. The bitstream structure for mesh geometry is shown in Figure 9.18.

I-MOP Coding: Uniform Triangulation When uniformly triangulated, the mesh geometry does not need to be coded, as the mesh geometry can be determined from the number of horizontal and vertical vertices and from the

Mesh type code ('01')	Number of horizontal mesh vertices	Number of vertical mesh vertices	Horizontal mesh rectangle size	Vertical mesh rectangle size	Triangle split code

(a)

Mesh type code ('10')	Total number of mesh vertices (N)	Number of boundary mesh vertices (N_b)	Vertex 0 (x)	Vertex 0 (y)	Delta (x) *	Delta (y) *

(b)

Fig. 9.18 Mesh geometry bitstream structure: (a) Uniform triangulation and (b) Delaunay triangulation.

horizontal and vertical interval sizes between vertices (see Figure 9.15). The 2D coordinate system for 2D mesh coding is similar to the one used for video, in which the y axis starts from the top and moves to the bottom and the x axis starts from the left and moves to the right. UT is useful mainly for rectangular shaped video objects, since it does not require sending any initial mesh geometry, except some header information with the parameters previously mentioned.

I-MOP Coding: Delaunay Triangulation   DT is useful in dealing with arbitrarily shaped video objects. In order to perform DT, the boundary vertices of the given mesh must be coded first. If boundary vertices are not distinguished in the given mesh, a unique DT cannot be guaranteed for arbitrarily shaped objects. Distinguishing boundary vertices from inside vertices, there will be $2N_i + N_b - 2$ triangles, where N_i denotes the number of interior vertices and N_b the number of boundary vertices in a given mesh. For a star-shaped mesh, the coding order of the vertices can be given as shown in Figure 9.19a. The coding of boundary vertices starts from the top-left most boundary vertex. The coding order of boundary vertices is counterclockwise. For the inside vertices, the coding order is determined by the distance between the last coded vertex and the current vertex to code. Hence, the first inside vertex would be chosen such that the distance between the last coded boundary vertex and the current inside vertex is the minimum. In Figure 9.19a, there are 10 boundary vertices ($N_b = 10$) and 2 interior vertices ($N_i = 2$), which yield 12 triangles.

Unlike UT, the geometry information needs to be transmitted if DT is used. A differential coding scheme has been adopted to efficiently compress the size of the mesh geometry. The previous x and y coordinates will be the

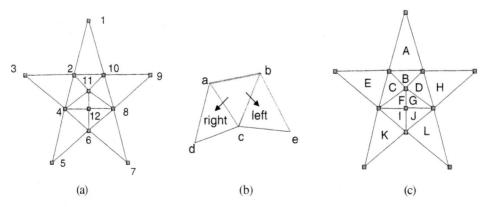

Fig. 9.19 (a) Coding order of the nodes after Delaunay triangulation, (b) example of mesh traversal with right and left marching, and (c) mesh traversal for the star-shaped mesh.

basis for the prediction of the coordinates of the next vertex: $x_n = x_{n-1} + dx_n$, $y_n = y_{n-1} + dy_n$, and $1 \leq n \leq N$, where dx and dy denote the delta values between the current and the previous x and y values. For the first x and y coordinates, there is no prediction, where the absolute coordinate values will be encoded as difference values.

9.3.2.3 Mesh Motion

Having a mesh geometry with uniform or Delaunay triangulation, it is possible to build a static mesh model eventually with some image or video mapped on it. The mesh motion data allows deforming the given mesh by changing the coordinates of the mesh vertices over time. Coding of mesh motion is done by coding the differential values between the current motion vectors and prediction motion vectors determined from the neighboring motion vectors. In order to increase the coding efficiency, the motion vectors are traversed in the mesh traversal order explained in the next section.

Mesh Traversal   The first vertex is the top left vertex of the mesh. The first edge connects the first vertex with the next one clockwise on the boundary of the mesh (edge [1,10] in Figure 9.19a). The top-left triangle becomes the *initial triangle* for mesh traversal (triangle <1,2,10> in Figure 9.19a). This triangle gets the lowest label ('A' in Figure 9.19c). The *current triangle* is the one that is currently being worked on. The *base edge* of a triangle is the edge that connects the current triangle to the already labeled neighboring triangle with the lowest label. In case of the initial triangle, the base edge is the first edge. The *right edge* of the current triangle is the next counterclockwise edge of the current triangle with respect to the base edge. The remaining edge of the current triangle is named *left edge* (see Figure 9.19b).

Starting with the initial triangle, the following breadth-first traversal of the mesh triangles is iterated:

1. Find the triangle with the lowest label and name it current triangle.
2. Identify base, left, and right edges for the current triangle.
3. If there is an unlabeled triangle adjacent to the current triangle sharing the right edge, label it with the next available lowest label.
4. If there is still an unlabeled edge, go to step 1.

As a result, the star-shaped mesh in Figure 9.19a has a traversal order, as shown in Figure 9.19c. Once this traversal order is determined, that order remains intact until the next I-MOP is decoded.

Mesh Motion Coding   Unlike motion for video objects, the motion vector associated to a certain vertex determines the movement of the triangles to which it belongs. For instance, the motion vector for vertex 11 in Figure 9.19a can affect the motion of its four neighboring triangles. Coding of vertex motion vectors, however, is very similar to video motion coding, as the coding operation is done on a 2D plane. It also is very likely that the motions associated to neighboring vertices are highly correlated.

Every vertex motion vector can be predicted based on two previously decoded motion vectors associated to two neighboring vertices, with the exception of the first two vertices. For the triangle $\triangle abc$ given in Figure 9.19b, the motion vector $\vec{V}_c$ of vertex c can be predictively coded using the previously coded motion vectors $\vec{V}_a = (Vx_a, Vy_a)^T$ and $\vec{V}_b = (Vx_b, Vy_b)^T$.

$$\vec{W}_c = 0.5 \times (floor(Vx_a + Vx_b + 0.5), floor(Vy_a + Vy_b + 0.5))$$

$$\vec{V}_c = \vec{W}_c + \vec{E}_c$$

where $\vec{W}_c$ and $\vec{E}_c$ denote the motion vector prediction and the delta motion vector that will be coded, respectively. The function $floor(x)$ returns the nearest integer value that is closer to zero (i.e., $floor[3.5] = 3$).

Only the nonzero motion vectors are encoded as in the case of video coding. To indicate whether the current motion vector is zero, there is a 1-bit code introduced in the syntax before each motion vector. This is particularly relevant when uniform triangulation is used and the movement of the mesh object is concentrated in a specific part of the object.

9.3.3 Example

Experiments on 2D mesh coding were conducted during the development of MPEG-4. Because of its close relation with video coding, the comparison was mostly made with video coding [TBTG98], notably between 2D mesh-based video coding and MPEG-4 block-based video coding. The major difference between the two coding tools is how motion is compensated and coded. A substantial bit saving in motion bits for mesh-based coding over block-based motion compensation was reported [TBTG98]. However, these savings are consumed for texture coding. As a result, video coding with mesh-based motion compensation performs comparably to regular block-based MPEG-4 video coding [WO98]. However, it must be noted that 2D mesh coding adds a new, important functionality: video manipulation through 2D mesh.

Figure 9.20 shows an example of 2D mesh coding with mapped video. From an original video sequence, a 2D mesh can be generated and tracked. Once the mesh is determined, a synthetic or natural texture can be overlaid on it as shown in Figure 9.20c.

Experimental results indicate that the coding of the vertex positions for an I-VOP takes about 12.3 bits/vertex. Motion vector coding requires between 2.3 bits/vertex and 3.1 bits/vertex for a head-and-shoulder scene with little motion. The fish shown in Figure 9.20 requires 6.5 bits/vertex to code the motion vectors [TBTG98].

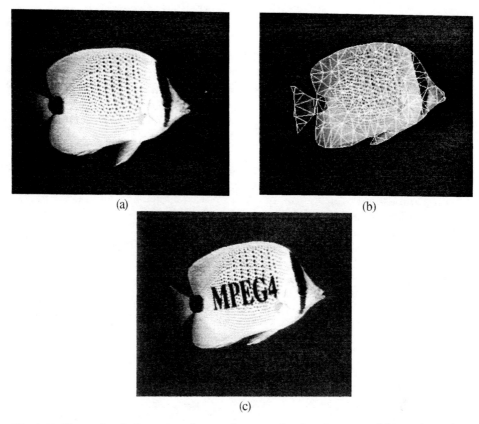

(a)

(b)

(c)

Fig. 9.20 Example of 2D mesh coding with mapped video: (a) original frame from the video sequence *Bream*, (b) mesh generated from the image in (a), (c) text overlaid on the video (and mesh).

9.4 3D MESH CODING

Three-dimensional mesh coding (3DMC) was introduced in MPEG-4 Visual, Version 2 [M4V2]. Because of its name, 3DMC might be confused with an extension to 2D mesh coding, which is not the case as 3DMC is a new SNHC technology to efficiently represent synthetic 3D models. Whereas 2D mesh coding was introduced to support video manipulation through 2D meshes, 3DMC provides a representation and compression tool for IndexedFaceSet nodes of 3D objects onto which images and video may be mapped. It is worth noting that 3DMC is to compress *static* 3D models, not their animation, which is yet another clear difference between 2D mesh coding and 3DMC. 3DMC provides additional functionalities—such as high compression, incremental rendering, and error resilience—that are useful to many applications.

3D (polygonal) meshes are not the only way to represent 3D models. There are many other well-known methods, such as NURBS and subdivision surfaces [Bloo97]. At the time of writing, these tools are under development in the context of MPEG-4 AFX. They will provide additional functionalities to support curved surfaces and subdivision surfaces that are popular in the games and animation industries. However, 3D mesh representation is one of the cheapest solutions to represent 3D models in software and hardware. Moreover, 3D meshes are a common way to exchange 3D models produced with different methods.

3DMC can be used as a stand-alone tool or together with other MPEG-4 tools, for example, for 3D home shopping in which the user can browse 3D products designed using 3D meshes. Even without compression, it is not difficult to find a Web home shopping or 3D gallery where the user can have 3D models. The major use of 3DMC, however, is as a component technology in graphics and multimedia applications. As mentioned before, 3D meshes are a popular representation method for some industries.

9.4.1 3D Mesh Object

The MPEG-4 3D mesh object is fully compatible with the `IndexedFaceSet` VRML/BIFS node (see Table 9.7). Major components in `IndexedFaceSet` are these:

☞ **Connectivity:** A 3D mesh consists of connected polygons. As for 2D meshes, the way to form a polygon from the given vertices is called *connectivity information*. Using a wireframe,[1] it is easy to see how polygons form a 3D mesh (see Figure 9.21a).

☞ **Geometry:** As for 2D meshes, the 3D coordinates of the nodes or vertices are called *geometry*. The coordinates are represented in the Cartesian coordinate system. A basic 3D model as shown in Figure 9.21a can be generated with only connectivity and geometry information. Hence, these two are the most important components in 3D mesh coding.

☞ **Photometry:** In designing the appearance of the 3D mesh object, one can add colors, normals, and texture (by allocating texture coordinates on the mesh) on top of the model. These properties are called *photometry information*, as they affect the rendered visual quality. Figure 9.21b shows a model with color applied and Figure 9.21c shows a model with texture and color applied. Figure 9.21d shows one of the texture maps used for Figure 9.21c. As can be seen from Figure 9.21, image and video data play an important supporting role in 3D model representation.

Table 9.7 shows the relationship between the fields of `IndexedFaceSet` and the three major components mentioned earlier. In the table, the fields

---

1. When a 3D object is visualized, a wireframe representation shows only vertices and edges and not surfaces. This is a way to render a 3D object quickly while showing the connectivity of the object.

Fig. 9.21 3D mesh model: (a) wireframe, (b) rendered model only with synthetic colors, (c) rendered model with texture map, and (d) texture map used for the model in (c).

Table 9.7 Relationship between `IndexedFaceSet` fields and 3D mesh components

`IndexedFaceSet` field name	3D mesh component
`set_colorIndex`	Photometry
`set_coordIndex`	Connectivity, Geometry
`set_normalIndex`	Photometry
`set_texCoordIndex`	Photometry
`color`	Photometry
`coord`	Connectivity, Geometry
`normal`	Photometry
`texCoord`	Photometry
`ccw`	Other property
`colorIndex`	Photometry

Table 9.7 Relationship between `IndexedFaceSet` fields and 3D mesh components (Continued)

`IndexedFaceSet` field name	3D mesh component
colorPerVertex	Photometry
convex	Other property
coordIndex	Connectivity, Geometry
creaseAngle	Other property
normalIndex	Photometry
normalPerVertex	Photometry
solid	Other property
texCoordIndex	Photometry

classified as *other property* deal with the general appearance of the 3D mesh object. The field `ccw`, meaning *counterclockwise*, is used to determine the visible face of the polygon.[2] The visible face of a polygon is determined by the counterclockwise order of the vertices. For example, the visible face of △ABC is the same as that of △BCA, whereas that of △ACB is the other side of the visible face of △ABC. If a face of a 3D mesh is made of a triangle, it is coplanar, which means that all points on the face reside in the same plane. If a 3D mesh can have polygons other than triangles (e.g., quadrilaterals, pentagons, hexagons), it is not always possible to assume that all points on the face of a polygon reside in the same plane. The field `convex` is a field that indicates if all faces are convex. A crease is a wrinkle or folded line on the edge. If two neighboring faces are close to coplanar, the crease should be invisible; otherwise, the crease will be shown on the edge. The `creaseAngle` is the threshold angle that makes the crease visible or not. If the `solid` field is TRUE, both sides of a face are visible; otherwise, the visible side will be determined by the field `ccw`. More details on these fields can be found in [MaCa97, Web3D].

It is important to highlight that 3DMC provides compression on top of `IndexedFaceSet`. However, compression is not the only advantage to using 3DMC. The following functionalities are supported by 3DMC:

☞ **Compression:** Near-lossless to lossy compression of 3D models is supported. Usually a VRML ASCII file can be compressed down to 2% to 4% of its original size without visual degradation.

☞ **Incremental rendering:** With 3DMC, there is no need to wait until the complete bitstream is received to start rendering it. With the incremental rendering capability, the decoder can begin building the model with

---

2. A polygon has two faces; if a face is chosen to be visible, the other face is automatically determined to be invisible.

just a fraction of the entire bitstream. This functionality is important when the latency is a critical issue, such as for home shopping.

☞ **Error resilience:** With a built-in error-resilience capability, 3DMC can suffer less from network errors, as the decoder can build a model from the partitions that are not corrupted by the errors.

☞ **Support of nonmanifold models:** Because of the compression characteristic using 3DMC topological surgery (see Section 9.4.2.1), only orientable and manifold models are supported. For nonorientable or nonmanifold models, a dedicated operation called *stitching* [GBTS99] is performed to support these models.[3]

☞ **Hierarchical buildup:** 3D mesh models can be quite complex, with millions of polygons. Depending on the viewing distance, the user may not need million-triangle accuracy, but may be satisfied with hundreds of triangles. A scalable bitstream allows building 3D models with different resolutions to serve such a case.

9.4.2 Coding Scheme

3DMC comprises three major coding blocks: topology analysis (data transformation); differential quantization of connectivity, geometry, and photometry information (quantization); and entropy coding. In Figure 9.22, the 3DMC decoder architecture is presented. As shown in this figure, connectivity data is

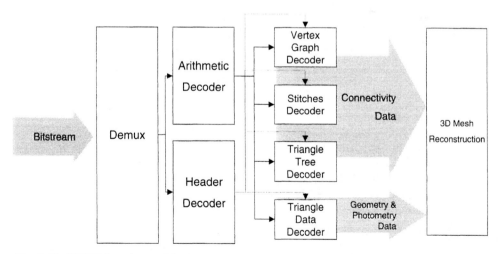

Fig. 9.22 3DMC decoder architecture.

---

3. A manifold model contains surfaces that can be locally approximated by a disk or half-disk. An orientable model does not contain any Mobius bands. Accordingly, a nonmanifold model cannot be approximated by a disk or half-disk and a nonorientable model contains at least one Mobius band [NB94].

decoded by three different modules and topologically assembled in the final reconstruction process. This is due to the topological analysis in the encoding process. Geometry and photometry data are decoded by the same module because of their similarity; they are properties per vertex or per face.

9.4.2.1 Topological Analysis

A 3D mesh includes vertices, edges, and triangles. An edge is made up of two vertices and a triangle is made up of three edges. In 3DMC, a 3D mesh is also called *triangular mesh*, as it is composed only of triangles. Other polygonal meshes also can be compressed using the MPEG-4 3DMC tool after decomposing the polygons into triangles. When decoded, the triangles will be restored to polygons. Connectivity information needs to be losslessly transmitted, as lossy compression implies a different topological model after compression. In 3DMC, connectivity information is encoded losslessly to preserve the topology of the given model.

In VRML, IndexedFaceSet describes the connectivity of the given 3D model through the set of vertices that form a polygon. For example, a polygon formed by the 1st, 3rd, 11th, and 37th vertices is represented as a sequence of "1, 3, 11, 37, –1." The last value, –1, indicates that vertex 37 is connected to vertex 1, building a polygon. This representation is easy and straightforward, but is not efficient enough. A vertex may be shared by multiple polygons, wherein it is obvious that the vertex must be written many times. For instance, vertex 1 from the previous example can be used to form polygons: "1, 3, 7, –1," "10, 1, 3, 8, –1," and so on.

Instead of treating a 3D mesh as a set of indexed faces (or polygons), a better way to represent a 3D mesh is to consider the 3D mesh as *connected polygons* or *connected components*. According to the current VRML (and BIFS) representation with IndexedFaceSet, consecutive polygons are not necessarily connected to one another. Once the 3D connectivity of polygons is identified, one can make a new vertex list (not in the same order as in IndexedFaceSet), starting from the first polygon. And there will be a 3D connectivity map for each connected component.

In 3DMC, topological analysis (also called *topological surgery*) is applied to 3D meshes[4] to collapse a 3D mesh connectivity and vertex list into a *2D mesh map, triangle tree,* and its *vertex graph*. This efficiently describes the connectivity information of 3D mesh structures (less than 2 bits/vertex). This is also called *dual graph representation*, as both triangle tree and vertex graph are complementary to each other to form a 3D mesh object. As shown in Figure 9.23a and b, the topological surgery operation consists in cutting through some edges to decompose the 3D mesh into a triangle tree (Figure 9.23c and d) and vertex graph (Figure 9.23e and f). The boundary edges of the triangle tree

---

4. At this time, a 3D mesh means a connected component, because topological analysis is done per connected component. If a 3D mesh is composed of multiple connected components (e.g., three dolphins in a scene), the topological analysis will be conducted three times, accordingly.

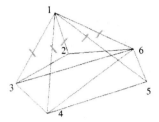

(a) topological surgery operation

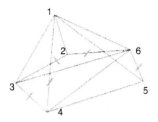

(b) topological surgery operation

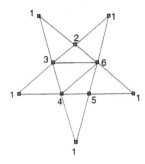

(c) triangle tree of (a)

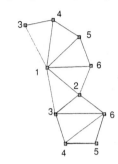

(d) triangle tree of (b)

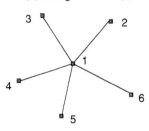

(e) vertex graph of (a)

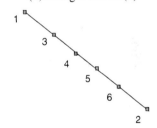

(f) vertex graph of (b)

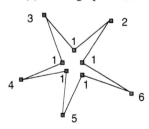

(g) bounding loop of (e)

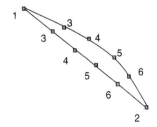

(h) bounding loop of (f)

Fig. 9.23 Examples of topological surgery for a 3D model. Notice that this 3D model is topologically the same as a sphere, as its Euler characteristic is two (genus 0) [FDFH96].

correspond to the edges of the vertex graph. In other words, the vertex graph is needed when reconstructing the 3D mesh connectivity from the triangle tree. In order to use the vertex graph to connect the triangle tree making it a 3D mesh, the vertex graph needs to be changed into a *bounding loop* as shown in Figure 9.23g and h [TaRo98].

Figure 9.23 depicts two different topological surgery operations on the same 3D model. There is no unique way to decompose a 3D model into triangle tree and vertex graph. Empirically, it is possible to state that the topological surgery in Figure 9.23b is better than the one in Figure 9.23a, since it has no branches[5] both in the triangle tree and vertex graph. The number of branches is an important factor that will determine the compression efficiency; the lesser the number, the better the compression efficiency.

3DMC supports manifold models. If the model is manifold, the topological surgery is straightforward as depicted in Figure 9.23. If the model is non-manifold or nonorientable, there are two ways to handle it. One is to simply treat the model as two manifold models and the other way is to apply *stitching* (see Section 9.4.2.4).

Even if the model is manifold, there can be some oddities (such as boundaries or holes within the model). Consider the case in which a triangle from the model shown in Figure 9.23 is deleted. If the triangle made by edges 4, 5, and 6 is deleted, the resulting triangle tree will look like the one in Figure 9.24a. In the triangle tree, a virtual triangle has been added to treat the model as if it was a regular manifold model without boundary edges. A *jump edge* is added in the vertex graph representation (Figure 9.24b) to form a bounding loop for the virtual triangle as shown in Figure 9.24c. In this way,

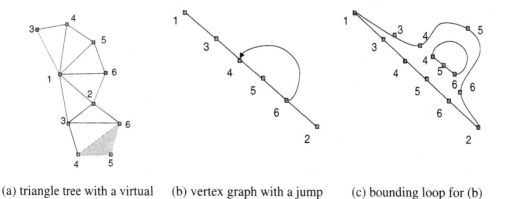

(a) triangle tree with a virtual   (b) vertex graph with a jump   (c) bounding loop for (b)
    triangle (Δ456)   edge

Fig. 9.24 Topological surgery decomposition for models with boundary edges or holes.

---

5. A branch is formed when two or more runs are required to represent the current graph; for example, a scissor has a branch and four runs.

manifold models with boundaries or holes also can be represented by topological surgery.

So far, only triangular meshes have been used as examples. This is because 3DMC inherently supports only triangular meshes. In order to support polygons other than triangles, the polygons must be divided into multiple triangles. In this case, additional information is encoded to distinguish whether the current triangle is part of a polygon.

Details on how topological surgery can be performed at the encoder side can be found in [Boss99, TaRo98, TaRo99, THLR98]. There are many different ways to conduct topological surgery, all of which can affect the compression efficiency. What is normative is the way to reconstruct the 3D model from the dual graph (triangle tree and vertex graph) created by topological surgery.

9.4.2.2 Connectivity Coding In Figure 9.22, connectivity data is decoded by the vertex graph decoder and triangle tree decoder.[6] Usually, triangle data—which contains the geometry coordinates, colors, normals, and also the texture coordinates—is the largest part of the bitstream. From Figure 9.22, it is clear that connectivity data is decoded separately from triangle data. In the 3DMC bitstream, connectivity data is packed separately from the triangle data and put into the bitstream before triangle data. This bitstream structure can bring some additional advantages:

☞ **Incremental rendering:** Once the decoder decodes the vertex graph and the triangle tree data, it has the full topology for the decoded 3D model. As such, it can render the 3D mesh object, as it begins to decode the geometry data. The latency time to view the 3D object is shortened in this way.

☞ **Error resilience:** In wired and wireless applications, network errors can have a severe impact, sometimes asking for the full retransmission of the bitstream. If the vertex graph and the triangle tree data are intact from the errors, it sometimes is possible to tolerate errors in the triangle data. In such a case, the decoder at least can form the 3D mesh structure with some missing geometry or photometry data. So, the user will be able to view, at least, a partly incorrect model; otherwise, no model is viewable. This is an important feature to localize errors and minimize their impact.

Vertex Graph Coding For a simple manifold model as shown in Figure 9.23, the vertex graph can be described with three fields: `last`, `run_length`, `leaf`. The `last` field is set to 1 if the current run[7] is the last branch of the

---

6. The stitches decoder is needed to encode nonmanifold models; details are explained in Section 9.4.2.4.

7. A run consists of a set of consecutive edges without branches inside. In other words, a run is a path between a leaf vertex and a branching vertex, between two branching vertices, or between two leaf vertices.

given branching node. A branching node (or vertex) is a node with three or more incident vertex graph edges, in which a normal node has only two incident vertex graph edges. For example, there are four branches with a branching vertex in Figure 9.23e[8] and no branches with no branching vertex in Figure 9.23f. For the first run of the vertex graph, and for the first run after the branching vertex, this field (`last`) is not coded. The `run_length` field serves to code the number of vertices in the run. The `leaf` field is set to 1 when the last node of the run is not the branching vertex, but a leaf vertex.[9] An example of vertex graph coding is given in Figure 9.25. From this figure, it can be concluded that no branch solution is better than one single branch in terms of compact representation since the nonbranch solution requires only two values (`run_length` and `leaf`; see Figure 9.25b).

For a manifold model with boundaries or holes, a different vertex graph[10] will be generated as shown in Figure 9.24. In this case, additional fields are added to keep track of the jump edges in the vertex graph: `last`, `forward_run`, `loop_index`, `run_length`, `leaf`, `loop`.[11] Other than the jump edge, the vertex graph would not have any branch node in Figure 9.24b. Because of the jump edge, there are now two branching vertex nodes: one with *opening* loop (vertex

last	run_length	leaf
(1)	2	0
(0)	1	1
0	1	1
0	1	1
1	1	1

last	run_length	leaf
(1)	6	1

(a) vertex graph coding for Figure 9.23e   (b) vertex graph coding for Figure 9.23f

Fig. 9.25 Vertex graph coding for the simple 3D mesh in Figure 9.23; values in () are not coded.

---

8. In Figure 9.23e, there are actually five runs from the branching vertex. A branch is counted after it encounters the branching vertex. There must be a run to reach the branching vertex, and this run is not counted as a branch.

9. A leaf vertex is a vertex at the end point of a run. A run can have one of two types of vertices at the end: leaf vertex or branching vertex.

10. A manifold model with boundaries or holes cannot be represented by a simple vertex tree. This is how the term *vertex graph* is coined, as there will be loops by jump edges to cover boundaries or holes with virtual triangles.

11. The term *loop* is introduced to represent the virtual triangle. In reconstructing the 3D model, the loops will generate the empty bounding loop that indicates the virtual (or empty) triangle.

6) and the other with *closing* loop (vertex 4). Coding these fields is not as easy as for simple manifold models, but it can follow the same principle with the assumption that loops are considered as branches. With this assumption, a run can be categorized as follows:

1. Run ended with a branching node
2. Run ended with a leaf node without a loop
3. Run ended with a leaf node opening a loop
4. Run ended with a leaf node closing a loop

The first two cases are the same as simple manifold models. The latter two cases are added to support the runs with loops. Depending on the run case, the encoded data fields are different as shown in Table 9.8. According to this table, the vertex graph in Figure 9.24 can be encoded as shown in Table 9.9. This table shows that there are five runs in the vertex graph including

Table 9.8 Encoded fields and corresponding values for vertex graph coding with loops (X: coded, 0: unset, 1: set)

Case	last[*]	forward_run[†]	loop_index[‡]	run_length	leaf	loop
1	X	1		X	0	
2	X	1		X	1	0
3	X	1		X	1	1
4	X	0	X			

[*]The field last is not coded if the run is the first run of a branch; a branch has two or more child runs.
[†]The field forward_run is not coded when there is no open loop.
[‡]The field loop_index is coded when there are two or more open loops.

Table 9.9 Example of vertex graph coding for the model in Figure 9.24

Run	Case	last	forward _run	run_length	leaf	loop	Status
[1,3,4]	2	1		"3"	0		depth = 0, open loop = 0
[4,5,6]	2			2	0		depth = 1
[6,2]	1			1	1	0	
[6]	3	1		0	1	1	depth = 0, open loop = 1
[4]	4	1	0				depth = −1, open loop = 0

loops. The ending condition of the vertex graph traversal is determined by the depth of the vertex graph. Whenever the encoder encounters a new branching vertex, the depth increases. As the encoder encodes the last run in the given branch, the depth decreases. If the value of depth reaches −1, it means that all of the vertices have been traversed. Once a loop is open (if loop is set), the number of open loops will increase. Once the number of open loops is 1 or more, the value of forward_run will be used to determine if the current run is a run closing a loop (case 4).

Triangle Tree Coding   Compared with vertex graph coding, triangle tree coding is rather simple as a triangle tree can be coded as a binary tree. Hence, only two fields are needed to code a triangle tree: run_length, leaf. If the edge (1,3) of the triangle (1,3,4) is taken as the starting edge, in Figure 9.23c and d, each and every triangle can be traversed, such that each triangle has up to two children: *left marching triangle* and/or *right marching triangle*. In the case of Figure 9.23c and d, the triangle (1,3,4) has only the left marching triangle. Triangle traversal is a similar problem to 2D meshes. However in 3DMC, the traversal order is determined differently. The breadth-first search in 2D mesh coding is not helpful to reconstruct a complex 3D object in an orderly and segmented fashion. The traversal of triangles in 3DMC is *left-most* search. The subtrees of the left marching triangle at the current branch will be traversed before the right marching triangle. Hence the triangle trees of Figure 9.23 can be traversed as shown by the alphabetical order in Figure 9.26; Figure 9.27 presents the run_length and leaf information for the two triangle trees. If run_length is greater than 1 as shown in Figure 9.27b, the way the 8 triangles are marched cannot be identified only with run_length and leaf information. So, additional information called *marching pattern* needs to be transmitted; this information is coded as geometry data. For instance, the marching pat-

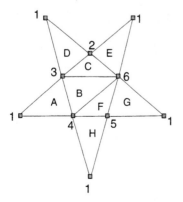

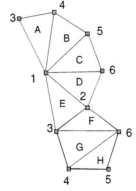

(a) traversal order for Figure 9.23c   (b) traversal order for Figure 9.23d

Fig. 9.26 Traversal order for the triangle trees in Figure 9.23.

run_length	leaf
2	0
1	0
1	1
1	1
1	0
1	1
1	1

run_length	leaf
8	1

(a) triangle tree coding for
Figure 9.23c

(b) triangle tree coding for
Figure 9.23d

Fig. 9.27 Triangle tree coding for the mesh in Figure 9.23.

tern for Figure 9.27b is [11111010], in which the symbol 0 stands for left marching and 1 for right marching.

9.4.2.3 Geometry and Photometry Coding After connectivity is coded with the vertex graph and the triangle tree, a data segment called *triangle data* follows. In the triangle data, geometry and photometry data (i.e., vertices coordinates, normals, colors, and texture coordinates) are encoded. Because the geometry and photometry data are provided per triangle, the decoder can render each triangle right after decoding its geometry and photometry.

Moreover, for each triangle a marching pattern and polygon edge information are also provided. The marching pattern indicates whether the current triangle is marched left or right from the previous triangle, whereas the polygon edge information determines if the current triangle is part of a polygon or a virtual edge that divides that polygon into triangles.

Differential Coding Both geometry and photometry are coded in a similar differential manner. Moreover, quantization is applied before differential coding; this process ensures that the collocated vertices belonging to different connected components (or models) are collocated even after compression. Properties such as geometry and photometry typically consist of two or three fields or coordinates (i.e., xyz, RGB). The maximum difference[12] for all of the fields or coordinates is used to normalize every field value. This normalized value is then quantized by the given Q_p bits.

---

12. If the minimum and maximum values of x, y, and z coordinates are 0, 0, 0, and 10, 100, 1,000, respectively, the maximum difference is 1,000.

To minimize the prediction error after quantization, three prediction modes exist for geometry and photometry coding:

1. **No prediction:** This means that the prediction error is equal to the value to code. This mode is used when there is no previous value available. The first traversed triangle of a given mesh is called the *root triangle*, and is typically coded with this mode.

2. **Parallelogram prediction (PP):** This is based on the previous three vertices to the current vertex position. The previous triangle information is used for the next triangle. Better prediction can be achieved with tree prediction, as it uses the surrounding vertices to predict the current vertex position. PP, however, is the preferred prediction method in 3DMC; particularly, only PP is allowed for geometry coding because it is less complex and not less efficient than tree prediction.

3. **Tree prediction:** For this mode, all of the topological information must be calculated to identify the surrounding vertices. The surrounding vertices will be used to compute the prediction value.

Adaptive Arithmetic Entropy Coding   The differential values computed for the geometry and photometry data are entropy coded using adaptive binary arithmetic coding [MPEG4-2]. Arithmetic coding is the wrapper for the entire 3DMC bitstream. The arithmetic coding scheme used is called *QF coder* [Boss99], and is similar to the one used for JBIG [Penn88] and JPEG [Penn93]. This coding is quite sensitive to errors, meaning that an error can result in the need to retransmit the entire bitstream. In order to be able to stream the 3DMC bitstream, arithmetic coded data is all byte-aligned along with start codes. This is useful in partitioning the bitstream for error resilience (see Section 9.4.2.5).

9.4.2.4 Stitching Mode   3DMC is inherently developed for manifold models. Thus nonmanifold or nonorientable models have to be divided into multiple manifold models for compression. This requires cutting a single nonmanifold model into two or more manifold models with some vertices being used multiple times instead of once. This would lead to the generation of more bits to encode the repetitive vertices and triangles. At the same time, the encoded representation would destroy the original connectivity, as it would generate topologically unconnected components.

The stitching mode in 3DMC allows synthesizing several manifold models back to a nonmanifold or nonorientable model. This is possible by sending a vertex clustering array (VCA) additionally, in which VCA is a mapping table that maps different vertices at the same coordinate into a single vertex.

In order to efficiently encode VCA data, 3DMC uses a stack-based coding method. In VCA, not all vertices are used for stitching. So, there is a 1-bit flag to determine if the current vertex is part of stitches. If the vertex is determined to be a stitching vertex, the encoder can use one of the following three stitching commands: PUSH, GET, or POP. Provided that a set of manifold meshes with more than 89 vertices exists, one of the stitching vertex arrays can be denoted as [2, 23, 57, 89], meaning that the four vertices need to be merged into a single anchor vertex [2] after stitching. One way to represent this array is to assign the label of the anchor vertex (Vertex 2, in this example) to every vertex as in Vertex[2] = Vertex[23] = Vertex[57] = Vertex[89]= 2.[13] Using the above stitching commands, the stitching array can be rewritten as [PUSH, GET [2], GET [2], POP [2]]. When encoding the first vertex of the stitching array, the PUSH operation will put the vertex index in the stack. When encoding the next vertex (23) of the same stitching array, it will use the GET command to get the vertex index (2) from the stack. The same happens for the following vertex [57]. If the next vertex (89) of the same stitching array is the last one, the POP operation will delete the vertex index (2) from the stack.

It also is likely that several manifold models resulting from a nonmanifold model can share not only vertices but also edges or polygons. In other words, if two manifold models share a triangle, a set of stitching vertex arrays may be sufficient instead of three.[14] This is called *variable length method* and is also applied normatively on top of the stack-based method to give further compression efficiency. With this variable length method, the incremental length will be included whenever each stitching command is encoded to indicate the length of the vertices that can be stitched automatically with the current vertex. More details on the stitching encoding and decoding procedures can be found in [GTLH98, GBTS99]. Methods for decomposing a nonmanifold model into multiple manifold models can vary depending on the implementation of the encoder. What is standardized in the stitching mode is the way stitching is done using stitching commands with variable lengths of stitches.

9.4.2.5 Error-Resilience Mode In order to minimize the impact of packet losses and burst errors on the 3DMC decoded data in wired or wireless environments, a data resynchronization tool is provided in 3DMC. Each *3DMC partition* or *packet* can be rendered with its vertex graph and triangle tree information independently. There are three modes for partitioning the 3D mesh bitstream as shown in Figure 9.28.

Partitioning in 3DMC is logical rather than physical, which means that the size of a partition might not be equal to the size of a packet.[15] However, a

---

13. Vertex 2 can be the first one to be coded, as numbering is likely to be performed in the coding order.

14. If a triangle (2,6,11) of one manifold model is the corresponding triangle (35,45,76) of the other manifold model, there will be three set of vertex arrays: {2,35}, {6,45}, and {11,76}. From the knowledge that the triangles are connected, such that 2→6→11 and 35→45→76, only one set {2,35} is enough to represent the stitching (instead of three sets).

15. A logical segment is called a partition, in which a packet denotes a physical segment with a fixed size. So, the size of a partition can be variable and packed in one or more packets.

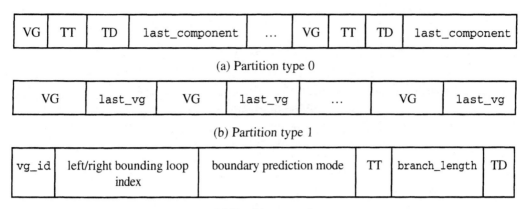

(a) Partition type 0

(b) Partition type 1

(c) Partition type 2

Fig. 9.28 Partition types for 3DMC error resilience.

partition can be designed to be a multiple of the packet size. Packing a whole partition into small-sized packets is not impossible, but would produce a lot of overhead because of the side information, the restart of arithmetic coding, and the like.

Without the error-resilience mode, the conventional bitstream is in the form of partition type 0 (a sequence of vertex graph [VG], triangle tree [TT], and triangle data [TD]). In the error-resilience mode, it is possible to separate VG data from TT and TD data into multiple partitions; usually, VG consumes 5 ~ 10% of the bit-rate budget of the entire bitstream. Without VG, however, it is impossible to reconstruct the 3D mesh. This explains why partition type 1 contains only VGs in the partition. This type of partitioning can be useful in prioritizing the partitions by their relative importance. The remaining TT and TD data will be contained in partition type 2, which means that a 3D mesh can have one type 1 partition and multiple type 2 partitions. In partition type 0, there can be one or more connected components.[16] The field `last_component` is a 1-bit flag to determine if the current connected component is the last one in the partition. A set of VG, TT, and TD data will make a connected component. In partition type 1, multiple VGs can be grouped into a partition. Pairing each VG with its TT and TD data will be done according to the decoding order in the bitstream; the first VG will be associated with the first TT and TD and so on. The field `last_vg` is used to determine if the current VG is the last one in the partition.

---

16. A connected component is a (manifold) model. A 3D mesh object can have multiple models (i.e., three dolphins that are apart from one another). In coding nonmanifold models, the model must be divided into multiple manifold models; this time, there will be multiple connected components, as well.

Before TT in partition type 2, some header information is attached to assist the traversal of the triangle tree (left/right bounding loop index) and to indicate the boundary prediction mode. If multiple type 2 partitions exist, this means that a connected component is cut into multiple pieces. This would generate a similar problem as with nonmanifold models (repetitive coding of the same vertices). There are two modes of coding vertices on the boundary between partitions:

☞ **Restricted boundary prediction mode:** This mode does not duplicate vertices between partitions. Because the vertices predicted in the previous partitions may not be available, prediction is done only with the available vertices that are predicted in the current partition.

☞ **Extended boundary prediction mode:** On the other hand, this mode duplicates the vertices shared with other partitions. Therefore, extended prediction would be less efficient in coding efficiency than restricted prediction. However, in the case of restricted prediction, when the previous partition is lost, the triangles that share vertices with previous partitions cannot be recovered.

If a connected component was cut on a branching triangle, an additional field, branch_length, gives the length of the subtree of the branching triangle and is decoded to reconstruct the branching triangle before the subtree of the branch is decoded (which will be in the next partition).

9.4.2.6 Progressive Transmission Mode The progressive transmission mode is similar to scalable video coding, as the full 3DMC data is structured into multiple layers: one base layer and one or more enhancement layers. The base layer coding uses the 3DMC coding scheme presented earlier. The enhancement layers provide *forest split* operations to increase the level of detail of the model by cutting some existing edges and creating new triangles in the opening as shown in Figure 9.29. Whereas the base layer may give a minimum quality, the enhancement layers gradually improve the quality of the 3D model. The enhancement layers will contain face forest, triangle tree, and triangle data, wherein face forest is a set of vertex trees (see Figure 9.29a). More details on the encoding and decoding operations for the 3DMC progressive transmission mode can be found in [M4V2, GuTa98, GTLH99].

9.4.3 Examples

In this section, some 3DMC examples are given to show some of its functionalities. The models used here were selected from the 300 test models used to perform the 3DMC core experiments [SNHC]. For simplicity, only connectivity and geometry data are coded.

During the 3DMC core experiments, it was found that 3DMC coding typically could compress VRML data 40 to 50 times more efficiently without

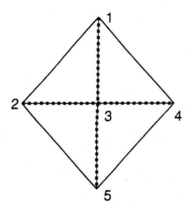

(a) cut through the edges of a vertex tree
(dotted line)

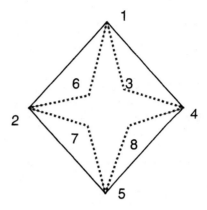

(b) opening forming a bounding loop while
allocating new indices for the newly created
vertices

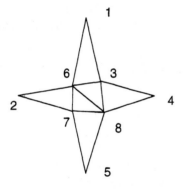

(c) triangulation on the opening to form a
triangle tree

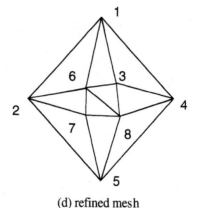

(d) refined mesh

Fig. 9.29 Forest split operation to refine a mesh based on the base mesh.

noticeable visual degradation. Coding examples are shown (in kByte) in Figure 9.30.

One useful 3DMC functionality is incremental buildup. As shown in Figure 9.31, the decoder can start rendering the 3D model while still decoding the remaining bitstream.

Figure 9.32 shows two different 3D models decoded taking benefit of the error-resilient mode using partitions 180 bytes long and a bitstream corruption with a packet loss rate of 10%. More specifically, the *Horse* model was

(a) *Cow* (VRML: 205 kB, 3DMC: 12 kB)

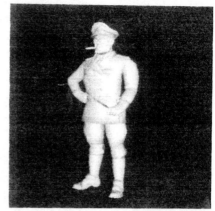

(b) *General* (VRML: 821 kB, 3DMC: 34 kB)

(c) *Pietá* (VRML: 232 kB, 3DMC: 13 kB)

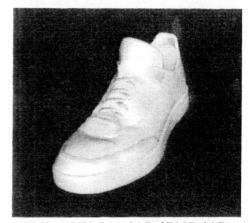

(d) *Shoe* (VRML: 112 kB, 3DMC: 8 kB)

Fig. 9.30 Examples of 3D mesh compression.

encoded with 199,088 bits without the error-resilience mode, and the bit-stream size increased to 296,880 bits when the 180-byte partition size was used (49% overhead). Typically, this is a large overhead because of the small partition size. In the experiment it was assumed that most users would get, in a broadcast scenario, the perfect model on the left of Figure 9.32. The experimental results from MPEG-4 core experiments showed that the typical overhead ranges from 17 and 50% [M4516]. If the partition size is bigger, the overhead decreases. However, the negative impact of network errors would increase as the size of undecodable data would become larger.

Fig. 9.31 Incremental buildup of a 3D mesh (*Horse*).

9.5 VIEW-DEPENDENT SCALABILITY

Scalability is one of the major functionalities in MPEG-4, as it can provide solutions to harmonize platforms with different capabilities, for example, in terms of bandwidth or computational power. In a 3D game scene, the user terminal might be burdened if it has to download the scene graph and all of its entities, especially when the user terminal is a mobile device. However, the virtual world, which the user can experience with MPEG-4, is bounded by the viewing position and field of view, just as human beings are bounded by eye positions and capabilities in the real world. View-dependent scalability is a tool for sending first to the decoding terminal only the visually needed data. If the user's viewing position or angle is changed, the server can update the

(a)   (b)

(c)   (d)

Fig. 9.32 Reconstructed 3D meshes with and without packet losses (10%).

visual data based on the new position and angle, but no useless data is sent (e.g., corresponding to angles where the user has never been).

The view-dependent scalability tool is useful when there is a back channel from the user to the server (see Figure 9.33). Hence, the major target applications for view-dependent scalability are network games and virtual reality applications in which terminals may vary in performance (e.g., computing power, memory size, and available bandwidth).

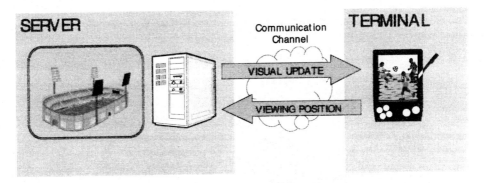

Fig. 9.33 Basic concept of view-dependent scalability.

9.5.1 View-Dependent Object

A view-dependent object is a 3D object with its geometry and texture information, in which the details of the object are updated depending on the viewer's position. Texture mapping on the 3D geometry for a view-dependent object is done for each triangle or quadrangle. In other words, one can assume there is a corresponding texture map for each polygon of the object.

For example, a view-dependent object may be a sphere object (as shown in Figure 9.34). Once the user terminal has the 3D geometry, the terminal needs to have the texture map that is inside the field of view (FOV) of the user. As the sphere rotates or the user turns around the sphere, the object texture must be updated. Moreover, as the user approaches the sphere, the texture quality for the same area can be improved using spatial scalability. In view-dependent scalability (VDS), the entire 3D geometry information is considered already available to the decoder. What is updated through time is the texture information depending on the viewer's position and viewing angle

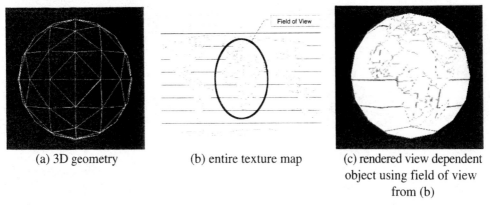

(a) 3D geometry   (b) entire texture map   (c) rendered view dependent
                                                              object using field of view
                                                              from (b)

Fig. 9.34 View-dependent object.

[HoJE97]. It is worth noting that the number of polygons in the mesh should be equal to the texture size in pixels divided by (8×8), as each quadrangle cell is designed to be 8 pixels wide in width and height. For example, an image with 512×512 luminance pixels resolution will have 4,096 quadrilaterals (or 8,192 triangles).

9.5.1.1 View-Dependent Scalability Parameters To reduce the amount of bits spent with texture data and render it efficiently, there are four criteria for texture data reduction: distance, rendering, orientation, and cropping:

- ☞ **Distance criterion (R$_d$):** If the viewpoint is far away from the view-dependent object, a downsampled texture can be used instead of the original texture, as shown in Figure 9.35.

- ☞ **Rendering criterion (R$_r$):** In rendering the view-dependent object, the following factors influence the final rendered texture: the distance between the viewpoint and the projection plane (p), and the ratio between the texture size and the projection plane size (q) in pixels. The rendering criterion parameter that will directly affect the size of rendered texture is calculated as follows:

$$R_r = \frac{p}{q}$$

- ☞ **Orientation criterion (R$_a$, R$_b$):** The final rendering of texture also can be affected by the vertical tilting (α) and the horizontal rotating (β) degrees. In the example shown in Figure 9.36, the vertical elevation (α) is 60 degrees. The smaller these parameters are, the more accurate the texture mapped must be. The orientation parameters are calculated as follows:

$$R_a = \cos \alpha,$$
$$R_b = \cos \beta$$

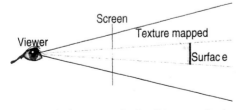

(a) rendering example for distance criterion

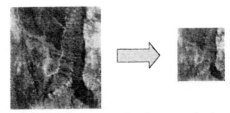

(b) scaled down texture for distance criterion

Fig. 9.35 Texture scaling for distant objects.

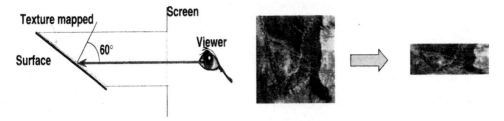

(a) rendering example for orientation criterion   (b) scaled down texture for orientation criterion

Fig. 9.36 Texture scaling for a tilted object.

☞ **Cropping criterion:** Depending on the 3D model geometry, some texture is hidden from the viewpoint. For instance, only one side of the moon is visible from the Earth. If the moon is a view-dependent object, the hidden texture does not need to be transmitted, as long as the viewpoint is located on the Earth.

9.5.2 Coding Scheme

The general decoding process for a view-dependent object is shown in Figure 9.37. The viewer's position and viewing angle will determine the field of view

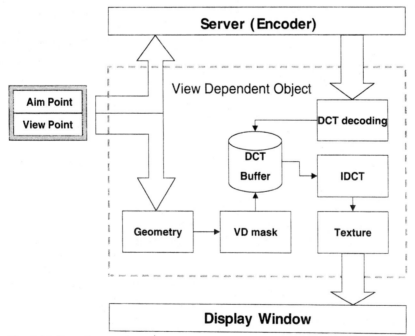

Fig. 9.37 Decoding architecture for a view-dependent object.

(FOV) of the given object, which will be transmitted as upstream data to the server (encoder). As a consequence, the server will transmit the necessary updated or new texture to the decoder; this includes updating or new DCT coefficients. In VDS, DCT-based coding is used for the texture. Wavelet-based coding currently is considered for the same purpose in the context of the SNHC AFX work. After partial decoding, the new DCT coefficients will be added to the DCT buffer where all of the transmitted DCT coefficients are kept. Given the FOV and the 3D geometry, a view-dependent mask (VD mask) can be generated by the decoder. The VD mask is used to determine which DCT blocks are visible in the display window. Only visible DCT coefficients, determined by the VD mask, will go through the inverse DCT process to yield the final texture to be mapped on the 3D model and shown on the display window.

9.5.2.1 View-Dependent Mask

The VD mask is a binary map of the texture image indicating if the corresponding DCT coefficients are needed, depending on the FOV and 3D geometry. The VD mask does not need to be transmitted or compressed as it can be calculated given the VDS parameters, according to the four criteria previously presented. Because the texture is mapped on top of 8×8 pixel blocks, the VD mask is processed on a block basis.

Overall, the VDS parameters are used to determine if a block is visible and how many DCT coefficients are needed for a given block according to the following:

☞ **Invisible blocks:** Some blocks will be invisible from the viewpoint (i.e., the backside of the moon). Using the cropping criterion (i.e., all of the vertices of those blocks are out of the field of view), the decoder terminal can distinguish the invisible blocks setting the corresponding 8×8 block VD mask to 0.

☞ **Visible blocks:** Using distance, orientation, and rendering criteria, the decoder can find out how many and which DCT coefficients in the given texture block are needed. When selecting DCT coefficients, if the rendered texture block is 8×2, the corresponding DCT coefficients also are cropped to be 8×2. In this case, only the first two rows are selected, corresponding to the lower frequency components. Because the texture block is 2D, the value of the VD mask (T_x, T_y) is determined as follows:

$$T_x = min\ (8, 8 \times R_r \times R_d \times R_b)$$
$$T_y = min\ (8, 8 \times R_r \times R_d \times R_a)$$
$$Mask(i, j) = \begin{cases} 1, & if \quad i \leq T_x, j \leq T_y \\ 0, & Otherwise \end{cases}$$

As the field of view changes over time, the encoder needs to transmit the new DCT coefficients to the decoder to update the rendered texture image. In the decoder architecture presented in Figure 9.37, the DCT buffer stores the transmitted DCT coefficients. Not all the DCT coefficients are needed for final rendering because of FOV. Whenever FOV is changed, the VD mask will be updated. Based on the updated VD mask, there are four categories of DCT coefficients: (1) transmitted and needed for rendering; (2) transmitted, but not needed anymore for rendering; (3) not transmitted, but needed for rendering; and (4) not transmitted and unneeded. Those coefficients that belong to the third category must be transmitted and updated. The DCT coefficients will be coded just like for the video intra coding mode (see Chapter 8).

9.5.3 Example

The view-dependent scalability saves texture information bandwidth by exploiting the viewer's position and field of view. Experimental results from the core experiments that were conducted during the MPEG-4 development showed 2× or 3× improved compression due to texture reduction. This is mainly due to the fact that during a typical session, not all texture needs to be transmitted in high resolution. More experimental results can be found in [HoJE97, JoEK98]. Yet the major advantage of this technology does not lie in the overall compression but in the streaming capability it provides. After an initial burst of bits, the server can maintain reasonably low bandwidth transmission with high-quality texture rendering. However, this kind of scenario places a high computational burden on the server. Therefore, the tool does not scale well and it was not included in any MPEG-4 visual object type.

9.6 PROFILES AND LEVELS

MPEG-4 is a toolbox standard, in which subsets of tools can be specified as object types and after profiles for different applications. Moreover, levels will define constraints associated to relevant parameters, building together the conformance points as profile@level combinations (see Chapter 13). There are four visual profiles defined including synthetic visual tools (i.e., object types defined using synthetic visual coding tools):

☞ **Simple face animation profile:** Using the FAPs, it is designed to animate a face model suitable for applications such as audiovisual presentations with an animated face.

☞ **Simple FBA profile:** This profile is an extension to the Simple face animation profile to include body animation capabilities and thus allow more complex applications including the animation of humanlike models.

☞ **Basic animated texture profile:** This allows the combination of facial animation objects with scalable still textures using the visual texture coding (VTC) tool described in Chapter 8, 2D meshes with uniform triangulation, and arbitrarily shaped textures.

☞ **Hybrid profile**: This supports both natural and synthetic visual object types targeting content-rich media applications including arbitrarily shaped video objects, scalable textures, 2D mesh objects, and face animation objects. This profile does not cover body animation or 3DMC, as the profile was designed with MPEG-4 Visual Version 1 tools (body animation and 3DMC were defined only in MPEG-4 Visual, Version 2).

View-dependent scalability is not yet part of any object type. Tools such as face and body animation also need the support of special nodes that are defined in the Systems part of MPEG-4 [MPEG4-1]. The text-to-speech interface (TTSI) that enables the control of talking faces with a speech synthesizer is defined in MPEG-4 Audio [MPEG4-3]. Details on the object types, profiles (like the Complete graphics profile), and levels can be found in Chapter 13 and Appendixes A to D.

9.7 SUMMARY

In this chapter, an overview of the MPEG-4 Visual SNHC tools was presented. SNHC work began in 1994 in the context of the MPEG-4 project, as a new type of activity in MPEG. MPEG-4 SNHC combines graphics, animation, compression, and streaming capabilities in a framework that allows for integration with (natural) audio and video. Face and body animation enables the integration of synthetic talking characters into an interactive media presentation. 2D mesh animation can be used to overlay an image or video over a movie (e.g., to integrate advertisements into the movie). 3D mesh coding allows for compressing a VRML ASCII file down to 2%–4% of its original size without visual degradation, for error resilient-transmission, and for incremental rendering of arbitrary 3D meshes. Profiles defined for 2D mesh animation allow for manipulating images and video and for face and body animation, which allows the efficient download and animation of synthetic faces and bodies.

The current SNHC activities target the AFX, which will enrich the SNHC vision and set of tools, notably by supporting advanced animation capabilities.

As MPEG-4 becomes *the standard* for natural audio- and video-coded data, it is expected that the strong commitment made by MPEG in terms of synthetic and hybrid data coding will provide a similar impact. Ongoing collaboration with the Web3D Consortium is a good sign that related standardization bodies and consortia recognize the work developed by MPEG and want to work together to achieve common goals.

9.8 ACKNOWLEDGMENTS

The authors would like to thank Olivia Ostermann for drawing Figure 9.7. Furthermore, we thank Fernando Pereira for his thorough review of this chapter.

9.9 REFERENCES

[AbPe99] Abrantes, G., and F. Pereira. "MPEG-4 Facial Animation Technology: Survey, Implementation and Results." *IEEE Transactions on Circuits and Systems for Video Technology*, 9 (2): 290–305. March 1999.

[ACM] ACM SIGGRAPH Video Review. *The Story of Computer Graphics.* Issue 137, *www.siggraph.org/movie*

[BKOS97] de Berg, M., M. van Kreveld, M. Overmars, and O. Schwarzkopf. *Computational Geometry—Algorithms and Applications.* Berlin: Springer-Verlag. 1997.

[Bloo97] Bloomenthal, J. (Ed.). *Introduction to Implicit Surfaces.* San Francisco: Morgan Kaufmann Publishers, Inc. 1997.

[Boss99] Bossen, Frank. "On the Art of Compressing Three-Dimensional Polygonal Meshes and Their Associated Properties." Thesis no. 2012. Ecole Polytechnique Federale de Lausanne. September 1999.

[BPO00] van Beek, P., E. Petajan, and J. Ostermann. "MPEG-4 Synthetic Video." *Advances in Multimedia: Systems, Standards and Networks* (A. Puri and T. Chen, Eds.). New York: Marcell Dekker, pp. 299–330. March 2000.

[BTZC99] van Beek, Peter, A. M. Tekalp, N. Zhuang, I. Celasun, and M. Xia. "Hierarchical 2-D Mesh Representation, Tracking, and Compression for Object-based Video." *IEEE Transactions on Circuits and Systems for Video Technology*, 9(2): 353–369. March 1999.

[CM93] Cohen, M. M., and D. W. Massaro. "Modeling Coarticulation in Synthetic Visual Speech." *Computer Animation '93* (M. Thalmann and D. Thalmann, Eds.). Tokyo: Springer-Verlag. 1993.

[CPO00a] Capin, T. K., E. Petajan, and J. Ostermann. "Efficient Modeling of Virtual Humans in MPEG-4." *ICME 2000*, New York: IEEE, pp. TPS9.1. 2000.

[CPO00b] Capin, T. K., E. Petajan, and J. Ostermann. "Very Low Bitrate Coding of Virtual Human Animation in MPEG-4." *ICME 2000.* New York: IEEE, pp. TPS9.2. 2000.

[ET97] Escher, M., and N. M. Thalmann. "Automatic 3D Cloning and Real-Time Animation of a Human Face." *Proc. Computer Animation '97.* Geneva: IEEE Press. 1997.

[Fari96] Farin, Gerald. *Curves and Surfaces for Computer Aided Geometric Design: A Practical Guide.* New York: Academic Press, Inc. 1996.

[FDFH96] Foley, J. D., et al. *Computer Graphics: Principles and Practice.* Reading, MA: Addison-Wesley. 1996.

[GBTS99] Gueziec, A., F. Bossen, G. Taubin, and C. Silva. *"Efficient Compression of Non-Manifold Polygonal Meshes,"* Visualization '99 Conference Proceedings. 1999.

[GTLH98] Gueziec, A., G.Taubin, F. Lazarus, and W. P. Horn. "Converting Sets of Polygons to Manifold Surfaces by Cutting and Stitching." *Visualization'98 Conference Proceeding.* Research Triangle Park, NC: Computer Science Press. October 1998.

[GTLH99] Gueziec, A., G. Taubin, F. Lazarus, and W. P. Horn. "A Framework for Streaming Geometry in VRML." *IEEE Computer Graphics & Applications.* March–April 1999.

[GuTa98] Gueziec, A., and G. Taubin. "A Framework for Memory-Efficient Levels of Detail." *Tenth Canadian Conference on Computational Geometry*, Montreal, Canada. August 1998.

[HoJE97] Horbelt, S., F. Jordan, and T. Ebrahimi. "View-Dependent Texture Coding for Transmission of Virtual Environments." *IEE Sixth International Conference on Image Processing and Its Applications (IPA97)*. Dublin, pp. 433–437. July 1997.

[Jang00] Jang, Euee S. "3D Animation Coding—Its History and Framework." *Proceedings of International Conference of Multimedia and Expo 2000*, New York. July 2000.

[JoEK98] Jordan, F., T. Ebrahimi, and M. Kunt. "View-Dependent Texture Coding Using the MPEG-4 Video Coding Scheme." *ISCAS'98*, (V.5), pp. 498–501. 1998.

[Leng99] Lengyel, J. "Compression of Time-dependent Geometry." *Symposium on Interactive 3D Graphics*, Atlanta, GA. 1999.

[LP99] Lavagetto, F., and R. Pockaj. "The Facial Animation Engine: Toward a High-Level Interface for the Design of MPEG-4 Compliant Animated Faces." *IEEE Transactions on Circuits and Systems for Video Technology*, 9(2): 277–289. 1999.

[M4516] Han, M.-J., M. Song, S. Kim, and E. S. Jang. *Results of M5 Core Experiment*, Doc. ISO/MPEG M4516, Seoul MPEG Meeting, March 1999.

[M4V2] ISO/IEC 14496-2:2000, Amendment 1. *Coding of Audio-Visual Objects—Part 2: Visual*, Version 2, 2000.

[MaCa97] Marrin, C., and B. Campbell. "Teach Yourself VRML in 21 Days." 1997. *Sams.net*

[MPEG4-1] ISO/IEC 14496-1:2001. *Coding of Audio-Visual Objects—Part 1: Systems*, 2d Edition, 2001.

[MPEG4-2] ISO/IEC 14496-2:2001. *Coding of Audio-Visual Objects—Part 2: Visual*, 2d Edition, 2001.

[MPEG4-3] ISO/IEC 14496-3:2001. *Coding of Audio-Visual Objects—Part 3: Audio*, 2d Edition, 2001.

[N4319] MPEG Requirements. *MPEG-4 Requirements*. Doc. ISO/MPEG N4319, Sydney MPEG Meeting, July 2001.

[NB94] Ni, X., and M. S. Bloor. "Performance Evaluation of Boundary Data Structures." *IEEE Computer Graphics and Applications*, 14(6): 66–77. 1994.

[OH97] Ostermann, J., and E. Haratsch. "An Animation Definition Interface: Rapid Design of MPEG-4 Compliant Animated Faces and Bodies." *International Workshop on Synthetic Natural Hybrid Coding and Three Dimensional Imaging*, Rhodes, Greece, pp. 216–219. September 5–9, 1997.

[Penn88] Pennebaker, W. B., et al. "An Overview of the Basic Principles of the Q-coder Adaptive Binary Arithmetic Coder." *IBM Journal of Research and Development* 32(6): 717–726. 1988.

[Penn93] Pennebaker, W. B., and J. Mitchell. *JPEG Still Image Data Compression Standard*. New York: Van Nostrand Reinhold. 1993.

[Perr01] T. S. Perry. "And the Oscar Goes To." *IEEE Spectrum*, pp. 42–49. April 2001.

[SNHC] MPEG SNHC Home Page, *www.sait.samsung.co.kr / snhc*

[TaRo98] Taubin, G., and J. Rossignac. "Geometry Compression through Topological Surgery." *ACM Transactions on Graphics*, pp. 84–115. April 1998.

[TaRo99] Taubin G., J. Rossignac, "Course on 3D Geometry Compression." *SIG-GRAPH'99*, Los Angeles 1999.

[TBTG98] Tekalp, A. M., P. van Beek, C. Toklu, and B. Gunsel. "Two-Dimensional Mesh-based Visual-Object Representation for Interactive Synthetic/Natural Digital Video." *Proceedings of the IEEE*, 86(6): 1029–1051. June 1998.

[TCWH99] H. Tao, H. H. Chen, W. Wu, T. S. Huang. "Compression of Facial Animation Parameters for Transmission of Talking Heads." *IEEE Transactions on Circuits and Systems for Video Technology*, 9(2): 264–276. 1999.

[THLR98] Taubin, G., W. P. Horn, F. Lazarus, and J. Rossignac. "Geometric Coding and VRML." *Proceedings of the IEEE*. July 1998.

[TO00] Tekalp, A. M., and J. Ostermann. "Face and 2-D Mesh Animation in MPEG-4." *Signal Processing: Image Communication*, 15 (4, 5): 387–421. 2000.

[TT01] Magnenat-Thalmann, T., and D. Thalmann (Eds.). *Deformable Avatars*. Boston: Kluwer Academic Publishers. 2001.

[VRML97] ISO/IEC 14772-1, "The Virtual Reality Modeling Language," 1997. *www.vrml.org/Specifications/VRML97*

[WaBS99] Walsh, A. E., and M. Bourges-Sevenier. *Core Web3D*. Upper Saddle River, NJ: Prentice Hall PTR. 1999.

[Web3D] Web3D Consortium Home Page, *www.web3d.org/*

[Web3D99] Web3D Working Group on Humanoid Animation. Specification for a Standard Humanoid, Version 1.1, August 1999.

[WL94] Wang, Y., and O. Lee. "Active Mesh—A Feature Seeking and Tracking Image Sequence Representation Scheme." *IEEE Trans. on Image Proc.*, 3(5): 610–624. Sept. 1994.

[WO98] Wang, Y., and J. Ostermann. "Evaluation of Mesh-based Motion Estimation in H.263 Like Coders." *IEEE Transactions on Circuits and Systems in Video Technology*, pp. 243–252. 1998.

[WOZ01] Wang, Y., J. Ostermann, and Y. Zhang. *Video Processing and Communications*. Upper Saddle River, NJ: Prentice Hall. 2001.

[X3DFAQ] X3D FAQ, *www.web3d.org/TaskGroups/x3d/faq/*

Speech Coding

by Masayuki Nishiguchi and Bernd Edler

Keywords: CELP, HVXC, LPC, ADPCM, VQ, speech coding, parametric coding, error protection

S peech coding technology has been developed for the purpose of saving the bandwidth and capacity of communication networks and storage systems when speech signals have to be transmitted or stored. During the last decade, the development of speech coding technology accelerated, especially in the area of quickly evolving personal mobile phone applications [RCR94, TIA99].

An object-based representation of audio signals, such as that provided by the MPEG-4 standards [MPEG4-1, MPEG4-3], requires a high degree of flexibility with respect to the kind of input material, the range of bit rates, and interaction capabilities supported. Therefore, MPEG-4 Audio [MPEG4-3] is not just another coding algorithm, but rather defines a system that allows the representation of complete audio scenes containing one or multiple natural or synthetic audio objects. In this context, *natural* objects are those which are derived from a sampled input waveform, whereas *synthetic* objects are usually only described at higher levels—for example, musical scores or written text. In conjunction with MPEG-4 Systems [MPEG4-1], the bitstream syntax [MPEG4-3] and the rendering process are specified, as illustrated by the decoder structure shown in Figure 10.1. Corresponding to their different characteristics, decoders for four audio signal types are included in the MPEG-4 Audio standard [MPEG4-3]. This chapter focuses on *natural speech*; the other

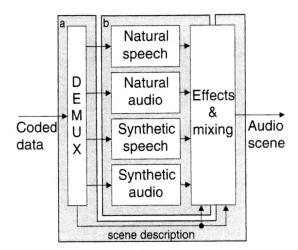

Fig. 10.1 MPEG-4 audio scene decoder including MPEG-4 Systems (a) and Audio (b) tools.

three types—*natural audio, synthetic speech*, and *synthetic audio*—are addressed in Chapters 11 and 12. The audio output signal is obtained by simply mixing the decoded data or by applying more complex effects, which also can be controlled by the bitstream.

10.1 INTRODUCTION TO SPEECH CODING

In contrast to general audio coders for *natural audio* signals, speech coders are usually expected to handle human voice as an input signal. This means that a model-based signal analysis and synthesis can be used in a highly efficient manner. The most common and widely used model-based speech analysis system is linear predictive (LP) analysis, also known as linear predictive coding (LPC). In LPC analysis, a speech signal is decomposed into two components: a set of LP coefficients and a prediction error signal (residue). The LP coefficients allow the construction of an analysis filter, which removes short-term correlation in the speech signal and provides a prediction error signal. In this analysis, a speech signal is modeled as a convolution of the glottal vibration with the human vocal tract response. The analysis filter has approximately an *inverse* characteristic of the human vocal tract response, and the prediction error signal represents the human glottal vibration. LPC analysis also can be seen as a whitening operation of the speech spectrum in the frequency domain.

The simplest speech coder that uses LPC analysis is an LPC vocoder [Rabi78]. In such an LPC vocoder, LPC residual signals are modeled by a pulse train or noise, which are switched depending on a voiced/unvoiced (V/UV) decision. In this way, an LPC vocoder uses model-based parameters not only for the spectral envelope but also for the residual signals to represent speech signals; therefore, an LPC vocoder can encode speech signals at very low bit rates, such as 800 to 1200 bit/s. Figure 10.2 shows the structure of the LPC vocoder synthesizer.

On the other hand, adaptive differential pulse code modulation (ADPCM) [Jaya84] is a coding method that does not use any modeling for the residual signal, although it uses, in principle, the same LPC modeling for the adaptive prediction of the speech samples. In ADPCM, the waveform of the LPC residual signal is transmitted using nonlinear scalar quantization. Because no modeling is used for the residual signal, ADPCM can basically encode and decode any kind of source signals of telephone bandwidth with reasonable quality. However, the operating bit rate of ADPCM is relatively high (in the range of 24 to 32 kbit/s).

The most prevalent speech coding technique is code excited linear prediction (CELP) [Schr85], in which a *codebook* representation of the residual signal is employed. In the *codebook*, several hundreds of typical residual waveforms (codevectors) are stored, and the most appropriate waveform is selected from the codebook for every 5- to 10-ms frame through an *analysis by synthesis* procedure in which the codevector generating the synthesized speech that best approximates the input speech is selected by waveform matching—that is, minimization of the perceptually weighted error. Because of the *codebook* structure, the variety of possible residual waveform shapes is less than that of ADPCM; however, it is more flexible for representing the residual waveform than the LPC vocoder, in which only the pulse train and noise residuals are allowed. The freedom in representing the LPC residual signal in CELP can be viewed as intermediate between that of the LPC vocoder and ADPCM. The typical operating bit rates of CELP range from 4 to 16 kbit/s for telephone bandwidth speech. When the bit rate is reduced to less than 4 kbit/s, however, the reproduced speech quality deteriorates due to the waveform matching mechanism of the CELP coder, in which the phase of the speech waveform as well as the speech spectral magnitudes are taken into account. For very low bit rates, however, it is desirable that the available bits are used to obtain better spectral magnitude reproduction rather than phase reproduction.

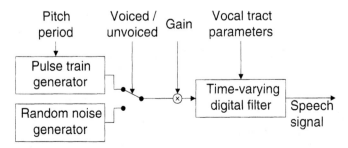

Fig. 10.2 Structure of the LPC vocoder synthesizer.

The next generation very-low-bit-rate coder made it possible to assign more bits to spectral magnitudes than phase by representing the LPC residual signal for voiced speech by a set of sinusoidal waveforms or harmonic waveforms [Nish95]. In this model, the residual signals are expressed by *parametric* representation, such as sinusoidal magnitudes and frequencies; this type of coders is referred to as *parametric coders*. By parameterizing the LPC residual waveform, phase information in the residual waveform can easily be discarded; hence the bit assignment for the spectral magnitudes can be increased. Although the bit rate is lower than in CELP coding, the spectral magnitudes at harmonic frequencies of the LPC residual are precisely and smoothly reproduced and therefore parametric coders can transmit good quality speech at very low bit rates such as 2 to 4 kbit/s.

In summary, LPC modeling of the spectral envelope (SE) of speech signals is a well-established technology and commonly used by most speech coders. However, the representation of the LPC residual signals varies depending on coder types and target bit rates, and it can be optimized for a variety of applications. Pulse train plus noise representation is used in LPC vocoders, waveform representation is used in ADPCM, codebook representation is used in CELP, and parametric representation is used in parametric coders to encode the LPC residual signals. In this chapter, both the MPEG-4 CELP speech coder and the MPEG-4 parametric speech coder based on harmonic vector excitation coding (HVXC) are presented.

10.2 OVERVIEW OF MPEG-4 SPEECH CODERS

MPEG-4 speech coding tools [MPEG4-3] consist of two different basic algorithms: CELP [Nomu98] and HVXC [Nish97a]. They provide not only high coding efficiency but also a variety of new functionalities, such as bit-rate scalability (also known as embedded coding); bit-rate controllability, which allows modification of the bit rate during encoding; variable bit-rate coding; and speed and pitch changes. Users can choose the best algorithm for their applications and requirements.

The CELP algorithm operates from 3.85 to 23.8 kbit/s for speech signals of 8- and 16-kHz sampling rate with two different excitation modes—that is, multipulse excitation (MPE) and regular pulse excitation (RPE). Parametric speech coding, HVXC, fits for even lower bit-rate coding of speech signals at 8-kHz sampling rate. HVXC uses a parametric representation of the harmonic spectral magnitudes of LPC residual signals for voiced segments, and a vector excitation coding (VXC) algorithm for unvoiced segments. Bit-rate scalable coding is enabled by a multistage structure of the vector quantization (VQ) scheme for LPC parameters and LPC residuals. In addition, speed and pitch changes are possible during decoding by manipulating the encoded parameters due to the parametric representation of the speech signals.

10.3 MPEG-4 CELP CODING

Like the other coding tools in the MPEG-4 standard, MPEG-4 CELP is defined by a normative bitstream syntax and a corresponding decoding process, whereas an example encoder is described in an informative annex [MPEG4-3]. This approach leaves room for application-specific encoder optimizations even after the finalization of the standard. In contrast to this solution, other standard CELP speech codecs—for example, the ITU-T G.723 series [G723]—also specify a normative encoder. MPEG-4 CELP is specified in a modular way and therefore fits well with the MPEG-4 toolbox approach. One benefit of this approach is high flexibility with respect to the range of covered bit rates and sampling frequencies. Thus, it can operate in a *narrowband mode*, with 8-kHz sampling frequency, and in a *wideband mode*, with 16-kHz sampling frequency. Furthermore, MPEG-4 CELP provides additional functionalities besides pure data compression, including bit-rate scalability (see Section 10.3.6). In the following section, a short overview on the general MPEG-4 CELP encoder structure is given, followed by a detailed description of the various decoding tools. Finally, two special modes for scalable coding and for the so-called *silence compression* are presented.

10.3.1 CELP Encoder

The general structure of a CELP encoder [Atal84], as shown in Figure 10.3, can also be applied for MPEG-4 CELP. This structure processes the audio signal in blocks (frames) consisting of a fixed number of samples and is optimized for representing speech, as explained earlier.

The key component in this system is the excitation generator shown in Figure 10.4, designed to produce an excitation signal, which in conjunction with the LPC synthesis filter resembles the input signal as closely as possible. Generation of periodic and quasi-random signals is supported by the weighted superposition of sample vectors from an adaptive codebook and from a fixed codebook. The adaptive codebook contains time-shifted versions

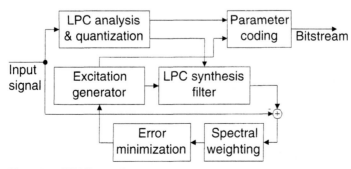

Fig. 10.3 CELP encoder.

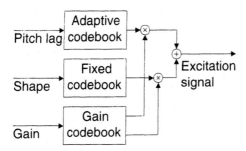

Fig. 10.4 Excitation generator.

of signal segments coded in previous frames and so is often referred to as long-term prediction (LTP). The fixed codebook is optimized for generating signal components, which cannot be derived from previous frames. Because these components often are of a random nature, the fixed codebook often is denoted as the *stochastic codebook*. The control parameters of the excitation generator are the indices for the adaptive codebook, which correspond to a pitch lag; the index for the fixed codebook, which controls the shape of the superimposed signal vector; and a gain index, which defines amplitude factors for the two signal components.

The excitation signal generated from a set of control parameters is fed through an LPC synthesis filter (see Figure 10.3) with a magnitude response resembling the SE of the input signal. The output of this filter then is compared to the input signal. The mean square error after filtering for spectral weighting according to perceptual criteria is used as a selection criterion for determining the optimum parameter set in an iteration loop. The excitation parameters and the LPC parameters are coded in the bitstream of a CELP audio object. Whereas the LPC parameters are updated once per frame (i.e., typically every 10 to 40 ms), the excitation parameters are updated in shorter intervals, which are called *subframes*.

10.3.2 CELP Decoder

As mentioned earlier, the general decoder structure of MPEG-4 CELP looks similar to that of a regular CELP system, as shown in Figure 10.5. However, the individual blocks are designed to be more flexible than those of previous CELP standard codecs. One indication of this flexibility is that two different types of excitation generators are supported:

1. **Multipulse excitation** (MPE) [Atal81] provides very high flexibility in terms of bit rates, bandwidth, and scalability, whereas

2. **Regular pulse excitation** (RPE) [Kroo86] is supported for enabling wideband speech coding applications with significantly lower encoder complexity at only slightly reduced compression efficiency.

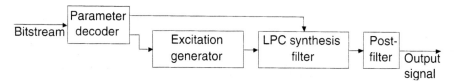

Fig. 10.5 CELP decoder.

Furthermore, the subframe lengths of both excitation modes can be varied, and the LPC synthesis filter and its parameter decoding are designed to operate for various frame-length configurations. Additionally, the postfilter is not normatively fixed by the MPEG-4 Audio standard, as it does not depend directly on elements of the bitstream syntax. Therefore, only an example implementation of a postfilter is described in an informative annex of MPEG-4 Audio [MPEG4-3].

In the following sections, the CELP normative components (i.e., the parameter decoding block and the excitation generators) are described in more detail.

10.3.3 Parameter Decoding

The parameter decoder has to parse the bitstream and to generate the control data for the excitation generator and the LPC synthesis filter. The excitation control parameters (such as pitch lag, shape, and gain) are coded differently for the MPE and RPE modes, as will be explained in the following.

The coding of the LPC parameters, on the other hand, is identical for both excitation modes. This coding is based on a representation as Line Spectrum Pairs (LSPs) [Itak75], also known as Line Spectrum Frequencies (LSFs). After decoding, the LSPs are interpolated and converted to LPC filter coefficients, as illustrated in Figure 10.6.

In the narrowband mode, the LPC filter order is 10, and the coding of the 10 LSPs is based on a two-stage vector quantizer as shown in Figure 10.7. The LSP decoder receives a codebook index for each of the two codebooks and an additional *predictor control* flag indicating whether the *interframe prediction* (described below) is active. If interframe prediction is deactivated, the outputs of the two stages are added and given to the LSP stabilizer, which enforces a minimum gap between neighboring LSP frequency values. For the frames with activated interframe prediction, the output of the first stage is averaged with the stabilized LSPs from the previous frame. The same VQ scheme is also used in HVXC [MPEG4-3, Tana96].

Fig. 10.6 LPC parameter decoder.

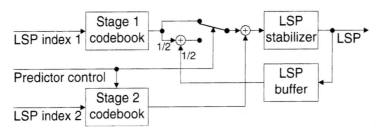

Fig. 10.7 Narrowband LSP decoder.

The LSP interpolator generates one set of LSPs per subframe by using the current decoded set for the last subframe and applying linear interpolation from the previous frame for all the other subframes. The interpolated LSPs are then converted to LPC filter coefficients, as shown in Figure 10.6.

In wideband mode, the LPC filter order is 20 and two LSP decoding modules are applied for the upper and lower 10 parameters, respectively. With the exception of the LSP stabilizer, both of these modules are identical to those used in the narrowband mode shown in Figure 10.7. The stabilization, the interpolation, and the LSP-to-LPC conversion are applied to merged parameter sets. Because the LSPs are ordered in frequency, the merging can be performed by simply appending the upper 10 parameters to the lower 10 parameters.

Another functionality of the LSP coding scheme is *fine rate control*, which allows disabling the parameter transmission in individual frames. In these frames, the LSP parameters are obtained by repetition from the previous frame or by interpolation, depending on control flags generated by the encoder. The use of repetition or interpolation is restricted to a maximum of one frame, after which LSP parameters have to be transmitted again. Using this mechanism, the bit rate can be controlled by selecting a percentage of transmitted LSP parameters between 50% and 100%.

10.3.4 Multipulse Excitation

The MPEG-4 CELP MPE generator [M694] can be operated in narrowband mode as well as in wideband mode (8- versus 16-kHz sampling frequency). The resulting (total) bit rate can be adjusted in small steps by varying the frame length, the subframe length, and the number of pulses generated by the fixed codebook per subframe [M1509]. The resulting bit rates range from 3.85 to 12.2 kbit/s in the narrowband mode (Table 10.1) and from 10.9 to 23.8 kbit/s in the wideband mode (Table 10.2).

The coded data for the excitation generator contains a V/UV mode indicator and an energy parameter, which are both transmitted only once per frame. The remaining parameters (such as pitch lag, gain, and shape) are present for every subframe. The individual subframe energies are obtained by linear inter-

Table 10.1 Narrowband MPE configurations

Bit rate (kbit/s)	Frame length (ms)	Subframe length (ms)	Pulses per subframe
3.85 ... 4.65	40	10	3 ... 5
4.90 ... 5.50	30	10	5 ... 7
5.70 ... 7.30	20	10	6 ... 12
7.70 ... 10.70	20	5	4 ... 12
11.00 ... 12.20	10	5	8 ... 12

Table 10.2 Wideband MPE configurations

Bit rate (kbit/s)	Frame length (ms)	Subframe length (ms)	Pulses per subframe
10.90 ... 14.30	20	5	5 ... 11
14.70 ... 21.10	20	2.5	3 ... 10
13.60 ... 17.00	10	5	5 ... 11
17.40 ... 23.80	10	2.5	3 ... 10

polation, in which each transmitted value is applied to the last subframe of the corresponding frame. The shape data conveys the positions and the signs of the individual pulses within the subframe, which all have equal magnitudes.

10.3.5 Regular Pulse Excitation

Although the MPEG-4 CELP RPE generator [M696] is restricted to wideband speech (i.e., 16-kHz sampling rate), it provides the flexibility to cover a range of bit rates, from 14.4 to 22.5 kbit/s. This is achieved by using different configurations of the frame length, the subframe length, and the pulse spacing (see Table 10.3).

The MPEG-4 CELP RPE is controlled by separately coded gain parameters for the adaptive and fixed codebooks. One signed gain value for the

Table 10.3 RPE configurations

Bit rate (kbit/s)	Frame length (ms)	Subframe length (ms)	Pulse spacing (samples)
14.4	15	2.5	8
16.0	10	2.5	8
18.7	15	15/8	5
22.5	15	1.5	4

adaptive codebook is coded per frame using nonuniform quantization. For the fixed codebook, the unsigned gain value of the first subframe is coded using nonuniform quantization, whereas the relative changes are coded for the other subframes. The generation of equidistant pulses with the spacing defined in Table 10.3 is controlled by the shape data, which contains the phase (i.e., the position of the first pulse) and the amplitude for each pulse. The only possible values for these amplitudes are –1, 0, and +1.

10.3.6 Scalability

As mentioned previously, MPEG-4 CELP not only supports a wide range of bit rates, it also allows scalable coding. This means that the bitstream is structured in a way that one or several lower rate streams are embedded, which can then be extracted and decoded, even if the full rate stream is not completely available. In other words, a full rate stream contains a base layer stream and one or several enhancement streams for improving the sound quality when decoded in conjunction with the base layer. The MPEG-4 scalable CELP is supported by the MPE mode, in which the enhancement layers can either reduce coding artifacts [M2083] or extend the bandwidth [M2486].

In the first case, the quality is improved by transmitting additional pulses in up to three enhancement layers. They can be added to any of the MPE configurations described earlier, each increasing the bit rate by 2 kbit/s in the narrowband mode or 4 kbit/s in the wideband mode. In the second case, the enhancement layer carries the information for extending the bandwidth from narrowband to wideband, thereby increasing the sampling rate from 8 kHz to 16 kHz. For this purpose, additional pulses are transmitted to increase the temporal resolution of the excitation. Furthermore, the LSPs need to be adjusted to the increased bandwidth. This is performed by the scalable LSP decoder shown in Figure 10.8. The LSP data from the base layer is decoded using the regular narrowband decoder. The resulting parameters are fed through a narrowband-to-wideband predictor with a set of fixed coefficients for all LSPs. The predictor output is added to the LSP enhancement parameters decoded from their representation as vector-quantized residuals from an additional moving average (MA) frame-to-frame predictor.

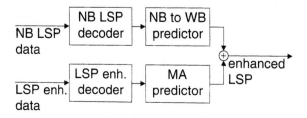

Fig. 10.8 Narrowband (NB) to wideband (WB) scalable LSP decoder.

The bit rate of the bandwidth extension layer depends on the narrowband configuration mode used for the base layer and ranges from 9 to 15 kbit/s.

10.3.7 Silence Compression

Silence compression is a special MPEG-4 CELP mode, which can be enabled to temporarily reduce the bit rate in frames with no or low voice activity. The basic idea is similar to that of HVXC variable bit rate coding, as described in Section 10.4.3.

For this purpose, the MPEG-4 CELP encoder is extended with an activity detector and a change detector. Although active frames are treated in the regular way, nonactive frames are classified according to changes in energy or spectral shape. According to this classification into a total of three nonactive modes, the transmission of frame energy (RMS) and LPC parameters (LSP) is enabled or disabled and the activity mode is signaled by the 2-bit parameter TX_flag, as shown in Table 10.4. For the first nonactive class (TX_flag = 0), no parameters are coded and the LPC parameters from the previous frame are used again in the decoder. For the second nonactive class (TX_flag = 3), only a new energy parameter is coded, although for the third nonactive class (TX_flag = 2) the LPC parameters are also updated. The LPC parameters are coded similarly to those of active frames as LSPs. This way the bit rate can be reduced down to 50 to 200 bit/s during periods without voice activity. The actual minimum rate depends on the frame configuration.

Based on the transmitted data, the decoder generates so-called *comfort noise* for the nonactive frames. For this purpose, a random excitation signal is generated using the same excitation generator as for the active frames, but with randomly created control parameters such as pulse positions or amplitudes. The structure of an MPEG-4 CELP decoder extended to include silence compression is shown in Figure 10.9.

10.4 MPEG-4 HVXC CODING

HVXC allows coding of speech signals at very low bit rates with reasonable quality—that is, communication quality to near toll quality.[1] In fact, two bit

Table 10.4 Frame classification for silence compression

TX_flag	Activity	Transmitted parameters
0	Nonactive	—
1	Active	CELP
2	Nonactive	LSP, RMS
3	Nonactive	RMS

---

1. Toll quality is the speech quality of public switched telephone networks, and communication quality is that of 2G cell phone networks.

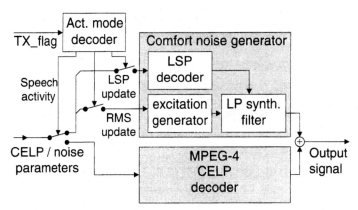

Fig. 10.9 CELP decoder with silence compression.

rates are accepted: 2 and 4 kbit/s, defining the so-called *2-* and *4-kbit/s modes*. This is possible due to the efficient representation of the LPC residuals, in which harmonic coding for voiced segments and VXC for unvoiced segments are employed. Major algorithmic features follow:

1. Perceptually weighted VQ of variable dimension spectral vectors for the harmonic coding
2. Fast harmonic synthesis algorithm by Inverse Fast Fourier Transform (IFFT)
3. Independent modification of speed and pitch by interpolation of coder parameters
4. 2- to 4-kbit/s scalable coding
5. Variable bit-rate coding for rates less than 2 kbit/s
6. Error-resilient syntax for error-prone transmission channels

Section 10.4.1 presents an overview on the encoder scheme and a detailed description of individual tools, including the weighted VQ of spectral vectors. Section 10.4.2 describes the decoder operation and key features such as a fast harmonic synthesis algorithm and speed/pitch change. Finally, a brief introduction to variable bit rate coding is given in Section 10.4.3, and error robustness is addressed in Section 10.5.

10.4.1 HVXC Encoder

In this section, an overview of the MPEG-4 HVXC encoder and a detailed description of some key features (such as the VQ of harmonic spectral vectors and the VXC algorithm) are given.

Figure 10.10 shows the overall structure of the HVXC encoder [Nish99]. Table 10.5 shows the bit allocation for 2- and 4-kbit/s coding (the parameters followed by enh are for the enhancement layer and used only in the 4-kbit/s

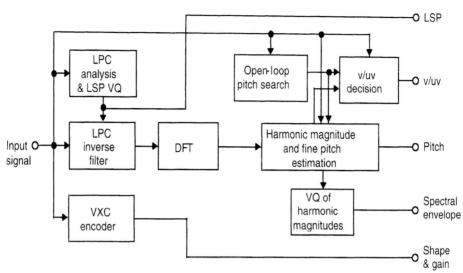

Fig. 10.10 HVXC encoder.

Table 10.5 Bit allocation for 2- and 4-kbit/s HVXC coding

	2 kbit/s		4 kbit/s	
	Voiced	**Unvoiced**	**Voiced**	**Unvoiced**
LSP	18 bits / 20 ms	18 bits / 20 ms	18 bits / 20 ms	18 bits / 20 ms
LSP(enh)			8 bits / 20 ms	8 bits / 20 ms
V/UV	2 bits / 20 ms	2 bits / 20 ms	2 bits / 20 ms	2 bits / 20 ms
Pitch	7 bits / 20 ms		7 bits / 20 ms	
Spectral shape	4+4 bits / 20 ms		4+4 bits / 20 ms	
Spectral gain	5 bits / 20 ms		5 bits / 20 ms	
Spectral shape(enh)			32 bits / 20 ms	
VXC shape		6 bits / 10 ms		6 bits / 10 ms
VXC gain		4 bits / 10 ms		4 bits / 10 ms
VXC shape(enh)				5 bits / 5 ms
VXC gain(enh)				3 bits / 5 ms
Total—2 kbit/s	**40 bits / 20 ms**	**40 bits / 20 ms**		
Total—4 kbit/s			**80 bits / 20 ms**	**80 bits / 20 ms**

mode). The operation of each signal-processing block in the encoder is described in the following.

LPC Analysis and LSP Vector Quantization   Speech input sampled at 8 kHz is divided into overlapped frames with a length and interval (offset) of 256 and 160 samples, respectively. Tenth-order LPC analysis is carried out using windowed input data over one frame. LPC parameters are converted to line spectral pair (LSP) parameters [Itak75] and vector quantized with a partial prediction and a multistage VQ scheme [Tana96]. LPC residual signals are computed by inverse filtering the input data using quantized and interpolated LSP parameters.

Open-Loop Pitch Search   The open-loop pitch value is estimated based on the peak values of the auto-correlation of the LPC residual signals; notice that this block computes the LPC residual signals independently, as different preprocessing has to be used from main path signal. Using estimated past and current pitch values, pitch tracking is conducted so as to generate a continuous pitch contour and increase the reliability of the pitch estimation. The V/UV decision of the previous frame also is used to improve the reliability of the pitch-tracking operation.

Harmonic Magnitude Estimation and Fine Pitch Search   The power spectrum of the LPC residual signal is then fed into the harmonic magnitude and fine pitch estimation block, wherein the harmonic SE of the LPC residual signal and fine pitch value are estimated as follows: An amplitude of a basis spectrum representing one harmonic spectrum is scaled and arranged with the spacing of the fundamental frequency obtained by the open loop pitch search. The amplitude scaling for each harmonic of the fundamental and the fundamental frequency are adjusted simultaneously, so that the difference between the synthesized power spectrum and the actual LPC residual spectrum is minimized. The harmonic SE for the voiced segment is then vector quantized.

Vector Quantization of Harmonic Magnitudes   In order to vector quantize an SE composed of a variable number of harmonics of the fundamental, the harmonic spectral vector is first converted to a fixed-dimension vector by band-limited interpolation [Nish93]. The fixed-dimension spectral vector is then vector quantized. The detailed operation of the VQ process is described in the following section.

Voiced/Unvoiced Decision   The V/UV decision is made based on the maximum auto-correlation of the LPC residual signals, the number of zero crossings, and the harmonic structure of the power spectrum of the LPC residual signals.

Vector Excitation Coding of Unvoiced Signals   For unvoiced segments, regular VXC coding is carried out, in which only stochastic codebooks are used. A 6-bit shape codebook of dimension 80 and a 4-bit gain codebook are used for the 2-kbit/s mode. For the 4-kbit/s mode, the quantization error of the

2-kbit/s mode is quantized again using a 5-bit shape codebook of dimension 40 and a 3-bit gain codebook at this additional stage. The detailed operation of the VXC encoder is presented in the following section.

10.4.1.1 Vector Quantization of Harmonic Spectral Magnitudes One of the most important HVXC algorithmic features is the coding of the harmonic spectral magnitudes, described in this section. Coding of harmonic spectral magnitudes is carried out in three major steps: computation of the perceptual weighting, dimension conversion, and VQ of the SE vectors.

Perceptual Weighting Filter and LPC Synthesis Filter Prior to the harmonic magnitude quantization, the transfer functions for the perceptual weighting filter $w(z)$ and the LPC synthesis filter $h(z)$ are defined, in order to allow the frequency responses of $w(z)$ and $h(z)$ to be used for the weighted VQ of the harmonic SE. The transfer function of the perceptual weighting filter $w(z)$ is as follows:

$$w(z) = \frac{\sum_{n=0}^{P} \alpha_n A^n z^{-n}}{\sum_{n=0}^{P} \alpha_n B^n z^{-n}}$$

where α_n are LP coefficients, $A = 0.9$, and $B = 0.4$. The α_n are computed by LPC analysis using a frame of input speech samples. The transfer function of the LPC synthesis filter $h(z)$ is as follows:

$$h(z) = \frac{1}{\sum_{n=0}^{P} \alpha_n z^{-n}}$$

The magnitude responses of $w(z)$ and $h(z)$ are computed at the frequencies required for spectral weighting, so that the resulting values can be used as the components of the diagonal weighting matrices **W** and **H**, as shown in the following section.

Dimension Converter The number of points that compose the SE varies, depending on the pitch value, because the SE is the set of the estimates of the magnitudes at each harmonic of the fundamental. The number of harmonics present in the spectrum ranges from about 9 to 70.

In order to vector quantize the SE, the coder must convert it to a constant number of spectral samples for a fixed-dimension VQ. A band-limited interpolation is used for this (spectral) sampling frequency conversion (i.e., resampling of the SE) to obtain the fixed-dimension spectral vectors. The number of spectral samples, which represent the shape of the SE, is modified without changing the shape. For this purpose, a dimension converter for an SE implemented as a combination of low pass filter and first-order linear interpolator is used, as shown in Figure 10.11. The coefficients for the 65th-order oversampling filter, $coef[i]$, are obtained from a *sinc (sinx/x)* multiplied by a raised cosine window, as shown below:

$$coef[i] = \frac{\sin \pi(i-32)/8}{\pi(i-32)/8}\left(0.5 - 0.5\cos 2\pi i/64\right) \qquad 0 \le i \le 64$$

The interpolation can be implemented efficiently using polyphase filters [Croc83]. For each offset k with respect to the original sampling grid, the coefficients of the corresponding polyphase filter of length 8 are shown here:

$$c_k[j] = coef[8 \times j + k] \qquad (1 \le k \le 7, \quad 0 \le j \le 7)$$

In the second stage, first-order linear interpolation is applied to the eight times oversampled data to compute the spectral magnitudes at fixed harmonic frequencies $n \times \omega_0$ $(1 \le n \le N)$; ω_0 is the fundamental frequency or frequency spacing of the components in the fixed dimension vector. The first-stage oversampling FIR filtering allows decimated computation in which only the points used at the second-stage linear interpolation are computed; explicitly, these are the left and right adjacent points of frequencies $n \times \omega_0$ $(1 \le n \le N)$. In this way, a spectral vector of fixed dimension N is obtained.

Scalable Vector Quantization Scheme for Harmonic Spectral Magnitudes   A fixed-dimension spectral vector is then quantized. Figure 10.12 shows the structure of the quantization scheme, and the VQ codebook dimensions and sizes are shown in Table 10.6. In order to reduce the memory requirements and search complexity while maintaining high performance, a two-stage VQ scheme is employed for the spectral shape together with a scalar quantizer for the gain—all this for 2-kbit/s coding.

The following weighted distortion measure D is used for the codebook search of both shape and gain:

$$D = \left\| \mathbf{WH}(\mathbf{x} - g(\mathbf{s}_0 + \mathbf{s}_1)) \right\|^2,$$

where $\mathbf{x}$ is the dimension-converted harmonic spectral magnitudes with fixed vector dimension (44), $\mathbf{s}_0$ is the output of the SE shape0 codebook, $\mathbf{s}_1$ is the out-

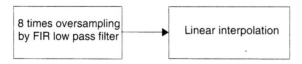

Fig. 10.11 Dimension conversion by a combination of oversampling filter and linear interpolation.

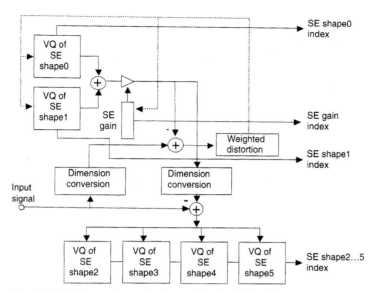

Fig. 10.12 Harmonic spectral magnitudes VQ scheme.

Table 10.6 Codebook size and dimension

VQ codebooks for 2-kbit/s mode	Sizes	4-bit + 4-bit shape and 5-bit gain		
	Dimension	44		
Split VQ codebooks for extension from 2- to 4-kbit/s mode	Sizes	7 bit	10 bit	9 bit	6 bit
	Dimension	2	4	4	4

put of the SE shape1 codebook, and g is the output of the SE gain codebook. The diagonal components of the matrices **H** and **W** are the frequency responses of the magnitudes of the LPC synthesis filter $h(z)$ and the perceptual weighting filter $w(z)$, respectively. For MPEG-4, the codebook design described in [Nish95] was used, and the fast search method for the two-stage VQ described in [Nish97b] is suggested as an efficient encoding strategy.

For the 4-kbit/s mode, another quantization stage is used in addition to the operation shown previously. In this stage, the quantized harmonic magnitudes from the 2-kbit/s mode with fixed dimension (44) are first converted to the dimension of the original harmonics by the same band-limited interpolation method as used before. The difference between the original harmonics and the dequantized and dimension-recovered harmonics is computed. This quantization error vector, having the original dimension, is then quantized by the additional vector quantizers. In this VQ, however, only the components corresponding to the lower 14 harmonics are quantized using split VQ [Gers92]. The codebooks used in this stage are SE shape2, SE shape3, SE shape4, and SE shape5, as shown in Figure 10.12. The dimensions and sizes of the VQ codebooks are shown in Table 10.6. For the codebook search of the split VQ, the same weights used in the first two-stage VQ are employed. This multi-stage structure allows the generation of scalable bitstreams, because the quantizer output of the 2-kbit/s mode is always used regardless of the use of vector quantizers in the additional stage for the 4-kbit/s mode.

10.4.1.2 Vector Excitation Coding for Unvoiced Segments   Another key feature of the HVXC encoder is the VXC algorithm. When a speech segment is unvoiced, the VXC algorithm is used. Figure 10.13 shows the overall structure of the VXC encoder.

Encoding Process   LPC coefficients α_n computed by the LPC analysis block are converted to LSP parameters and quantized, and the quantized LSPs are converted back to LPC coefficients; quantized LPC parameters $\hat{\alpha}_n$ are obtained in this manner. The perceptually weighted LPC synthesis filter $H_w(z)$ is formed as shown below:

$$H_w(z) = \frac{W_{vxc}(z)}{A_q(z)}$$

where $1/A_q(z)$ is the transfer function of the LPC synthesis filter with $\hat{\alpha}_n$, and $W_{vxc}(z)$ is a perceptual weighting filter derived from the LPC coefficients.

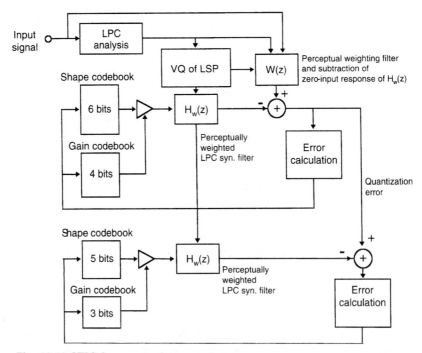

Fig. 10.13 VXC for unvoiced segments.

Let $x_w(n)$ be the perceptually weighted input signal. Subtracting the zero-input response $z(n)$ from $x_w(n)$, a reference signal $r(n)$ for the analysis-by-synthesis procedure of the VXC encoder is obtained. Optimal shape and gain vectors are searched using the distortion measure E:

$$E = \sum_{n=0}^{N-1} [r(n) - g \times syn(n)]^2$$

where $syn(n)$ is the zero-state response of $H_w(z)$, driven by a VXC shape vector $s(n)$, which is an output of the VXC shape codebook; g is a gain, which is an output of the VXC gain codebook; N is the vector dimension of the shape codebook. The codebook search process for the VXC coding consists of two steps:

1. Search $s(n)$ that maximizes $E_s = \dfrac{\displaystyle\sum_{n=0}^{N-1} r(n) \times syn(n)}{\sqrt{\displaystyle\sum_{m=0}^{N-1} syn(m)^2}}$

2. Search g that minimizes $E_g = (g_{ref} - g)^2$ where $g_{ref} = \dfrac{\displaystyle\sum_{n=0}^{N-1} r(n) \times syn(n)}{\displaystyle\sum_{m=0}^{N-1} syn(m)^2}$

The quantization error $e(n)$ is then computed as shown below:

$$e(n) = r(n) - g \times syn(n)$$

When the bit rate is 4 kbit/s, one more stage is used for the quantization of unvoiced segments, and $e(n)$ is used as the reference input to the second stage VQ. The operation of the second-stage VQ is the same as that of the first-stage VQ.

The 2-kbit/s coder uses 6-bit shape and 4-bit gain codebooks for the unvoiced excitation, every 10 ms. The 4-kbit/s scheme adds 5-bit shape and 3-bit gain codebooks, every 5 ms, to quantize the quantization error of the 2-kbit/s mode. Both configurations are shown in Table 10.7.

Table 10.7 VXC codebook configurations

1st stage	(dimension 80, 6-bit shape + 4-bit gain) / 10 ms
2d stage	(dimension 40, 5-bit shape + 3-bit gain) / 5 ms

10.4.2 HVXC Decoder

In this section, an overview of the MPEG-4 HVXC decoder and its key features (such as fast harmonic synthesis and pitch/speed control algorithm) is given. Figure 10.14 shows the overall structure of the HVXC decoder. The basic decoding process is composed of four steps: dequantization of parameters, generation of excitation signals for voiced frames by sinusoidal synthesis (harmonic synthesis) and noise component addition, generation of excitation signals for unvoiced frames by codebook look-up, and LPC synthesis. To enhance the synthesized speech quality, a postfilter is used.

For voiced frames, a fixed-dimension harmonic spectral vector (obtained by dequantization of the spectral magnitude) is first converted to a vector having the original dimension, which varies frame by frame in accordance with the pitch value. This is done by the dimension converter in which a band-limited interpolator generates a set of spectral magnitude values at harmonic frequencies, without changing the shape of the SE [Nish93]. The structure of the dimension converter used in the decoder is the same as the one used in the encoder (see Figure 10.11). Using the spectral magnitude values at harmonic frequencies, a time domain excitation signal is generated by the fast harmonic synthesis algorithm, using an IFFT [Nish95]. In order to make the synthesized speech sound natural, an additional noise component is used. A spectral component of Gaussian noise, covering the frequency range of around 2 to 3.8 kHz, is colored in accordance with the harmonic spectral magnitudes in the frequency domain, and its Inverse Discrete Fourier Transform (IDFT) is added to voiced excitation signals in the time domain. The amount and bandwidth of this additive noise is controlled by the 2-bit V/UV value set by the encoder, which is derived based on the normalized maximum auto-correlation of the LPC residual signal. The harmonic excitation signal for voiced segments, including the added noise, is then fed into the LPC synthesis filter, followed by the postfilter.

For unvoiced segments, the usual VXC decoding algorithm is used, in which an excitation signal is generated by multiplying the gain value by the

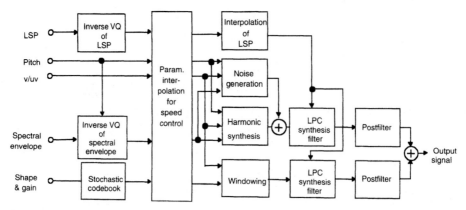

Fig. 10.14 HVXC decoder.

shape codevector. The result is then fed into the LPC synthesis filter followed by the postfilter. Finally, the synthesized speech components for voiced and unvoiced segments are added in the time domain to form the output signal.

10.4.2.1 Fast Harmonic Synthesis by IFFT
One drawback of harmonic coding is the high complexity of the synthesizer. If the voiced output, $v(n)$, is computed directly from this equation:

$$v(n) = \sum_m A_m(n) \cos[\theta_m(n)] \qquad (0 \le m < M, \ 0 \le n < N_0), \qquad (10.1)$$

with interpolated magnitudes $A_m(n)$ and phases $\theta_m(n)$ [Grif88], in which n is a discrete time index and m is a harmonic index, then the complexity of the implementation is of the order of $\gamma N_0 M$; γ is a constant related to the interpolation of magnitude and phase; N_0 is a frame interval (in samples); and M is the maximum number of harmonics. Typically, $N_0 = 160$, $M = 64$, and $\gamma = 5$. In MPEG-4 HVXC [MPEG4-3], in order to reduce the complexity of this process, a fast synthesis method [Nish95] using an IFFT and sampling rate conversion is specified, in which $A_m(n)$ and $\theta_m(n)$ are linearly interpolated.

Suppose that at the kth frame there is a spectrum with M_1 harmonics, with the magnitude of each being $A_m(0 \le m < M_1)$. The pitch lag expressed in terms of the number of samples is $2M_1$. Appending zeros to this array $A_m(n)$ yields a new array, with 2^b components ranging from 0 to π. The number b can be arbitrarily chosen, so that $M \le 2^b$; typically, $b = 6$ is used. The same processing is done on the array of phase data. The phase data used here are generated from those of the previous frame, assuming that the fundamental frequency is linearly interpolated; this is described as phase prediction in [Grif88]. A 2^{b+1}-point IFFT is applied to these arrays of magnitude and phase data, with the constraint that the results be real numbers.

With this operation, an oversampled version of the time domain waveform over a one-pitch period is obtained. Let this be $w_1(i)$ $(0 \le i \le 2^{b+1})$. Thus, 2^{b+1} points are used to express the one-pitch period of the waveform, whereas the actual pitch is $2M_1$, as the oversampling ratio ov_1 is

$$ov_1 = 2^b / M_1.$$

Similarly, another one-pitch period of the waveform at the $k + 1^{th}$ frame may be obtained, which has an oversampling ratio of

$$ov_2 = 2^b / M_2$$

where the pitch lag is $2M_2$. Let this waveform be $w_2(i)$ $(0 \le i \le 2^{b+1})$. Here the function $f(n)$, which maps the time index n from the original sampling version to the oversampled version, is defined under the condition that the pitch is linearly interpolated to be

$$f(n) = \int_0^n (ov_1 \frac{N_0 - t}{N_0} + ov_2 \frac{t}{N_0}) dt.$$

The number of oversampled data needed to reconstruct a waveform of length N_0 at the original sampling rate is at most L:

$$L = nint[f(N_0)] = nint[\frac{N_0}{2}(ov_1 + ov_2)],$$

where $nint(x)$ returns the nearest integer to x. Cyclically extending $w_1(i)$ and $w_2(i)$, the waveforms $\tilde{w}_1(l)$ and $\tilde{w}_2(l)$ of length L are obtained:

$$\tilde{w}_1(l) = w_1[\mod(l, 2^{b+1})] \qquad (0 \le l < L),$$

$$\tilde{w}_2(l) = w_2[\mod(\textit{offset} + l, 2^{b+1})] \quad (0 \le l < L),$$

where

$$\textit{offset} = 2^{b+1} - \mod(L, 2^{b+1}),$$

and $\mod(x, y)$ returns the remainder of x divided by y.

These two waveforms, $\tilde{w}_1(l)$ and $\tilde{w}_2(l)$, from the spectra of the kth and $k + 1^{\text{th}}$ frames, have the same *pseudo* pitch (2^{b+1}) and are aligned. So, simply adding these two waveforms using appropriate weights produces the result $w(l)$:

$$w(l) = \frac{L-1}{L}\tilde{w}_1(l) + \frac{1}{L}\tilde{w}_2(l) \quad (0 \le l < L),$$

where each A_m is linearly interpolated between the adjacent frames. Finally, $w(l)$ has to be resampled so that the resulting waveform can be expressed at the original uniform sampling grid. This operation brings the waveform back from the *pseudo* pitch domain to the real pitch domain, as well. In principle, the resampling operation is just:

$$v'(n) = w(f(n)) \quad (0 \le n < N_0),$$

Usually $f(n)$ does not return integer values. So, $v'(n)$ is obtained by linearly interpolating $w[\lceil f(n) \rceil]$ and $w[\lfloor f(n) \rfloor]$, where $\lceil x \rceil$ and $\lfloor x \rfloor$ denote the smallest integer greater than or equal to x, and the largest integer less than or equal to x, respectively.

$v'(n)$ is a good approximation of $v(n)$ in Eq. (10.1), reducing the complexity of the process to the order of magnitude of $\alpha 2^b(b + 1) + \beta N$. α and β are constants related to the IFFT and linear interpolation, respectively; typically, $\alpha = 7$ and $\beta = 12$ and thus the complexity is less than one-tenth of the direct synthesis method.

10.4.2.2 Time Scale Modification One of the most important HVXC features is its speed-control capability. The HVXC decoder has a scheme for parameter interpolation to generate the parameters at any arbitrary time instant. A sequence of encoded parameters with modified intervals is applied to the speech synthesizer to generate speech with a modified time scale. The operation of this time scale modification block is described below [Nish99].

The arrays of original and interpolated parameters are denoted as $param[n]$ and $mdf\_param[m]$, respectively, wherein n and m are the time indices (frame number) before and after the time scale modification. The frame intervals are both 20 ms. Parameters represented here are pitch, LSP, and residual spectrum. Let us define the ratio of speed change as spd:

$$spd = N_1/N_2$$

where N_1 is the duration of the original speech and N_2 is the duration of the speed-controlled speech, with $0 \leq n < N_1$ and $0 \leq m < N_2$. The time-scale-modified parameters are expressed as follows:

$$mdf\_param[m] = param[m \times spd] \tag{10.2}$$

In general, however, $m \times spd$ is not an integer number. So it is necessary to define

$$\begin{cases} fr_0 = [m \times spd] - 1 \\ fr_1 = fr_0 + 1 \end{cases}$$

to generate parameters at a time index $m \times spd$ by linearly interpolating the parameters at time (frame) indices fr_0 and fr_1. In order to execute the linear interpolation, let us define:

$$\begin{cases} left = m \times spd - fr_0 \\ right = fr_1 - m \times spd \end{cases}$$

Then equation (10.2) can be approximated this way:

$$mdf\_param[m] = param[fr_0] \times right + param[fr_1] \times left \tag{10.3}$$

Because the excitation signal at the VXC decoder is a time domain waveform, the interpolation method proposed by equation (10.3) cannot be used. Thus, one has to take one frame (160 samples) of excitation samples from the original parameters $param[n]$ centered around the time $m \times spd$ and compute the energy over the frame (160 samples); this energy is denoted as E. Gaussian noise consisting of 160 samples is then generated, and its gain is adjusted, so that its energy is equal to E. This gain-adjusted Gaussian noise sequence is used for the time-scale-modified VXC excitation as $mdf\_param[m]$.

Depending on the V/UV decision at fr_0 and fr_1, the interpolation strategy varies in the following way:

☞ **voiced -> voiced**

All of the parameters are interpolated as in equation (10.3).

☞ **unvoiced -> unvoiced**

All of the parameters are interpolated as in equation (10.3); the VXC excitation centered around $m \times spd$ is generated as previously described.

☞ **voiced -> unvoiced**

If *left* < *right*: All of the parameters of frame fr_0 are used.

If *left* ≥ *right*: All of the parameters of frame fr_1 are used; the VXC excitation centered around fr_1 is generated in the same manner as described above.

☞ **unvoiced -> voiced**

If *left* < *right*: All of the parameters of frame fr_0 are used; the VXC excitation centered around fr_0 is generated in the same manner as described above.

If *left* ≥ *right*: All of the parameters of frame fr_1 are used.

In this manner, all of the necessary parameters for the HVXC decoder are generated. By applying these modified parameters, $mdf\_param[m]$, to the speech synthesizer in the same way as in the usual decoding process, the time-scale-modified output is obtained. Apparently, when $N_2 < N_1$, speed-up decoding results, and when $N_2 > N_1$, speed-down decoding results. Power spectrum and pitch are not affected by this speed control, meaning that good-quality speech for speed-control factors of about $0.5 < spd < 2.0$ may be obtained.

10.4.2.3 Pitch Modification   Another important HVXC feature is the pitch-modification functionality, in which the pitch of the synthesized speech can be altered during decoding. In the regular decoding process, the fixed dimension harmonic spectral vector is converted to the one having original dimension by the dimension converter, as described in Section 10.4.1.1. Pitch modification is carried out by simply modifying the target pitch frequency for the dimension conversion. Figure 10.15 shows an example of the pitch-up decoding, in which the spacing between the harmonics (pitch frequency) is widened with respect to normal decoding without altering the shape of the SE.

10.4.3 Variable Bit-Rate Coding

HVXC supports a variable bit-rate coding mode in which the average bit rate is reduced by using the varying voice activity of the input speech signal. Bit assignment is varied depending on V/UV decisions, and bit-rate saving is

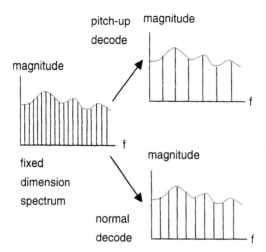

Fig. 10.15 Dimension conversion of harmonic spectral magnitudes.

obtained mostly by reducing the bit assignment for background noise and unvoiced speech frames. When the unvoiced mode is selected, it is checked whether the frame is *real unvoiced speech* or *background noise* (i.e., non-speech). If the frame is declared to be *background noise,* then bit assignment to the frame is further reduced. In the *background noise* mode, only the two V/UV mode bits (signaling a voiced, unvoiced, mixed voiced, or background noise decision) or a *noise update frame* is transmitted, depending on the change of the background noise characteristics. In the variable bit-rate coding mode, the interpretation of two V/UV mode bits indicating V/UV decision and amount of additive noise (see Section 10.4.2) is altered so that one of the additive noise modes is used to indicate *background noise.* Using this variable rate mode, the average bit rate is reduced to between 50% and 85% of the fixed bit-rate mode, depending on the input speech signal. Details of the operation of the HVXC variable rate coding are given in [MPEG4-3].

10.5 ERROR ROBUSTNESS

One of the important new features of the MPEG-4 Audio standard [MPEG4-3] is that it provides error-robustness tools that allow improved performance on error-prone transmission channels. These tools comprise the following:

☞ Error-resilience coding tools for some *natural speech* and *natural audio* codecs,

☞ Error-resilient (reordered) bitstream payload syntax,

☞ An error-protection tool, and

☞ Error-concealment tools (informative).

The error-resilience tools and the error-resilient bitstream syntax are defined normatively for the corresponding audio object types (see Chapter 13); the error-protection tool is defined in a normative way, but its use is optional; the error-concealment tools are described in the informative part of the MPEG-4 Audio standard and thus are not normative. A brief introduction to these MPEG-4 Audio tools is given in the following sections.

Error-Resilience Coding Tools   The virtual codebooks tool (VCB11), the reversible variable length coding tool (RVLC), and the Huffman codeword reordering (HCR) for advanced audio coding (AAC; see Chapter 11) are typical examples of codec-specific error-resilience tools [MPEG4-3]. These tools improve the perceived audio quality of the decoded audio signal in case of corrupted bitstreams, which may occur for noisy transmission channels. For speech coders, quantizer codebooks are designed to be robust to transmission errors; thus, no special additional tools are required.

Error-Resilient Bitstream Reordering   Error-resilient bitstream reordering allows the effective use of advanced channel coding techniques, such as unequal error protection (UEP), which can be perfectly adapted to the needs of the different source coding tools. The basic idea is to rearrange the general audio or speech frame coded data, depending on its error sensitivity in one or more segments belonging to different error-sensitivity categories (ESCs). This rearrangement can be either data element-wise or even bit-wise, depending on the distribution of the error sensitivity over the coded data. An error-resilient frame is built by concatenating these segments. This new arranged bitstream is channel-coded with, for example, the EP tool presented in the following section, taking benefit of the existence of bitstream segments with different error sensitivity, transmitted and channel decoded. Prior to general audio or speech decoding, the bitstream is rearranged to its original order.

Error-Protection Tool   The error-protection (EP) tool provides UEP for MPEG-4 Audio streams (general audio and speech) in conjunction with the error-resilient reordered bitstream payload. UEP is an efficient method to improve the error robustness of source coding schemes. The UEP encoder receives several classes of bits from the audio and speech coding tools, and then applies forward-error-correction codes (FEC) and/or cyclic redundancy codes (CRC) to each class, giving better protection to more error-sensitive bits. Main features of this tool are these [MPEG4-3]:

- ☞ Providing a set of error-correcting/detecting codes with performance and redundancy selectable in a wide range;
- ☞ Providing a generic and bandwidth-efficient error protection framework, which covers both fixed-length frame bitstreams and variable-length frame bitstreams;
- ☞ Providing a UEP configuration control with low overhead.

Error-Concealment Tools To further increase the perceived audio quality of decoded signal with channel errors, optional (non-normative) error-concealment tools can be used in the decoders. The concealment strategies depend on the source coding algorithm; therefore, they are optimized for each of the source coders. An example of a concealment algorithm for HVXC is described in Section 10.5.1.3 [MPEG4-3].

10.5.1 Error-Resilient HVXC

This section describes the HVXC error-resilient reordered syntax, an example configuration of the EP tool for 2-kbit/s HVXC source coding, and an error-concealment tool for HVXC source coding; this syntax and these tools compose the error-robustness tools for HVXC. In the error-robust HVXC scheme, some of the more perceptually important bits are protected by FEC (allowing some degree of correction) and some others are checked by CRC to judge whether erroneous bits were included in the bitstream (allowing only some degree of error detection). When a CRC error is detected, error concealment is applied to reduce the perceptible degradation.

10.5.1.1 Error-Sensitivity Categories In order to apply a UEP scheme, HVXC-encoded bits are reordered and classified into several categories, according to their error sensitivity, with the most error-sensitive bits coming first. Each of these categories is referred to as an ESC, and the reordered bitstream syntax is sometimes called *error-resilient syntax*. For reference, Table 10.8 shows the bit assignments for the coder parameters at 2 kbit/s. The number of bits for each ESC at the 2-kbit/s mode is shown in Table 10.9 and Table 10.10, for voiced and unvoiced frames, respectively. For example, one can read from Table 10.9 that 5 bits of LSP1 are grouped into ESC0, 2 bits of LSP2 are grouped into ESC0, and 5 bits of LSP2 are grouped into ESC3 for voiced frames. For the ESC bit assignments at 4 kbit/s, see [MPEG4-3].

Table 10.8 Bit assignments for 2 kbit/s

Parameter name	Contents	Number of bits
LSP1	LSP index 1	5
LSP2	LSP index 2	7
LSP3	LSP index 3	5
LSP4	LSP index 4	1
VUV	V/UV flag	2
Pitch	Pitch lag index	7
SE_gain	Spectral gain index	5
SE_shape1	Spectral shape index 0	4

Table 10.8 Bit assignments for 2 kbit/s (Continued)

Parameter name	Contents	Number of bits
SE_shape2	Spectral shape index 1	4
VX_shape1[0]	VXC shape index 0	6
VX_shape1[1]	VXC shape index 1	6
VX_gain1[0]	VXC gain index 0	4
VX_gain1[1]	VXC gain index 1	4

Table 10.9 Number of ESC bits for voiced frames (2 kbit/s)

Parameters	ESC0	ESC1	ESC2	ESC3	Total
LSP1	5	—	—	—	**5**
LSP2	2	—	—	5	**7**
LSP3	1	—	—	4	**5**
LSP4	1	—	—	—	**1**
VUV	2	—	—	—	**2**
Pitch	6	—	—	1	**7**
SE_gain	5	—	—	—	**5**
SE_shape1	—	4	—	—	**4**
SE_shape2	—	—	4	—	**4**
Total	**22**	**4**	**4**	**10**	**40**

Table 10.10 Number of ESC bits for unvoiced frames (2 kbit/s)

Parameters	ESC0	ESC1	ESC2	ESC3	Total
LSP1	5	—	—	—	**5**
LSP2	4	3	—	—	**7**
LSP3	2	1	2	—	**5**
LSP4	1	—	—	—	**1**
VUV	2	—	—	—	**2**
VX_gain1[0]	4	—	—	—	**4**
VX_gain1[1]	4	—	—	—	**4**
VX_shape1[0]	—	—	2	4	**6**
VX_shape1[1]	—	—	—	6	**6**
Total	**22**	**4**	**4**	**10**	**40**

10.5.1.2 EP Tool Setting The MPEG-4 Audio EP tool can provide a variety of combinations of multiple (unequal) error protection and detection capabilities, optimized to the different error sensitivity of the bits in single-source coder bitstreams. This is implemented by partitioning the source coder bits into several groups, based on the error sensitivity. The EP tool then applies multiple convolutional coding rates for the FEC, and multiple CRC checkers corresponding to the multiple groups of source coder bits. Each of these groups is called *EP class* and is used to denote the EP classes used by the EP tool. Usually, each ESC is assigned to one EP class, or several ESCs are combined to form a single EP class, as described in the following section. Figure 10.16 shows the functional block diagram of the EP encoder.

Table 10.11 shows an example of bit assignment for the various EP classes, FEC coding rate (ratio of source and channel coding bits), and number of CRC bits when using the EP tool. In this example, a bit-rate setting for a total of 3.5 kbit/s using 2-kbit/s source coding is shown, in which two consecutive source coder frames are processed as one set to form the bits for CRC. The mapping from ESCs to EP classes for this example is as follows:

☞ **EP class 1:** ESC0 bits of both frames
☞ **EP class 2:** ESC1 bits of the first frame

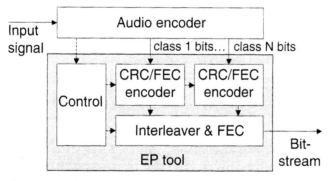

Fig. 10.16 Block diagram of the EP encoder.

Table 10.11 Example of bit assignments when using the EP tool

	EP class 1	EP class 2	EP class 3	EP class 4	EP class 5	EP class 6
Source coder bits	44	4	4	4	4	20
CRC parity bits	6	1	1	1	1	0
FEC coding rate	8/16	8/8	8/8	8/8	8/8	8/8
Total bits in class	100	5	5	5	5	20
Total of bits for all classes	140 (3.5 kbit/s)					

☞ **EP class 3:** ESC2 bits of the first frame

☞ **EP class 4:** ESC1 bits of the second frame

☞ **EP class 5:** ESC2 bits of the second frame

☞ **EP class 6:** ESC3 bits of both frames

A single 6-bit CRC is assigned to EP class 1, and FEC coding also is implemented. A 1-bit CRC is assigned for each of the EP classes 2, 3, 4, and 5. Finally, EP class 6 is protected neither by a CRC nor by FEC.

10.5.1.3 Error Concealment When a CRC error is detected, error-concealment processing (*bad frame masking*) may be carried out to improve the decoded speech quality. Although error concealment is not normative in the MPEG standards in general, the HVXC error-concealment algorithm described here is included as an informative annex to the MPEG-4 Audio standard [MPEG4-3]. This tool is implemented based on the state transition diagram shown in Figure 10.17. The state is updated according to the decoded CRC result for the EP class 1 bits. The initial state is state = 0. An arrow with a character 1 denotes the transition in the case a CRC error is detected, whereas one with a character 0 denotes the transition in the case of no CRC error. The concealment processing for the HVXC LSP parameters and excitation signals using the transition diagram suggested is described in the following section.

LSP Parameters   LSP-concealed parameters are generated depending on these states:

☞ **State 0:** All of the received bits are used, without any concealment processing.

☞ **States 1–6:** The received LSP parameters are replaced with those from the previous frame.

☞ **State 7:** When interframe prediction is not used, the current LSP parameters are used; when interframe prediction is used, LSP parameters are generated by mixing the LSP parameters of the previous frame and those obtained by the output of the first-stage VQ of the current frame. The mixing ratio is controlled depending on the previous states (1–6). The larger the previous state number, the smaller the ratio with which the previous LSP parameters are used.

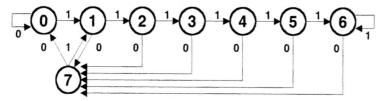

Fig. 10.17 State diagram for HVXC error concealment.

Excitation Signal Depending on the state, the gain of the excitation signal is controlled in seven steps. For higher state numbers, a reduced gain factor multiplies the excitation signal (generated by using the latest error-free parameter values). For state 6, a gain factor equal to 0 is used to completely mute the excitation signal. If the latest error-free frame is voiced, harmonic spectral magnitudes from that frame are used; if the latest error-free frame is unvoiced, the VXC shape vectors are generated by randomly creating index values, and the gain is computed based on the frame gain of the latest error-free frame.

SE_shape1 and SE_shape2 in voiced frames are assigned to separate ESCs (ESC 1 and 2), as can be seen in Table 10.9. In the example described in Section 10.5.1.2, the ESCs from the two consecutive frames are directly mapped (individually, not together) to EP classes 2 to 5. According to Table 10.11, these EP classes are individually checked using a one-bit CRC. In states 0 or 7, when CRC errors in both of the EP classes containing the SE shape data of a frame are detected at the same time, the magnitudes of the lower frequency components of the quantized harmonics are suppressed.

10.5.2 Error-Resilient CELP

For MPEG-4 CELP, error resilience is supported by a normative assignment of the bitstream syntactic elements to five ESCs (ESC0–ESC4) with decreasing significance, so that UEP can be applied. The individual protection parameters for each ESC, however, are not fixed in the standard, in order to allow optimization for different channel conditions. The detailed assignment depends on the CELP operating mode, as can be seen in Table 10.12, Table 10.13, and Table 10.14. Bitstream elements with entries in multiple columns are distributed among the indicated ESCs, with the significance of the bits or components decreasing from left to right. As an example for the narrowband

Table 10.12 ESC assignment for the narrowband MPE mode

Bitstream element	ESC0	ESC1	ESC2	ESC3	ESC4
Fine rate control	x				
V/UV indicator		x			
Frame energy	x		x		x
LSP		x	x	x	x
Subframe pitch lag	x	x	x	x	
Subframe gain			x	x	x
Pulse positions					x
Pulse signs				x	

Table 10.13 ESC assignment for the wideband MPE mode

Bitstream element	ESC0	ESC1	ESC2	ESC3	ESC4
Fine rate control	x				
V/UV indicator		x			
Frame energy	x		x		x
LSP	x	x	x	x	x
Subframe pitch lag		x	x	x	x
Subframe gain			x	x	x
Pulse positions					x
Pulse signs				x	

Table 10.14 ESC assignment for the wideband RPE mode

Bitstream element	ESC0	ESC1	ESC2	ESC3	ESC4
Fine rate control	x				
LSP	x	x	x	x	x
Subframe pitch lag		x	x	x	x
Subframe gain	x		x		x
Pulse data (phase and ampl.)					x

MPE mode, the most significant bit of the subframe pitch lag gets the strongest protection with ESC0, whereas its least significant bits get relatively weak protection with ESC3. Additional assignments for CELP modes using silence compression are defined in the MPEG-4 Audio standard [MPEG4-3].

10.6 SUMMARY

This chapter has described the MPEG-4 tools for coding of *natural speech*— that is, representations derived from sampled speech signals. The required flexibility with respect to bit-rate and quality ranges, speech bandwidth, and other functionalities lead to the support of two coding algorithms: CELP and HVXC.

The results of the verification tests presented in Chapter 16 show that MPEG-4 CELP at 6 kbit/s provides competitive quality to ITU-T G723.1 [G723] at 6.3 kbit/s, and higher quality at higher bit rates that is competitive to the existing standards for the same bit-rate range of telephone band speech, allowing flexible bit-rate controllability and scalability up to 12 kbit/s. Wideband speech coding with 16-kHz sampling rate is supported at 10 to 24 kbit/s, in which quality comparable to ITU T G.722 [G722] 48/56 kbit/s is obtained at around 18 kbit/s [N2424]. Bandwidth scalability (in addition to bit-rate scalability) is supported, in which telephone band decoding is possible using a subset of the bitstream for wideband speech. The operating bit-rate range of the CELP coder, however, starts at about 4 kbit/s—as, for lower bit rates, the quality deteriorates because of the nature of the waveform matching used in the CELP algorithm.

For bit rates of up to 4 kbit/s, HVXC provides near toll-quality speech at 4 kbit/s, and communication-quality speech at 2 kbit/s, outperforming FS1016 CELP [Camp91] at 4.8 kbit/s [N2424]. When the variable bit-rate mode is used, the average bit rate can go down to about 1.2 kbit/s. HVXC also has the functionality of bit-rate scalability, in which 2-kbit/s decoding is possible not only using 2-kbit/s bitstream but also using a subset of a 4-kbit/s bitstream. Additionally, playback speed and pitch can be modified independently, without the need of complex postprocessing operations. Until MPEG-4, there has been no speech-coding standard including such new functionalities, although they are quite useful for fast speech database search, browsing, and some other applications.

Together with the tools for coding of *natural audio* signals, for the representation of *synthetic speech and audio* signals, and for scene description, MPEG-4 provides all components necessary for a powerful object-based representation of audio.

10.7 REFERENCES

[Atal81] Atal, B. S., and J. Remde. "A New Model of LPC Excitation for Producing Natural-Sounding Speech at Low Bit Rates." *Proc. ICASSP-81*, 1981.

[Atal84] Atal, B. S., and M. R. Schroeder. "Stochastic Coding of Speech Signals at Very Low Bit Rates." *Proc. IEEE Int. Conf. on Communications*. Amsterdam, May 1984, p. 48.1.

[Camp91] Campbell, J. P., V. C. Welch, and T. Tremain. "The DOD 4.8 kbps Standard (Proposed Federal Standard 1016)," *Advances in Speech Coding* (Atal et al., Eds.). Dordrecht, Holland: Kluwer Academic Publishers, 1991.

[Croc83] Crochiere, R. E., and L. R. Rabiner. *Multirate Digital Signal Processing*. Englewood Cliffs, NJ: Prentice Hall, 1983.

[G722] ITU-T Recommendation G.722. *7 kHz Audio Coding within 64 kbit/s*. 1988.

[G723] ITU-T Recommendation G.723.1. *Dual Rate Speech Coder for Multimedia Communications Transmitting at 5.3 and 6.3 kbit/s*. 1996.

[Gers92] Gersho, A., and R. M. Gray. *Vector Quantization and Signal Compression*. Boston, MA: Kluwer Academic Publishers, 1992.

[Grif88] Griffin, D. W., and J. S. Lim, "Multiband Excitation Vocoder." *IEEE Trans. ASSP*, 36: 1223–1235. Aug. 1988.

[Itak75] Itakura, F. "Line Spectral Representation of Linear Predictive Coefficients of Speech Signals." *Journal of the Acoustical Society of America*, 57: S35. 1975.

[Jaya84] Jayant, N. S., and P. Noll. *Digital Coding of Waveforms: Principles and Applications to Speech and Video*. Englewood Cliffs, NJ: Prentice Hall, 1984.

[Kroo86] Kroon, P., E. F. Deprettere, and R. J. Sluyter. "Regular-Pulse Excitation: A Novel Approach to Effective and Efficient Multipulse Coding of Speech." *IEEE Trans. ASSP*, 34(5): 1054–1063, 1986.

[M1509] Nomura, T., et al. *Proposal of Compression Algorithm with Rate Control for MPEG-4/Audio Core Experiments*. Doc. ISO/MPEG M1509, Maceió MPEG Meeting, November 1996.

[M2083] Nomura, T., et al. *A Bitrate Scalable Tool for the Narrow Band CELP Coder of the MPEG-4/Audio VM*. Doc. ISO/MPEG M2083, Bristol MPEG Meeting, April 1997.

[M2486] Nomura, T., et al. *An Extension of the Narrow-Band CELP VM Coder to a Bandwidth Scaleable CELP Coder*. Doc. ISO/MPEG M2486, Stockholm MPEG Meeting, July 1997.

[M694] Nomura, T., et al. *Technical Description of the 6 kbps Compression Algorithm for MPEG-4/Audio Submission from NEC*. Doc. ISO/MPEG M694, Florence MPEG Meeting, March 1996.

[M696] Wuppermann, F., and F. de Bont. *Detailed Technical Description for an MPEG-4 Audio Codec with a Bit Rate of 16 kbit/s at a Reference Sampling Frequency of 16 kHz*. Doc. ISO/MPEG M696, Florence MPEG Meeting, March 1996.

[MPEG4-1] ISO/IEC 14496-1:2001, *Coding of Audio-Visual Objects—Part 1: Systems*, 2d Edition, 2001.

[MPEG4-3] ISO/IEC 14496-3:2001, *Coding of Audio-Visual Objects—Part 3: Audio*, 2d Edition, 2001.

[N2424] MPEG Audio. *Report on the MPEG-4 Speech Codec Verification Tests*. Doc. ISO/MPEG N2424, La Baule MPEG Meeting, October 1998.

[Nish97b] Nishiguchi, M., K. Iijima, and J. Matsumoto. "Low Bit Rate Speech Coding by Harmonic Vector Excitation Coding." *Proc. ASJ 1-2-4*, September 1997.

[Nish93] Nishiguchi, M., J. Matsumoto, S. Ono, and R. Wakatsuki. "Vector Quantized MBE with Simplified V/UV Division at 3.0 kbps." *Proc. ICASSP-93*, pp. II-151–154, April 1993.

[Nish95] Nishiguchi, M., and J. Matsumoto. "Harmonic and Noise Coding of LPC Residuals with Classified Vector Quantization." *Proc. ICASSP-95*, pp. I-484–487, May 1995.

[Nish97a] Nishiguchi, M., K. Iijima, and J. Matsumoto. "Harmonic Vector Excitation Coding of Speech at 2.0 kbps." *IEEE Workshop on Speech Coding*, September 1997.

[Nish99] Nishiguchi, M., A. Inoue, Y. Maeda, and J. Matsumoto. "Parametric Speech Coding—HVXC at 2.0–4.0 kbps." *IEEE Workshop on Speech Coding*, June 1999.

[Nomu98] Nomura, T., M. Iwadare, M. Serizawa, and K. Ozawa. "A Bit Rate and Bandwidth Scalable CELP Coder." *Proc. ICASSP-98*, pp. I-341–344, May 1998.

[Rabi78] Rabiner, L. R., and R. W. Schafer. *Digital Processing of Speech Signals*. Englewood Cliffs, NJ: Prentice Hall, 1978.

[RCR94] *Personal Digital Cellular Telecommunication System RCR Standard (RCR STD27)*. Association of Radio Industries and Businesses, Japan. 1994.

[Schr85] Schroeder, M. R., and B. S. Atal. "Code-Excited Linear Predictive (CELP): High-Quality Speech at Very Low Bit Rates." *Proc. ICASSP-85*, pp. 937–940, 1985.

[Tana96] Tanaka, N., et al. "A Multi-mode Variable Rate Speech Coder for CDMA Cellular Systems." *Proc. IEEE VTC*, pp. 198–202, April 1996.

[TIA99] *Mobile Station-Base Station Compatibility Standard for Wideband Spread Spectrum Cellular Systems (ANSI / TIA / EIA-95-B-99)*. Telecommunications Industry Association, 1999.

General Audio Coding

by Jürgen Herre and Heiko Purnhagen

Keywords: natural audio coding, audio coding tools, advanced audio coding, AAC, TwinVQ, T/F coding, transform coding, PNS, LTP, low-delay audio coding, parametric audio coding, HILN, BSAC, scalable audio coding, error resilience, error robustness

*T*his chapter introduces the concepts and tools behind the *MPEG-4 General Audio* coding technology—that is, the coding algorithms within the MPEG-4 natural audio framework that are not targeted at specific types of audio signals but aim at the faithful reproduction of all types of input audio signals. This implies that encoding has to be done in a flexible way rather than relying on a specific source model.

Originally, the term *general audio coding* was created to refer to MPEG-4 audio coders based on the coding of spectral components derived from an analysis filterbank, the so-called *time/frequency (T/F) coders.* In a wider sense, the MPEG-4 parametric audio coder also can be considered a general audio coder, as it aims at the good reproduction of arbitrary input audio signals by means of a flexible decomposition of the input signal into distinct sound components. This chapter will cover both technologies—T/F and parametric coders—in terms of their basic concepts and the actual specification in the context of the MPEG-4 Audio standard [MPEG4-3].

Starting with T/F-based coding, the basic concepts underpinning this type of coder will be discussed briefly. Because most parts of the MPEG-4 T/F coder are based on MPEG-2 advanced audio coding (AAC) technology, this coder will be explained. Building on the MPEG-2 AAC technology, MPEG-4 defines a number of extensions to enhance compression performance (perceptual noise substitution, long-term prediction) and enable operation at extremely low bit rates (TwinVQ), very low delays (low-delay AAC), and under error-prone transmission conditions (error-resilience tools). The principles and algorithms involved in these extensions will be described one by one. One major novel functionality that was not supported by any previous MPEG Audio standard is *bit-rate scalability*. This aspect will be covered in a separate section by discussing the basic principles behind scalability and describing the two different approaches *MPEG-4 Audio offers for realizing this concept—that is, large-step* scalable audio coding and bit-sliced arithmetic coding. Finally, an introduction into parametric audio coding techniques and a section on the MPEG-4 Harmonic and Individual Lines plus Noise (HILN) parametric audio coder will conclude the description of MPEG-4 general audio coding technology.

11.1 INTRODUCTION TO TIME/FREQUENCY AUDIO CODING

The term *T/F coder* was chosen in MPEG-4 to refer to coders that adhere to the traditional paradigm of perceptual audio coding by coding a spectral (frequency domain) representation of the input signal rather than the time domain signal itself. This type of coding technology has made tremendous progress in the past 10 years and has become the coder type of choice for music distribution in broadcasting, over the Internet, and on other media. This may be explained by the fact that the T/F coder framework combines both redundancy reduction and exploitation of the potential provided by irrelevancy removal. Coding of a spectral representation is an efficient way of exploiting linear correlation between subsequent samples of an input signal (i.e., the signal's spectral *unflatness*) [JayN84]. Further redundancy removal can be achieved through entropy coding of the spectral coefficients. A further advantage of a spectral representation of the input signal is that it makes possible good first-order modeling of the limits of human auditory perception. Once the threshold of perceptibility (audibility) is estimated, the precision of the coding process can be easily adapted to match that threshold by adjusting the quantization distortion of the corresponding spectral coefficients.

Although the development of T/F based coding dates back all the way to the 1970s, an important step in its evolution was marked by the development of the MPEG-1 and MPEG-2 Audio standards [MPEG1-3, MPEG2-3]; these standards prepared the way for the large-scale application of this technology. The upper part of Figure 11.1 depicts the basic block diagram of a generic monophonic T/F audio coder:

☞ The input signal is mapped to a subsampled spectral representation using various types of *analysis filterbanks*. For reasons of coding efficiency, modern coding schemes typically employ filterbanks with critical sampling (i.e., same number of input samples and spectral coefficients) and overlapping analysis windows between subsequent analysis frames.

Examples include the Modified Discrete Cosine Transform (MDCT) [PrJB87], polyphase filterbanks [Rotw83], or hybrid structures [BEHE92]. A tutorial on common filterbanks for perceptual audio coding can be found in [Bosi99].

☞ The signal's time- and frequency-dependent threshold of perceptibility (masking threshold) is estimated by a *perceptual model*. This threshold describes the maximum quantization error that can be introduced into the audio signal while still maintaining perceptually unimpaired signal quality. Typically, perceptual models for T/F audio coding consider psychoacoustic effects such as frequency domain masking (inter- and intra-band masking), temporal masking effects (postmasking), and the *asymmetry of masking* between tonal and noiselike stimuli [Helm72]. Also, modeling typically is carried out considering the intrinsic frequency selectivity of the human auditory system, as described by the concept of *critical bands* and the *Bark* scale [Moor89, ZwiF90]. Typical modeling strategies can be found, for example, in Annex D of [MPEG1-3] or in the literature on perceptual measurements [BraS92, BeeS92].

☞ The spectral values are *quantized* and *coded* with a precision corresponding to the masking threshold estimate. In this way, the quantization noise is hidden (masked) by the respective coded signal and is thus not perceptible after decoding. Both uniform and nonuniform scalar quantization have been used in the context of perceptual audio coding [BSDJ92]. Additional reduction of redundancy can be achieved by employing entropy coding techniques, such as Huffman coding. Usually, the quantization/coding kernel of an audio encoder operates under two constraints: Although the estimated masking threshold defines a target for the minimum precision of the coded signal representation, coders frequently are operating under the limitations of a fixed bit rate selected by the user. As it may not be possible to satisfy both constraints at the same time, a good encoding algorithm has to produce an encoding solution representing an acceptable compromise with respect to perceived audio quality in such cases. Besides the perceptual modeling aspect, the strategies for *bit allocation* or *noise allocation* are essential elements of an optimized audio coder. A tutorial on issues of quantization/coding strategies can be found in [Herr99].

☞ Finally, all relevant information (i.e., the coded spectral values and additional side information) is packed into a bitstream and transmitted to the decoder.

Correspondingly, the processing steps appear in reverse order in the decoder (see the lower part of Figure 11.1). The bitstream is decoded and parsed into coded spectral data and side information. The inverse quantization of the quantized spectral coefficients is then carried out. Finally, the spectral values are mapped back to a time domain representation using a synthesis filterbank.

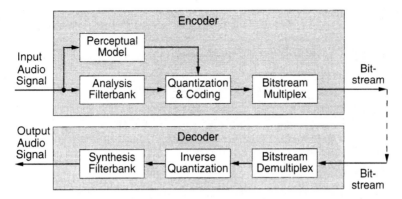

Fig. 11.1 Generic architecture of a perceptual T/F audio encoder and decoder.

Although almost all common T/F based audio coders are based on this generic structure, practical coders are extended by a number of optional building blocks (*coding tools*) that provide further improvements in coding performance for specific coding configurations. These include provisions for joint coding of stereo signals and enhanced exploitation of redundancy (e.g., by means of predictors) or irrelevance (by enhancing the coder's noise shaping abilities).

11.2 MPEG-2 ADVANCED AUDIO CODING

After the completion of the MPEG-2 multichannel audio standard in 1994 [MPEG2-3], the MPEG Audio subgroup started another standardization effort to define a multichannel coding standard allowing a higher quality than is achievable while requiring MPEG-1 backward compatibility, as was the case for the first MPEG-2 multichannel audio standard. The aim of this development, initially called the MPEG-2 *Non-Backward Compatible* (NBC) coding scheme [BBQF97], was to reach *indistinguishable quality* according to the EBU definition [N1419] at bit rates of 384 kbit/s or lower for five full bandwidth channel signals. The standardization process was successfully completed in 1997, renamed MPEG-2 Advanced Audio Coding (MPEG-2 AAC) subsequently, and became Part 7 of the MPEG-2 standard (ISO/IEC 13818-7) [AAC].

In formal verification tests, MPEG-2 AAC demonstrated near-transparent subjective audio quality at a bit rate of 256 to 320 kbit/s for five channels [KirW97] and at 96 to 128 kbit/s for stereophonic signals [N2006]. Although originally designed for near-transparent audio coding, testing inside MPEG revealed that the coder exhibits excellent performance also at very low bit rates down to 16 kbit/s. As a result, MPEG-2 AAC was adopted as the core of the MPEG-4 General Audio (T/F) coder [MPEG4-3].

11.2.1 Coder Overview

The MPEG-2 AAC algorithm [BBQF97, AAC] is a good example of a modern audio coding algorithm that is equipped with a number of coding tools, including joint stereo coding and different kinds of predictive coding. Figure 11.2 provides an overview of an MPEG-2 AAC encoder and its building blocks, the *coding tools*.

In terms of overall approach and structure, some commonalities between the MPEG-2 AAC coder and the MPEG-1/2 Layer-3 coder can be observed in that both schemes employ a switched filterbank providing a high-frequency resolution, a nonuniform power-law quantizer, and Huffman code–based entropy coding. Beyond these commonalities, the MPEG-2 AAC coder includes a considerable number of novel coding tools to increase the coder flexibility and performance. Furthermore, the MPEG-2 AAC standard defines three profiles, which correspond to different configurations of the basic coding scheme providing different trade-off options between coding performance and complexity:

☞ **Low-Complexity (LC) profile:** Defines a baseline coder that is both efficient in coding and has moderate complexity (no interframe prediction is used, the maximum temporal noise shaping [TNS] filter order is limited to 12).

☞ **Main profile:** Does not carry the preceding restrictions and delivers somewhat higher compression performance at the expense of higher memory and computational demands. Because the Main profile is a true superset of the LC profile, all LC profile bitstreams can be decoded by a Main profile decoder.

☞ **Scalable Sampling Rate (SSR) profile:** Can provide decoder configurations with even lower complexity than the LC profile; if not, the entire audio bandwidth is decoded. This is achieved by using a preprocessing stage (including a first filterbank and the gain control stage) in combination with a filterbank of modified length. Only partial compatibility is achieved with the LC profile.[1]

The following sections of this chapter will discuss the most important building blocks of the AAC coder. This includes gain control, filterbank, prediction, quantization, noiseless coding, TNS, prediction, mid/side (M/S) stereo coding, intensity stereo coding, and bitstream multiplexing.

11.2.2 Gain Control

The gain control (preprocessing) module is used exclusively by the MPEG-2 AAC SSR profile as an additional block of the input stage of the encoder. The

---

1. LC profile content can be decoded by an SSR profile decoder, but the bandwidth of the decoded signal will be limited to approximately 5 kHz, corresponding to the nonaliased portion of the lowest band of the first filterbank.

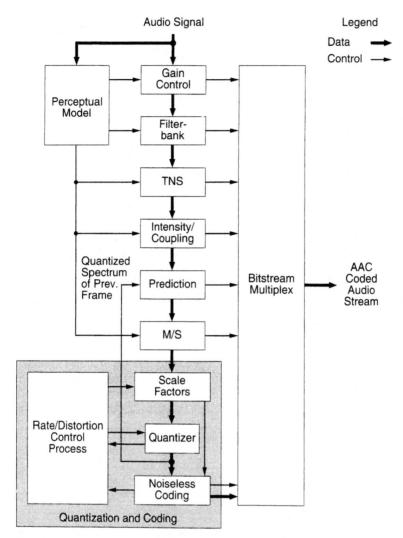

Fig. 11.2 Overview of the MPEG-2 AAC encoder [AAC].

module includes a polyphase quadrature filterbank (PQF), gain detectors, and gain modifiers. Each audio channel input is split into four frequency bands of equal bandwidth (for a sampling rate of 48 kHz, this corresponds to the bands of 0–6 kHz, 6–12 kHz, 12–18 kHz, and 18–24 kHz). The signals in these bands are examined for rapid changes in signal energy by the gain detectors. Based on the result of this analysis, adjustments of the signal amplitude over time are conducted by the gain modifiers in order to compress the dynamics of the signal. Each preprocessed signal is subsequently passed on to an MDCT filterbank to produce 256 spectral coefficients, resulting in a total of 1,024 spectral coefficients for each input frame of 1,024 samples.

The postprocessing (inverse gain control) in the SSR decoder uses the same components as the encoder preprocessing but arranged in reverse order. Each subset of 256 spectral coefficients is mapped back to the subband representation by four Inverse Modified Discrete Cosine Transform (IMDCT) stages. For each of the four bands, the inverse gain change is carried out to restore the original signal dynamics and the outputs are combined by the inverse PQF filterbank to form the output signal.

The pre/postprocessing stage is designed to reduce the temporal spread of the quantization noise for transient input signals (pre-echo). Because higher PQF bands comprise a smaller perceptual bandwidth (number of critical bands), the effectiveness of the gain control algorithm increases with the number of the PQF bands, allowing a finer noise control at higher frequencies. For this reason, gain control is not used in the lowest band. Instead, the TNS tool is employed for this band, achieving a similar function with the flexible choice of the target frequency range.

The MPEG-2 AAC SSR profile permits the definition of decoders with lower maximum signal bandwidth and complexity by discarding the signal processing for higher PQF bands. In this way, 1-band, 2-band, and 3-band decoders achieve maximum audio bandwidths of 6 kHz, 12 kHz, and 18 kHz, respectively (assuming a sampling rate of 48 kHz). In this way, decoder complexity can be reduced as output bandwidth is decreased. Due to the fact that TNS is used for the lowest band rather than gain control, a 1-band SSR bitstream (limited to a bandwidth of 6 kHz) can be decoded by an LC decoder, and vice versa.

11.2.3 Filterbank

The MPEG-2 AAC encoder employs a high-frequency resolution filterbank to map the time domain input samples to a subsampled spectral representation. More specifically, an MDCT is used which is a perfect reconstruction filterbank relying on the concept of *Time Domain Aliasing Cancellation* (TDAC) [PrJB87]. This filterbank type has been successfully used for audio coding in a number of previous systems [BHJM91, BraS92, John96] due to its favorable properties, such as high-frequency resolution and coding gain, critical subsampling, and the absence of blocking effects. There is an overlap of 50% of the window size between subsequent analysis windows. Due to the critical sampling of the filterbank, the result of mapping back each portion of spectral data by an IMDCT contains temporal aliasing components. The final output signal is then obtained by means of an overlap/add (OLA) operation between adjacent windows, which then leads to the cancellation of these aliasing terms.

In standard operation mode (ONLY_LONG_SEQUENCE[2]), the AAC encoder analyzes input windows of 2,048 samples with a shift length of 1,024 samples

---

2. The terms ONLY_LONG_SEQUENCE, LONG_START_SEQUENCE, EIGHT_SHORT_SEQUENCE, and LONG_STOP_SEQUENCE denote the four different states of the filterbank (i.e., specific window shapes and filterbank resolutions).

between subsequent windows. As a result, the filterbank produces 1,024 spectral coefficients, representing 1,024 uniformly spaced filterbank channels with a frequency resolution of 23.4 Hz (assuming a sampling rate of 48 kHz). This high-frequency resolution allows for a very fine spectral shaping of the quantization noise, which is particularly important in the lower frequency range, where the critical bands are narrower. Also, a high degree of redundancy extraction and coding gain is achieved for stationary tonal signals, for which individual tonal components are resolved and coded efficiently [John96].

On the other hand, coding of transient signal portions with a long window size tends to cause temporal smearing of the quantization noise within the whole window size in such a way that the conditions for temporal masking are not met any longer (the *pre-echo* effect) [JohB92, HerJ96]. This can be alleviated by adaptive switching of the MDCT window size [Edle89] depending on the stationary or transient character of the input signal. In the case of the AAC coder, the filterbank resolution can be switched by an 8:1 ratio, that is, either one set of 1,024 or eight sets of 128 spectral coefficients each are produced per frame (filterbank states ONLY_LONG_SEQUENCE and EIGHT_SHORT_SEQUENCE). Switching between the two resolutions while maintaining perfect reconstruction is achieved via two transition windows (LONG_START_SEQUENCE and LONG_STOP_SEQUENCE) that combine half of a long and a short window, the latter being extended by flat window segments toward a frame size of 1,024 spectral coefficients. Due to the window size, the EIGHT_SHORT_SEQUENCE state also is frequently referred to as *short windows*, whereas all other states are referred to as *long windows*. Figure 11.3 shows a typical window sequence, consisting of the states ONLY_LONG_SEQUENCE,

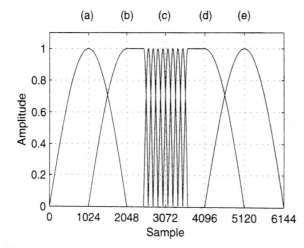

Fig. 11.3 Typical window switching sequence: ONLY_LONG_SEQUENCE (a), LONG_START_SEQUENCE (b), EIGHT_SHORT_SEQUENCE (c), LONG_STOP_SEQUENCE (d), ONLY_LONG_SEQUENCE (e).

LONG_START_SEQUENCE, EIGHT_SHORT_SEQUENCE, LONG_STOP_SEQUENCE, and ONLY_LONG_SEQUENCE.

Another degree of flexibility for the AAC filterbank is provided by the concept of *adaptive window shape selection*. For each frame, the shape of the transform window can be adaptively selected between a sine window and a so-called *Kaiser-Bessel-derived* (KBD) window [FBDD96]. Compared to the sine window, the KBD window provides improved far-off rejection of the resulting filter response at the expense of a somewhat wider main lobe. Figure 11.4 shows the shape and corresponding frequency response of both window types.

It is possible for the encoder to select the optimum window shape, depending on the characteristics of the input signal. In order to maintain perfect reconstruction while changing window shapes, the shape of the left half of each window is always required to match the shape of the right half of the preceding window [Edle89]. Thus, a new window shape is chosen by introducing it as a new right half. Window shape selection is possible for both the long and short window coding modes.

11.2.4 Quantization

This section will focus on describing three normative aspects of the AAC quantization scheme: nonuniform quantization, noise shaping by using scalefactors, and encoding of scalefactors.

11.2.4.1 Nonuniform Quantization AAC uses a nonuniform power-law quantization scheme according to the following equation,

$$ix(i) = \text{sign}(x(i)) \cdot \text{nint}\left[\left(\frac{|x(i)|}{\sqrt[4]{2}^{scale\_factor}} \right)^{0.75} - \alpha \right], \qquad (11.1)$$

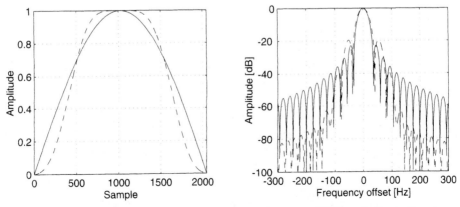

Fig. 11.4 AAC window shapes and frequency responses; solid lines indicate the sine window, whereas dashed lines indicate the KBD window.

where $x(i)$ and $ix(i)$ denote unquantized/quantized values, $sign(\cdot)$ returns the sign $(-1, 0, 1)$ of its argument, $nint(\cdot)$ returns the nearest integer value to the real-valued argument, α is a small constant, and `scale_factor` relates to a quantization resolution parameter (see below).

The principle behind this quantization scheme was inherited from the MPEG-1/2 Layer-3 coder and has the advantage of an intrinsic noise shaping capability even within a group of spectral coefficients sharing a common quantizer setting. Due to the compressive characteristics of the power-law quantizer, smaller values are quantized finer and larger values are represented coarser in comparison to a uniform quantizer. Thus, the quantization noise tends to be shaped under the spectral regions containing larger coefficients, where masking is more effective. The range of quantized values is limited to absolute values not exceeding 8,191.

11.2.4.2 Noise Shaping by Using Scalefactors   A crucial aspect of the perceptual encoding process is its ability to achieve a dynamic shaping of the quantization noise according to perceptual requirements as assessed by the psychoacoustic model. In general, the way to achieve this functionality in MPEG-2 AAC (as well as in MPEG-1/2 Layer-3) is to vary the effective quantizer step size by means of *scalefactors*.

The basic idea is to scale the spectral coefficients prior to passing them through a quantizer with a fixed quantization characteristic and then perform an inverse scaling operation at the decoder end. In AAC, the resolution of the scaling operation can be controlled in increments of 1.5 dB, thus enabling very fine control of the amount of quantization noise introduced. In order to achieve frequency-dependent noise shaping, all 1,024 spectral coefficients are grouped into *scalefactor bands,* with one scalefactor per band. The scalefactor bands were chosen to provide a high resolution of approximately 1/2 Bark, under the constraint that each band would hold an integer multiple of four spectral coefficients (to accommodate entropy coding of quadruples in the noiseless coding stage; see Section 11.2.5) and comprise a maximum number of 32 spectral coefficients. The scalefactor of the first scalefactor band, in relation to which all other scalefactors are expressed, is called `global_gain`.

If the filterbank is switched to higher time resolution for optimum encoding of transient signal portions (using short windows), spectral coefficients are grouped not only in frequency but also in time (i.e., among several of the eight subsequent analysis windows). This window grouping mechanism allows arbitrary numbers of subsequent windows to share scalefactors and thus helps to keep the amount of side information low for the processing of transient signals.

Please note that the meaning of scalefactors is defined with different polarity in MPEG-2 AAC and MPEG-1/2 Layer-3. Whereas larger scalefactor values indicate larger signals in AAC, the opposite is true for the MPEG-1/2 Layer-3 coder.

11.2.4.3 Encoding of Scalefactors
All scalefactors are differentially encoded—that is, the difference between scalefactor bands adjacent in frequency is calculated. The result is coded by a dedicated Huffman code table. Due to this coding scheme, a manipulation of the first scalefactor will effectively change the level of the decoder output signal in steps of 1.5 dB. Consequently, this scalefactor, called global_gain, is coded by means of an extra fixed-length 8-bit PCM field to allow convenient access and change of its value. In order to avoid the transmission of redundant side information, scalefactors belonging to scalefactor bands for which all-zero spectral coefficients have been signaled are omitted in the bitstream.

11.2.5 Noiseless Coding

The *noiseless coding* stage encodes the set of 1,024 quantized spectral coefficients as efficiently as possible by exploiting statistical redundancy without any further reduction in precision [Quac97]. As one simple method of noiseless dynamic range reduction, up to four coefficients can be coded separately by replacing them with a value of +/–1 in the quantized coefficient array, thus carrying the sign information. The magnitudes of these coefficients are coded separately, together with the corresponding frequency indices. This method is not available for use with the high time resolution filterbank mode (short windows).

The general coding mechanism of the noiseless coding stage, however, is based on a flexible entropy coding framework using *sectioning* and *Huffman coding* techniques.

11.2.5.1 Sectioning
As the quantized coefficient statistics vary with frequency and depend on the signal type, a good entropy coding kernel needs to be able to adapt to these conditions. This is addressed by providing a set of 11 Huffman code tables optimized for various statistics (plus a pseudo-table indicating all-zero coefficients).

In principle, an individual Huffman codebook selection could be used to code the coefficients of each scalefactor band. However, in order to reduce the side information overhead, several adjacent scalefactor bands can be grouped to form a *section* and share a common Huffman codebook. Accordingly, a section is described in the bitstream by the number of grouped scalefactor bands plus the index of the Huffman codebook used. In this way, entropy coding of all active scalefactor bands is defined by choosing sectioning parameters. Because this sectioning process provides a high level of freedom, an encoder will search for sectioning parameters that result in an overall minimization of the bit demand needed to encode both the quantized coefficients and the sectioning side information.

11.2.5.2 Huffman Coding
Multidimensional Huffman coding is used to represent n-tuples of quantized spectral coefficients with the Huffman codebook

selected from 1 of 12 predefined codebooks. Both 2-dimensional and 4-dimensional codebooks are available for coding. Table 11.1 shows the index of the Huffman codebook together with its n-tuple size and the largest absolute value that can be encoded with each codebook. For a given largest absolute value, there may be more than one codebook available in order to fit different distribution statistics (codebooks #1 and #2, codebooks #3 and #4, etc.). Furthermore, there are codebooks that encode signed values and codebooks representing only absolute values. For the latter case, the sign information is simply appended to the codeword for each nonzero coefficient in the encoded n-tuple. In order to save codebook storage memory (ROM) at both the encoder and decoder, the larger codebooks are defined using unsigned encoding of pairs of coefficients.

Among the Huffman codebooks listed previously, two tables exhibit special characteristics:

☞ **Codebook #0** signals that all coefficients within this section have a value of zero. Thus, no bits are needed to transmit these coefficients (by Huffman codewords) other than the codebook index. This can be usefully applied, for example, to efficiently code band-limited audio signals. The resulting high-frequency spectral coefficients are zero and thus can be encoded efficiently by a single section using Huffman codebook #0. No scalefactors are transmitted for the scalefactor bands belonging to such sections.

☞ **Codebook #11** is not restricted to coding maximum absolute values of 16 (see Table 11.1), but is able to invoke an escape mechanism to repre-

Table 11.1 AAC Huffman code tables

Codebook index	n-Tuple size	Largest absolute value	Signed / unsigned
0	–	0	
1	4	1	Signed
2	4	1	Signed
3	4	2	Unsigned
4	4	2	Unsigned
5	2	4	Signed
6	2	4	Signed
7	2	7	Unsigned
8	2	7	Unsigned
9	2	12	Unsigned
10	2	12	Unsigned
11	2	16 (ESC)	Unsigned

sent still larger values that can be necessary for certain tonal signals (or at high bit rates). The basic magnitude of the coefficients is limited to 16, and the corresponding 2-tuple is Huffman coded. The sign bits, as needed, are appended to the codeword. For each coefficient magnitude greater than or equal to 16, an escape code is appended, representing the excess value portion of the codeword. In this way, absolute values up to a value of 8,191 can be covered.

11.2.5.3 Grouping/Interleaving

If the filterbank is switched to the EIGHT_SHORT_SEQUENCE state (i.e., to high time resolution mode), then its 1,024 output values consist of a matrix of 8×128 spectral coefficients representing the signal's time-frequency characteristic over the duration of the eight short analysis windows. These coefficients are quantized and need to be efficiently encoded by the noiseless coding stage. As this is done by using section-based entropy coding, it is advantageous to arrange the order of the quantized spectral coefficients such that minimum sectioning overhead is incurred.

As outlined previously, a grouping operation is performed on the sequence of short windows to reduce the demand for side information (scalefactors). Similarly, the grouping scheme is used to interchange the order of the scalefactor bands and windows in the spectral coefficient array before noiseless coding/sectioning is applied. This increases the probability of having coefficients of similar magnitude sorted into contiguous regions of the coefficient array and thus enhances coding efficiency. Also, all zero portions within each group due to band limiting are combined into one section.

11.2.6 Temporal Noise Shaping

TNS represents a novel concept in perceptual audio coding and was first introduced in MPEG-2 AAC [HerJ96]. This coding tool is motivated by the fact that the handling of transient and pitched input signals still presents a challenge for T/F based coding schemes. More specifically, coding of such signals is difficult because of the temporal mismatch between masking threshold and quantization noise (pre-echo problem) [JohB92]. This is due to the fact that the quantization noise will be evenly distributed within each filterbank window, whereas the actual time-dependent masking threshold may vary considerably within that time period. The TNS tool addresses these concerns by allowing a fine temporal shaping of the coder's quantization noise.

11.2.6.1 Principle

The concept underpinning the TNS technique can be characterized by the following two main aspects:

☞ **T/F duality:** The concept of TNS exploits the duality between the time and frequency domains to further extend the known predictive coding techniques. From traditional signal-processing applications, it is well known that signals with a *nonflat* spectral envelope show correlation between subsequent signal samples. Thus, they can be coded efficiently either by directly coding spectral values, *transform coding*, or by applying predictive coding methods to the time signal [JayN84]. Consequently, the corresponding dual statement leads to the coding of signals with an *unflat* time structure—that is, transient signals. Efficient coding of transient signals thus can be achieved either by directly coding time domain values or by applying predictive coding methods to the spectral data. Such a predictive coding of spectral coefficients over frequency, in fact, constitutes the dual concept to the intrachannel prediction tool described later in this chapter. Although intrachannel prediction over time increases the coder's temporal scope, prediction over frequency enhances its temporal resolution.

☞ **Noise shaping by predictive coding:** If a time signal is coded by means of an open-loop (forward) predictive coding technique, the quantization error in the final decoded signal is known to be adapted in its power spectral density (PSD) to the PSD of the input signal [JayN84]. As a dual to this, if predictive coding techniques are applied to spectral data over frequency, the temporal shape of the quantization error signal will appear adapted to the temporal shape of the input signal at the output of the decoder. This effectively adapts the temporal envelope of the quantization noise to the temporal envelope of the signal and, in this way, avoids problems of temporal masking, either in transient or in pitched signals. Therefore, this type of predictive coding of spectral data is referred to as the TNS technique. Because TNS processing can be applied either to the entire spectrum or to only part of the spectrum, the time-domain noise control can be applied in any necessary frequency-dependent fashion. If necessary, it is, for example, possible to use several predictive filters operating on different frequency (coefficient) regions.

A more extensive discussion on the theoretical background of TNS can be found in [HerJ97].

Compared to the standard architecture of a generic perceptual encoder and decoder, the predictive encoding/decoding process over frequency can be realized by adding one building block. This concept is shown, for the encoder, in Figure 11.5. Subsequent to the analysis filterbank, an additional block, called *TNS filtering*, is inserted, which implements an in-place filtering operation on the spectral values—that is, it replaces the set of spectral coefficients to which TNS is applied with the corresponding prediction residual. Filtering with both increasing and decreasing frequency order is possible. The TNS prediction coefficients are determined via a linear predictive coding (LPC) analysis of the spectral coefficients and communicated to the decoder.

Accordingly, the TNS decoding process is achieved by inserting one additional processing block, called *inverse TNS filtering*, immediately before the synthesis filterbank (see Figure 11.6). An inverse in-place filtering operation is performed on the residual spectral values such that the target spectral

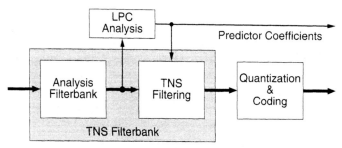

Fig. 11.5 TNS filtering at the encoder.

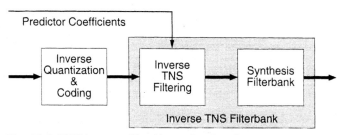

Fig. 11.6 TNS inverse filtering at the decoder.

coefficients are replaced with the decoded spectral coefficients by means of the inverse prediction filter.

The combination of the encoder filterbank and the adaptive (TNS) prediction filter can be interpreted as a compound continuously signal adaptive filterbank. In fact, the behavior of this type of adaptive composite filterbank dynamically shifts between the characteristics of a high-frequency resolution filterbank (for stationary signals) and a high-time resolution filterbank (for transient signals). This technique thus approaches the optimum filterbank structure for a given signal [HeJ97b]. Note that this allows a soft adaptation of the T/F resolution to the properties of the input signal in contrast to the *switched* window switching approach.

11.2.6.2 Specifics For signaling the TNS operating mode to the decoder, the following information is conveyed as side information within the bitstream:

- ☞ A global on/off flag signaling whether the TNS technique is active in the current frame/filterbank window or not;
- ☞ The number of TNS filters being used (maximum of one filter per filterbank window for short windows; otherwise up to three filters);
- ☞ A target frequency range for each TNS filter, together with the filter parameters, including filter order and filter coefficients.

The filter coefficients are represented as reflection coefficients that are quantized by a nonuniform (`arcsin`) quantization. Depending on the profile, a maximum filter order of 12 (MPEG-2 AAC LC and SSR profiles) or 20 (MPEG-2 AAC Main profile) is possible.

For the MPEG-2 AAC coder, TNS has been shown to provide significant gain in coding quality, in particular for speech signals and other input signals with transient characteristics (up to one grade on the ITU-R five-grade impairment scale). Furthermore, the necessity for TNS becomes increasingly important when the coder is run at lower sampling rates and the effective duration of the filterbank window increases.

11.2.7 Prediction

Predictive coding techniques have been used for a long time in traditional speech coding schemes as a way of exploiting redundancy by removing correlated parts of the input signal. As is known from coding theory [JayN84], redundancy also can be exploited by coding spectral (subband) data and taking advantage of the nonflat spectral distribution of the signal energy. Note that the latter mechanism is intrinsically built into the basic framework of a perceptual audio coder (owing to the reliance on the coding of spectral data).

Although both approaches are equivalent in theory (i.e., they will provide the same asymptotic gain for stationary signals), the exploitation of periodic and stationary signal components by a perceptual audio coder is limited by the size of its filterbank window. Stationary components that have a very close frequency spacing cannot be resolved as individual spectral *lines* and thus will not provide transform gain. This limitation for very long-term periodic/stationary signals can be overcome by incorporating predictive coding techniques into the framework of a perceptual audio coder.

11.2.7.1 Principle   The general principle of the prediction tool in MPEG-2 AAC can be explained as follows: Based on the quantized spectral coefficients of the previous frame(s), an estimation of their current value is calculated by means of a *predictor*. A bit-rate saving is achieved by quantizing and coding the prediction residue (error/difference signal) rather than the actual spectral coefficients. Because the prediction of new values is based on previously quantized and reconstructed data, this corresponds to a *backward prediction* scheme [JayN84].

The prediction is carried out separately for each spectral coefficient (subband), based on the value of the corresponding spectral coefficient (subband) for the two preceding frames. More specifically, a second-order lattice predictor scheme is employed for each spectral coefficient.

To achieve fast adaptation and tracking of the prediction coefficients, a least mean squares–based adaptation algorithm is used. The adaptation process is carried out simultaneously at the encoder and decoder, leading to a *backward-adaptive* prediction scheme [JayN84]. This differs from previous prediction schemes in MPEG-2 audio multichannel coding [MPEG2-3].

As an important consequence of the backward-adaptive scheme, no prediction coefficient side information has to be transmitted from the encoder to the decoder (only the prediction errors themselves are transmitted). On the other hand, the update of the prediction coefficients needs to take place in both encoder and decoder, which implies both computation and the storage of the adaptation algorithm's state variables.

The AAC prediction tool implements prediction over time (effectively increasing the temporal scope of the coder), whereas the TNS tool addresses prediction over frequency (and effectively provides an enhanced temporal coding resolution). Thus, both tools are dual counterparts in the T/F sense and complement each other in terms of functionality.

11.2.7.2 Specifics To achieve seamless and efficient operation of the prediction-based scheme, a number of practical aspects have been addressed in the definition of the AAC prediction tool:

☞ **Enable/disable signaling:** Due to the use of a backward-adaptive prediction scheme, no costly transmission of prediction coefficients is required. There are, however, situations in which the coding of the prediction residual signal would need more bits than the coding of the actual (unpredicted) spectral coefficients caused by a nonoptimum adjustment of the predictors. Thus, a one-bit flag is transmitted in the bitstream for each scalefactor band, indicating whether prediction is used for that band (i.e., whether the residual signal is quantized and coded instead of the original spectral coefficients). In the case of joint coding of a stereo signal pair, each flag bit determines the use of prediction for both the left and the right channels for a given scalefactor band.

☞ **Predictor reset:** Because encoder and decoder prediction algorithms may be running on different processor/computing platforms with different numerical behavior, it is important to assure that coefficient adaptation in both parts does not diverge beyond tolerable limits. To this end, a *predictor reset* mechanism has been included that allows the encoder and decoder predictors to be reset to a well-defined initial state synchronously. Groups of predictors belonging to specific spectral coefficients are addressed for reset by definition of the so-called *predictor reset groups*. By selecting one predictor reset group at a time (i.e., within a frame), a sliding reset of all predictor instances across frequency can be achieved over time without noticeable change in the overall prediction gain.

☞ **Handling of short filterbank windows:** Because the prediction of spectral coefficients (subband data) over time is only meaningful between values with the same T/F interpretation, no prediction can be achieved when the coder has selected the short window filterbank state. In this case, all predictor state variables are reset and residual calculation is switched off. Note that the use of short windows usually indicates the presence of transient/nonstationary signal parts for which achieving a prediction gain is unlikely.

☞ **Complexity:** As a consequence of the backward-adaptive prediction scheme, both the encoder and decoder have to carry the burden of the actual prediction operation and the coefficient adaptation algorithm (storage, prediction computation, and update of state variables). As this results in considerable additional complexity compared to the basic decoder parts, a number of measures have been taken. First, prediction is not applied beyond a certain upper frequency limit. This limit is defined for each encoding sampling rate and amounts to around 16 kHz for a sampling rate of 48 kHz. Furthermore, the memory demand for the prediction calculation is reduced by 50% by storing the associated state values as a truncated (16-bit) IEEE floating point format rather than the full floating point format (32-bit).

Because MPEG-2 AAC prediction is still associated with a significant complexity penalty, only MPEG-2 AAC `Main` profile decoders support this tool. A perceptible increase in subjective sound quality (up to one grade on the ITU-R five-grade impairment scale) was encountered for very stationary input signals, such as the *Harpsichord* or *Pitch Pipe* test signals [N2006].

11.2.8 Joint Stereo Coding

For coding of high-quality stereophonic (or multichannel) audio signals at low bit rates, joint coding techniques have proven to be extremely valuable. They provide mechanisms that account for binaural psychoacoustic effects while frequently reducing the required bit rate for stereophonic signals to rates significantly below the rate for separate coding of the input channels. Figure 11.7 illustrates how the basic scheme of a monophonic perceptual audio coder (see Figure 11.1) is extended by means of joint stereo coding techniques (for clarity, the perceptual model block has been ignored here). An additional block is inserted after the analysis filterbank, allowing the joint processing of the two (or more) channel signals in the spectral domain.

Because MPEG-2 AAC was initially conceived as a multichannel audio coder, the MPEG-2 AAC coding scheme also includes two techniques for joint stereo coding of audio signals: M/S stereo coding (also known as *sum/difference coding*) and intensity stereo coding. Both stereo coding tools can be combined by

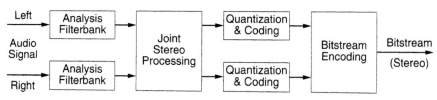

Fig. 11.7 Generic architecture of a joint stereo coder.

selectively applying them to different frequency regions. By using M/S stereo coding, intensity stereo coding, and left/right (L/R) independent coding as appropriate, it is possible to avoid the expensive overcoding caused by binaural masking level depression (BMLD) [Blau83], to account for noise imaging, and often to achieve a significant saving in data rate. The concept of joint stereo coding in the MPEG-2 AAC framework is discussed in greater detail in [JHDG96].

11.2.8.1 M/S Stereo Coding   The technique of M/S coding of stereo signals has been used in perceptual audio coding for a considerable time (just as in FM radio broadcasting and microphone recording) [JohF92]. Instead of coding the left and right signals independently, the sum and difference signals are handled; these signals are referred to as the middle (M) and the side (S) channels, respectively. With respect to Figure 11.7, the joint stereo processing stage is implemented by a sum/difference (M/S) matrix and its reverse processing at the decoder side.

Clearly, M/S stereo coding is a very efficient coding method for the case of near monophonic signals, in which both the difference signal is very weak and the spatial attributes of the coding noise, as compared to the spatial attributes of the original signal, become important. In particular, this technique is capable of addressing the issue of BMLD [Blau83, Moor89], in which a signal at lower frequencies (below 2 kHz) can show up to 20-dB difference in masking thresholds depending on the phase of the signal and noise present (or lack of correlation in the case of noise).

In the context of the MPEG-2 AAC framework, M/S stereo coding is employed within each channel pair of the multichannel signal—that is, between a pair of channels that are arranged symmetrically on the left/right listener axis. In this way, imaging problems due to spatial unmasking are avoided to a large degree. M/S stereo coding can be used in a flexible way by selectively switching in time (on a frame-by-frame basis), as well as in frequency (on a scalefactor band basis), as required by the signal. The switching state (M/S stereo coding *on* or *off*) is transmitted to the decoder as an array of signaling bits.

11.2.8.2 Intensity Stereo Coding   The second joint stereo coding tool adopted in MPEG-2 AAC relates to the concept of *intensity stereo coding* [WaaV91, HeBD94]. This idea was used in MPEG-1 and MPEG-2 audio stereophonic and multichannel coding as well as in other coders, being known under various names such as *dynamic crosstalk* and *channel coupling*.

Intensity stereo coding attempts to exploit the fact that the perception of high-frequency sound components mainly relies on their energy-time envelopes [Blau83]. Thus, it is possible for certain types of stereo (or multichannel) signals to transmit a single set of spectral values that is shared among the audio channels with virtually no loss in sound quality. The original energy-time envelopes of the coded channels are approximately preserved by means of a scaling operation such that each channel signal is reconstructed with its original level after decoding. With respect to Figure 11.7, the joint stereo processing stage is implemented by combining the spectral data of the single channel signals and by extracting their original energies (see Figure 11.8). At the decoder, the combined signal is rescaled independently for each signal channel to match the average envelope (or signal energy) for the corresponding coder frame. The scaling information is calculated and coded once for each frame and scalefactor band.

As in the case of M/S stereo coding, MPEG-2 AAC provides both time and frequency selective intensity stereo coding using two mechanisms:

☞ **AAC intensity stereo coding:** The first mechanism is based on the *channel pair* concept as used for M/S stereo coding and implements an easy-to-use coding concept that covers most of the needs without introducing noticeable signaling overhead into the bitstream. For simplicity, this mechanism is referred to as the AAC *intensity stereo coding* tool. Although this tool only implements joint coding within each channel pair, it may be used for coding of both two-channel and multichannel signals. The use of the intensity stereo coding tool is signaled by means of two pseudo-codebook numbers in the noiseless coding side information (INTENSITY_HCB=15; INTENSITY_HCB2=14), which indicate in-phase and out-of-phase intensity stereo coding for the corresponding section, respectively (see Section 11.2.5.1). The shared spectral coefficients are transmitted in the left channel part of the channel pair element. Instead of scalefactors, the right channel carries *intensity stereo position* values, indicating the right channel scaling as compared to the left channel.

☞ **AAC coupling channels:** An alternative, more sophisticated mechanism is available that is not restricted by the channel pair concept. This mechanism is called the AAC *coupling channel* and provides two functionalities. Coupling channels may be used to implement generalized intensity stereo coding in which channel spectra can be shared across channel boundaries (including sharing between different channel pairs). Coupling channels can furthermore perform a downmix of additional sound objects into the

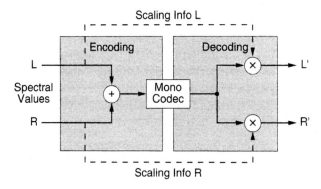

Fig. 11.8 Intensity stereo processing principle.

stereo image so that, for example, a commentary channel can be added to an existing multichannel program (voice-over). The coupling channel is implemented as a dedicated syntax element (see Section 11.2.9 on syntax elements), which basically consists of a single channel of audio information plus a list of target channels onto which the audio data will be added. For each target channel, separate scaling/gain information is specified. Depending on the profile, certain restrictions apply with regard to the consistency between coupling channels and target channels in terms of window sequence and window shape parameters.

11.2.9 Bitstream Multiplexing

In contrast to previous MPEG Audio standards, MPEG-2 AAC (MPEG-2 Part 7) distinguishes between the *transport layer* (AUDIO_DATA_INTERCHANGE_FORMAT [ADIF] or AUDIO_DATA_TRANSPORT_STREAM [ADTS]) and the *compression layer* (raw audio bitstream data) [AAC]. Whereas the latter carries the actual compressed audio data, the transport layer may provide functions such as synchronization, signaling of data packetization, and entry points for resynchronization (see Chapter 7). As no specific transport layer can be appropriate for all applications, only the coded audio bitstream format is normatively defined within the MPEG-2 AAC standard (and can, in fact, be decoded on its own).

A second difference with respect to earlier related MPEG audio coders is the departure from the constant rate header paradigm. The MPEG-1/2 Layer-3 bitstream consists of header portions that occur at a regular rate corresponding to the nominal bit rate. In order to allow for local variations of the bit rate (*bit reservoir*), the main coded information is located somewhere in between the header time grid, with a variable offset depending on the actual fill level of the reservoir. The MPEG-2 AAC bitstream format departs from this scheme by simply concatenating header and body portions of the bitstream. This results in a straightforward bitstream structure with a variable distance between subsequent header portions (depending on the actual bit-rate usage for the frame). For the sake of simplicity, each header will start on a byte boundary (byte alignment). Fast-forward and fast-reverse functionality have to be provided by means of the transport layer, which may contain appropriate pointers into the bitstream for direct access to future or past frames.

Another novel concept in bitstream definition is represented by the AAC modular bitstream syntax. Contrary to former MPEG Audio standards, which are based on a monolithic bitstream definition, the AAC bitstream consists of a number of syntactic elements that can be extracted, rearranged, and stripped off to be decoded on their own or form new bitstreams. Each syntactic element is identified by a unique element ID. In general, there may be more than 1 (up to 16) instance of each syntactic element type. Table 11.2 provides an overview of the various AAC syntactic element types and their functionality.

Further discussion on these bitstream elements can be found in [JHDG96].

Table 11.2 MPEG-2 AAC syntactic elements

Syntactic element name	Symbol	Purpose
Single_channel_element	SCE	Represents a single audio channel
Channel_pair_element	CPE	Represents a stereo audio channel pair
Coupling_channel_element	CCE	Multichannel coupling
Lfe_channel_element	LFE	Represents a low frequency effects channel
Data_stream_element	DSE	Carries associated data
Program_config_element	PCE	Defines program/channel configuration
Fill_element	FIL	Adjusts bit rate for constant rate channels
Terminator	TERM	Signals end of frame

11.2.10 Other Aspects

Beyond the functions already described, some other aspects of MPEG-2 AAC should be noted here briefly.

The AAC coder offers a high degree of flexibility in terms of the range of supported sampling rates. All necessary coder parameters are defined from 8 kHz to 96 kHz (nominal sampling rates are 8,000; 11,025; 12,000; 16,000; 22,050; 24,000; 32,000; 44,100; 48,000; 64,000; 88,200; and 96,000 Hz). AAC also allows using other sampling rates, in which case the coder parameter set for the nearest nominal sampling rate will come into effect.

In order to accommodate the desire for *dynamic range control* (DRC)—that is, the modification of the decoded signal's dynamic range according to the needs of the reproduction environment or listener preferences—a bitstream format for this functionality was defined for MPEG-2 AAC. Although there is no normative or mandatory definition of the actual dynamic range processing algorithm, this bitstream format ensures interoperability between different encoders of DRC information and optional dynamic range processors at the decoder end. The DRC information is embedded into fill elements, thus providing both full backward compatibility with non-DRC-enhanced bitstreams without occupying any of the data stream elements available to the user.

11.3 MPEG-4 ADDITIONS TO AAC

Besides the building blocks provided by MPEG-2 AAC, the MPEG-4 T/F coder includes several extensions in order to enhance coding efficiency and offer new functionalities. Examples of additional building blocks include *perceptual noise substitution* (PNS), *long-term prediction* (LTP), and the *transform-*

domain weighted interleave vector quantization (TwinVQ) coding kernel. The *low-delay AAC* (AAC-LD) coder and the *error-resilience* (ER) provisions illustrate how new functionalities can be supported by variation/modification of the basic AAC technology. The capability for *scalable audio coding* deserves particular attention in the context of MPEG-4 (see Section 11.4). Scalability is achieved by introducing concepts of large-step scalable coding into the AAC coding framework, resulting in a flexible toolboxlike solution that combines coding elements with different characteristics into a single, unified framework. Alternatively, the *fine grain scalability* (FGS) mode based on bit-sliced arithmetic coding (BSAC) may be used.

Although some of these extensions were added only in the 1st Amendment to MPEG-4 Audio Version 1 (the so-called Version 2 [MP4V2-3]), such as ER provisioning, low-delay T/F coding, and the BSAC scalability mode, no distinction will be made between Version 1 and Version 2 tools in the following discussions.

11.3.1 Perceptual Noise Substitution

So far, all the coding schemes discussed (with the exception of intensity stereo coding) constitute techniques for coding the waveform of the input signal— that is, they lead to the perfect reconstruction of the input signal in the absence of quantization. As known from speech coding (*vocoder principle*), relaxing the requirement for waveform coding can lead to very efficient coding at low bit rates and intermediate quality. Related coding approaches usually aim at producing a perceptually equivalent output signal rather than reproducing the waveform of the input signal.

11.3.1.1 Principle   The PNS tool [HerS98] is an example of such a nonwaveform coding method. It is designed to represent noiselike components of the input signal with a very compact parametric representation and, in this way, further increases the compression efficiency of the AAC coder for suitable input signals.

The PNS technique is based on the fact that the subjective sensation stimulated by a noiselike signal is not determined by its actual waveform, but by its spectral and temporal fine structure. This phenomenon is exploited in the context of the MPEG-4 filterbank-based audio coder by extension of the basic AAC coder concepts as follows (see Figures 11.1 and 11.9):

☞ The input signal is analyzed in the encoder to determine noiselike signal components for each frame and scalefactor band.

☞ If a particular scalefactor band is considered noiselike, the corresponding set of spectral coefficients is not quantized and coded as usual but is omitted from this process. Instead, a noise substitution flag is transmitted to the decoder, together with the total power of the substituted set of spectral coefficients.

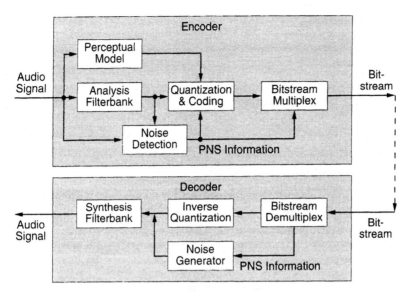

Fig. 11.9 PNS coding principle.

☞ At the decoder, the transmitted signaling information is analyzed; for scalefactor bands for which noise substitution has been signaled, pseudo-random numbers are inserted instead of actually coded/transmitted spectral coefficients, with a total noise power equal to the transmitted level.

Because only signaling and energy information is transmitted, once per scalefactor band (no spectral coefficients), this approach results in a highly compact representation of noiselike signal components.

11.3.1.2 Specifics Within the MPEG-4 T/F coder, the PNS tool is designed to integrate smoothly into the existing architecture of the AAC coder. More specifically, the following AAC mechanisms are reused in the context of PNS (see also Section 11.2.8.2 on signaling of intensity stereo coding):

☞ **PNS signaling:** The signaling of the noise substitution status for particular scalefactor bands is realized by using one of the available Huffman codebook numbers (pseudo-codebook number NOISE_HCB=13)—that is, noise substitution is assumed for all scalefactor bands covered by a NOISE_HCB section (see Section 11.2.5.1).

☞ **Transmission of noise energies:** For noise-substituted scalefactor bands, the associated energy values are transmitted in the place of scale-factors using a similar type of encoding (Huffman coding of differential log values with a resolution of 1.5 dB).

☞ **L/R correlation:** In order to account for some aspects of binaural perception, the correlation status of the PNS-generated pseudo-noise components for the left and right channels of a channel pair can be signaled in the bitstream. If a particular scalefactor band is signaled to be coded via PNS in both channels of a channel pair, the M/S bit (which is not meaningful in this context) is used to switch between uncorrelated and correlated pseudo-noise generation modes in both channels. This capability is important to successfully handle both stereophonic and near-monophonic stereo signals without introducing artifacts into the stereo image.

As a consequence of this tight integration of the PNS tool into the AAC framework, the extended bitstream syntax is downward compatible with the MPEG-2 AAC syntax—that is, an MPEG-2 AAC decoder will be able to decode the extended MPEG-4 bitstream format as long as the PNS feature is not invoked. Based on the three profiles of MPEG-2 AAC (see Section 11.2.1), three corresponding configurations of the MPEG-4 T/F coder with the PNS tool are defined: AAC Main, AAC LC, and AAC SSR. Such configurations are referred to as *audio object types* (see Chapter 13).

The PNS tool has been shown to provide some boost in coding efficiency for complex musical signals (both classical and pop) in the range of (nontransparent) coding at low bit rates and is associated with little additional decoder complexity, both in terms of computational and memory requirements.

11.3.2 Long-Term Prediction

Another technique well known from speech coding has been incorporated into the AAC coding framework under the name LTP. In speech coding, this technique is commonly used to exploit redundancy in speech signals that is related to the periodicity of the voiced signal portions (i.e., the signal pitch).

11.3.2.1 Principle Although common speech coders apply LTP within a time-domain coder, the MPEG-4 Audio LTP tool has been integrated into the framework of a generic perceptual audio coder—that is, quantization and coding are performed on a spectral representation of the input signal. Figure 11.10 illustrates how both concepts are combined into one coding scheme [OjaV99], in which the LTP tool produces a prediction of the input signal based on the quantized values (here, spectral values) of the preceding frames (backward prediction [JayN84]):

☞ The quantized spectral values are mapped back to a time domain representation by the synthesis filterbank and the associated inverse TNS filtering operation to form the reconstructed quantized time signal.

☞ By matching the reconstructed time signal to the actual input signal, the optimum parameters for delay (*pitch lag*) and amplitude scaling (*gain*) are determined to form the predicted signal. The LTP parameters are transmitted to the decoder (forward adaptation [JayN84]).

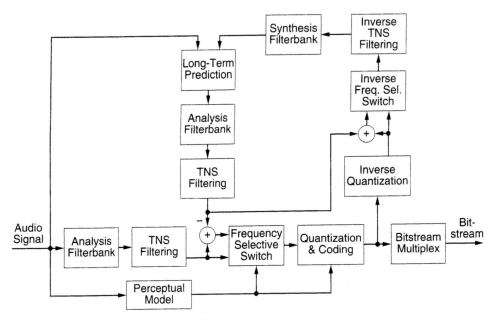

Fig. 11.10 LTP Integration into a T/F coder.

☞ Both the input and the predicted signals are then mapped to a spectral representation by means of an analysis filterbank and a forward TNS filtering operation.

☞ Both representations are subtracted from each other to form a residual signal. In order to enable the frequency selective use of the predicted signal (and thus the LTP tool), either the difference (residual) signal or the original signal is used for coding, whichever is more favorable to minimize the bit-rate demand.

The choice between the residual and the original signals for each scalefactor band is implemented by means of a so-called *frequency selective switch* (FSS), which is also employed in the context of the MPEG-4 large-step scalable coders (see Section 11.4.1).

11.3.2.2 Specifics   Because the LTP tool is based on a forward-adaptive scheme, a number of prediction parameters are transmitted to the decoder as part of the side information. In particular, this includes the prediction lag and gain values and flags indicating whether LTP is in use or not. As the LTP tool can generally also be used with the EIGHT_SHORT_SEQUENCE filterbank state (in contrast to the MPEG-2 backward-adaptive prediction tool), a flag bit for each filterbank window is transmitted to indicate the use of the LTP tool in this case. For other filterbank states, one bit per scalefactor band is sent, indicating whether the residual signal is used for encoding or not.

For one particular coder configuration (the so-called MPEG-4 Audio AAC LTP object type; see Chapter 13) that does not support scalable coding, the use of the LTP tool is restricted to long windows and, in this way, achieves backward compatibility with the bitstream syntax of the MPEG-2 AAC LC profile—that is, MPEG-2 AAC LC bitstreams can be decoded with an MPEG-4 AAC LTP decoder, and MPEG- 4 AAC LTP bitstreams are decodable by MPEG-2 AAC LC decoders, if LTP is not used.

As would be expected from the underlying principle, the LTP tool provides good coding gain for stationary harmonic signals as well as some gain for nonharmonic tonal signals (such as polyphonic tonal instruments).

In comparison with the rather complex MPEG-2 AAC prediction tool, the MPEG-4 LTP tool requires only approximately one-half of the resources, both in terms of computational complexity and RAM storage.

11.3.3 TwinVQ

In addition to the MPEG-2 AAC quantization/coding process, MPEG-4 includes an alternative coding kernel (i.e., an alternative to the gray block *Quantization and Coding* in Figure 11.2), the so-called TwinVQ [IwMM95, IwaM96], which is mainly for use with the MPEG-4 scalable T/F audio coder (see Section 11.4.1 on large-step scalable audio coding). This TwinVQ tool is designed to provide good coding performance at extremely low bit rates (down to 6 kbit/s) for general types of audio signal, including music. Contrary to the MPEG-2 AAC quantization/coding modules, it is based on vector quantization (VQ) techniques [JayN84]. During the standardization process, the originally proposed TwinVQ coding scheme has been harmonized to fit into the unified MPEG-4 T/F coding framework and operate on the spectral representation provided by the AAC filterbank [HABD98].

11.3.3.1 Principle In the TwinVQ kernel, quantization/coding of the spectral coefficients is carried out in two steps. A first stage—*spectral normalization*—normalizes the spectral coefficients to a specified target amplitude range; these are then quantized/coded in a second phase by means of a weighted VQ stage—*weighted VQ*.

Spectral Normalization The normalization (*flattening*) process provides a rescaling of the spectral coefficients' amplitude to a desired range and extracts characteristic signal parameters. As a result, the spectral coefficients are flattened across the frequency axis. The parameters associated with this normalization process are quantized and transmitted as side information to the decoder.

The spectral normalization proceeds in three stages:

☞ **LPC spectral estimation:** The overall coarse spectral envelope of the input signal is estimated using an LPC model and used to normalize the amplitude of the spectral coefficients. The LPC envelope parameters are coded efficiently by means of the line spectral pair (LSP) representation.

☞ **Periodic component coding:** For frames using long filterbank windows, *periodic peak* components (corresponding to harmonic peaks in the spectrum) are coded. To this end, the fundamental signal frequency (pitch) is estimated, and a number of periodic peak components are extracted from the preflattened coefficients. The pitch value and the average gain of the components are quantized and coded as side information.

☞ **Bark-scale envelope coding:** Further flattening on the resulting coefficients is carried out by making use of a spectral envelope based on the Bark-related AAC scalefactor bands. These envelope values are quantized by means of a vector quantizer with interframe prediction.

Weighted Vector Quantization   In a second phase, the coding of the flattened spectral coefficients is carried out. This process consists of the following steps:

☞ **Interleaving of spectral coefficients:** Before VQ, the flattened spectral coefficients are interleaved and divided into subvectors (see Figure 11.11). If the subvectors were constructed from spectral coefficients that are consecutive in frequency, the subvectors corresponding to the lower frequency range would require much finer quantization (more bits) than the subvectors with the higher frequencies. In contrast, interleaving of the spectral coefficients in frequency leads to subvectors with comparable properties in terms of required accuracy and thus allows keeping a constant bit allocation for each subvector. It is possible to achieve perceptual shaping of the quantization noise by using an adaptive weighted distortion measure that is controlled by a perceptual model (employing the spectral envelope/flattening parameters [IwaM96]).

☞ **Vector quantization:** After interleaving, the subvectors are vector quantized by means of a VQ scheme. The quantization distortion is minimized by determining the best codebook indices. For each subvector, a weighted

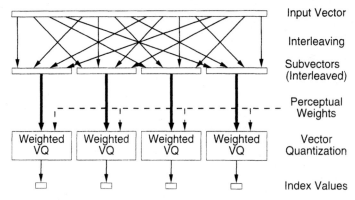

Fig. 11.11 TwinVQ interleaving and VQ encoding.

distortion measure is applied during the codebook index selection. In this way, perceptual control of the quantization distortion is achieved.

Due to the nature of the interleaved VQ scheme, no adaptive bit allocation is necessary for individual quantization indices—that is, an equal number of bits is spent for each of the quantization indices. The main part of the coded information consists of the selected codebook indices.

11.3.3.2 Specifics Aside from several tools available within the AAC coding framework, the MPEG-4 TwinVQ coding kernel uses two dedicated (de)coding tools that resemble its general algorithmic structure (as outlined previously):

☞ **Interleaved VQ:** This tool combines the steps of inverse VQ and deinterleaving. The interleaved subvectors are vector quantized using a two-channel conjugate structure with two sets of codebooks. Thus, the optimum quantization is determined by the best combination of codebook indices when two code vectors are added. One bit of polarity information is coded for each of the two code vectors. In order to support scalable coding (see Section 11.4), the frequency range to which a TwinVQ-based enhancement layer bitstream contributes can be selected from four possible choices.

☞ **Spectrum normalization:** This tool includes the decoding and denormalization steps for global gain information, Bark-scale envelope, periodic peak components, and LPC spectral envelope. The global gain information is quantized with a quasi-logarithmic (μ-law) characteristic [JayN84]. Bark-scale envelope, periodic peak components, and LPC spectral envelope are coded by means of VQ techniques. In order to allow some control over the trade-off between coded audio bandwidth and coding distortion, the bitstream syntax permits signaling a band-limiting operation of the decoded spectral coefficients. In this way, the encoding vector codebook search can improve the precision of the output signal by finding the codebook indices that represent the best approximation within the reconstructed frequency range.

The TwinVQ coder kernel codes the values of the AAC filterbank and may be used in conjunction with the LTP and TNS tools. This coder configuration (audio object type TwinVQ) operates at bit rates between 6 kbit/s/ch and 16 kbit/s/ch and is frequently employed in the scalable configurations of the MPEG-4 T/F audio coder (see Section 11.4.1).

11.3.4 Low-Delay AAC (AAC-LD)

Although the standard MPEG-4 T/F coder provides very efficient coding of general audio signals at low bit rates, it exhibits an algorithmic (i.e., minimum theoretic) delay of up to several hundred ms and is thus not well-suited for applications requiring low coding delay, such as real-time bidirectional communications. As an example, the T/F coder operating at 24 kHz sampling rate and 24 kbit/s has an algorithmic coding delay of about 110 ms, with up to an additional 210 ms for the bit reservoir. In contrast to this, traditional speech coding schemes provide good quality coding at low delay only for a narrow class of signals (i.e., speech).

To enable coding of general audio signals with an algorithmic delay down to 20 ms, MPEG-4 Audio Version 2 specifies a so-called *low-delay audio coding mode*, which is derived from the MPEG-4 AAC LTP audio object type [AGHS99] with the following modifications:

☞ It operates at sampling rates up to 48 kHz and uses a frame length of 512 or 480 samples, compared to the 1,024 or 960 samples used in standard MPEG-4 AAC. Accordingly, the size of the window used in the analysis and synthesis filterbank is reduced by a factor of two.

☞ No window switching is used to avoid the *look-ahead* delay, which is incurred in the encoder due to the window switching decision. To reduce pre-echo artifacts for the case of transient signals, only TNS is employed in conjunction with window shape adaptation.

☞ Although for the nontransient parts of the signal a sine window is used, a so-called low overlap window is applied for the case of transient signals in order to achieve optimum TNS performance, thus reducing the effects of temporal aliasing as a result of the MDCT filterbank. The low overlap window consists of the combination of a LONG_START_SEQUENCE and a LONG_STOP_SEQUENCE and effectively replaces the standard KBD window (see Figure 11.4 and Figure 11.12).

☞ Furthermore, the use of the bit reservoir is minimized at the encoder in order to reach the desired target delay. In the extreme case, no bit reservoir is used at all.

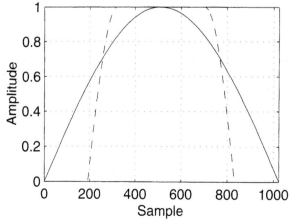

Fig. 11.12 AAC-LD: sine and low overlap windows.

Despite the limitations in coding strategy owing to the low target delay values, the reduction in coding delay is reached at a moderate cost of only circa 8 kbit/s/ch bit-rate increase, compared to standard MPEG-4 AAC Main audio object type (see Chapter 16). As a further point of reference, the MPEG-4 AAC-LD coder compares favorably with the widely used MPEG-1/2 Layer-3 (well-known as MP3) coder at a bit rate of 64 kbit/s/ch [AGHS99].

Compared to CELP coders, which by principle do not perform well for non-speech signals, the MPEG-4 AAC-LD coder is capable of coding both music and speech signals with good quality. Unlike speech coders, furthermore, the coding quality achieved scales up with bit rate for all types of signals. All stereo and multichannel capabilities have been inherited from the standard AAC coder with a view toward future high-quality bidirectional communication applications. This coder configuration is referred to as audio object type ER AAC LD.

11.3.5 Error Robustness

Another important functionality addressed by MPEG-4 Audio Version 2 is the error robustness of coded audio bitstreams, in order to provide improved performance on error-prone transmission channels, such as wireless channels (see also the section on error robustness in Chapter 10). More specifically, the error-robustness provisions of MPEG-4 Audio can be divided into codec-specific *error-resilience* (ER) measures and a common *error-protection* (EP) tool. In addition, *error-concealment* techniques can be employed in the decoder, but these are not within the scope of the normative part of the MPEG-4 standard. Subsequently, the codec-specific ER measures for MPEG-4 Audio will be discussed.

Improved error robustness for AAC is provided by a set of ER tools that reduce the perceived degradation of the decoded audio signal caused by corrupted bits in the bitstream. Three tools are provided to improve the error robustness of an MPEG-4 AAC frame:

☞ **Virtual codebook (VCB11) tool:** The VCB11 tool enables the detection of serious errors within the AAC spectral data by extending the sectioning information. In particular, VCB11 enhances the error resilience for scalefactor bands that contain large spectral coefficients (with absolute values > 16) and, therefore, have to be coded using the *Escape Huffman codebook* ESC_HCB=11 (see Section 11.2.5 on AAC Noiseless Coding). Because bit errors in these parts of the bitstream can lead to obtrusive signal distortions, virtual codebooks are used to limit the largest possible absolute value within these scalefactor bands. While referring to the same codes as codebook 11, the 16 virtual codebooks introduced by this tool provide 16 different limitations of the spectral values belonging to the corresponding section. In this way, large spectral coefficients that result from bit errors can be detected and concealed appropriately.

☞ **Reversible variable length coding (RVLC) tool:** The RVLC tool enhances the error resilience of Huffman-coded DPCM scalefactors in the

AAC bitstream by using symmetric codewords, thus enabling both forward and backward decoding of the scalefactor data. In order to make backward decoding possible, both the length of the scalefactor data and the value of the last scalefactor are transmitted. As an additional benefit, some degree of error-detection capability is provided, as not all conceivable RVLC codewords are used (see [TakW95] for construction of RVLCs).

☞ **Huffman codeword reordering (HCR) tool:** The HCR tool for AAC spectral data exploits the idea that some Huffman codewords can be placed at known regular positions within the bitstream so that proper synchronization is guaranteed for them, independently of potential preceding bit errors. This is done by defining a regular grid of positions in the bitstream, each of which is the start of a *priority codeword* (PCW). The remaining non-PCWs are filled into the gaps between the PCWs. In this way, error propagation is avoided for the most important spectral coefficients. A more extensive description of HCR can be found in [Sper00].

Improved error robustness for MPEG-4 BSAC is provided by the SBA tool, as described in Section 11.4.2. Because of its VQ-based design and the use of fixed-length codewords (i.e., no VLC), MPEG-4 TwinVQ intrinsically provides good error robustness, and no additional error-robustness tools are required.

Furthermore, improved error resilience is achieved by the so-called *ER bitstream payload syntax*. The basic idea is to rearrange (reorder) the conventional bitstream payload depending on its error sensitivity and thus provide an interface for using adapted channel coding techniques according to the *error-sensitivity category* (ESC) of each bitstream element. This concept is referred to as *unequal error protection* (UEP) [MasW67]. All object types added by MPEG-4 Audio Version 2 [MP4V2-3] use such an ER bitstream syntax: Object types added in Version 2 were directly defined this way (ER BSAC, ER AAC LD, ER HILN, ER Parametric), whereas for object types from Version 1 a reordered bitstream syntax was defined (ER AAC LC, ER AAC LTP, ER AAC Scalable, ER TwinVQ). The corresponding Version 2 object types for MPEG-4 speech coding (ER CELP, ER HVXC) and the MPEG-4 Audio EP tool for UEP are described in Chapter 10.

11.4 MPEG-4 SCALABLE AUDIO CODING

According to the traditional approach to perceptual audio coding (see Section 11.1), the bit rate used for the compressed representation of the audio signal is specified at the time of encoding. This may, however, not be appropriate if the bitstream is to be distributed subsequently via transmission channels not known beforehand or channels with varying transmission capacity (e.g., Internet, wireless transmission). This problem is addressed by the concept of MPEG-4 scalable audio coding, which enables the transmission and decoding of the bitstream with a bit rate that can be adapted to dynamically varying requirements, like the instantaneous transmission channel capacity (or, in some cases, the decoder computational resources). Although certain types of scalability were already available for video coding in previous MPEG standards, MPEG-4 is the first standard to provide this core functionality in the context of MPEG Audio standards.

11.4.1 Large-Step Scalable Audio Coding

Scalability within the MPEG-4 Audio standard is mostly achieved by using the concept of *hierarchical embedded coding*—that is, scalable audio bitstreams consisting of several partial bitstreams that can be decoded on their own and are combined to form a meaningful decoding result. Thus, the decoding of a hierarchical subset of the full bitstream will lead to a valid decoded signal, although at a lower quality. Combining the concepts of scalability and tool-based structure, MPEG-4 Audio has defined a flexible framework in which the existing MPEG-4 audio coding schemes have been integrated into a combined scalable coding structure.

11.4.1.1 Principle Generally, the key concept of hierarchical scalable coding can be described as follows (see Figure 11.13):

☞ A first coder (the *base layer* coder, coder #1) codes the input audio signal, producing the base layer part of the scalable bitstream.

☞ Subsequently, a coding error is determined by local decoding and subtraction from the original signal. This residual signal is coded by the next layer coder (an *enhancement layer* coder, coder #2), producing the next part of the composite scalable bitstream.

The process of repeated refinement of the signal can be continued as often as appropriate, with each partial coding process producing one further enhancement layer of bitstream information. Due to the inherent hierarchy of the representation, (partial) decoding of an M-layer bitstream always consists of combining all contributions of layers 1 to N (n = 1 . . . N, N ≤ M). As N grows, the decoded signal quality increases, starting from the basic quality level provided by the base layer coder and increasing with each additional enhancement layer. Within MPEG-4 Audio, this concept provides coders with a limited number of layers (typically 2 to 4) and is referred to as *large-step scalability*.

The general architecture of an MPEG-4 large-step scalable audio coder is depicted in Figure 11.14. This configuration (optionally) comprises a non-T/F base layer coder and several enhancement layers based on T/F coding modules. The non-T/F base layer coder (e.g., an MPEG-4 CELP narrow band speech coder) operates at a lower sampling rate than the subsequent

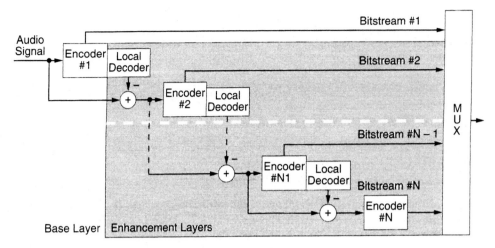

Fig. 11.13 Embedded hierarchical coding principle.

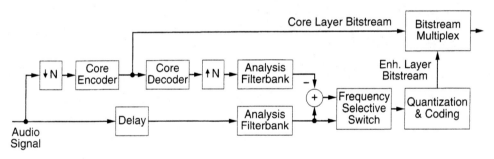

Fig. 11.14 Architecture of an MPEG-4 large-step scalable audio encoder.

enhancement layers and is called a *core coder* [Gril97]. The scalable combination of these component coders works as follows:

☞ The input audio signal is downsampled and encoded by the core coder. The produced bitstream constitutes the base layer portion of the scalable bitstream. It is decoded locally and upsampled to match the sampling rate of the T/F-based enhancement layers and passed through the T/F analysis filterbank (MDCT). Note that, although the downsampling operation has to include explicit lowpass filtering, the upsampling operation can be implemented simply by inserting zero values into the time signal and band limiting the spectral coefficients after the filterbank computation in order to remove redundant spectral replicas.

☞ In a second signal path, the delay-compensated input signal is passed through the T/F analysis filterbank (MDCT) and used to compute the residual coding error signal.

☞ The residual signal is passed through a *frequency selective switch* (FSS) tool that permits one to fall back to the original signal on a scalefactor band basis, if this allows a more efficient coding (see Sections 11.3.2 and 11.2.7 on the MPEG-4 LTP and the MPEG-2 prediction tools, respectively).

☞ The spectral coefficients are quantized/coded by an AAC coding kernel, leading to an enhancement layer bitstream.

Further stages of refinement (enhancement layers) by recoding the residual coding error signal could follow, as explained before. This is done efficiently by calculating subsequent residual coding error signals directly in the spectral domain (rather than in the time domain) and recoding them through an AAC coding kernel.

If no core coder is used, the base layer coder is a filterbank-based coder (i.e., AAC or TwinVQ), just like the enhancement layer coders, and the preceding scheme simplifies, as no sampling rate conversion is necessary and all quantization and coding are performed on a common set of spectral coefficients [HABD98].

Figure 11.15 illustrates the structure of a large-step scalable decoder. The composite bitstream is split into the individual coding layers. Decoding of the core stream is then performed; its output signal may be presented via an optional postfilter stage. In order to use the core decoder signal within the scalable decoding process, it is upsampled to the sampling rate of the scalable coder, delay compensated with respect to the other layers, and decomposed by the coder analysis filterbank (MDCT).

Higher layer bitstreams are decoded by applying the standard AAC noiseless decoding, inverse quantization, and summing all spectral coefficient contributions. An FSS tool combines the resulting spectral coefficients with the contribution from the core layer by selecting either the sum of them or only the coefficients originating from the enhancement layers as signaled by the encoder. Finally, the result is mapped back to a time domain representation by the synthesis filterbank (IMDCT).

The general scheme of quality scaling by adding subsequent enhancement layers may be used to obtain both improvements in coding precision

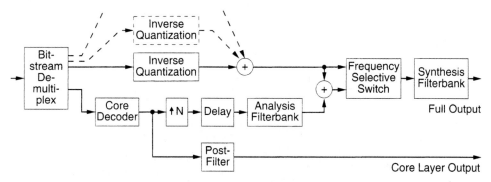

Fig. 11.15 Architecture of an MPEG-4 large-step scalable audio decoder.

(SNR) and extended signal bandwidth as more layers are included. It applies to both the coding of monophonic and (two-channel) stereophonic signals, including the use of intensity and M/S joint stereo coding as well as other coding tools.

In addition to the SNR/bandwidth scalability concept described previously, the MPEG-4 scalable coder also offers the possibility of enhancing sound reproduction from a monophonic to a stereophonic signal when going to the next highest layer [GriT98]. This *mono/stereo scalability* capability is unique so far among the known audio coders, including previous MPEG audio schemes, and is achieved in the following way: The lower layers encode a mono downmixed version of the stereo signal that can be used as an approximation of the mid (sum) signal of the M/S coded stereo channel pair. A subsequent stereo layer may then make use of this approximation when choosing M/S stereo coding or fall back to an L/R representation, if this is more efficient.

11.4.1.2 Specifics   Although the generic principle of large-step scalable coding allows a huge number of conceivable combinations of coders, a number of restrictions have been imposed in order to increase the regularity and simplicity of the MPEG-4 large-step scalable system:

☞ The MPEG-4 narrow-band CELP coder (see Chapter 10) can only play the role of the core (i.e., non-T/F base layer) coder.

☞ The TwinVQ coding kernel can be used both as a base layer coder and as an enhancement layer coder, provided that the coding of the previous layer is based on TwinVQ as well (this means that if the base layer is not TwinVQ-based, no other layer will be).

☞ The AAC coder provides the greatest flexibility in the context of scalable coding, as it can act both as a base layer and as an enhancement layer coder for various core coders. In order to accommodate combinations with coders using a basic granularity of multiples of 10 ms (such as speech coders), MPEG-4 scalable AAC also supports a frame size of 960 samples in addition to the standard 1,024-sample configuration.

In practice, a particularly interesting configuration consists of the combination of an MPEG-4 narrow-band CELP coder together with several AAC-based enhancement layers, which will provide good speech quality even at the output of the lowest layer. If emphasis is put on the acceptable reproduction of general audio and music signals at the lowest layer, TwinVQ may be used as the base layer coder instead of CELP.

As a consequence of the scalable encoding paradigm, the bitstream syntax of the MPEG-4 scalable coding scheme is, to some degree, different from the standard MPEG-2 style (nonscalable) AAC syntax:

☞ The bitstream syntactic elements that are common to all layers are arranged to be transmitted within the base layer bitstream only.

☞ The MPEG-4 scalable AAC syntax does not support the multichannel capabilities of the MPEG-2 AAC coder (e.g., coupling channel elements and related functions of the program configuration element). Thus, MPEG-4 scalable bitstreams convey either mono or (two-channel) stereo material.

☞ In view of the more general concept for transport that is provided by MPEG-4 Systems (see Chapter 7), the MPEG-4 audio scalable syntax departs from the MPEG-2 AAC concept of modular bitstream syntax (embodied by bitstream syntactic elements, as described in Section 11.2.9) and relies on appropriate transport and multiplexing mechanisms.

In order to account for the special coding functionalities provided by some MPEG-4 AAC coding modes, the previously shown simple concept for adding spectral contributions with each layer is extended to consider the coding modes (e.g., intensity stereo coding, PNS) of both preceding and actual layer in order to ensure a meaningful decoding result. For each combination, a particular way of combining the lower layer output signal with coefficients of the current layer is defined. As an example, an enhancement layer using regular coding of spectral coefficients will override (rather than add to) the spectral coefficients of a preceding layer that is encoded using PNS. Coder configurations employing AAC-based large-step enhancement layer(s) are referred to as audio object type AAC scalable.

11.4.2 Bit-Sliced Arithmetic Coding

The large-step scalable audio framework described in the previous section is a very efficient system for scalable configurations with a few enhancement layers. Due to the side information carried in each layer, however, the use of small scalability steps—that is, a low bit rate per enhancement layer—results in a decreased overall efficiency of this system at the full bit rate.[3] Hence, MPEG-4 Audio (as MPEG-4 Visual) provides an alternative tool for applications requiring FGS. This tool, named BSAC [PaKS97, KiPK01], is specified in MPEG-4 Audio Version 2 [MP4V2-3] and builds on the MPEG-4 AAC tools, replacing the noiseless coding of the quantized spectral data and the scalefactors (i.e., replacing the gray block *Quantization and Coding* in Figure 11.2).

BSAC provides scalability in steps of approximately 1 kbit/s per audio channel. This FGS is obtained by applying bit-sliced coding, also known as bit-plane coding, to the quantized spectral data. For this, a sign/magnitude format is used for the quantized spectral coefficients, and the magnitudes (i.e., abso-

---

3. Although scalability always has a quality price when compared with a single layer coding at the same bit rate, this quality price may increase significantly if the number of enhancement layers is high—for example, as would be desirable for an Internet transmission.

lute values) are represented as binary integers with up to 13-bit length. The bits of the magnitudes are processed in slices according to their significance, starting from the most significant bit (MSB) plane and progressing to the least significant bit (LSB) plane. Within a slice, the bits are processed in the original order of the spectral data from low to high frequencies. The sign bit of a spectral coefficient is *inserted* in this reordering process directly after the first "1" bit encountered in the magnitude of the coefficient. Figure 11.16 gives an example of bit-sliced coding, showing the processing order of the bits.

These bit-slices are then encoded by using a BSAC scheme to perform entropy coding with minimal redundancy. Several arithmetic coding models are provided to cover the different statistics of the bit-slices. For each model, the context-dependent probability of a bit in the bit-slice being either "0" or "1" is defined, in which the context is given by bits from the same slice and from the more significant bit-slices of the same frequency band that are already known to the decoder. Arithmetic coding also is employed for the efficient representation of the scalefactors.

The BSAC base layer carries the side information, the scalefactors and the first bit-slices of spectral data up to a certain audio bandwidth, which depends on the base layer bit rate. With an increasing number of enhancement layers used by the decoder, quantized spectral data is refined by providing more LSB information, which means decoding more bit-slices. At the same time, the audio bandwidth increases as more scalefactors and spectral data bits (up to the current bit-slice) in higher frequency bands become available to the decoder. In this way, quasi-continuous scalability is possible, as the decoder may stop to decode almost at any bit in a bitstream frame.

Because each enhancement layer carries only about 20 to 60 bits per AAC frame (typically 20 to 30 ms), such small data packets can result in a significant packetization overhead. To reduce this effect, the data packets of consecutive frames within each layer can be grouped to form larger payload packets. BSAC error resilience is improved by using the optional segmented

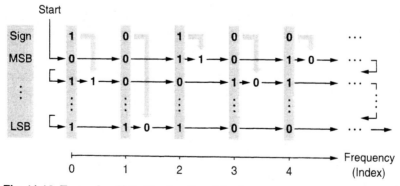

Fig. 11.16 Example of bit-sliced coding (bit-plane coding) for spectral coefficients in sign/magnitude representation.

binary arithmetic (SBA) coding tool. This tool forms several segments by grouping adjacent enhancement layers into segments. The arithmetic coder is initialized at the start of each segment and terminated at its end, so that the potential error propagation in the decoder is limited to such a segment.

BSAC FGS (audio object type ER BSAC) works well for high bit rates (about 48 to 64 kbit/s/ch), where it comes at only marginal loss in efficiency for full decoding of all layers when compared to a nonscalable AAC coder. When it is scaled down to lower bit rates, however, the loss in coding efficiency increases progressively (see Chapter 16).

11.5 INTRODUCTION TO PARAMETRIC AUDIO CODING

The design of an audio coding system is based on model assumptions for the signal source and signal perception [PaiS00]. A coding system designed for these models is able to exploit redundancy and irrelevancy in the audio signal to obtain a compact coded representation of the signal, which can be efficiently transmitted or stored.

Two examples help to illustrate the possible range of abstraction for representing audio signals. From a physical point of view, the temporal sound pressure function $x(t)$ (i.e., the waveform) is an obvious way to represent an audio signal x. On the other hand, the score of a piece of music provides a rather abstract representation of an audio signal. Although there are many ways to transform such a musical score into an audio signal, this ambiguity in interpretation is acceptable for many applications.

Abstract representations usually permit a compact description of an audio signal compared to, for example, the original representation as a sampled waveform. Thus, they offer a promising approach to multimedia applications that require efficient coding of audio signals—namely, by minimizing transmission bandwidth or storage requirements, as well as enabling new functionalities that are elicited by a semantic description of multimedia content.

For audio material available as a musical score, as a MIDI file, or in similar high-level formats, the structured audio part of the MPEG-4 Audio standard [MPEG4-3] provides both a score language (SASL) and an orchestra language (SAOL) to define a scorelike representation and a procedure to transform it into an audio signal (see Chapter 12 and [VeGS98]). A symbolic (scorelike) representation of music can be either directly generated by the content creator (e.g., a composer) or obtained as a result of automatic transcription from real-world signals. However, a severe limitation occurs in this latter case in the sense that the automatic transcription of an audio signal into a musical score is only feasible for a limited class of audio material [Moor77, Sche00].

A general audio coding system, however, should be able to cope with a wide range of real-world signals. Hence, it can only employ signal representations that can be derived automatically from real-world signals. T/F coders use an orthogonal set of base functions to represent the waveform of the audio signal, for instance, by the OLA MDCT paradigm. In contrast, the parametric audio coding paradigm uses a signal representation that is based on parameterized models, in which the actual signal is described by means of model parameters (such as the frequency and amplitude of a pure tone).

Parametric modeling approaches have long been used in musical instrument analysis/synthesis and music synthesizers [RisM69] as well as in vocoders for speech [Dudl39]. Sinusoidal modeling, a specific type of parametric modeling, has been successfully applied to speech signals [McaQ86] and audio signals [GeoS92]. It has been extended with noise modeling for use with musical sounds [Serr89, Serr97]. Recently, the parametric modeling paradigm has been employed for low-bit-rate coding of general audio signals [PuEF98, Levi98, OomB99]. In MPEG-4, parametric speech coding is provided by HVXC (see Chapter 10), and parametric audio coding is provided by the HILN tools described in Section 11.6.

11.5.1 Source and Perceptual Models

Parametric audio coding heavily relies on the availability of source models that allow the description of signals or signal components with a small number of parameters. A brief overview of the applicable models is given below, and more details can be found in [Purn99].

☞ **Physical models:** The sound generation of most musical instruments can be described by an excitation (e.g., pulselike, periodic, or noiselike) in conjunction with multiple coupled resonances [FleR91]. An example of such a system is a plucked string instrument, where the plucking is a pulselike excitation that triggers the periodic vibration of the string. This sound is further shaped by the resonances of the instrument's body. The automatic estimation of the parameters of such physical models often is feasible from recordings of a single known instrument. This approach, however, becomes difficult when the recognition of the instrument or even the separation of several instruments in an orchestra is required. Therefore, its application in coding natural signals is currently restricted to the area of speech coders, in which a voiced or unvoiced excitation and the vocal tract are modeled.

☞ **Sinusoidal models:** Many natural audio signals are composed of several fairly steady tones. A relatively simple mathematical approach to model such a tonal signal, $x(t)$, is to treat it as the superposition of N individual sinusoidal components. Each of the components is described by slowly varying parameters for the amplitude $a_i(t)$, the frequency $f_i(t)$, and a start phase φ_i, thus forming a sinusoidal trajectory.

$$\hat{x}(t) = \sum_{i=1}^{N} a_i(t) \cdot \sin\left(\varphi_i + 2\pi \int_0^t f_i(\tau)\, d\tau\right) \tag{11.2}$$

Sinusoidal modeling was first used for the analysis/synthesis of musical instruments [RisM69] as its parameters can be easily interpreted and modified. It was later applied to speech [McaQ86] and audio signals [Serr89, GeoS92, Serr97].

To adapt to the time-variant properties of an audio signal, the signal is commonly processed as a sequence of short time intervals (frames) of duration T. Hence, the model parameters are sampled once per frame such that the n-th frame is represented by the parameters $a_i(nT)$ and $f_i(nT)$.

☞ **Transient models:** In principle, transient signals (for instance, those from percussion instruments) can also be described by a sinusoidal model. This, however, requires the capability to cope with rapid variations of the model parameters. Although an increased temporal resolution of the parameters can be accomplished by using shorter frames, it can be more efficient to apply an additional temporal envelope function to the amplitude of appropriate sinusoidal components [EdPF96]. Similar to music synthesizers, such an envelope can be described by its temporal position and the attack and decay rates. An alternative approach approximates the DCT spectrum of the transient by means of a sinusoidal model [VeLM97]. Techniques that are also known from transform or wavelet coding can be used to represent a transient [Levi98].

☞ **Noise models:** The waveform of a noiselike signal unfortunately does not contain much redundancy that could be exploited by a source model. However, with respect to perception, it is not necessary to maintain the waveform of noiselike signals. Hence, such signals can be adequately modeled by a random noise with appropriate spectral and temporal envelopes. The spectral shaping of the noise can be efficiently performed by filter structures as they are used in LPC-based speech coders. Temporal shaping can be accomplished by time-variant filter coefficients. The use of the so-called *reflection coefficients* enables an easy adaptation of the filter order to the required level of spectral detail. Alternative models are based on a piecewise linear noise spectrum approximation or Bark-band/ERB noise modeling [Good97].

☞ **Harmonic models:** Because the frequencies of the partials of a harmonic tone are multiples of its fundamental frequency f_0, the sinusoidal modeling can be extended accordingly [PuEF98]. Thus, only the model parameter f_0 and the amplitudes a_i of the partials are required. The number of parameters can be further reduced by modeling the spectral shape of a harmonic tone using LPC models similar to those applicable to noise modeling [PurM00]. To account for slight inharmonicities of the higher partials—as, for example, caused by a stiff string—the model can be extended by a *stretching* parameter.

Many of these source models can only be applied to a limited class of signals. In order to allow the efficient representation of arbitrary audio signals, different source models have to be combined to form a hybrid model. The audio signal is decomposed into several components, each of which can be adequately modeled by one of the available source models.

To achieve high coding efficiency, a parametric audio coding scheme also needs to take into consideration the perception of the decoded signal by the listener. For this purpose, two types of perceptual models are of interest:

☞ **Perceptual models for parameter quantization:** The efficient coding of model parameters requires quantizers that are designed to take into account perceptual criteria. These criteria might even depend on the current signal content. All quantizers for frequency and amplitude parameters should hence be adjusted to the audibility thresholds for deviations known as *just noticeable differences* (JNDs). For frequency parameter quantizers, the step size should be approximately proportional to the frequency-dependent critical bandwidth (i.e., the Bark scale). For amplitude parameter quantizers, a logarithmic characteristic is appropriate. Furthermore, the quantization of the parameters for spectral and temporal envelopes needs to relate to the sensitivity of the human ear in detecting deviations.

Subjective evaluations have shown that the relevancy of sinusoid phase parameters is generally so low that they do not need to be transmitted [PuEF98]. However, in this case the temporal structure of transients must be maintained by using an appropriate transient model. Moreover, phase continuity of sinusoidal trajectories spanning several frames must be ensured.

☞ **Perceptual models for component selection:** For very low bit-rate coding applications, only the parameters of a few signal components can be conveyed in the bitstream. Hence, it is important to select those components that are most relevant for the perceived subjective quality of the decoded signal.

11.5.2 Parametric Encoding and Decoding Concepts

The optimal decomposition of an audio signal into components suitable for parametric representation minimizes the number of bits required by the coded model parameters to reconstruct a signal from these parameters at a given perceptual quality. To obtain such an optimal decomposition, all source and perceptual models used in a parametric coder need to be taken into account.

Signal Decomposition and Parameter Estimation   The decomposition of the signal into components and the estimation of model parameters are two interdependent tasks. An efficient approach that takes into account these interdependencies is the iterative component extraction and parameter estimation, leading to an analysis-by-synthesis scheme. For each signal component, an appropriate source model is selected and then the model parameters are estimated. The extracted signal component is then resynthesized from these parameters, and all resynthesized components are accumulated. By using all these accumulated components, a residual is calculated, containing only those signal components that have not been modeled.

Special attention is required for signal components that can be represented by different alternative source models. For example, modeling the partials of a harmonic tone as individual sinusoids should be avoided if a source model for a harmonic tone is available. Another problem is the proper discrimination between sinusoidal and noiselike signal components.

Parameter Quantization and Coding   To convey the model parameters in a bitstream of a given capacity (i.e., bit rate), the perceptually most relevant components should be selected and their model parameters should be quantized according to perceptual criteria. Correlation between the model parameters within one frame and between consecutive frames can be exploited by parameter prediction. The quantized parameters are finally entropy-coded and multiplexed to form a bitstream.

Parameter Decoding and Signal Synthesis   At the decoder, the model parameters can be reconstructed from the bitstream by decoding and dequantization. Each signal component then can be synthesized according to its source model, and finally all synthesized components are added to compose the output signal of the audio decoder. The signal components of consecutive time frames can be either combined by OLA or synthesized by using model parameter interpolation.

Bit-Rate Scalability and Signal Modification   Besides providing a compact signal representation, the abstract models used for parametric audio coding allow further interesting functionalities:

☞ Bit-rate scalability can be achieved by transmitting the parameters of the perceptually most important signal components in a base layer bitstream and the parameters of further signal components in additional enhancement bitstreams [FeSF98, Verm00]. In the case of limited-transmission bandwidth, only the base layer is received and decoded. If a higher bandwidth is available, one or more enhancement layers are also received. Together with the base layer information, the audio signal can then be reconstructed at a higher quality.

☞ Time-scaling and pitch-shifting can easily be implemented in a parametric audio decoder. By varying the length of the synthesized time frames, the playback speed of the signal can be modified in the decoder without affecting the pitch of the signal. On the other hand, it is possible to multiply all frequency parameters by a given factor prior to synthesis to alter the pitch of the decoded signal without any influence on the playback speed.

11.6 MPEG-4 HILN PARAMETRIC AUDIO CODING

The MPEG-4 parametric audio coding tools (called HILN) are defined in Version 2 of the MPEG-4 Audio standard [MP4V2-3]. By means of a parametric signal representation, they allow the coding of general audio signals at very low bit rates, down to about 4 kbit/s [PurM00]. Based on the general principles introduced in Section 11.5, the specific architecture and design of the HILN tools will now be described.

It should be noted that, as usual for MPEG standards, only the bitstream format and the decoding process are defined in the normative part of the MPEG standard, whereas for the encoding process only an example is given in an informative annex of the standard. This approach allows the optimization of the encoding process as well as its adaption to special requirements (e.g., low encoder complexity or transmission over error-prone channels) without affecting the interoperability of encoders and decoders conformant to the standard.

In spite of the fact that MPEG standards do not define a normative encoding process, it is beneficial to start the description of the HILN tools with an overview of the techniques applied in a typical encoder. This is followed by a description of the normative aspects—namely, the bitstream format, which defines the coded representation of the quantized parameters and the decoding process.

11.6.1 HILN Parametric Audio Encoder

Figure 11.17 shows the block diagram of an HILN parametric audio encoder as used during the development of the MPEG-4 Audio standard [MPEG4-3]. HILN encoding is performed on frames of input samples, using overlapping window functions. A typical frame length (hop size) for signals sampled at 16 kHz is 32 ms. Each frame of the input signal is decomposed into different components, and the model parameters for the components' source models are estimated. The following three types of components are used in HILN, listed here with the major parameters of their corresponding source models:

☞ **Individual sinusoids** are described by their frequencies and amplitudes.

☞ A **harmonic tone** is identified by its fundamental frequency, amplitude, and the spectral envelope of its partials.

☞ A **noise** component is marked by its amplitude and spectral envelope.

Sinusoidal components that live for more than one frame are treated as sinusoidal trajectories to ensure phase continuity at frame boundaries. Although the phase parameters of sinusoidal components are estimated as well, they usually are not conveyed in the bitstream, thus exploiting the low phase sensitivity of the human ear. The modeling of transient signals is improved by optional parameters describing the components' temporal envelope. To represent the spectral envelope of the harmonic tone and of the noise component, spectral modeling (as known from the theory of LPC for speech signals) is employed. The frequency response of an all-pole LPC synthesis filter is used as a spectral envelope, and the number of LPC parameters used to describe the spectral envelope is adapted to provide the desired level of spectral detail.

As shown in Figure 11.17, the signal decomposition and parameter estimation can be implemented as a three-step process:

☞ **Sinusoidal extraction:** The sinusoidal components are extracted from the current frame of the input signal.

☞ **Harmonic tone detection:** If several sinusoids share a common fundamental frequency, they are grouped as a single harmonic tone to permit efficient coding, whereas all remaining sinusoids are handled as individual sinusoids.

☞ **Noise definition:** After extracting the sinusoidal components, the magnitude spectrum of the residual signal is used to find the parameters of the noise component.

Due to the very low target bit rates of typically 6 to 16 kbit/s, only the parameters for a small number of components can be transmitted. Therefore, a perceptual model is employed to select those components that are most important for the perceptual quality of the signal.

11.6.1.1 Signal Decomposition and Parameter Estimation
The three-step process of signal decomposition and parameter estimation is now explained in greater detail, starting with the windowing of the input signal and the estimation of temporal envelope parameters for transients.

Windowing   Normally, the input signal is processed in frames obtained by means of a window function with 50% overlap—that is, the window of each frame reaches up to the centers of both neighboring frames. However, if the energy of the signal components in the overlap regions outside the actual

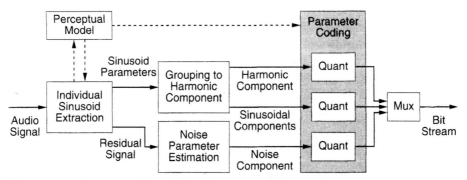

Fig. 11.17 Architecture of an HILN parametric audio encoder.

frame in relation to the energy within the frame exceeds a given threshold, a window with shorter overlap is selected. As a result, the influence of strong transients on neighboring frames is reduced [EdPF96].

Estimation of Temporal Envelope   For the sinusoidal modeling of signal segments containing strong temporal amplitude variations, parameters approximating the temporal envelope are estimated for the full input signal prior to decomposition. This envelope estimation should be focused on the strongest variation within a frame, as usage of the temporal envelope can be signaled individually for each sinusoidal component.

To find the temporal amplitude envelope of the input signal, an FIR Hilbert transformer can be used to construct an analytic signal corresponding to the positive frequency components in the current frame. The magnitude of this complex signal is normalized with respect to its maximum and then approximated by a triangular shape. The approximation is specified by parameters for the time of its maximum t_{max} and the angles for left slope r_{atk} (attack rate) and right slope r_{dec} (decay rate). These parameters can be obtained by using regression techniques with appropriate weighting, depending on the amplitude and the distance from the maximum [EdPF96]. For example, Figure 11.18 shows the original envelope and the approximated envelope as specified by the obtained envelope parameters for one frame of the signal *Castanets* containing a strong transient *(clapp)* [SQAM].

Extraction of Individual Sinusoids   To decompose the current frame of the input signal into sinusoidal components and to estimate their parameters,

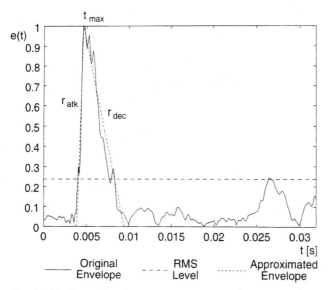

Fig. 11.18 Normalized temporal amplitude envelope $e(t)$ and envelope parameters t_{max}, r_{atk}, and r_{dec} for one frame of the signal *Castanets*.

an iterative procedure based on an analysis/synthesis loop as shown in Figure 11.19 can be used. This is the most complex task to be performed at the encoder.

Prior to the start of the iterative procedure, the magnitude spectrum $|X(f)|$ of the input signal is calculated with a fast Fourier transform (FFT). Furthermore, the accumulator collecting the resynthesized versions of the extracted sinusoids is reset; that is, $s_0(t) = 0$. For each cycle i of the loop, the magnitude spectrum $|S_{i-1}(f)|$ of the synthesized signal $s_{i-1}(t)$ containing all $i-1$ previously extracted sinusoids is calculated first. This spectrum is subtracted from the original spectrum $|X(f)|$, and the resulting difference is limited to positive values and squared to obtain a power spectrum $|E_i(f)|$, indicating how much the original spectrum $|X(f)|$ exceeds the synthesized spectrum $|S_{i-1}(f)|$.

The next step is the detection of the maximum ratio of $|E_i(f)|$ over an estimated masked threshold $|M_{i-1}(f)|$, which characterizes the psychoacoustic effect of simultaneous masking produced by the synthesized signal $s_{i-1}(t)$. The location of this maximum is used as a coarse frequency estimate, $f_{c,i}$, of the i-th sinusoid.

The masked threshold $|M_{i-1}(f)|$ can be calculated quite efficiently by employing a parametric psychoacoustic model that makes direct use of the frequency and amplitude parameters of the extracted sinusoids and thus avoids an additional spectral analysis of the synthesized signal $s_{i-1}(t)$. By including such a perceptual model in the analysis/synthesis loop, it is possible to extract the perceptually most relevant sinusoids first.

The accuracy of $f_{c,i}$ is determined by the FFT frequency resolution, which is in the order of 15 Hz for a typical window length. The frequency resolution of the human ear, however, is much higher—in fact, on the order of 3 Hz for tones below 500 Hz. In order to obtain a higher resolution for the frequency parameter f_i and improve the modeling of sinusoids with slowly varying frequencies and amplitudes, a high-accuracy parameter estimation is performed on the residual $r_i(t) = x(t) - s_{i-1}(t)$ in a frequency interval surrounding $f_{c,i}$. This

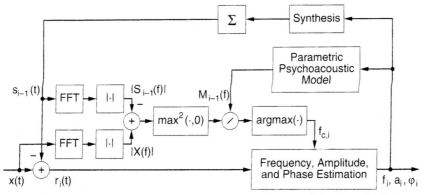

Fig. 11.19 Analysis/synthesis loop for the extraction and parameter estimation of individual sinusoids.

can be accomplished by using a narrow bandpass filter centered around $f_{c,i}$ followed by a regression-based tracking of the instantaneous phase of the analytic (i.e., complex) bandpass signal [EdPF96]. Assuming the phase is a parabolic function of time, the mean frequency as well as the rate of linear frequency variation over time (*sweep rate*) can be estimated.

Once these frequency parameters, *mean frequency* and *sweep rate*, are obtained, amplitude and phase are estimated. For this purpose, a complex sinusoid is synthesized according to the estimated frequency parameters. Amplitude and phase are then determined as the magnitude and phase of a complex correlation coefficient between the complex sinusoid and the residual $r_i(t)$. The same procedure is applied to the complex sinusoid additionally multiplied by the temporal envelope function generated from the estimated envelope parameters. The best match is selected to determine whether the temporal envelope should be applied to the current sinusoid. Based on the estimated parameters, f_i, a_i, φ_i, and the *sweep rate*, the extracted sinusoidal component is resynthesized in preparation for the next cycle of the analysis/synthesis loop.

The iterative procedure can be terminated when all relevant sinusoids have been extracted, or when the desired bit rate does not allow conveying the parameters of further components.

To identify sinusoidal trajectories that span several frames, the frequency and amplitude parameters of the sinusoids extracted for the current time frame can be compared to those of the previous frame. If the parameters of the best potential predecessor are sufficiently similar, it is assumed that this sinusoidal component is continuing from the previous frame. Otherwise it is handled as a new sinusoidal component (see Section 11.6.2).

Grouping as Harmonic Tone   The parameter estimation for a harmonic tone can be based on the parameters of the extracted individual sinusoids by searching for patterns of frequencies that are multiples of a common fundamental frequency. The goal of this grouping is an efficient representation that uses only the fundamental frequency and the amplitudes of the partials.

Noise Component Modeling   The final residual signal from the iteration loop for extracting sinusoids can be regarded as noise component. Hence, its amplitude and the LPC parameters describing its spectral envelope are estimated. The LPC-based spectral modeling approach can be strongly influenced by sinusoids not extracted in the loop. Therefore, the noise modeling requires a careful selection of the number of iterations of the loop for extracting sinusoids. Particularly at very low bit rates, the number of extracted sinusoids usually exceeds significantly the number of sinusoids that can be transmitted as harmonic and individual lines in order to obtain a suitable noise component.

11.6.1.2 Aspects of Encoder Optimization   As explained previously, the most critical and complex task in the HILN encoder is the decomposition of the input signal into the perceptually most relevant components and the estimation of their component parameters. Initially, the HILN encoder development focused primarily on maximum coding efficiency in order to prove the merit of the HILN parametric audio coding tools in the MPEG-4 standardization process. Although this has led to rather high computational complexity of the encoder implementation, later work aimed at complexity reduction so as to speed up the encoding process significantly.

For the encoder, as presented in Figure 11.17, the analysis/synthesis loop for the extraction of sinusoidal components—which is by far the most complex module—is implemented in the time domain. However, as the sinusoidal signal components are highly localized in frequency, a frequency domain implementation can offer major advantages in terms of computational complexity. With an optimized encoder based on this frequency domain approach, real-time encoding on common PCs is easily possible with only a minor penalty in terms of coding efficiency [PuME00].

11.6.2 HILN Bitstream Format and Parameter Coding

The estimated parameters for the components selected in the encoder are quantized, coded, and multiplexed to form a bitstream.

11.6.2.1 Bitstream Format Overview

The bitstream is structured as a sequence of access units (AU), each carrying the compressed data of one frame of the audio signal. In addition, the general setup information for the decoder (such as audio bandwidth, frame length, etc.) is conveyed out-of-band in the so-called decoder-specific configuration (see Chapter 3).

An HILN AU starts with main control data—that is, flags signaling the presence of a harmonic tone or a noise component and the number of individual sinusoids in the current frame. The *continue flags* signal which components are continued from the previous frame. If parameters for the optional temporal envelope are transmitted, there is also an *envelope flag* for each component. Most of the bits in an AU convey the quantized and entropy-coded parameters for the components.

The sequence of AUs as described previously form a nonscalable bitstream, which can also be used as base layer for bit-rate scalable configurations. In the latter case, one or more enhancement layer bitstreams are used, each of which carries the parameters of further individual sinusoids (not yet transmitted in lower layers) as well as the corresponding control data, like the *continue flags*.

To enable the use of UEP for improved performance in the case of error-prone transmission, each bitstream element in an HILN (base layer) bitstream AU is assigned to a certain error-sensitivity category (ESC). The bitstream format is defined in a way that an AU starts with the most important bits (i.e., the bitstream elements in ESC0), which are followed by the bitstream elements in ESC1, ESC2, ESC3, and ESC4 that are less error-sensitive. More details on the definition of the five ESCs for HILN can be found in [PuEM01].

11.6.2.2 Parameter Quantization and Subdivision Coding Nonuniform quantization is used for amplitude and frequency parameters to take into account the perception of just noticeable differences. Hence, the amplitudes of all components are quantized on a logarithmic scale with 1.5 dB step size. Frequencies of individual sinusoids are quantized on a Bark scale with 1/32 Bark step size—that is, approximately 3 Hz steps below 500 Hz and approximately 10 cent[4] steps above 500 Hz. For the fundamental frequency of a harmonic tone, a logarithmic scale with approximately a 5-cent step size is used instead.

For the parameters of components continued from the previous frame, differential coding (i.e., interframe prediction) is employed for the quantized parameters. Because these changes in frequency and amplitude of a continued component are not uniformly distributed, a variable length code is used for entropy coding.

To encode the quantized parameters of the new individual sinusoids (also called lines) in a frame, a sophisticated technique is used that is referred to as subdivision coding (SDC) [PurM00]. It is motivated by the observation that it is not required to convey the frequency/amplitude parameter pairs of new individual lines in a specific order. By allowing any arbitrary permutation of this set of N parameter pairs, a redundancy of $\log_2 (N!)$ bits per frame can be exploited. To accomplish this, first the parameter sets are sorted according to their frequencies f_i. Then each f_i is entropy coded according to its probability distribution, which depends on the value of the previous parameter f_{i-1} and the number of remaining parameters $N-i$. The coding of the corresponding amplitude parameter a_i is described below.

Assuming, for example, that the N frequencies f_i are uniformly distributed in the interval $[0, F]$, the probability density function (PDF) of the first frequency f_1 (which is also the lowest) is given by $p_1(f_1) = N/F \cdot (1 - f_1/F)^{N-1}$.

To implement SDC, the quantized frequencies f'_i are sorted to achieve $0 \leq f'_1 \leq f'_2 \leq \ldots \leq f'_N < F'$ and then coded in this order. For each frequency f'_i, the range of possible values is limited by the previously coded frequency f'_{i-1} (0 for the first one) and the highest possible frequency $F'-1$—that is, the audio bandwidth of the coded signal. This interval $[f'_{i-1}, F'-1]$ is divided into two partitions of equal probability: $P = 1/2$. One bit is transmitted to indicate in which of these two partitions the actual frequency f'_i is located. This partition becomes the new interval for the frequency, which is then divided again. This subdivision step is repeated recursively until the remaining frequency interval consists of only a single possible value.

Because the real probability distribution of the f'_i differs somewhat from the uniform distribution assumed in the example above, empirically collected data were used to design the boundary tables for the subdivision process in the standard [MPEG4-3]. These tables specify the exact positions of the

---

4. A semitone interval corresponds to 100 cent—that is, 1 cent denotes a frequency ratio of $1{:}2^{(1/1200)}$.

boundaries between partitions. As these positions differ according to the number $N-i$ of frequency parameters that follow, the boundaries are stored for zero up to seven following parameters. If more than seven frequencies follow, the boundary table for seven is used. The boundaries are stored in a normalized form and have to be scaled to the actual width $F'-f'_{i-1}$ of the initial interval when used. Only the boundaries for the first five subdivision steps are stored. If further subdivision is necessary, the boundary is located in the middle of the remaining frequency interval, assuming a locally uniform distribution.

For each frame, the maximum amplitude of the components is transmitted as *global gain* with 6 dB resolution. The amplitudes of new components are then coded in relation to this maximum. For new individual lines, these relative amplitudes usually are quantized using a 3-dB step size, and SDC with a special boundary table is used to achieve entropy coding.

Normally, a random start phase is used for new individual lines and the partials of a harmonic tone. Nonetheless, it is optionally possible to transmit the start phase for selected new lines in the bitstream to achieve deterministic behavior of the decoder. For such phase data, uniform quantization in the interval $[0, 2\pi]$ with 5-bit resolution is applied.

11.6.2.3 Spectral Model Parameter Coding   To quantize and code the LPC parameters describing the spectral envelopes of the harmonic and noise components, a logarithmic area ratio (LAR) representation is adopted [PalK95]. Unlike the line spectrum frequencies (LSFs) representation commonly used in speech coding (see Chapter 10), LARs allow changes of the LPC filter order from frame to frame while still permitting efficient interframe prediction of LPC parameters. This interframe prediction uses a predictor coefficient of 0.75 or 0.5 and takes into account the mean LARs of *default* lowpass spectra.

The LARs for the noise and harmonic spectrum are quantized uniformly with empirically determined quantizer step sizes of approximately 0.1 (harmonic) or approximately 0.3 (noise). The interframe prediction error of the LARs has a PDF that is well approximated by a Laplacian distribution. After quantization with the desired step size, the probability of the quantized values decreases approximately by a factor of one-half when the absolute value is increased by one. This property permits entropy coding with an algorithmic code[5] instead of a Huffman code without any loss of coding efficiency [PurM00].

In order to enable proper decoding of the LPC parameters, the filter order (i.e., the number of LPC parameters) has to be transmitted first. In the case of a harmonic tone, the actual number of partials is also transmitted to enable the reconstruction of their amplitudes from the spectral envelope.

---

5. An algorithmic code is one in which the codewords are defined by a simple algorithm so that no codebook table is required for encoding and decoding.

11.6.3 HILN Parametric Audio Decoder

The block diagram of the HILN parametric audio decoder (audio object type ER HILN) is shown in Figure 11.20. First, the parameters of the components are decoded, and then the component signals are resynthesized according to the transmitted parameters. By combining these signals, the output signal of the HILN decoder is obtained.

11.6.3.1 Parameter Decoding and Signal Synthesis
The procedure for decoding and dequantization of the component parameters follows directly from the description of the bitstream format and parameter coding. In this way, the parameters of all components conveyed in the base layer bitstream are decoded. If one or more enhancement layers are available to the decoder, the parameters of additional individual lines conveyed with these bitstreams are decoded. Finally, all parameters are passed to the corresponding synthesis tools. If, however, time-scaling or pitch-shifting is desired, the involved parameters (like frame length or the components' frequencies) are modified accordingly.

All sinusoids (i.e., all individual and harmonic lines as decoded from the bitstream) are synthesized in the same way. Overlapping synthesis windows are used to obtain smooth fade-in and fade-out of sinusoids that begin or end in the current frame. For a sinusoid continuing from one frame to the next, the OLA procedure is replaced by synthesis of a single sinusoidal trajectory to guarantee phase continuity. In this case, amplitude and frequency parameters are linearly interpolated between the two frame centers. Usually the start phase of a new (born) sinusoid is not conveyed in the bitstream; so a random start phase is used, instead. To avoid phase discontinuity for a sinusoid that is coded as individual line in one frame and as partial of a harmonic tone in the adjacent frame, the sinusoid is reconnected in the decoder if the corresponding frequencies and amplitudes are similar to each other. The sinusoidal components that have an envelope flag set are multiplied with the temporal amplitude envelope reconstructed from the envelope parameters encoded in the bitstream.

To synthesize the noise component, a random number generator producing a white noise signal is used. This white noise is filtered by an IIR all-pole

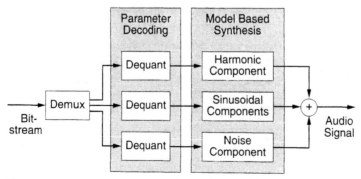

Fig. 11.20 Architecture of an HILN parametric audio decoder.

filter (i.e., an LPC synthesis filter) to shape its spectrum according to the LPC spectral envelope parameters decoded from the bitstream. This output is then multiplied by a gain factor calculated from the noise amplitude parameter conveyed in the bitstream. To obtain smooth transitions between frames, a windowed OLA scheme is used for the noise components synthesized in neighboring frames.

11.6.3.2 Integrated Parametric Coder

In MPEG-4, the parametric audio and speech coders—namely, HILN and HVXC (see Chapter 10)—can be combined to form an integrated parametric coder covering a wider range of signals and bit rates (audio object type ER parametric). This integrated coder can operate in *switched mode* or *mixed mode* and supports speed and pitch change functionalities. The switched mode provides the ability to alternate dynamically between HILN and HVXC. By means of a speech/music classification tool in the encoder, the HVXC and HILN coders can be selected automatically for speech and music signals, respectively. Such automatic HVXC/HILN switching has been successfully demonstrated [M2481], and the classification tool is described in an informative annex of the MPEG-4 Audio standard [MPEG4-3]. The mixed mode provides for simultaneous use of both coders and combines their output signals. In this way, a speech signal that is coded with HVXC at 2 kbit/s, for example, can be combined with background music coded with HILN at 4 kbit/s. Although this mode can be easily used if the speech and music signals are available separately, the automatic decomposition of a mixed signal into a speech and a music component is a difficult task and the subject of ongoing research [EdlP98].

11.7 SUMMARY

This chapter described the technology the MPEG-4 Audio standard provides for the coding of general audio signals (general audio coding), as opposed to specific signal classes, such as speech signals. Most of the MPEG-4 coding tools for general audio coding adhere to the paradigm of filterbank-based (T/F) coding of spectral coefficients in general and MPEG AAC in particular. Introduced within MPEG-2 Audio, the AAC coder is a powerful multichannel coder with a rich set of coding tools that allow the coder to adapt flexibly to arbitrary input signals. Building on this core technology, MPEG-4 defines a number of extensions to further enhance compression performance and support new modes of operation and functionalities.

The PNS tool achieves a compact parametric coding of noiselike signal components. LTP exploits the redundancy of stationary signals and effectively replaces the MPEG-2 AAC prediction tool at about one-half of its complexity. The TwinVQ coding kernel can be used to replace the standard AAC quantization/coding tools to enable operation at extremely low bit rates (down to 6 kbit/s/ch), which is of particular interest in the context of scalable audio coding.

The AAC-LD coder enables the coding of multichannel general audio signals with algorithmic delays as low as 20 ms, thus fulfilling the requirements for high-quality two-way communication applications. Improved performance under error-prone transmission conditions is achieved by using the ER tools.

One major novel functionality, which was not supported by any previous MPEG Audio standard, is bit-rate scalability—that is, the ability to meaningfully decode subsets of the bitstream and thus to enable the subsequent distribution via transmission channels not known beforehand or channels with varying transmission capacity (e.g., Internet or wireless transmission). MPEG-4 Audio offers two different approaches to bit-rate scalability. The large-step scalability approach allows the combination of existing coding tools to build coders with several layers that may gradually enhance coding precision, reproduction bandwidth, and the number of reproduction channels. Further interesting combinations arise if a coder other than AAC is used as base layer coder (e.g., narrow-band CELP or TwinVQ). On the contrary, the FGS mode is based on BSAC of the AAC spectral coefficients and permits quasi-continuous bit-rate scalability for this type of coder, yielding good results for high enough bit rates.

Finally, the MPEG-4 HILN parametric audio coder is considered to be a general audio coder in the wider sense, as it aims at the good reproduction of arbitrary input audio signals by means of a flexible decomposition of the input signal into distinct sound components—namely, a harmonic tone, individual sinusoids, and a noise component. The coder can operate on very low bit rates (down to about 4 kbit/s), provides bit-rate scalability, and can be combined with the MPEG-4 HVXC parametric speech coder. Due to the underlying principle of parametric signal representation, the HILN coder provides the pitch change and speed change functionalities without requiring noticeable extra computation.

11.8 ACKNOWLEDGMENTS

The authors wish to express their sincere thanks to Viti Bramigk of Fraunhofer IIS-A for her valuable assistance in preparing the text of this chapter as well as to Ian Burnett, Bernd Edler, Aníbal Ferreira, and Fernando Pereira for reviewing the manuscript of this chapter and for their valuable suggestions for improvements.

11.9 REFERENCES

[AAC] ISO/IEC 13818-7:1997. *Generic Coding of Moving Pictures and Associated Audio Information—Part 7: Advanced Audio Coding*, 1997.

[AGHS99] Allamanche, E., R. Geiger, J. Herre, and T. Sporer. *MPEG-4 Low Delay Audio Coding Based on the AAC Codec*. 106th AES Convention, Preprint 4929. Munich, 1999.

[BBQF97] Bosi, M., K. Brandenburg, S. Quackenbush, L. Fielder, K. Akagiri, H. Fuchs, M. Dietz, J. Herre, G. Davidson, and Y. Oikawa. "ISO/IEC MPEG-2 Advanced Audio Coding." *Journal of the AES*, 45(10): 789–814. October 1997.

[BeeS92] Beerends, J., and J. Stemerdink. "A Perceptual Audio Quality Measure Based on a Psychoacoustic Sound Representation." *Journal of the AES*, 40(12): 963–978. 1992.

[BEHE92] Brandenburg, K., E. Eberlein, J. Herre, and B. Edler. "Comparison of Filterbanks for High Quality Audio Coding." *Proc. IEEE ISCAS*, San Diego, 1992.

[BHJM91] Brandenburg, K., J. Herre, J. Johnston, Y. Mahieux, and E. F. Schroeder. *ASPEC: Adaptive Spectral Perceptual Entropy Coding of High Quality Music Signals*. 90th AES Convention, Paris, Preprint 3011. 1991.

[Blau83] Blauert, J. *Spatial Hearing*. Cambridge, MA: MIT Press. 1983.

[Bosi99] Bosi, M. "Filterbanks in Perceptual Audio Coding." *Proc. of the 17th International AES Conference on High Quality Audio Coding*, Florence, 1999.

[BraS92] Brandenburg, K., and T. Sporer. "NMR and Masking Flag: Evaluation of Quality Using Perceptual Criteria." *Proc. of the 11th International AES Conference on Audio Test and Measurement*, Portland, 1992.

[BSDJ92] Brandenburg, K., G. Stoll, Y. Dehéry, J. Johnston, L.v.d. Kerkhof, and E. Schroeder. *The ISO/MPEG-Audio Codec: A Generic Standard for Coding of High Quality Digital Audio*. 92nd AES Convention, Vienna, Preprint 3336. 1992.

[Dudl39] Dudley, H. "The Vocoder." *Bell Labs Rec.*, 17: 122. 1939.

[Edle89] Edler, B. "Codierung von Audiosignalen mit überlappender Transformation und adaptiven Fensterfunktionen." *Frequenz*, 43: 252–256 (in German). 1989.

[EdlP98] Edler, B., and H. Purnhagen. *Concepts for Hybrid Audio Coding Schemes Based on Parametric Techniques*. AES 105th Convention, San Francisco, Preprint 4808. 1998.

[EdPF96] Edler, B., H. Purnhagen, and C. Ferekidis. *ASAC—Analysis/Synthesis Audio Codec for Very Low Bit Rates*. 100th AES Convention, Copenhagen, Preprint 4179. 1996.

[FBDD96] Fielder, L., M. Bosi, G. Davidson, M. Davis, C. Todd, and S. Vernon. "AC-2 and AC-3: Low-Complexity Transform-Based Audio Coding," *Collected Papers on Digital Audio Bit-Rate Reduction* (N. Gilchrist and C. Grewin, Eds.), pp. 54–72. New York: AES, 1996.

[FeSF98] Feiten, B., R. Schwalbe, and F. Feige. *Dynamically Scalable Audio Internet Transmission*. 104th AES Convention, Amsterdam, Preprint 4686. 1998.

[FleR91] Fletcher, N., and T. Rossing. *The Physics of Musical Instruments*. New York: Springer. 1991.

[GeoS92] George, E. B., and M. J. T. Smith. "Analysis-by-Synthesis/Overlap-Add Sinusoidal Modeling Applied to the Analysis and Synthesis of Musical Tones." *Journal of the Audio Engineering Society*, 40(6): 497–516. June 1992.

[Good97] Goodwin, M. *Adaptive Signal Models: Theory, Algorithms, and Audio Applications*. Ph.D. thesis, University of California, Berkeley, 1997.

[Gril97] Grill, B. *A Bit Rate Scalable Perceptual Coder for MPEG-4 Audio*. 103rd AES Convention, New York, Preprint 4620. 1997.

[GriT98] Grill, B., and B. Teichmann. *Scalable Joint Stereo Coding*. 105th AES Convention, San Francisco, Preprint 4851. 1998.

[HABD98] Herre, J., E. Allamanche, K. Brandenburg, M. Dietz, B. Teichmann, B. Grill, A. Jin, T. Moriya, N. Iwakami, T. Norimatsu, M. Tsushima, and T. Ishikawa. *The Integrated Filterbank Based Scalable MPEG-4 Audio Coder.* 105th AES Convention, San Francisco, Preprint 4810. 1998.

[HeBD94] Herre, J., K. Brandenburg, and D. Lederer. *Intensity Stereo Coding.* 96th AES Convention, Amsterdam, Preprint 3799. 1994.

[HeJ97b] Herre, J., and J. Johnston. "A Continuously Signal-Adaptive Filterbank for High-Quality Perceptual Audio Coding." *Proc. IEEE Workshop on Applications of Signal Processing to Audio and Acoustics,* Mohonk, 1997.

[Helm72] Hellman, R. P. "Asymmetry of Masking between Noise and Tone." *Perception and Psychophysics,* 11: 241–246. 1972.

[HerJ96] Herre, J., and J. D. Johnston. *Enhancing the Performance of Perceptual Audio Coders by Using Temporal Noise Shaping (TNS).* 101st AES Convention, Los Angeles, Preprint 4384. 1996.

[HerJ97] Herre, J., and J. D. Johnston. *Exploiting Both Time and Frequency Structure in a System that Uses an Analysis/Synthesis Filterbank with High Frequency Resolution.* 103rd AES Convention, New York, Preprint 4519. 1997.

[Herr99] Herre, J. "Temporal Noise Shaping, Quantization and Coding Methods in Perceptual Audio Coding: A Tutorial Introduction." *Proc. of the 17th International AES Conference on High Quality Audio Coding,* Florence, 1999.

[HerS98] Herre, J., and Donald Schulz. *Extending the MPEG-4 AAC Codec by Perceptual Noise Substitution.* 104th AES Convention, Amsterdam, Preprint 4720. 1998.

[IwaM96] Iwakami, N., and T. Moriya. *Transform Domain Weighted Interleave Vector Quantization (TwinVQ).* 101st AES Convention, Los Angeles, Preprint 4377. 1996.

[IwMM95] Iwakami, N., T. Moriya, and S. Miki. "High-Quality Audio-Coding at Less Than 64 kbit/s by Using Transform-Domain Weighted Interleave Vector Quantization (TWINVQ)." *Proc. IEEE ICASSP,* pp. 3095–3098. Detroit, 1995.

[JayN84] Jayant, N., and P. Noll. *Digital Coding of Waveforms.* Englewood Cliffs, NJ: Prentice-Hall. 1984.

[JHDG96] Johnston, J. D., J. Herre, M. Davis, and U. Gbur. *MPEG-2 NBC Audio: Stereo and Multichannel Coding Methods.* 101st AES Convention, Los Angeles, Preprint 4383. 1996.

[JohB92] Johnston, J., and K. Brandenburg. "Wideband Coding Perceptual Considerations for Speech and Music," *Advances in Speech Signal Processing* (S. Furui and M. M. Sondhi, Eds.). New York: Marcel Dekker. 1992.

[JohF92] Johnston, J. D., and A. J. Ferreira. "Sum-Difference Stereo Transform Coding." *Proc. IEEE ICASSP,* pp. 569–571. 1992.

[John96] Johnston, J. D. "Audio Coding with Filter Banks," *Subband and Wavelet Transforms* (A. N. Akansu and M. J. T. Smith, Eds.), pp. 287–307. Norwell, MA: Kluwer Academic Publishers. 1996.

[KiPK01] Kim, S.-W., S.-H. Park, and Y.-B Kim. *Fine Grain Scalability in MPEG-4 Audio.* 111th AES Convention, New York, 2001.

[KirW97] Kirby, D., and K. Watanabe. *Formal Subjective Testing of the MPEG-2 NBC Multichannel Coding Algorithm.* 102nd AES Convention, Munich, Preprint 4418. 1997.

[Levi98] Levine, S. *Audio Representations for Data Compression and Compressed Domain Processing.* Ph.D. thesis, Stanford University, 1998.

[M2481] Purnhagen, H., B. Edler, Y. Maeda, K. Iijima, and M. Nishiguchi. *Proposal for the Integration of Parametric Speech and Audio Coding Tools Based on an Automatic Speech/Music Classification Tool.* Doc. ISO/MPEG M2481, Stockholm MPEG Meeting. July 1997.

[MasW67] Masnick, B., and J. Wolf. "On Linear Unequal Error Protection Codes." *IEEE Trans. Information Theory,* IT-13: 600–607. October 1967.

[McaQ86] McAulay, R., and T. Quatieri. "Speech Analysis/Synthesis Based on a Sinusoidal Representation." *IEEE Trans. ASSP,* 34(4): 744–754. August 1986.

[Moor77] Moorer, J. A. "On the Transcription of Musical Sound by Computer." *Computer Music Journal,* 1(4): 32–38. 1977.

[Moor89] Moore, B. C. J. *Introduction to the Psychology of Hearing,* 3rd ed. New York: Academic Press. 1989.

[MP4V2-3] ISO/IEC 14496-3:1999/Amd.1:2000. *Coding of Audio-Visual Objects—Part 3: Audio, Amendment: Audio Extensions,* Version 2. 2000.

[MPEG1-3] ISO/IEC 11172-3:1992. *Coding of Moving Pictures and Associated Audio for Digital Storage Media at up to about 1.5 Mbit/s—Part 3: Audio,* 1992.

[MPEG2-3] ISO/IEC 13818-3:1994. *Generic Coding of Moving Pictures and Associated Audio Information—Part 3: Audio,* 1994.

[MPEG4-3] ISO/IEC 14496-3:1999. *Coding of Audio-Visual Objects—Part 3: Audio,* Version 1, 1999.

[N1419] MPEG Audio. *Report on the Formal Subjective Listening Tests of MPEG-2 NBC Multichannel Audio Coding.* Doc. ISO/MPEG N1419, Maceió MPEG Meeting. November 1996. Available at http://mpeg.telecomitalialab.com/

[N2006] MPEG Audio. MPEG-2 AAC Stereo Verification Test Results. Doc. ISO/MPEG N2006, San Jose MPEG Meeting, February 1998. Available at *http://mpeg.telecomitalialab.com/*

[OjaV99] Ojanperä, J., and M. Väänänen. *Long Term Predictor for Transform Domain Perceptual Audio Coding.* 107th AES Convention, New York, Preprint 5036. 1999.

[OomB99] Oomen, A. W. J., and A. C. den Brinker. "Sinusoids Plus Noise Modelling for Audio Signals." *Proc. of the 17th International AES Conference on High Quality Audio Coding,* pp. 226–232. Florence, 1999.

[PaiS00] Painter, T., and A. Spanias. "Perceptual Coding of Digital Audio." *Proc. IEEE,* 88(4): 451–513. April 2000.

[PaKS97] Park, S.-H., Y.-B. Kim, and Y.-S. Seo. *"Multi-Layered Bit-Sliced Bit-Rate Scalable Audio Coding.* 103rd AES Convention, New York, Preprint 4520. 1997.

[PalK95] Paliwal, K. K., and W. B. Kleijn, "Quantization of LPC Parameters," *Speech Coding and Synthesis* (K. K. Paliwal and W. B. Kleijn, Eds.). New York: Elsevier. 1995.

[PrJB87] Princen, J., A. Johnson, and A. Bradley. "Subband/Transform Coding Using Filter Bank Designs Based on Time Domain Aliasing Cancellation." *Proc. IEEE ICASSP,* pp. 2161–2164. 1987.

[PuEF98] Purnhagen, H., B. Edler, and C. Ferekidis. *Object-Based Analysis/Synthesis Audio Coder for Very Low Bit Rates.* 104th AES Convention, Amsterdam, Preprint 4747. 1998.

[PuEM01] Purnhagen, H., B. Edler, and N. Meine. *Error Protection and Concealment for HILN MPEG-4 Parametric Audio Coding.* 110th AES Convention, Amsterdam, Preprint 5300. 2001.

[PuME00] Purnhagen, H., N. Meine, and B. Edler. *Speeding up HILN: MPEG-4 Parametric Audio Encoding with Reduced Complexity.* 109th AES Convention, Los Angeles, Preprint 5177. 2000.

[PurM00] Purnhagen, H., and N. Meine. "HILN: The MPEG-4 Parametric Audio Coding Tools." *Proc. IEEE International Symposium on Circuits and Systems (ISCAS),* Geneva, 2000.

[Purn99] Purnhagen, H. "Advances in Parametric Audio Coding." *Proc. IEEE Workshop on Applications of Signal Processing to Audio and Acoustics (WASPAA),* pp. 31–34. Mohonk, 1999.

[Quac97] Quackenbush, S. "Noiseless Coding of Quantized Spectral Components in MPEG-2 Advanced Audio Coding." *Proc. IEEE Workshop on Applications of Signal Processing to Audio and Acoustics.* Mohonk, 1997.

[RisM69] Risset, J.-C., and M. V. Matthews. "Analysis of Musical Instrument Tones." *Physics Today,* 22: 22–30. February 1969.

[Rotw83] Rothweiler, J. "Polyphase Quadrature Filters: A New Subband Coding Technique." *Proc. IEEE ICASSP,* pp. 1280–1283. Boston, 1983.

[Sche00] Scheirer, E. *Music-Listening Systems.* Ph.D. thesis, MIT Media Lab, 2000.

[Serr89] Serra, X. *A System for Sound Analysis / Transformation / Synthesis Based on a Deterministic plus Stochastic Decomposition.* Ph.D. thesis, Stanford University, 1989.

[Serr97] Serra, X. "Musical Sound Modeling with Sinusoids plus Noise." *Musical Signal Processing* (C. Roads et al., Eds.). Lisse, The Netherlands: Swets and Zeitlinger. 1997.

[Sper00] Sperschneider, R. *Error Resilient Source Coding with Variable Length Codes and Its Application to MPEG Advanced Audio Coding.* 109th AES Convention, Los Angeles, Preprint 5271. 2000.

[SQAM] EBU. *SQAM: Sound Quality Assessment Material Recordings for Subjective Tests.* Audio-CD, Brussels, Belgium: EBU, 1988.

[TakW95] Y. Takishima, M. Wada, and H. Murakami. "Reversible Variable Length Codes." *IEEE Trans. Comm.,* 43(2/3/4): 158–162. 1995.

[VeGS98] Vercoe, B., W. Gardner, and E. Scheirer. "Structured Audio: Creation, Transmission, and Rendering of Parametric Sound Representations." *Proc. IEEE,* 86(5): 922–940. May 1998.

[VeLM97] Verma, T., S. Levine, and T. Meng. "Transient Modeling Synthesis: A Flexible Analysis/Synthesis Tool for Transient Signals." *Proc. International Computer Music Conference (ICMC),* 1997.

[Verm00] Verma, T. *A Perceptually Based Audio Signal Model with Application to Scalable Audio Compression.* Ph.D. thesis, Stanford University, 2000.

[WaaV91] Waal, R. G. V. D., and R. N. J. Veldhuis. "Subband Coding of Stereophonic Digital Audio Signals." *Proc. IEEE ICASSP,* pp. 3601–3604. 1991.

[ZwiF90] Zwicker, E., and H. Fastl. *Psychoacoustics, Facts and Models.* Berlin: Springer Verlag. 1990.

SNHC Audio and Audio Composition

by Riitta Väänänen and Jyri Huopaniemi

Keywords: synthetic audio, SNHC audio, structured audio, text-to-speech synthesis, audio composition, AudioBIFS, spatial audio, 3D audio

*T*he audio tools in MPEG-4 are divided into *coding* and *scene description* tools, as illustrated in Figure 12.1. The coding tools are further classified as *natural* and *synthetic* audio coding. Whereas speech and natural audio coding tools were presented in Chapters 10 and 11, synthetic audio coding and audio composition are the major focus of this chapter.

Synthetic audio coding refers to techniques by which sounds are described and generated (or synthesized) through an algorithm with the aid of a computing device. In MPEG-4, this means that the synthetic audio data that is transmitted in a bitstream contains parametric descriptions of the sound instead of a compressed version of recorded sound, which is the case in speech and natural audio coding (as described in Chapters 10 and 11).

The composition of audio is a process by which different decoded audio streams are mixed together before the sound is finally reproduced—that is, made audible to the user. In MPEG-4, this process is defined in the semantics of the AudioBIFS nodes that are part of the BIFS tools defined in the Systems part of the standard ([MPEG4-1], see Chapter 4). The AudioBIFS nodes can be used for mixing, postprocessing, and spatial presentation of decoded audio streams. An *audio subtree* constructed with these nodes is used for hierarchical composition of streams locally at the terminal. This composition process also can be affected via local interactivity through the interaction mechanisms defined in BIFS.

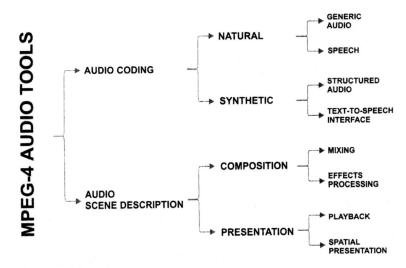

Fig. 12.1 Audio tools in MPEG-4.

This chapter is divided into four sections. In the first, the concepts associated with *synthetic-natural hybrid coding* (SNHC) for audio will be addressed. In the SNHC framework, both synthetic and natural audio as well as digital sound effects can be included in a single MPEG-4 presentation. Then the synthetic audio coding technologies that are included in the MPEG-4 Audio standard [MPEG4-3] are discussed. The main tools for synthetic audio coding are *structured audio* (SA) and the *text-to-speech interface* (TTSI). Finally, in the two last sections of the chapter, the tools related to the composition and presentation of sound that are defined in the Systems part of MPEG-4 [MPEG4-1] will be presented. These tools are included in the specification of AudioBIFS and its extension, Advanced AudioBIFS. Whereas AudioBIFS nodes are used mainly for composition of sounds, Advanced AudioBIFS nodes are intended for enhanced spatial presentation of sound, including the modeling of room acoustics.

The application areas of the synthetic and SNHC audio tools defined in the MPEG-4 standard are wide, ranging from digital sound synthesis and text-to-speech (TTS) synthesis in multimedia applications to postprocessing and mixing of natural and synthetic audio, and, finally, to interactive 3D audiovisual environments for virtual and augmented reality. The SNHC audio toolset combined with the AudioBIFS presentation tools represents the state-of-the-art technology in synthetic audio, and its flexibility ensures a wide range of applications.

12.1 SYNTHETIC-NATURAL HYBRID CODING OF AUDIO

The SNHC of audio is enabled by MPEG-4 Systems through the AudioBIFS architecture and can be used to combine sound streams that are decoded with different (synthetic or natural) audio decoders, as depicted in Figure 12.2. In this figure, it can be seen that at the BIFS level the decoded sounds are processed in a similar way, regardless of the decoder used. The hierarchical AudioBIFS subtree mixes the decoder outputs, and these mixed sounds are associated with the rest of the scene with the aid of the topmost nodes of the AudioBIFS graphs. In Figure 12.2, the topmost node is Sound, which can be given a 3D spatial location and a simple, directive sound radiation pattern. Thus, the AudioBIFS graph describes the *audio signal flow* from the output of the audio decoders through the processing subtree defined in AudioBIFS, and finally to the playing of the sound. The final output at the terminal is the sum of the outputs of the different AudioBIFS subtrees in the overall BIFS scene graph. If spatial properties are given to the topmost sound nodes of the audio subtrees, they can be thought of as representing real-world sound sources in a virtual scene [ScVH99, ScLY00]. This effect can be further enhanced if these nodes are associated with visual objects.

This SNHC framework allows the encoding of sounds at different sampling rates and the production of audio bitstreams at different bit rates to be used in the same MPEG-4 session. When the streams are composed according to the AudioBIFS descriptions, they are converted to the same sampling rate before the mixing or summing operations. The AudioBIFS specification defines that the sampling rate of sound, which is an output of the composition process, must be the highest of the elementary sound streams. In this manner, it is guaranteed that the quality of the elementary audio streams is not decreased when composing the sounds.

As an overview, MPEG-4 SNHC of audio allows for:

1. **Integration of synthetic and natural audio streams:** Any type of decoded MPEG-4 audio elementary stream can be combined in a single

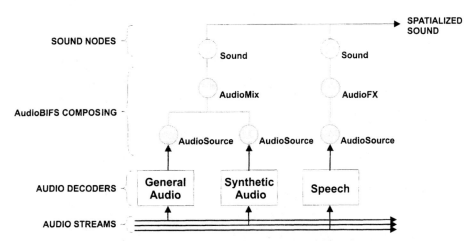

Fig. 12.2 Hybrid coding of synthetic and natural sounds: AudioBIFS nodes define how the different streaming and decoded sound objects are composed together.

MPEG-4 scene with the help of the various processing functionalities provided by AudioBIFS.

2. **Efficient, object-based coding of sounds:** This means the audio streams (like any other MPEG-4 stream) are handled in an object-based manner. In other words, some sounds can be encoded with methods particularly designed for those types of sound—for example, speech coding and synthetic audio coding of music. The total audio bit rate using this selective coding approach may be lower than when encoding with the wideband general audio coding method (for the same quality). Moreover, the object-based approach allows for the possibility of including different types of sounds in a single application; dealing with them as separate sources of sound; coding them, taking into account their specific characteristics; and also receiving sounds from multiple servers.

3. **Digital audio effects:** SNHC tools enable the addition of synthetic digital filtering effects to single sound streams or combinations of multiple sound streams. These filtering effects can be transmitted in a BIFS bitstream (using a node called AudioFX, as it will be explained later in this chapter).

4. **Spatial presentation of sound:** If a composed soundtrack is given spatial properties, it is processed according to the current viewing point and sound location information before reproducing the sound. The result is that the sound is heard coming from its virtual position. This processing is done in real time, locally at the terminal, and it also takes into account movements of the listening point and the source position. This functionality will be explained in detail when the Sound, Sound2D, and DirectiveSound nodes are presented. This feature is useful, for example, in virtual reality applications but also in creating spatial effects at the terminal for the decoded sounds. In this framework, the sounds also can be associated with visual objects with the help of the usual BIFS grouping nodes (such as Group and Transform) to form audiovisual objects. Thus, regardless of the encoding method of the sound (synthetic or natural), the decoded stream can be given synthetic 3D effects.

The presented SNHC audio framework allows the application developer to fully use the different capabilities offered by the MPEG-4 Audio standard. In the following sections, the synthetic audio and audio scene description tools are discussed in greater detail.

12.2 STRUCTURED AUDIO CODING

The concept of SA, as presented in [VeGS98], refers to the semantic and symbolic description of sound events, which enables very low bit-rate transmission, configurable and scalable synthesis, and flexible control of sound. In SA, the sound waveform itself is not coded, but parametric models are defined and used to generate *descriptions* of the sound algorithms and events.

Structured audio in the MPEG-4 standard is used as a novel concept of coding and transmission of digital sound effects and synthetic music. The required bandwidth for SA depends on the type of synthesis or processing algorithm and control method used, but data rates of less than 1 kbit/s for simple music can easily be achieved [SchR98]. In this section, the MPEG-4 SA coding principles and the tools available for algorithmic and wavetable synthesis are presented [MPEG4-3].

The SA paradigm in MPEG-4 enables the creation and transmission of extremely low bandwidth—even if high quality—synthetic sound or processing algorithms. The MPEG-4 SA specification does not specify particular sound synthesis methods, but allows for the creation of arbitrary signal processing algorithms using a specialized processing language and specifies methods for the control of the signal processing routines. In the following, the main elements of the SA toolset are listed (see [MPEG4-3]):

1. **Structured audio orchestra language (SAOL)** [SchV99]: SAOL is a digital signal processing language that is normatively defined in the MPEG-4 standard. SAOL is based on the CSound computer music programming language [Verc95], belonging to the family of the so-called *Music-N* languages.

2. **Score and control languages:** The structured audio score language (SASL) is a score and control language that is used in conjunction with SAOL to produce audible output from the SA bitstream. An alternative method that can be used (in conjunction with or instead of SASL) for controlling SA is Musical Instrument Digital Interface (MIDI) [MIDI96], which has been widely used in the music industry since its adoption in 1983.

3. **Structured audio sample bank format (SASBF):** The sample bank format is used for the transmission of audio sample banks for wavetable synthesis and associated simple processing algorithms.

4. **Scheduler description:** The normative scheduler is a runtime element of the SA decoding process, mapping the control specified by SASL or MIDI to real-time events produced using the algorithms described in SAOL.

From the sound synthesis point of view, the SA specification can be divided into two main categories of tools: algorithmic and wavetable synthesis [SchR98]. These will be presented in the following sections.

12.2.1 Algorithmic Synthesis and Processing

Unlike for MPEG-4 natural audio coding techniques, in algorithmic sound synthesis, the sound is not recorded and compressed, but instead is created as sound descriptions that define what the decoded sound should be like. In structured audio, these descriptions are programmed in SAOL, which includes C-like syntax and basic operators and a library of functions that can be used to describe *instruments* (music synthesis algorithms) as networks of digital signal processing (DSP) routines. This library is composed of approximately 100 *core opcodes*.

The SAOL does not define any specific sound synthesis method, but it can be used to implement any type of sound synthesis that takes advantage of the sound generation and sound processing routines defined in SAOL.[1] The SAOL opcodes include sets of mathematical functions, pitch converters, operations for the manipulation of tables, signal generators, noise generators, DSP filters and transforms, gain control functions, delay functions, sound effects, and functions for speed changes. The functionalities of the SAOL opcodes are not described in detail here, but the reader is referred to the SA specification [MPEG4-3] for further information.

The SA bitstream includes the following data:

☞ Header information, consisting of instruments, sample data, and predefined score data; and

☞ Actual bitstream, containing streaming score and sample data.

The SAOL code that defines the music synthesis algorithms (called *instruments* in the SA terminology) is transmitted in the SA bitstream *header* and compiled at the terminal. The output of the compiling process is executed by the *synthesis engine* of the SA decoder. The synthesis engine is reconfigurable in the sense that its execution is described case-by-case by the corresponding compiled (static) SAOL code. In summary, the instruments define the algorithms (but not what the decoded sound will be), and the associated score information gives the time-dependent control input to play the synthetic instruments. The block in the bitstream header that contains the instrument definitions is called the *orchestra file*. In addition to the instruments, the orchestra file contains:

1. A global block that defines common parameters for the instrument definitions following it;

2. Opcode declarations that can be used to better structure the code of instruments by user-defined subroutines; and

3. Template declarations that can be used to define multiple instruments with small variations.

The global block of an SAOL orchestra contains data such as the output number of channels and the audio sampling rate of the decoded stream, the

---

1. It should be noted that the SAOL opcodes are *normatively defined* (to be implemented as specified in the standard). This means that all signal processing algorithms are built based on the defined functions. This also guarantees that a particular SAOL bitstream always produces the same sound when played on a MPEG-4 conformant decoder.

control rate of the orchestra, the sequencing of instruments, and the bus send/ route statements that define where the output of each instrument is sent.

12.2.1.1 Controlling the Sound Synthesis in Structured Audio
The SAOL instruments can be instantiated and controlled by the SASL. This means that time-dependent information can be sent in the SA header or bitstream for producing sound output from the decoder. The control language allows triggering instantiations of SAOL instruments, sending control events to variables of an instrument, changing the tempo of the decoding process, dynamically creating or deleting wavetables to be used by an SAOL orchestra, and finally ending the orchestra—that is, stopping the decoding process.

In addition to SASL, MIDI score events can be used for controlling the playing of SAOL instruments. This form of control is included in SA to enable backward compatibility with MIDI-based synthesis (which will be discussed in more detail later in this section). When used in SAOL-based synthesis, the MIDI events are converted to SAOL orchestra control events.

Both the MIDI and the SASL control information can be transmitted in the SA stream header as well as in the bitstream following it. The control events in the header (in either a MIDI or an SASL file) must have timing information, which is used to register each event with the *scheduler* of the decoder to be used later in the decoding process. The MIDI or SASL events in the bitstream following the header may or may not have a time stamp telling when the event takes place. If an event has a time stamp that indicates a later time than that of the event arriving at the decoder, it is registered in the scheduler to be used later. If there is no time stamp associated with the event, the event is triggered immediately when it is received at the decoder. There is also a priority bit that indicates if a late event should be started or skipped. MPEG-4 SA also allows interactive playing of the synthetic sound (SAOL instruments or wavetable synthesis) at the terminal, but it does not normatively define the interface between the user and the SA decoder.

A graphical overview of the SA decoder is shown in Figure 12.3. The decoder consists of the scheduler, the synthesis engine, and the sample data storage elements.

In the decoding process, the SA bitstream is first demultiplexed and the score data, the SAOL instruments, and the wavetable data are parsed and passed on to the scheduler, the synthesizer, and the sample storage, respectively. The header information is used for initializing the signal-processing network for algorithmic synthesis, for storing the wavetables in the sample storage, and for registering the predefined SASL data with the scheduler. The streaming data typically includes score and sample data for control and synthesis, respectively. The synthesized sound is then passed on to the MPEG-4 Systems layer for audio composition using AudioBIFS. From the Systems layer, there can be various controls to the SA decoder. For example, audio effects implemented in SAOL can be called and controlled using the AudioFX node, and decoded samples can be provided for synthesis using the

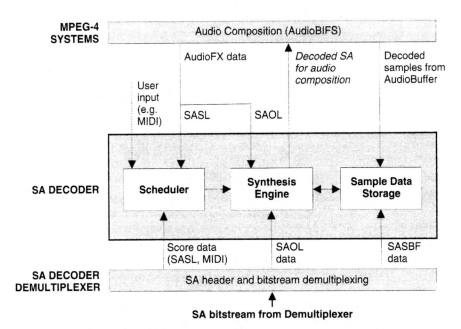

Fig. 12.3 Structured audio decoder.

`AudioBuffer` nodes (discussed in the next section). User interactivity can also be enabled (e.g., for playing an SAOL instrument), but this is not normatively specified (marked with a dashed line in Figure 12.3).

12.2.2 Wavetable Synthesis

Wavetable synthesis refers to the generation of sound from processed and looped samples of sound. This is a widely used and simple way of performing sound synthesis, and it is included as a part of MPEG-4 structured audio [MPEG4-3] to guarantee high-quality wavetable synthesis, especially when MIDI is used. A wavetable-bank format (SASBF) can be used to transmit wavetables in the bitstream. A wavetable consists of sound samples (waveforms), which can be used in the synthesis process. Typically, the music synthesis from these wavetables can be controlled with MIDI instructions, but this can also be done with SASL when the wavetable synthesis is controlled within the SAOL instrument code.

The SASBF is based on *MIDI Downloadable Sounds 2* format (DLS2), which, like MIDI, is also specified by the MIDI Manufacturers Association (MMA) [MIDI99]. The purpose of this format is to guarantee the quality of the synthesized sound and the compatibility between different decoders. General MIDI specifies only the mapping between the MIDI instructions and the music instruments; it does not normatively define the quality of the music synthesis. MIDI alone enables a very low bit-rate transmission of sound, but it is entirely dependent on the synthesizer concerning what the output will sound like. The downloadable sound concept in SASBF, on the other hand, is used to transmit the wavetables along with the bitstream, providing a normative way of controlling the quality of the played samples. In addition to the waveform data and the associated looping information, DLS2 contains parameters for real-time filtering and manipulation (articulation) of the waveform. Apart from SASBF, wavetables also can be transmitted or synthesized using standard SAOL table generators.

Any wavetable, either SASBF or custom (user-defined or some other commonly used wavetable format), can be referred to by SAOL instruments or included in them. This way, for example, postprocessing effects written in SAOL can be added to wavetable instruments. In addition, the wavetables can be used by the SAOL instruments, for example, as excitation signals of instrument code or as impulse responses that are convolved with the instrument output (e.g., for adding a reverb effect to an instrument by representing the acoustic impulse response with a wavetable).

Wavetables cannot be linked directly to other MPEG-4 audio decoders and transmitted in compressed format. However, there is an MPEG-4 Systems functionality that makes it possible to transmit MPEG-4 encoded (compressed) sound to be used in the wavetable synthesis. This will be discussed in more detail when addressing the interface between the synthetic audio coding and the MPEG-4 Systems layer, in the context of the AudioBIFS nodes.

12.2.2.1 Structured Audio Object Types

In MPEG-4 SA, there are four different object types (see Chapter 13) that can be associated with a particular SA bitstream (see [MPEG4-3]). An *object type* defines the set of coding tools that can be used to represent the object being considered. As such, object types provide a mechanism to generate bitstreams with different coding tools and thus also with different complexities. The object type is defined in the bitstream header. In the following, the four SA object types are presented:

1. **General MIDI**: With the General MIDI object type, only MIDI files and MIDI events are transmitted in the bitstream. This means that the decoding uses non-normative ways to generate sound (as explained previously), and the mapping between the MIDI instruments and the synthesis is done according to the patch mappings defined in the general MIDI standard. Thus, a bitstream of this object type is completely backward compatible with the MIDI specification, but it does not provide normative quality of the decoded sound, as this is not part of the MIDI specification.

2. **Wavetable synthesis**: With the Wavetable synthesis object type, MIDI files and SASBF wavetables can be transmitted in the bitstream, and MIDI events are used to control the playing of the instruments.

3. **Algorithmic Synthesis and AudioFX**: With the Algorithmic Synthesis and AudioFX object type, the synthetic instruments are defined only with SAOL statements and variables. SASBF is not supported; only

SASL score events can be used to control the sound synthesis process. This object type (or the next one) must be supported if the AudioFX node is to be used in a BIFS scene. AudioFX is an AudioBIFS node (defined in MPEG-4 Systems) that can be used to apply SA effects to any decoded MPEG-4 audio streams.

4. **Main Synthetic**: With the Main Synthetic object type, all SA tools—including SASBF and the use of AudioFX node in BIFS, as well as MIDI files and MIDI events—are allowed.

12.2.3 Interface Between Structured Audio Coding and AudioBIFS

Now the algorithmic and wavetable synthesis in the MPEG-4 SA specification have been presented. In addition to being used as a sound generator (i.e., an audio decoder), SA is also closely related, in two main ways, to the audio composition tools that are specified in MPEG-4 Systems [MPEG4-1]. First, AudioBIFS nodes are part of the MPEG-4 scene description tools that define the normative composition of decoded audio streams. AudioBIFS gives the possibility of applying custom postprocessing effects to the same audio streams (both synthetic and natural audio); this is achieved through the AudioFX node. Second, AudioBIFS also defines a way of transmitting MPEG-4 encoded sounds to be used as SAOL wavetables, as will be explained in more detail below. In the following, only some of the AudioBIFS nodes are discussed (namely, those having a more direct relationship with SA). A more thorough discussion of AudioBIFS is given in Section 12.4.

The AudioFX node can be used to create sound processing (filtering) effects in SAOL as a part of the BIFS bitstream. In this case, the SAOL code written in the orch field of the AudioFX node is stored in the SA decoder, and corresponding SAOL instruments are initialized. When processing the input sound of the AudioFX node, this sound is routed to the input bus of the SAOL orchestra (defined by the orch field) of the AudioFX node, and the output of the orchestra is made the output of the AudioFX node. Alternatively, the SASL code can be written in the score field of the AudioFX node; this data then can be used to control the parameters of the SAOL instruments in the orch field.

The AudioBuffer node of AudioBIFS can be used to store clips of sound from the input sound that is provided by its children AudioBIFS nodes. When an AudioBuffer node is a child of an AudioSource node that refers to an SA stream, these clips of decoded MPEG-4 sounds can be used as wavetables in the SAOL decoding process. Through this mechanism, both uncompressed and compressed wavetables can be used by an SA decoder.

12.2.4 Structured Audio Applications

The versatile MPEG-4 SA toolset enables a rich variety of applications. The SNHC concept itself allows for the combination of synthetic and natural audio content with synthetic and natural visual data. Some example applications of the SA toolset organized in terms of the major SA functionalities are presented in the following:

1. **Algorithmic synthesis:** The MPEG-4 SA algorithmic synthesis capability offers high-quality user-definable sound synthesis, including expressive control. Because any signal processing routine can be written with SA, the application area is very wide. Current applications include generic sound synthesis and computer music, video games, low bit-rate Internet delivery of music, virtual reality models, and entertainment [SchR98].

2. **Wavetable synthesis:** The MPEG-4 SA wavetable synthesis engine offers a bounded-complexity sound synthesis implementation, enabling the implementation on low-complexity decoders. Envisaged applications may include karaoke systems and musical backgrounds for Web pages [SchR98].

3. **Audio effects processing:** The SAOL descriptions can be used as custom effects processing modules for natural and synthetic audio. Example custom digital effects are reverb, flanger, phaser, and time-scale modification of sound.

4. **Generalized audio coding:** This concept refers to a wider (theoretical) application area of SA. In [SchK99], the use of MPEG-4 SA for generalized audio coding is described. This application is based on the fact that the SA decoder can be used to emulate the behavior of natural audio coders. In essence, any decoder can be implemented using SA tools, as shown in [SchK99]. The SA toolset, however, may not provide the computationally optimal tools to achieve the general audio coding principle in all cases.

12.3 TEXT-TO-SPEECH INTERFACE

The Text-To-Speech Interface (TTSI) is used for the generation of synthetic speech from textual data (either text or phoneme). In this framework, the transmission of speech data is enabled at very low bit rates (200 bps to 1.2 kbit/s). Speech synthesis, in general, is useful in various kinds of multimedia applications, and thus MPEG-4 defines flexible means for its use in different situations. MPEG-4 TTSI allows the following additional information to the plain text:

- ☞ Speaker-related information (speech rate, age, and gender of the speaker);
- ☞ Prosody (e.g., time-dependent variation of pitch);
- ☞ Language code, or lip shape information when used for video dubbing; and
- ☞ Face animation-related parameters when used in synchronization with an animated face.

In MPEG-4 TTSI, no normative speech synthesis method is used; only the interface (the bitstream syntax and its semantics) between the textual data and the speech synthesizer is defined. Consequently, no TTS synthesis methods will be presented in this section; but a brief description of the different components of an MPEG-4 TTS bitstream will be presented.

The TTSI is defined as an audio tool in MPEG-4 Audio [MPEG4-3]. Because it is considered as a part of both the speech coding tools and the synthetic audio tools, it is included in both the speech and synthetic audio profiles of the standard (see Chapter 13).

The TTS bitstream contains *sentences* that carry the text to be converted to speech at the decoder. The sentences also may contain information characterizing certain properties of the speaker, such as the gender and the age, as these are useful features for distinguishing among different speakers in a TTS application. *Prosodic* information (meaning properties related to the tempo, energy, and the frequency contour of the speech) also can be associated with the speech sentences. These properties give an impression of the speaker's emotional or mental state. In MPEG-4, the synthetic speech also can be used for video dubbing, in which case information that is used for synchronizing the speech with the visual information can be added. Furthermore, TTS can be used for controlling the lip positions of animated faces. In this case, the phonemes of the speech or the face animation parameter (FAP) bookmarks that can be added to the text part of the sentence are converted to FAPs. These are defined in the MPEG-4 Visual standard [MPEG-2], and they serve to animate 3D faces, as explained in Chapter 9.

A TTS *sequence* that is one of the TTS bitstream components is used for identifying common information for all the TTS sentences following this TTS sequence. A TTS sequence has a sequence ID and a language code. The language is expressed as one of ISO 639 Language Code values [ISO89], or it can be "00," indicating that the speech is given by phonemes as defined in the International Phonetic Alphabet (IPA; [IPA99]). The TTS sequence also contains several Boolean variables that can be used to enable or disable certain properties of the sentences, such as the selection of gender or age, adjusting the speech rate, adding prosody to the sentence, video dubbing, and adding lip shape data to be converted to facial animation data [MPEG4-3; Section 6].

The MPEG-4 TTSI decoder is presented in Figure 12.4. The architecture of the decoder can be presented as a collection of normative interfaces. Therefore, the normative behavior of MPEG-4 TTSI is described by the interfaces, not by the actual sound or animated output. Five TTSI interfaces are defined:

- ☞ Interface between the demultiplexer and the syntactic decoder
- ☞ Interface between the syntactic decoder and the speech synthesizer
- ☞ Interface from the speech synthesizer to the compositor
- ☞ Interface from the compositor to the speech synthesizer
- ☞ Interface between the speech synthesizer and the phoneme-to-FAP converter

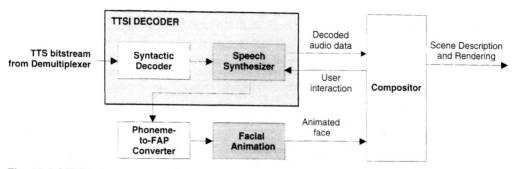

Fig. 12.4 MPEG-4 text-to-speech decoder.

In the decoding process, the MPEG-4 TTS syntactic decoder passes the following information to the speech synthesizer:

☞ Text and phoneme information, including bookmarks to be used in face animation;

☞ Input type (pure TTS or synchronized TTS with facial animation decoding);

☞ Sequence control information containing general information for the TTS sentences;

☞ Prosody information;

☞ Lip shape information; and

☞ User interaction mode.

The shaded blocks in Figure 12.4—namely, the speech synthesizer and the facial animation block—are not normatively described in the MPEG-4 standard and operate in a terminal-dependent manner. When TTS is used to control the facial animation, the phoneme-into-FAP bookmark information is passed further to the phoneme-to-FAP converter, where it is converted to FAPs to be used by the facial animation module. The output from this process is a 3D animated face, with the lips synchronized with the phonemes. The synthesized speech is passed to the compositor, where it is treated as a regular audio stream. Simultaneously, an animated face can be output to the compositor to be presented in synchrony with the synthetic speech. In addition to previously discussed tools, the TTSI in MPEG-4 defines a way to control the speech synthesis process locally at the terminal. This interface (between the compositor and the speech synthesizer) can be used to define the start and stop times of the speech synthesis; to change the rate, pitch, or the dynamic range of the pitch of the speech; and to select the gender or age of the speaker.

12.4 Audio Composition

This section addresses the audio composition and presentation capabilities specified in the BIFS part of the MPEG-4 Systems standard [MPEG4-1]. The tools providing these functionalities are a subset of the BIFS tools called AudioBIFS (together with its extension, Advanced AudioBIFS, which was specified in the first amendment to the MPEG-4 Systems standard [AMD1-1]). The nodes included in AudioBIFS and Advanced AudioBIFS act as an interface between the systems and the audio components of MPEG-4 content. Audio composition is a process by which decoded audio streams (corresponding to different audio objects, possibly encoded with different coding methods) are combined at the terminal to form one or several audio tracks. Audio presentation, on the other hand, includes the processing related to playing back the composed soundtrack and the actual reproduction of the sound via headphones or loudspeakers. The presentation stage also can include the spatial processing of sound that is carried out according to the relative positions of a sound source and a listening point.

In this section, the VRML sound model [VRML97] is reviewed briefly as a background to the object-based audio scene composition developed for MPEG-4. Furthermore, it is explained why the 3D sound model and the sound composition capabilities provided by the VRML sound model are not sufficient for most MPEG-4 applications. After that, the functionality of each AudioBIFS node with relevance regarding the composition and presentation of audio in MPEG-4 will be presented. In the following section, the details on how the spatial properties of audio scenes can be enhanced with the support of Advanced AudioBIFS nodes to produce immersive audiovisual scenes in 3D applications will be presented. Also, the production and control of advanced spatial sound processing effects in the MPEG-4 terminal will be covered.

12.4.1 VRML Sound Model in BIFS

The sound nodes specified in the Virtual Reality Modeling Language (VRML, see [VRML97])—namely, the Sound and AudioClip nodes—are also a part of the BIFS specification (see Chapter 4 for the relationship between VRML and MPEG-4). As an introduction to the MPEG-4 audio presentation tools, the principles of the VRML sound model will be reviewed. This sound model has two different tasks: One is to attach a sound to a scene, and the other is to spatially present it in the local coordinate system of the terminal. The textual node interfaces of Sound and AudioClip are presented in Figure 12.5. This type of description shows the fields for the nodes and their default values [MPEG4-1].

To associate sounds to a scene, VRML enables only the interactive downloading of sound clips that are referred to in the url field of an AudioClip node. However, in MPEG-4 it is necessary to be able to include streamed and decoded sounds into a scene and in applications for which interactive download commands are not possible or available. Therefore, a node called Audio-Source has been added to the MPEG-4 BIFS specification, implementing the link between a streamed sound and the BIFS scene.

The task of the Sound node is to attach a decoded sound to a 3D scene and place it in a specified location. In VRML, it is also the only node that can be

```
Sound {                          AudioClip {
    direction    0, 0, 1             description      ""
    intensity    1.0                 loop             FALSE
    location     0, 0, 0             pitch            1.0
    maxBack      10.0                startTime        0
    maxFront     10.0                stopTime         0
    minBack      1.0                 url              []
    minFront     1.0                 duration_changed
    priority     0.0                 isActive
    source       NULL            }
    TRUE
}
```

Fig. 12.5 Textual descriptions of the Sound and AudioClip nodes.

used to make audible the sound referred to by an AudioClip node. Thus, also in MPEG-4, this node acts as a topmost node of an audio subtree, where the leaf nodes of this subtree contain a reference to the encoded sound and the intermediate nodes (those between the leaf nodes and the topmost Sound node) determine the composition of the sound streams.

The spatial properties of the Sound node are defined by the fields location and direction, specifying the 3D position and orientation of the sound source, and by the fields maxBack, minBack, maxFront, and minFront, which specify the directivity and distance-dependent attenuation pattern of the source (see Figure 12.6). In Figure 12.6, the sound level is constant, inside the inner ellipsoid (defined by minBack and minFront); the sound attenuates to −20 dB from the original value, within the region defined by maxFront and maxBack; and outside the outer ellipsoid, the sound is not heard. The spatialize field defines whether the effect of direction of sound arriving at the listener is rendered or not. No specific technique for implementing this spatialization is defined in the standard, as the sound reproduction system is not included as a normative part of the decoder. For reference about different spatialization techniques, see [Bega94].

In VRML, only downloadable clips of sound can be included in the source field, but MPEG-4 extends the semantics of the source field of the Sound node so that it can contain a reference to an audio stream or a composed soundtrack (i.e., an audio subtree). A detailed description of the Sound node is given in [VRML97].

The AudioClip node is used to associate the actual source of sound (e.g., MPEG-4 encoded sound stream) with the Sound node in a time-dependent way. Its fields, startTime and stopTime, are used to trigger and stop the playing (and downloading when retrieved from a remote location) of the sound. When the current time of the decoder is bigger than the value of the startTime field and smaller than the value of the stopTime field, the node is active—that is, it is passing sound to the node above it. AudioClip can be used to add simple sound feedback to scenes (for example, as a feedback to user input) and thus increase the feeling of interaction with the scene. A detailed description of the semantics of this node is available in [VRML97].

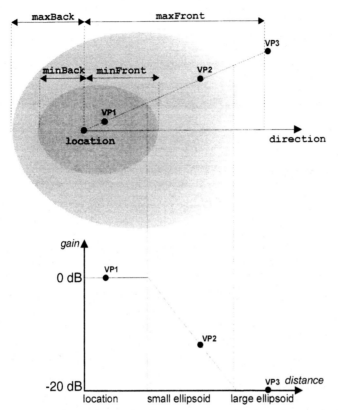

Fig. 12.6 Spatial sound radiation pattern for the Sound node (the nested ellipsoids on top of the picture) and the attenuation curve of sound in the region between the inner and outer ellipsoids of the Sound node.

12.4.2 Other AudioBIFS Nodes

AudioBIFS consists of a set of nodes that are used for mixing, effects processing, interactive playing, and simple 2D or 3D spatial presentation of sound. In addition to the VRML nodes Sound and AudioClip, eight audio nodes that are specific to MPEG-4 were defined: AudioSource, AudioMix, AudioSwitch, AudioFX, AudioDelay, AudioBuffer, ListeningPoint, and Sound2D. These nodes and their fields are listed in Figure 12.7. A more detailed specification of these nodes can be found in the MPEG-4 Systems standard [MPEG4-1], including the types of the various fields.

As the sound composition nodes can be used to hierarchically define an audio subtree, each of them has a field called children defining the input sound to that node. This means that the input audio channels of the nodes in the audio subtree are the same as the output channels of their children nodes. Thus, for each of the AudioBIFS nodes, the children field (except in the case of AudioSource, as will be explained later) may contain one of the following nodes: AudioSource, AudioDelay, AudioMix, AudioSwitch, AudioFX, Audio-

```
AudioSource {                          AudioFX {
    addChildren                            addChildren
    removeChildren                         removeChildren
    children        []                     children        []
    url             []                     orch            ""
    pitch           1.0                    score           ""
    speed           1.0                    params          []
    startTime       0                      numChan         1
    stopTime        0                      phaseGroup      []
    numChan         1                  }
    phaseGroup      []
}                                      AudioBuffer {
                                           loop  FALSE
AudioMix {                                 length          0.0
    addChildren                            pitch           1.0
    removeChildren                         startTime       0
    children        []                     stopTime        0
    numInputs       1                      children        []
    matrix          []                     numChan         1
    numChan         1                      phaseGroup      [1]
    phaseGroup      []                     duration_changed
}                                          isActive
                                       }
AudioSwitch {
    addChildren                        Sound2D {
    removeChildren                         intensity       1.0
    children        []                     location        0,0
    whichChoice     []                     source          NULL
    numChan         1                      spatialize      TRUE
    phaseGroup      []                 }
}
                                       ListeningPoint {
AudioDelay {                               set_bind
    addChildren                            jump            TRUE
    removeChildren                         orientation     0, 0, 1, 0
    children        []                     position        0, 0, 10
    delay           0                      description     ""
    numChan         1                      bindTime
    phaseGroup      []                     isBound
}                                      }
```

Fig. 12.7 Textual descriptions of the AudioBIFS nodes and their fields.

Clip, or AudioBuffer. The fields addChildren and removeChildren are used to add or remove the audio nodes to the children field.

Fields that are common to the leaf nodes and intermediate AudioBIFS *nodes (those above the leaf nodes but below the topmost sound node) are the* numChan and phaseGroup nodes. They are used to specify the number of input sound channels to the node and to indicate whether there are relationships between the different input channels that should be kept unaffected in the composition and presentation process. numChan is the total number of output channels of the children audio nodes, and phaseGroup indicates whether there are phase relationships between the audio channels in the associated sound stream. This information is used when the topmost node of the audio subtree to which this node belongs performs spatialization to its children sounds. The topmost node of an audio subtree is always one of the following: Sound, Sound2D, or DirectiveSound (the latter of which is an Advanced AudioBIFS node that will be explained in the following section). If the different channels of the input audio stream of a topmost audio node contain mutual dependencies that should be maintained, no spatialization should be done to those channels; instead, the channels should be passed through the audio subtree without any spatial processing. Otherwise (if the spatialize flag of the topmost node of this audio subgraph is set to true), the various channels of a multichannel sound are summed to form a monophonic sound, and this monophonic sound is spatialized as defined in the semantics of the Sound, Sound2D, or DirectiveSound.

12.4.2.1 AudioSource As mentioned, AudioSource is a node that can be used to include streaming sounds in MPEG-4 scenes. One decoded audio stream can be associated to each AudioSource node, which is used as a leaf node in a scene. Thus, the output of each AudioSource node is a decoded sound stream that can be further composed or processed with other AudioBIFS nodes.

The reference to the actual sound stream is given in the url field of this node. The fields pitch and speed can be used to change the pitch of the stream without affecting the playback speed of the stream, and the speed without changing the pitch, respectively. However, these fields can be applied only to sound streams that have been encoded with low- and high-level parametric audio coding methods (HVXC and HILN; see Chapters 10 and 11, respectively) and to SA streams. This is because the parametric and structured decoders contain built-in functionalities for changing the speed and pitch, regardless of each other. If the pitch or speed of another type of sound stream has to be changed, this could be done using the AudioFX node (explained later in this section), through a specific *opcode* for its associated orchestra.

As in the case of AudioClip, the activity of this node is defined by the fields startTime and stopTime. Thus, by manipulating these fields, it is possible to switch on and off the sound that is passed through this node.

The children field of AudioSource differs from other audio nodes in that it can contain only AudioBuffer nodes when the source sound of this node is an output of an SA decoder. Thus, the semantics of this field is different compared to the other AudioBIFS nodes. This functionality will be explained in more detail in the context of the AudioBuffer node. A more detailed description of the semantics of the AudioSource node is available in [MPEG4-1].

12.4.2.2 AudioMix The AudioMix node can be used to mix an arbitrary number of input channels to an arbitrary number of output channels. Any of the input channels in the children field of this node can be mixed in proportions defined by the mixing matrix of this node.

The mixing rules for forming the output channels as a combination of the input channels (from the children audio nodes) are defined in the matrix field.

The number of rows of the matrix is the same as the number of input channels, and the number of columns equals the number of output channels of this
node. See [MPEG4-1] for details about how the mixing matrix is built up. The
field numInputs gives the number of all input channels of the children nodes,
and numChan gives the number of output channels of this node.

12.4.2.3 AudioSwitch This node is a simplified version of the AudioMix node,
enabling one to choose a specified set of input channels as an output of this
node. Thus AudioSwitch does not affect the gains of the input channels, but
passes the selected ones through this node unaffected.

The selected input channels are defined by the whichChoice field of this
node, which consists of zeros and ones, indicating which channels in the children sound channels are selected as an output of this node. (A more detailed
description of the functionality of this node is provided in [MPEG4-1].)

12.4.2.4 AudioFX AudioFX is a node that can be used to add sound filtering
effects to its children sounds, performed locally at the decoding terminal.
These effects are expressed in the SAOL, which allows building up arbitrary
sound filtering (signal processing) effects; in addition, it includes various built-
in filtering routines that can be efficiently used to process sounds. Also, the
SASL (for controlling the SAOL orchestra) can be used in the context of this
node to change the sound filtering parameters locally at the terminal.

This node is useful, for example, in applications in which it is necessary
to add and vary sound effects interactively (such as studio applications); in
some cases, this sound filtering scheme may also reduce the total bit rate of
the transmitted sound. To clarify this effect, consider a situation in which
anechoic (or dry) speech is transmitted to the decoding terminal, where a
reverberation effect is added, in order to make it sound as it was heard in a
natural acoustic environment (such as a room). In this case, it is likely that
the speech can be efficiently encoded with one of the MPEG-4 speech coding
techniques (and, therefore, the bandwidth it requires may be small). The
reverberation algorithm may be transmitted only once at the beginning of the
session using BIFS and thus does not add much to the required bandwidth.
On the other hand, if the speech was recorded in reverberant conditions, it is
possible that the speech coding algorithms would perform poorly (after all,
they rely on speech produced by a single speaker without a considerable
amount of noise or reverberation). In that case, the general audio codec might
be needed—and it would probably require a higher bit rate than the encoded
speech and the audio effect together.

It must be emphasized that, in most cases, synthetic audio (produced by
algorithmic, wavetable, or TTS synthesis) is anechoic,[2] and often some room

---

2. Anechoic means that the sound does not contain any acoustic effect caused by wall
   reflections or other obstacles in the environment.

acoustic effects are needed to make it sound more natural and pleasant. In MPEG-4 SA, these effects can be encoded in SAOL as a part of the SA bitstream, but another option is to use the BIFS AudioFX node, which also allows the effects to be interactively controlled at the terminal. Because AudioFX can also be applied to any MPEG-4 decoded sound, the use of synthetic audio effects is not restricted to synthetic audio streams.

The filtering of the children sounds of this node is defined by its orch and score fields. The orch field contains a description of the filtering routines that are to be applied to the children sounds of this node. This description is written in SAOL and contains orchestra header and instrument definitions, which form the actual signal processing block that is applied to the input sound of this node. The score field, on the other hand, is an optional field of this node; it can contain code written in SASL. This code then can be used to control the parameters of the SAOL code in the orch field.

For a more detailed explanation of the use of AudioFX, see [MPEG4-1], the corresponding part in the MPEG-4 Audio standard [MPEG4-2], and [ScVH99, SchV99, ScLY00].

12.4.2.5 AudioDelay AudioDelay is used to delay the output sounds of its children audio nodes with the same amount of time. The main purpose of this node is to enable fine-tuning of synchronization between the output sound of this node and other streams or scene events. The delay that is applied to the input sound channels of this node is defined by the delay field of this node. The normative description of this node is given in [MPEG4-1].

12.4.2.6 AudioBuffer The AudioBuffer node provides a similar functionality as the AudioClip node, as it also can be used to include sound clips in a scene. Unlike the case of AudioClip, in which the clips are interactively retrieved from the server, AudioBuffer is used to *cut* clips of sound from incoming audio streams for interactive use. The main difference between the two nodes is that AudioBuffer can be used in one-way applications (e.g., broadcasting applications) in which no back-channel download command can be sent to the server. An audio clip of a specified length is taken from the input sound of this node, which can be played in an interactive and time-dependent manner, as for the source sound of AudioClip.

The length (in seconds) of the clipped sound from the children sounds of this node is defined by the value of the length field. This clip is taken from the start of the input sound; or, whenever the value of the length field changes, the new clip is taken from the input sound starting at the current time. Similarly to AudioClip and AudioSource, the fields startTime and stopTime define whether the node is active (i.e., producing sound as its output). Also, pitch is used to increase or decrease the playback speed (and the pitch, accordingly) similarly as with the AudioClip node.

This node can be used for transmitting MPEG-4 encoded sound to be used in SA wavetable synthesis. When this node is used in the children field of an `AudioSource` node and the `url` of the `AudioSource` points to an SA bit-stream, the sound clips that are the output of the `AudioBuffer` node can be used by the SA decoder as SASBF wavetables. This functionality makes it possible to transmit wavetables in a compressed form, instead of using the uncompressed SASBF or SA table format. The complete normative description of the `AudioBuffer` node can be found in [MPEG4-1].

12.4.2.7 Sound2D This node is especially useful for 2D applications as well as other applications not requiring any 3D sound properties. It is used similarly to the `Sound` node, as it acts as a topmost node of an audio subtree and thus attaches a sound stream or a composed soundtrack to a scene. This node is functionally the simplest of the three sound nodes available for similar purpose: `Sound2D`, `Sound`, and `DirectiveSound` (to be explained in the context of Advanced `AudioBIFS`). Even if 2D applications typically do not require any spatial sound capabilities, simple 2D spatialization capabilities allowing one to place the sound on a 2D plane are provided by the semantics of this node.

The viewing area in 2D applications is restricted to an area of 2 meters in the horizontal and 1.5 meters in the vertical plane [MPEG4-1]. The virtual viewpoint is assumed to face this plane from a 1-meter distance on an axis that cuts the plane perpendicularly at its center point. The *location* of `Sound2D` is given as 2D coordinates on this plane.

When the `spatialize` field is set to `TRUE`, the source sound should be perceived as coming from the direction of the defined 2D location. Thus, unlike `Sound`, `Sound2D` does not provide the spatial effects (ellipsoidal directivity pattern and distance attenuation) that depend on the relative distance between the source and the listener, and on the orientation of the source with respect to the listener. When `spatialize` is set to `FALSE`, the input sound to this node is passed through unaffected. Thus, this node offers the simplest way of attaching decoded sounds to scenes without any spatial effects. A more detailed description of this node is given in [MPEG4-1].

12.4.2.8 ListeningPoint While addressing the `Sound` node, it was explained that the spatial presentation of the sound depends on the relative positions and orientations of the `Sound` source object and the `Viewpoint` of the scene. This means that the visual view and the viewpoint from which the sounds are heard coincide, which is typically the desired case in audiovisual virtual world applications. However, sometimes it may be useful to define a sound scene that does not dynamically change in synchronization with the visual scene. The `ListeningPoint` node provides the possibility of having a stable spatial sound scene, for example, where the listening point does not change even if the viewpoint changes. Alternatively, the `ListeningPoint` can be animated irrespective of the `Viewpoint` movement. This node also can be useful in audio-only applications, in which no visual view is defined.

The fields of `ListeningPoint` have the same semantics as those of `Viewpoint` (presented in Chapter 4), but this node is applicable only to the spatial rendering of the sound scene. For the full description of this node, see [MPEG4-1].

12.4.2.9 AudioBIFS **Scene Graph: An Example**   The scene graph example presented below was written using the textual format interfaces of the nodes (i.e., in a similar manner as the nodes described in [MPEG4-1, AMD1-1, VRML97]). It contains the same AudioBIFS nodes that were used to build up the scene graph illustrated in Figure 12.2.

In the AudioMix (at line 15) node referred to in SOUND1 (lines 7 to 36), two sound streams are mixed, both containing two audio channels. The mixing matrix (line 19) in this example sums the first channels of the input streams to the first output channel, and the second channels of the input streams to the second output channel (see further details in [MPEG4-1]). The SOUND2 Sound node (lines 38 to 59) refers to an audio stream that is routed through an AudioFX node, adding a filtering effect to the sound that is decoded from an SA stream.

```
 1   Group {
 2    children [
 3     Viewpoint {
 4       position 0 0 3
 5       orientation 0 1 0 0
 6     }
 7    DEF SOUND1 Sound{
 8      spatialize FALSE
 9      location 1 0 0
10      minBack 1
11      minFront 2
12      maxBack 3
13      maxFront 10
14
15      source AudioMix {
16        numinputs 4
17        numChan 2
18        phaseGroup [1 1]
19        matrix [1 0 0 1 1 0 0 1]
20        children [
21         AudioSource {
22          url 10 // reference to, e.g. General
23                 // Audio stream
24          startTime 0
25          stopTime -1
26          numChan 2
27         }
28         AudioSource {
29          url 11 // reference to, e.g. Synthetic
30                 // Audio stream
31          startTime 0
32          stopTime -1
33          numChan 2
34         }
35      ]
```

```
36   }
37
38   DEF SOUND2 Sound {
39     spatialize TRUE
40     location -1 0 0
41     direction 0 0 1
42     minBack 1
43     minFront 2
44     maxBack 3
45     maxFront 10
46     source AudioFX {
47       numChan 2
48       phaseGroup [1 1]
49       children [
50        AudioSource {
51          startTime 0
52          stopTime -1
53          numChan 1
54          url 12 // reference to, e.g. Speech stream
55        }
56      orch "..." // e.g. Reverberator written in SAOL
57
58     }
59    }
60    ]
61   }
```

12.4.3 Enhanced Modeling of 3D Audio Scenes in MPEG-4

The first amendment made to the MPEG-4 Systems standard [MPEG4-1], also known as MPEG-4 Systems Version 2 [AMD1-1], adds a powerful toolset for audio rendering complementing the tools already included in MPEG-4 Systems Version 1. Whereas the concepts of sound rendering in the first version of MPEG-4 were mostly adopted from VRML [VRML97], MPEG-4 Systems Version 2 includes two complementary approaches for virtual acoustics rendering: the *physical* and *perceptual* approaches.

In this section, important concepts of virtual audio rendering are presented first. Efficient parameterization of room acoustics for multimedia is discussed, and the two approaches adopted for Advanced AudioBIFS in MPEG-4 Systems Version 2 are presented [AMD1-1]. The section concludes with details on the node structure and example scenes.

Modeling of 3D sound scenes or virtual acoustics is normally divided into three main parts [Bega94, SHLV99]:

1. **Source modeling:** In the case of MPEG-4, this refers to decoded natural or synthetic audio bitstreams, which are given acoustical characteristics such as directivity.

2. **Transmission medium modeling:** In room acoustics, there are various methods for the modeling and simulation of the sound behavior in rooms

or other enclosed spaces (see [Kutt91] for details). In the context of MPEG-4, the standardized methods, following *physical* and *perceptual* approaches (discussed in more detail below), provide capabilities for real-time rendering of room acoustical effects.

3. **Receiver modeling:** The receiver modeling in a virtual acoustic system corresponds to the form in which the sound is presented to the end user (taking into account the directional hearing characteristics of the listener). In the case of MPEG-4, the receiver modeling is not defined in the standard. The reason for this is simply that it is up to the receiving device and end user to decide what type of reproduction method (loudspeakers, headphones, multichannel) is going to be used for presenting MPEG-4 audio.

In virtual room acoustics, the impulse response (shown in Figure 12.8) of the system is the main element of modeling and parameterization. A room impulse response normally is divided, in the time domain, into direct sound, early reflections, and late reverberation [Jot99, SHLV99] and may contain the previously mentioned features of the source, transmission medium, and receiver. The physical and perceptual approaches for interactive virtual acoustics have many differences, motivated by the different application areas for which they are most suitable. It should be noted that, although detailed modeling of sound transmission has an important role, many applications do not necessarily target the simulation of a reverberating, enclosed space such as a room or concert hall. It also may be desirable to simply model the sound transmission in the air (including effects such as distance-dependent attenuation, air absorption, or Doppler effect); or effects such as the occlusion caused by obstructing

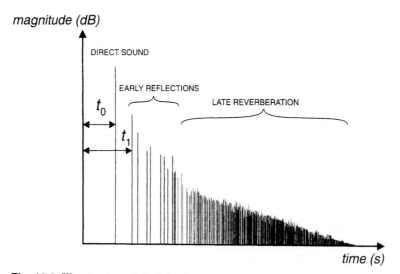

Fig. 12.8 Illustration of the impulse response of a room as it is often considered when synthetically reproduced (in virtual room acoustics).

objects between the sound and the listening point; or individual reflections, echoes, or reverb. This is specifically the case in modern, real-time virtual reality applications, such as computer games—for which there may be dynamic movement of the sound source and the listener, for example, that results in time-varying acoustics. For these types of applications, it may not be necessary to hear detailed room acoustics, but just simple environmental acoustic effects that increase the degree of immersiveness of an audiovisual application.

Conceptually, a division can be made between a physical modeling approach and a perceptual modeling approach:

☞ **Physical modeling:** This approach aims at capturing both the macroscopic (reverberation time, room volume, absorption area, etc.) and the microscopic (reflections, material absorption, air absorption, directivity, diffraction, etc.) room acoustical features by a geometrical acoustics approach, and at diffusing late reverberation using statistical means.

☞ **Perceptual modeling:** The goal here, on the other hand, is to find an orthogonal set of parameters using which a virtual acoustic rendering algorithm can be controlled to produce a desired auditory sensation. Therefore, the output is user-controllable (through the possibility of interactively accessing the room acoustic parameters), not environment-controllable, as in the physical approach. The perceptual approach uses physical properties for the direct sound, macroscopic room acoustical features and a static image-source method for directional early and diffuse early reflections, and a statistical late reverberation module for late reverberation [Jot99].

This division clearly highlights the two major functional goals and application domains for virtual acoustics modeling. The first approach is more suitable for accurate simulation of spaces such as concert halls and auditoria, and for auditorily accurate audiovisual rendering, such as modeling of environmental audio effects that are consistent with the visual objects in virtual reality scenes. The latter approach provides the possibility to create a high-quality room acoustic response based on intuitive, perceptually controlled parameters. Therefore the perceptual rendering scheme better suits applications aiming at creating room acoustic effects, such as postprocessing of sound or music performances and teleconferencing.

The physical and perceptual approaches form the basis of the Advanced AudioBIFS specification, which will be presented in the next section.

12.4.4 Advanced AudioBIFS for Enhanced Presentation of 3D Sound Scenes

Advanced AudioBIFS is a set of nodes extending AudioBIFS and aiming at modeling 3D sound propagation, taking into account acoustic phenomena as described above. Advanced AudioBIFS includes four nodes: DirectiveSound, AcousticScene, AcousticMaterial, and PerceptualParameters. The first three are used in the physical approach to sound scene modeling, meaning it is possible with them to define scenes in which geometrical surfaces cause sound reflections or obstruct the sound when they appear in the direct path between the source and the listener. When the DirectiveSound node is used together with the PerceptualParameters node, the modeling approach is considered as perceptual, as the room acoustic response is controlled by the fields of the PerceptualParameters node. Figure 12.9 shows the textual node interfaces of each of the Advanced AudioBIFS nodes.

In the next section, the functionality of each of the Advanced AudioBIFS nodes will be described. For a more detailed description, see [AMD1-1].

12.4.4.1 DirectiveSound The DirectiveSound node is used in both the physical and the perceptual room acoustics modeling approaches as a 3D sound source object. Similarly to the Sound and Sound2D nodes, it acts as a topmost node of an audio subgraph—that is, it is used to attach a sound stream or a composed combination of sound streams to a scene. Many functionalities associated with this node are similar for the physical and perceptual modeling approaches. The main difference between these two approaches is in the rendering of the room acoustic response, which also depends on the other Advanced AudioBIFS nodes.

```
DirectiveSound {
        angles                  0
        directivity             1
        frequency               []
        speedOfSound            340
        distance                100
        useAirabs               FALSE
        direction               0, 0, 1
        intensity               1
        location                0, 0, 0
        source                  NULL
        perceptualParameters    NULL
        roomEffect              FALSE
        spatialize              TRUE
}

AcousticMaterial {
        reffunc             0
        transfunc           1
        refFrequency        []
        transFrequency      []
        ambientIntensity    0.2
        diffuseColor        0.8, 0.8, 0.8
        emissiveColor       0, 0, 0
        shininess           0.2
        specularColor       0, 0, 0
        transparency 0
}
```

```
AcousticScene {
        center          0 0 0
        size            -1 -1 -1
        reverbTime      0
        reverbFreq      1000
        reverbLevel     0.4
        reverbDelay     0.5
}

PerceptualParameters {
        sourcePresence      1.0
        sourceWarmth        1.0
        sourceBrilliance    1.0
        roomPresence        1.0
        runningReverberance 1.0
        envelopment         0.0
        lateReverberance    1.0
        heavyness           1.0
        liveness            1.0
        omniDirectivity     1.0
        directFilterGains   1.0, 1.0, 1.0
        inputFilterGains    1.0, 1.0, 1.0
        refDistance         1.0
        freqLow             250.0
        freqHigh            4000.0
        timeLimit1          0.02
        timeLimit2          0.04
        timeLimit3          0.1
        modalDensity        0.8
}
```

Fig. 12.9 Textual descriptions of the Advanced AudioBIFS nodes.

Some functionalities of the DirectiveSound node are the same as for the Sound node: The location and the orientation of the source are still given with the location and direction fields. Thus, the relative positions of the Viewpoint (or the ListeningPoint) and of the DirectiveSound node also are taken into account in the same way as for the Sound node when the sound is being processed. Also, the spatialize field of DirectiveSound is interpreted as for Sound: It either allows or disallows the rendering of the incident angle of the sound that arrives at the listener.

The fields angles, directivity, and frequency are used to define a directivity pattern for the sound source. The angles field allows the inclusion of an arbitrary number of directions (with respect to the direction given by the direction field of DirectiveSound). For each angle, the fields directivity and frequency can be used together to define frequency-dependent filtering. In the simplest case, the source is omnidirectional (i.e., radiates sound equally to all directions). Also, simple frequency-independent directivity can be obtained by associating a pure scalar value (in the directivity field) with each of the angles (given by the angles field). Figure 12.10 illustrates the rendering of the

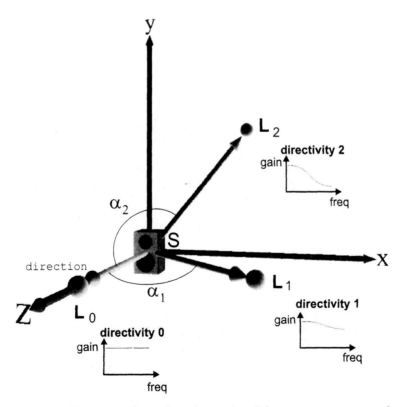

Fig. 12.10 Frequency-dependent directivity of the DirectiveSound node as a function of the angle with respect to the reference direction.

frequency-dependent directivity of DirectiveSound. Three angles are defined: 0 degrees, α_1, and α_2; for each of them, a transfer function is defined, performing an increasing low-pass filtering to the sound as a function of the increasing angle.

The effects related to the DirectiveSound node that depend on the distance between the source and the listener are the propagation delay, air absorption, and distance-dependent attenuation (see previous section for explanations of these effects). The propagation delay can be affected (or disabled) by the speedOfSound field, which can be given different values for controlling the strength of this effect. When sound reflections are present and the physical approach is adopted, this field also affects the delays of the reflections with respect to the direct sound. The distance-dependent attenuation also can be controlled with the distance field. For the physical approach, this field defines a distance toward which the sound attenuates to –60 dB from the original level at the location of the source; after this distance, the sound is not heard. For the perceptual approach, this field is used only for defining the radius of the audibility region around the source location, and the source-Presence field of the PerceptualParameters node defines the distance attenuation effect. Finally, the distance-dependent low-pass filtering caused by air absorption is enabled by the useAirabs field. This filtering is carried according to distance-dependent air absorption curves given in [ISO9613].

The rendering of the room acoustic effect (sound reflections and reverberation, to be explained in the sections on AcousticScene, AcousticMaterial, and PerceptualParameters nodes) can be enabled or disabled by the room-Effect field. This way it is possible to build up scenes in which some of the sound sources are anechoic—that is, they sound like they would when listened to in free-field conditions (with no sound reflecting or obstructing objects), whereas others are processed with a room acoustic effect. This property gives the scene author the freedom to design scenes in which he or she controls the processing costs required by different sound source objects. A more detailed description of the DirectiveSound node is given in [AMD1-1].

12.4.4.2 AcousticScene   The AcousticScene node is used only in the physical approach and has three functionalities:

1. **Rendering region of sound:** AcousticScene defines a region (a 3D box) in the scene, inside of which the sound is heard when both the source and the listener are within the same area.

2. **Late reverberation:** AcousticScene adds late reverberation to DirectiveSound sources when they are inside this region.

3. **Combining acoustic surfaces to form complete rooms:** AcousticScene can be used to group geometrical surfaces together in one room acoustic modeling process. This is useful for scenes in which there are several geometrical room configurations and the purpose is to associate a single acoustic room to one sound processing event. By grouping the

walls of one room with one `AcousticScene` node, and by defining the 3D box so that it covers the expected listening area in the room, it is possible to restrict the processing inside that box so that it takes into account only the surfaces belonging to that room.

The 3D rectangular box is defined by the fields `center` and `size`. `center` is the center point of the 3D box, whereas `size` is a 3D vector giving the lengths of the edges of the box in the x-, y-, and z- directions.

The fields `reverbTime`, `reverbFreq`, `reverbLevel`, and `reverbDelay` together define characteristics of the late reverberation that can be added to sounds when they are in the 3D `AcousticScene` region. The fields `reverbTime` and `reverbFreq` together are used to define a frequency-dependent reverberation time (i.e., the time during which the impulse response of the reverberation decays to –60 dB from the original value). This decay time can be given a different value at different frequencies to simulate different types of rooms or halls. `reverbDelay` defines the delay of the first output of the reverberator impulse response, and `reverbLevel` defines the level of that output. The delay can be used to roughly adjust the start of the late reverberation with respect to the direct sound and early reflections, so that they do not overlap in time in the room acoustic impulse response. Thus, this delay should be proportional to the size of the simulated room. These fields also can be used to give a reverberating effect to sounds (within the `AcousticScene` region), even when there are no sound-reflecting geometrical surfaces in the scene. A more detailed description of this node is given in [AMD1-1].

12.4.4.3 `AcousticMaterial` `AcousticMaterial` is a material node (see Chapter 4 about different node types) defined in Advanced `AudioBIFS` that, in addition to giving visual properties (as the `Material` node) to a geometrical shape object, can be used to associate acoustic properties with it. These acoustic properties are sound reflectivity (describing the size of the portion of the sound reflected by the object) and sound transmission (describing how much of the sound passes through the object). Following are descriptions of the functionalities of `AcousticMaterial` used for the purpose of associating acoustic properties to an object.

This node is used in the `material` field of an `Appearance` node, giving visual properties to a geometrical shape by binding the appearance and geometry together with a `Shape` node. Because of the real-time constraints of a sound scene rendering, the sound reflectivity in BIFS is defined only for flat, polygonal surfaces. In these conditions, the specular sound reflections can be calculated with an *image source* method [Bori84, AllB79]. This method allows real-time tracing of the reflections as well as smooth changes in the delays and directions of each reflection of a surface in dynamic conditions—that is, when the listener, sound source, or one of the reflective surfaces changes its position [SHLV99]. Therefore, the semantics of the `AcousticMaterial` node allow associating it only with an `IndexedFaceSet` node, which is used to form polygons that may contain an arbitrary number of vertices. The reflectivity and transmission properties of surfaces can be given as frequency-modifying coefficients (like the directivity in the case of DirectiveSound node). The reflectivity is expressed with the fields reffunc and refFrequency, whereas the transmission function is expressed with the fields transfunc and transFrequency. As for the directivity, these properties also can be given as simple, scalar coefficients that do not modify the frequency proportions in the sound that is reflected or transmitted through the surface. (For a more detailed description of the AcousticMaterial functionalities, see [AMD1-1].)

12.4.4.4 PerceptualParameters As explained earlier, the perceptual modeling approach involves controlling the characteristics of the acoustic effect with a set of parameters describing how the room effect is perceived instead of defining the geometry of the room. In BIFS, this is made possible through the PerceptualParameters node that is included as a field of the DirectiveSound node. Thus, the room acoustic response is defined for each sound source separately. Furthermore, the response does not relate to scene components other than the relative positions of the sound source and the listener. This is an important difference compared to the geometrical approach, in which the positions of sound-reflecting or sound-obstructing surfaces also are taken into account. With the perceptual approach, scenes can be created in which the sound sources can feature different room acoustic effects, even when they exist in the same 3D space.

With the perceptual modeling approach, the energy relations between the various parts of the room impulse response are controlled in both the time and frequency domains. As can be seen in Figure 12.11, the time-domain impulse response for the perceptual approach is divided into four sections: *direct sound, directional early reflections, diffuse early reflections,* and *late reverberation.* The perceptual room acoustic parameters are used to control the time limits between these different sections and the relative energies of the various parts of the response in three frequency bands.

The field values of the PerceptualParameters node are used to control the way the user perceives the presence of the sound source, the relative energies of the source and the room acoustic response, and the acoustic energy caused by the room only. These parameters are mapped to the energies of the four different time sections of the impulse response in three frequency bands (*low-, mid-,* and *high-frequency* bands; see [AMD1-1] for a detailed definition of this mapping). The temporal time limits (t0 to t3 in Figure 12.11) have an effect on the perceived room size. The boundary frequencies are set between the low and middle frequencies, and between the middle and high frequencies. Additionally, three different DSP filters can be defined to modify the direct sound, the rest of the room response (without the direct sound), and the input sound (equivalent to filtering both the direct sound and the room response). These filters also are defined in terms of gains at three frequency bands.

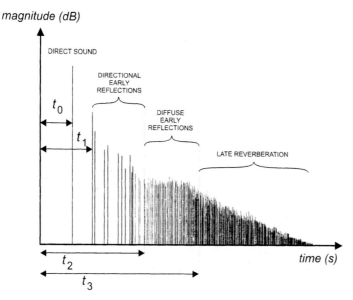

Fig. 12.11 Different sections of a time-domain room impulse response as defined for the perceptual approach to modeling room acoustics.

The various fields of the `PerceptualParameters` node are briefly described (see also Figure 12.9) here:

☞ **Time-domain division of the room impulse response:** The fields `timeLimit1`, `timeLimit2`, and `timeLimit3` are the starting times of the three sections of the room response (directional reflections, diffuse early reflections, and late reverberation). According to Figure 12.11, this means that `timeLimit1` = t_1-t_0, `timeLimit2` = t_2-t_0, and `timeLimit3` = t_3-t_0. The delay of the direct sound (t_0) is defined by the distance between the source and the listener, and it depends on the `speedOfSound` field of the parent `DirectiveSound` node in the same way as for the physical approach.

☞ **Frequency domain division of the response:** The fields `freqLow` and `freqHigh` define the boundary frequencies for the low, middle, and high frequencies of the DSP filters used to modify the frequency content of the four time sections of the room impulse response. Thus, these fields act together with the fields `sourceWarmth`, `sourceBrilliance`, `heaviness`, `liveness`, `directFilterGains`, `inputFilterGains`, and `omniDirectivity` in defining the magnitude spectra of the filters applied to the different time-domain sections of the response.

☞ **Modal density:** Modal density of a room response regards the density of resonances (modes) in the frequency-domain response. This property typically depends on the size and the geometrical configuration of a room. The field `modalDensity` of the `PerceptualParameters` node can be used as a guideline for the renderer to implement the room response so that the modal density it produces matches the desired one as closely as possible.

☞ **Energy of the early room response:** In the `PerceptualParameters` node, three fields are included for controlling the properties of the direct sound and the directional early reflections. These fields are `sourcePresence`, `sourceWarmth`, and `sourceBrilliance`. They define the frequency-dependent *energy of the direct sound and the directional early reflections*: `sourcePresence` is the total early energy, and `sourceWarmth` and `sourceBrilliance` are used to emphasize the low and high frequencies of this part of the response, respectively. These frequency modifications are applied to the low- and high-frequency bands defined by the `freqLow` and `freqHigh` fields.

☞ **Effect of the source-listener distance in the perceived room response:** The field `refDistance` of the `PerceptualParameters` node defines the distance between the source and the listener at which the values of the fields that define the energy of the various parts of the response are valid. For example, in a 3D scene in which the listener can navigate, the actual distance between the user and the sound source changes—and this change is taken into account by modifying the relative energies of the four sections of the response. This corresponds to a situation in a natural environment in which the perception of the source relative to the room effect decreases as a function of the distance from the source. This change is carried out by automatically computing a distance-dependent value for the *source presence* parameter (used instead of the value given in the `sourcePresence` field), which affects all the four parts of the response, as defined in [AMD1-1].

☞ **Controlling the late reverberation:** The late reverberation part of the response is controlled in terms of its energy and frequency-dependent reverberation time, meaning the time during which an exponentially decaying late reverberation decreases to –60 dB from 0 dB. The field `roomPresence` defines the energy of the late reverberation, `lateReverberance` defines the reverberation time at middle frequencies, and the fields `heavyness` and `liveness` are used to modify the reverberation times at low and high frequencies, respectively.

☞ **Envelopment:** `envelopment` is a field that defines the relative energies of the directional early reflections and the direct sound.

☞ **Early decay time:** Early decay time of a room response is the time during which the sound decreases to –10 dB from the level of the direct sound. This room acoustic parameter is given in the `runningReverberance` field of the `PerceptualParameters` node.

☞ **Additional filtering of the direct sound and room effect:** There are three fields in the `PerceptualParameters` node that allow DSP filtering (frequency and magnitude modification) of different parts of the room response. These fields are `directFilterGains`, `inputFilterGains`, and `omniDirectivity`. They are used to control the filtering of the direct sound, the whole sound (including the four sections of the impulse response), and the room part of the response (early reflections, diffuse reflections, and late reverberation only), respectively. The filtering of the direct sound (defined by the `directFilterGains` field) corresponds to simulating an *occlusion* effect, meaning a situation in which the sound source is behind a sound-obstructing object, in which case the direct sound is attenuated but the response of the room is still heard. The use of the field `inputFilterGains`, on the other hand, affects the whole sound, which can be thought of as simulating a situation in which the sound is heard from outside the room where the sound source is located. Finally, `omniDirectivity` is a field that enables establishing the average directivity of the sound source, which modifies the spectrum of the reflections and the late reverberation. These filters are given as magnitude gains at the three frequency bands defined by the fields `freqLow` and `freqHigh`.

12.4.4.5 Examples of the Use of Advanced `AudioBIFS` Nodes

A simple example of a scene in which room acoustics modeling is done according to the physical approach is presented in the following. In this scene, there is one `DirectiveSound` source (defined at lines 7–20) and two `AcousticScene` nodes, the first of which (lines 25–30) adds late reverberation to the sound when the source is inside the rendering region defined by the `AcousticScene` node. The second `AcousticScene` node (lines 37–42) is associated with a set of acoustically responding polygons (`IndexedFaceSet` at lines 47–64, the acoustic properties of which are given at lines 68–69 in the `AcousticMaterial` node). These polygons produce reflections to the sound when the source and the listener are inside the rendering region of that `AcousticScene` node. This node also adds a short reverb to the sound with the aid of the `reverbTime` and `reverbFreq` fields. In both `AcousticScene`s, the reverberation time is given at two frequencies. In this example, the reverberation is clearly shorter at the high frequencies than at the low frequencies, which is often the case with natural reverberating spaces.

```
 1    Group{
 2     children [
 3      Viewpoint {
 4       Position
 5       Orientation
 6      }
 7     DirectiveSound {
 8      source AudioSource {
 9       url 10
10       startTime 0
11       stopTime -1
12      }
13      location 2.5 2.5 -3
14      spatialize TRUE
```

```
15      roomEffect TRUE
16      speedOfSound 340
17      distance 100
18      directivity [1 0.5 0.30 0.1]
19      angles [0 1.05 2.09 3.14]
20      }
21
22   // 3D rectangular region where sounds are added reverberation
23   // When the source and the listener are inside of this region
     the acoustic surfaces
24   // inside of the second AcousticScene region are ignored.
25   AcousticScene {
26    center -5 2.5 -3
27    size 10 10 12
28    reverbTime [2.5 0.5]
29    reverbFreq [0 11025]
30          }
31
32     // Group which contains one AcousticScene that groups together
       sound
33     // reflecting surfaces, and restricts the region where the
       (sound and the)
34     // sound reflections are rendered.
35   Group {
36    children [
37     AcousticScene {
38       center 2.5 2.5 -3
39       size 10 10 12
40      reverbTime [1.0 0.2]
41       reverbFreq [0 11025]
42      }
43
44   // "Surfaces of the box"
45   // (Note that these are grouped under the same Group node as the
     AcousticScene)
46    Shape {
47     geometry IndexedFaceSet {
48      coord Coordinate {
49       point [0 0 0,
50             5 0 0,
51             5 5 0,
52             0 5 0,
53             0 0 -6,
54             5 0 -6,
55             5 5 -6,
56             0 5 -6]
57      coordinateIndex [1, 5, 6, 2, -1,
58                       5, 4, 7, 6, -1,
59                       0, 3, 7, 4, -1,
60                       0, 1, 2, 3, -1,
61                       0, 1, 5, 4, -1,
62                       3, 2, 6, 7, -1]
63      }
64     }
65      appearance Appearance {
```

```
66        material AcousticMaterial {
67          emissiveColor 0 0 1
68          reffunc 0.5
69          transfunc 0.2
70          }
71        }
72      }
73    ]
74  }
```

The next example includes a DirectiveSound (lines 2–32) node that is associated with a PerceptualParameters node. Thus, the sound scene rendering is done according to the perceptual approach to room acoustics modeling. The position and directivity of this sound source are defined as for the previous example. The room acoustic effect is enabled by the PerceptualParameters node (given at lines 21–31). Lines 22 through 24 define the source-presence-related properties (through the energies of the direct sound and the directional early reflections), and lines 25 through 30 define the properties more related to the effect of the room (as explained in context of the PerceptualParameters node).

```
1   Group {
2     Viewpoint {
3       position 0 0 2
4       orientation 0 1 0 0
5     }
6
7     DirectiveSound {
8       source AudioSource {
9       url 3
10      startTime 0
11      stopTime -1
12      }
13      location 0 0 0
14      direction 0 0 1
15      spatialize TRUE
16      useAirabs TRUE
17      distance 100
18      roomEffect TRUE
19      directivity [1 0.5 0.30 0.1]
20      angles [0 1.05 2.09 3.14]
21      perceptualParameters PerceptualParameters {
22        sourcePresence 1.0
23        sourceWarmth 3.0
24        sourceBrilliance 0.5
25        roomPresence 0.5
26        runningReverberance 0.1
27        envelopment 0.1
28        lateReverberance 3.5
29        heavyness 2.0
30        liveness 0.5
31      }
32    }
33  }
```

12.5 SUMMARY

In this chapter, the MPEG-4 audio SNHC and composition tools were presented. The SNHC framework enables both synthetic and natural audio, as well as digital sound effects, to be included in a single MPEG-4 presentation. The main tools for synthetic audio coding, structured audio (SA), and the text-to-speech interface (TTSI), were presented. Finally, the tools related to the composition and presentation of sound that are defined in the Systems part of MPEG-4 were discussed. These tools are known as AudioBIFS and its extension, Advanced AudioBIFS.

The application areas of the MPEG-4 SNHC audio and audio composition tools range from digital sound synthesis and TTS synthesis to postprocessing and mixing of natural and synthetic audio, and to interactive 3D audiovisual environments for virtual and augmented reality.

12.6 REFERENCES

[AllB79] Allen, Jont B., and David A. Berkley. "Image Method for Efficiently Simulating Small-Room Acoustics." *Journal of the Acoustical Society of America*, 65(4). April 1979.

[AMD1-1] ISO/IEC 14496-1:2000, Amendment 1, *Coding of Audio-Visual Objects—Part 1: Systems*. 2000.

[Bega94] Begault, Durand R. *3-D Sound for Virtual Reality and Multimedia*. Cambridge, MA: AP Professional. 1994.

[Bori84] Borish, J. "An Extension to Image Model to Arbitrary Polyhedra." *Journal of the Acoustical Society of America*, 75(6): 1827–1836. 1984.

[IPA99] International Phonetic Association: *Handbook of the International Phonetic Association: A Guide to the Use of the International Phonetic Alphabet*. Cambridge, UK: Cambridge University Press. 1999.

[ISO639] ISO 639:1988. *Code for the Representation of the Names of Languages*. 1988.

[ISO9613] ISO 9613-1. *Acoustics: Attenuation of Sound During Propagation Outdoors, Part 1: Calculation of the Absorption of Sound by the Atmosphere*. 1993.

[Jot99] Jot, Jean-Marc. "Real-Time Spatial Processing of Sounds for Music, Multimedia, and Interactive Human-Computer Interfaces." *Multimedia Systems*, 7: 55–69. 1999.

[Kutt91] Kuttruff, H. K. *Room Acoustics* (3rd ed.). Essex, UK: Elsevier Science. 1991.

[MIDI96] MIDI Manufacturers Association (MMA). *The Complete MIDI 1.0 Detailed Specification*, v. 96.2. 1996.

[MIDI99] MIDI Manufacturers Association (MMA). *The MIDI Downloadable Sounds Specification*, Level 2, v. 1.0c. 1999.

[MPEG4-1] ISO/IEC 14496-1:1999. *Coding of Audio-Visual Objects—Part 1: Systems*, Version 1, December 1999.

[MPEG4-2] ISO/IEC 14496-2:1999. *Coding of Audio-Visual Objects—Part 2: Visual*, Version 1, December 1999.

[MPEG4-3] ISO/IEC 14496-3:1999. *Coding of Audio-Visual Objects—Part 3: Audio*, Version 1, December 1999.

[SchK99] Scheirer, Eric D., and Youngmoo E. Kim. *Generalized Audio Coding with MPEG-4 Structured Audio*. 17th Audio Engineering Society (AES) International Conference, Florence, Italy. Preprint 5–3. September 1999.

[SchR98] Scheirer, Eric D., and Lee Ray. Algorithmic and Wavetable Synthesis in MPEG-4 Multimedia Standard. 105th Audio Engineering Society (AES) Convention, San Francisco. Preprint No. 4811. September 1998.

[SchV99] Scheirer, Eric D., and Barry L. Vercoe. "SAOL: The MPEG-4 Structured Studio Orchestra Language." *Computer Music Journal*, 23(2): 31–51. 1999.

[ScLY00] Scheirer, Eric D., Youngjik Lee, and Jae-Woo Yang. "Synthetic and SNHC Audio in MPEG-4." *Signal Processing: Image Communication*. Special Issue on MPEG-4 (15): 445–461. 2000.

[ScVH99] Scheirer, Eric D., Riitta Väänänen, and Jyri Huopaniemi. "AudioBIFS: Describing Audio Scenes with the MPEG-4 Multimedia Standard." *IEEE Transactions on Multimedia*, 1(3): 237–250. September 1999.

[SHLV99] Savioja, Lauri, Jyri Huopaniemi, Tapio Lokki, and Riitta Väänänen. "Creating Virtual Acoustic Environments," *Journal of the Audio Engineering Society*, 47(9): 675–705. September 1999.

[VeGS98] Vercoe, Barry L., William G. Gardner, and Eric D. Scheirer. "Structured Audio: Creation, Transmission, and Rendering of Parametric Sound Representations." *Proceedings of the IEEE*, 86(5): 922–940. May 1998.

[Verc95] Vercoe, Barry L. *Csound: A Manual for the Audio Processing System*. Cambridge, MA: MIT Media Laboratory. 1995.

[VRML97] ISO/IEC 14772-1. *VRML97 Standard. Information Technology—Computer Graphics and Image Processing—The Virtual Reality Modeling Language (VRML), Part 1: Functional Specification and utf-8 encoding*. April 1998.

Profiling and Conformance:
Approach and Overview[1]

by Rob Koenen and Fernando Pereira

Keywords: MPEG-4, profile, level, conformance, MPEG-4 Video,
MPEG-4 Audio, MPEG-4 Systems, interoperability, interworking

Profiles and levels in MPEG-4 are standardized in order to give users a number of well-defined and well-chosen conformance points. They serve two main purposes: (a) ensuring interoperability between MPEG-4 implementations and (b) allowing conformance to the standard to be tested. Profiles exist not only for the Audio and Visual parts of the standard (audio profiles and visual profiles) but also for the Systems part of the standard, in the form of graphics profiles, scene graph (or scene description) profiles, MPEG-J profiles, and an object descriptor (OD) profile. Different profiles are created for different application environments. They allow manufacturers to use a subset of the (large) MPEG-4 toolbox. The policy for defining profiles is that they should enable as many applications as possible while keeping the number of different profiles low. Having too many different profiles adversely affects interoperability. MPEG has defined quite a few profiles for MPEG-4, but some more may be added when the need becomes apparent.

---

1. The introductory sections of this chapter are adapted from *Signal Processing: Image Communication*, 15 (4–5) (2000), Rob Koenen, pp. 463–478, "Profiles and Levels in MPEG-4: Approach and Overview.," Copyright 2000, with permission from Elsevier Science.

MPEG will be restrictive in defining any new profiles, listening carefully to what its industrial users have to say.

An oft-heard criticism of the MPEG-4 standard concerns its size and complexity. The complaint is that the standard is too big and unwieldy to implement. This would have been true without the existence of profiles and levels (see [KoPC97, Koen99] and the other chapters in this book for an overview of the MPEG-4 technologies). This chapter will describe MPEG-4's *profiles* and *levels* and the philosophy behind their definitions.

Profiles are known from MPEG-2 Video [MPEG2-2], wherein the most used profiles are Main in end-user systems and, more recently, 4:2:2 for professional purposes. Both in MPEG-2 and in MPEG-4, profiles limit the tool set that needs to be implemented; they are created for users who wish to use only a part of the standard. Such users are usually companies and industrial consortia, rather than end users. In fact, profiles can be regarded as a compromise between maximizing interoperability and minimizing the cost incurred by implementation overhead. The essence of profiling the MPEG-4 standard is, as this chapter will argue, about finding optimal balances and making the right trade-offs. The larger and more diverse a standard like MPEG-4 becomes, the more difficult it is to make these choices and to arrive at a transparent and usable division into subsets while maintaining interoperability between systems.

MPEG-4 defines not only visual (as in MPEG-2 Video) and audio profiles but also graphics profiles, scene graph (or scene description) profiles, MPEG-J profiles, and OD profiles. The need to limit the complexity applies to these various technology dimensions as it does to visual and audio tools. Visual profiles are logically defined in the Visual part of the standard (Part 2) [MPEG4-2]; audio profiles in the Audio part (Part 3) [MPEG4-3]; and the other four types of profiles can be found in the Systems part of MPEG-4 (Part 1) [MPEG4-1]. The DMIF part of the MPEG-4 standard [MPEG4-6] does not have profiles; they simply are not needed, as implementing the whole Part does not incur significant complexity over only implementing a subset.

In Section 13.1, this chapter will first explain the most important concepts. Section 13.2 will describe the procedure and policy that was developed for choosing the needed profiles. Finally, Section 13.3 will describe the profiles that are currently defined and their envisioned application areas.

13.1 PROFILING AND CONFORMANCE: GOALS AND PRINCIPLES

The goal of defining profiles and levels is twofold. The first goal is to *ensure interoperability*. Implementations of a profile at a certain level result in a decoder that behaves in a predictable way. Content encoded (e.g., by a real-time encoder) or authored (e.g., for streaming from a server) for such a combination will work on any decoder implementation that conforms to that combination. The second goal is to allow *conformance testing* to take place.

In theory, it would be possible to signal, at the beginning of a communication, the tools that a decoder has on board, so that the server or encoder could use only these tools—and profiles wouldn't be necessary. There are a number of reasons why this does not work in practice. First, content often is pre-encoded, and you want to be 100% sure that all target decoders can consume it. In low-complexity devices (often communication terminals) users want to be sure that their terminals will be able to communicate, and not leave that to the chance that the tool sets overlap enough. Just "getting new tools online" is often impossible and cumbersome at best. Exchanging capabilities before exchanging content will at least incur unwanted delays and requires a bidirectional connection that does not always exist.

Profiles by themselves do not constitute a so-called *conformance point* for the standard. To define a conformance point (a precisely defined specification at which different implementations can interoperate), a *level* is needed. Whereas the profile restricts the tool set, the level defines the bounds of complexity that can be expected in the bitstream for a particular profile. Without a level definition, the complexity that needs to be handled by, say, a video decoder could still be arbitrarily complex, as a profile does not specify, for example, maximum bit rates or spatial resolution. Note that a profile and level combination is usually referred to as *profile@level* (pronounced as a profile at a level).

A profile and level combination gives a well-defined conformance point. For such a conformance point, tests can be devised to determine whether implementations of the standard really operate as the standard specifies. Typically, such tests define input (bitstreams) and expected decoder output (e.g., waveforms or pixel values for decoded audio and video objects, respectively). Note that conformance testing is specified in a separate part of the MPEG-4 standard (Part 4) [MPEG4-4]. This part of MPEG-4 specifies procedures for testing the conformance of bitstreams and decoders that follow the specifications found in MPEG-4 Parts 1 (Systems), 2 (Visual), 3 (Audio), and 6 (Delivery Multimedia Integration Framework; DMIF). Bitstream and decoder conformance (also called *compliance*) may be defined like this:

☞ **Bitstream conformance:** A bitstream or set of bitstreams compliant with a given profile@level shall only contain the allowed syntactic elements for that profile, and any parameter value shall not exceed the allowed values for that profile@level. Additionally, the set of bitstreams shall not violate the complexity restrictions defined for the profile@level in question.

☞ **Decoder conformance:** A decoder compliant with a particular profile@level shall be able to interpret all allowed values of all allowed syntactic elements for that profile@level (*static compliance*) and shall have enough resources to perform all allowed decoding operations according to the decoding semantics for the syntax supported by that profile@level at the required pace (*dynamic compliance*).

Part 4 also gives guidelines on how to construct bitstream test suites to verify decoder conformance. Some test bitstreams generated according to these guidelines are provided as an electronic annex to this part of the standard.

The bounds set by the level need to be observed by both the encoder and the decoder. The encoder needs to make sure it doesn't exceed bitstream complexity bounds when encoding; the decoder needs to be built such that it can at least handle the most complex bitstream possible under the level definition. Hence, the level gives minimum decoder implementation bounds. For decoding *hardware*, the profile@level combination gives minimum performance constraints to be observed at design and manufacture time. For decoding *software*, the combination also may imply resource availability to be monitored at run time.

The MPEG-4 standard, like MPEG-1 and MPEG-2, only defines the decoding process and the syntax and semantics of the bitstream. The encoding process is not specified, with the restriction that a valid bitstream (or possibly a set of elementary streams [ESs]) must result which observes the complexity limits imposed by the level definition. In spite of MPEG only defining the decoder, profile@level combinations do impact encoding systems, as mentioned previously. To encoders, regardless of whether implemented in hardware or software, a profile@level combination gives implicit bounds to observe while encoding a bitstream. The encoder must constantly check whether the output ES or streams are still within the limits defined by the level. In practice, for *encoder,* one should actually read the more general term *authoring system*, because MPEG-4 content may well be created from different pre-encoded objects, perhaps in combination with real-time objects. Summarizing, a profile@level is an upper bound on the complexity of the bitstream (to be observed by the encoder) and a lower bound on the capabilities of the decoder.

MPEG-4 is an object-based standard, and audiovisual scenes are composed of different objects. Audio and visual profiles@levels do *not* define the maximum complexity per individual MPEG-4 object but, rather, give bounds on the *total of all objects* in the scene. How many objects in the scene need to be decoded simultaneously is important in determining the complexity of the decoder. Content authors will require the freedom to spend the limited decoder resources they need to work with flexibly. They need to be able to choose between one very difficult object and, perhaps, five easier ones—as long as they observe decoder resource limits. After giving the relevant definitions, a more detailed explanation will be provided on why this approach was chosen.

Before going into more detail, this section will introduce the relevant definitions and terminology used in MPEG. These are largely extracted from the MPEG-4 Requirements document [N4319].

☞ **Object type:** An object type defines the syntax of the bitstream for one single object, which can represent a meaningful entity in the (audio or visual) scene. An object type corresponds to a set of tools. The object type does not define any complexity boundaries. There are audio object types and visual object types.

☞ **Profile:** A profile defines the set of a certain type of tools that can be used in a certain MPEG-4 terminal. There are media profiles (audio, visual, graphics) and systems profiles (scene graph,[2] MPEG-J, and OD). Audio and visual profiles are defined as a set of (audio or visual) object types.

☞ **Level:** A level is a specification of the constraints and performance criteria on an audio, visual, graphics, scene description, MPEG-J, or OD profile and thus on the corresponding tools. (The MPEG-J dimension currently has no levels associated with it. The OD dimension doesn't have such levels either.)

☞ **Conformance point:** A conformance point is a specification of a particular audio, visual, graphics, scene description, MPEG-J, or OD profile at a certain level at which conformance may be tested (profile@level). Conformance points are normatively defined within the MPEG-4 standard.

Audio and visual profiles are more than just lists of tools. They define the kinds of audio and visual objects the MPEG-4 terminal needs to be able to decode and, hence, give a list of admissible object types (ES types; see Chapter 3 for a discussion on ESs in MPEG-4).

In MPEG-2, (video) profiles can be thought of as containing only a single, rectangular object. In MPEG-4, scenes can contain more than one object, and the objects can be of different natures. Therefore, the concept of an object type is introduced as an intermediate level of definition between tools and profiles. Object types not only define which tools are available to code an object in the scene, they also give restrictions on how they can be combined, as they also specify the syntax and the semantics. When a profile would only consist of a list of tools, many more combinations of tools would be possible than are now allowed in the predefined object types. Not all of these combinations would make sense, and some of them would be hard to implement. Hence, object types are a required step in the definition of an audio or visual profile.

Graphics profiles define, in terms of BIFS nodes (see Chapter 4), which graphical elements can be used in the scene. Scene graph profiles define the scene description capabilities required in the terminal also in the form of allowed BIFS nodes in the bitstream. Note that there are two different types of BIFS nodes. The first type (*media nodes*) is used to create objects in the scene, or to refer to ESs associated with media objects. This is the type of node found in the graphics profiles. The second type of nodes is used to build the scene structure and to define object and user interactions. These are called *scene graph nodes* and are found in the scene graph profiles.

---

2. Note that *scene graph* profiles is the official term used in the MPEG-4 standard. *Scene description* would probably have been a better term, as it cannot be as easily confused with the *graphics* profiling dimension.

MPEG-J profiles define subsets of the MPEG-J Application Programming Interfaces (APIs; see Chapter 5). They restrict the power of the virtual machine a device needs to have under its hood.

Finally, the OD profiles define required terminal capabilities in terms of OD and synchronization layer (SL) tools (see Chapter 3 for a detailed explanation of these tools).

The audio, visual, and graphics profiles can be called *media profiles* as they govern the media elements in the scene. Note that MPEG has chosen not to prescribe which combinations of audio, visual, and graphics profiles are allowed in a terminal. The same applies to the systems profiles (scene graph, OD, and MPEG-J). MPEG wants to let the market decide these combinations, but it does take care that matching profiles exist to create, for example, a handheld device. Creating these combinations has been discussed, but ultimately decided against. There would simply be too many possible combinations with the six profiling dimensions that currently exist, and MPEG did not feel it had enough knowledge of the future market for MPEG-4 to prescribe such combinations. This is again a compromise—it would have helped interoperability if MPEG could prescribe, normatively, combinations of profiles. Reality, however, teaches that it is even difficult to make sure industry implements singular profiles completely according to the specifications; in the MPEG-2 case, industrial consortia have introduced additional restrictions in many occasions (e.g., the U.S. Advanced Television Standards Committee did this in the context of the U.S. digital TV system).

From the preceding, it can be concluded that profiles for which levels are defined only make sense in combination with a certain level. In terms of compliance statements, it is meaningless to just name the profile and not give a level. Some profiles, however, currently have only one level defined, in which case mentioning the (default) level could be omitted. As of February 2002, the MPEG-J and OD profiles have no levels, but this could change in the future.

13.2 PROFILING POLICY AND VERSION MANAGEMENT

The policy in defining MPEG-4 profiles and levels is aimed at obtaining a minimum number of profile@level combinations that are as widely usable as possible [N2565]. This means both a low total number (giving a "global" optimum) and a low number of different conformance points that address roughly the same application type (giving a "local" optimum); for example, there are quite a few profiles that address slightly different guises of Web applications. The profiling policy should be understood together with the approach to version management. This chapter will first discuss version management before returning to the profiling policy.

The fact that the MPEG-4 standard is delivered in versions necessitates the existence of version management procedures [N2200]. Different versions are issued in the form of amendments to the standard. Versions of MPEG-4

are meant for major improvements and enhancements of the tool set. New tools are added only if they bring new functionalities or significant gains in performance for the functionalities already provided. This implies that a marginal increase in, for example, coding efficiency is not enough reason for adding a tool, which would make the standard more complex, more expensive to implement, and less stable but not significantly more powerful and useful.

A new version of a Part of the standard, by definition, extends the standard in a backward-compatible way. This compatibility is preserved by adding new profiles to the existing set, notably in the area of (audio and visual) coding tools. These *may* be supersets of existing profiles, but they do not have to be. It is important to note that an existing profile will not be modified in a new version, because doing so would render existing products noncompliant with the standard. That would clearly be highly undesirable. The version management procedures as documented within MPEG define this as follows: "New versions, notably in the area of coding tools, are managed by adding new profiles, significantly different from existing ones. New versions will not make changes to existing profiles" [N2200]. To illustrate this, Version 2 (e.g., of MPEG-4 Visual) includes all Version 1 profiles and adds new ones. Profile E could be a superset of D, but it could also be a subset of an existing profile with new tools added. E could even be a subset of D, which means that only Version 1 tools are used in this profile defined in Version 2, as seen in Figure 13.1.

Now that versioning is clear, it is good to revisit the profiling policy. In the first MPEG meetings when potential profiles for MPEG-4 were discussed, there were many requests for different profiles. Sometimes the proposals were very close in terms of tools to be included. With these many requests, it was clear that a strict policy was needed in order to limit the number of different profiles, especially in the Visual part of the standard. Having too many

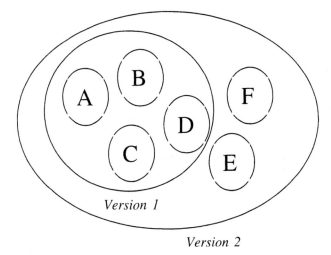

Fig. 13.1 Adding profiles in versions.

profiles confuses the market place and is unhelpful in achieving interoperability. MPEG has defined a number of rules to keep profile proliferation under control:

☞ There should be a demonstrated need for the new profile in terms of applications that will use it. These applications need to be described by proponents.

☞ A new profile needs to be significantly different from existing ones in terms of tools and functionality. This must be backed by evidence (e.g., proof of additional functionality, proof of error-resilience gains, or proof of significantly better compression efficiency). When a new subset of existing tools is proposed that aims at reducing complexity, there should be no existing profile that (perhaps at the expense of some slight overhead in terms of unnecessary tools) can already support the applications addressed by that new profile.

☞ There must be significant company or consortium support: A statement that plans are in existence to implement and deploy the profile is required.

☞ Collectively, supporting companies must commit to delivering conformance bitstreams.

☞ In the case of audio and visual profiles, supporting companies must collectively commit to performing verification tests (usually subjective tests that prove the performance of the profile in question).

Note that, in spite of these requirements, MPEG never discusses business models and never requires business information to be supplied by proponents.

When requests for profiles are on the table, MPEG tries to see if the requirements for the new profile can be met with an existing profile. When multiple requests are discussed, MPEG tries to harmonize these and merge two or more profile proposals. This is not an easy process. Typically, MPEG experts have rather strong opinions about what subset of MPEG-4 technology they want to support in their products, and although the MPEG community understands that the number of profiles needs to be limited, it is hard to merge different proposals. Similar requirements apply before new object types and levels can be defined, but the strongest requirements are placed on the profiles.

Whereas profiling information is an integral part of the standard, defining a profile or a level is very different than defining a tool. A new profile in itself does not affect the interworking of different tools or extend the syntax of the bitstream (with the exception that a new value needs to be defined for the field that signals the profile in the OD; see Chapter 3). Although some technical checks need to be carried out to confirm that the profile definition is solid, this implies there is no hurry in defining profiles and that they can be added at a relatively late stage. This is in contrast to the tools themselves, which need to be much more rigorously tested and cross-checked with other tools. In other words, the tools in the standard need to anticipate future needs, whereas the profiles can be defined when their requirement becomes apparent, which means that MPEG can take a conservative attitude in defining profiles. Note that the addition of profiles or levels still requires the standard to be amended.

The same strategy is applied in the definition of levels for a profile: Only define levels that are going to be used; additional ones can always be added at a later stage. Also, their parameters need to differ enough from existing levels. As far as producing the standard is concerned, adding levels is a matter of merely *adding a line to a table*. This can always be done when the need becomes apparent; although the formal process is the same as for any amendment, it requires far less technical work than adding tools. Thus MPEG is not concerned with completely defining all possible levels within a given profile, but only with those that are envisaged to be used at a certain moment in time.

Although, in principle, all the work in MPEG is driven by the requirements, it may turn out that tools in the standard have not found a "home" in any of the profiles, at least not yet.[3] As long as the number of these tools is low, this is not a problem. Work on tools is often started in advance of real market need, based on anticipated requirements. When, later in the process, the exact market requirements take shape, some of the tools may not be as useful as originally thought. Note that the converse also happens: Needs become apparent that are unsupported by tools. In such a case, quick action must be taken, especially if the technology to fulfill the requirements exists. In MPEG-4, this was the case for the MPEG-4 file format and the management and protection of intellectual property (IPMP).

The profiles in MPEG-2 Video are organized as an almost complete hierarchical structure. This implies that decoders for the higher profiles can, by definition, also understand the ones "below" them, which use a subset of tools. In MPEG-4, profile hierarchy is implemented when possible, but it is not pursued at all costs. This policy was adopted because keeping a strict hierarchy in MPEG-4 is much harder than in MPEG-2, as there are more types of tools and hence more dimensions. Also, there are more application domains that need to be accommodated. As an example, consider MPEG-4 Visual. A strict hierarchy could be maintained for all of the profiles addressing compression efficiency for rectangular objects. Different application areas, however, have different needs for choosing combinations of natural and synthetic object types, and using those with or without scalability, error resilience, or shape representation. Thus, a strict hierarchy across all the visual and audio profiles could not be maintained.

---

3. In February 2002, this was still the case for the overlapped block motion compensation (OBMC) tool in the Visual standard [MPEG4-2].

13.3 OVERVIEW OF PROFILES IN MPEG-4

This section gives an overview of the profiles in MPEG-4 that were defined as of October 2001. It first lists the media profiles—visual, audio, and graphics—and then the nonmedia profiles. These nonmedia profiles are defined in the Systems part of the standard, as are the graphics media profiles [MPEG4-1]. The audio and visual media profiles can be found in the Audio [MPEG4-3] and Visual [MPEG4-2] parts of the standard, respectively. The media profiles govern the usage of object types and BIFS elements that can be used to create the objects in the MPEG-4 scene.

Figure 13.2 provides a graphical representation of the profile structure in MPEG-4. Note that the figure gives examples and does not list all profiles, nor does it reflect hierarchical relationships.

The next sections will first describe the media profiles and then discuss the nonmedia profile dimensions.

13.3.1 Visual Profiling

This section addresses the visual object types and profiles, as defined in MPEG-4 Visual [MPEG4-2]. It gives the status as of October 2001. Level information will be supplied without too much detail; for the full details on levels definition see Appendix A.

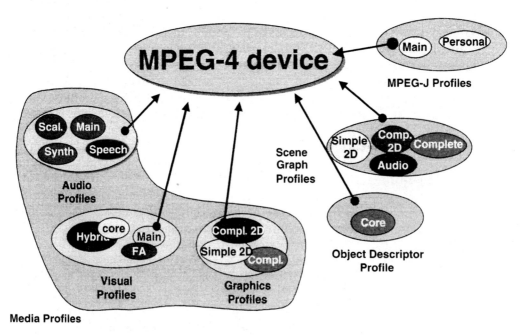

Fig. 13.2 Overview of the profile structure in MPEG-4.

13.3.1.1 Visual Object Types

As the visual profiles are defined using the visual object types, it is necessary to discuss these before discussing the profiles themselves. For details on the visual coding tools, see Chapters 8 and 9. There are a number of object types for representing natural visual (video) information [MPEG4-2].

Rectangular Video

☞ The **Simple** object type defines an error-resilient, rectangular natural video object of arbitrary height-to-width ratio. It uses relatively simple and inexpensive coding tools, based on intra (I) and predicted (P) video object planes (VOPs), the MPEG-4 term for frames.

☞ The **Advanced real-time simple (ARTS)** object type, a superset of the Simple object type, targets real-time coding situations. It employs a back channel to monitor throughput, adapt resolution and reduce error propogation. It also adds dynamic resolution conversion to the Simple object type tool set.

☞ The **Advanced simple** object type also defines a rectangular video object, but adds tools that enhance compression efficiency, such as 1/4 pel motion estimation, global motion estimation, and B frames.

☞ The **Fine granularity scalable (FGS)** object type [N3904] looks like Advanced simple, but includes temporal and fine-granular SNR scalability.

☞ The **Simple scalable** object type is a scalable extension of Simple, which gives temporal and spatial scalability, using Simple as the base layer. The enhancement layer is still rectangular.

Arbitrarily Shaped Video

☞ The **Core** object type uses a tool superset of Simple, giving better quality through the use of bidirectional interpolation (B-VOPs), and has binary shape coding. It supports temporal scalability based on sending extra P-VOPs. Note that binary shape can include a constant transparency but excludes the variable transparency offered by gray-scale shape coding.

☞ The **Core scalable** object type adds rectangular temporal and spatial scalability to the Core object type as well as object-based spatial scalability.

☞ The **Main** object type adds to Core, coding of gray-scale shape, sprites, and interlaced content in addition to the coding of progressive material.

☞ The **Advanced coding efficiency (ACE)** object type is similar to Main, but includes some extra coding efficiency tools, notably 1/4 pel motion compensation and global motion compensation (GMC); but it does not include sprites.

☞ The **N-bit** object type is equal to the Core object type but it can vary the pixel depth from 4 to 12 bits for the luminance as well as the chrominance planes.

☞ The **Simple studio** object type [N3898] defines an I-frame-only object for very high quality and bit rates. It has arbitrary shape and multiple alpha planes. It differs from the Simple object type in the syntax, which is closer to MPEG-2 to allow easy transcoding, and in the fact that P-frames are not supported.

☞ The **Core studio** object type [N3898] adds P-frames to the Simple studio object type, which makes it more complex but also more efficient.

The Simple object type uses a subset of the tools in Core, and Core in return uses a subset of the tools in Main. The tools in the Simple scalable object type are a superset of the tools in Simple, whereas the N-bit object type is a superset of Core (and hence also of Simple).

Still Visual

☞ The **Scalable texture** object type defines an arbitrary-shaped still image that uses wavelet coding for scalability and incremental download and build-up.

☞ The **Advanced scalable texture** object type supports (in addition to what the Scalable texture object type can do) error resilience, better scalable shape coding, and partial decoding of the bitstream.

Synthetic Visual

The following object types use synthetic coding tools, some of them in combination with natural video or still picture coding tools.

☞ The **Basic animated 2D texture** object type allows mesh animation with arbitrary-shaped still images (according to the Scalable texture object type; see preceding). There are two types of meshes supported in MPEG-4: *uniform* and *Delaunay*. In this object type, only the uniform mesh type can be used. (See Chapter 9 for further explanation on the different types of mesh coding tools in MPEG-4.)

☞ The **Animated 2D mesh** object type combines the synthetic mesh tool (either rectangular or Delaunay topology) with natural video. The natural video coding uses the same tools as the Core object type. This video can be mapped onto the mesh and deformed by moving the vertices in the mesh. This gives interesting animation possibilities. Note that this object can be of arbitrary (binary) shape.

☞ The **Simple face animation** object type has the tools for facial animation. This object type does not define what the face looks like, just its animation; the animation can be applied to any local model of choice. Note that MPEG-4 does include tools to download a predefined 3D facial model to the decoder, but these tools are not mandatory in the Simple face animation object type.

☞ The **Simple face and body animation** object type adds body animation to the Simple face animation object type.

Table 13.1 lists the tools against the object types; for further details, see [MPEG-4-2], [N3898], and [N3904].

Table 13.1 Visual tools versus visual object types

Visual tools \ Visual object types	Simple	Simple Scalable	Advanced Simple	ARTS	Core	Core Scalable	Main	ACE	N-bit	FGS	Simple Studio	Core Studio	Animated 2D Mesh	Basic Animated Texture	Advanced Scalable Texture	Scalable Texture	Simple Face Animation	Simple FBA
I-VOP	•		•	•	•	•	•	•	•	•	•	•	•					
P-VOP	•	•	•	•	•	•	•	•	•	•		•	•					
B-VOP		•	•		•	•	•	•	•	•			•					
P-VOP with OBMC (Texture)																		
Basic Tools • AC/DC Prediction • 4-Motion Vectors • Unrestricted Motion Vectors	•	•	•	•	•	•	•	•	•	•	•		•					
Basic Tools (Studio Object Types) • Frame/Field Structure • Slice Structure • Studio DPCM Block • Studio Binary Shape • Studio Grayscale Shape											•	•						
Error Resilience • Slice Resynchronization • Data Partitioning • Reversible VLC	•	•	•	•	•	•	•	•	•	•			•					
Short Header	•		•	•	•	•	•	•	•	•			•					
Method 1/Method 2* Quantization			•		•	•	•	•	•	•			•					
P-VOP Based Temporal Scalability • Rectangular • Arbitrary Shape				•	•	•	•	•	•				•					

Table 13.1 Visual tools versus visual object types (Continued)

Visual tools ↓ / Visual object types →	Simple	Simple Scalable	Advanced Simple	ARTS	Core	Core Scalable	Main	ACE	N-bit	FGS	Simple Studio	Core Studio	Animated 2D Mesh	Basic Animated Texture	Advanced Scalable Texture	Scalable Texture	Simple Face Animation	Simple FBA
Binary Shape					•		•	•	•				•	•	•			
Gray Shape							•	•										
Interlace			•				•	•		•	•	•						
Sprite							•											
Temporal Scalability (Rectangular)		•				•												
Spatial Scalability (Rectangular)		•				•												
N-Bit									•									
Scalable Still Texture													•	•	•	•		
2D Dynamic Mesh with Uniform Topology														•	•			
2D Dynamic Mesh with Delaunay Topology														•				
Dynamic Resolution Conversion				•														
NewPred				•														
Global Motion Compensation			•					•										
Quarter-pel Motion Compensation			•					•										
SA-DCT								•										
Error Resilience for Visual Texture Coding															•			
Wavelet Tiling															•			

Table 13.1 Visual tools versus visual object types (Continued)

↓ Visual tools / Visual object types →	Simple	Simple Scalable	Advanced Simple	ARTS	Core	Core Scalable	Main	ACE	N-bit	FGS	Simple Studio	Core Studio	Animated 2D Mesh	Basic Animated Texture	Advanced Scalable Texture	Scalable Texture	Simple Face Animation	Simple FBA
Scalable Shape Coding for Still Texture															•			
Object Based Spatial Scalability						•												
Fine Granularity Scalability (FGS)										•								
FGS Temporal Scalability										•								
Facial Animation Parameters																	•	•
Body Animation Parameters																		•

°A mark for Method 1/Method 2 Quantization indicates that both quantization methods are supported by the corresponding visual object type; where there is no mark, only Method 2, also called H.263 quantization, is supported (see Chapter 8).

13.3.1.2 Visual Profiles The visual profiles govern which visual object types can be present in the scene, thereby determining which coding tools can be used to code these objects and, hence, what their ESs look like. Visual profiles are thus defined as lists of admissible object types. Quite a few of the profiles have names that correspond to the most complicated object type they support. Following is a list of the visual profiles defined as of October 2001 [MPEG4-2, N3898, N3904], mentioning some of the application areas they address. Note again that these are only suggestions and that profiles are not intended to address specific applications. This is why their names are generic and refer to tools rather than applications or services.

Rectangular Video

☞ The **Simple** profile accepts only objects of type Simple and was created with low-complexity applications in mind. The first usage is mobile (audio)visual services, and the second is putting very-low-complexity video on the Internet. Also small camera devices recording moving video to, for example, disk or memory chips can make good use of this profile. There were four levels for the `simple` profile with bit rates from 64 kbit/s in Level 0 to 384 kbit/s in Level 3.

Levels 1 and higher support up to four objects in the scene with, at the lowest level, a maximum total surface of a QCIF picture.[4] Level 0 was defined at the request of the 3rd Generation Partnership Project (3GPP). It is similar to Level 1 in bit rate, but includes some extra restrictions; for instance, there is only one object allowed in the scene. The levels also define the maximum total surface for the objects and the amount of macroblocks per second that the decoder needs to be able to decode. Further, they define the size of various (hypothetical) buffers needed for decoding. Although the maximum total object size is defined, the aspect ratio is not prescribed. This gives maximum creative freedom. It could be used, for instance, in a personal computer screen, on which a very wide or a very tall object could be created or several smaller objects could be put in various places on the screen, not confined to a typical rectangular area (e.g., QCIF). The same level philosophy is followed for restricting the complexity of the natural video objects in all the visual profiles. For all details on the definition of levels for visual profiles, see Appendix A.

☞ The **Advanced simple** profile accepts object types `Simple` and `Advanced simple`. It is useful in Internet streaming applications and other applications that have fairly low bandwidth available, but it scales to television-size pictures and TV quality. There are six levels, from 0 to 5. Levels 0 to 3 have bit rates from 128 kbit/s to 768 kbit/s. Levels 0 and 3 correspond to the same levels in `Simple`. Support for interlaced coding is added for Levels 4 and 5, with bit rates of 3 Mbit/s and 8 Mbit/s. This represents a compromise between the desire to support interlace at higher bit rates and the desire to keep the number of profiles manageable. Normally new tools such as interlaced are not introduced in levels, but require a different profile. It is likely that another level, termed 3b, will be added; compared to Level 3, it will allow higher bit rates without having to support interlaced coding.

☞ The **Fine granularity scalability (FGS)** profile [N3904] defines how to add FGS layers of scalability to either `Simple` or `Advanced simple`, which can both function as base layers.

☞ The **Simple scalable** profile can supply scalable coding in the same operational environments as foreseen for the `Simple` profile, but now in a scalable fashion. `Simple scalable` has two levels, which build on Levels 1 and 2 of `Simple` profile. A Level 0 corresponding to `Simple` Level 0 was in the process of being defined at the time of writing.

☞ The **Advanced real-time simple (ARTS)** profile is useful in real-time coding situations, taking advantage of the back channel and the adaptive

---

4. 144×176 pixels for the luminance and 72×88 pixels for each chrominance.

encoding to create higher resilience to errors and better performance under changing bandwidth conditions.

Arbitrarily Shaped Video

☞ The **Core** profile accepts Core and Simple object types. It is useful for higher-quality interactive services, combining good quality with limited complexity and supporting arbitrarily shaped objects. Also, mobile broadcast services can be supported by this profile. The maximum bit rate is 384 kbit/s in Level 1 and 2 Mbit/s in Level 2. Although the levels do not prescribe the visual session size, they are created with a certain session size in mind, called the *typical visual session size*. For Simple and Core, these are QCIF and CIF, depending on the level. The maximum amount of macroblock surface that can be used to create objects is usually chosen such that a scene using this typical session size can have overlapping objects and still be "filled."

☞ The **Core scalable** profile is a superset of the Simple, Simple scalable, and Core profiles. It adds scalability to Core, according to the Core scalable object type.

☞ The **Advanced core** profile combines the natural video coding provided by the Core object type with the possibilities of the Advanced scalable texture object type. Advanced core has two levels, for 384 kbit/s and 2 Mbit/s, respectively.

☞ The **Main** profile was created with broadcast services in mind, addressing progressive as well as interlaced material. It combines the highest quality with the versatility of arbitrarily shaped objects using grayscale coding. The highest of three levels accepts up to 32 objects (of Simple, Core, or Main type) for a maximum total bit rate of 38 Mbit/s.

☞ The **Advanced coding efficiency (ACE)** profile starts with the Simple and Core object types and includes some extra tools for coding efficiency, such as global motion compensation and 1/4 pel motion compensation, through the ACE object type. Four levels exist, with bit rates from 384 kbit/s to 38.4 Mbit/s.

☞ The **N-bit** profile is useful for applications that use thermal images, such as surveillance applications. Also, medical applications may want to use the enhanced pixel depth, giving a larger dynamic range in color and luminance. It accepts objects of type Simple, Core, and N-bit. Currently only one level is defined: Level 1 at 2 Mbit/s with 16 objects.

☞ The **Simple studio** profile [N3898] includes only the Simple studio object type. It is meant for editing video in the studio and other professional applications needing similar quality. Four levels exist, with bit rates from 180 Mbit/s to 1800 Mbit/s.

☞ The **Core studio** profile [N3898] also is intended for editing uses in the studio. It includes the Simple and Core studio object types. Also, four levels are defined, the highest with a bit rate up to 900 Mbit/s. Note that this is lower than for `Simple studio`, because the P frames add compression efficiency.

Still Visual

☞ The `Scalable texture` profile is meant for audiographic applications: applications including only audio and still visual material. It was requested by companies that want to build rather simple terminals—for example, mobile devices—which combine sound with synchronously displayed pictures and possibly BIFS-based graphics. Three levels restrict the complexity, and the maximum surface ranges from about 400k pixels to about 6M pixels.

☞ The `Advanced scalable texture` profile includes solely the `Advanced scalable texture` object type, and has three levels similar to the `Scalable texture` profile.

Synthetic and Hybrid Natural/Synthetic Visual

☞ The `Simple face animation` profile accepts only objects of type `Simple face animation`. Depending on the level, either one or a maximum of four faces can appear in the scene—for example, for a virtual meeting. Bit rates remain very low; even for the second level, 32 kbit/s is more than adequate for driving a maximum of four faces.

☞ The `Simple face and body animation (FBA)` profile does the same as the `Simple face animation` profile, but now for face and body animation. Similar level restrictions apply (i.e., one or four FBA objects in Levels 1 and 2, with 32 kbit/s or 64 kbit/s, respectively).

☞ The `Hybrid` profile allows combining natural and synthetic objects in the same scene while keeping complexity reasonable. On the natural side, it compares to the `Core` profile, whereas on the synthetic side, it adds animated meshes, scalable textures, and animated faces—a rich set of tools for creating attractive hybrid natural and synthetic content. This profile can be used to place "real" objects into a synthetic world and also to do the opposite: add synthetic objects to a natural environment. There are two levels, corresponding to `Core` profile at Levels 1 and 2 and giving matching restrictions on the synthetic objects.

☞ The `Basic animated texture` profile allows animation of still pictures using meshes and facial animation. Attractive content can be created at very low bit rates. Two levels are defined, corresponding to the levels in the `Simple face animation` profile, giving restrictions on the meshes as well.

A partial hierarchy exists in the visual profiles, the same hierarchy that was described for the corresponding object types. This means that `Main` is a superset of `Core`, which is a superset of `Simple`. `N-bit` is a superset of `Core`. `Simple scalable` is a superset of `Simple`, in such a way that the `Simple` profile can decode the base layer of a `Simple scalable` bitstream. `Advanced simple` is a superset of `Simple`, and `FGS` can work with both `Simple` and `Advanced simple` as a base layer.

Table 13.2 shows the relations between visual object types and visual profiles. For all details on the visual levels, see Appendix A.

Table 13.2 Visual profiles versus visual object types

Visual profiles ↓ \ Visual object types →	Simple	Advanced Simple	Fine Granularity Scalable	Simple Scalable	Advanced Real-Time Simple	Core	Core Scalable	Main	Advanced Coding Efficiency	N-Bit	Scalable Texture	Advanced Scalable Texture	Animated 2D Mesh	Basic Animated Texture	Simple Face Animation	Simple FBA	Simple Studio	Core Studio	Number of levels
Simple	•																		4
Simple Scalable	•			•															2
Advanced Simple	•	•																	6
FGS	•	•	•																6
ARTS	•				•														4
Core	•					•													2
Core Scalable	•			•		•	•												3
Advanced Core	•					•						•							2
Main	•					•		•			•								3
ACE	•					•			•										4
N-Bit	•					•				•									1
Scalable Texture											•								3
Advanced Scalable Texture												•							3
Simple Face Animation															•				2
Simple FBA																•			2
Hybrid	•					•						•	•	•					2
Basic Animated Texture												•	•	•					2
Simple Studio																	•		4
Core Studio																	•	•	4

13.3.2 Audio Profiling

This section will first describe the audio object types and then show how they are grouped into profiles. Although there are quite a few object types in the Audio part of the standard (13 plus the Null object type), the number of profiles is only eight [MPEG4-3]. Again, level information will be presented only in broad lines, but the full details are available in Appendix B.

13.3.2.1 Audio Object Types   There are different audio object types for general audio, speech, speech and general audio, synthetic audio, and synthetic speech. (For details on the coding tools, see Chapters 10, 11, and 12.)

General Audio
For general audio coding, MPEG-4 includes the advanced audio coding (AAC), twin vector quantization (TwinVQ), bit-sliced arithmetic coding (BSAC), and harmonic and individual lines plus noise (HILN) algorithms. With these basic coding tools, the following object types were defined.

☞ The **AAC main** object type is similar to—and compatible with—the AAC main profile that is defined in MPEG-2 [MPEG2-3]. MPEG-4 AAC adds the perceptual noise shaping (PNS) tool. The object type has multi-channel capability, to give five full channels plus a separate, low-frequency channel in one object. An MPEG-2 AAC decoder can parse and decode the bitstream for this object type, albeit at somewhat lower quality, as it does not understand the PNS tool.

☞ The **AAC low complexity** (AAC LC) object type is a low-complexity version of the AAC main object type.

☞ The **Error-resilient AAC low complexity** (ER AAC LC) object type adds error resilience to the MPEG-4 AAC LC object type.

☞ The **AAC scalable sampling rate** (AAC SSR) object type is the counterpart to the MPEG-2 AAC scalable sampling rate profile, again adding PNS.

☞ The **AAC long-term prediction** (AAC LTP) object type is similar to the AAC main object type, with the long-term predictor replacing the MPEG-2 AAC predictor. This gives the same efficiency with significantly lower implementation cost.

☞ The **Error-resilient AAC long-term prediction** (ER AAC LTP) object type is the error-resilient version of the AAC LTP object type.

☞ The **Error-resilient AAC low-delay** (ER AAC LD) object type includes the low-delay, PNS, and LTP tools. It also supports syntax for error resilience. It allows general low-bit-rate audio coding in applications requiring a very low delay of the encoding/decoding chain (e.g., full-duplex real-time communications). Note that there is no *nonerror-resilient* counterpart to this object type.

☞ The **TwinVQ** object type is based on fixed-rate vector quantization instead of the Huffman coding used in AAC. It operates at lower bit rates than AAC, supporting mono and stereo sound.

☞ The **Error-resilient TwinVQ** (ER TwinVQ) object type is the error-resilient counterpart to the MPEG-4 TwinVQ object type.

☞ The **Error-resilient bit-sliced arithmetic coding** (ER BSAC) object type provides FGS audio coding and also supports error resilience. A large number of scalable layers can be used, giving 1 kbit/s per channel enhancement layers.

☞ The **Error-resilient for harmonic and individual lines plus noise** (ER HILN) object type uses HILN parametric coding to code general audio signals suitable for very low bit rates (4 kbit/s to 16 kbit/s). Parametric coding means that speed and pitch changes while decoding are easy to do. Bit-rate scalability is also supported.

Speech

MPEG-4 specifies two different algorithms for speech coding—code excited linear prediction (CELP) and harmonic vector excitation coding (HVXC)—each operating at different bit rates; these algorithms are used in the following object types:

☞ The **CELP** object type is based on the CELP coder and supports 8 kHz and 16 kHz sampling rates at bit rates from 4 kbit/s to 24 kbit/s. CELP bitstreams can be coded in a scalable way using bit-rate scalability and bandwidth scalability.

☞ The **Error-resilient CELP** (ER CELP) object type is the error-resilient version of the CELP object type, also supporting silence compression for greater efficiency.

☞ The **HVXC** object type is based on the HVXC coder providing a parametric representation of 8 kHz, mono speech at fixed bit rates between 2 kbit/s and 4 kbit/s. One can even go below 2 kbit/s when the variable bit-rate mode is used. With HVXC, it is possible to change pitch and speed during decoding.

☞ The **Error-resilient HVXC** object type is the error-resilient version of the HVXC object type.

Speech and General Audio

☞ The **AAC scalable** object type allows a large number of scalable combinations, including combinations with TwinVQ and CELP coder tools, as the core coders. It supports only mono or two-channel stereo sound.

☞ The **Error-resilient AAC scalable** (ER AAC scalable) object type is the error-resilient version of AAC scalable object type.

☞ The **Error-resilient parametric** object type includes the HILN and HVXC parametric coding tools. This integrated parametric coder com-

bines the functionalities of the ER HILN and the ER HVXC object types.
Only 8 kHz sampling rate and mono audio channel are supported.

Synthetic Audio

☞ The **Main synthetic** object type collects all MPEG-4 structured audio
tools. It supports flexible, high-quality algorithmic synthesis using the
Structured Audio Orchestra Language (SAOL) music-synthesis language
and efficient wavetable synthesis, with the *Structured Audio Sample-
Bank Format* (SASBF), and enables high-quality mixing and postproduction with the audio BIFS tool set. Sound can be described in the MPEG-4
structured audio format even from 0 kbit/s (meaning that sound continues without input—until it is stopped by an explicit command) to 3 kbit/s
to 4 kbit/s for extremely expressive sounds.

☞ The **Algorithmic synthesis and AudioFX** object type provides SAOL-
based synthesis capabilities for very low-bit-rate terminals. The sample
bank tool is not included in this object type. (Note that FX stands for
effects.)

☞ The **Wavetable synthesis** object type is a subset of the Main synthetic
object type, making use of only the SASBF format and MIDI[5] tools. It
provides relatively simple sampling synthesis.

☞ The **General MIDI** object type supports interoperability with existing
MIDI content (see Chapter 12). Unlike the Main synthetic or Wavetable
synthesis object types, it does not give completely predictable (i.e., normative) sound quality and decoder behavior.

Synthetic Speech

☞ The **Text-to-speech interface** (TTSI) object type offers an extremely
low-bit-rate phonemic representation of speech. The actual text-to-
speech synthesis is not specified; only the interface is defined. Bit rates
range from 0.2 kbit/s to 1.2 kbit/s. The synthesised speech can be synchronized with a facial animation object.

   Finally, the **Null** object type provides the possibility to feed raw PCM
data directly to the MPEG-4 audio compositor to allow the mixing in of
local sound at the decoder. This means that support for this object type is
in the compositor, not in the decoder.

   Table 13.3 gives a list of all the audio object types and the tools they use
[MPEG4-3]. A complete explanation of the MPEG-4 Audio tools can be found
in Chapters 10, 11, and 12 and, of course, in the MPEG-4 Audio standard itself
[MPEG4-3].

---

5. MIDI means Musical Instrument Digital Interface, a popular wavetable format in
   wide use. (See [MIDI]).

Table 13.3 Audio tools versus audio object types

↓ Audio object types / Audio tools →	MPEG-2 ACC Main	MPEG-2 LC	MPEG-2 SSR	PNS	Long Term Pred.	TL SS	Twin VQ	Low Delay AAC	CELP	Silence Compression	HVXC	HVXC 4kbs VB	BSAC	HILN	Error Robustness	SA Tools	SASBF	MIDI	TTSI	Superset of ...
AAC main	•			•																AAC LC
AAC LC		•		•																
ER AAC LC		•		•											•					
AAC SSR			•	•																
AAC LTP		•		•	•															AAC LC
ER AAC LTP		•		•	•										•					ER AAC LC
AAC Scalable		•		•	•	•														
ER AAC Scalable		•		•		•									•					
ER AAC LD				•	•			•							•					
TwinVQ					•		•													
ER TwinVQ							•								•					
ER BSAC				•									•		•					
ER HILN														•	•					
CELP									•											
ER CELP									•	•					•					
HVXC											•									
ER HVXC											•	•			•					
ER Parametric											•	•		•	•					
Main Synthetic																•	•	•		Wave-table and Algor. Synthesis
Algorithmic Synthesis and AudioFX																•				

Table 13.3 Audio tools versus audio object types (Continued)

↓ Audio object types	MPEG-2 ACC Main	MPEG-2 LC	MPEG-2 SSR	PNS	Long Term Pred.	TL SS	Twin VQ	Low Delay AAC	CELP	Silence Compression	HVXC	HVXC 4kbs VB	BSAC	HILN	Error Robustness	SA Tools	SASBF	MIDI	TTSI	Superset of ...
Wavetable Synthesis																		•	•	General MIDI
General MIDI																		•		
TTSI																			•	
Null																				

13.3.2.2 Audio Profiles   There are fewer profiles in MPEG-4 Audio than there are object types. Which audio profiles there are is explained in this section. Remember that claims for conformance of decoders and bitstreams cannot be made against object types, but only to profiles at a certain level.

Before explaining the audio profiles in more detail, it is helpful to say a few words on the level definition. Some of the audio profiles have levels that are defined in terms of complexity units. There are two different types of complexity units: processor complexity units (PCU), specified in millions of operations per second, and RAM complexity units (RCU), specified in terms of number of kwords of memory. The standard also specifies the complexity units required for each object type. In this way, authors have maximum freedom in choosing the right object types and allocating resources among them. An example makes this clear. The profile could contain Main AAC and Wavetable synthesis object types. A level could specify a maximum of two objects of each. This would prevent the resources reserved for the Main AAC objects to be used for a third and fourth Wavetable synthesis object, even though it would not break the decoder. With the complexity units, the author is completely free to use decoder resources for any combination of objects, as long as their types are supported by the profile. This is a similar approach to what the video buffering verifier does for visual profiles, see Appendixes A and B.

Note that in the Scalable audio profile, this type of level definition was used only for the highest level, because the other three levels can be expressed in simpler ways. Here, only for the highest level will the decoder be complex enough to benefit from the type of flexibility offered by the resource-based way of defining levels. For further details on audio levels, see Appendix B and [MPEG4-3].

One note on error resilience: In MPEG-4 Audio Version 2, error-resilient object types were added, often as extensions (supersets) of Version 1 object types; Version 2 profiles predominantly use the error-resistant versions of the appropriate object types: an example is the MAUI profile described in the following section.

General Audio

Only one pure general audio MPEG-4 audio profile has been defined:

☞ The **Mobile audio internetworking (MAUI)** profile combines the ER AAC LC, ER ACC LD, and ER AAC Scalable object types with the ER TwinVQ and ER BSAC ones, but it does not contain speech coders. This profile was designed to add general, high-quality audio to devices that already include (non-MPEG-4) speech coders. An example is a mobile phone, which will already include a speech coder prescribed by an appropriate standards body.

Speech and General Audio

A number of profiles address speech and general audio:

☞ The **Scalable** profile was defined to allow good quality, reasonable complexity, low-bit-rate audio on the Internet, an environment in which bit rate varies from user to user and from one minute to the next. Scalability allows making optimal use of available and even dynamically changing bandwidth while having only to encode and store the material once. The Scalable profile was not defined exclusively for the Internet, however. For example, in broadcast situations scalability can be a desirable feature. The Scalable profile has four levels that restrict the number of objects in the scene, the total number of channels, and the sampling frequency. The highest level employs the novel concept of complexity units, as explained above. The object types in this profile support both speech and general audio.

☞ The **High-quality audio** profile contains tools for high-quality natural audio coding. The CELP object type is included for speech, and the AAC LC and ACC LTP object types are included for general audio. Scalable coding is included through the AAC scalable object type. The profile also includes error-resilient versions of all these object types, and can be used in an error-prone environment.

☞ The **Natural audio** profile contains all natural audio coding tools in MPEG-4, including speech, AAC, and parametric object types. It also contains the corresponding error-resilient object types. A bit stranger in the collection of object types is TTSI. This object type was included in all four Version 1 audio profiles because it is only an interface and the text-to-speech decoder itself is not normative.

Synthetic Audio

There is only one purely synthetic audio profile:

☞ The **Synthetic** profile groups all of the synthetic object types. The main application areas are found where good quality sound is needed at very low data rates, while the sound source is usually not a microphone. There are three levels that define the amount of memory for data, the sampling rates, the amount of TTSI objects, and some further processing restrictions.

Hybrid Natural/Synthetic Audio

There are three hybrid natural/synthetic audio profiles:

☞ The **Main** profile includes all object types of MPEG-4 Audio Version 1. It is useful in environments in which processing power is available to create very rich, highest-quality audio scenes that may combine microphone-recorded sources with synthetic ones. Example application areas are DVD and multimedia broadcast. This profile has four levels, defined in terms of complexity units.

☞ The **Speech** profile addresses speech applications, as its name suggests. Two levels are defined, determining whether either 1 or a maximum of 20 objects can be present in the (audio) scene. The supported object types are CELP, HVXC, and TTSI (all of the speech-related object types that were defined in MPEG-4 Audio Version 1).

☞ The **Low-delay audio** profile is designed for situations in which a low delay is important, such as two-way communications. It contains the HVXC and CELP object types and their error-resilient versions, the ER AAC LD object type and the text-to-speech interface TTSI.

Table 13.4 lists all of the MPEG-4 audio profiles and the object types they include.

Table 13.4 Audio profiles versus audio object types

↓ Audio object types / Audio profiles →	Main	Scalable	Speech	Synthetic	High-Quality Audio	Low-Delay Audio	Natural Audio	MAUI
AAC main	•						•	
AAC LC	•	•			•		•	
ER AAC LC					•		•	•
AAC SSR	•						•	
AAC LTP	•	•			•		•	
ER AAC LTP					•		•	
AAC Scalable	•	•			•		•	
ER AAC Scalable					•		•	•
ER AAC LD						•	•	•
TwinVQ	•	•					•	

Table 13.4 Audio profiles versus audio object types (Continued)

↓ Audio object types / Audio profiles →	Main	Scalable	Speech	Synthetic	High-Quality Audio	Low-Delay Audio	Natural Audio	MAUI
ER TwinVQ							•	•
ER BSAC							•	•
ER HILN							•	
CELP	•	•	•		•	•	•	
ER CELP					•	•	•	
HVXC	•	•	•			•	•	
ER HVXC						•	•	
ER Parametric							•	
Main Synthetic	•			•				
Algorithmic Synthesis and AudioFX	•			•				
Wavetable Synthesis	•			•				
General MIDI	•			•				
TTSI	•	•	•	•		•	•	
Null								
Number of Levels	4	4	2	3	8	8	4	6

13.3.3 Graphics Profiling

Graphics profiles define which of the graphics and textual elements can be used to build a scene. They are expressed in terms of BIFS nodes. Although these profiles are defined in the Systems part of the standard [MPEG4-1], they are really media profiles like the audio and visual ones, as they are used to produce visible or audible objects in the scene.

Four graphics profiles are defined in MPEG-4: Simple 2D, Complete 2D, Complete, and 3D audio. They differ in the BIFS nodes to be supported at the decoder.

☞ The **Simple 2D** profile provides the basic functionalities needed to create a visual scene with visual objects, without giving additional graphic ele-

ments. It includes only the nodes needed to put visual objects in the
scene. One level has been already defined.

☞ The **Complete 2D** profile allows elements like bitmaps, backgrounds, circles, boxes, and lines all in a flat (2D) space to be used. These elements
have characteristics like line width and color. Note that BIFS commands
exist to change, for example, the color of a circle; such commands are also
included in this profile. There are no levels defined yet for this profile.

☞ The **Complete** profile contains the full set of Version 1 BIFS graphics
nodes with which complete and elaborate 3-dimensional graphics can be
created. It adds to Complete 2D, for example, the sphere, the cone, 3D
boxes, and so on, and directional lighting as well. Note that *flat* video
material can be projected into this 3D space on an arbitrary plane. This
is a complex profile to implement. It is likely that subsets will be
deployed first, in the form of new profiles. Some of these profiles were
being defined when this chapter was written (see below). No levels have
been defined yet for this profile.

☞ The **3D Audio** profile includes graphics tools that are required to define
the acoustical properties of the scene, such as geometry, acoustics absorption, diffusion, and acoustical transparency of the material. There are no
levels defined yet for this profile.

Additional Graphics Profiles

At the time of writing, several more graphics profiles were either included in
draft amendments or under study. It was apparent that Simple 2D is limited
to devices that were indeed (just as it says) simple and that, on the other end
of the spectrum, the Complete 2D and Complete profiles were too complex for
most cases. Because it is highly likely that the new profiles would obtain status of International Standard before the end of 2001 or early in 2002, they
have been included in this chapter. Please note that their inclusion in the
final amendment was not 100% certain, and the technical details or the
names may have changed in the final stage of the processing of the corresponding amendment.

☞ The **Simple 2D+text** profile is much like Simple 2D, but also contains the
BIFS nodes to display text, possibly colored or transparent. Like Simple
2D, this profile is useful for low-complexity audiovisual devices. One level
is proposed, which restricts some of the fields in the supported BIFS
nodes.

☞ The **Core 2D** profile includes support for relatively simple 2D graphics
and text: picture-in-picture, video warping animated ads, logos, and so
on. It is aimed at devices such as set-top boxes. Two levels are being
defined for this profile.

☞ The **Advanced 2D** profile provides a set of tools for advanced 2D graphics.
Complex and streamed graphics animations, cartoons, advanced graphi-

cal user interfaces, and games can all be implemented using Advanced 2D. Two levels are proposed for this profile.

☞ The **X3D Interactive** profile is the only 3D profile currently poised to be added to MPEG-4. X3D Interactive is designed to be compatible with Web3D's X3D Interactive profile under development [Web3D], which provides a rich environment for 3D applications such as virtual worlds and games.

Most of the graphics profiles have a counterpart in the form of a scene graph (or scene description) profile with the same name (see the next section).

Table 13.5 lists the BIFS nodes that must be implemented to comply with each of the graphics profiles [MPEG4-1]. For a better understanding of the table, see the relevant chapters in this book. Note that the levels for the graphics profiles are still under development; for details see Appendix C.

Table 13.5 BIFS nodes in the graphics profiles

↓ BIFS nodes / Graphics profiles →	Simple 2D	Simple 2D + Text	Core 2D	Advanced 2D	Complete 2D	3D Audio	Complete	X3D Interactive
AcousticMaterial						•		•
Appearance	•	•	•	•	•	•	•	
Background							•	
Background2D		•	•	•	•		•	•
BAP								
BDP								
Bitmap	•	•	•	•	•		•	•
Body								
BodyDefTable								•
BodySegment ConnectionHint								
Box							•	
Circle			•	•	•		•	
Color			•	•	•		•	•
Cone							•	
Coordinate						•	•	

Table 13.5 BIFS nodes in the graphics profiles (Continued)

Graphics profiles → ↓ BIFS nodes	Simple 2D	Simple 2D + Text	Core 2D	Advanced 2D	Complete 2D	3D Audio	Complete	X3D Interactive
Coordinate2D			•	•	•		•	
Curve2D				•	•		•	
Cylinder							•	
DirectionalLight							•	
ElevationGrid							•	
Expression							•	
Extrusion							•	
Face							•	
FaceDefMesh							•	
FaceDefTable							•	
FaceDefTransform							•	•
FAP							•	
FDP							•	•
FIT							•	
Fog							•	
FontStyle		•	•	•	•		•	•
IndexedFaceSet						•	•	
IndexedFaceSet2D			•	•	•		•	
IndexedLineSet							•	
IndexedLineSet2D				•	•		•	
LineProperties				•	•		•	•
Material							•	
Material2D		•	•	•	•		•	
MaterialKey				•				•
MatteTexture				•				
Normal						•	•	

Table 13.5 BIFS nodes in the graphics profiles (Continued)

Graphics profiles → ↓ BIFS nodes	Simple 2D	Simple 2D + Text	Core 2D	Advanced 2D	Complete 2D	3D Audio	Complete	X3D Interactive
PixelTexture			•	•	•		•	
PointLight							•	•
PointSet							•	
PointSet2D					•		•	
Rectangle		•	•	•	•		•	
Shape	•	•	•	•	•	•	•	
Sphere							•	
SpotLight							•	
Text		•	•	•	•		•	
TextureCoordinate				•	•		•	
TextureTransform				•	•		•	
Viseme							•	

13.3.4 Scene Graph Profiling

The scene graph profiles define what types of scene description capabilities need to be supported by the terminal. As in the case of the graphics profiles, these profiles are defined in terms of the BIFS nodes the decoding terminal will understand. Examples are translations, (3D) rotations, and elements like input sensors with which interactive behavior can be created. The scene graph profiles follow a structure similar to the graphics profiles, with one profile added for audio-only scenes. Although many names are similar to the graphics profiles, and indeed their matching names correspond to matching technology sets, there is no hard and fast rule that these are more adequate combinations. Other combinations of standardized profiles are allowed just as well—this is all at the discretion of users of the MPEG-4 standard.

Levels for many of the scene graph profiles still need to be defined; in some cases they already exist. For more details on levels for scene graph profiles, see Appendix D. The newer profiles, especially those still under construction (see below), were all developed with one or more levels. In fact, it is now a requirement in MPEG that profiles should have at least one level before they can be accepted into the standard. The following is a list of scene graph profiles, as they were defined in October 2001:

☞ The **Audio** profile includes the nodes to build sophisticated audio-only MPEG-4 content. It does not include nodes for displaying and transforming visual elements. It can be used for relatively simple applications like radio but also for more complex applications requiring audio effects. This profile has four levels under study.

☞ The **Simple 2D** profile contains the BIFS scene graph elements necessary to place one or more audiovisual objects in a scene. The audiovisual content can be presented and the scene graph can be updated, but there are no interaction capabilities. The **Simple 2D** scene graph profile supports applications such as classic broadcast television. One level had been defined and another was under ballot at the time this chapter was written.

☞ The **Complete 2D** profile contains all the two-dimensional scene graph nodes from MPEG-4 Version 1. It is a rather complex profile to implement, with strong support for complicated and highly interactive applications. It is likely that, in the short term, application builders will want to choose subsets of this profile. Some of these subsets were under consideration to become new profiles at the time of this writing (see below). This profile is a superset of **Simple 2D**.

☞ The **Complete** profile includes all the scene graph nodes from Version 1, both 2D and 3D. This is a superset of **Complete 2D**. The **Complete** scene graph profile can enable applications like dynamic virtual 3D worlds and games, but also in this case it is likely that subsets will be defined, in the form of profiles, that can be implemented more easily. One such subset, stemming from the VRML world (the **X3D Interactive** profile), was under definition when this chapter was written (see below).

☞ The **3D Audio** profile has the BIFS scene graph elements for audio-only applications in which advanced 3D rendering of audio is employed. The 3D audio nodes were added in MPEG-4 Systems after the scene graph **Audio**-only profile was defined. This profile has four levels under study; they restrict, for example, the number of spatialized sources per scene, the number of reverberations, and the complexity of filters used to create the 3D effects.

Additional Scene Graph Profiles

The situation for scene graph profiles is very much like that for graphics profiles. Also here, more profiles were being considered at the time this chapter was written. These are the scene graph profiles that were either under study or already in draft amendments under ballot at the time of writing.[6]

☞ The **Basic 2D** profile is meant to provide basic 2D composition for very simple scenes with only audio and visual elements. It includes basic 2D

---

6. This list refers to October 2001; the **Advanced Main 2D** profile mentioned in Appendix D became a profile under study in December 2001.

composition and audio and video node interfaces: nodes that can include an audio or a video object in the scene. One level is proposed, limiting the use of the nodes included in the profile.

☞ The `Core 2D` profile has nodes for creating scenes with audio and visual elements using basic 2D composition, 2D texturing, local interaction, local animation, BIFS updates, quantization, and access to Web links and subscenes. It also includes tools for interactive services such as `Server-Command`, `MediaControl`, and `MediaSensor`, to be used in video-on-demand services. Two levels are under ballot.

☞ The `Advanced 2D` profile is a superset of the `Basic 2D` and `Core 2D` profiles, adding advanced 2D composition, advanced local interaction, streamed animation (`BIFS-Anim`), scripting, advanced audio, and `PROTO` nodes. Again two levels are being balloted.

☞ The `Main 2D` profile extends `Core 2D` with the FlexTime model, all input sensors, the `Layer2D`, and `WorldInfo` nodes. `Main 2D` gives an interoperability point with SMIL (see [SMIL01]), and provides rich functionality for highly interactive applications, for example, on the World Wide Web. This name is tentative and may change in the final specification.

☞ The `X3D Interactive` profile provides a common interoperability point with the Web3D specifications [Web3D] and the MPEG-4 standard; this means the same profile is being defined as a Web3D specification. The profile is a subset of nodes that allows the implementation of 3D applications on a low-footprint engine (e.g., a Java applet or small browser plug-in). It also addresses limitations of software renderers.

For a few BIFS nodes, their presence is not just inferred from the scene graph profile but follows logically from the audio and visual profiles the terminal implements. For instance, any audio profile would require `AudioClip` and `AudioSource` to be present; another example is the `Core` visual profile, which will require the `Texture`, `Background2D`, `Background`, and `MovieTexture` BIFS nodes. The presence of these nodes can always be inferred from the chosen audio and visual profiles, and the Systems part of the MPEG-4 standard provides the tables with the required nodes. Table 13.6 gives an example of such a table, taken from [MPEG-4-1].

Table 13.6 BIFS nodes inferred from the visual profiles

Visual profile	BIFS nodes available at the terminal
Simple	ImageTexture, Background2D, Background, MovieTexture
Simple Scalable	ImageTexture, Background2D, Background, MovieTexture
Core	ImageTexture, Background2D, Background, MovieTexture
Main	ImageTexture, Background2D, Background, MovieTexture
Simple Scalable	ImageTexture, Background2D, Background, MovieTexture

Table 13.6 BIFS nodes inferred from the visual profiles (Continued)

Visual profile	BIFS nodes available at the terminal
N-Bit	ImageTexture, Background2D, Background, MovieTexture
Hybrid	ImageTexture, Background2D, Background, MovieTexture, Face, Expression, FAP, FDP, FIT, FaceDefMesh, FaceDefTable, FaceDefTransform, Viseme
Basic Animated Texture	ImageTexture, Background2D, Background, Face, Expression, FAP, FDP, FIT, FaceDefMesh, FaceDefTable, FaceDefTransform, Viseme
Scalable Texture	ImageTexture, Background2D, Background
Simple Face Animation	Face, Expression, FAP, FDP, FIT, FaceDefMesh, FaceDefTable, FaceDefTransform, Viseme

Table 13.7 lists the BIFS nodes for the scene graph profiles. This table was based on [MPEG4-1], new amendments to MPEG-4 Systems, and the MPEG-4's Profiles under Consideration document [N4321]. For an explanation on the BIFS nodes, see Chapter 4; for more details on the levels for scene graph profiles, see Appendix D.

Table 13.7 BIFS nodes in the scene graph profiles

Scene graph profiles → ↓ BIFS nodes	Basic 2D	Simple 2D	Core 2D	Advanced 2D	Complete 2D	Audio	3D Audio	Complete	Main 2D	X3D Interactive
AcousticScene							•			
AnimationStream				•	•		•	•	•	•
Anchor			•	•	•		•	•		
AudioBuffer				•	•	•	•	•		
AudioDelay				•	•	•	•	•		
AudioFX				•	•	•	•	•		
AudioMix				•	•	•	•	•		
AudioSwitch				•	•	•	•	•		
Billboard							•	•		
ColorInterpolator		•	•	•				•		
Collision							•			

Table 13.7 BIFS nodes in the scene graph profiles (Continued)

Scene graph profiles → ↓ BIFS nodes	Basic 2D	Simple 2D	Core 2D	Advanced 2D	Complete 2D	Audio	3D Audio	Complete	Main 2D	X3D Interactive
CompositeTexture2D					•			•		
CompositeTexture3D								•		
Conditional			•	•	•		•	•	•	•
CoordinateInterpolator2D			•	•	•			•	•	•
CoordinateInterpolator							•	•	•	
CylinderSensor								•		
DirectiveSound							•			
DiscSensor				•	•			•		
Form					•			•		
Group				•	•	•	•	•		•
Inline			•	•	•		•	•	•	
InputSensor			•	•						
Layer2D				•	•			•	•	
Layer3D								•		
Layout					•			•	•	
ListeningPoint					•		•	•	•	
LOD							•	•		•
MediaBuffer				•					•	
MediaControl			•	•						•
MediaSensor			•	•					•	•
NormalInterpolator								•	•	
NavigationInfo								•	•	
OrderedGroup	•	•	•	•	•			•	•	
OrientationInterpolator							•	•		
PerceptualParameters							•			
PlaneSensor2D			•	•				•	•	

Table 13.7 BIFS nodes in the scene graph profiles (Continued)

Scene graph profiles → ↓ BIFS nodes	Basic 2D	Simple 2D	Core 2D	Advanced 2D	Complete 2D	Audio	3D Audio	Complete	Main 2D	X3D Interactive
PlaneSensor								•	•	
PositionInterpolator							•	•	•	
PositionInterpolator2D			•	•	•		•	•		•
ProximitySensor							•	•		
ProximitySensor2D				•	•		•	•	•	
QuantizationParameter			•	•	•		•	•		
ScalarInterpolator			•	•	•			•		•
Script				•			•	•	•	
Sound							•	•		
Sound2D	•	•	•	•	•	•	•	•		•
ServerCommand			•	•					•	
SphereSensor								•		
Switch			•	•	•		•	•	•	
TemporalTransform								•		
TemporalGroup								•		
TermCap				•	•		•	•		
TimeSensor			•	•	•		•	•		
TouchSensor			•	•	•		•	•	•	•
Transform							•	•	•	
Transform2D		•	•	•	•			•		
Valuator			•	•	•		•	•	•	
Viewpoint							•	•		
VisibilitySensor							•	•	•	
WorldInfo				•	•		•	•	•	
Node Update			•	•	•		•	•	•	
Route Update			•	•	•		•	•		

Table 13.7 BIFS nodes in the scene graph profiles (Continued)

Scene graph profiles → ↓ BIFS nodes	Basic 2D	Simple 2D	Core 2D	Advanced 2D	Complete 2D	Audio	3D Audio	Complete	Main 2D	X3D Interactive
Scene Update		•	•	•	•	•	•	•	•	
ROUTE			•	•	•		•	•	•	
PROTO				•						•

13.3.5 Object Descriptor Profiling

According to MPEG-4 Systems [MPEG4-1], the OD profiling dimension speci-fies the allowed configurations of the OD and SL tools. The OD contains all descriptive information, whereas the SL tool provides the syntax to convey, among other things, timing information for ESs (see Chapter 3). The main rea-son for wanting to subject the OD to profiling lies in reducing the amount of asynchronous operations and the necessary permanent storage. Currently, only one default OD profile exists, termed Core. It was created to allow the cre-ation of levels, which in turn is necessary to reduce, for example, the number of different time bases a system needs to support simultaneously. However, no levels have been defined yet.

13.3.6 MPEG-J Profiling

MPEG-J stands for MPEG-Java (see Chapter 5). Two MPEG-J profiles exist: Personal and Main. MPEG-J profiles have been defined in MPEG-4 Systems Version 2. There are no levels defined for these profiles; as for the other pro-files without levels, this may still happen in the future.

☞ The **Personal** profile provides a lightweight set of MPEG-J functions for personal devices, constrained in processing power and memory. Exam-ples of such devices are personal digital assistants (PDAs), mobile audio-visual telephones, and hand-held gaming devices. This profile includes the following packages of MPEG-J APIs:
  ✗ Terminal
  ✗ Network
  ✗ Scene
  ✗ Resource
  ✗ Decoder

☞ The **Main** profile includes all the MPEG-J APIs. It is meant for a multitude of consumer devices such as set-top boxes, but also for desktop multimedia players. Main is a superset of the Personal profile, adding the following packages to personal:

✗ Service information and section filtering

13.4 SUMMARY

MPEG-4 profiles and levels are meant for interoperability and conformance checking. MPEG-4 profiles start from MPEG-2 Video principles, consistently applied to all parts of the MPEG-4 standard. The fact that an MPEG-4 scene is potentially composed of multiple objects implies that profiles and levels must give bounds for the total of objects in the scene rather than for individual ones. While implementations of MPEG-4 reach a more mature stage, so does the understanding of industry's needs for interoperability points. It is not unlikely that more profiles need to be added in the future, and it is probable that some profiles will never be used. This means that the set of profiles is less elegant than it could have been, but this does not pose a major problem. What really counts is that industry sectors concentrate on a few profiles and levels, and that these are well defined, so that multiple independent parties can build interoperable systems. Early developments indicate that this is happening: Simple and Advanced simple are the preferred visual profiles, for instance, in the initial phases of MPEG-4 deployment.

MPEG-4 has been designed as a future-proof multimedia standard. This means that although MPEG-4 is largely completed, new elements are being added as the need arises. For instance, there is work on a specification for more interoperable intellectual property management and protection (IPMP), and MPEG is also studying new coding methodologies that will allow MPEG-4 to keep delivering state-of-the art technology. This work will result in new profiles and very likely (for IPMP) even new profiling dimensions.

13.5 ACKNOWLEDGEMENTS

The authors would like to thank the MPEG community for the interesting discussions and meetings that led to this chapter.

13.6 REFERENCES

[Koen99] Koenen, Rob. "MPEG-4: Multimedia for our Time." *IEEE Spectrum*, 36(2). February 1999.

[KoPC97] Koenen, Rob, Fernando Pereira, and Leonardo Chiariglione. "MPEG-4: Context and Objectives." *Image Communication Journal*, 9(4). May 1997.

[MIDI] MIDI Manufacturers Organization. Home Page, *www.midi.org*

[MPEG2-2] ISO/IEC 13818-2:2000. *Generic Coding of Moving Pictures and Associated Audio Information—Part 2: Video*, 2000.

[MPEG2-3] ISO/IEC 13818-3:1994. *Generic Coding of Moving Pictures and Associated Audio Information—Part 3: Audio*, 1994.

[MPEG4-1] ISO/IEC 14496-1:2001, *Coding of Audio-Visual Objects—Part 1: Systems*, 2d Edition, 2001.

[MPEG4-2] ISO/IEC 14496-2:2001, *Coding of Audio-Visual Objects—Part 2: Visual*, 2d Edition, 2001.

[MPEG4-3] ISO/IEC 14496-3:2001, *Coding of Audio-Visual Objects—Part 3: Audio*, 2d Edition, 2001.

[MPEG4-4] ISO/IEC 14496-4:2001, *Coding of Audio-Visual Objects—Part 4: Conformance Testing*, 2d Edition, 2001.

[MPEG4-6] ISO/IEC 14496-6:1999, *Coding of Audio-Visual Objects—Part 6: Delivery Multimedia Integration Framework (DMIF)*, Version 1, December 1999.

[N2200] MPEG Requirements, *MPEG-4 Version Management Procedures*, Doc. ISO/IEC MPEG N2200, Tokyo MPEG Meeting, March 1998.

[N2565] MPEG Requirements, *MPEG-4 Profiling Policy*. Doc. ISO/IEC MPEG N2565, Rome MPEG Meeting, December 1998.

[N3898] MPEG. *Studio Profiles*. Final Draft Amendment. Doc. ISO/MPEG N3898, Pisa MPEG Meeting, January 2001.

[N3904] MPEG. Final Draft Amendment. *Streaming Video Profiles*. Doc. ISO/MPEG N3904, Pisa MPEG Meeting, January 2001.

[N4319] MPEG Requirements. *MPEG-4 Requirements*. Doc. ISO/IEC N4319, Sydney MPEG Meeting, July 2001.

[N4321] MPEG Requirements. *New MPEG-4 Profiles Under Consideration*. Doc. ISO/IEC MPEG N4321, Sydney MPEG Meeting, July 2001.

[SMIL01] W3C. *Synchronized Multimedia Integration Language (SMIL 2.0)*. W3C Recommendation. August 2001. *www.w3.org/TR/smil20/*

[Web3D] Web3D Consortium. Home Page, *www.web3d.org*

Implementing the Standard: The Reference Software

by Zvi Lifshitz, Gianluca Di Cagno, and Matthew Leditschke

Keywords: reference software, conformance, normative software, informative software, MPEG-4 player, 2D compositor, 3D compositor, plug-ins, scene graph

*T*he reference software activity has always been an important part of the standard-making process in MPEG and a significant contribution to the success of MPEG standards. The reference software serves three main purposes: verification, clarification, and promotion of the standard, as explained next.

Implementing the standard during the specification development helps to verify that the technical elements are properly specified, that the various tools in the standard are consistent, and that the standard, when put together, serves the goal it is intended to serve.

When writing an international standard, many words are debated as if it was a legal document, and efforts are extended to ensure clarity and remove ambiguity. Nevertheless, it is almost impossible for a complex technical document compiled out of contributions of tens of parties to be perfectly clear and clean. Whereas legal documents in dispute come before a court, the reference software is the judge for conflicting interpretations of the technical standard specifications. Obviously, this judge cannot reverse the text of the standard, but it can be consulted when the text is helplessly ambiguous. It should be emphasized that in any case of inconsistency between the text and the software (which may occur due to human errors), it is the text that prevails.

As for the promotion, the reference software makes it easier to develop complying hardware and software, therefore encouraging engineers to implement products based on the standard. When they do so, the software can assist them in two different ways. First, the reference software is an open-source project. The code is usually not optimized, but nevertheless may be helpful as a basis for implementation. Second, developers can check interoperability between their applications and the reference software.

Regarding this last point, there is a need to mention an important concept in MPEG standards, and actually in any kind of standard—conformance. A product complies with a standard at a certain conformance point if it meets the conditions associated to that point. More than one conformance dimension may exist, as explained in Chapter 13. MPEG-4 defines conformance in terms of visual, audio, graphics, scene graph, object descriptor (OD), and MPEG-J profiles and levels. A conformance point is a stage during the execution of an application claiming conformance to the standards, when some results are expected, and these results (final or intermediate) can be compared to the results produced at that stage by an application that is known to be conformant to the standard, such as the reference software.

For instance, the decoding of a compressed video sequence produces a sequence of uncompressed images. These images can be represented as a sequence of bits in some known format—luminance and chrominance (YUV) bitmaps, for instance—and the sequence can be compared to another sequence produced by a conformant application out of the same compressed video sequence. In this case, the YUV values of the decoded images are a *conformance point*. However, the visual result of the rendering of the decoded image on a screen cannot be considered as a conformance point, because the colors might look a bit different on different devices or to different eyes.

It is not unexpected that conformance points are defined only at the decoding side. It is the nature of compressing media that there would be unlimited numbers of possibilities to encode raw media, whereas the decoding process is deterministic to ensure interoperability. For instance, a video sequence can be compressed more or less efficiently, preserving the quality of the original media better or less. However, there is only one way to decode the compressed sequence and convert it to a sequence of YUV bitmaps, guaranteeing that whatever the encoder behavior, all the compliant decoders will create the same decoded sequence (at least within very narrow boundaries) [MPEG4-4].

Still, regarding interoperability and conformance, this is the place to mention that there are reference software modules of two kinds: normative and informative. A normative software module is one that takes input in some format and produces output in another format, following strict rules defined in the standard. This means that every implementation complying with the standard that processes the same input and uses the same format for output will produce exactly the same output (sometimes allowing a small variation) [MPEG4-4]. An example is the reference software implementation of a video decoder, which takes an encoded video stream as input and produces raw video as a series of bitmaps. Every complying implementation that tries to do the same will produce the same output, notably the one produced by the reference software for each sample input [MPEG4-5].

Non-normative reference software modules are still useful for the purposes described above, but they are only informative because the normative specification of their behavior is not essential to guarantee interoperability. A good example for this is a reference video encoder. It gets a series of bitmaps and converts it to an encoded video stream. However, unlike the reference decoder, there are many different ways to encode video, all valid, and other implementations are not required to follow the same operation or even to produce the same output (e.g., with the same quality, encoding decisions). There are good encoders and bad encoders, but as long as they produce valid MPEG-4 encoded streams, they are all *MPEG-4 encoders*. Consequently, a software module is normative in its behavior at conformance points, and this is unrelated to other results it may produce.

The reference software activity took place like any other standardization activity in MPEG: Members contributed code and accompanying documents, contributions were discussed over emails and in face-to-face meetings, and editors did integration work. The MPEG-4 reference software was published as Part 5 of the standard [MPEG4-5]; this software can be bought from ISO or from various national standardization bodies.

14.1 REFERENCE SOFTWARE MODULES

For each part of the standard (see Chapter 1) there exists one or more reference software modules [MPEG4-5]. The various modules in the reference software are listed in Table 14.1. As mentioned in the introduction to this chapter, some of the modules are normative whereas others are only informative. In this table, the normative modules are marked with an (N) symbol.

14.2 SYSTEMS REFERENCE SOFTWARE

The rest of this chapter is dedicated to the Systems part of the reference software, or, more precisely, to the part that connects all the tools in the standard to create a complete MPEG-4 player. Unlike most other parts, this part of the reference software does not process streams in a linear manner, but accepts input from many sources in parallel, including various media streams and user interaction, creating rich audiovisual experiences based on a complex state machine. Understanding this part, or at least the general architecture behind it, is extremely helpful for understanding the standard itself and the entire MPEG-4 vision.

Table 14.1 Reference software modules

MPEG-4 Part	Reference software modules	Description
Systems (ISO/ IEC 14496-1)	MPEG-4 players	An MPEG-4 player that receives a complete MPEG-4 presentation (systems, audio, and visual streams), decodes the corresponding streams, synchronizes them, and renders the complete composition.
		The software contains hooks for video and audio decoder plug-ins, but these decoders are not part of the Systems reference software.
		Two players exist—one restricted to 2-dimensional content (2D) and the second a 3D player. A third *pseudo-player* (**N**) processes the streams as the real players but, instead of rendering them, sequentially logs its activity in a textual file. The three players use a common *core* code (**N;** see below).
		All the players are written in C++.
	MPEG-J	MPEG-J decoder integrated in the 2D player (**N**). MPEG-J encoder software. Programming language is C++.
	BIFS/OD compiler	An application that converts a textual file in VRML format (extended to describe BIFS updates and ODs) to binary BIFS and OD streams. The tool is written in C++.
	MP4 library	A C-language source file library that includes routines for MP4 file format access—create, store, and retrieve (**N**). Programming language is C.
	MP4 multiplex	An application that converts individual MPEG-4 streams of all types to a multiplexed MP4 file. A textual script file describes the input and output of this utility. The tool is written in C++.
Visual (ISO/ IEC 14496-2)	Video decoders (**N**)	Offline nonoptimized video decoders that receive MPEG-4 compressed video streams and convert them to raw video format (bitmaps). There are two decoder implementations, one in C (MoMuSys) and another in C++ (Microsoft).
	Video encoders	Offline nonoptimized video encoders that compress a sequence of frames or VOPs in raw video format (bitmaps) into an MPEG-4 stream. There are two encoder implementations, one in C (MoMuSys) and another in C++ (Microsoft).

Table 14.1 Reference software modules (Continued)

MPEG-4 Part	Reference software modules	Description
	Synthetic video decoders	Face and body animation decoders (**N**); mostly in C, but part in C++.
		2D and 3D mesh decoders (**N**). Programming language is C++.
		Visual texture coding (VTC) decoder (**N**). Programming language is C++.
Audio (ISO/ IEC 14496-3)	Audio decoders (**N**)	Offline nonoptimized natural audio decoders that decode MPEG-4 audio streams and produce uncompressed audio samples in raw audio format.
		There are decoders for the various audio and speech coding algorithms. Programming language is C.
	Audio encoders	Offline nonoptimized audio encoders that convert uncompressed audio samples in raw audio format to compressed MPEG-4 audio streams.
		There exist encoders for the various audio and speech coding algorithms. Programming language is C.
	Synthetic audio decoders	A structured audio decoder (**N**) in C/C++.
		Audio composition software C/C++.
		A text to speech decoder in C.
	Synthetic audio encoders	A structured audio encoder in C/C++.
		A text to speech encoder in C.
DMIF (ISO/ IEC 14496-6)	DMIF plug-ins	Components that plug into the MPEG-4 player and feed input streams to it. There exists a plug-in to process local MP4 files, and a plug-in that demonstrates the implementation of remote retrieval.
		Programming language is C++.

This chapter will continue with a description of the general architecture of the MPEG-4 player in the reference software. This description will be followed by detailed discussions on some components that are especially tricky, notably the scene graph handling, synchronization, and OD management. A section describing the player plug-ins comes next. Finally, two sections describing the 2D and 3D compositors conclude the chapter.

The reader of this chapter is assumed to be acquainted with the MPEG-4 standard, especially the Systems Part. Understanding Chapters 3, 4, and 7 of this book is a prerequisite for a good comprehension of this chapter. Notably, the terms OD (described in Chapter 3) and scene graph (described in Chapter 4) will be frequently referenced throughout this chapter.

14.3 MPEG-4 PLAYER ARCHITECTURE

The major keyword in the architecture of the reference software player is modularity. Although modularity is a general goal in software architecture, in this case it is the *raison d'être* of the project. The player reflects the standard, which is the result of collaborative work. The player project was made possible through the definition of modules that could be developed independently and still work tightly together. In addition, the MPEG-4 player developed as a part of the Systems reference software was used as a platform for the integration of software modules developed by other MPEG subgroups, such as the video and audio decoders.

The MPEG-4 player is written in C++. When this chapter references the player code, it often uses C++ terms like *class* and *object*. This use of the term *object* should not be confused with the term *object* as used in the MPEG-4 standard, which refers to audiovisual objects in the scene. In this chapter, the later will always be called *media objects* or *audiovisual objects*, in order to distinguish between the two. Note that one of the C++ classes mentioned in this chapter is called MediaObject. This is still a C++ object, which represents, not accidentally, a media object.

The modules comprising the kernel part of the player were integrated at source level, whereas audio, video, and Delivery Multimedia Integration Framework (DMIF) plug-ins were developed and integrated as Dynamic Link Libraries (DLLs).

Figure 14.1 illustrates the modular structure of the MPEG-4 player. Its main modules are the core, the compositor, the frame application, the DMIF plug-ins, and the decoders.

At the basis of the player is the core module. This is the heart of the player—the part that receives the input streams, processes the synchronization information, parses the scene and ODs, constructs the memory image of the scene graph, and manages the flow of data among all other modules [M3111].

The core module consists of several submodules. The most important ones are these:

1. **Executive:** Creates and connects program components (C++ objects) that process the input streams.

2. **Buffer and time management:** Manages the decoding and composition buffers and ensures that access and composition units are retrieved from the buffers right on time to be properly synchronized.

3. **BIFS/OD decoders:** Parses the BIFS and OD streams and constructs the scene graph.

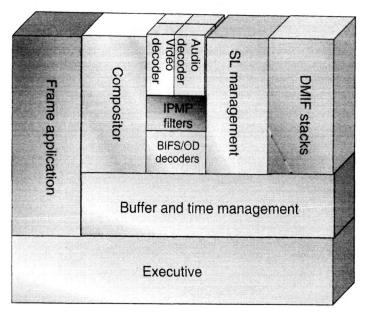

Fig. 14.1 MPEG-4 player modules.

4. **Synchronization layer (SL) management:** Parses SL packets, extracts timing information, and initiates the synchronization process.

Figure 14.2 shows other modules that are outside of the core but use the core services as illustrated in Figure 14.3. The modules are these:

1. **Compositor:** Scans the memory image of the scene graph, fetches the decoded media units from the composition buffer, and renders the composed scene on the output devices (screen and speakers).

2. **Frame application:** Wraps all the player modules into a single application and provides the application user interface.

3. **DMIF stacks:** Network stacks that handle input media streams arriving from various sources—local or remote. Each delivery scenario is handled by a different DMIF plug-in. The plug-ins interface with the player through the DMIF application interface (DAI), which hides the network specifics from the player.

4. **Media decoder plug-ins:** Fetch access units (AUs) from decoding buffers, decode them, and place the result in the composition buffers.

5. **Intellectual property management and protection (IPMP) filter plug-ins:** Optional components that are similar to decoder plug-ins, in the sense that their role is also to process the media streams—converting them to another format or extracting information from them. AUs pass through a set of decoders and IPMP translators, together called *filters*, before arriving in the composition buffers, ready for presentation.

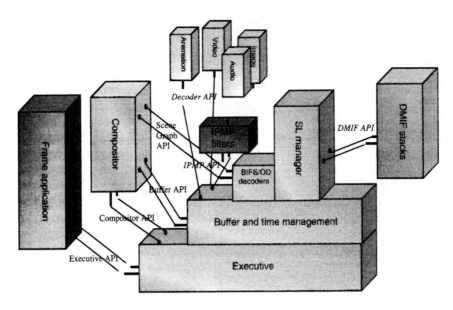

Fig. 14.2 MPEG-4 player APIs.

This modular architecture enables a cooperative development process through published APIs, which define the connection between the modules. These APIs are presented in Figure 14.2.

The APIs are divided in two groups:

☞ **Source-level APIs:** Provide the infrastructure over which developers can cooperate in developing the player application. Source code can be contributed by different parties and compiled together into one application.

☞ **Binary-level APIs:** Define the interfaces between the player application and DLLs, also called *plug-ins*. Plug-ins can be developed independently and easily integrated at the end-user platform. Usually multiple plug-ins can use the same API, run in parallel, or replace each other. Thus, for instance, multiple decoders can be plugged into the player, and alternate versions may exist for each decoder. In Figure 14.2, the binary-level API names are in *italics*: Decoder API, IPMP API, and DMIF API.

The main source-level APIs are these:

1. **Executive API:** Interfaces between the frame application and the core, providing the connection between the user interface and the player behavior.

2. **Compositor API:** Used by the executive part of the core to control the compositor—that is, to activate, deactivate, and delegate user interaction events.

3. **Scene graph API:** Used by the compositor to access the scene graph, which is generated and updated by the core.

4. **Buffer API:** Because media units flow between the player components through first-in-first-out (FIFO) buffers, the buffer API is used to store and retrieve units from these buffers and to control the scheduling of these operations.

The binary-level APIs are as follow:

1. **DMIF API:** Interfaces between the core and the DMIF plug-ins.

2. **Decoder API:** Interfaces between the core and the decoder plug-ins.

3. **IPMP API:** Interfaces between the core and the IPMP plug-ins.

14.3.1 Player Structure

Figure 14.3 shows how the modules described above are organized inside the player. The flow of data starts on the left-hand side of the diagram, where the

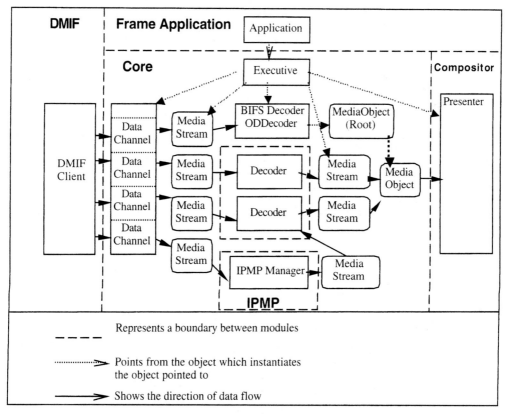

Fig. 14.3 Player structure.

media streams arrive from the network or files are processed by DMIF plug-ins. The DMIF plug-ins demultiplex the input channels and feed elementary streams (ESs) to the core.

At this time, the core starts to process the streams. The SL manager instantiates a `DataChannel` object to handle each stream. This object extracts the synchronization information from each AU and places it in a decoding buffer, along with its time attributes.

The `MediaStream` objects manage the buffers. These objects appear in the diagram wherever media are transferred between two modules.

At the right-hand side of the diagram, connected to `MediaStream` objects and to the BIFS decoder on one side and to the compositor on the other side, are the `MediaObjects`. These objects have a major role in the architecture of the player, as they actually comprise the memory image of the scene graph. Each node in the scene graph has a corresponding `MediaObject`. These objects are instantiated and connected to each other by the BIFS decoder. The compositor then traverses the scene graph—that is, it goes through each of the `MediaObject` objects and processes them, creating the audiovisual experience represented by the collection of the objects. By doing this continuously several times per second, the entire presentation is realized.

Media nodes (i.e., nodes that consume media streams) have associated `MediaStream` objects attached to them, which are used to fetch composition units from the composition buffers.

The `Executive` object controls the entire process. The frame application, implemented as an `Application` object, wraps the entire application and handles its user interface.

This chapter continues by focusing on certain aspects of this architecture, explaining in more detail some of the topics mentioned briefly above.

14.4 SCENE GRAPH

The previous section mentioned the `MediaObject` class as the programming entity that manages the scene graph. The implementation details of this class are tricky and worth further elaboration.

A simplified structure of the scene graph implementation is illustrated in Figure 14.4. `MediaObject` is the base class for scene nodes of every type. Each node type has a corresponding class that extends `MediaObject`. Each node in the scene graph is represented by an instance of the appropriate class.

Scene graph nodes contain fields. The node fields in `MediaObject`-derived objects are represented by objects of type `NodeField`. As explained in Chapter 4, there are several types of fields: scalar fields and vector fields, fields of various data type (i.e., integers, strings), and so on. Therefore the fields are of types `SFInt`, `SFString`, `MFInt`, and so on. Fields also can be of type `SFNode` or `MFNode`—that is, they can contain child nodes. Consequently, the `NodeField` class is a base class for classes that represent type-specific fields.

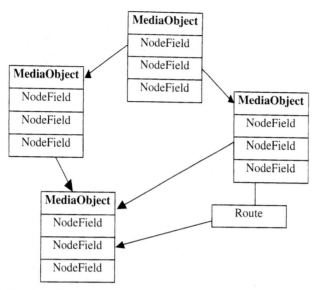

Fig. 14.4 Scene graph structure and objects.

The graph of the scene is created when nodes are contained as SFNode or MFNode fields of other nodes. In the reference software, classes that implement SFNode contain a member field that is a pointer to a MediaObject object. Classes that implement MFNode contain a member field that is an array of such pointers. Multiple node fields may reference a single instance of a node, when the USE feature is used in the scene graph.

Another class worth mentioning is the Route class, which implements ROUTEs. A Route object connects two node fields, so that a change in the source field is immediately reflected in the destination field. To achieve this, Node-Field objects contain a member field that is a pointer to a linked list of the ROUTEs originating from that field. NodeField classes include definitions of overloading operators, so that each assignment or manipulation of the field triggers the activation of a routine that copies the changed value to ROUTE destinations.

The example in Figure 14.4 shows several MediaObject objects that represent scene nodes. Each of these nodes contains member variables of type NodeField, representing node fields. Some of these fields are of type SFNode (not shown in the figure) and therefore point to other MediaObject objects. Also, one Route object is shown, connecting a NodeField in one MediaObject to a NodeField in another.

14.4.1 Parsing the Scene Graph

The parsing code in the reference software is probably its trickiest part. There are more than 100 node types, each with a different set of fields. In addition,

SFNode and MFNode fields are parsed differently, depending on the node type expected in each field. To complicate matters further, the code needs to handle the initial scene description, as well as several types of update commands, and the syntax of each update command is still context-dependent.

Scene parsing in the reference software is implemented with very compact code, but at the price of making the code less legible. The text below clarifies some tricks and shticks that were used.

The code relies heavily on C++ polymorphism. Each program entity—node, field, or route—has a Parse function, which it uses to parse itself.

MediaObject :: Parse is a nonvirtual function that commonly parses all scene nodes. However, each node type has its own *parsing table*. A parsing table is an array of BifsFieldTable entries, with each entry corresponding to one node field. The format of the BifsFieldTable structure is shown in Table 14.2.

The m_offset field is the memory offset of the member field (a NodeField object) in MediaObject, representing the node field. This is used to locate the field object when its serial number in the node (and therefore its corresponding BifsFieldTable entry) is known. The code in Table 14.3 shows the trick.

The m_function field points to a routine that is executed whenever the value of an eventIn field is modified. This is how incoming events are executed.

The m_bChildren field contains a non-0 value for children fields, which may be updated by update commands with special syntax (as explained in Chapter 4).

Chapter 4 also explains the field enumeration system used in BIFS parsing, when each field has a context-dependent serial number, depending if it is a field, exposedField, eventIn, or eventOut or if it is referenced in anima-

Table 14.2 BifsFieldTable structure

```
Struct BifsFieldTable
{
    short m_offset;
    EventHandler m_function;
    BYTE m_bChildren;
    short m_defID, m_inID, m_outID, m_dynID;
    char m_nQuantizationType, m_quantParams;
    char m_AnimationType;
    BYTE m_nDimension;
    float m_fMin [3], m_fMax [3];
    LPCSTR m_szName;
};
```

Table 14.3 MediaObject :: GetField implementation

```
NodeField *MediaObject :: GetField
(const BifsFieldTable *pEntry)
{
    return (NodeField *) (((char *) this) + pEntry ->   m_offset);
}
```

tion streams. These serial numbers are stored in the m_defID, m_inID, m_outID, and m_dynID fields in the BifsFieldTable structure.

The rest of the fields, excluding the last, are all used in animation streams, or with Quantization. For the sake of simplicity and conciseness, they will not be described here (see [MPEG4-5]).

The last field, m_szName, points to a character string containing the symbolic field name (as referenced in Chapter 4). Although symbolic names are not part of the BIFS, they appear in the code for the use of utilities that perform text-to-binary and binary-to-text BIFS conversion.

It should now be straightforward to understand how parsing tables are used as one-code-fits-all mechanisms to parse nodes and node fields, whether the latter appear as part of a complete node syntax, in update commands, or in ROUTEs. In addition to the MediaObject :: Parse function mentioned above, the MediaObject class includes functions such as ReplaceValue, InsertValue, and DeleteValue that are called when executing BIFS update commands. In each of these functions, the first thing the function does is find the parsing table of the node. MediaObject :: GetFieldTable is a virtual function that returns the address of the table for each node object. Similarly, MediaObject :: GetFieldCount returns the length of the table for each node. Once these are known, the parsing code can determine how many bits are used to represent the field number in the corresponding context, parse the field number from the stream, and—using the GetField function whose code was shown above—find the address of the NodeField object representing that field. The next step is to call on the field object and ask it to parse its own value from the stream. Because there are few different field types, NodeField :: Parse is a virtual function whose implementation is overloaded for each NodeField-derived class.

This leads to another prickly issue—the parsing of SFNode and MFNode fields. When parsing a node field, it is necessary to first determine the type of node that follows, then to parse the node itself. The SFNode syntax is described using the Syntactic Description Language (SDL, as described in Chapter 3), as in Table 14.4.

It can be seen in Table 14.4 that, in order to determine the node type, the parser must know which nodes are allowed for the particular SFNode field. This information is called nodeDataType, which is actually an identifier of a corresponding *node coding table*. Each SFNode (or MFNode) field can be of a certain data type, such as SF2DNode, SF3DNode, or SFGeometryNode, and each node data type has its own coding table.

The reference software handles this mechanism by first deriving SFGenericNode and MFGenericNode classes from NodeField, and then extending these further to classes that represent SF2DNode, SF3DNode, SFGeometry-Node, and so on. For each of these classes, a static member field m_nodeTable is defined as an array of integer values. The array contains the list of the nodes in the corresponding coding table, each internally represented by a unique number called nodeType.

Table 14.4 SFNode definition

```
Class SFNode(int nodeDataType) {
    bit(1) isReused ;
    if (isReused)
        bit(BIFSConfig.nodeIDbits) nodeID;
    else {
        bit(GetNDTnbBits(nodeDataType)) localNodeType;
        nodeType =
            GetNodeType(nodeDataType,localNodeType);
        bit(1) isUpdateable;
        if (isUpdateable)
            bit(BIFSConfig.nodeIDbits) nodeID;
        bit(1) MaskAccess;
        if (MaskAccess)
            MaskNodeDescription
            mnode(MakeNode(nodeType));
        else
            ListNodeDescription
            lnode(MakeNode(nodeType));
    }
}
```

With this information, SFNode and MFNode objects are ready to parse themselves. The virtual Parse function defined for each field class uses its m_nodeTable field to determine the length in bits of the nodeType value. The nodeType is then parsed and used to instantiate the appropriate node object. This operation is executed by yet another programming stunt—factory objects. Factory objects are C++ objects that instantiate other objects. These are usually implemented with templates, as in the example in Table 14.5.

Instances of this template do two things. When instantiated, such as in the following,

```
static BifsNodeFactory<TouchSensor> bifsFactoryTouchSensor
```

the object constructor stores the object address in the appropriate entry of a factory object table. When it is needed to instantiate a node object, nodeType is

Table 14.5 BifsNodeFactory class

```
template <class T> class BifsNodeFactory :
public BifsFactory
{
    virtual MediaObject *CreateNode () {return new T;}
public:
    BifsNodeFactory ()
    {
        SFGenericNode ::
        StoreFactory (T :: GetCode (), this);
    }
};
```

used to index that entry in the table, and the `CreateNode` function of the corresponding factory object is called. As one can see, this function creates the appropriate `MediaObject` object and returns its address.

Much of the complex code described so far is hidden behind *parsing macros*. C++ macros are pieces of source code that expand at compilation time into other, usually much lengthier code. Often, macros are called with arguments, such as run-time subroutines. The expansion at compilation time enables operations like planting a macro argument as part of a symbol name in the expanded code. The reference software uses this feature in several places. The text below will include one example. Looking at the node class definitions (object derived from `MediaObject`), one can see that the first line in each definition, preceding the series of node field definitions, is a call to the `BIFS_DECLARE_NODE` macro. This macro expands to declarations of virtual functions and static member fields that are common to each node—such as the parsing table and the `GetFieldTable` and `GetFieldCount` functions mentioned above.

Another source file contains the implementation of the nodes. There are a few macro calls per node. The example in Table 14.6 shows what a complete node implementation looks like, in this case for the `ScalarInterpolator` node type.

`BIFS_IMPLEMENT_NODE_START` is expanded to code that includes the implementation of the variables and functions declared in `BIFS_DECLARE_NODE`. The macro code starts with the following lines:

```
#define BIFS_IMPLEMENT_NODE_START(className) \
char *className:: GetName () {return #className;} \
```

This is an example of the source code expansion feature mentioned earlier. With the example in Table 14.6, the first line of the macro code expands to this:

```
char *className :: GetName () {return "ScalarInterpolator";}
```

That is, it becomes a function that returns a character string of the node name. The rest of the macro code, not shown here, assigns a numeric value to the node type, instantiates a factory object for the node, and starts the values assignment to the parsing table, with the following line:

```
const BifsFieldTable className :: m_fieldTable [] = {
```

Table 14.6 Parsing macros for `ScalarInterpolator`

```
BIFS_IMPLEMENT_NODE_START(ScalarInterpolator)
 BIFS_EVENT(ScalarInterpolator,set_fraction,0,-1,0,0)
 BIFS_FIELD1(ScalarInterpolator,key,0,1,0,-1,8,0,0,1)
 BIFS_FIELD1(ScalarInterpolator,keyValue,1,2,1,-1,0,0,0,0)
 BIFS_FIELD(ScalarInterpolator,value_changed,-1,-1,2,-1,0,0)
BIFS_IMPLEMENT_NODE_END(ScalarInterpolator)
```

The BIFS_FIELD (or BIFS_FIELDx or BIFS_EVENT) macros assign values to the rows of the parsing table for each node field. BIFS_EVENT is used for eventOut fields, whereas several flavors of the BIFS_FIELD macro exist, because different Quantization schemes require different parameters. The BIFS_IMPLEMENT_NODE_END macro closes the definition by terminating the parsing table assignment and implementing functions like this:

```
int className :: GetFieldCount () const
{return sizeof m_fieldTable / sizeof(BifsFieldTable);}
```

With these hurdles behind, it is time for an overview of the code at the other side of the scene graph objects—the rendering.

14.4.2 Rendering the Scene Graph

One of the nice aspects of the reference software architecture is the complete separation between the layers—the parsing (or, in a broader sense, scene manipulation) layer and the rendering layer. Different implementers, using the MediaObject and its derived classes as an interface, implemented the two modules almost independently.

The rendering code traverses the scene graph in continuous iterations, rendering each node it passes through. During each scene traversal iteration, the scene is locked to prevent racing conditions between the scene manipulation and the rendering code. Scene updates can take place between rendering iterations, and this happens completely transparently to the rendering layer.

The method by which parsing and rendering are separated, even though they use the same objects, requires explanation. It has been explained before how all nodes are treated by the parser in a generic way, using the parsing tables that are defined for each node. This, of course, cannot work for rendering, which depends on the semantics of the nodes. Therefore, another C++ feature was used here: *class overloading*. Class overloading is the mechanism in C++ (and other object-oriented languages) by which a class definition is derived from a base class, and the derived class redefines some of the methods that were defined at the base class. If the redefined methods are declared as *virtual*, a call to the method of the base class results in the actual execution of the method of the derived class. The rendering part of the players was implemented by overloading the bare-boned node classes, adding semantics to the derived classes by putting content in the Render (virtual) function of each node.

This is the place to mention another trick. The mechanism the parser uses to create scene class nodes via factory objects was described above. This works well for classes that are known to the parser, but it jeopardizes the principle of layer separation if the classes are overloaded by the rendering layer. Fortunately, the factory object mechanism provides a clean solution to this problem. Because nodes are instantiated by factory objects that are stored in a table indexed via nodeType, all that is needed is to replace the factory objects in the table with objects that will create the proper node objects. Again using a macro, the compositor derives its own version of the node classes from the original node classes—for example, `TimeSensorImp` from `TimeSensor`. Then, somewhere in the code, a line similar to the following will appear:

```
OVERLOAD_NODE (TimeSensorImp);
```

This macro is defined as this:

```
#define OVERLOAD_NODE(className) \
static BifsNodeFactory<class className> \
bifsFactory##className
```

Therefore, for the above example, it would expand to this:

```
static BifsNodeFactory<class TimeSensorImp>
bifsFactoryTimeSensorImp;
```

Recalling what happens when `BifsNodeFactory` is instantiated, this line inserts a pointer to the factory object that creates `TimeSensorImp` into the factory object table, at index `TimeSensorImp :: GetCode()`. Because `TimeSensorImp` is derived from `TimeSensor`, `GetCode` would return the same value for both, and therefore the pointer to `bifsFactoryTimeSensorImp` will substitute the pointer to `bifsFactoryTimeSensor` in the table.

This puts in place the entire infrastructure for scene parsing, construction, manipulation, and rendering. On each rendering iteration, the compositor calls the `Render` function on the scene root object, which renders itself and its child nodes, again calling `Render` on the respective objects. The two last sections of this chapter describe the 2D and 3D players and elaborate on the rendering.

One more issue to clarify before concluding this topic is the implementation of ROUTEs, briefly mentioned earlier. During rendering, the player may change some field and `eventOut` values. If there are ROUTEs originating from these fields, the value will be copied to other fields and affect the rendering accordingly. Because it is desirable to keep the activation of the ROUTEs transparent to the rendering code, operator overloading is used to implement seamless activation of ROUTEs. Remember that node fields are implemented as classes derived from `NodeField`. The implementation of these classes overloads the assignment operator and provides utility functions to manipulate the fields. Every time a node field is changed, the ROUTEs originating from that field (if there are any) are activated, and the new field value is copied to the target node fields.

14.5 PROTOS

PROTO is a way to define new nodes by combining the functionality of predefined nodes. Scenes that use PROTOS contain a section of PROTO definitions, which can then be instantiated in the scene and used like normal nodes.

The implementation of PROTO was done with the following objectives:

1. Efficient implementation with minimal changes to the reference code, in order not to compromise the efficiency and stability of the code.
2. Restriction of the changes to the scene manipulation layer so that no modifications will be required at the rendering layer.

To achieve that, the following problems needed to be solved:

1. Both PROTO definition and PROTO instantiation need to be implemented— two tasks that seem fairly complex.
2. Although a PROTO can be used as a node, its structure is different; nodes have fixed fields that are described by hard-coded node tables, whereas PROTO fields are defined at run time.
3. PROTOS, being a decoding construct in BIFS rather than representing real media objects, need to be hidden from the compositor. Without PROTOS, the compositor traverses the scene graph by walking from fields that point to nodes down to the nodes themselves. With PROTOS, a field might point to a PROTO instance instead of to a node of a known type. Requiring the compositor to handle this would mean assigning decoding tasks to a component that is supposed to perform only rendering tasks.

So, the first step in implementing PROTOS is defining a class for a PROTO— the Proto class. The trick is to inherit this class from MediaObject so that it behaves like a regular scene node. This is illustrated in Figure 14.5. It can be seen that the polymorphism capability of C++ is used to hide the differences between regular node fields (static and predefined) and PROTO fields (defined dynamically). It also can be seen that a Proto object includes two parts that do not exist in regular nodes—a list of nodes and ROUTEs that together construct the PROTO code. Moreover, the technique of ROUTEs is used to associate the PROTO fields with the corresponding node fields in the PROTO code. Therefore, any modification to a PROTO field is automatically mirrored in the associated node field, and vice versa.

The second step is the implementation of a PROTO instantiation—that is, when an instance of the PROTO is used within the scene graph. This is done, as illustrated in Figure 14.6, by a *cloning* mechanism. A Clone function is implemented in each of the classes Proto, MediaObject, NodeField, and Route in such a way that each object clones itself and calls the Clone functions of all the objects it contains.

The last thing that needs to be solved in implementing PROTOS is PROTO handling at the rendering layer. As explained above, the compositor is implemented as an independent module. It is desirable to hide the existence of PRO-TOS from the compositor, so that this module will not need to be changed. Figure 14.7 refers to a branch in the scene graph that represents a 3D model of a person. The person consists of an audio node and an IndexedFaceSet node, and both are bound under a Group node. The figure contains three charts:

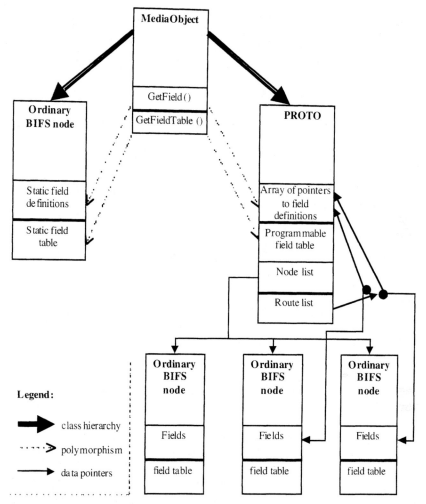

Fig. 14.5 PROTO as extension of MediaObject.

1. The left chart shows how the person would be represented in a scene graph without the use of PROTO. In this case, when the compositor traverses the scene it encounters the Group node and its children and renders them appropriately.

2. The middle chart shows the same, but here the person's branch is encoded as a PROTO (i.e., a PROTO node containing an AudioSource node and an IndexedFaceSet node under a Group), it is declared beforehand, and a reference to this PROTO exists in the scene instead of the entire person branch. In this case, when the compositor traverses the scene it encounters the Proto object. The compositor then needs to know how to handle a Proto object.

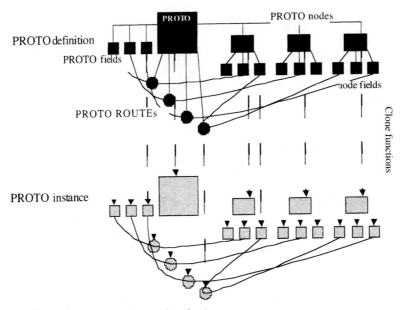

Fig. 14.6 PROTO instantiation by cloning.

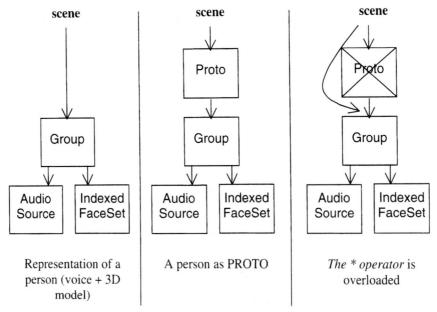

Fig. 14.7 Using operator overloading to hide PROTOs.

3. The right chart shows the procedure for rendering Proto objects by a compositor that is not aware of the existence of PROTOs. Here comes for rescue the C++ feature of operator overloading. At some point in the scene, there is a pointer that would point to the Group object in the left chart and to the Proto object in the middle chart. In order to hide the existence of the Proto object from the compositor, the *pointing* operator (*) is overloaded so that a pointer that points to a Proto object will actually behave as if it points to the first object of the PROTO code, in this case the Group node. The chart illustrates the effect of this operation—as if the Proto object disappeared and the pointer that was pointing to it is redirected to point to the Group node, just as in the chart on the left-hand side.

14.6 SYNCHRONIZATION

Another responsibility of the core module is the handling of stream synchronization. This involves two tasks—the extraction of synchronization information attached to incoming ESs and the maneuvering of the stream data so that it arrives at the compositor on time.

14.6.1 Sync Layer

Chapter 3 describes the MPEG-4 SL and its role. The SL is implemented in the reference software through DataChannel objects. A DataChannel object is instantiated for each open ES and handles all the DMIF events (see Chapter 7) occurring for this stream (as illustrated in Figure 14.3).

The DataChannel :: OnPacketReceived function is called for each incoming SL packet. Depending on the calling DMIF instance, the packet may include a packet header that contains the SL information for that packet. Alternatively, the DMIF layer can parse this information by itself and pass it to DataChannel in a separate structure. OnPacketReceived then performs two separate tasks— packet loss handling and calculation of timestamps. Packet loss can be detected by the underlying DMIF layer, in which case OnPacketReceived is notified via a function argument, or the function itself can detect it by analyzing SL header parameters such as packetSequenceNumber. The behavior in the case of data loss depends on the implementation, because the MPEG-4 specification assumes a theoretical error-free case and thus no error behavior is standardized. The reference software discards all data following a lost packet until the next random access point of that stream. More sophisticated implementations can deliver all data, and the notification about the loss, to the decoding layer for better error resilience by means of sophisticated error-concealment solutions.

The processing of time stamps includes the alignment of the various time stamps—composition time stamps (CTSs), decoding time stamps (DTSs), and object clock references (OCRs)—potentially delivered with different resolution and different modulus basis, on a common timeline. Once this is done, OnPacketReceived uses MediaStream objects to perform the actual buffering and synchronization.

14.6.2 MediaStream Objects

A MediaStream object has two roles: to manage the buffers through which the stream units flow and to control the timing of the flow so that synchronization is achieved at presentation time. For each stream, there is one or more objects of this class, connecting between submodules that process the stream. For instance, in the case of a video stream, one object connects the SL (the DataChannel object) to the video decoder, implementing the *decoding buffer*, and another connects the decoder to the compositor, implementing the *composition buffer*. If an IPMP filter is required for the stream processing, whether before or after decoding, more MediaStream objects are instantiated, as in Figure 14.8.

Two main parameters are used for instantiating a MediaStream object: the size of the controlled buffer and a pointer to a ClockReference object. The buffer size of decoding buffers is calculated according to the DecodingBuffer-Size value conveyed in the DecoderConfigDescriptor associated with the stream. The buffer size of the composition buffer is calculated according to the type of the (decompressed) media. The MediaStream constructor allocates the required memory for a buffer that will be handled as a member variable. The ClockReference object is basically an object that keeps the timeline of the stream. Because different streams may have different timelines, multiple ClockReference objects may exist. However, objects that share the timeline will use a common ClockReference object. Hence, the MediaStream objects that handle the decoding and composition buffers of the same stream will use

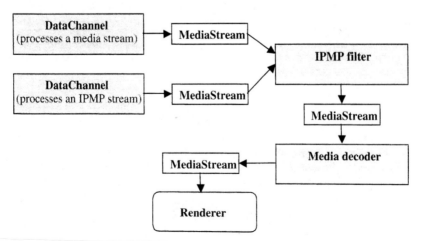

Fig. 14.8 MediaStream connections.

a common ClockReference object, as will buffers used by different ESs that use the same timeline.

A ClockReference is initialized when the first OCR of the corresponding OCR stream is received and is subsequently updated as more OCRs are received. When OCRs are not used, as is the case when the media is accessed from local files, the player device's native system time is used.

The process of establishing MediaStream connections is as follows. Before a connection is established, two conditions must be met:

1. The scene graph must contain an active node that consumes the media stream.

2. An OD describing the media stream must be received.

Once these two conditions are met, the player (or, more precisely, the code that is part of the implementation of all media-consuming nodes; see next section) opens up an appropriate DMIF channel and instantiates a corresponding DataChannel and media decoding objects as well as the necessary MediaStream objects to create the path for the data flow. Each software component that needs to access a MediaStream object is given a handle to the object, and it proceeds to open the object. There are always two sides that access such objects—the sender side and the receiver side. The sender side opens the object by calling the Open function, and the receiver side does the same with the Start function.

At this point, the sender side is ready to store data in the buffer controlled by the MediaStream object, and the receiver side is ready to fetch it from there. Four main functions exist for this purpose: Allocate and Dispatch for the sender, Fetch and Release for the receiver.

The sender calls Allocate to request a contiguous block of memory on the buffer, where it will later store the media unit (AU or composition unit). The function returns a pointer to an unused block. If a block of the requested size is not yet available in the buffer, the calling thread is suspended until the receiver retrieves data from the buffer and frees enough memory. The allocated block is kept protected and cannot be accessed by any other component until Dispatch is called. There is one twist: If at some point the sender realizes that the block it requested was not large enough, it calls a variant function, Reallocate, to resize the original block (this may result in shifting the block and its content to another memory location). Once the sender finishes preparing a full unit, it calls the Dispatch function to specify the exact length of the unit, so that unused space may be released and to signal that the unit is available for processing by the next component. The function also is used to specify the composition time of the unit that is saved in the buffer along with the data.

At this point, the data is available for retrieval. To retrieve data from the buffer, the receiver side calls the Fetch or FetchNext function. An important difference between the two is that the execution of Fetch depends on the time stamps attached to the unit. A mechanism implemented within the MediaStream class, which uses the ClockReference object associated with each class object, ensures that Fetch always gets the unit appropriate for composition at that time. Different parameters of the function signal whether the calling thread is blocked until the unit is ready (or return without retrieving data), and whether expired units can be skipped. FetchNext, on the other hand, immediately retrieves the next available unit and if necessary waits until one exists or end-of-stream is signaled by the sender. FetchNext is used by components that do not care about composition time, usually the media decoders. Fetch is used when synchronization is required—that is, when the compositor requests a decoded block for presentation.

Fetch and FetchNext return a pointer to the data without physically removing it from the buffer. The data remains locked in the buffer until it is no longer needed by the receiver side, at which point the receiver calls Release. This signals that the block is available as free memory.

To summarize, the crucial tasks of buffering and synchronization of content in the reference MPEG-4 player software are handled by one class, MediaStream. The code implementing this class is rather compact, but still flexible enough to be reused all across the application. MediaStream constructs the artery system of the player.

14.7 Object Descriptors

Chapter 3 describes the OD framework in MPEG-4 as a powerful and flexible framework that serves the goal of separating the information about the content of ESs from the information on how to compose that content in the scene. The description of MPEG-4 streams can be very complex, ranging from SL and decoder configuration information to the definition of alternative and scalable content, use of IPMP tools, and object content information (OCI).

14.7.1 Syntactic Description Language

Because it is important for the discussion in this chapter, a brief recap of the essentials of SDL is given here (see also Chapter 3). SDL is the language that was invented by MPEG [MPEG4-1] to describe the layout of data on bit-streams using syntax borrowed from C++.[1]

SDL describes data structures made of fields, which can be scalar or aggregated. Structures are described as *classes*. Aggregated fields are arrays, subclasses, or both (i.e., arrays of subclasses). Conditional expressions are used to indicate that the syntax of some fields depends on the value of fields preceding them in the stream or on other context values.

The class inheritance mechanism of C++ is also imitated. A base class can be defined once and extended in multiple places, which means that the extensions use the fields defined in the base class plus some more.

---

1. Note that MPEG SDL is different than ITU's SDL (*www.tdr.dk/public/SDL/whatis.html*).

The base class for all ODs is `BaseDescriptor`. `BaseDescriptor` is an *expandable* empty class, which means that any class derived from it uses an extension mechanism. This mechanism allows for the addition of private fields to the classes, so that applications that do not recognize the extensions can safely skip them. Basically, each class starts with a tag field identifying the class type, followed by a length field that contains the total size of the class in bytes. If an application does not recognize the class by its tag, it can use the length field to skip that class. This also enables fields that are variable-length arrays of subclasses. The length of such arrays is implicit. By parsing the class tags, the parser can determine how many consecutive classes of the same type exist; and by knowing the total size of the container class, the parser knows when to stop field parsing. For instance, the `ObjectDescriptor` structure is defined in Table 14.7.

`ObjectDescriptor` is an expandable class whose tag equals `Object-DescrTag` (defined in the MPEG Systems specification [MPEG4-1]). Its first field is a 10-bit `ObjectDescriptorID`, followed by a 1-bit `URL_Flag` field, then by a 5-bit reserved field with a preassigned value. The rest of the structure depends on the value of the `URL_Flag` field and is fairly straightforward.

This example also uses arrays of subdescriptors of undefined length. There can be 1 to 30 `ES_Descriptor` objects, 0 to 255 `OCI_Descriptor` objects, and so on. A numeric tag that identifies the descriptor type precedes each descriptor.

14.7.2 Parsing the OD Stream

Parsing is conceptually easy, but the code that parses the myriad of classes and options can be excessively long and complex. As with the equally stuffed scene description syntax, a few tricks can make the code more compact.

Table 14.7 `ObjectDescriptor` definition

```
Class ObjectDescriptor extends BaseDescriptor :
bit(8) tag=ObjectDescrTag
{  bit(10) ObjectDescriptorID;
   bit(1) URL_Flag;
   const bit(5) reserved=0b1111.1;
   if (URL_Flag)
   {   bit(8) URLlength;
       bit(8) URLstring[URLlength];
   }
   else
   {   ES_Descriptor esDescr[1 .. 30];
       OCI_Descriptor ociDescr[0 .. 255];
       IPMP_DescriptorPointer ipmpDescrPtr[0 .. 255];
   }
   ExtensionDescriptor extDescr[0 .. 255];
}
```

As with scene graph handling, there is one base class from which all kinds of descriptors are derived. This is the `BaseDescriptor`. Also similar to the `MediaObject` used to implement scene nodes, `BaseDescriptor`-derived classes (*descriptors* hereinafter) have fields, parsing tables to describe the fields, and virtual functions like `GetFieldCount` and `GetField`.

The `EncDec` function decodes ODs, similar to the `Parse` function that was mentioned in the previous section, which decodes BIFS nodes, or `Media-Objects`. It is called `EncDec`, short for encode-decode, because the function can also work the other way—encoding the memory structure of a descriptor into a series of bits. An extra parameter in the function declaration tells whether the function is called for decoding or for encoding.

SDL fields are represented by member variables in descriptors. The base field object is `SDLField`, extended by specific field types such as `SDLInteger`, `SDLFloat`, and `SDLIntegerArray`. Here also each class implements a virtual `EncDec` function, which every field object uses to decode (and encode) itself. Fields that are descriptors or descriptor arrays by themselves are represented by classes derived from `SDLBaseClass` and `SDLBaseClassArray`, respectively. The C++ template mechanism is used to extend these classes. For instance, a template class `SDLClass<T>` is derived from `SDLBaseClass`, and a specific field type is defined as `SDLClass <ObjectDescriptor>`.

For example, `ObjectDescriptor` (whose SDL syntax was shown in Table 14.7) is defined in Table 14.8.

By following these explanations about scene graph parsing, one may already have guessed the role of `OD_DECLARE_DESCRIPTOR`. It is a macro that expands to functions and variable declarations common to all descriptors. And

Table 14.8 `ObjectDescriptor` class

```
class ObjectDescriptor : public BaseDescriptor
{
    OD_DECLARE_DESCRIPTOR
public:
    SDLInt<10> ObjectDescriptorID;
    SDLInt<1> URL_Flag;
    SDLInt<5> reserved;
    SDLString URLstring;
    SDLClassArray<class ES_Descriptor, 1, 255> esDescr;
    SDLClassArray<class OCI_Descriptor, 0, 255> ociDescr;
    SDLClassArray<class IPMP_DescriptorPointer, 0, 255>
            ipmpDescrPtr;
    SDLClassArray<class ExtensionDescriptor, 0, 255> extDescr;
public:
    ObjectDescriptor () : reserved (0x1f) {}
protected:
    virtual void DescEncDec (ODCoder *pODCoder,OPTYPE op);
};
```

here too there are parsing tables and macros. The parsing table is of type
`ODFieldTable`, which is simpler than in the case of BIFS:

```
struct ODFieldTable
{
    WORD m_offset;
    LPCSTR m_szName;
};
```

That is, all that is needed to know is the offset of the field within the
descriptor object. The symbolic field name is also worth keeping, whether for
text-to-binary encoding or for binary-to-text analysis.

The tables for all the descriptors are coded with parsing macros, as in
Table 14.9.

`OD_IMPLEMENT_DESCRIPTOR_START` defines a few functions and starts
assigning values to the field table, `OD_IMPLEMENT_FIELD` adds entries to the
table, and `OD_IMPLEMENT_DESCRIPTOR_END` closes the table assignment.

From the implementation point of view, there are two kinds of descriptors: simple and complex. Simple descriptors have a series of fields, with no
conditional or loop statements. This is, for example, the situation with
`DecoderConfigDescriptor`, shown in Table 14.10.

Table 14.9 Parsing macros for `ObjectDescriptor`

```
OD_IMPLEMENT_DESCRIPTOR_START (ObjectDescriptor)
    OD_IMPLEMENT_FIELD
        (ObjectDescriptor, ObjectDescriptorID)
    OD_IMPLEMENT_FIELD (ObjectDescriptor, URL_Flag)
    OD_IMPLEMENT_FIELD (ObjectDescriptor, reserved)
    OD_IMPLEMENT_FIELD (ObjectDescriptor, URLstring)
    OD_IMPLEMENT_FIELD (ObjectDescriptor, esDescr)
    OD_IMPLEMENT_FIELD (ObjectDescriptor, ociDescr)
    OD_IMPLEMENT_FIELD (ObjectDescriptor, ipmpDescrPtr)
    OD_IMPLEMENT_FIELD (ObjectDescriptor, extDescr)
OD_IMPLEMENT_DESCRIPTOR_END (ObjectDescriptor)
```

Table 14.10 `DecoderConfigDescriptor` class

```
class DecoderConfigDescriptor extends BaseDescriptor :
bit(8) tag=DecoderConfigDescrTag {
    bit(8) objectTypeIndication;
    bit(6) streamType;
    bit(1) upStream;
    const bit(1) reserved=1;
    bit(24) bufferSizeDB;
    bit(32) maxBitrate;
    bit(32) avgBitrate;
    DecoderSpecificInfo decSpecificInfo[0 .. 1];
}
```

A generic `EncDec` function can easily encode and decode this kind of descriptor, with the following short code:

```
int nCount = GetFieldCount ();
for (int i = 0; i < nCount; i++)
    GetField (i) -> EncDec (pCoder, this, op);
```

There is no way, however, to handle complex cases in such a generic way. The solution is to have a generic virtual function that is used for parsing all of the simple descriptors, and to overload this function in each complex case. More specifically, there is one `EncDec` function that starts the encoding or decoding for all descriptors, and then calls a virtual `DescEncDec` function. The default implementation of this function contains the three lines of code quoted above. Individual implementations for complex descriptors follow the SDL description of the descriptors, as in Table 14.11.

One can immediately see the similarity between this code and the SDL syntax of the `ObjectDescriptor` entity as specified in the standard [MPEG4-1]. Because it is so similar, it is easy to implement automatic tools that convert the text in the standard specification into the C++ code of the reference software.

There are, of course, many more details in this implementation. Only the essentials of OD parsing are summarized here. Before concluding this description, a buzzword the reader should be familiar with by now will be mentioned again—factory objects. Here, too, there are factory objects and the factory object table. While parsing fields that are descriptors (i.e., of type `SDLClass<T>` or `SDLClassArray<T>`), there is a need to instantiate descriptor objects by knowing their tags. So there is a table of objects of type `ODFactory`, and the table is filled up once during application initialization. The `OD_IMPLEMENT_ DESCRIPTOR_START` macro mentioned above also contains the following definition,

```
static ODFactory<class className> odFactory##className;
```

Table 14.11 `ObjectDescriptor::DescEncDec` implementation

```
void ObjectDescriptor::DescEncDec(ODCoder *pCoder,OPTYPE op)
{
   ObjectDescriptorID.EncDec (pCoder, this, op);
   URL_Flag.EncDec (pCoder, this, op);
   reserved.EncDec (pCoder, this, op);
   if (URL_Flag)
   {
      URLstring.EncDec (pCoder, this, op);
   }
   else
   {  esDescr.EncDec (pCoder, this, op);
      ociDescr.EncDec (pCoder, this, op);
      ipmpDescrPtr.EncDec (pCoder, this, op);
   }
   extDescr.EncDec (pCoder, this, op);
}
```

which, as for `bifsFactory`, instantiates a static factory object for the class `className` and inserts a pointer to it in the OD factory object table. And, of course, table entries can be overridden with factory objects that create overloading descriptor objects, a fact that leads us to the following topic.

14.7.3 Execution of OD Objects

Having parsed an OD frame, the player now needs to execute it. For this purpose, a special class is derived from `BaseDescriptor`—`BaseCommand`. This is the base class for all OD command descriptors such as `ObjectDescriptor-Update` and `ES_DescriptorRemove`. `BaseCommand` adds one virtual function: `Commit`. OD command descriptors overload this function if they want to act.

The mechanism of separating parsing from execution is repeated here. The basic OD command classes do only parsing (and encoding), but OD execution code can be easily added by deriving appropriate classes and using the `OVERLOAD_OD_CLASS` macro to replace the corresponding factory object in the OD factory table. So, for instance, one can find a definition of `ObjectDescriptorUpdateImp` that is derived from `ObjectDescriptorUpdate`, with an overloaded implementation of the function `Commit`. The application initialization code includes the line

```
OVERLOAD_OD_CLASS (ObjectDescriptorUpdateImp);
```

and so on for the other command descriptors.

More interesting is the action that takes place inside the `Commit` functions. As explained in Chapter 3, an OD basically describes a stream (or streams) that is consumed by nodes in the scene. Not all node types consume streams. Those that do, such as `MovieTexture`, are called media nodes. Media nodes contain a `url` field, which is actually an object descriptor ID (`OD_ID`) referencing an associated OD. In the reference software, the renowned `Media-Object` has a derivative called `StreamConsumer`. All media nodes are derived from `StreamConsumer` instead of from `MediaObject` directly. This gives them free access to the bunch of code implemented in `StreamConsumer` that handles everything that has to do with streams, from the handling of OD events through the fetching of decoded media from the composition buffer.

Of course, it takes two to tango, and thus two entities must exist in an MPEG-4 player to make a media stream dance: a media node and a corresponding OD. The player cannot assume that one of these entities will come first, so the first thing it does when a media node is inserted into the scene graph, or when an OD arrives in the player, is to register these objects. The `OD_ID` is qualified by a namespace, and a new namespace is created for each `Inline` node, so the place to register these objects is on the `Inline` object. And, indeed, the code for `InlineImp`, a class derived from `Inline` to implement its functionality, includes code for handling the list of registered media nodes and ODs.

Coming back to the `ObjectDescriptorUpdateImp :: Commit` function, this function determines the appropriate `InlineImp` object, and notifies it upon reception of the new ODs. The descriptors are registered, and a search takes place for a matching registered media node. An analogous process is performed if the media node registers last. When a match is found, `StreamConsumer :: SetStream` is called, requesting the media node to process the OD conveyed in the function argument.

This is the place where the full setup of the stream is performed. A request to add a channel is sent to DMIF; a channel object is created (a.k.a. `DataChannel`); media decoders are instantiated, configured, and set up; and the entire data path is connected through `MediaStream` objects. At the end of this procedure, if completed successfully, a pointer to the `MediaStream` object representing the composition buffer is placed on the `StreamConsumer` object. The rendering code can use the `GetStream` function defined for this class to get a handle to the `MediaStream` object and start fetching and rendering the units.

`Commit` functions for other OD commands follow this pattern in a fairly straightforward way. This concludes the discussion of the core part of the player. The rest of this chapter deals with components that are considered peripheral to the core—the plug-ins and the compositors.

14.8 PLUG-INS

A *plug-in* in the context of the reference software is a software component that is part of the application, but is developed, compiled, and linked independently, usually as a DLL, and communicates with the principal part of the application via binary-level APIs. Three types of plug-ins are used in the MPEG-4 reference software: DMIF plug-ins, decoder plug-ins, and IPMP plug-ins.

14.8.1 DMIF Plug-Ins

DMIF was discussed in detail in Chapter 7. It is the stack that implements content access for diverse networks and platforms [MPEG4-6]. The MPEG standardization effort has gone as far as defining a conceptual DMIF API, or the DAI, which hides the specifics of the content access from the principal player part. It also provides a unified interface to the access procedures for all the scenarios—local file, broadcast, and remote retrieval [M6124].

Two main entities are defined in DMIF: *service* and *channel*. A service can be loosely defined as a collection of media managed centrally. This can be a file, a media server, a component that tunes into digital television channels in set-top boxes, and so on. From an implementation point of view, a server needs to be able to locate specific media streams and hand them to the terminal. A channel is a logical path for a media stream. Note that, although in network technology a channel is associated with network resources and multiple logical streams can often be multiplexed in one channel, in the narrow context of the DAI one channel is opened for each ES. All together, an MPEG-4 session requires the creation of one or more services, with a set of channels for each service.

The DAI is based on a set of *primitives*. There are *request, confirmation, indication,* and *response* primitives. The application sends requests to the DMIF stack and receives confirmations. This means that the DAI is based on nonblocking calls—that is, the application sends a request without waiting for the answer to that request, and the DMIF asynchronously sends a reply (called confirmation even though some requests may not be confirmed) after executing the request. Indications are sent from the DMIF stack to the application to notify about events that are not direct replies to DMIF requests; the application answers these indications by issuing response primitives.

The DMIF stack is implemented in the reference software as a DLL. The DLL is built around a main class, called `DMIFClientFilter`. The core loads the DLL dynamically and calls a `CreateInstance` function implemented in the DLL, which returns a pointer to a `DMIF_Filter` object. After creating this object, its `SetSink` function is called, passing it a sink object that the DMIF stack uses to call the application. The sink object happens to be the `Executive` object, which controls the player execution. Thus, the application calls member functions of the `DMIF_Filter` to pass requests and responses, and the `DMIF_Filter` calls member functions of the `Executive` to pass confirmations and indications.

The main DAI primitives were described in Chapter 7, so only a few of them will be briefly repeated here:

1. `DAI_ServiceAttach_Req`: Called by the application to log into a new service.

2. `DAI_ServiceDetach_Cnf`: Sent by the DMIF stack in reply to `DAI_ServiceAttach_Req`.

3. `DAI_ChannelAdd_Req`: Called by the application to add channels to an established service.

4. `DAI_ChannelAdd_Cnf`: Sent by the DMIF stack to confirm a `DAI_ChannelAdd_Req` request; this opens the way for the data to flow from DMIF to the application.

5. `DAI_DataReceive_Ind`: Sent by the DMIF stack to signal the arrival of data packets.

6. `DAI_ChannelDelete_Req`: Called by the application to close channels that are no longer needed.

7. `DAI_ServiceDetach_Req`: Called by the application to remove the entire service when it is no longer required.

The DMIF filter is a general stack that provides services for all scenarios and networks. The specific content access procedure for each case is implemented as a separate DLL. Parallel to the DAI that connects the application and the DMIF filter, another API connects the DMIF filter to individual plug-ins. This interface is called DMIF Plug-in Interface (DPI). The DPI is almost completely analogous to the DAI—even function names are the same, except that they now start with `DPI_` rather than with `DAI_`. Indeed, for many of them the only task of the DMIF filter is to map DAI primitives to DPI primitives, and vice versa. Still, the role of the filter is not negligible. The DMIF filter performs an important job when DAI_ServiceAttach_Req is called. This is the place where the filter decides which plug-in to load, depending on the URL that is passed to the function as a parameter. The URL identifies the service—its location and its type, whether this is a file or a Web site, and so on. It is the service type that determines which DMIF plug-in to load, and once the appropriate one is loaded, its DPI_ServiceAttach_Req is called, again receiving the URL, to locate the service and start accessing it.

The DMIF filter is also capable of handling *FlexMux* [MPEG4-1]—that is, the standard MPEG-4 multiplex scheme used to interleave several ESs in one network channel. So, in this case, the DPI channel is actually a FlexMux channel—a collection of interleaved ESs.

The following walkthrough summarizes the setup procedure for the DMIF plug-ins:

1. The player is initialized, the DMIF filter DLL is loaded, and its Create-Instance entry point is called to return a pointer to a DMIFClientFilter object.

2. The DMIFClientFilter object is used to instantiate DMIF plug-ins and delegates DAI calls to them.

3. DMIFClientFilter :: DAI_ServiceAttach_Req is called to attach a new service, identified by a URL.

4. DMIFClientFilter uses the URL type to determine which DMIF plug-in to load, loads the appropriate DLL, and calls its CreateInstance entry point.

5. The CreateInstance function returns a pointer to a Service object. DMIFClientFilter sets itself as the sink object of the Service object.

6. Once the plug-in infrastructure is in place, the player passes DMIF requests to DMIFClientFilter, which delegates the request to the appropriate Service. Then, the plug-in sends confirmations back to DMIFClientFilter, which handles them back to the player.

Altogether, these create a mechanism to attach different modules that handle different input scenarios to an MPEG-4 player, in a way that would make all scenarios handled transparent by the player.

14.8.2 Decoder Plug-Ins

Media decoders form another type of plug-ins. They carry out the vital role of decoding compressed audio [MPEG4-3] and visual [MPEG4-2] streams. Essentially, they take compressed AUs from the decoding buffers and place the decompressed composition units in the composition buffers [M6123].

A variety of media decoders exist to handle assorted media types. The decoder type is conveyed through the OD framework, more specifically in the DecoderConfigDescriptor contained in an ES_Descriptor that is part of an ObjectDescriptor. This descriptor determines which plug-in to use. The entire procedure works as follows:

1. As explained in the previous section about the OD's implementation, it all starts when the player receives an OD and the associated Stream-Consumer object (a derivation of MediaObject) exists in the memory image of the scene graph.

2. This StreamConsumer object receives a SetStream call with an Object-Descriptor object as a parameter—or, more precisely, ObjectDescriptorImp, which extends ObjectDescriptor with execution code.

3. The MediaObject object calls ObjectDescriptorImp :: SetStream, asking the object to set up the delivery chain for the new media stream. This function examines the ES_Descriptors descendants and determines which of them to use for accessing the media.

4. The ObjectDescriptorImp object calls ES_DescriptorImp :: SetDecoder. This function looks at the DecoderConfigDescriptor and decides which DLL to load as the decoder.

5. A decoder plug-in needs to implement the Decoder interface. A decoder DLL must have a CreateInstance entry point that returns a pointer to an object derived from Decoder. Having obtained this object, SetDecoder calls Decoder :: Setup, passing on the DecoderSpecificInfo as parameters.

6. If SetDecoder is executed successfully, Decoder :: SetFormat is called to tell the decoder which uncompressed media format the player knows to play (e.g., YUV or RGB, in case of a video stream).

7. If up to this point everything is still going well, the player proceeds to open the associated DMIF channel.

8. When all of these components are in place, they need to be connected to each other with MediaStream objects. To connect the decoder to the chain, Decoder :: SetInputStream and Decoder :: SetOutputStream are called. The functions get as parameters the MediaStream objects of the decoding buffer and composition buffer, respectively. The output Media-Stream is also the return value of the SetDecoder function, which propagates it back to the StreamConsumer object. The latter now uses this object to fetch decoded data from the composition buffer.

9. At some point during its initialization, the Decoder plug-in spawns an execution thread to perform its decoding task. This thread enters a loop of fetching units from the input MediaStream, decoding them and dispatching decoded units to the output MediaStream.

10. The loop described in step 9 goes on until somebody calls Decoder :: Terminate and terminates the process.

Altogether these create a mechanism to plug different decoders that handle different media types into an MPEG-4 player in a way that would make all streams handled transparent by the player.

14.8.3 IPMP Plug-Ins

IPMP plug-ins are, as their name suggests, components that handle protected content. This process can take place before media decoding (for instance, when encoded media is encrypted) or after (as is the case with watermarks). The IPMP modules required to process a stream are specified in the ODs. They are loaded and initialized together with the media decoders, so that when protected data enters the player, the complete path of decoding and IPMP processing is already in place [M3860].

The similarity to media decoders is reflected by the similarity in the APIs. In fact, decoder and IPMP plug-ins share so much functionality that they are implemented through classes derived from a common class. The `Decoder` class previously described and the `IPMPManager` class that handles IPMP are both derived from `Filter`. They both use `MediaStream` objects to fetch the data they need to work on and dispatch the processed data, and therefore the functions `SetInputStream` and `SetOutputStream` are found also in `IPMPManager`. But there are a few differences:

1. In order to process the streams, IPMP modules usually require some external information—the security data. For instance, keys are required for decryption. As explained in Chapter 3, this data can arrive in two forms—in ESs, called *IPMP streams*, or in *IPMP descriptors*. The `IPMPManager` interface includes virtual functions that handle both: `SetIPMPStream` and `SetDescriptor`. The first is called during the initialization of the plug-in, when the OD framework indicates that an IPMP stream is required. The second is called when the `ObjectDescriptor` or the `ES_Descriptor` of the media stream includes an `IPMP_DescriptorPointer`, and a descriptor referenced by that pointer enters the terminal.

2. The `IPMPManager` interface has an `IsPostDecoding` function. This function returns a Boolean value that is `false` if the IPMP plug-in processes the media before decoding, and `true` if after. The player calls this function during initialization of the plug-in to determine where to link it in the data flow chain.

14.9 2D COMPOSITOR

This section describes some of the most important issues regarding the implementation of the 2D MPEG-4 compositor, a software implementation that had as a target the `Complete 2D` scene graph and graphics profiles (see Chapter

13). Because object composition is not normative in MPEG-4, this module is not included in the reference software as a normative part but as an informative one. Still, the code described here was useful to validate the specification, as only the technology that was implemented in the context of the reference software was included in the standard.

Being a 2D compositor, only a subset of the MPEG-4 BIFS nodes has been implemented (a list of the nodes implemented can be found in [MPEG4-5]). As a consequence, the problems that had to be faced were reduced. In particular, lights, navigation, and, of course, 3D primitives were left out of the scope of this implementation. This section will illustrate how the core framework was used, along with the implementation details of the efficient rendering of 2D media objects and the synchronization of ESs. This description does not cover all of the implementation issues of an MPEG-4 terminal, but illustrates some implementation choices made when implementing a compositor on top of the core reference software classes.

14.9.1 Using the Core Framework

The core set of classes does not include the implementation of the rendering capabilities of an MPEG-4 terminal. The problem of adding 2D graphics has been solved by implementing *proxy objects*. The proxy objects are a design pattern useful for implementing a specific functionality on a framework of classes. The idea is to add to each BIFS node class a proxy class that encapsulates the rendering functionality of that node. In this way, a separation between the rendering part and the node semantics is introduced. This is particularly useful when implementing an MPEG player while the standard specification is evolving, because when cosmetic changes of the specification occur, they do not necessarily affect the implementation of the proxy object. Also, this architecture mirrors the task allocation in the player implementation, wherein people working on the rendering are not necessarily the same as those working on the core framework. The last important advantage of this approach is that it confines the platform-dependent part of the player implementation to the proxy objects, so that the porting of the player to another OS or graphic library implies only changes in the proxy objects. There is, however, one drawback to using this design pattern: The proxy nodes are traversed before scene rendering, at each tick of the simulation. BIFS nodes fields are cached in proxy objects data structures. This implies that at each scene traversal a proxy object has to test whether or not the associated node has changed.

14.9.2 DEF and USE Handling

One important feature of the BIFS language lets content authors define a portion of a scene as reusable in other parts of the same scene. This derives from the VRML specification [VRML97], and the reason is to improve performance by caching the geometric computation done for the first occurrence and merely rendering for subsequent occurrences. Because scene traversal is done on the proxy objects, a proxy object has been designed to render more than one instance of a BIFS node during each simulation tick. This is achieved by creating a *drawable context*, a data structure containing all the information (bounding rectangle, clipping rectangle, visible/not visible, pointers to its appearance, geometry, event handler nodes) a renderer needs to render the node. A drawable context is taken from a pool of free contexts at each visit to a node. If a node is visited twice during a single tree traversal, this means that it is being *used*; hence, two contexts are used to render the media object in different positions. The visual renderer has a list of used drawable contexts and uses this list to draw the media objects. The next paragraph illustrates the method used to display the media objects.

14.9.3 Rendering Optimization

The rendering of visual objects is a process that basically implies two steps: object composition in a memory surface and copy of the memory surface into the frame buffer. Figure 14.9 shows an MPEG-4 scene composed of several visual objects providing different views of a cycling event. BIFS text nodes are rendered on top of the videos, displaying information about the event. The composition process starts when the BIFS and OD decoders have fin-

Fig. 14.9 A screen shot of the MPEG-4 2D player (courtesy of TDF/ENST).

ished decoding the respective AUs. It is a cyclic process done at a specific rate (typically 25 times per second). First, the BIFS tree is traversed to collect information about the nodes to draw. Media objects are drawn in the order in which they appear in the BIFS scene, unless an OrderedGroup node is in the scene. If this is the case, the drawing priority field indicates which media object should be drawn first. A list of objects to be drawn is created. Objects are then drawn using an algorithm that minimizes the number of objects to draw.

Drawing all of the media objects at each tick is not required and would be highly inefficient. For instance, a rectangle that is painted at the beginning of the scene and does not change in appearance during the rest of the simulation does not require a 25-times-per-second drawing. At each tick of simulation, a media object is drawn if

1. its appearance or geometry nodes have changed (or, in the case of a MovieTexture node, there is a new frame to display);
2. its position in the scene has changed; or
3. its bounding box intersects the area that needs to be redrawn. In that case, only the intersection is drawn.

After composition, the memory surface is copied to the frame buffer.

14.9.4 Synchronization

This section describes how ES synchronization is achieved. The 2D compositor implementation uses the reference software mp4 file format implementation for the playback of mp4 streams. As a consequence, this section applies only to local file playback and does not cover all of the issues concerning the synchronization of ESs in a push scenario. In a pull scenario, such as local file playback, ES synchronization may be achieved without taking into account OCRs, as there are no problems of clock skew between client and server terminals.

The core framework provides a set of classes that associate a clock to each ES. If the authors of the MPEG-4 scene require two or more streams to be synchronized, they must signal in the OD of each stream at authoring time whether that stream has to have its own clock or shares the clock with other streams. Synchronization between streams hence is achieved by sharing the same clock between ESs.

Consider the case when a terminal has to synchronize two ESs: one carrying audio, the other video. Two different threads are used to render audio and video. The audio thread fetches audio frames produced by the audio decoder and copies them into the audio board buffer. The audio thread is set to a higher priority than the video one, because failure in copying the audio samples into the audio board buffer on time (for instance, in the case the CPU is loaded) heavily affects the quality of the presentation. Video frames, on the other hand, can eventually be skipped without seriously affecting the overall quality. The audio and video renderers access composition buffers to get composition units. Core media buffers expose a set of methods to access the composition units in sequence, or to get the most mature units (i.e., units whose composition time matches the composition time of the simulation). Video frames are fetched from the composition buffer using a method that compares the object clock to the composition time and gives back the most appropriate valid frame. As a result, video frames may be skipped if the terminal is late during the presentation. Audio frames, on the other hand, are fetched one after the other, as no skipping is desirable when rendering audio. Audiovisual synchronization is achieved using the `AdjustClock` method of the core. When ESs share the same clock, this method sets the clock to the value indicated as an argument. After initial audio buffering required to smoothly play audio, the audio rendering starts; then the composition time of the audio sample that is being played is used as an argument in the `AdjustClock` function, causing all of the ESs sharing that same clock to be in sync.

14.10 3D COMPOSITOR

This section describes some of the most important issues regarding the implementation of the MPEG-4 3D compositor. This implementation interfaces to the core classes to visually render a scene graph containing 3D and 2D BIFS nodes, ensuring that objects in the scene are drawn at the correct location in the correct order.

The 3D compositor is part of a complete application, often referred to as the MPEG-4 3D player, encompassing all of the MPEG-4 Systems' functionality (the core code, plug-ins, and 3D compositor) along with a frame application providing windows and menus. Although the 3D compositor and frame application are separate modules, they have been developed and maintained together. Figure 14.10 shows a sample screen shot of the MPEG-4 3D player.

The development of this 3D compositor began at the same time as the other MPEG-4 Systems reference software, in particular the core code. The intention was to implement the 3D and 2D BIFS nodes accepted in the standard, providing a way to validate the standard as it evolved.

The major aim of this section is to provide an introduction into how the 3D compositor part of the MPEG-4 Systems reference software works. It is intended that this material be used in conjunction with the source code itself [MPEG4-5], as it tries to provide some background information to help understand the code. It is hoped this text will answer some questions that arise when looking at the source code. At the time of this writing, the latest version of the 3D player, including the 3D compositor, was version 3.4.0.

The 3D rendering of the scene is achieved using the *OpenGL* 3D graphics library [SgAk99]. OpenGL is available for both Unix and Windows, but the latest version of the 3D compositor (v3.4.0) has been tested only on the Windows

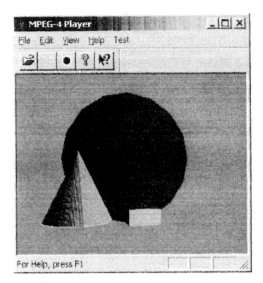

Fig. 14.10 Screen shot of the MPEG-4 3D player.

platform. Earlier versions of the 3D compositor were also able to compile and run on Unix.

14.10.1 Differences Between the 2D and 3D Compositors

The 3D compositor was developed independently of the 2D compositor described in the previous section. Although both the 2D and 3D compositors make use of the same core classes, the independent development of the two compositors has led to differences in architecture and different approaches being adopted in places.

A particular difference is the way the 2D and 3D compositors render a scene. The 2D compositor renders the objects in a scene into an off-screen buffer, calculating the order to draw objects and how to minimize the amount of redrawing that needs to be done. The 3D compositor uses the OpenGL graphics library to render the objects. The 3D compositor only needs to provide a list of the objects to be drawn and their 3D coordinates. The OpenGL graphics library determines in what order to draw the objects and how to improve rendering performance by only drawing visible objects.

Another important variation relates to the difference between a 2D-only scene and a 3D scene. In a 2D scene, the compositor knows the width and the height of the scene, and objects in the scene are placed at particular coordinates within this scene. The 2D compositor needs to ensure that objects are drawn in the correct order. In a 3D scene, the position of an object is given by 3D coordinates and, more importantly, the location of the viewer also has a 3D position and orientation. This allows the viewer to move through the scene, looking at it from different angles. The 3D player provides a way for the user to navigate through a scene, as described in more detail in Section 14.10.6.

14.10.2 Using the Core Framework

As with the 2D compositor, the 3D compositor relies on the core classes to provide the functionality to receive MPEG-4 scenes and streams of decoded data. The 3D compositor adds to these core classes the ability to visually render the 3D and 2D nodes contained in a scene and to interact with this scene.

In order to render a 3D scene, the 3D compositor needs to be able to add render-specific information to the nodes in the scene graph. Instead of modifying the definition of the core classes, the 3D compositor makes use of proxy objects to associate rendering information with each node. Thus, whenever the core creates a node in the scene graph, a proxy object is created for this node that will contain relevant information for the 3D compositor to be able to render this node.

14.10.3 Overview of the Key Classes

The 3D compositor is implemented using a large number of classes. These classes implement the functionality required for the BIFS nodes, as well as the general functionality to support the operation of the 3D graphical rendering process. This section will focus on those classes that contribute to the fundamental operation of the 3D compositor and frame application and their relationship to the core module of the MPEG-4 Systems reference software.

Figure 14.11 shows the hierarchy of the key classes in the 3D compositor and related core classes. This shows how classes from the core code are extended (using inheritance) to add rendering capabilities. Thus the OpenGL-Presenter, OpenGLVisualRenderer, Executive3D, and OpenGLApplication are classes that work with the rest of the core classes to make a complete operational MPEG-4 player. In Figure 14.11, classes in italics belong to the core; other classes are part of the 3D compositor and player.

The large majority of the classes that comprise the 3D compositor inherit from the MediaObjectProxy class, with only a fraction of these classes shown in Figure 14.11 for brevity. These classes contain the code that makes the necessary OpenGL function calls to render each of the BIFS nodes.

14.10.4 3D Rendering Process

To help understand the 3D rendering process, this section provides a high-level walkthrough of the steps taken to render a scene in the 3D compositor. The main task of graphically rendering a 3D scene is done in the OpenGLVisualRenderer :: Paint method. In this function, the following steps are performed:

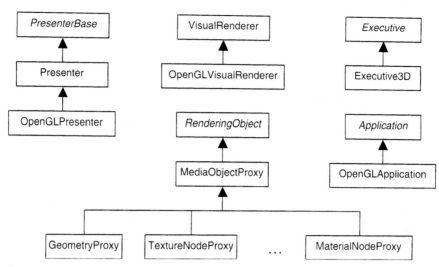

Fig. 14.11 Class hierarchy in the 3D compositor.

1. Obtain a pointer to the root node of the current scene graph maintained by the core code. Each of the nodes in this scene graph will have proxy objects (derived from `RenderingObject`) attached to them, and the tree of these proxy objects is used for the rendering.

2. Scan through all time-based sensors, determining if their states need to be changed.

3. If a mouse button has been pressed, pass this information on to the media object in the scene currently under the mouse. This involves mapping the 2D mouse coordinates into the 3D space and determining which media object in the scene graph should receive this event. This allows users to interact with objects (particularly touch sensors) in the scene.

4. Adjust the current viewpoint based on user navigation (discussed in more detail in Section 14.10.6). This allows users to move through the scene, changing their view of the scene.

5. Prepare the viewing transformation for the current viewpoint. This is the 3D transformation that maps the coordinates of the objects in the scene into the 2D view the user sees.

6. Render the scene's background media object, if it exists.

7. Activate any lighting media objects in the scene.

8. Call the `CullOrRender` method of the root node's proxy object. This recursively moves through the scene graph of proxy objects, only drawing those media objects that are partially or entirely visible. The drawing of each node in the scene is achieved by calling the appropriate OpenGL function. In this way, the OpenGL libraries take care of depth, occlusion, and visibility issues.

14.10.5 Support for 2D Nodes

The 3D player can also render 2D BIFS nodes, such as `Circle`, `Rectangle`, and `Transform2D`. The 3D compositor can render these objects at the same time as 3D BIFS nodes, drawing the 2D BIFS nodes in the X–Y plane. Thus, a 2D object such as a rectangle is treated as a flat object lying in the X–Y plane, rendered interspersed between the objects laid out in 3D coordinates. Because users can change their viewpoint, it will be possible to look at these 2D nodes from angles other than directly front-on, as is the case with the 2D compositor.

14.10.6 User Navigation

The frame application of the 3D player allows users to navigate through the scene, changing their position and orientation. This allows users to view the objects in the scene from different *vantage points*. This navigation is achieved by changing the users' viewpoint in the scene.

The viewpoint is composed of where the user currently is (as a 3D coordinate), in what direction he or she is looking (how much to the left/right and up/down), and his or her orientation (how much the head tilted to the left or right). So there are six parameters involved: 3D coordinates (3), direction looking left/right (1) and up/down (1), and orientation to the left or right (1). Navigation through the scene is achieved by translating the location of the user and rotating the angles of his or her direction and orientation.

The 3D player allows users to change their viewpoint to navigate through the scene by using the mouse. Different combinations of mouse buttons and mouse movements affect the viewpoint parameters as follows:

1. **Press and hold down the left mouse button**. Moving the mouse up and down will translate the viewpoint forward and backward. Moving the mouse left and right will rotate the viewpoint to the left and right.

2. **Press and hold down the right mouse button**. Moving the mouse up and down will translate the viewpoint up and down. Moving the mouse left and right will translate the viewpoint to the left and right.

3. **Press and hold down both mouse buttons**. Moving the mouse up and down will rotate the viewpoint up and down. Moving the mouse left and right will roll the viewpoint clockwise and counterclockwise.

Figure 14.12 presents this navigation control in a different manner. This shows how the viewpoint location, direction, and orientation can be changed, and what combination of mouse buttons and movement is required.

The user navigation is implemented by having the frame application (the `CMmiView` class) passing mouse button presses and mouse movements on to the visual renderer. Inside the main rendering loop, these mouse movements are converted into adjustments (translations and/or orientation changes) to the current viewpoint.

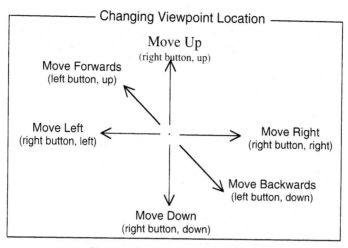

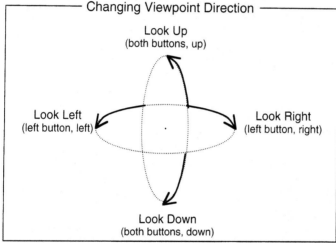

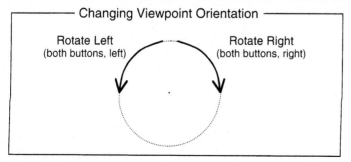

Fig. 14.12 Viewpoint navigation in the 3D player. This shows what button and mouse movements allow the user to change viewpoint location, direction, and orientation.

14.11 SUMMARY

This chapter has illustrated how the MPEG-4 standard specifications, explained in the previous chapters, were used in the real world to implement an actual player that could demonstrate the MPEG-4 experience. This chapter emphasized the role of MPEG-4 Systems as one of the most innovative parts of the MPEG-4 standard, as was introduced in Chapters 3, 4, and 7. The chapter also has shown how the audiovisual decoders, introduced in Chapters 8 through 12, could be integrated into the player.

After presenting the concept of reference software in MPEG-4, this chapter elaborated on the software architecture of the MPEG-4 player. The chapter was written by members of the implementation group of this specific software and, as such, does not do full justice to the reference software effort as a whole. Nevertheless, MPEG-4 Systems, as the part that most differentiates MPEG-4 from other media representation standards and the core of the player implementation, well deserves this focus. A special attention was given to the trickiest parts of the implementation, hoping that the reader who tries to understand the reference software, designs his or her own implementation, or just wishes to deepen his or her knowledge of the standard will find these explanations useful.

14.12 REFERENCES

[M3111] Lifshitz, Zvi. *APIs for Systems Software Implementation*. Doc. ISO/MPEG M3111, San Jose MPEG Meeting, February 1998.

[M3860] Lifshitz, Zvi. *IPMP Development Kit*. Doc. ISO/MPEG M3860, Atlantic City MPEG Meeting, October 1998.

[M6123] Lifshitz, Zvi. *Decoder Development Kit for IM1*, Version 2.0. Doc. ISO/MPEG M6123, Beijing MPEG Meeting, July 2000.

[M6124] Lifshitz, Zvi, "*DMIF Development Kit*," Version 2.2. Doc. ISO/MPEG M6124, Vancouver MPEG Meeting, July 1999.

[MPEG4-1] ISO/IEC 14496-1:2001. *Coding of Audio-Visual Objects—Part 1: Systems*, 2d Edition. 2001.

[MPEG4-2] ISO/IEC 14496-2:2001. *Coding of Audio-Visual Objects—Part 2: Visual*, 2d Edition. 2001.

[MPEG4-3] ISO/IEC 14496-3:2001. *Coding of Audio-Visual Objects—Part 3: Audio*, 2d Edition. 2001.

[MPEG4-4] ISO/IEC 14496-4:2001. *Coding of Audio-Visual Objects—Part 4: Conformance Testing*, 2d Edition. 2001.

[MPEG4-5] ISO/IEC 14496-5:2001. *Coding of Audio-Visual Objects—Part 5: Reference Software*, 2d Edition. 2001.

[MPEG4-6] ISO/IEC 14496-6:2000. *Coding of Audio-Visual Objects—Part 6: Delivery Multimedia Integration Framework (DMIF)*, 2d Edition. 2000.

[SgAk99] Segal, Mark, & Kurt Akeley. *The OpenGL Graphics System: A Specification*, Version 1.2.1. April 1999. *www.opengl.org*

[VRML97] ISO/IEC 14772-1. *The Virtual Reality Modeling Language*. 1997. *www.vrml.org / Specifications / VRML97*

Video Testing for Validation

by Laura Contin, Vittorio Baroncini, and
Fernando Pereira

Keywords: perceived video quality, benchmarking, subjective
assessment methods, verification tests

*I*n its standards development approach, MPEG may adopt formal subjective evaluation procedures at two stages: at the beginning and at the end of the specification process (see Chapter 1). At the beginning it evaluates the proposals answering to a specific call for proposals, and at the end it evaluates and verifies the performance of the tools specified in fulfilling the initially defined requirements. In particular in this second evaluation phase, MPEG also may compare the performance of the new standard with respect to available ones and evaluate the adequacy of defining specific profiles (see Chapter 13). For this reason, these tests are commonly called *verification tests*. These verification tests are an important instrument for MPEG to show the industry how good its standards are in addressing the requirements initially identified and thus to deploy new and improved applications.

The MPEG-4 verification testing activities are based on the following principles [N1707]:

☞ Verification tests have the purpose of assessing the state of the work, getting feedback, and demonstrating the potential of MPEG-4.

☞ Verification tests can address audio or video technologies, as well as their combination, or still audio, video, and systems together.

☞ Each test should address a specific (MPEG-4) functionality that can be evaluated in the context of particular profiles.

☞ The normative tools used to produce test conditions must be considered stable by the relevant MPEG subgroups (e.g., video coding tools by the video subgroup).

☞ The non-normative tools used to produce test conditions must be optimized (as much as possible at that stage); for example, rate control used in video testing must be optimized.

☞ MPEG-4 performance may be compared with the performance of existing standards when this is meaningful and useful.

This chapter addresses the verification tests that were performed (as of December 2001) to validate MPEG-4 video technology [MPEG4-2]. The formal subjective tests run at the beginning of MPEG-4 to evaluate the video proposals answering the initial MPEG-4 call for proposals [N998] are described in [AlBa97, PeAl97]. Through December 2001, the following video verification tests were performed or planned to be performed in the context of MPEG-4:

1. Error resilience [N2604]
2. Scalability in Simple Scalable profile [N2605]
3. Content-based coding [N2711]
4. Coding efficiency for low and medium bit rates [N2826]
5. Advanced Coding Efficiency profile [N2824]
6. Temporal scalability in Core profile [N2823]
7. Advanced Real-Time Simple profile [N2825]
8. Fine granularity scalability [N4456][1]
9. Studio profiles [N4457]

Because of the large number of verification tests performed, in this chapter only a few will be addressed. The tests selected cover most of test methods adopted and the most important MPEG-4 functionalities.

The following sections illustrate the rationale behind the test design, the applied assessment methods, the test conditions, and the test results. In particular, Section 15.1 deals with general aspects related to the test methodology, and Section 15.2 illustrates the assessment methods used for the tests. Each of the remaining sections is devoted to a particular video verification test, providing details about the coding tools and corresponding settings, test material, experimental design, and test results.

15.1 GENERAL ASPECTS

This section addresses some general aspects related to video subjective testing—in principle, independent of the assessment method—notably, the crite-

---

1. Planned to be performed in 2002.

ria and procedure to select the test material and the test subjects, the definition of the laboratory setup, the specification of the test plan, and the procedures to train the subjects before the test.

In order to guarantee the validity of the results, a group of experts in formal subjective testing was in charge of designing and performing the tests, according to rigorous assessment methods. Moreover, the tests were (as much as possible) conducted in independent laboratories—that is, in laboratories not directly involved in the development of the tools under test. When independent laboratories were not available, the tests were duplicated to cross-check results.

15.1.1 Selection of Test Material

Adequate test material should be selected among a set of sequences that are representative of the application or functionality under examination. Moreover, the criticality of the sequences should be carefully evaluated in order to guarantee that the encoded material covers a wide range of quality.

MPEG has gathered a wide library of video and audiovisual sequences to be used in subjective evaluation [N1740]. The selection of these sequences was based on the criticality and the structure of the content. Concerning criticality, sequences with different levels of spatial detail and amount of movement were included, ranging from videophone to high-motion sports sequences. Concerning the structure of the scene, three types of sequences were included: segmented sequences, traditional frame-based sequences, and blue-screen sequences that can be easily segmented. In the first case, an alpha channel was associated to the video sequences.

15.1.2 Selection of Test Subjects

When subjective tests are aimed at estimating the evaluation of a typical user, the subjects who participate in a viewing test should not be directly concerned with video quality as part of their normal work; they also should not be experienced assessors. In short, they should be *nonexperts*.

On the contrary, when a particular type of artifact has to be evaluated (e.g., blurring or blockiness), expert viewers should perform the test. This is particularly important when high-quality signals are evaluated.

The number of nonexpert viewers participating in a test should be higher than or equal to 15 [BT500, P910]. When viewers are experts, a lower number may be sufficient.

In general, nonexpert viewers were selected to participate in the MPEG-4 verification tests, whereas experts participated in the initial competition tests [AlBa97, PeAl97].

Prior to the test, subjects must be screened for visual acuity by using, for example, a Monoyer Optometric Table. Besides, testing for normal color vision should be performed using Ishihara's tables [Ishi95]. These tables are designed to detect color blindness or strong color vision deficiencies; this means they check the ability of observers in discriminating colors. Subjects who do not pass these tests must be discarded.

15.1.3 Laboratory Setup

The subjective assessment environment used in the MPEG-4 tests consisted of one *control room* and one or more *test rooms*. The control room hosted the systems to generate the test items (i.e., D1 or other videotape recorders [VTRs]) and the control system to synchronize presentations and scoring. From this room the experimenter also was able to check the correct running of the test and to monitor, through a closed-circuit TV system, the subjects while they were performing the test in the test room(s).

The setup of the subjective test laboratory (the test rooms) was based on Recommendation ITU-R BT.500 [BT500], which specified the viewing conditions in terms of parameters such as luminance, brightness and contrast of the monitor, viewing distance, luminance and chromaticity of the background, and room illumination. In general, professional studio monitors were used, with a size between 19 and 21 inches. These were calibrated using PLUGE signal [BT814] instruments, such as luminance meter and spectrophotometer.

In general, the test rooms were arranged to host more than one subject at a time: In some cases, they were all viewing the same monitor; in others, each subject was assigned a different monitor.

Often, votes were collected with automatic PC-based systems. In other cases, the scores were recorded on voting sheets.

15.1.4 Test Plan

The first step in the organization of a subjective test is the selection of sufficient test sequences and the definition of the test conditions to be used. If the effect of parameter variation is of interest, it is necessary to choose a set of parameter values that cover the impairment grade range in a number of roughly equal steps.

Next, the experimenter must design the test in terms of number and type of test items,[2] number of test sessions, duration of each session, distribution of test items throughout the sessions and among subjects,[3] repetition of test items within the test, and presentation order.

---

2. The most common way for generating test conditions is to include all combinations of sequence and test condition, but in some cases only some subsets might be selected for the test.

3. In general, in the MPEG-4 verification tests, all the subjects evaluated all the test items, but a test can be arranged in order to make different groups of subjects evaluate different test items (e.g., same test conditions, but different sequences).

The test design is closely related to the statistical analysis that will be performed and is defined by considering a number of hypotheses [Myer79]. Some general rules, aimed at reducing bias due to presentation order and listeners' fatigue, are widely adopted. Some of them are listed here:

1. Two subsequent presentations of the same sequence should be avoided, even if the treatment is different.
2. For different groups of subjects, different presentation orders should be used.
3. Whenever test conditions are repeated, different presentation orders should be adopted.
4. Duration of a test session should be less than 30 minutes.

15.1.5 Training Phase

At the beginning of each test, written instructions should be provided to the subjects, to ensure that all of them receive exactly the same information. The instructions should include explanations about what the subjects are going to see, what they must evaluate (i.e., differences in quality), and how they are to express their opinion. Any question from the subjects should be answered to avoid, as much as possible, any opinion bias from the test administrator.

After the instructions, a demonstration session should be run. In this way subjects are acquainted both with the voting procedures and with the type of impairments.

Finally, a mock test should be run, in which a number of representative conditions are shown. The sequences should be different from those used in the test, and they should be played one after the other without any interruption.

When the mock test is finished, the experimenter should check that the subjects have understood the task; this can be obtained by verifying that no inversion of votes was recorded. If this happens, the experimenter should repeat both the explanation and the mock test.

15.2 TEST METHODS

The verification tests basically were carried out applying the test methods described in Recommendations ITU-R BT.500 [BT500] and ITU-T P.910 [P910]. In some cases, slight modifications to the standard test procedures were applied, to improve the reliability of test results; in other cases, important modifications were introduced, to properly evaluate new MPEG-4 functionalities. In the latter case, these modifications were proposed to the ITU groups in charge of developing subjective test methodology; they were included in the above-mentioned recommendations.

The test methods used for the MPEG-4 video verification tests are listed below:

- ☞ Single stimulus (SS)
- ☞ Double stimulus impairment scale (DSIS)
- ☞ Double stimulus continuous quality scale (DSCQS)
- ☞ Simultaneous double stimulus for continuous evaluation (SDSCE)

Depending on the goal of the test, most of these methods can be applied in different ways: For example, using different types of scoring scales, repeating the presentation of test material, using different presentation orders, and so on. In the following sections, the test methods are described as they were applied for the MPEG-4 verification tests. A more detailed presentation of all possible variants can be found in the relevant recommendations [BT500, P910].

15.2.1 Single Stimulus Method

In the SS method [BT500], the subjects are presented with a single picture or picture sequence; after each presentation, a midgray (i.e., $Y = U = V = 128$) frame is displayed for some time. During the midgray period, the subjects are requested to evaluate the quality of the picture or sequence they have seen just before the midgray period. The combination of a presentation and the subsequent scoring period is called an *assessment trial*. The duration of an assessment trial varies with the subject task, content, and factors considered, but 20 seconds (10 seconds for the stimulus plus 10 seconds for the scoring) is a typical time. Figure 15.1 illustrates the SS method protocol.

The test session consists of a series of assessment trials. Because the evaluation might depend on the presentation order—as, implicitly, any subject is comparing the stimulus that is evaluating with previous stimuli—the assessment trials will be presented in random order and, preferably, in a different random order for each observer.

Regarding the scoring, ITU recommendations specify a number of different scales, either categorical or numerical, evaluating different factors, such

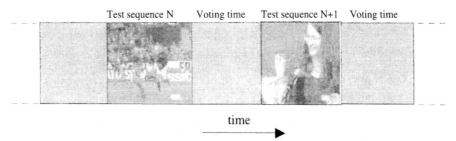

Fig. 15.1 SS method protocol.

as quality and degradation. In MPEG-4, the 11-grade scale shown in Figure 15.2 was adopted for the SS method, as it provides the highest sensitivity and stability to this evaluation method.

15.2.2 Double Stimulus Impairment Scale Method

In the DSIS method, the subjects are presented with pairs of sequences, as illustrated in Figure 15.3. The first stimulus presented in each pair is the source reference, whereas the second stimulus is the same source processed through one of the systems (e.g., coding schemes) under test. Then, the subjects are asked to evaluate the degradation of the second stimulus regarding the first one (this means the reference), and so on.

Pairs including the reference condition both in the first and second place also are introduced. This is done for two reasons: to provide a lower limit for the impairment range and to check the reliability of subjects.

The range of impairments should be chosen so that the majority of observers use all grades and the grand mean score (averaged overall judgments made in the experiment) is close to the middle of the scale.

The method uses an impairment scale like the one shown in Figure 15.4. For this scale, the stability of the results usually is greater for small impairments than for large ones.

	11
Excellent	10
	9
Good	8
	7
Fair	6
	5
Poor	4
	3
Bad	2
	1

Fig. 15.2 The 11-grade quality scale for the SS method.

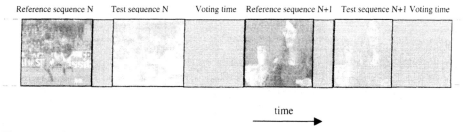

Reference sequence N   Test sequence N   Voting time   Reference sequence N+1   Test sequence N+1 Voting time

time

Fig. 15.3 DSIS method protocol.

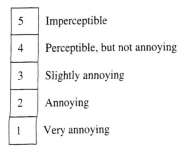

5	Imperceptible
4	Perceptible, but not annoying
3	Slightly annoying
2	Annoying
1	Very annoying

Fig. 15.4 The 5-level impairment scale for the DSIS method.

Concerning the duration of the sequences, experience suggests that extending the periods of exposition of original and processed (e.g., coded) material beyond 10 seconds does not improve the subjects' ability to grade the pictures or sequences. The time between reference and test conditions should be about 2 seconds, and the voting time should be about 8 to 10 seconds.

Both the content and the test conditions should be presented using pseudorandom order, covering all required combinations. The same test picture or sequence should never be presented on two successive occasions with the same or different levels of impairment.

A session should not last more than roughly half an hour, including the training phase; the test sequence could begin with a few pictures indicative of the range of impairments to be presented; scores for these pictures should not be taken into account in the final results. At the end of the series of sessions, the mean score for each test condition and test picture is calculated.

15.2.3 Double Stimulus Continuous Quality Scale Method

The DSCQS method is especially useful when it is not possible to provide test stimulus that exhibit the full range of quality. Thus, it has been widely applied for evaluating high-quality TV pictures. The test protocol for the DSCQS method is illustrated in Figure 15.5.

The subjects are presented with a series of pairs of pictures or sequences, each from the same source; one is processed by the system/scheme under evalu-

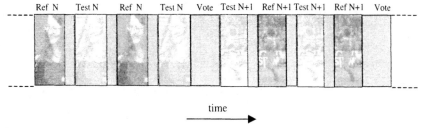

Fig. 15.5 DSCQS method protocol.

ation (test condition), and the other is just the source (reference). Within each pair the presentation order of the test condition and the corresponding reference are random and not disclosed to the subjects. The duration of each sequence, reference or processed, is about 10 seconds; the time between two sequences of the same pair is about 2 seconds; and the voting time is about 10 seconds.

Each pair is presented twice, using the same internal order (reference–test or test–reference). Already during the second presentation, subjects are asked to assess the overall picture quality of each sequence in the pair by inserting a mark on a vertical scale. The vertical scales are printed in pairs to consider the double presentation of each test sequence. The scales provide a continuous rating system to avoid quantization errors, but they are divided into five equal lengths, which correspond to the usual five-point quality scale, as illustrated in Figure 15.6. The associated terms categorizing the different levels are included for general guidance.

The pairs of assessments (reference and test) for each test condition are linearly converted from measurements of length on the score sheet to normalized scores in the range of 0 to 100. Then, the differences between the assessment of the reference and of the test condition are calculated. Finally, statistics such as mean and standard deviation or 95% confidence interval are calculated for each test condition.

Experience has shown that the scores obtained for different test sequences are dependent on the criticality of the test material used. A more complete understanding of codec performance can be obtained by presenting results for different test sequences separately, rather than only as aggregated averages across all the test sequences used in the assessment.

15.2.4 Simultaneous Double Stimulus for Continuous Evaluation Method

The test methods described in the previous sections are based on the evaluation of short sequences and are not fully adequate for evaluating the time-

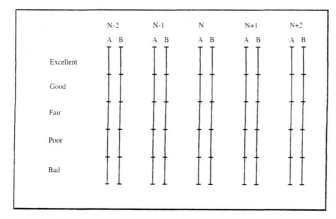

Fig. 15.6 Example of score sheet for the DSCQS method.

varying artifacts produced by digital coded content subject to transmission errors [ADGH95a]. ITU-R introduced an improvement in this direction by recommending the single stimulus continuous quality evaluation (SSCQE) method [BT500], which is based on long sequences, lasting up to many minutes. In this method, subjects are asked to assess the quality of the sequence continuously, while they are viewing it. To express their opinions, they use a slider connected to a PC that converts the position of the slider into a vote.

This method is useful to trace the overall quality of a sequence, under typical home viewing conditions, but it is not adequate to evaluate the *fidelity* of the sequence. Using an object-based coding scheme, it is possible that, because of a transmission error, an object on the scene is not updated. This might be not perceived as an impairment if the test condition is not directly compared to a suitable reference (e.g., the source or the error-free decoded sequence).

Because the goal of the MPEG-4 error-robustness verification tests was to compare the error-free decoded sequences against the same decoded sequences corrupted by transmission errors, the MPEG subjective test experts modified the SSCQE method to design a new, more adequate method called simultaneous double stimulus for continuous evaluation (SDSCE). Before applying it to the MPEG-4 verification tests, the SDSCE method was validated at CCETT in France and TILAB (formerly CSELT) in Italy in the framework of the ACTS European project TAPESTRIES [M1604]. Afterward, this new test method was endorsed by both ITU-R [BT500] and ITU-T [P910].

This method will be described in more detail, as it was purposely designed by MPEG experts for the MPEG-4 verification tests. In this method, the panel of subjects watches two sequences at the same time: One is the reference, the other is the test condition. If the format of the sequences is CIF[4] or smaller, the two sequences can be displayed side by side on the same monitor (see Figure 15.7); otherwise, two aligned monitors should be used.

Subjects are asked to evaluate the differences between the two sequences and to judge the fidelity of the video information by moving the slider of a handset voting device. When the fidelity is perfect, the slider should be at the top of the scale range (coded 100); when the fidelity is null, the slider should be at the bottom of the scale (coded 0).

Subjects are aware of which is the reference, and they are asked to express their opinions while viewing the sequences throughout their whole duration.

A multistep statistical analysis is then performed:

☞ **Step 1:** Means and standard deviations are calculated for each point of vote (the moment when a vote is collected) by accumulation of the observers.

☞ **Step 2:** Averages and standard deviations of the averages calculated at the previous step are calculated for each second.[5] The results of this step can be represented in a temporal diagram, as shown in Figure 15.8.

---

4. Common Intermediate Format (CIF) corresponds to a spatial resolution of 288×352 pixels for the luminance and 144×176 pixels for the chrominances.

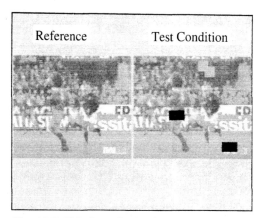

Fig. 15.7 Side-by-side display on the same monitor.

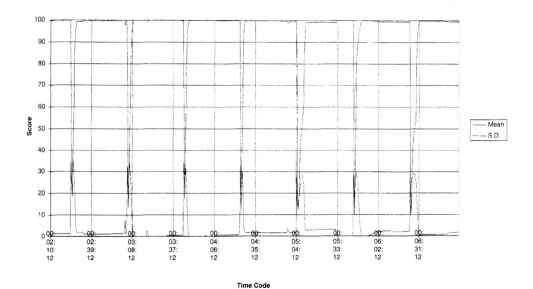

Fig. 15.8 Temporal diagram with mean (about 100) and standard deviation (about 0) for the SDSCE method.

5. In general the calculation of averages and standard deviations in this step might include votes collected during a longer period, up to 10 seconds. This is possible because no recency or forgiveness effects impact the assessment of sequences that last up to 10 seconds [SeGP92, ADGH95b]. In MPEG testing, the period was short in order to keep track of sudden impairments due to transmission errors.

☞ **Step 3**: The statistical distribution of the means calculated in the previous step and their frequency of appearance are determined. This calculation usually is based on a quantization of the rating scale in a suitable number of steps. If, for example, the scale is quantized in 20 steps, the histogram indicates the frequency of votes between 1 and 5, the frequency of votes between 6 and 10, and so forth, until the last frequency of votes between 96 and 100. In order to avoid recency effects,[6] the first 10 seconds of votes for each test condition are rejected.

☞ **Step 4:** A global annoyance characteristic is calculated by accumulating the frequencies calculated in the previous step. The confidence intervals should be taken into account in this calculation.

The reliability of the subjects can be qualitatively evaluated by checking their behavior when reference–reference pairs are shown. In these cases, subjects are expected to give evaluations very close to 100. This proves that at least they understood their task and are not giving random votes.

A more precise analysis of the reliability of subjects is based on the following two parameters:

☞ **Systematic shifts:** During a test, a subject may be too optimistic or too pessimistic, or may even have misunderstood the voting procedure (e.g., the meaning of the voting scale). This can lead to a series of votes systematically more or less shifted from the average series, if not completely out of range.

☞ **Local inversions:** As in other well-known test procedures, observers sometimes vote without taking much care in watching and tracking the quality of the sequence displayed. In this case, the overall vote curve can be relatively within the average range but local inversions nevertheless can be observed.

These two undesirable effects (atypical behavior and inversions) can be avoided. Training of the participants is important, of course. The use of a tool allowing one to detect and, if necessary, discard inconsistent observers also is possible. A proposal for a two-step process allowing such a filtering of unreliable subjects is described in [BT500].

15.3 ERROR-RESILIENCE TEST

Universal access is one of the most important MPEG-4 functionality (see Chapter 1); this functionality refers to the provision of acceptable quality even

---

6. Recency effect is the bias introduced by the sequence previously observed. If, for example, the quality level of the previous sequence is very poor, the votes associated to the new sequence are in general higher than usual. After 10 seconds, the voting can be considered stabilized.

for error-prone channels such as mobile environments (burst-error environment) or the Internet (packet-loss environment). MPEG-4 video error-resilience tools [MPEG4-2] are the solution for this problem. The error-resilience test performed in June 1998 had the objective of verifying the MPEG-4 capabilities in terms of error resilience under realistic error-prone conditions [N2165, N2604]. It is expected that the error-resilient MPEG-4 video profiles[7] will be used for existing and future wireless and mobile networks (such as GSM, DECT, GPRS, and IMT-2000), with bit rates ranging from 24 kbit/s to 2 Mbit/s. In this context, possible applications are personal real-time conversational services (e.g., videotelephony), video retrieval (e.g., video on demand), surveillance, and remote monitoring. It should be noted that the transmission delay and the total bit rate are important for these applications, especially in mobile environments; video error resilience can help in this context by allowing the decoder to gracefully recover from residual bit errors in the transmission.

15.3.1 Test Conditions

To perform the error-resilience test, the wireless system model shown in Figure 15.9 was used. This model is a typical example of a wireless multimedia communication system and consists of the following four layers:

- ☞ **Application layer:** MPEG-4 video encoder/decoder and MPEG-4 audio encoder/decoder.
- ☞ **Access unit (AU) layer:** A component of the MPEG-4 Systems layer [MPEG4-1].
- ☞ **TransMux layer:** H.223/Mobile, which is a mobile extension of ITU-T's standard for low bit-rate multimedia communications [H.223].
- ☞ **Physical layer:** 10-ms burst errors were adopted as a typical wireless channel condition.

The FlexMux layer (see Chapter 3) was skipped, except for the AU layer, so as not to loose error resiliency in the multiplexing layer. As the simplest model of the AU layer, the following assumptions were adopted [N2604]:

1. One AU corresponds to one video packet (see Chapter 8).
2. One AU is mapped into one adaptation layer-protocol data unit (AL-PDU).
3. No AL-PDU header is used.

H.223/Mobile mode 2 (H.223 with its Annex B) [H223] was selected as the TransMux layer (see Chapter 7) among a variety of H.223 solutions and its extensions. The header information in multiplexed packets was strongly pro-

---

7. All MPEG-4 video (natural visual) object types with the exception of the studio object types include error-resilience tools.

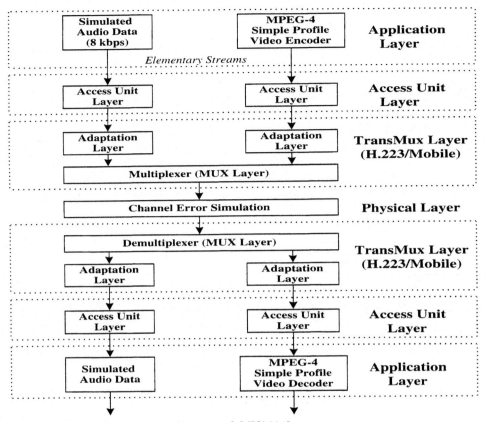

Fig. 15.9 Wireless system simulation model [N2604].

tected, but its payload (i.e., video packets and dummy audio packets) was not protected at all. Other parameters regarding the TransMux configuration can be found in [N2604].

Video coding tools were selected to comply with the simple profile and thus with the simple object type (see Chapter 13; remember that the simple object type includes only rectangular objects). Forced intra refreshment was used for all the test sequences, with a number of macroblocks in each VOP being transmitted in intra coding mode. Two (non-normative) techniques were used for intra refreshment: cyclic intra refresh (CIR) and adaptive intra refresh (AIR) [MPEG4-2]. CIR performs cyclic intra refreshment of the frame by intra coding sequential sets of N macroblocks each time the frame is coded. This way, full refreshment will be obtained after a certain number of frames is coded. The AIR technique intra refreshes more often in high-motion areas, allowing a quicker recovery from the errors in the motion areas, typically more difficult to conceal. The video data were encoded and multiplexed and stored on a hard drive offline (i.e., in nonreal time). Summarizing, the following video coding tools were used (see Chapter 8):

☞ I- and P-modes

☞ AC/DC prediction

☞ Reversible VLC

☞ Slice resynchronization

☞ Data partitioning

☞ Header extension code (HEC)

☞ CIR

☞ AIR

The video coding bit rate was specified taking the audio bit rate and multiplex overhead into account. A fixed quantization step value was chosen for the video coding, which allows reaching the specified video bit rate as an average value over the whole test sequence; thus, no rate control mechanism was applied. The refresh time was equal to two frames, and the HEC was introduced after the first resynchronization marker (RM). Table 15.1 presents the coding parameters used for the error-resilience test [N2604].

In the decoder, error concealment was used when errors were detected. The error-detection method was identical to the method used in the MPEG-4 video error-resilience core experiments [N1646]. No additional information provided by the demultiplexer (e.g., CRC check results and frame border indication) was used at all.

As audio bitstreams to be multiplexed with video bitstreams, dummy data were used for multiplexing purposes, assuming 8 kbit/s speech coding[8] as an example of MPEG-4 Audio [MPEG4-3]; a constant audio frame length of 20 ms was adopted. As a consequence, the demultiplexed audio data was simply discarded.

Concerning the simulation of error conditions, an initial error-free period of 100 bytes to protect the video object (VO) header (4 byte) and the video

Table 15.1 Coding parameter values for the error-resilience test

Channel bit rate (kbit/s)	Video bit rate (kbit/s)	Packet size[*]	Frame resolution	Frame rate (fps)	Quantization step	AIR refresh rate	CIR refresh rate
32	20.4	480	QCIF	7.5	15	3	1
128	116.4	600	CIF	10	10	4	4
384	360	1440	CIF	15	6	7	3

[*]Determines the introduction of RMs.

8. In this test, audio data were not introduced because at the time of this test the error-resilience tools for audio were not sufficiently stable. These tools were tested separately in a test addressed in Chapter 16.

object layer (VOL) header (12 byte) was chosen, because errors on this header information may cause fatal results in the decoding.

Error patterns were generated using software supplied by NTT DoCoMo. As a typical example of wireless and mobile transmission channels, 10-ms burst errors were used. Error conditions were defined as follow:

☞ **Typical error condition:** 10^{-4} bit error rate for 10-ms burst errors

☞ **Critical error condition:** 10^{-3} bit error rate for 10-ms burst errors

☞ **Initial error-free period:** 100 bytes

The critical error condition corresponds to the "Robustness to Information Errors and Loss" requirement included in the MPEG-4 Requirements document [N4319]. For current wireless systems, the critical error condition corresponds to the worst cases, which occur at the edge of radio service area, whereas the typical error condition corresponds to the average situation under which users usually enjoy the communication.

15.3.2 Test Material

Three test sequences were selected for the error-resilience test (see Table 15.2), one for each bit rate. The criterion for selecting the sequences was criticality: The higher the bit rate, the higher the amount of movement and spatial details in the sequence.

The source format adopted was ITU-R BT.601 [BT601], but the sequences were down-sampled to the input format indicated in Table 15.2. The duration of each original sequence was 1 minute. In order to study the error robustness for a long period, each of the sequences was repeated three times to get sequences 3 minutes long. These down-sampled and repeated sequences were delivered to each of the encoding sites to ensure that the same source material was used by all testing sites.

15.3.3 Test Method and Design

Because the occurrence and characteristics of transmission errors are strongly time-varying and randomly distributed, the impact of the transmission errors

Table 15.2 Test material for the error-resilience test

Sequence name	Content description	Source format (lines/fps)	Input format	Target bit rate
Artbit	Virtual studio	625/50	CIF	384 kbit/s
Australia	News	625/50	CIF	128 kbit/s
Overtime	Videotelephony	525/60	QCIF	32 kbit/s

had to be observed and evaluated in periods of time considerably longer than the traditional 10 seconds commonly used for video quality evaluation. Moreover, in order to evaluate the impact of transmission errors only, it was important to compare sequences that were compressed but not affected by transmission errors with sequences that were compressed and affected by transmission errors. Thus, the SDSCE method (described in Section 15.2.4) was chosen for the error-resilience test. The subjects were required to evaluate the annoyance of impairments in compressed sequences with transmission errors against the same sequences without transmission errors. The two sequences were displayed simultaneously on two windows on the same monitor (see Figure 15.7).

This new test method developed for this verification test was validated by MPEG through a pilot test [M3600] that was able to verify any possible bias of the results due to particular test conditions (sites, source format, etc.). The pilot test was carried out in three different laboratories (CCETT in France and TILAB and FUB in Italy), at three bit rates (32, 128, and 384 kbit/s), with three different test sequences (each 3 minutes long), at both 50- and 60-Hz frame rates, changing the order of presentation of the test sequences and showing to the subjects three types of test sequences: with no errors (the reference), with small errors, and with large amounts of transmission errors (simulating *typical* and *critical* error conditions, respectively). The pilot test demonstrated that there was no bias of the results due to the laboratories, to the source format, to the order of presentation, or to the test material. Moreover, the results showed that the subjects easily discriminated the conditions not affected by transmission errors.

Based on the results of the pilot test, it was decided to run the final test with one test session, designed according to the following rules:

☞ A single presentation order
☞ No inclusion of the "decoded without transmission errors" condition in the test[9]
☞ All test conditions presented in the same session
☞ 60-Hz frame rate
☞ A total of 33 subjects
☞ Two laboratories (CCETT and FUB)

The setup of the test laboratories was done according to Recommendation ITU-R BT.500 [BT500]. One or more 19-inch professional monitors (e.g., Sony BVM1910) were used to display the sequences. The subjects were located in front of the monitor at a distance equal to 4 times the height of the screen. Each test at each test site was performed using a panel of nonexpert subjects, organized in groups of three in front of each monitor. The sequences were

---

9. This means that no side-by-side comparison of two *without transmission errors* sequences was performed.

recorded on a D1 tape, one after the other, without any interval between two successive sequences.

15.3.4 Data Analysis

The statistical analysis was performed according to the procedures described in Section 15.2.4. In the next section, only the results of step 2 of this analysis are presented, because these are the results that show the (local) behavior in terms of transmission errors of the systems under test. Of particular relevance were the number of errors perceived by the observers, the severity of the impairments, and the recovery time.

15.3.5 Test Results

Some results for the error-resilience test are reported in Figure 15.10, Figure 15.11, and Figure 15.12. The thick lines at the top of the charts represent the average scores obtained for all subjects for a certain sequence. The values close to 100 indicate that no (or few) transmission errors were detected; lower values indicate higher degradation of the images due to transmission errors. The thin lines at the bottom represent the behavior of the associated standard deviation. As usual, higher averages generally are associated with lower standard deviations. This is a consequence of the different use of the scale: Some subjects react to any impairment with larger movements of the slider than others. However, when test conditions are close to the fidelity, most subjects keep the slider at the upper end position.

At 32 kbit/s, typical and critical error conditions both perform very close to the maximum of the evaluation scale (see Figure 15.10). This means that

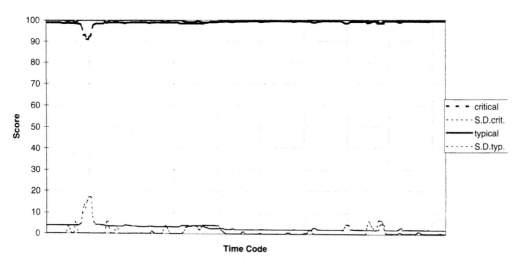

Fig. 15.10 Error-resilience test results for *Overtime* at 32 kbit/s.

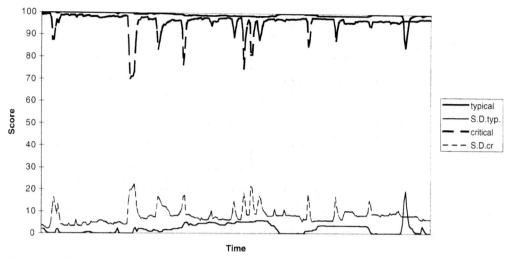

Fig. 15.11 Error-resilience test results for *Australia* at 128 kbit/s.

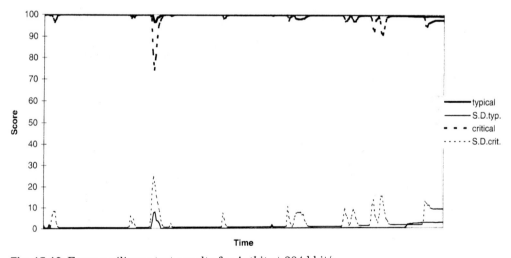

Fig. 15.12 Error-resilience test results for *Artbit* at 384 kbit/s.

the errors were hardly perceived. This can be due to a considerable masking effect because of the presence of visible coding artifacts; thus, transmission errors do not increase the annoyance due to coding artifacts.

At 128 kbit/s, the impairments introduced by the transmission errors were more visible but the scores were not too low and the recovery time was quite short (see Figure 15.11). Finally, at 384 kbit/s, the critical error condition got rather high scores (see Figure 15.12). This may be due to the higher redundancy available because of the higher bit rate.

Summarizing, it can be concluded that the MPEG-4 video error-resilience tools perform well for all cases tested and provide a satisfactory level of quality for the video signal, even in the presence of critical transmission errors. Moreover, the recovery time from errors is short and the decoder quickly restores the full quality of the signal. These results were achieved with low overhead—less than that usually associated with the GOP coding structure used in MPEG-1 and MPEG-2 [N2604].

The test results show that the MPEG-4 error-resilience tools can provide quality comparable to error-free quality under the typical error condition with burst errors at a bit error rate of 10^{-4}. They also have demonstrated that the quality level can be classified as *usable* even under critical error conditions; thus, they meet the MPEG-4 requirements [N4319].

The overall results confirm that the SDSCE method is particularly well-adapted to evaluate the effect of sparse impairments such as transmission errors on the fidelity of visual information in video sequences already degraded by a low bit-rate encoding.

15.4 CONTENT-BASED CODING TEST

To fulfill the MPEG-4 functionalities presented in Chapter 1, typical video frame-based coding was no longer enough; thus, the object-based data model was adopted for MPEG-4. Because encoding a video sequence as a composition of objects requires coding not only the texture but also shape data (see Chapter 8), it is important to know at which cost in terms of subjective quality come the content-based functionalities made possible by the independent coding of the video objects. This section describes the content-based coding test that had the target of comparing the subjective quality performance of MPEG-4 object-based coding regarding MPEG-4 frame-based coding.

15.4.1 Test Conditions

The MPEG-4 content-based coding test, performed in February 1999, targeted the comparison in terms of coding efficiency of an MPEG-4 video object-based encoder with an MPEG-4 video frame-based encoder, under the conditions specified in Table 15.3 [N2602, N2711]. Two ranges of bit rates were considered: The lower range was up to 384 kbit/s and the higher range was up to 1.15 Mbit/s. Table 15.3 lists some of the coding parameters used to produce the test sequences.

The encoding sites were free to choose the optimal encoding options, as long as the general conditions depicted in Table 15.3 were followed. Concerning the VOP structure, intra coding was used only for the first frame, and no B frames were used (M = 1). Finally, the rate control suggested by MPEG was applied for all bit rates.

Table 15.3 Coding conditions for the content-based coding test

	Low bit rate		High bit rate		
Sequences	*News, Children, Coastguard*		*Singer, Dancer, Stefan*		
Resolution	CIF		CIF		
Bit rate	256 kbit/s	384 kbit/s	512 kbit/s	768 kbit/s	1.15 Mbit/s
Input frame rate	10 Hz	10 Hz	25 Hz	25 Hz	25 Hz
I-period	1st VOP only	1st VOP only	1st VOP only	1st VOP only	1st VOP only
P-period	M = 1	M = 1	M = 1	M = 1	M = 1
Rate control	MPEG-4	MPEG-4	MPEG-4	MPEG-4	MPEG-4

15.4.2 Test Material

Two sets of alpha plane sequences were selected for this test. The first set included critical sequences (i.e., *Singer, Dancer*, and *Stefan*) and was characterized by fast movements and considerable amounts of spatial details. These sequences were used to test the higher range of bit rates (from 512 kbit/s to 1.15 Mbit/s). The second set of sequences included less critical sequences (i.e., *News, Children*, and *Coastguard*) and was used for the lower range of bit rates (from 256 to 384 kbit/s).

The decoded video sequences were displayed in CIF format and inserted in the center of a midgray ITU-R BT.601 frame [BT601].

15.4.3 Test Method and Design

The MPEG-4 content-based coding test was performed using the DSIS test method (see Section 15.2.2). The DSIS test method was designed in ITU-R [BT500] to test video signals with a broad range in terms of quality, as it was expected for this test taking into account the bit rates and sequences selected.

The setup of the test laboratory was performed according to Recommendation ITU-R BT.500 [BT500]: 19-inch professional monitors were used (e.g., Sony BVM1910) to display the sequences. The test subjects were located in front of the monitor at a distance equal to 4 times the height of the screen. Each run was done using a panel of nine nonexpert subjects. Three of them were located in front of each monitor.

The test was organized into two sessions according to the range of bit rate; as such, they will be identified as the low and the high bit-rate sessions, respectively. The former included all test conditions encoded at 256 and 384 kbit/s. The latter included all test conditions encoded at 512, 768, and 1,150 kbit/s.

Each session was made up of all the possible combinations of sequences and coding conditions, including both the frame- and object-based modes; furthermore, five dummy combinations were added at the beginning of each session, to allow stabilization of the opinions of the subject. The votes collected during this stabilization phase were discarded before the data analysis.

15.4.4 Data Analysis

For each test condition and each sequence, mean, standard deviation (SD), and the 95% confidence interval (CI) were calculated.

15.4.5 Test Results

The results obtained for the content-based coding test are reported in Table 15.4 and Table 15.5, where *OB* stands for object-based and *FB* for frame-based. The tables report in the rightmost column the overall mean scores, averaged across subjects and sequences. The three columns on its left list the mean scores obtained for each sequence.

Table 15.4 Results for the low bit-rate test session

Bit rate (kbit/s)	Mode (OB/FB)	Mean [95% CI] *News*	Mean [95% CI] *Children*	Mean [95% CI] *Coastguard*	Overall mean [95% CI]
384	OB	4,389 [0.359]	3,444 [0.507]	3,278 [0.630]	3.704 [0.318]
384	FB	4.167 [0.396]	3.667 [0.549]	3.056 [0.582]	3.630 [0.316]
256	FB	4.389 [0.393]	2.333 [0.549]	3.167 [0.396]	3.296 [0.342]
256	OB	4.333 [0.354]	2.944 [0.681]	2.611 [0.393]	3.296 [0.346]

Table 15.5 Results for the high bit-rate test session

Bit rate (kbit/s)	Mode (OB/FB)	Mean [95% CI] *Dancer*	Mean [95% CI] *Singer*	Mean [95% CI] *Stefan*	Overall mean [95% CI]
768	FB	4.833[0.177]	4.944 [0.109]	3.722 [0.589]	4.5 [0.253]
768	OB	4.778[0.338]	4.944 [0.109]	4.222 [0.561]	4.648 [0.233]
1150	FB	4.722 [0.347]	4.667 [0.317]	4.111 [0.632]	4.5 [0.268]
1150	OB	4.833 [0.238]	4889 [0.218]	4.167 [0.455]	4.63 [0.203]
512	FB	4.889 [0.218]	4.833 [0.238]	3.667 [0.672]	4.463 [0.287]
512	OB	4.778 [0.198]	3.944 [0.488]	3.444 [0.658]	4.056 [0.313]

Considering the results for particular sequences, it can be observed that at the lower bit rates the least critical sequence is *News*. This is not surprising, as a very low activity characterizes this sequence. On the contrary, at the higher bit rates, the most critical sequence is *Stefan*, which is characterized by fine spatial details and high amounts of movement, including global variations (panning) and local displacement (the tennis player is continuously moving).

The results obtained for the two test sessions clearly demonstrate that the object-based functionality is provided by MPEG-4 with no overhead or loss in terms of visual quality, with respect to the frame-based coding solution. In fact, the means are very close to each other and the confidence intervals are (in general) overlapping.

15.5 CODING EFFICIENCY FOR LOW AND MEDIUM BIT-RATE TEST

Although not the only driving force, coding efficiency was a major requirement [N4319] for MPEG-4 as for previous video coding standards. The MPEG-4 Requirements document explicitly states in the "Object Quality and Fidelity" requirement:

> Quality in MPEG-4 shall be as high as possible. This means that MPEG-4 shall provide a subjective video quality that is better than the quality achieved by the available or emerging standards in similar conditions.

In this section the results of the formal subjective verification test made to evaluate the performance of MPEG-4 video coding tools in terms of coding efficiency regarding MPEG-1 Video [MPEG1-2] at low and medium bit rates will be presented [N2710, N2826]. In this coding efficiency test, only frame-based coding was studied.

15.5.1 Test Conditions

The MPEG-4 coding efficiency test, performed in May 1999, targeted the comparison in terms of coding efficiency of an MPEG-4 video encoder, including all the coding efficiency tools available in the MPEG-4 verification model (see Chapter 1) versus an MPEG-1 video encoder, under the conditions specified in Table 15.6 [N2826]. This table also specifies the VOP (periodic) structure used: N is the parameter indicating the number of VOPs between two I-VOPs plus one, whereas M is the parameter indicating the number of VOPs between two P-VOPs plus one (see Chapter 8): M = 1 means that no B-VOPs were coded.

Two ranges of bit rates were considered for this test: a low bit-rate range between 40 and 256 kbit/s, and a medium bit-rate range between 384 and 768 kbit/s.

The coding tools included all of the video coding efficiency tools in MPEG-4 Visual Version 1 (thus, no global motion compensation and no 1/4-pel, which were added in Version 2) except sprites and overlapped block motion compensation (OBMC). The encoding sites were free to choose the optimal encoding options, as long as the general conditions presented in Table 15.6 were followed.

The MPEG-1 Video standard was used instead of the MPEG-2 Video standard for this test because, for progressive sequences, these two standards perform identically, except that MPEG-1 uses less overhead and so is globally more efficient. The test used typical test sequences, encoded with the same rate control for both MPEG-1 and MPEG-4 (TM5[10] from MPEG-2) to compare the coding algorithms without the impact of different rate control schemes.

15.5.2 Test Material

Two sets of sequences were selected for this test, according to their criticality. The most critical sequences (i.e., *Dancer*, *Stefan*, and *Table Tennis*) were used for the higher bit-rate range, and less critical sequences (i.e., *Carphone*, *Foreman*, and *Coastguard*) were used for the lower bit-rate range.

The decoded video sequences in CIF/QCIF format were upsampled to the ITU-R BT.601 frame format using a filter previously made available [N2826].

15.5.3 Test Method and Design

For this coding efficiency test, the SS test method (described in Section 15.2.1) was chosen. The SS test method was designed by ITU-R [BT500] to test video signals with a large quality range. In an SS test session, the subjects are asked to evaluate each sequence independently. All of the viewing conditions were compliant with Recommendation ITU-R BT 500 [BT500].

Table 15.6 Coding conditions for the coding efficiency test

	Low bit rate				**Medium bit rate**		
Sequences	*Carphone, Foreman, Coastguard*				*Dancer, Stefan, Table Tennis*		
Resolution	QCIF		CIF		CIF		
Bit rate	40 kbit/s	64 kbit/s	128 kbit/s	256 kbit/s	384 kbit/s*	512 kbit/s	768 kbit/s
Input frame rate	7.5 Hz	7.5 Hz	7.5 Hz	10 Hz	25 Hz	25 Hz	25 Hz
I-period	1st VOP only	1st VOP only	1st VOP only	1st VOP only	N = 24	N = 24	N = 24
P-period	M = 1	M = 1	M = 1	M = 1	M = 1	M = 1	M = 3

*In the prescreening tests, it was pointed out that MPEG-1 Video could not code *Stefan* at 384 kbit/s and thus this bit rate was replaced by 440 kbit/s for this sequence but only for MPEG-1 [N2710].

---

10. TM5 refers to the rate control algorithm used in MPEG-2 Test Model (TM), Version 5.

The test was performed at CCETT. Six groups of three nonexpert observers participated in the tests, organized in two sessions: the low bit rate and the medium bit rate. Each session was made up of all possible combinations of sequences and coding conditions; furthermore, five dummy combinations were added at the beginning of each session, to allow stabilization of the opinions of the subjects. The votes collected during the stabilization phase were discarded.

To limit the effect of the contextual effect (easily present in an SS test), the sequences under test were presented twice in each session; the subjects were not aware of this solution and thus repeated their evaluation for the same sequences.

15.5.4 Data Analysis

For each test condition, mean and 95% CIs were calculated. Moreover, a complete *Student* statistical analysis was performed, in order to calculate the probability of equivalence for each pair of test conditions. This allows one to conclude whether the difference between means is statistically significant. When the difference between two (or more) test conditions is not statistically significant, nothing can be concluded (based on this statistical analysis) about the test conditions' relative performance, as there is not sufficient precision in the statistical estimate of the mean opinion scores to allow relative ranking.

15.5.5 Test Results

Table 15.7 and Table 15.8 show the test results for the low and medium bit-rate tests, respectively, where Mx_y stands for MPEG-x (MPEG-1 or MPEG-4 in this case) and y identifies the bit rate. In these tables, one row is used for each test condition, and one column is used for each test sequence. The name

Table 15.7 Results for the low bit-rate coding efficiency test

Coding condition	Mean [95% CI] Carphone	Mean [95% CI] Coastguard	Mean [95% CI] Foreman	Overall mean [95%CI]	NSSD
M4_256	4,19 [0,53]	3,22 [0,49]	3,06 [0,52]	3,49 [0,31]	M4_128
M1_256	3,42 [0,58]	3,14 [0,58]	2,89 [0,43]	3,15 [0,31]	M4_128
M4_128	2,72 [0,42]	1,75 [0,33]	1,72 [0,35]	2,06 [0,23]	M4_64
M1_128	1,94 [0,35]	1,97 [0,42]	1,53 [0,30]	1,81 [0,21]	M4_64
M4_64	1,58 [0,33]	1,25 [0,24]	1,36 [0,28]	1,40 [0,17]	M1_64
M1_64	1,19 [0,28]	0,97 [0,29]	0,67 [0,21]	0,94 [0,15]	M1_40
M4_40	0,89 [0,34]	1,06 [0,30]	0,56 [0,21]	0,83 [0,17]	M1_40
M1_40	0,08 [0,09]	0,22 [0,16]	0,25 [0,14]	0,19 [0,08]	

Table 15.8 Results for the medium bit-rate coding efficiency test

Coding condition	Mean [95% CI] *Dancer*	Mean [95% CI] *Table Tennis*	Mean [95% CI] *Stefan*	Overall mean [95% CI]	NSSD
M4_768	4,86 [0.64]	4,47 [0.55]	3,25 [0.53]	4,19 [0.35]	M1_768
M1_768	3,97 [0.58]	4,11 [0.56]	2,81 [0.43]	3,63 [0.32]	M4_384
M4_512	4,25 [0.54]	3,08 [0.44]	2,31 [0.42]	3,21 [0.31]	M4_384
M4_384	3,14 [0.40]	2,56 [0.35]	2,19 [0.35]	2,63 [0.22]	M1_384
M1_512	3,00 [0.48]	2,97 [0.48]	1,72 [0.40]	2,56 [0.29]	M1_384
M1_384	2,06 [0.36]	2,19 [0.44]	1,39 [0.39]	1,88 [0.24]	

of the test condition appears in the first column of each table, and the name of the sequence appears at the top of each column. In each cell of the tables, the mean and the 95% CI on the mean opinion score are indicated. The tables also give the performance for each codec being tested, averaged over all test sequences. The codecs are ordered in the tables in decreasing order of their mean opinion score. The far-right column in each table gives, for each proposal, the next codec in the table with a statistically significantly different mean opinion score. Referring to Table 15.7, the next statistically significant different (NSSD) condition with respect to MPEG-4 at 256 kbit/s (M4_256) is MPEG-4 at 128 kbit/s. In other words, M4_256 is not NSSD from M1_256 (MPEG-1 at 256 kbit/s), but it is NSSD with respect to M4_128. In the same way, M4_64 performed statistically significantly better than M1_64. Results are graphically presented in Figure 15.13 and Figure 15.14.

The results for the coding efficiency test show a general superiority of MPEG-4 video frame-based coding regarding MPEG-1 video (always frame-based) coding for both the low and medium bit rates, whatever the criticality of the sequence.

15.6 ADVANCED REAL-TIME SIMPLE PROFILE TEST

As described in Chapter 13, the Advanced Real-Time Simple (ARTS) object type is a superset of the Simple object type, which targets real-time coding situations by employing a back-channel to monitor the video throughput in order to adapt the temporal resolution and reduce error propogation. In this context, the ARTS profile (which includes the Simple and ARTS object types) provides two major functionalities that were tested with two separate tests: *error robustness* and *temporal resolution stability*. In fact, although targeting similar applications to the Simple profile, the ARTS profile provides some improvements, notably these [N2825]:

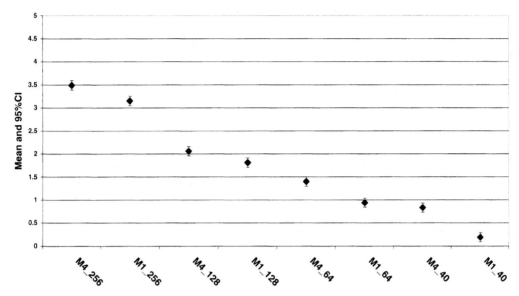

Fig. 15.13 Overall mean and 95% CI for the low bit-rate coding efficiency test.

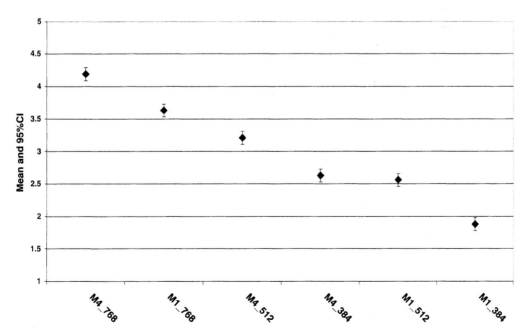

Fig. 15.14 Overall mean and 95% CI for the medium bit-rate coding efficiency test.

☞ **Improved error robustness:** The ARTS profile can provide the same quality of service level regarding the Simple profile under about 10× worse error conditions. This allows the expansion of the applications as well as the improvement of the performance already supported by the Simple profile if some conditions are fulfilled—notably, the existence of an upstream (back) channel.

☞ **Quick error recovery:** Quick error recovery from *erred states* can be provided while keeping a high coding efficiency. Artifacts are removed even faster in the case of scene changes.

☞ **Transmission buffering delay and temporal resolution stability:** The transmission buffering delay can be stabilized by minimizing the jitter of the number of bits per VOP. Large frame skips also are prevented by imposing some restrictions on the buffer, and the encoder can control the temporal resolution even for highly active scenes. This is important for low-delay applications.

15.6.1 Test Conditions

As mentioned previously, two functionalities were separately tested in the context of the ARTS profile test: *error robustness* and *temporal resolution stability*.

ARTS Error-Robustness Test

The ARTS error-robustness test targeted the evaluation of NEWPRED, a new error-resilience tool supported by the ARTS object type. Here the ARTS profile is to be compared with the Simple profile in terms of error-resilience performance. NEWPRED is a technique in which the reference picture for interframe coding changes adaptively, according to the upstream messages from the decoder ([MPEG4-2]; see Chapter 8).

Test conditions were similar to those used for the Version 1 error-resilience test presented in Section 15.3. For the ARTS error-robustness test, the MPEG-4 video encoder/decoder was combined with the wireless system model shown in Figure 15.15. This model is similar to the model used for the Version 1 error-resilience test shown in Figure 15.9. In this system, Recommendation ITU-T H.223/Annex B [H223] was used in the TransMux layer, and dummy data were used as audio data. The audio bit rate was 8 kbit/s for all test conditions. The back-channel messages generated by the video decoder in the context of the ARTS profile were multiplexed with H.223 and transmitted to the video encoder. The transmission channel errors were applied to the back-channel messages as well as to the forward video and audio data.

The coding tools used for the ARTS error-robustness test are presented in Table 15.9 and the coding parameters in Table 15.10.

The ARTS profile has higher coding efficiency than the Simple profile, because the NEWPRED tool allows recovering from erred states without forced intra MBs as used for the Simple profile (see Chapter 8). To evaluate this feature, the same quantization step values were used both for the ARTS

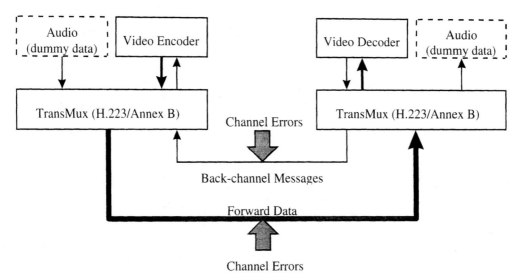

Fig. 15.15 Wireless system simulation model [N2825].

Table 15.9 Coding tools for ARTS and Simple profiles

Tools	ARTS profile*	Simple profile†
I/P-VOP	✔	✔
AC/DC prediction	✔	✔
Slice resynchronization	✔	✔
Data partitioning	✔	✔
Reversible VLC	✔	✔
HEC	✔	✔
NEWPRED	✔	
AIR and CIR‡		✔

*The ARTS profile includes the Simple and ARTS object types and thus the coding tools correspond to the coding tools in these two object types.

†The Simple profile includes only the Simple object type and thus the coding tools correspond to the tools in the Simple object type.

‡AIR and CIR are extra coding refreshment non-normative tools.

and for the Simple profiles. Therefore, the ARTS profile video coding rates were smaller than those of the Simple profile; stuffing bits were inserted to keep the same channel bit rate in the TransMux layer. As the errors on the stuffing bits do not cause any degradation, the total number of errors on the video data was smaller (in general) for the ARTS profile bitstream. Quantization step values were not changed along the sequence to evaluate only the degradation due to

Table 15.10 Coding conditions for the ARTS error-robustness test

	ARTS profile			Simple profile		
	32 kbit/s	64 kbit/s	128 kbit/s	32 kbit/s	64 kbit/s	128 kbit/s
Video coding rate	19 kbit/s	48 kbit/s	67 kbit/s	20 kbit/s	51 kbit/s	112 kbit/s
Resolution	QCIF	CIF	CIF	QCIF	CIF	CIF
Target frame rate	6 fps	6 fps	10 fps	6 fps	6 fps	10 fps
Quantization step value	10	12	12	10	12	12
Rate control	Frame skip (fixed QP)			Frame skip (fixed QP)		
Video packet size	11 MBs	44 MBs	44 MBs	480 bits	600 bits	600 bits
HEC insertion	All video packets			2nd video packet only		
I-VOP period	1st VOP only			1st VOP only		
Forced intra MBs (AIR/CIR)	0/0	0/0	0/0	3/1	10/2	40/4
Round trip delay	300 ms			-		

the errors, although rate control was used in this test. Frame skipping was used when necessary to prevent the overflow of the buffer.

The assumed round-trip delay, which is the time from sending the video packet to receiving the back-channel message, was about 300 ms for the ARTS profile. This value is appropriate for typical mobile systems.

The error conditions used for this test were as follows:

☞ **Critical error condition:** 10^{-3} bit error rate for 10-ms burst errors

☞ **Very critical error condition:** 10^{-3} bit error rate for 1-ms burst errors

W-CDMA is one of the most promising candidates for the next generation of mobile communication systems; its burst error length is said to be shorter than that of conventional systems [N2825]. Assuming the same bit error rate, the number of burst errors in the short burst error system is larger than that of the long burst error system, and the packet error rate of the short burst error system is higher than that of the long burst error system. Therefore, the short burst error conditions were evaluated as *very critical error conditions* in this test.

Error patterns were generated using software supplied by NTT DoCoMo. As the image degradation due to errors depends on the error pattern, 25 runs of error patterns were simulated for each condition in this test. The typical error pattern was automatically selected as the one producing the PSNR that was the nearest to the average PSNR over all error patterns. The typical error pattern sequence was the one used for the subjective evaluations.

ARTS Temporal Resolution Stability Test

The temporal resolution stability test targeted the evaluation of the *reduced resolution coding* tool [MPEG4-2] supported by the ARTS object type in comparison to the Simple object type, under conditions similar to those used for the error-resilience test (see Section 15.3). The reduced resolution VOP coding tool allows one to encode highly active scenes with higher VOP rate by using a macroblock size of 32×32 pixels for motion compensation and a block size of 16×16 pixels.

The coding tools used for the ARTS temporal resolution stability test are presented in Table 15.11 and the coding parameters in Table 15.12 [N2825]. An I-VOP was included only at the beginning of each sequence; all other VOPs were encoded as P-VOPs. Concerning rate control, the macroblock-based rate control suggested by MPEG was applied and frames were skipped when necessary to prevent buffer overflow.

15.6.2 Test Material

ARTS Error-Robustness Test

For the ARTS error-robustness test, two test sequences (each 1 minute long[11]) were selected: *Australia* and *Overtime*. The sequences were down-sampled into

Table 15.11 Coding tools for the ARTS temporal resolution stability test

Tools	ARTS profile	Simple profile
I/P-VOP	✔	✔
AC/DC prediction	✔	✔
Dynamic resolution conversion (DRC)	✔	

Table 15.12 Coding conditions for the ARTS temporal resolution stability test

Sequences	Australia, Overtime, Foreman, Crowd		
Resolution	CIF		
Bit rate	64 kbit/s	96 kbit/s	128 kbit/s
Target frame rate	10 Hz for *Australia* and *Overtime*	15 Hz for *Australia* and *Overtime*	15 Hz for *Australia* and *Overtime*
	5 Hz for *Foreman* and *Crowd*	5 Hz for *Foreman* and *Crowd*	7.5 Hz for *Foreman* and *Crowd*
I-period	1st VOP only		
Rate control	MPEG-4 MB-based rate control with frame skipping [MPEG4-2]		

---

11. To make the test less tiring to the subjects, it has been decided to reduce each video sequence length to 1 minute.

suitable input formats, and both were handled at 60 Hz. At 32 and 128 kbit/s, only one sequence was used. The sequences and their parameters are summarized in Table 15.13.

ARTS Temporal Resolution Stability Test

For the ARTS temporal resolution stability test, four test sequences were selected: *Australia*, *Overtime*, *Foreman*, and *Crowd* (background of *Akiyo* with *Crowd* sequence). In this case, the *Australia* and *Overtime* sequences were obtained extracting a 10-second spot of highly active scene from the original long sequences used for the error-robustness test. For the *Australia* sequence, a scene was selected in which a man stands up and goes to the map. For the *Overtime* sequence a scene in which a man comes back and sits down was selected. All the sequences were converted into CIF.

15.6.3 Test Method and Design

ARTS Error-Robustness Test

The ARTS error-robustness test was performed using the SDSCE test method (see Section 15.2.4). This test method was proposed to MPEG to evaluate error robustness at very low bit rates, and it had been already successfully used for the Version 1 error-resilience test (see Section 15.3). The decoded video sequences with and without errors (reference) were displayed in CIF format, side by side, at the center of a midgray ITU-R BT.601 frame (60 Hz). The reference was located on the left-hand side, and the sequences decoded in presence of errors were located on the right-hand side, as shown in Figure 15.16. The subjects were asked to judge the difference in degradation between the left (reference) and the right windows. The display format was always CIF; the QCIF sequences were up-sampled into CIF format.

The subjective evaluation was done by asking the subjects to move a slider along a scale ranging from *imperceptible* (i.e., top of the scale, coded 100) to *very annoying* (i.e., bottom of the scale, coded 0). The votes were sampled every 500 ms from the status of the slider. The collection of data was done automatically using a PC connected to the slider. A typical SDSCE test session was made up of one or more test sequences, each lasting for at least 1 minute.

The test was performed in two laboratories: CCETT and FUB. The general setup of the test laboratories was done according to Recommendation ITU-R BT.500 [BT500]. Three subjects were located in front of each monitor at

Table 15.13 Test material for the Arts error-robustness test

Sequence name	Content description	Source format (BT.601)	Input format	Target bit rate
Australia	News show	625/50	CIF	64/128 kbit/s
Overtime	Video-telephony	525/60	QCIF/CIF	32/64 kbit/s

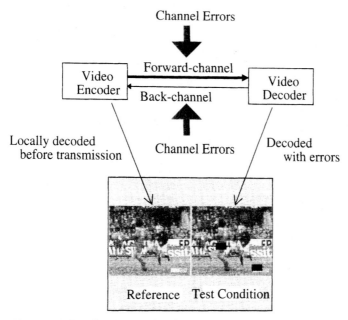

Fig. 15.16 Display format for the error-robustness test.

a distance equal to 4 times the height of the screen. Each test site used a panel of 18 nonexperts.

ARTS Temporal Resolution Stability Test

The temporal resolution stability test was performed using the SS test method. The decoded video sequences were displayed in CIF format and inserted at the center of a midgray ITU-R BT.601 frame.

The laboratory setup was designed according to Recommendation ITU-R BT.500 [BT500]. The test was performed at NTT-AT test laboratories, using a total of 18 subjects distributed in six groups of three nonexpert observers, organized into three sessions.

The test sessions were made up using all possible combinations of sequences and coding conditions; furthermore, five dummy combinations were added at the beginning of each session, to allow stabilization of the opinions of the subjects. The votes collected during this stabilization phase were discarded.

15.6.4 Data Analysis

The statistical analysis was performed according to the procedures described in Section 15.2.1 for the SS method and in Section 15.2.4 for the SDSCE method. In the next section, only the results for step 2 of this analysis are presented, because they clearly show the (local) behavior of the systems under test—in particular, the number of errors perceived by the observers, the severity of the impairments, and the recovery time.

15.6.5 Test Results

ARTS Error-Robustness Test
The results obtained for the ARTS error-robustness test are shown in Figures 15.17 through 15.20. Figures 15.17 and 15.18 show the results at the two test sites for the *very critical error condition*, whereas Figures 15.19 and 15.20 show the results at the two test sites for the *critical error condition*. Each

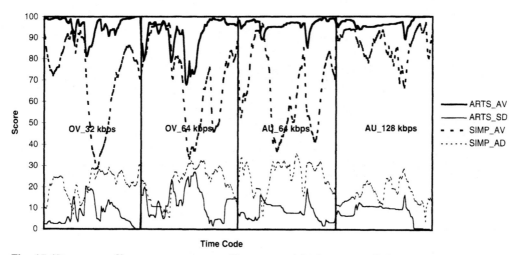

Fig. 15.17 ARTS profile versus Simple profile—very critical error condition (test site A).

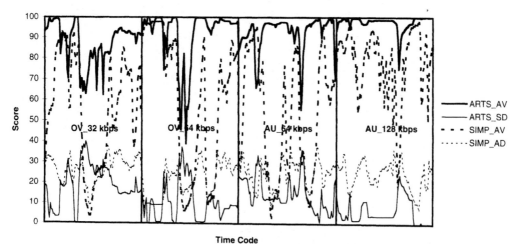

Fig. 15.18 ARTS profile versus Simple profile—very critical error condition (test site B).

chart reports the results averaged over subjects, but separate for each sequence and bit rate: for example, OV_32 refers to *Overtime* at 32 kbit/s and AU_64 refers to *Australia* at 64 kbit/s. In all the four charts, the upper curves represent the behavior of the ARTS profile and the lower curves the behavior of the Simple profile (regarding the average scores). The values close to 100 indicate that no (or few) transmission errors have been detected; lower values indicate higher degradation of the images due to transmission errors. Dotted lines at the bottom of the charts represent the behavior of the associated *standard deviation*. As usual, higher averages are (in general) associated to lower standard deviations.

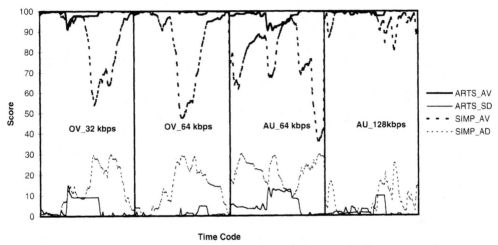

Fig. 15.19 ARTS profile versus Simple profile—critical error condition (test site A).

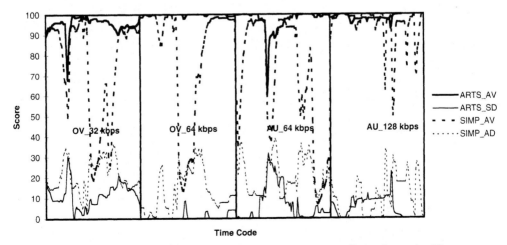

Fig. 15.20 ARTS profile versus Simple profile—critical error condition (test site B).

Figures 15.17 through 15.20 show a clear superiority of the ARTS profile over the Simple profile in terms of error robustness, notably error-recovery time. This superiority is clear for both test sites involved and for both error conditions (critical and very critical). For the critical error condition, the ARTS profile provides results that (most of the time) are close to a complete transparency, whereas the Simple profile is still severely affected by errors.

ARTS Temporal Resolution Stability Test

Figure 15.21 shows the results for the ARTS temporal resolution stability test. The results show evidence of a quality improvement for the ARTS profile regarding the Simple profile. Furthermore, the results show that the ARTS profile at 64 kbit/s outperforms the Simple profile at 96 kbit/s, and that the ARTS profile at 96 kbit/s performs as well as the Simple profile at 128 kbit/s. In conclusion, the results report a clear superiority of the ARTS profile when compared with the Simple profile in terms of subjective quality for similar bit rates.

15.7 SUMMARY

Verification tests are an important step in the MPEG-4 process of developing standards. These tests are aimed at verifying that the new standard has reached the goals specified by the initially defined requirements. The results

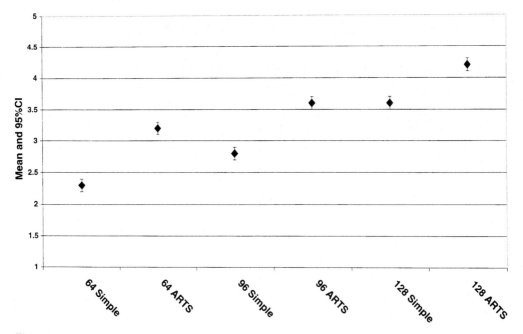

Fig. 15.21 ARTS profile versus Simple profile for the temporal resolution stability test.

of these tests are suitable to estimate the improvement introduced by the new technology and to determine a minimum threshold in terms of achievable performance by systems based on that standard, as improvements certainly can be introduced by optimizing the implementations. Verification tests are one of the elements the industry can take into account in deciding whether to bet on the new standard.

In the case of the MPEG-4 Visual standard [MPEG4-2], several verification tests were performed to test the functionality provided by tools and profiles. In this chapter, a selection of these has been analyzed in detail, including error resilience, content-based coding, coding efficiency for low and medium bit rates, and `Advanced Real-Time Simple` profile. The study of these tests allows a good review of the conditions and methodologies used for the MPEG-4 video verification tests. The test results presented in this chapter show that the evaluated tools fulfill the requirements initially specified for the standard.

15.8 REFERENCES

[ADGH95a] Aldridge, R. J. Davidoff, M. Ghanbari, D. Hands, and D. Pearson. "Measurement of Scene-Dependent Quality Variations in Digitally Coded Television Pictures." *IEE Proceedings on Vision, Image and Signal Processing*, 142(3): 149–154. 1995.

[ADGH95b] Aldridge, R., J. Davidoff, M. Ghanbari, D. Hands, and D. Pearson. *Recency Effect in the Subjective Assessment of Digitally-Coded Television Pictures.* Fifth IEE International Conference on Image Processing and its Applications, pp. 336–339. Edinburgh, UK. July 1995.

[AlBa97] Alpert, T., V. Baroncini, D. Choi, L. Contin, R. Koenen, F. Pereira, and H. Peterson. "Subjective Evaluation of MPEG-4 Video Codec Proposals: Methodological Approach and Test Procedures." *Signal Processing: Image Communication*, 9(4): 305–325. May 1997.

[BT500] Rec. ITU-R BT.500-10. *Methodology for the Subjective Assessment of the Quality of Television Pictures.* March 2000.

[BT601] Rec. ITU-R BT.601. *Studio Encoding Parameters of Digital Television for Standard 4:3 and Wide-Screen 16:9 Aspect Ratios.* July 1995.

[BT814] Rec. ITU-R BT.814-1. *Specifications and Alignment Procedures for Setting of Brightness and Contrast of Displays.* July 1994.

[H223] ITU-T Recommendation H.223. *Multiplexing Protocol for Low Bit Rate Multimedia Communication.* March 1998.

[Ishi95] Ishihara, S. *Ishihara's Tests for Color-Blindness.* Tokyo: Kanehara & Co. 1995.

[M1604] Alpert, T., and L. Contin. *DSCQE (Double Stimulus using a Continuous Quality Evaluation) Experiment for the Evaluation of the MPEG-4 VM on Error Robustness Functionality.* Doc. ISO/MPEG M1604, San Jose MPEG Meeting, February 1997.

[M3600] Contin, L., M. Quacchia, S. Pefferkorn, and V. Barboncini. *Report of the Formal Verification Tests on MPEG-4 Video Error Resilience.* Doc. ISO/MPEG M3600, Dublin MPEG Meeting, July 1998.

[MPEG1-2] ISO/IEC 11172-2:1993. *Information Technology: Coding of Moving Pictures and Associated Audio for Digital Storage Media at Up to About 1.5 Mbit/s—Part 2: Video*, 1993.

[MPEG4-1] ISO/IEC 14496-1:2001. *Coding of Audio-Visual Objects—Part 1: Systems*, 2d Edition, 2001.

[MPEG4-2] ISO/IEC 14496-2:2001. *Coding of Audio-Visual Objects—Part 2: Visual*, 2d Edition, 2001.

[MPEG4-3] ISO/IEC 14496-3:2001. *Coding of Audio-Visual Objects—Part 3: Audio*, 2d Edition, 2001.

[Myer79] Myers, J. L. *Fundamentals of Experimental Design*. Boston: Allyn & Bacon. 1979.

[N1646] MPEG Video. *Description of Error Resilient Core Experiments*. Doc. ISO/IEC N1646, Bristol MPEG Meeting, April 1997.

[N1707] MPEG Test. *Principles and Work Plan for Verification Tests*. Doc. ISO/MPEG N1707, Bristol MPEG Meeting, April 1997.

[N1740] MPEG Test. *List of Video and Audiovisual Sequences for Use by MPEG in Audiovisual Coding Standard Development*. Doc. ISO/IEC N1740, Stockholm MPEG Meeting, July 1997.

[N2165] MPEG Test. *Revised Work Plan for Formal Verification Tests on Video Error Resilience*. Doc. ISO/MPEG N2165, Tokyo MPEG Meeting, March 1998.

[N2602] MPEG Test. *Revised Test Conditions and Test Plan for Video Verification Test on Content-Based Coding*. Doc. ISO/MPEG N2602, Rome MPEG Meeting, December 1998.

[N2604] MPEG Test. *Report of the Formal Verification Tests on MPEG-4 Video Error Resilience*. Doc. ISO/MPEG N2604, Rome MPEG Meeting, December 1998.

[N2605] MPEG Test. *Report of the Formal Verification Tests on Temporal Scalability in Simple Scalable Profile*. Doc. ISO/MPEG N2605, Rome MPEG Meeting, December 1998.

[N2710] MPEG Test. *Revised Test Conditions and Test Plan for Video Verification Test on Coding Efficiency*. Doc. ISO/MPEG N2710, Seoul MPEG Meeting, March 1999.

[N2711] MPEG Test. *Report of the Formal Verification Tests on MPEG-4 Content-Based Coding*. Doc. ISO/MPEG N2711, Seoul MPEG Meeting, March 1999.

[N2823] MPEG Test. *Report of the Formal Verification Tests on MPEG-4 Temporal Scalability in Core Profile*. Doc. ISO/MPEG N2823, Vancouver MPEG Meeting, July 1999.

[N2824] MPEG Test. *Report of the Formal Verification Tests on Advanced Coding Efficiency Profile*. Doc. ISO/MPEG N2824, Vancouver MPEG Meeting, July 1999.

[N2825] MPEG Test. *Report of the Formal Verification Tests on MPEG-4 Advanced Real Time Simple Profile*. Doc. ISO/MPEG N2825, Vancouver MPEG Meeting, July 1999.

[N2826] MPEG Test. *Report of the Formal Verification Tests on MPEG-4 Coding Efficiency for Low and Medium Bitrates*. Doc. ISO/MPEG N2826, Vancouver MPEG Meeting, July 1999.

[N4319] MPEG Requirements. *MPEG-4 Requirements*. Doc. ISO/MPEG N4319, Sydney MPEG Meeting, July 2001.

[N4456] MPEG Test/Video. *MPEG-4 Visual Fine Granularity Scalability Tools Verification Test Plan*. Doc. ISO/MPEG N4456, Pattaya MPEG Meeting, December 2001.

[N4457] MPEG Test/Video. *Results of the MPEG-4 Visual Studio Profile Verification Test*. Doc. ISO/MPEG N4457, Pattaya MPEG Meeting, December 2001.

[N998] MPEG AOE. *Proposal Package Description (PPD), Revision 3*. Doc. ISO/MPEG N998, Tokyo MPEG Meeting, July 1995.

[P910] Rec. ITU-T P.910. *Subjective Video Quality Assessment Methods for Multimedia Applications*. September 1999.

[PeAl97] Pereira, F., and T. Alpert. "MPEG-4 Video Subjective Tests Procedures and Results." *IEEE Transactions on Circuits and Systems for Video Technology,* 7(1): 32–51. February 1997.

[SeGP92] Seferedis, V., M. Ghanbari, and D. Pearson. "The Forgiveness Effect in the Subjective Assessment of Packet Video." *Electronics Letters*, 28(21): 2031–2014. 1992.

Audio Testing for Validation

by Laura Contin

Keywords: perceived audio quality, benchmarking, subjective assessment methods, verification tests

*T*his chapter addresses the subjective listening tests that were performed during the final self-assessment phase for the MPEG-4 Audio standard. Previously (by the end of 1995), other listening tests were performed for evaluating the submissions answering the call for proposals issued at the beginning of the MPEG-4 standard development phase. These tests are described in [N999, N1144, CEMS97].

As for video, audio verification tests were planned with great care to avoid any type of bias due to incorrect design or inadequate statistical analysis. A group of experts in formal subjective testing was charged with the design and performance of the tests. In general, the assessment methods were selected among those recommended by ITU, but in one case no suitable method was available because of the new functionality under evaluation. In that case, the test experts proposed a new method based on psychometric principles [Guil54]. The same experts also were asked to analyze the results and draft the conclusions. To increase the credibility of the conclusions, the tests have been conducted, whenever possible, in laboratories that were not directly involved in the development of the tools under test. When this was not possible, the tests were duplicated for cross-checking the results.

The MPEG-4 Audio standard [MPEG4-3] includes a number of tools for compressing *natural audio* signals. These tools have been optimized for the type of signals to be encoded (e.g., music or speech), range of bit rate, and supported functionalities (e.g., low delay, scalability, speed variation). This makes the MPEG-4 Audio standard suitable for a wide range of application areas.

The MPEG-4 audio verification tests were designed to evaluate the performance of the standard in the most promising application areas, including *speech communication, digital radio broadcasting,* and *audio on the Internet.* *These test cases covered different ranges of bit rates and included most of the supported functionalities. To this end, representative test conditions were* selected and suitable test methods were applied.

The tests to verify the performance of MPEG-4 audio tools were run at an early stage, so that the test results were available before MPEG-4 Audio reached the status of final draft international standard (FDIS). Because MPEG only standardizes the decoding process, most of the encoders used to produce the MPEG-4 audio test bitstreams were not yet fully optimized. Thus, the results of the verification tests provide an indication of the minimum performance that can be achieved by using the MPEG-4 solutions. With time, it is expected that better MPEG-4 audio encoders will appear in the market and higher performance will be reached.

This chapter illustrates the rationale behind the test design, the applied assessment methods, the test conditions, and the audio test results. Section 16.1 deals with general aspects related to the test methodology. Section 16.2 illustrates the assessment methods used for the tests. Sections 16.3 through 16.7 are devoted to a particular audio verification test. In these sections, details about coding tools and corresponding settings, test material, experimental design, and test results are provided. Finally, a summary concludes the chapter.

16.1 GENERAL ASPECTS

This section addresses some general aspects related to any listening test, including the criteria and procedure to select the test material and the test subjects, the definition of the laboratory setup, the specification of the test plan, and the procedures to train the subjects before the test.

16.1.1 Selection of Test Material

MPEG has gathered a large library of audio excerpts to be used for developing and testing its standards [M3598]. During the preparation of the verification tests, whenever the experts recognized a lack of suitable material, either in terms of the type of content or in terms of the level of criticality, they issued a call for new test material [N2168], specifying the desired characteristics. In this way, a considerable number of excerpts were collected after each call. To identify the items to be used in the tests, a two-step selection was performed. First, the audio experts listened to the source material to eliminate the content that was not relevant for the test or that was *a priori* considered either too critical or too simple regarding its encoding. Then, the excerpts were encoded with the codecs under test, and the experts made a blind selection of the material based on some predefined criteria, including *criticality, type of content, some characteristics of the signals*, and *type of use.*

Concerning *criticality*, a distinction between critical and typical signals often was made. Critical signals are somehow *difficult* signals, showing more annoying artifacts than other signals when processed by the encoders in a given test. Typical signals are signals that show the average perceivable distortion when processed by the encoders in a given test.

The *type of content* often was classified as speech, single instrument, pop, classic, and complex sound. This last was either a combination of two of the previous types of content or a combination of speech with noise (e.g., babble noise and car noise).

The relevant *characteristics of the signals*, such as the sampling rate and the number of channels (e.g., monophonic, stereophonic, or multichannel), were strictly related to the codec under test.

As far as it concerns the *type of use*, the excerpts were classified as training items or test items: The former were used for explaining and demonstrating the testing procedures and the range of possible artifacts, and the latter were used for actually testing the codecs.

Finally, the selection panel also could recommend some preprocessing for part of the excerpts, such as loudness adjustment, if necessary.

16.1.2 Selection of Test Subjects

The characteristics of the sample of listeners participating in a subjective test are an important point in the test design. Among the selection criteria most commonly used (such as age, gender, and profession), the most important one is the level of expertise in audio quality assessment. For example, usually nonexpert listeners participate in speech tests, because the typical telephone user is not an expert. On the contrary, expert listeners are preferred for audio quality evaluation, because they can better recognize coding impairments. This is particularly important when high-quality signals are being evaluated.

In general, the number of nonexpert listeners participating in a test should be higher than or equal to 15. When listeners are experts, a lower number may be sufficient.

A further important requirement for the listeners is their native language, which should be the same as for the speech test items. An exception to this rule was made in the MPEG-4 audio verification tests, when a test was run at different test sites located in different countries. In those cases, to cross-check the results, it was important that a common set of excerpts was played in all test sites, and it was difficult to find native speakers for the language, typically English, used for the common set of test items.

16.1.3 Laboratory Setup

ITU recommendations describing test methodologies [BS1116, P800] also specify the characteristics of the test room in terms of the minimum requirements for the volume of the room, reverberation time, background noise, and so on. These requirements obviously are more restrictive in the case of sound played through loudspeakers than in the case of sound played through headphones.

Concerning the playing system, test items can be presented either via a digital player (e.g., DAT) or by means of a PC-based presentation system, connected either to loudspeakers or to headphones. In the case of MPEG-4 audio verification tests, test items were always played through high-quality headphones, such as STAX LAMBDA NOVA. With this approach, more than one listener can perform the test at a time because there is no impact from the listener's position.

16.1.4 Test Plan

The test plan defines the overall organization of the test. Generally, this includes the presentation order of the test items; the number of repetitions, if any; and the timing of each test session. The definition of the test plan is closely related to the statistical analysis that will be performed and generally is defined by a test expert, considering a number of hypotheses and the goal of the test [Myer79]. However, some general rules, aimed at reducing bias due to presentation order and listeners' fatigue, are widely adopted. Some of them are listed below:

1. Two subsequent presentations of the same item should be avoided, even if the treatment is different.
2. For different groups of listeners, different presentation orders should be used.
3. Whenever test conditions are repeated, different presentation orders should be adopted.
4. Duration of a test session should be less than 30 minutes.

16.1.5 Training Phase

The goal of the training phase is to make the listeners familiar with the quality levels presented during the test and grading procedures. The training phase generally includes three steps:

1. **Instructions:** The test administrator gives instructions to the subjects. The instructions should be written or played from an audio recording system to guarantee that all subjects receive exactly the same explanation. The instructions should include explications about the goal of the test, the presentation of the stimuli, and the grading procedures. Unclear

points should be discussed and clarified with the test administrator before the test starts.

2. **Training listening:** The subjects listen to the training items in order to become familiar with the nature of the artifacts. They can discuss the perceived artifacts, but subjects are not allowed to talk about specific grades, in order to avoid bias in individual grading.

3. **Mock testing:** A mock test is run, using the training items; usually these are different from the items presented during the test, but they have the same kind and intensity of artifacts. In this way, the subjects become familiar with the grading procedures.

16.2 TEST METHODS

Test methods must be defined according to the goal of the test. Usually this is done by selecting one of the methods recommended by the International Telecommunication Union (ITU). In fact, both the T sector (ITU-T, formerly CCITT) and the R sector (ITU-R, formerly CCIR) have been very active in defining subjective assessment methodologies since the 1970s. ITU-T mainly addresses methodologies for speech quality evaluation both in analog and digital systems, whereas ITU-R mainly addresses methodologies for audio quality evaluation.

When ITU methods were considered not suited to the goal of the test and type of excerpts to be evaluated by MPEG, new test methods were developed based on psychometric principles [Guil54].

In the MPEG-4 audio verification tests, three test methods were used: the *absolute category rating* (ACR) method for speech and the *paired comparison* (PC) and *multistimulus test with hidden reference and anchors* (MUSHRA) methods for audio.

16.2.1 Absolute Category Rating Method

The ACR method is one of the subjective assessment methods recommended by ITU-T for evaluating speech quality [P800]. The test procedure is quite simple: Excerpts are played in sequence, with a period of silence of a few seconds in between. During this period, subjects are asked to evaluate the quality level of the last heard excerpt. To express their judgment, the subjects use a five-grade quality scale, as illustrated in Table 16.1.

Every test should include reference conditions, produced, for example, by means of the modulated noise reference unit (MNRU). This system is used for impairing speech items by applying controlled multiplicative noise, as specified in [P810], and for producing predictable quality levels. When the signal path has unity gain, the generation of output speech-plus-modulated-noise can be expressed by:

$$y(i) = x(i)[1 + 10^{-Q/20}N(i)]$$

Table 16.1 ACR quality scale

Score	Category
5	Excellent
4	Good
3	Fair
2	Poor
1	Bad

where $y(i)$ is the output speech-plus-modulated-noise, $x(i)$ is the input speech, Q is the ratio of speech power to modulated noise power, and $N(i)$ is the random noise. The specification of both wideband and narrowband MNRU is given in [P810].

To analyze the test results, a statistical analysis is performed to obtain the average and standard deviation of scores for each test condition. Further analysis, such as the analysis of variance, can be used to investigate aspects such as the reliability of subjects and dependency of test laboratory.

16.2.2 Paired Comparison Method

ITU-R has issued two recommendations specifying audio subjective evaluation methods: BS.1116 [BS1116] and BS.1284 [BS1284]. BS.1116 describes the *double-blind triple-stimulus with hidden reference* method, which is suited for evaluating signals affected by barely perceptible impairments. For example, this method was used by MPEG to evaluate advanced audio coding (AAC) at high bit rates [N1419, N2006]. BS.1284 includes a few test methods suited for signals characterized by lower quality levels. Among those methods, MPEG chose the paired comparison method for the following MPEG-4 audio verification tests: narrowband audio broadcasting, audio on the Internet, and Version 2 coding efficiency evaluation.

In the paired comparison method, the test includes a number of trials. Each trial consists of the presentation of a pair of stimuli: The first stimulus is called *reference*, the second is called *test stimulus*. In the context of this test method, the *reference* is the uncompressed, often band-limited signal: Typically, it is the input to the system under test. The *test stimulus* usually is the reference after undergoing some type of processing, for example, coding with the technology under testing. Usually, at least one of the test stimuli is produced by using a well-known codec. These codecs are sometimes called *anchors*, because they are introduced to make an indirect comparison with the systems under test.

Each pair of stimuli can be presented either once or twice in a trial, depending on the experimental design. When the pair of stimuli is presented twice in a trial, a short interval of about 2 seconds is inserted between the two presentations. It is preferable to present the pair of stimuli twice when their differences are small, or, in other words, when the quality level of the test condition is quite high and close to the quality level of the reference.

Listeners are asked to evaluate the quality level of the test condition, with respect to its reference, by using a scale similar to that illustrated in Table 16.1. However, in this case the scale is intended as a continuous scale, with a resolution of 1 decimal place. This means, for example, that *excellent* corresponds to the vote 5.0 (not 5, as in Table 16.1), *good* corresponds to the vote 4.0, and subjects may use all one-decimal place resolution numbers between 1.0 and 5.0.

The statistical analysis is similar to that described for the ACR method.

16.2.3 MUSHRA Method

The MUSHRA method was developed by the European Broadcasting Union (EBU) project, Group B/AIM, to measure the perceived audio quality in a range of quality lower than *high fidelity*. MUSHRA was proposed to ITU-R for standardization in 2000 [BS1534]. At the time of some of the MPEG-4 audio verification tests described below, this method was not yet standardized; but a test conducted by EBU [EBU00] had already shown that this method could provide valid and reliable results for the evaluation of intermediate audio quality levels. Thus, the MPEG test experts decided to apply this method in the test on audio error resilience [N3075]. This method can provide more accurate evaluation, because the items can be directly compared with one another and each item can be played by the listener as many times as needed.

As explained above, MUSHRA stands for multistimulus test with hidden reference and anchors. The expression *multistimulus* refers to the fact that for each trial the subjects can listen to an excerpt under a number of different conditions. These conditions are classified as test conditions, reference, and anchors. In the context of this test method, a *test condition* is generated by processing an original excerpt through one of the systems under test: These systems can be codecs, emulators of a transmission link, and so on. The *reference* is the unprocessed original excerpt with full bandwidth; this should be considered as the upper limit of the quality range presented during the trial. An *anchor* is a low-pass filtered reference;[1] anchors are included to guarantee that the whole range of quality is spanned in each trial.

During a trial (for an excerpt), the subjects are given the reference and a number of items, which in general include again the reference, two anchors,

---

1. Notice that in the context of the MUSHRA test method, an anchor is a low-pass filtered original while for most other test methods, notably the paired comparison, an anchor is a test stimulus produced with a relevant system, e.g., a codec, for comparison purposes.

and a number of test conditions. Subjects are not made aware which are the anchors and the reference in the group of items to evaluate; for this reason, the name of the method includes the expression *hidden reference and anchors*. The reference and the items available in a trial are different versions of the same original excerpt, and all of the tested versions of an excerpt must be collected in one trial. Each subject can individually decide

1. whether to listen to the reference or to an item under test (which also may be the reference, although the subject does not know);
2. which item to play;
3. how many times to play an item; and
4. when to vote.

In particular, the subject can decide whether to compare an item against either the reference or another item. This is particularly important when the quality level of a test condition is considerably different from the quality level of the reference; in general, the comparison between items provides more accurate evaluations. Subjects also are allowed to keep changing their evaluations for the various items, until they are convinced about all of them. The evaluations are given through continuous quality scales like those shown in Figure 16.1.

The subjects might individually control the stimuli presentation order and vote insertion, for example, by means of a computer-based playing system through a user interface, like the one illustrated in Figure 16.1.

The MUSHRA test is divided into three phases:

1. **Instructions:** During the instructing phase, the experimenter explains the purpose of the test, the organization of the test, and the procedures for voting.
2. **Training:** During the training phase, the subjects listen to all items to become familiar with the range of quality and the type of impairments they will listen to during the trials. The training also includes a mock

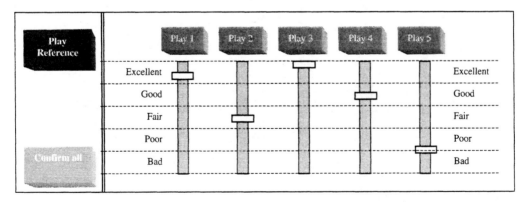

Fig. 16.1 Example of user interface to be used for individual test controlling.

test to familiarize the subjects with the user interface for voting (see Figure 16.1) and to check that they correctly understood the instructions.

3. **Trials:** Finally, the trials are the actual evaluation phase. For each trial, the subjects listen to and evaluate a number of items, which correspond to all versions of one excerpt. Each item in a trial might be the original, unprocessed excerpt (reference) or a processed version of the original, through either one of the systems under test (test item) or a low-pass filter (anchor).

Data analysis is performed according to the following procedure:

1. First, the evaluations expressed in a graphic form with the continuous quality scale are linearly converted into a number between 0 and 100, corresponding to the bottom and top end of the scale, respectively.

2. Next, the difference between the vote associated to the hidden reference and the vote associated to each item is computed.

3. Finally, averages across subjects and corresponding 95% confidence intervals (CIs) are derived for each item. Additional analysis—such as averages across excerpts, averages across system parameters, factorial analysis of variance (ANOVA), and so on—may be calculated, depending on the goal of the test and according to the experimental design.

16.3 NARROWBAND DIGITAL AUDIO BROADCASTING TEST

The test described in this section was aimed at evaluating the suitability of MPEG-4 audio technology for an amplitude modulation (AM) digital audio broadcasting application, as described in [N2724]. Although so-called wideband codecs were included in this test, the test was called *narrowband digital audio broadcasting* (NADIB) [N2276], because AM uses 10- to 20-kHz bandwidth on the radio channel, whereas other digital audio broadcast schemes use more (e.g., IBOC-FM uses 200 to 300 kHz, and Eureka-DAB 1.5 MHz).

This test was conducted in collaboration with the NADIB consortium, a Eureka project (EU 1559) with the mandate of defining a worldwide usable standard for improving the existing analog service in the AM modulated bands, such as HF, MF, and LF.

With the intent of identifying a suitable digital audio broadcast format to provide improvements over the existing AM services, several coding configurations involving the MPEG-4 Audio code excited linear prediction (CELP), TwinVQ, and AAC tools (see Chapters 10 and 11), covering a range of bit rates from 6 to 24 kbit/s, were compared to a reference AM system.

The benchmarking was focused mainly on scalability. In fact, the evaluation of the performance differences between the scalable and nonscalable codecs was of great interest, as a scalable broadcasting system can offer, under good reception conditions, a good quality at full bit rate (i.e., base + enhancement layers) whereas, under bad error conditions, it can still provide meaningful output at lower bit rate (i.e., only base layer).

The comparison between scalability and simulcast also was investigated in terms of quality. In the simulcast mode, the number of stand-alone bitstreams that are broadcast in parallel is equal to the number of supported bit rates. This solution makes the decoding process less complex because only one decoder at a time has to be used to get the desired quality. On the other hand, in the simulcast mode, higher bit-rate streams do not take advantage from the information transmitted in the lower bit-rate streams, as happens in the scalable mode with the enhancement and base layers. Thus, the coding process is expected to be globally less efficient in the simulcast mode.

16.3.1 Test Conditions

In the NADIB test, the following topics were investigated: comparison of different base layer codecs, comparison between scalable and nonscalable codecs at the same bit rate, and comparison between scalable and simulcast solutions at the bit rate of the enhancement.

A realistic condition for bit rates suitable for narrowband digital audio broadcasting is 6 kbit/s for the base layer stream and a total bit rate of 24 kbit/s for the full bitstream. This means the bit rate of the enhancement layer is 18 kbit/s.

Because music, speech, and their combinations can be broadcast, both speech and general audio codecs were tested.

These considerations led to the selection of the following codecs for testing, grouped in terms of investigation topic:

☞ **Comparison of different base layer codecs:** Only narrowband codecs (i.e., codecs with an audio bandwidth of approximately 3.5 kHz) were tested as base layer codecs—namely, *MPEG-4 narrowband CELP* (NB-CELP) as a speech coder and *TwinVQ* as a general audio coder. The anchor condition was produced with the codec specified in Recommendation ITU-T G.723.1 [G723]. All the codecs were operating at approximately 6 kbit/s, and they were tested in the *narrowband test*.

☞ **Comparison between scalable and nonscalable codecs:** To keep the size of the test within reasonable dimensions, only the most interesting conditions were selected; in particular, only wideband codecs were evaluated. The one-layer (i.e., nonscalable) AAC codec was compared against two AAC scalable versions: one using TwinVQ as the base layer (AAC + T-VQ), and another using the narrowband CELP as the base layer (AAC + NB-CELP). Both scalable codecs were operating at the same total bit rate of 24 kbit/s, with the base layer operating at 6 kbit/s.

☞ **Comparison of scalability against simulcast:** Also for this case, only wideband codecs were tested. The scalable codecs at 24 kbit/s were compared to the AAC one-layer codec at 18 kbit/s. These codecs were tested together with the codecs indicated above for comparison between scalable and nonscalable codecs. The conditions addressed in this and the previous topic were evaluated in the *wideband test*.

All coders operated at fixed bit rate, in monophonic mode. A maximum short-time buffer of 6,144 bytes was allowed, corresponding to the maximum bit reservoir of AAC.

The anchors for the wideband conditions were MPEG-2 Layer 3 and Perfect AM. Perfect AM was simulated using a band-pass filter with the following characteristics: –50 dB at 24 Hz, –3 dB at 73 Hz, –3 dB at 2400 Hz, and –50 dB at 5300 Hz. This characteristic was based on a measurement of consumer AM receivers [N2276]. No further distortions, which might be caused by the AM modulation scheme, were taken into account in this simulation.

The codec settings used for the narrowband and wideband conditions are listed in Table 16.2 and Table 16.3, respectively.

16.3.2 Test Material

Considering that music, speech, and their combinations can be broadcast, all of these classes of items were used in this test.

Half of the selected excerpts were speech items and the other half were music or complex items. The set of speech excerpts included items either in English or in the native language of each of the test sites (i.e., French and Swedish). For each language, both male and female voices were included.

Overall, 12 monophonic items were selected, including 3 music items, 3 complex items, 2 English speech items, 2 French speech items, and 2 Swedish speech items.

At each test site, respectively located in France and Sweden, the speech items that were neither English nor the native language were discarded. In this way, at each test site 10 items were used: 8 of them were common to both test sites, and 2 were specific for the language of the country in which the test was run.

Table 16.2 Test conditions for the narrowband test

Codec	Sampling rate (kHz)	Total bit rate (kbit/s)	Estimated bandwidth (kHz)	Frame length (ms)	Delay (ms)
NB-CELP mode VIII[*]	8	6	3.5	20	25
TwinVQ	24	6	3.5	43	107
G.723.1[†]	8	6.3	3.5	30	37.5

[*]Using vector quantization and MPE excitation module (see Section 16.5.1).
[†]Anchor codec.

Table 16.3 Test conditions for the wideband test

Codec	Sampling rate (kHz)	Total bit rate (kbit/s)	Estimated bandwidth (kHz)
Wideband CELP (WB-CELP) mode III[*†]	16	18.2	7.5
AAC 18 Low complexity[*]	16	18	6.5
AAC 24 Low complexity[‡]	24	24	7.5
Scalable AAC (AAC + NB-CELP)	24	24[§]	7
Scalable AAC (AAC + TVQ)	24	24[§]	6
MPEG-2 Layer 3 (MP2/L3)[**]	24	24	6
Perfect AM	N.A.	N.A.	−3 dB at 2.4 −50 dB at 5.3

[*]For scalability versus simulcast comparison.
[†]Using scalar quantization and RPE excitation module (see Section 16.5.1).
[‡]For scalability versus nonscalability comparison.
[§]6 kbit/s for the base layer + 18 kbit/s for the enhancement layer.
[**]Anchor codec.

The same set of excerpts was used for both the wideband and narrowband test conditions, and all items presented some audible impairment after encoding.

A set of training items was also selected, including speech, music, and complex items. Additional details about this test material can be found in [N2276].

16.3.3 Assessment Method and Test Design

The test was divided into two separate subtests: the narrowband and the wideband tests described above.

For each of the subtests, the paired comparison evaluation method was used. Each pair *reference–test item* was repeated in each presentation. The 8-kHz and 24-kHz sampling rate original items were used, respectively, as references for the narrowband and wideband tests. Two test sites ran both the narrowband and the wideband tests: TERACOM in Sweden and CCETT in France. The former applied the set of items including the 8 common excerpts and 2 Swedish items, whereas the latter used the 8 common excerpts and 2 French items.

To prevent bias due to listener fatigue, each of the two subtests was split into sessions of approximately 20 to 25 minutes. This resulted in two sessions for the narrowband test and five sessions for the wideband test, meaning seven sessions per listener.

Both test sites applied four different pseudorandomizations per test to evenly distribute the codec solutions and items across sessions. Three trials were repeated within each session, to check the scoring consistency of the subjects.

16.3.4 Data Analysis

The main goal of each test was to rank the performance of the codecs under test. Additional investigations were performed, such as the evaluation of the reliability of the subjects and the dependency on the test site. The Massachusetts Institute of Technology (MIT) performed the statistical analysis.

The data analysis was organized in the following steps:

1. **Data collection and sanity check:** The scores coming from the two test sites were collected in one data sheet and cross-checked to detect editing mistakes.

2. **Check of listener reliability:** The listener scoring consistency was checked on those trials that were repeated within each session. Let SCORE1 and SCORE2 be the two scores associated by a listener to the repeated presentation of the same stimulus; for each subject, the mean and 95% CI of the score difference (SCORE1 − SCORE2) was calculated. If, for any subject, the CI extended beyond [−1,1], that subject was eliminated. Based on this criterion, scores of 4 subjects out of 58 were discarded.

3. **Check of test site dependency:** An ANOVA was computed, including the following factors: test site, codec, and item. As expected, the results showed a significant effect of test site. The differences between the results from the different laboratories could be due to many factors, notably cultural differences, translation of the scaling labels, translation of the instructions, and differences between listeners' samples. Typically, the differences do not affect the ranking of the codecs, but they introduce a bias that is reflected in a general shift of the results of one laboratory with respect to the results of another. The ANOVA also showed that there was a significant interaction between test site and codec. This also is an expected result, for similar reasons. However, there was no significant interaction between the test site and item, and no three-way interaction. Thus, the analysis was conducted separately by test site for the codec-by-codec evaluation; for the item-by-item evaluation, data coming from different laboratories were pooled.

4. **Calculation of mean opinion scores and CI:** The statistics were calculated over subjects, for each codec and item. Scores also were averaged over items. Based on the ANOVA results, scores from different laboratories were grouped only in the case of item-by-item analysis.

16.3.5 Test Results

Means and 95% CI for each of the codecs at each of the test sites were computed, to evaluate their overall performance. These results are shown in Figure 16.2 and Figure 16.3.

Additionally, *Student* tables were computed for both sites and both tests. These tables summarize the results of the *Student* analysis that was used to verify whether the difference between mean opinion scores was statistically significant. The *Student* tables provide a simple representation of the ranking of the tested codecs.

Further analyses—for example, to evaluate the item-by-item codec performance—were also conducted. All details can be found in [N2276].

Before commenting on the results shown in Figure 16.2 and Figure 16.3, it should be noted that this was the first MPEG-4 Audio verification test, performed when MPEG-4 Audio was still in final committee draft (FCD) stage; hence, most of the tested encoders have been further optimized since this test and their performance has improved.

The results from the two test sites provided the same ranking for the codecs under test, although it was not always the case that differences which were statistically significant at one test site were statistically significant at the other, and vice versa.

The results for the narrowband test showed that G.723.1 and NB-CELP were superior to TwinVQ at both test sites. G.723.1 performed slightly better

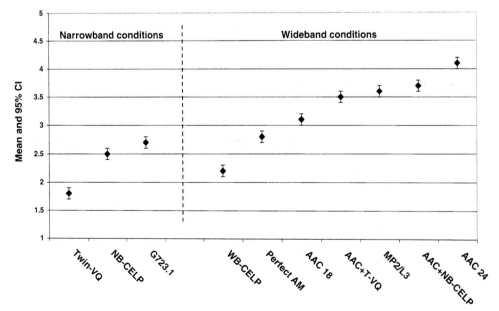

Fig. 16.2 Codec-by-codec results at CCETT.

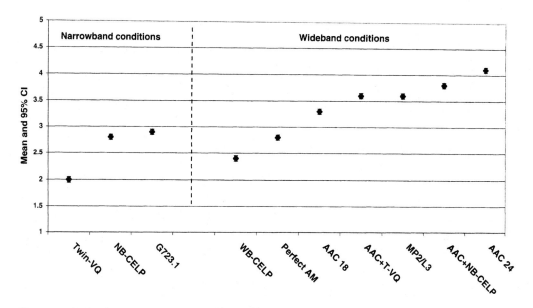

Fig. 16.3 Codec-by-codec results at TERACOM.

than NB-CELP, although they resulted statistically equivalent at CCETT and statistically different only at TERACOM.

Concerning the wideband test, the nonscalable AAC at 24 kbit/s (AAC 24) was the best codec. At TERACOM, the scalable AAC + NB-CELP codec was second in position, performing better than the scalable AAC + T-VQ codec. AAC + T-VQ was statistically equivalent to MPEG-2 Layer 3 (MP2/L3). Then, MPEG-2 Layer 3 was followed, in order of performance, by the nonscalable AAC at 18 kbit/s (AAC 18), Perfect AM, and WB-CELP, with statistically significant differences at each step.

At CCETT, the scalable AAC + NB-CELP codec and MPEG-2 Layer 3 were statistically equivalent, but worse than AAC 24. MPEG-2 Layer 3 and AAC + T-VQ were statistically equivalent. Then, AAC + T-VQ was followed, in order of performance, by AAC 18, Perfect AM, and WB-CELP, with statistically significant differences at each step.

Summarizing, the test results led to the following conclusions:

☞ Nonscalable encoding provided the best performance for a given net bit rate.

☞ If two levels of quality of service are targeted, one for normal reception conditions and another for bad reception conditions, the results proved that a layered scalable codec offers better quality than simulcast.

☞ For the scalable mode and when programs contained music, TwinVQ performed better than NB-CELP, as expected. However AAC + NB-CELP performed somewhat better than AAC + T-VQ.

☞ For the scalable mode, AAC + NB-CELP gave the best results when programs contained speech. More generally, this appeared to be the best compromise for generic audio broadcasting among all the coders in this test.

☞ For the material used in this test, the tested TwinVQ at 6 kbit/s performed worse than both G.723.1 and NB-CELP, with the exception of some music items. A possible reason for that was the used sampling rate: In fact, the tested TwinVQ directly quantized input signals sampled at 24 kHz. This was done for simplifying the connection to the AAC scalable system. It can be assumed that TwinVQ can achieve higher quality at the same bit rate if a lower sampling rate is used in the nonscalable mode.

☞ For the speech items, the results of WB-CELP and Perfect AM were statistically equivalent,[2] whereas for music items Perfect AM performed better. The NB-CELP test results were statistically equivalent to those of ITU-T G.723.1 at CCETT, and slightly worse at TERACOM. It must be noted that the NB-CELP coder operated at a lower bit rate and with a shorter delay.

The overall conclusion was that digital audio in AM bands using 24 kbit/s audio coding demonstrated the potential to offer better quality compared to existing analog modulation techniques.

16.4 AUDIO ON THE INTERNET TEST

Considering the great success of the MP3 format (i.e., MPEG-1/2 Layer 3), and also that MPEG-4 provides new functionalities and includes new coding tools aimed at increasing audio coding efficiency, one of the most promising applications for the MPEG-4 Audio standard is music online.

The test described in this section was aimed at evaluating the performance of the MPEG-4 Audio tools in this context. For this reason, this verification test was called *audio on the Internet* [N2425].

The following subsections provide information about what was evaluated, how the testing procedures were applied, and what were the most important results and conclusions of the test.

16.4.1 Test Conditions

The starting point in defining the test conditions for the audio on the Internet test was the speed of connection to Internet, through either modem or ISDN. Thus, the bit rates for audio coding were divided into three groups, corresponding to an overall bit rate of 28.8, 33.6, and 64 kbit/s. These bit rates are considerably lower than those considered in previous MPEG audio tests [N1419, N2006], in which test signals were perceptually very close to the corresponding source signals. For this reason, the paired comparison test method

---

2. This was not computed but concluded from the figures.

was used instead of the double-blind triple-stimulus with hidden reference method, which was used in previous MPEG-2 audio tests.

For each bit-rate group, one or more audio bit rates and suitable codecs were selected from those included in the MPEG-4 Audio standard [MPEG4-3]. For example, for the 28.8 kbit/s group, two bit rates were selected (6 and 16 kbit/s), and for each bit rate two codecs were chosen: HILN and TwinVQ for 6 kbit/s, and HILN and AAC for 16 kbit/s.

At medium and high bit rates, the attention was focused on scalability; thus, scalable AAC was tested. The AAC base layer operated on monophonic signals at 24 kbit/s, whereas the availability of the first or both 16 kbit/s enhancement layers enabled the decoder to provide a stereophonic signal. The resulting bit rates for this codec were 24 kbit/s (monophonic) for the base layer, 40 kbit/s (stereophonic) in the case of the base layer + one enhancement layer, and 56 kbit/s (stereophonic) in the case of the base layer + two enhancement layers. At the high bit rates, besides the scalable AAC, the AAC coder with a small-step-scalable bit-sliced arithmetic coding (BSAC) noiseless coder was evaluated. Different from the scalable AAC, this algorithm is not based on a mono/stereo scalable system; therefore, the results for BSAC could not be compared with the results of the scalable AAC.

In addition to the test conditions, one anchor condition was included in each test, in order to compare the performance of the new standard with existing standards (e.g., MP3, G.722) or to compare the performance of MPEG-4 codecs providing new functionalities (e.g., scalability) against the base MPEG-4 audio codecs.

For each test condition, the optimal sampling rate and mode (i.e., monophonic/stereophonic) were chosen. Test conditions are summarized in Table 16.4 through Table 16.6.

16.4.2 Test Material

For this test, new audio excerpts were called for first [N2168], and from the 90 collected excerpts, audio experts selected those to be used for the test and for the training.

Table 16.4 Test conditions for the 28.8-kbit/s bit-rate group

Very Low Bit Rate (VLBR)				Low Bit Rate (LBR)			
Codec	Mode	Sampling rate (kHz)	Bit rate (kbit/s)	Codec	Mode	Sampling rate (kHz)	Bit rate (kbit/s)
HILN	Mono	8	6	HILN	Mono	16	16
TwinVQ	Mono	16	6	AAC	Mono	16	16
MP3[*]	Mono	8	8	G.722[*]	Mono	16	48

[*]Anchor condition.

Table 16.5 Test conditions for the 33.6-kbit/s bit-rate group

Medium Bit Rate (MBR)			
Codec	Mode	Sampling rate (kHz)	Bit rate (kbit/s)
AAC	Mono	24	24
AAC/sc	Mono	24	24
MP3*	Mono	16	24

*Anchor condition.

Table 16.6 Test conditions for the 64-kbit/s bit-rate group

High Bit Rate (HBR)			
Codec	Mode	Sampling rate (kHz)	Bit rate (kbit/s)
AAC/sc 40	Stereo	24	40
BSAC 40	Stereo	24	40
AAC 40*	Stereo	24	40
MP3 40*	Stereo	24	40
AAC/sc 56	Stereo	24	56
BSAC 56	Stereo	24	56
AAC 56*	Stereo	24	56
MP3 56*	Stereo	24	56

*Anchor condition.

Based on the content type, the excerpts were divided into five categories: speech, single instrument, pop, classic, and complex sound. For each content type and for each class of bit rate, the experts selected one training excerpt and two test excerpts, using criticality as the selection criterion. The training items allowed them to illustrate the full range of quality variation and all of the typical artifacts. For each content category, the test items included one *critical excerpt* and one *typical excerpt* [N2425].

16.4.3 Assessment Method and Test Design

Test conditions were divided into four groups (according to the bit rates) corresponding to four separate tests. In the following, the four tests are identified as very low bit-rate (VLBR), low bit-rate (LBR), medium bit-rate (MBR), and high bit-rate (HBR) tests. (This classification is also shown in Table 16.4 through Table 16.6.)

For each of these tests, the paired comparison method was applied. The reference condition was the uncompressed signal, down-sampled according to the coding sampling rate.

For the VLBR and LBR tests, the pairs *reference–test condition* were played only once in each presentation. For the other tests, the pairs *reference–test condition* were repeated within each presentation. The repetition was introduced because, at bit rates higher than or equal to 24 kbit/s, the detection and evaluation of some artifacts is not as straightforward as at lower bit rates.

For the VLBR and LBR tests, the total duration of the test was less than 21 minutes, whereas for the MBR and HBR tests, the total duration of the test was longer than 30 minutes. To prevent bias due to fatigue effects, the MBR test was organized into two test sessions, lasting approximately 18 minutes each, and the HBR test was organized into four sessions, lasting about 26 minutes each. Within each test, the presentation order was pseudorandomized.

Four laboratories participated in the tests: Sony (Japan), Mitsubishi Electric, America (U.S.), NTT (Japan), and Samsung AIT (Korea).

16.4.4 Data Analysis

The main goal of each test was to rank the performance of the codecs under test, but additional investigations were performed, such as the evaluation of the reliability of the subjects and the dependency on the test site. The statistical analysis was performed by MIT for all the laboratories.

The data analysis was organized in the following steps:

1. **Data collection and sanity check:** The scores coming from the four test sites were collected onto one data sheet and cross-checked to detect editing mistakes.

2. **Check of listener reliability:** Two rejection criteria were used. One criterion was simply the capacity of distinguishing between the reference and the coded items. To this aim, *t-tests* were performed for each listener, over the listener's aggregate responses to all coded items, to check that the scores were significantly different from the maximum of the scale, corresponding to the score to be given to the reference. No listeners were rejected on this criterion. The other criterion was the capacity of consistently distinguishing between codecs. In this case, for each subject a one-way ANOVA was performed for each of the tests in which he or she participated. Because of this second criterion, four subjects were rejected.

3. **Check of test-site dependency:** An overall comparison among results coming from different test sites was conducted to determine whether the scores from different test sites could be merged in a same data set. As explained in Section 16.3.4, usually a bias is observed in the results from different laboratories. Surprisingly, in the case of the MBR and HBR tests, the laboratory was not a significant source of variation; therefore, the results from different laboratories could be merged into two data sets, cor-

responding to the two classes of bit rates. By contrast, analysis for the VLBR and LBR test data was made for each test laboratory separately.

4. **Calculation of mean opinion scores and CIs:** The statistics were calculated over subjects, for each test condition and each test item. Scores from different laboratories were grouped, when appropriate. Scores also were averaged over items.

16.4.5 Test Results

The detailed results for this test are available in [N2425]. In this section, only the mean score and the 95% CI for each codec are reported.

Figure 16.4 illustrates the results for the VLBR and LBR tests. As mentioned above, data coming from different test sites were analyzed separately. Comparing the results from different laboratories, it was observed that the ranking did not change, although most of the mean opinion scores of one laboratory were shifted with respect to the mean opinion scores of another. As this does not affect the conclusions of the analysis, Figure 16.4 shows the results of only one laboratory.

The results show that the difference between TwinVQ at 6 kbit/s and MP3 at 8 kbit/s is not statistically significant. On the contrary, HILN at 6 kbit/s showed a significantly worse average quality than both TwinVQ and MP3. Further investigations [M4087] showed that the HILN quality was highly dependent on the test material, and it was better than the quality of TwinVQ

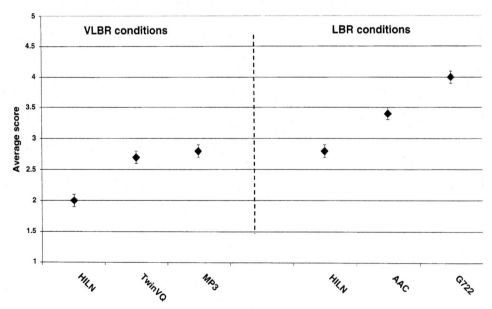

Fig. 16.4 Mean and 95% CI for the VLBR and LBR tests.

for some items. The LBR test results showed that AAC at 16 kbit/s performed slightly worse than G.722 at 48 kbit/s. This was interpreted as a good result, considering that AAC was operating at one-third of the bit rate used for G.722. HILN at 16 kbit/s performed equal to or worse than AAC at 16 kbit/s for almost all items. The test results showed that for this case as well, the quality of HILN was highly dependent on the test material. The conclusion, therefore, was that more work on HILN was required to improve the coding quality for critical material. To allow this, HILN was moved to Version 2 of MPEG-4 Audio and tested again in a second verification test (see Section 16.6).

Figure 16.5 shows results for the MBR and HBR tests. As highlighted in Section 16.4.4, for these two tests the scores coming from different test sites were pooled.

AAC always performed better than MP3 at the same bit rate. Concerning the comparison between the large-step-scalable system (AAC/sc) and the non-scalable AAC, at the base layer (i.e., 24 kbit/s), AAC/sc provided almost the same quality as AAC; at higher layers (i.e., 24 + 16 kbit/s and 24 + 32 kbit/s), it showed about 0.4–0.5 grades worse quality. Still, all layers performed slightly better (highest layer) or significantly better (lower and middle layer) than MP3 for the same bit rate. Therefore the scalable system showed good performance compared to older standards, while providing the additional functionality of mono/stereo scalable coding.

The small-step-scalable system (BSAC) performed well at the higher tested bit rate. At 40 kbit/s, however, BSAC performed worse than expected.

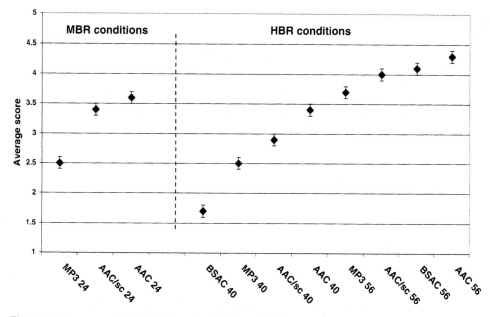

Fig. 16.5 Means and 95% CIs for the MBR and HBR tests.

Although this algorithm was mainly designed for bit rates in the range of 40 to 64 kbit/s, mono, at 48-kHz sampling rate, the BSAC tool should provide reasonably good performance when going from 56-kbit/s stereo to 40-kbit/s stereo, at 24 kHz sampling rate. The conclusion, therefore, was that the integration of BSAC in the MPEG-4 Audio framework needed further investigation to check whether the integration was incomplete or needed changes. To permit the improvement of BSAC, it was moved to Version 2 of MPEG-4 Audio and tested again in a second verification test (see Section 16.6).

16.5 SPEECH COMMUNICATION TEST

An important application area covered by the MPEG-4 Audio standard is speech communication. The algorithms standardized for speech coding were designed for transmitting voice over different networks, including PSTN, IP, and wireless.

Other standardization bodies (such as ITU-T and the European Telecommunication Standards Institute [ETSI]) have standardized speech coding algorithms for similar applications [G722, G723, G729, ETSI97].

In the test described in this section, the performance of the MPEG-4 algorithms for speech communications is compared with the performance of other standards addressing the same type of applications [N2424].

16.5.1 Test Conditions

MPEG-4 speech coding can operate for a bit-rate range of 2 to 24 kbit/s, by means of two basic coding schemes: the harmonic vector excitation coding (HVXC) and CELP algorithms. HVXC is a parametric coding scheme, which provides fixed bit-rate coding at 2 and 4 kbit/s. In addition, it can operate in a variable bit-rate mode with lower average bit rates, down to 1.2 kbit/s. HVXC supports pitch and speed change functionalities. CELP can support both narrowband and wideband speech, corresponding to signal bandwidths of 3.4 kHz and 7 kHz, respectively. More details about these codecs are given in Chapter 10.

In the MPEG-4 speech communication test, the performances of both the parametric and the CELP coding solutions were evaluated. Parametric coding was evaluated for the supported fixed bit rates (i.e., 2 and 4 kbit/s). The reference condition for this case was a U.S. federal standard called FS1016 [CaWT91]. In addition, four MNRUs (see Section 16.2.1) were introduced as anchors. Test conditions are summarized in Table 16.7.

CELP coding was evaluated for both narrowband and wideband conditions. The differences between the various CELP algorithms tested were mainly related to the type of quantization, the type of excitation module that drives the LPC filter, and the supported functionality (see Chapter 10). Concerning quantization, both scalar quantization (SQ) and vector quantization

Table 16.7 Test conditions for the narrowband parametric (NB-P) test

Codec	Sampling rate (kHz)	Bit rate (kbit/s)	Frame length (ms)	Delay (ms)
Parametric 2	8	2	20	36
Parametric 4	8	4	20	36
FS1016*	8	4.8	30	37.5
MNRU10*	8	N.A.	N.A.	N.A.
MNRU20*	8	N.A.	N.A.	N.A.
MNRU30*	8	N.A.	N.A.	N.A.
MNRU40*	8	N.A.	N.A.	N.A.

*Anchor condition.

(VQ) were considered. Then, regarding the excitation module, both regular pulse excitation (RPE) and multipulse excitation (MPE) were tested. The combination of SQ and RPE was identified as *CELP Mode III*, whereas the combination of VQ and MPE was identified as *CELP Mode VIII*.

Furthermore, bandwidth scalability was tested by using 6 kbit/s for the narrowband signal (8 kHz sampling rate) plus 10 kbit/s for the bandwidth extension, resulting in a wideband signal (16 kHz sampling rate).

For NB-CELP, two coding modes were considered: mode VIII multirate and mode VIII scalable. Three different bit rates were considered: 6, 8, and 12 kbit/s. For each bit rate, a reference condition was introduced. For the 6 and 8 kbit/s, two speech coders recommended by ITU-T for multimedia communications were adopted: G.723.1 [G723] and G.729 [G729]. For 12 kbit/s, the enhanced full rate (EFR) encoder, recommended by ETSI for wireless communications, was used as a reference [ETSI99]. Finally, four MNRUs (see Section 16.2.1) were introduced as references.

Table 16.8 summarizes the coding parameters used for producing NB-CELP test conditions.

Concerning the wideband test conditions, the performance of four different MPEG-4 CELP algorithms was evaluated. Table 16.9 summarizes the coding parameters used for producing wideband CELP test conditions.

As before, some reference conditions were introduced. At the time of designing the speech communication test, no wideband algorithms were standardized for speech coding; thus, generic coding algorithms, such as MPEG-2 Layer 3 (MP2/L3) and ITU-T G.722, were used as references. Because neither of these coding algorithms was optimized for speech signals, considerably higher bit rates were assigned to both. Also for this test, MNRU conditions (see Section 16.2.1) were included to cover a wide range of quality and to introduce references with well-known quality levels.

Table 16.8 Test conditions for the NB-CELP test

Codec	Quantiz. mode	Excitation	Sampling rate (kHz)	Bit rate (kbit/s)	Frame length (ms)	Delay (ms)
CELP mode VIII multirate (CELP-MR 6)	VQ	MPE	8	6	20	25
ITU-T G.723.1*	VQ	MPE	8	6.3	30	37.5
CELP mode VIII multirate (CELP-MR 8.3)	VQ	MPE	8	8.3	20	25
CELP mode VIII scalable (CELP-SC 8)	VQ	MPE	8	8†	20	25
ITU-T G.729*	VQ	Algebraic codebook	8	8	10	15
CELP mode VIII multirate (CELP-MR 12)	VQ	MPE	8	12	10	15
CELP mode VIII scalable (CELP-SC 12)	VQ	MPE	8	$12^{\dagger}$	20	25
GSM-EFR*	VQ	Algebraic codebook	8	12.2	20	20
MNRU10*	N.A.	N.A.	8	N.A.	N.A.	N.A.
MNRU20*	N.A.	N.A.	8	N.A.	N.A.	N.A.
MNRU30*	N.A.	N.A.	8	N.A.	N.A.	N.A.
MNRU40*	N.A.	N.A.	8	N.A.	N.A.	N.A.

*Reference condition.
†6 kbit/s for the base layer + 2 kbit/s for each of the 3 enhancement layers.

16.5.2 Test Material

It is well known that the performance of a speech encoder depends on the content of the excerpts to be coded. Thus, to obtain results that are representative of different conditions, it is important to use test items in different languages, including both male and female voices, and in some cases additional environmental noises, such as car and babble noises. In the MPEG-4 speech communication tests, a wide range of content type was used: four languages (i.e., Japanese, English, German, and Swedish), male and female speaking, speech with environmental noise (i.e., babble noise and car noise), speech with music

Table 16.9 Test conditions for the WB-CELP test

Codec	Quantiz. mode	Excitation	Sampling rate (kHz)	Bit rate (kbit/s)	Frame length (ms)	Delay (ms)
CELP Mode VIII scal. (CELP-SC)	VQ	MPE	16	16 (6+10)	20	30
Optimized VQ +MPE	VQ	MPE	16	17.9	20	25
Optimized VQ+RPE	VQ	RPE	16	18.1	15	33.75
CELP Mode III (SQ+RPE)	SQ	RPE	16	18.2	15	18.75
MPEG-2 Layer3 (MP2/L3)[*]	psycho-acoustical	N.A.	16	24	1152	210
G.722-48[*]	ADPCM	N.A.	16	48	1 sample	1.5
G.722-56[*]	ADPCM	N.A.	16	56	1 sample	1.5
MNRU10[*]	N.A.	N.A.	16	N.A.	N.A.	N.A.
MNRU20[*]	N.A.	N.A.	16	N.A.	N.A.	N.A.
MNRU30[*]	N.A.	N.A.	16	N.A.	N.A.	N.A.
MNRU40[*]	N.A.	N.A.	16	N.A.	N.A.	N.A.

[*]Reference condition.

background, and pure music. This last type of content was introduced in consideration of services such as audio on demand through communication networks.

16.5.3 Assessment Method and Test Design

Considering the range of bit rates and sampling rates adopted, three separate subtests were organized for evaluating MPEG-4 speech codecs: parametric, NB-CELP, and WB-CELP. The corresponding test conditions are illustrated in Table 16.7 through Table 16.9. The ACR evaluation method (described in Section 16.2.1) was used for each test.

Three laboratories ran the tests, two of them in Europe (Nokia and Fraunhofer Institute) and one in Japan (NTT). All test sites performed all of the tests. Considering the languages spoken in the countries hosting the speech communication tests, two sets of speech items were used: one including only Japanese speech, and another including only European languages (i.e., English, German, and Swedish). Within each language group, items were associated to each test according to their criticality: The least critical items were used in the parametric codec test, and the most critical items were used in the WB-CELP test.

16.5.4 Data Analysis

Different from previous tests, for the speech communication test each test site analyzed its own scores. The detailed analysis is provided in [N2424]. All laboratories calculated the following:

- ☞ Mean opinion scores and 95% CIs for each codec, averaged over subjects and items.
- ☞ Mean opinion scores and CIs for each codec and each item, averaged over subjects; for the sake of conciseness, only the conclusions of this analysis are reported in this chapter.

Nokia performed two additional investigations:

- ☞ Mean opinion scores and CIs for each codec and each European language.
- ☞ *Student*'s statistical analysis to check whether differences between mean opinion scores averaged over subjects and items were statistically significant.

16.5.5 Test Results

The results of all tests showed a distribution of scores for MNRUs aligned with the usual evaluations associated to these conditions. This can be considered as proof of the reliability of the test results and of the validity of the test method.

Figure 16.6 shows results for the narrowband parametric (NB-P) test. MPEG-4 HVXC at both 2.0 (Parametric 2 in the figure) and 4.0 kbit/s (Parametric 4) outperformed the reference codec FS1016 at 4.8 kbit/s. The *Student* analysis conducted by Nokia confirmed that the difference between the reference codec and the parametric codecs was statistically significant. This result was considered positive also because the HVXC coder provides more functionality, such as pitch and speed change and bit-rate scalability. Concerning the criticality of languages, based on descriptive statistics, the most critical language appeared to be German, whereas the least critical seemed to be Swedish.

Figure 16.7 shows results for the NB-CELP test. The MPEG-4 NB-CELP coder with a bit rate ranging from 6 to 12 kbit/s provided quality levels that are comparable to the quality levels provided by speech coding standards, such as G.723.1 and G.729, that are optimized for specific bit rates. Furthermore, the tested MPEG-4 CELP coders support bit-rate scalability, so that speech quality can be improved step-by-step by adding enhancement layers on top of the base layer.

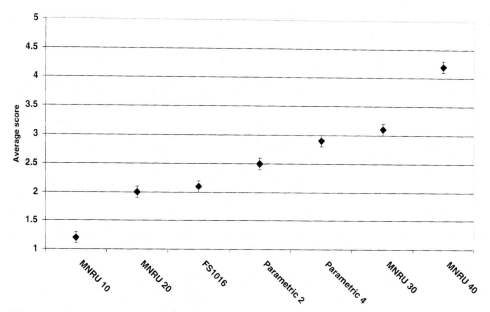

Fig. 16.6 Overall means and 95% CIs for the NB-P test.

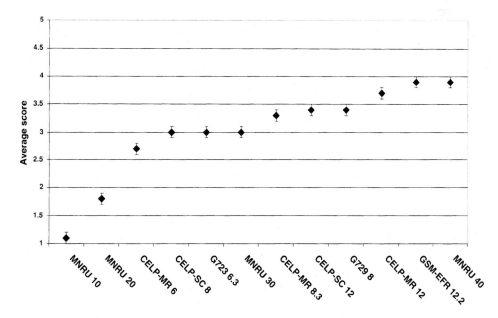

Fig. 16.7 Overall means and 95% CIs for the NB-CELP test.

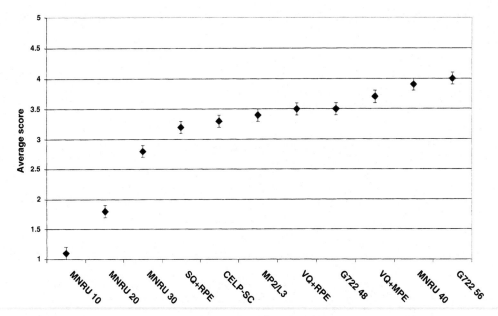

Fig. 16.8 Overall means and 95% CIs for the WB-CELP test.

Also in this case, the lowest scores generally were those associated to the German items, and the highest scores those associated to the Swedish items.

Figure 16.8 shows results for the WB-CELP test. For speech items, MPEG-4 WB-CELP coders at 18 kbit/s provided competitive quality compared with G.722 at 48 kbit/s and MPEG-2 Layer 3 at 24 kbit/s, bringing additional functionality, such as bit rate, bandwidth, and complexity scalability. Because CELP using scalar quantization performed clearly worse than CELP using vector quantization, the MPEG Audio subgroup decided not to include scalar quantization in the final standard.

Concerning the differences between languages, the results did not show a clear trend; therefore, it was not possible to derive any general conclusion.

16.6 VERSION 2 CODING EFFICIENCY TEST

MPEG-4 Audio Version 1 includes a considerable number of tools for representing audiovisual information, either in natural or in synthetic form, supporting most of the functionalities described in the initial MPEG-4 Proposal Package Description [N998] and in the MPEG-4 Requirements document [N4319] (see Chapter 1). Nevertheless, MPEG experts felt that MPEG-4 could be further improved by including new technologies and additional functionalities. This fact motivated the decision to work for MPEG-4 Audio Version 2, developing new tools with the constraint that they would be complementary and not a substitution for the tools already included in the MPEG-4 Audio Version 1. The set of new tools was nicknamed within MPEG as *MPEG-4 Audio Version 2* and it was included in the MPEG-4 Audio standard as an amendment [MP4V2-3].

The set of new tools included in MPEG-4 Audio Version 2 includes the following (see Chapters 10 and 11):

☞ **New codecs**: HILN, AAC with low delay (AAC LD), and BSAC: These audio object types were directly defined with an error-resilient bitstream syntax and thus will be referred to as ER HILN, ER AAC LD, and ER BSAC (see Chapter 13)

☞ **Error-robustness features:**

✗ Error-resilient bitstream syntax for all Version 1 audio object types, with the exception of AAC main, AAC in scalable sampling rate configuration (AAC SSR), and synthetic audio

✗ Error-protection (EP) tool (see Chapter 10)

✗ Additional error-resilience tools for ER AAC low complexity (ER AAC LC), ER AAC long-term prediction (ER AAC LTP), ER AAC scalable, and error-resilience mode for ER BSAC (by including segmented binary arithmetic coding)

☞ **MPEG-4 Version 1 codec extensions**: Silence-compression tool for ER CELP and variable rate coding for ER HVXC at 4 kbit/s

Considering the number of new tools introduced in MPEG-4 Audio Version 2, a subset was selected for verification, including the following error-resilient codecs: ER HILN, ER BSAC, ER AAC LD, ER AAC LC, and ER TwinVQ. Two functionalities were evaluated in the MPEG-4 Audio Version 2 tests:

1. **Coding efficiency:** Three subtests were performed to evaluate the coding efficiency of ER HILN, ER BSAC, and ER AAC LD, respectively. The goal of this test was to estimate the performance of these codecs against other codecs, either MPEG-4 Audio Version 1 codecs or other standard codecs. In this case, transmission errors were not taken into account.

2. **Error robustness:** Two subtests were performed to evaluate the error robustness of ER AAC LC and ER TwinVQ. In this case, bitstreams were corrupted with simulated transmission error patterns.

This section presents the coding efficiency test, and the next section will present the error-robustness test.

16.6.1 Test Conditions

Three subtests were carried out for evaluating the coding efficiency of three error-resilient MPEG-4 audio object types (codecs)—namely, ER HILN, ER BSAC, and ER AAC LD.

Table 16.10 shows the settings of the codecs evaluated in the HILN (ER HILN) subtest. The scalable ER HILN codec was configured for a bit rate of 6 kbit/s for the base layer and 10 kbit/s for the enhancement layer. In this test, the anchor conditions were TwinVQ and AAC.

In the audio on the Internet verification test (described in Section 16.4) an earlier, nonerror-resilient, version of the HILN coder was assessed and compared to TwinVQ and AAC, at 6 and 16 kbit/s, respectively. The results of that test showed that HILN had a worse performance with respect to the other codecs, especially for critical material. Therefore, it was decided to continue the work on the HILN algorithm for Version 2, to improve its compression efficiency, and to add both the error-resilience and the scalability functionalities. So, this test was mainly aimed at verifying the improvements over the previous HILN version and at evaluating the impact of scalability on the perceived quality. For this reason, the test conditions used in this subtest were close to those used in the audio on the Internet test.

For the evaluation of the error-resilient fine-step bit-rate scalable codec (ER BSAC), the codec was set to operate on stereophonic excerpts, sampled at 32 kHz. The tested bit rates were set with steps of 8 kbit/s, in a range between 64 and 96 kbit/s. This algorithm was evaluated earlier in the audio on the Internet test (a nonerror-resilient version), for a bit-rate range between 40 and 56 kbit/s. As for the previous test, the anchor was AAC configured to produce bitstream payloads corresponding to the AAC main object type (see Chapter 11). AAC was set to operate at the two ends of the bit-rate range selected for the ER BSAC codec. Table 16.11 summarizes the test conditions used in the BSAC subtest.

For the evaluation of the ER AAC LD codec, two bit rates were considered: 32 and 64 kbit/s. Table 16.12 summarizes the settings of the codecs evaluated in this test.

Table 16.10 Test conditions for the ER HILN subtest

Codec	Mode	Sampling rate (kHz)	Bit rate (kbit/s)	Frame length (ms)	Bit reservoir size (bit I ms)
ER HILN 6	Mono	16	6	32	384 I 64
ER HILN sc 6[*]	Mono	16	6	32	No bit reservoir
TwinVQ[†]	Mono	16	6	64	No bit reservoir
ER HILN 16	Mono	16	16	32	1024 I 64
ER HILN sc 16	Mono	16	16[‡]	32	No bit reservoir
AAC[†]	Mono	22.05	16	43	>5400 I 340

[*]Base layer.

[†]Anchor conditions.

[‡]6 kbit/s (base layer) + 10 kbit/s (enhancement layer).

Table 16.11 Test conditions for the ER BSAC subtest

Codec	Mode	Sampling rate (kHz)	Bit rate (kbit/s)
AAC*	Stereo	32	96
ER BSAC	Stereo	32	96
ER BSAC	Stereo	32	88
ER BSAC	Stereo	32	80
ER BSAC	Stereo	32	72
ER BSAC	Stereo	32	64
AAC*	Stereo	32	64

*Anchor conditions.

Table 16.12 Test conditions for the ER AAC LD subtest

Codec	Mode	Sampling rate (kHz)	Bit rate (kbit/s)	Delay (ms)
G.722*	Mono	16	64	1.5
ER AAC LD 64	Mono	48	64	20
AAC 56*	Mono	44.1	56	146
ER AAC LD 32	Mono	32	32	30
AAC 24*	Mono	24	24	323
WB-CELP*	Mono	16	24	15

*Anchor conditions.

16.6.2 Test Material

For the ER HILN subtest, a panel of experts selected the excerpts from those used for the audio on the Internet test. The set of excerpts included two music items, two complex items, two single instrument items, and one speech item. Additionally, the panel chose three training items, belonging to the speech, complex, and music content categories.

For the ER BSAC subtest, the attention was focused on music content. The set of excerpts included two single instrument items, three music items, one complex item, and one speech item. For training, only music and complex items were used.

The excerpts for the ER AAC LD subtest were selected from those used for the NADIB test (described in Section 16.3). For this case, the attention was focused on speech items, because the low-delay functionality was introduced mainly for interpersonal real-time communications. The set of the selected excerpts included one single instrument item, two speech items, two music items (one of them was only vocal), and one complex item, which was a combination of male speech and music. For the training, one speech and one complex (speech + music) item were used.

16.6.3 Assessment Method and Test Design

For the MPEG-4 Audio Version 2 coding efficiency test, the paired comparison test described in Section 16.2.2 was adopted. This method was chosen to allow the results to be somehow comparable to those obtained for the MPEG-4 Version 1 tests.

The tests were performed without repetition—that is, each pair reference–test stimulus was presented only once in a trial. All tests were preceded by a training session. Each test was divided into sessions of approximately 20 minutes.

Concerning the playing system, the HILN subtest was performed with a DAT player, whereas the other two subtests were performed with a PC-based presentation system. For all subtests, the STAX LAMBDA NOVA headphones were used.

For the HILN subtest, two listeners at a time participated in the test, whereas for the other subtests, four listeners at a time participated in the test. The pseudorandom presentation order was different for each group of listeners.

The HILN and AAC LD subtests were both divided into two sessions, according to the bit rate. For example, in the case of the HILN subtest, one session was devoted to the 6 kbit/s condition and the other to the 16 kbit/s condition.

Each test was performed at one laboratory only: The ER HILN subtest was performed at Samsung, whereas the ER BSAC and ER LD AAC subtests were performed at NTT.

16.6.4 Data Analysis

For this test, each test site analyzed its own scores; the detailed analysis is provided in [N3075]. All laboratories calculated the following:

- ☞ Mean opinion scores and CIs for each codec, averaged over subjects and items; and
- ☞ Mean opinion scores and CIs for each codec and each item, averaged over subjects; for the sake of conciseness, only the conclusions of this analysis are reported in this chapter.

16.6.5 Test Results

Figure 16.9 shows the mean scores and associated two-sided 95% CIs obtained for the HILN subtest. Considering the overlapping of the CIs, it was concluded

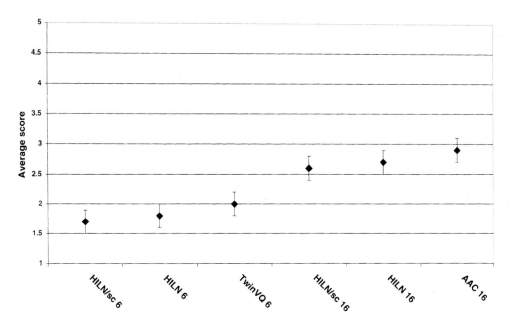

Fig. 16.9 Mean opinion scores and 95% CIs for the ER HILN subtest.

that neither the differences between the 6-kbit/s codecs nor the differences between the 16-kbit/s codecs were generally perceived by the subjects as significant. This was confirmed by the analysis performed item-by-item, which showed that both at 6 and 16 kbit/s, for only one item out of seven did the anchor perform better than both ER HILN and scalable ER HILN.

Therefore, it was concluded that the difference in quality between HILN and the reference codecs observed in the audio on the Internet test (see Section 16.4.5) was recovered in this second test. Additionally, besides error-resilience capabilities, ER HILN also provides speed and pitch change capabilities, as described in Chapter 11. Finally, the scalable HILN provides a perceived quality level similar to that provided by other nonscalable MPEG-4 codecs operating at the same bit rate.

Figure 16.10 presents the mean opinion scores for the ER BSAC subtest. Considering the overlapping of the CIs for the BSAC conditions, it was concluded that, in general, listeners were able to distinguish between different bit rates. Only for the step between 80 and 88 kbit/s was there a small overlapping of the CIs. In any case, it was observed that the ER BSAC performance for the tested set of bit rates was monotonic with the bit rate (i.e., an incrementally higher rate resulted in an incrementally higher performance).

Considering the comparison between BSAC and AAC, it was evident that there were no significant differences at 96 kbit/s. On the contrary, at 64 kbit/s BSAC performed worse than AAC at the same bit rate. At 72 kbit/s, BSAC was nearly comparable to AAC at 64 kbit/s—because, although the CIs calculated

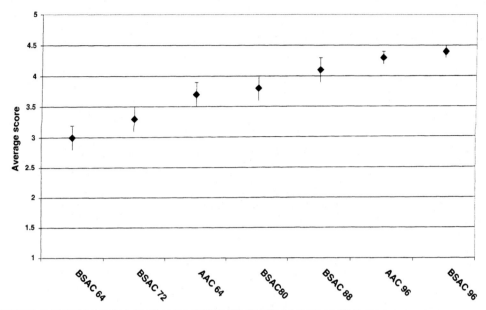

Fig. 16.10 Mean opinion scores and 95% CIs for the ER BSAC subtest.

over all the items were not overlapping, the item-by-item analysis showed that for all items except one, the CIs for BSAC at 72 kbit/s overlapped the CIs for AAC at 64 kbit/s. This suggests that the efficiency overhead introduced by scalability at the low end of the tested set of bit rates is approximately 12.5%.

The results for this test and those for the audio on the Internet test (Section 16.4.5) are not directly comparable, because they covered two different ranges of bit rates. Nevertheless, it was observed that for the audio on the Internet test at a lower bit rate (40 kbit/s), BSAC performed worse than AAC, whereas at a higher bit rate (56 kbit/s), it performed as well as AAC. A second consideration related to the outcome of the two tests concerns the performance of BSAC with music items: The main BSAC problem in the previous test was its strong degradation in sound quality while using its scalable feature. The Version 2 audio test showed that, for a reduced number of enhancement layers, the degradation in sound quality is rather moderate.

Concerning the ER AAC LD subtest, for the higher bit-rate range the overall averages were quite close to one another (4.338 for ER AAC LD at 64 kbit/s and 4.34 for AAC at 56 kbit/s). Thus, the mean opinion score and the 95% CI are shown item by item in Figure 16.11, to provide a more detailed analysis of these codecs. The left-most six scores are related to the ER AAC LD at 64 kbit/s, and the right-most six scores are for AAC at 56 kbit/s. The CIs obtained for the two codecs for each item always overlap. Thus, it was not possible to conclude that there were significant differences in terms of performance between them. This led to the conclusion that, at the higher range of the tested bit rates, a reduction in delay of 86% (one-way delay reduced from 146 ms for AAC to 20 ms for ER AAC LD) came at a cost of an increase in bit rate of approximately 14% (increased from 56 kbit/s to 64 kbit/s).

Figure 16.12 represents the mean opinion scores and CIs calculated over all items for the lower bit-rate conditions tested in the ER AAC LD subtest.

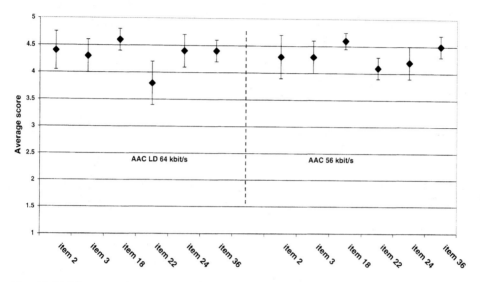

Fig. 16.11 Mean opinion scores and 95% CIs, calculated item by item, for the ER AAC LD subtest at higher bit rates.

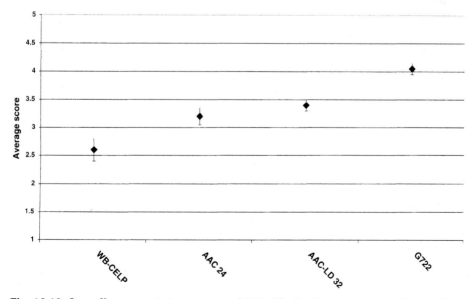

Fig. 16.12 Overall mean opinion scores and 95% CIs for the ER AAC LD subtest at lower bit rates.

Regarding the overlapping of the CIs for the ER AAC LD and AAC cases, it was concluded that a reduction in delay of 91% (one-way delay reduced from 323 ms for AAC to 30 ms for ER AAC LD) came at a cost of an increase in bit rate of approximately 33% (increased from 24 kbit/s to 32 kbit/s).

Moreover, the G.722 codec at 64 kbit/s and the WB-CELP codec at 24 kbit/s showed a performance that is significantly different from both AAC codecs; these codecs were positioned, respectively, at the top-end and the bottom-end of the range of quality variation.

Because one of the anchors (WB-CELP) was a speech-coding algorithm, a further analysis was performed by processing the scores for music items separately from those for speech items. Considering only the music items, G.722 at 64 kbit/s had statistically better performance than all other systems in this subtest, although the difference with respect to both the AAC codecs was less than 0.4 on a 5-level scale. Moreover, ER AAC LD at 32 kbit/s was not statistically different from AAC at 24 kbit/s. Finally, CELP at 24 kbit/s had statistically worse performance than all other codecs in this subtest.

Considering only speech items, G.722 at 64 kbit/s had statistically better performance than all other systems in this test. Moreover, CELP at 24 kbit/s had statistically better performance than both ER AAC LD at 32 kbit/s and AAC at 24 kbit/s. Finally, ER AAC LD at 32 kbit/s was not statistically different from AAC at 24 kbit/s. This led to the conclusion that, for applications restricted to speech signals only, the MPEG-4 WB-CELP coder has a higher performance, expressed by a lower delay (15 ms versus 30 ms) and a lower bit rate (24 kbit/s versus 32 kbit/s), than the ER AAC LD coder.

16.7 VERSION 2 ERROR-ROBUSTNESS TEST

As explained in the previous section, the MPEG-4 Audio Version 2 verification tests included three coding efficiency subtests and two error-robustness subtests. In Section 16.6, the coding efficiency test was addressed. This section deals with the error-robustness test.

The evaluation of error robustness implied a comparison between the perceived quality of a stream corrupted by transmission errors against the perceived quality of the same stream when uncorrupted. For this, it was essential to select relevant channel conditions.

16.7.1 Test Conditions

All natural speech and audio coding tools included in the MPEG-4 Audio standard allow for error-robust operation by means of their *error-resilient bitstream payload syntax* (see Chapter 11). To keep the size of the error-robustness test within manageable dimension, it was decided to focus the test on those coding tools that had the most advanced error-robustness capabilities (i.e., ER AAC LC/LTP/LD and ER TwinVQ) and error conditions representative of mobile communications (i.e., burst errors with two different burst lengths), due to the importance of this type of transmission. In the following, the term *system* is used to identify a combination of coding scheme and channel condition.

The two codecs, ER AAC LC and ER TwinVQ, were evaluated in two separate subtests, under the conditions illustrated in Table 16.13 and Table 16.14. One full-bandwidth uncompressed condition was included as hidden reference (identified as HR). In addition, three band-limited uncompressed conditions were added as anchors; they are identified as *A 70*, the 7-kHz bandwidth; *A 35*, the 3.5-kHz bandwidth; and *A 17*, the 1.7-kHz bandwidth. This last was included only for the lower bit-rate condition.

Concerning the generation of ER AAC LC streams, a Version 1 to Version 2 AAC transcoder was used to translate bitstream payloads from the AAC LC object type to those of the ER AAC LC object type. The transcoder was config-

Table 16.13 Test conditions for the error-robustness test at 16 kbit/s

Codec/System	Mode	Sampling rate (kHz)	Bit rate (kbit/s)	Error condition
ER TwinVQ crit	Mono	32	16	Critical
ER TwinVQ verycrit	Mono	32	16	Very Critical
ER TwinVQ no err	Mono	32	16	No error
HR[*]	Mono	48	N.A.	No error
A 70[†]	Mono	16	N.A.	No error
A 35[†]	Mono	8	N.A.	No error
A 17[†]	Mono	4	N.A.	No error

[*]Hidden reference.
[†]Anchor.

Table 16.14 Test conditions for the error-robustness test at 96 kbit/s

Codec/System	Mode	Sampling rate (kHz)	Bit rate (kbit/s)	Error condition
ER AAC LC crit	Stereo	32	96	Critical
ER AAC LC verycrit	Stereo	32	96	Very Critical
ER AAC LC noerr	Stereo	32	96	No error
HR[*]	Stereo	48	N.A.	No error
A 70[†]	Stereo	16	N.A.	No error
A 35[†]	Stereo	8	N.A.	No error

[*]Hidden reference.
[†]Anchor.

ured to apply noiseless AAC error-resilience tools (see Chapter 11). The error-protection tool developed in Version 2 was used to produce unequal error-protected bitstream payloads. Details about its configuration are given in [N3075]. A combination of noise substitution and prediction in conjunction with energy interpolation was used as a concealment technique at the decoder [LauS01]. The selection of the appropriate concealment method depended on the signal characteristics. A delay of one audio frame was inserted, due to the concealment process. If a multiple frame loss occurred, the reconstructed spectra were attenuated.

ER TwinVQ test streams were produced according to the following steps: source encoding, header removing, EP tool encoding, error insertion/Mux-demux, EP tool decoding, header insertion, source decoding, and concealment. Details about error-concealment techniques can be found in [N3075]. Error insertion and multiplexing/demultiplexing were applied to error-protected bitstreams. Based on the frame erasure information and the CRC information, concealment processes were carried out, and there was no additional delay due to the concealment.

For the simulation of channel errors, all the excerpts were concatenated prior to encoding. Then a multiplex layer inserted the EP data, to produce a bitstream ready for error insertion. The error pattern was applied to this bitstream. After decoding, the sequence was split again into the original items, which were then separately presented to the listeners.

To produce the error patterns, the Gilbert Model, which is a two-state Markov model, was used [Gilb60]. Considering the typical error conditions expected in mobile communications, bit errors occurred only within the error burst, during which the bit error rate is 50%. The probability of making a transition from a burst interval to a clear channel interval and back is this:

Probability of BAD to GOOD (P_BADtoGOOD) =
1.0/ average burst length (in bits)

Probability of GOODtoBAD =
AverageBER × P_BADtoGOOD × (0.5 − AverageBER)

Following the MPEG-4 Requirements document [N4319], the average bit error rate (BER) was set to 10^{-3}, whereas for the error burst length, two values (i.e., 10 ms and 1 ms) were selected as critical and very critical conditions. For the current wireless systems, the critical error condition corresponds to the worst case occurring at the edge of the radio service area, and the very critical condition is so bad that it should not be observed for a real transmission channel in normal operation.

Twenty-five different error patterns were generated, by feeding the error pattern generator with different random seeds. Among them, the error pattern that provided the signal-to-noise ratio (SNR) nearest to the average SNR over all error patterns was selected as the most representative and used to corrupt the streams.

16.7.2 Test Material

For the error-robustness test, the test items were selected from the items used in the audio on the Internet test. The set of test items included two items for each of the following content categories: speech, single instrument, and complex. In addition, one item of classic music and one item of pop music were included.

16.7.3 Test Method and Experimental Design

For the error-robustness test, the MUSHRA test method (as described in Section 16.2.3) was used. This method was selected because some of the systems under test were expected to output audio signals in a rather low quality range. The test was divided into two parts: one for testing the 16-kbit/s conditions, and another for evaluating the 96-kbit/s conditions. For each subtest, anchor conditions were introduced, as indicated in Tables 16.13 and 16.14.

The error-robustness test was conducted at two test sites: NTT DoCoMo (Japan) and Fraunhofer Institute (Germany). For each test, the appropriate set of items was used, according to the native language of the test site. Audio presentation and grading were done by using a computer-based system, connected to STAX LAMBDA PRO headphones. One listener at a time participated in the test.

16.7.4 Data Analysis

Each test site analyzed its own scores, according to the following steps:

1. **Postscreening of subjects:** A few subjects were discarded, either because they were not following the testing rules (e.g., they did not associate the maximum of scores to the full-band hidden reference) or because they demonstrated a hearing sensitivity significantly lower than average (e.g., they did not associate low scores to very impaired test items).

2. **Calculation of mean opinion scores and CIs for each codec**, averaged over subjects and items.

3. **Calculation of mean opinion scores and CIs for each codec and each item**, averaged over subjects. For the sake of conciseness, only the conclusions of this analysis are reported in this chapter.

For the detailed analysis of the results, see [N3075].

16.7.5 Test Results

The goal of this test was to evaluate how transmission errors affect the perceived quality of audio streams encoded with the ER TwinVQ codec at 16 kbit/s and the ER AAC LC codec at 96 kbit/s. Therefore, the comparison was between audio data decoded from streams corrupted by channel errors and audio data decoded from the same streams but now uncorrupted. Reduced bandwidth anchors also were taken into account.

Figure 16.13 and Figure 16.14 illustrate the results obtained for the MPEG-4 Version 2 audio error-robustness test.

Concerning the comparison between clean and corrupted conditions, it was observed that in all tests and test sites, except for the data collected at Fraunhofer for 96 kbit/s, the results for error-prone systems were not statistically different from one another. Thus, it seems that the ER tools may be able to address a wide variety of channel error characteristics with only a modest overhead in bit rate. For the ER AAC LD test at 96 kbit/s, the total overhead was 9.5% (2% for ER and 7.5% for EP); for the ER TwinVQ test at 16 kbit/s, the total overhead was 17% (EP only).

Concerning the comparison of the results for the systems under error-prone conditions against the results for the anchors, the test results showed that at 16 kbit/s and at both test sites the 95% CIs of both error-prone channel systems were between the 95% CIs of the 3.5-kHz bandwidth and the 1.7-kHz bandwidth anchors. At 96 kbit/s, and for both test sites, the 95% CIs of both error-prone channel systems were between the 95% CIs of the full-bandwidth reference and the 7.0-kHz anchor. As expected, at both bit rates and for both test sites, the clear channel system had statistically better performance than both error-prone channel systems.

As further validation of the new test method (MUSHRA), it was observed that, at both bit rates and for both test sites, all anchor signals had quality

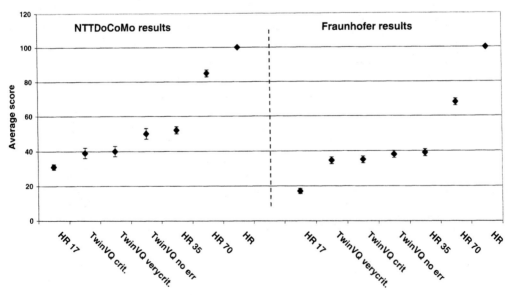

Fig. 16.13 Overall mean opinion scores and 95% CIs for 16 kbit/s.

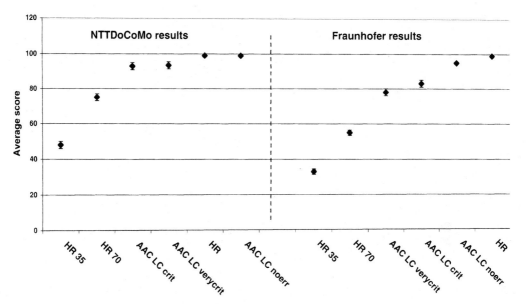

Fig. 16.14 Overall mean opinion scores and 95% CIs for 96 kbit/s.

scores that were monotonically decreasing with a decreasing bandwidth. This is confirmed by the comparison between conditions, performed item by item; for example, the full-bandwidth items were always judged better than the corresponding 7-kHz bandwidth-limited items, and these last were always judged better than the 3.5-kHz bandwidth-limited items.

16.8 SUMMARY

Verification tests are part of the standards development approach adopted by MPEG. These tests are aimed at verifying that the tools developed can be used to assemble the target systems and to provide the desired functionalities with an adequate level of performance.

In the case of the MPEG-4 Audio standard, two rounds of verification tests were performed: one for Version 1 and another for Version 2.

For Version 1, the following target applications were considered: narrow-band audio broadcasting, audio on the Internet, and speech communications. For each of these cases, suitable functionalities were taken into account. The test results confirmed that the MPEG-4 Version 1 audio tools are suitable for the above-mentioned applications and generally outperform the existing standards. In very few cases, it came out that some specific tools needed further optimization. These tools were improved during the development of MPEG-4 Audio Version 2 and were tested during the second round of verification tests.

For Version 2, the attention was focused on the following functionalities: scalability associated to speed and pitch change, fine granularity scalability, low delay, and error resilience. For all cases, the tested tools effectively supported the above-mentioned functionalities.

The test results presented in this chapter show that the evaluated tools fulfill the requirements specified for the standard.

16.9 REFERENCES

[BS1116] ITU-R Recommendation BS.1116. *Methods for the Subjective Assessment of Small Impairments in Audio Systems Including Multichannel Sound Systems.* 1994.

[BS1284] ITU-R Recommendation BS.1284. *Methods for the Subjective Assessment of Sound Quality: General Requirements.* 1996.

[BS1534] ITU-R SG10-11 WP10-11Q. *Draft New Recommendation ITU-R BS.1534: Method for the Subjective Assessment of Intermediate Audio Quality.* Doc. 6/106-E. May 2001.

[CaWT91] Campbell, P., V. C. Welch, and T. Termain. "The DOD 4.8 kbps Standard (Proposed Federal Standard 1016)." *Advances in Speech Coding* (B. S. Atal, V. Cupermann, and A. Gersho, Eds.), pp. 121–133. Dordrecht, Holland: Kluwer Academic. 1991.

[CEMS97] Contin, L., B. Edler, D. Meares, and P. Schreiner. "Tests on MPEG-4 Audio Codec Proposals." *Signal Processing: Image Communication,* 9(4): 327–342. May 1997.

[EBU00] ITU-R. *Multi Stimulus Test with Hidden Reference and Anchors (MUSHRA), EBU Method for Subjective Listening Tests of Intermediate Audio Quality.* ITU-R delayed contribution, Document 10-11Q/62-E 1. February 2000.

[ETSI97] ETSI. *Digital Cellular Telecommunications System, Phase 2+ (GSM). Half Rate Speech: ANSI-C Code for the GSM Half Rate Speech Codec GSM. ETSI EN 300 723,* Version 6.0.1. Release 1997.

[ETSI99] ETSI. *Digital Cellular Telecommunications System, Phase 2+ (GSM). Enhanced Full Rate (EFR) Speech Transcoding GSM. ETSI EN 300 726,* Version 7.0.2. Release 1998.

[G722] ITU-T Recommendation G.722. *7 kHz Audio Coding Within 64 kbit/s.* 1988.

[G723] ITU-T Recommendation G.723.1. *Dual Rate Speech Coder for Multimedia Communications Transmitting at 5.3 and 6.3 kbit/s.* 1996.

[G729] ITU-T Recommendation G.729. *Coding of Speech at 8 kbit/s Using Conjugate-Structure Algebraic-Code-Excited Linear-Prediction (CS-ACELP).* 1996.

[Gilb60] Gilbert, E. N. "Capacity of a Burst-Noise Channel." *Bell Systems Technology Journal,* p. 1253. September 1960.

[Guil54] Guilford, J. P. *Psychometric Methods.* New York: McGraw-Hill. 1954.

[LauS01] Lauber, P., & R. Sperschneider. *Error Concealment for Compressed Digital Audio.* 111th AES Convention, New York. November 2001.

[M3598] Contin, L. *List of Audio, Video and Audiovisual Excerpts That Have Been Released to MPEG.* Doc. ISO/MPEG M3598, Dublin MPEG Meeting, July 1998.

[M4087] Purnhagen, H., and B. Edler. *On HILN and TwinVQ Performance in the Audio on Internet Verification Test.* Doc. ISO/MPEG M4087, Atlantic City MPEG Meeting, October 1998.

[MP4V2-3] ISO/IEC 14496-3. Amendment 1. *Coding of Audio-Visual Objects—Part 3: Audio,* Version 2, 2000.

[MPEG4-3] ISO/IEC 14496-3:2001. *Coding of Audio-Visual Objects—Part 3: Audio,* 2d Edition, 2001.

[Myer79] Myers, J. L. *Fundamentals of Experimental Design.* Boston: Allyn & Bacon. 1979.

[N1144] MPEG Audio and Test. *MPEG-4 Audio Test Results (MOS Tests).* Doc. ISO/MPEG N1144, Munich MPEG Meeting, January 1996.

[N1419] MPEG Audio and Test. *Report on the Formal Subjective Listening Tests of MPEG-2 NBC Multichannel Audio Coding.* Doc. ISO/MPEG N1419, Maceió MPEG Meeting, November 1996. Available at *http://mpeg.telecomitalialab.com/*

[N2006] MPEG Audio and Test. *Report on the MPEG-2 AAC Stereo Verification Tests.* Doc. ISO/MPEG N2006, San Jose MPEG Meeting, February 1998. Available at *http://mpeg.telecomitalialab.com/*

[N2168] MPEG Implementation and Test. *Call for New Audio and Audiovisual Test Sequences for MPEG-4.* Doc. ISO/MPEG N2168, Tokyo MPEG Meeting, March 1998.

[N2276] MPEG Audio and Test. *Report on the MPEG-4 Audio NADIB Verification Tests.* Doc. ISO/MPEG N2276, Dublin MPEG Meeting, July 1998. Available at *http://mpeg.telecomitalialab.com/*

[N2424] MPEG Audio and Test. *Report on the MPEG-4 Speech Codec Verification Tests.* Doc. ISO/MPEG N2424, Atlantic City MPEG Meeting, October 1998. Available at *http://mpeg.telecomitalialab.com/*

[N2425] MPEG Audio and Test. *MPEG-4 Audio Verification Test Results: Audio on Internet.* Doc. ISO/MPEG N2425, Atlantic City MPEG Meeting, October 1998. Available at *http://mpeg.telecomitalialab.com/*

[N2724] MPEG Requirements. *MPEG-4 Applications.* Doc. ISO/MPEG N2724, Seoul MPEG Meeting, March 1999.

[N3075] MPEG Audio and Test. *Report on the MPEG-4 Audio Version 2 Verification Test.* Doc. ISO/MPEG N3075, Maui MPEG Meeting, December 1999. Available at *http://mpeg.telecomitalialab.com/*

[N4319] MPEG Requirements. *MPEG-4 Requirements.* Doc. ISO/MPEG N4319, Sydney MPEG Meeting, July 2001.

[N998] MPEG AOE. *Proposal Package Description (PPD): Revision 3.* Doc. ISO/MPEG N998, Tokyo MPEG Meeting, July 1995.

[N999] MPEG AOE. *MPEG-4 Testing and Evaluation Procedures.* Doc. ISO/MPEG N999, Tokyo MPEG Meeting, July 1995.

[P800] ITU-T Recommendation P.800. *Methods for Subjective Determination of Transmission Quality.* 1996.

[P810] ITU-T Recommendation P.810. *Modulated Noise Reference Unit (MNRU).* 1996.

Levels for Visual Profiles

by Fernando Pereira and Paulo Nunes

*Keywords: visual profile, visual level, conformance, video buffering
verifier, video rate buffer verifier, video complexity verifier, video
memory verifier*

*A*s described in Chapter 13, the MPEG-4
Visual standard defines (as of October 2001) 18 visual object types and 19
visual profiles.

Nine visual profiles were defined in MPEG-4 Visual Version 1 [MPEG4-2]:
Simple, Simple Scalable, Core, Main, N-bit, Scalable Texture, Simple Face
Animation, Basic Animated Texture, and Hybrid.

Six additional visual profiles were defined in MPEG-4 Visual Version 2
[MPEG4-2]: Core Scalable, Advanced Core, Advanced Coding Efficiency,
Advanced Real-Time Simple, Advanced Scalable Texture, and Simple FBA.

Moreover, two additional profiles were defined in the 1st amendment to
the 2d Edition of the MPEG-4 Visual standard [N3898], Simple Studio and
Core Studio, and two more were defined in the 2d amendment to the 2d Edition of the MPEG-4 Visual standard [N3904]: Advanced Simple and Fine
Granularity Scalability.

Tables 13.1 and 13.2 show which visual coding tools include each visual
object type and which visual object types include each visual profile.

In the following, the mechanism specified to define video levels, the Video
Buffering Verifier, as well as the visual levels defined for all visual profiles are
presented.

A.1 VIDEO BUFFERING VERIFIER MECHANISM

The idea of using a Video Buffering Verifier mechanism to bound the decoding complexity of a given set of bitstreams is not new, and it was already adopted in MPEG-1 Video [MPEG1-2] and MPEG-2 Video [MPEG2-2]. In these standards, the major purpose of the Video Buffering Verifier mechanism was to set some restrictions on the maximum variability of the number of bits per picture, especially in the case of constant bit-rate operation, and thus on the complexity of the encoded video streams.

Generically, the complexity of the encoded video is directly related to the encoded bit rate and to the decoded video data rate that the decoder generates (e.g., measured in terms of the number of MB/s). For frame-based video coding (e.g., MPEG-1 and MPEG-2 video), the decoded video data rate is typically constant since the frames have fixed dimensions and are usually encoded at fixed frame rates. This is not the general case for object-based video coding, as in MPEG-4, since the several video objects composing a scene may vary in size along time and may be encoded at different VOP rates. Therefore, the amount and type[1] of MB/s that a given object-based video decoder has to process may vary over time in comparison with frame-based coding solutions [Nunes].

In the MPEG-4 context, to limit the decoding complexity of a set of bitstreams corresponding to a video scene it is necessary to set some limits on the variability of the number of decoded MB/s, and their complexity, as well as on the picture memory required to store the decode data. This constitutes the major novelty of the MPEG-4 Video Buffering Verifier mechanism, relative to the previous MPEG standards, since it not only bounds the bitstream buffer memory but also the MB decoding capacity and the MB picture memory.

The MPEG-4 Video Buffering Verifier mechanism [MPEG4-2, Annex D] consists of three normative models (see Figure A.1), each one defining a set of rules and limits to verify that the amount required for a specific type of decoding resource is within the values allowed by the corresponding profile and level specifications. Table A.1 provides details of the specifications.

1. **Video Rate Buffer Verifier (VBV):** This model is used to verify that the bitstream memory required at the decoder(s) does not exceed the values specified for the corresponding profile and level. The model is defined in terms of the VBV buffer sizes for all the VOLs corresponding to the objects building the scene. Each VBV buffer size corresponds to the maximum amount of bits that the decoder can store in the bitstream memory for the corresponding VOL. There is also a limitation on the sum of the VOL VBV buffer sizes. The bitstream memory is the memory where the decoder puts the bits received for a VOL while waiting to be decoded.

---

1. For an arbitrarily shaped video object, three types of MBs may exist: transparent, opaque, and boundary.

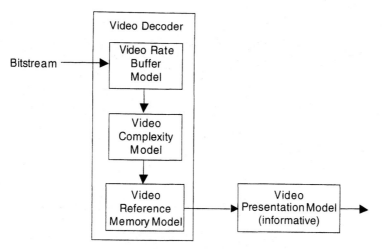

Fig. A.1 Video buffering verifier model [MPEG4-2].

2. **Video Complexity Verifier (VCV):** This model is used to verify that the computational power (processing speed), defined in terms of MB/s, required at the decoder does not exceed the values specified for the corresponding profile and level. The model is defined in terms of the VCV MB/s decoding rate and VCV buffer size and is applied to all MBs in the scene. If arbitrarily shaped VOs exist in the scene, an additional VCV buffer and VCV decoding rate are also defined, to be applied only to the boundary MBs.

3. **Video Reference Memory Verifier (VMV):** This model is used to verify that the picture memory required at the decoder for the decoding of a given scene does not exceed the values specified for the corresponding profile and level. The model is defined in terms of the VMV buffer size, which is the maximum number of decoded MBs that the decoder can store during the decoding process of all VOLs corresponding to the scene.

The Video Presentation Model (VPM) is not a normative part of the MPEG-4 Visual specification [MPEG4-2]. It is an algorithm for checking that the set of bitstreams corresponding to a scene does not require an amount of presentation memory higher than a given amount of memory (expressed in units of MB). It is also used to constrain the speed of the compositor to a maximum number of MB/s. The Video Presentation Verifier (VPV) operates in the same way as the VCV in terms of occupancy dynamics [MPEG4-2].

In order that the set of visual elementary streams corresponding to a given scene be considered compliant with a given profile and level, the encoder must guarantee that none of the previously mentioned buffers overflows and, additionally, it must guarantee that, in certain circumstances, the VBV buffer never underflows.

A.1.1 Video Rate Buffer Verifier Definition

The MPEG-4 VBV model defines a set of rules and limits for examining a video elementary bitstream with a delivery rate function, $R(t)$. This model simulates the occupancy of the decoder bitstream buffer in order to control the amount of bitstream memory required at the decoder. Its purpose is to guarantee that the bitstream memory required is less than the specified buffer size (i.e., to verify that the decoder bitstream buffer occupancy never goes beyond the limits of the specified buffer size for the relevant profile@level). In the case of visual scenes composed by multiple VOs, each with one or more VOLs, the MPEG-4 Visual standard specifies that the video rate buffer model must be applied independently to each VOL (using a particular buffer size and rate function for each VOL). Additionally, the maximum total bitstream buffer size (defined as the sum of all VOL bitstream buffer sizes) for the given profile and level must not be exceeded (see Table A.1). Notice that the bit rate and buffer size allocation, among the several VOs—and for each VO among the several VOLs—is a non-normative issue. Still, it can significantly determine the performance of object-based video encoders and thus deserves careful attention.

The VBV applies to video data encoded as a combination of I-, P-, B-, and S-VOPs, using several coding tools organized in terms of video object types (see Chapter 13). Face animation, still texture, and mesh objects are not constrained by the VBV model. The coded video bitstreams shall be constrained to comply with the requirements of the VBV specified in the following sections.

A.1.1.1 VBV Model Parameters
The VBV model for a given elementary stream (ES) is defined by the three following parameters: vbv_buffer_size, vbv_occupancy, and bit_rate. These parameters have to be defined for all the ESs corresponding to the various objects in a scene. These parameters can be specified at video level (i.e., through the video ES), or by means of systems-level configuration information [MPEG4-1]. In the first case, the VBV model parameters are specified in the VOL header, when the one-bit flag vbv_parameters is set to 1. In the second case, the VBV model parameters are conveyed to the video decoder through the Object Description Information (more precisely, through the DecoderConfigDescriptor field of the ES_Descriptor associated with the ES in question).

When the vbv_buffer_size and vbv_occupancy parameters are specified by systems-level configuration information, the bitstream shall be constrained according to the specified values, and these values shall not be part of the video ES. It may happen, however, that these parameters are not explicitly specified; in this case, it is assumed that the ES is constrained according to the default values of the corresponding profile and level combination.[2]

---

2. Except for the short video header case.

VBV Buffer Size The VBV buffer size for a VOL specifies the minimum bit-stream memory required at the decoder to properly decode the corresponding VOL ES. The VBV buffer size for a VOL is defined by the 18-bit `vbv_buffer_size` field in units of 16384 bits (the value zero is forbidden). The maximum VBV buffer size in bits, vbv_{BS}, is then given by

$$vbv_{BS} = 16384 \times vbv\_buffer\_size$$

The `vbv_buffer_size` value is bounded by *Max VOL VBV buffer size* (see Table A.1), which specifies the levels' constraints, and the sum of all these values for all VOLs is bounded by *Max total VBV buffer size.*

The default value of `vbv_buffer_size` for a VOL is the maximum value of `vbv_buffer_size` allowed for the profile@level combination in question (called *Max VOL VBV buffer size*). Still, it must be checked that the sum of the `vbv_buffer_size` default values does not exceed *Max total VBV buffer size.*

In terms of the levels' specifications shown in Table A.1, two constraints are defined: *Max VOL VBV buffer size*, which sets the limit for each VOL, and *Max total VBV buffer size*, which sets the limit on the sum of all the VOL buffer sizes.

VBV Occupancy The VBV occupancy for a VOL specifies the initial occupancy of the VBV buffer for that VOL; this means the occupancy that the VBV must reach in order the decoding process may start with the removal of the first VOP following the VOL header. This parameter, together with the `bit_rate` parameter, establishes the initial decoding delay, the so-called VBV latency. The VBV occupancy is defined by the 26-bit `vbv_occupancy` field in units of 64 bits.[3]

The default value of `vbv_occupancy` for a VOL, in 64-bit units, is given by $170 \times$ `vbv_buffer_size` (for that VOL), where `vbv_buffer_size` is in 16,384-bit units (of course, the maximum value of `vbv_occupancy` is `vbv_buffer_size` for the corresponding VOL). This corresponds to an initial occupancy (before the removal of the first VOP from the buffer) in bits, vbv_{0^-}, of approximately two-thirds of the defined buffer size:

$$vbv_{0^-} = 64 \times vbv\_occupancy$$
$$= 64 \times \left(170 \times \frac{vbv_{BS}}{16384}\right) \approx \frac{2}{3} \times vbv_{BS} \quad [\text{bit}]$$

---

3. For basic sprites, the `vbv_occupancy` field specifies the initial VBV occupancy before decoding the first S-VOP in the elementary stream (i.e., not the very first VOP in a basic sprite, which must be an I-VOP, but the subsequent VOP—an S-VOP). Low-latency sprites, which allow the transmission of large sprites progressively (both spatially and in terms of quality), are treated as any other VOL.

Note that there is no explicit limitation on vbv_occupancy in terms of levels' definitions.

Bit Rate   When present for a VOL, the bit-rate parameter, bit_rate, defined by the 30-bit *bit_rate* field in units of 400 bit/s (value zero is forbidden), specifies the ES peak bit rate for VOL$_{ij}$,[4] such that

$$R_{VOL_{ij}}(t) \le 400 \times bit\_rate$$

where $R_{VOL_{ij}}(t)$ is defined as the instantaneous VOL channel bit rate for VOL$_{ij}$ (in bit/s), counting only the visual syntax.

If the channel, with a total instantaneous channel rate, $R(t)$, is a serial time multiplex of several streams (e.g., as defined by MPEG-4 Systems [MPEG4-1]), then $R_{VOL_{ij}}(t) = R(t)$ for the time instants where the channel is occupied by the relevant VOL$_{ij}$ bits; otherwise, it is zero [MPEG4-2]:

$$R_{VOL_{ij}}(t) = \begin{cases} R(t) & \Leftarrow \quad t \in \left\{ \text{bit time interval from VOL}_{ij} \right\} \\ 0 & \text{otherwise} \end{cases}$$

Notice that the purpose of the bit-rate parameter is to provide an upper bound on the VOL ES bit rate rather than a precise value of the actual VOL bit rate since MPEG-4 Visual does not specify any temporal window to measure the actual ES bit rate.

In terms of the levels' specifications shown in Table A.1, only the sum of the bit rate for all the VOLs for all the objects in the scene is bounded, assuming that this total bit rate can be shared among the VOLs at an author's wishes (signaled using the bit_rate field for each VOL).

A.1.1.2 VBV Occupancy Dynamics   The VBV occupancy dynamics specify when the bitstream bits enter the VBV buffer and when they are removed from it to be decoded (i.e., the process by which the VBV buffer is filled and drained). This process is mainly driven by the time instants at which the VOP bits are removed from the VBV.

VBV Buffer Filling   The VBV buffer for each ES is initially empty and it is filled as coded data arrives, until it reaches the value specified in the vbv_occupancy field or the first VOP decoding time arrives. The first bit that is put in the VBV buffer is the first bit of the elementary stream (the VOL header bits are not taken into account since they are not considered to be part of the elementary stream data; see [MPEG4-2]).

VBV Buffer Draining   The VBV buffer is instantaneously emptied at the VOP decoding times (see Figure A.2, which shows the VBV occupancy for a

---

4. VOL$_{ij}$ corresponds to VOL j of VO i.

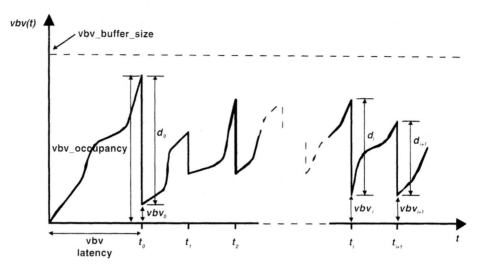

Fig. A.2 Dynamics of the VBV occupancy for one VOL [Nunes].

VOL, $vbv[t]$, as a function of time). This instantaneous removal property distinguishes the VBV buffer model from a real bitstream buffer. This way, the model accommodates the worst-case scenario (i.e., the case where the decoder stores all the encoded data for the current VOP in its bitstream buffer until it starts decoding it).

VOP Decoding Time Computation   In order to keep a good estimate of the decoder bitstream buffer occupancy, the encoder needs to know when the encoded data will be removed from the VBV buffer (i.e., the VOP decoding times). Since the VOP time information carried in the VOP ES is the VOP composition time, the encoder needs to compute the corresponding VOP decoding time from this information. In MPEG-4 Visual [MPEG4-2], the time at which each VOP must be available in the composition memory for composition is given by this VOP composition time plus a fixed delay: *VCV Latency* (see the description in Section A.1.2). This delay sets the minimum latency of the decoding process.

The usage, in some profiles, of B-VOPs, which may be coded using more than one prediction (i.e., may be predicted from preceding I- or P-VOPs [*forward prediction*] and from upcoming I- or P-VOPs [*backward prediction*], as explained in Chapter 8), implies that the VOP decoding order and the VOP composition order are different for these cases. In fact, some VOPs must be decoded in advance—before their natural composition order—because they are needed for the prediction of other VOPs. In terms of decoder operation, this implies additional delay and VOP memory for the decoding and storage of the backward predictions.

MPEG-4 Visual clearly defines the time instants at which a given VOP has to be available at the bitstream buffer (all its bits) for decoding; these time instants have to be computed by the encoder in order to track the occupancy of the decoder bitstream buffer. For further details, see [MPEG4-2, Annex D].

A.1.1.3 VBV Model Constraints   This section applies to all the cases considered in the VBV model except for basic sprites, which have a special treatment. The first I-VOP of a sprite VO is divided into N sections of 396 MBs and each section is treated as a different VOP. The remaining S-VOPs are treated as any other VOP.

Constraints on VBV Occupancy   The main constraint imposed to the VBV model is that each VOL VBV buffer never overflows or underflows. The VBV buffer occupancy for a VOL, immediately following the removal of VOP i from the bitstream buffer, vbv_i, as shown in Figure A.2, can be iteratively defined by equation (A.1)[5]

$$vbv_0 = vbv_{0^-} - d_0$$
$$vbv_{i+1} = vbv_i + \int_{t_i}^{t_{i+1}} R_{vol}(t)dt - d_{i+1} \quad \text{for } i \geq 0 \tag{A.1}$$

where vbv_0 is the initial VBV occupancy just before the removal of the first VOP from the buffer, d_0 is the number of bits for the first VOP in the ES (VOP 0), and d_i is the number of bits for VOP i.

The condition that the VBV buffer should never overflow or underflow can then be expressed by

$$\begin{cases} vbv_i + d_i \leq vbv_{BS} \\ \quad vbv_i \geq 0 \end{cases} \quad \text{for all } i,$$

where vbv_{BS} is the buffer size in bit units for the relevant VOL.

Constraints on VOP Coded Size   The VBV occupancy constraints for a VOL require that the coded VOP size must always be less than the VBV buffer size (i.e., $d_i < vbv\_buffer\_size$ for all i).

Annex D of MPEG-4 Visual includes the VBV model restrictions that apply in the case the short video header is in use [MPEG4-2].

A.1.2 Video Complexity Verifier Definition

The MPEG-4 VCV model defines a set of rules and limits for examining a set of ESs building a visual scene to control if the required amount of decoder pro-

---

5. To avoid accumulating errors, the MPEG-4 Visual standard specifies that real-valued arithmetic should be used to compute vbv_i.

cessing power is less than the maximum complexity specified for the given profile and level, both measured in MBs per second (see Table A.1). This model is applied to all MBs of all ESs of the scene together.

The VCV applies to video objects encoded as a combination of I-, P-, B- and S-VOPs.[6] A separate VCV model applies to still texture objects [MPEG4-2]. Face animation and mesh objects are not constrained by this model.

The coded video bitstreams for a certain scene will be constrained to globally comply with the requirements of the VCV defined in the following sections.

A.1.2.1 VCV Model Parameters The VCV model consists of two virtual buffers accumulating the number of MBs in the encoded data:

1. The VCV Buffer, which accumulates all MBs of all VOLs for the scene.
2. The Boundary MB VCV Buffer (B-VCV),[7] which accumulates only boundary MBs.

Notice that boundary MBs (i.e., MBs which are not totally transparent or totally opaque) are included in both the VCV and the B-VCV buffers.

The VCV model is defined by the size of the buffers mentioned previously, the corresponding draining rates (i.e., the VCV and B-VCV decoding rates), and the latency of the VCV model, which depends on the VCV buffer size and VCV decoding rate.

VCV Buffer Sizes and VCV Decoding Rates Each VCV buffer can be seen as a queue, instantaneously filling with all the MBs of each VOP at the VOP decoding time and delivering MB encoded data to the decoding process at a constant rate.

The size of each VCV buffer, respectively `vcv_buffer_size` and `boundary_vcv_buffer_size`, defines the maximum number of MBs that a given decoder can instantaneously have in the decoding queue to process (i.e., the maximum occupancy of the VCV buffers in MB units). In the current MPEG-4 Visual specification [MPEG4-2], the two buffers always have the same maximum dimension for all profile@level combinations.

These MBs are consumed by the decoder, from each buffer, at a given VCV decoding rate, in MB/s, as specified for each profile@level. The VCV decoding rate, H, specifies the draining rate of the VCV buffer while the B-VCV decoding rate, H_B, specifies the draining rate of the B-VCV buffer. Together they define the maximum speed of the decoding process. As can be seen in Table A.1, the B-VCV decoding rate, H_B, is typically half the VCV decoding rate, H.

---

6. For sprites, a hypothetical number of MBs is defined for each S-VOP [MPEG4-2].

7. The B-VCV is only defined for profiles supporting arbitrarily shaped video objects.

For each profile@level combination, MPEG-4 Visual defines the maximum VCV buffer size (the same size for both VCV and B-VCV buffers) and the draining rates for the VCV and B-VCV buffers.

VCV Latency   The VCV latency, L, is defined as the time it takes to decode a full VCV buffer and thus is given by the following equation

$$L = \frac{vcv\_buffer\_size}{H} \tag{A.2}$$

This parameter imposes a minimum latency in the decoding process, as explained in Section A.1.1. Notice that, by definition, the latency of the VCV model is imposed by the VCV buffer and not by the B-VCV buffer. Since the B-VCV decoding rate, H_B, is typically half the VCV decoding rate, H, this means that it is not possible to decode a full B-VCV during a time interval of L because the two buffers have the same size. This implies that at full decoding rate, the amount of boundary MBs in the scene cannot exceed 50% of the total number of MBs.

A.1.2.2 VCV Occupancy Dynamics   The VCV dynamics simulate the VOP decoding process. At the VOP decoding times, the VOP encoded data is added to the VCV buffers; it is removed from these buffers as the decoding process progresses. The time instant at which a given VOP is completely decoded depends on the amount and type of MBs to be decoded, the occupancy of the VCV buffers at the VOP decoding time, and the maximum decoding speed specified through the VCV decoding rates for the profile@level in question.

VCV Buffer Filling   Let M_i be the total number of MBs in VOP i and M_{Bi} be the number of boundary MBs in the same VOP. For S-VOPs, M_i is given by $MB_{S\text{-}VOP}$, the hypothetical number of MBs in an S-VOP, as specified in [MPEG4-2, Annex D].

Each VCV buffer is empty at the start of decoding and is filled instantaneously with encoded data at VOP decoding times as the decoding process advances. At the VOP decoding time, t_i, M_i is added to the VCV buffer occupancy, $vcv(t)$, and simultaneously M_{Bi} is added to the B-VCV buffer occupancy, $b\text{-}vcv(t)$.

VCV Buffer Draining   The VCV buffers' occupancies decrease linearly at rates H and H_B, respectively for the VCV buffer and for the B-VCV buffer, until occupancy is zero or until the next VOP decoding time, t_{next}, where t_{next} is the earliest VOP decoding time greater than t_i for any VOP of any ES of the scene.

If the occupancy of the VCV buffers becomes zero, the VCV model decoder becomes idle and remains idle until t_{next}, as exemplified in Figure A.3.

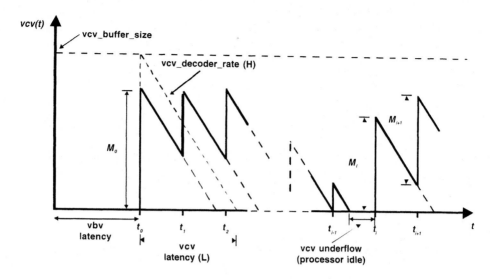

Fig. A.3 Dynamics of the VCV occupancy [Nunes].

VOP Decoding Duration In order to avoid the violation of the VCV model, each VOP must be decoded in time. The interval of time where VOP i is being decoded extends from s_i to e_i, which are defined by equation (A.3)

$$
\begin{aligned}
s_i &= t_i + \max\left[\frac{vcv(t_i)}{H}, \frac{b\text{-}vcv(t_i)}{H_B}\right] \\
e_i &= t_i + \max\left[\frac{(vcv(t_i) + M_i)}{H}, \frac{(b\text{-}vcv(t_i) + M_{Bi})}{H_B}\right]
\end{aligned}
\tag{A.3}
$$

where $vcv(t_i)$ is the VCV occupancy before the MBs representing VOP i, M_i, are added to $vcv(t)$, H is the VCV decoding rate, $b\text{-}vcv(t_i)$ is the B-VCV occupancy before the boundary MBs of VOP i, M_{Bi}, are added to $b\text{-}vcv(t)$, and H_B is the B-VCV decoding rate.

A.1.2.3 VCV Model Constraints Compliance regarding the VCV model can only be guaranteed if the set of ESs building a scene fulfills the constraints imposed by the VCV model relative to the occupancy of the VCV buffers and the VOP decoding duration defined in the following paragraphs.

Constraints on VCV Occupancy A given set of visual ESs building a scene conforms to a given profile@level with respect to the VCV model if they never overflow the VCV buffers.

When the VCV buffers are empty, the decoder simply remains idle and the VCV buffer occupancies, $vcv(t)$ and $b\text{-}vcv(t)$, remain unchanged during the idle period; this is illustrated in Figure A.3, which shows the occupancy of a VCV buffer, $vcv(t)$, as a function of time.

Constraints on VOP Decoding Duration   In addition to not overflowing the VCV buffer, the decoding of each VOP i must be completed by $\tau_i + L$ (composition time plus the latency of the VCV decoding process). Notice that the latency L of the VCV decoding process is constant for all VOPs.

A.1.3 Video Reference Memory Verifier Definition

To examine the set of ESs building a visual scene, the MPEG-4 VMV model defines a set of rules and limits to control whether the required amount of decoder picture memory, measured in MB units, is less than the maximum memory specified for the chosen profile and level (see Table A.1). The VMV models the memory requirements of all VOLs of all VOs in the scene (this model assumes a common memory space, shared by all VOLs of all VOs).

The VMV applies to video objects encoded as a combination of I-, P-, B-, and S-VOPs, and still texture objects. Face animation, mesh objects, and I-VOPs in basic sprite sequences are not constrained by this model.

The coded video bitstreams will be constrained to comply with the requirements of the VMV defined in the following sections.

A.1.3.1 VMV Model Parameters
The VMV model consists of an MB buffer that accumulates all the decoded MBs of all VOPs and stores them until they are no longer needed for the prediction of other VOPs. The VMV model is defined by the size of this buffer, `vmv_buffer_size`, declaring the maximum amount of decoded MBs that the decoder can store at any time instant (see Table A.1).

A.1.3.2 VMV Occupancy Dynamics
The VMV dynamics simulates the decoded VOP memory allocation and deallocation process. As each VOP is being processed, the decoder needs to allocate memory to store the decoded data. This data remains in the decoder memory until it is no longer needed (e.g., for prediction). At this point in time, the memory allocated to store this data is instantaneously released and can be used again.

VMV Buffer Filling   The VMV buffer is initially empty; it is filled with decoded data as each MB is decoded (see Figure A.4). For I-, P-, and B-VOPs, the amount of picture memory required for the decoding of the i-th VOP is defined as the number of MBs in the VOP, M_i. This memory, called reference memory in the MPEG-4 Visual standard [MPEG4-2], is consumed at the same constant rate specified for the VCV buffer (i.e., H MB/s) as the decoding process occurs. This solution contemplates the worst-case scenario in terms of

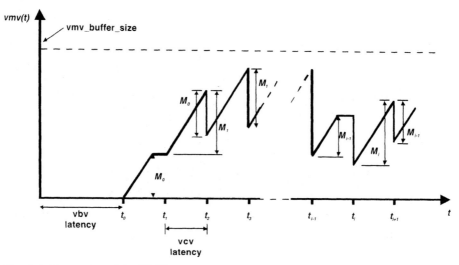

Fig. A.4 Dynamics of the VMV occupancy [Nunes].

memory consumption because the VCV has the highest decoding rate (consumes memory faster than the B-VCV) and accumulates all the MBs (consumes all the needed memory).

For S-VOPs, the amount of picture memory required for the decoding of the VOP is defined as the number of MBs in the reconstructed VOP. The memory used for storing the sprite is not constrained by the VMV model.

The decoding duration of VOP i, T_i, is identical in the VCV and VMV models, and it starts at s_i and ends at e_i, as defined in Section A.1.2.2.

VMV Buffer Draining The VMV draining depends on the coding type of the VOP being decoded, as explained in the following paragraphs [MPEG4-2].

☞ **I- and P-VOPs**

At the VOP composition time (or presentation time in a decoder without a compositor) plus VCV latency, $\tau_i + L$, the total memory allocated to the previous I- or P-VOP in the decoding order is instantaneously released.

☞ **B-VOPs**

At the VOP composition time (or presentation time in a decoder without a compositor) plus VCV latency, $\tau_i + L$, the total memory allocated to the current B-VOP is instantaneously released.

A.1.3.3 VMV Model Constraints A given set of visual ESs building a scene conforms to a given profile@level, with respect to the VMV model, if it never overflows the VMV model buffer.

A.1.4 Interaction Between the VBV, VCV, and VMV Models

A given set of ESs building a visual scene is considered compliant with a given profile and level if it fulfills all the constraints defined by the three Video Buffering Verifier models. Bitstream compliance with a given profile@level guarantees that the resources required at the decoder do not exceed a certain defined amount corresponding to the relevant profile@level. Moreover, it defines strict timing for completion of decoding and composition of VOPs as explained in the following:

1. The VBV model defines the time at which the coded bits for each VOP are available for decoding and the time at which they should be removed from the VBV buffer—the coded bits for each VOP should be removed from the VBV buffer at the VOP decoding times, t_i, computed from the composition time information in the video ES or conveyed by systems decoding time stamps.

2. The VCV model defines the decoding speed of the MB data, and, thus, the time at which each VOP is available for composition—a given VOP should be available for composition, at most, at the VOP composition time plus the VCV latency (i.e., at the time it is supposed to be available to the compositor).

3. The VMV model defines the amount of picture memory allocated at each time instant and the time it should be released—a given VOP should be removed from the VMV buffer at its composition time plus the VCV latency (B-VOP) or at the composition time plus the VCV latency of the next P- or I-VOP.

The various models are independent but interact with each other as described in the following paragraphs.

The Decoder Cannot Arbitrarily Decode in Advance   From a decoding point of view, it could be advantageous to process the incoming data as far in advance as possible; however, this is constrained by two factors:

1. The decoder can only start decoding if the bits are available for decoding —*a constraint imposed by the VBV model.*

2. As the decoder decodes the incoming data, it generates macroblocks that consume picture memory. If the decoder decodes too fast, it may not have enough picture memory to store the decode data—*a constraint imposed by the VMV model.*

The Decoder Cannot Decode Too Late   If the decoder starts decoding too late, it may not be able to complete the decoding in time, and the following situations may occur:

1. The VOP bits may be removed from the bitstream buffer before they can be decoded.

2. The composition time for the current VOP may arrive without the VOP having been completely decoded.

3. The time to release the picture memory required for the prediction of the current VOP may arrive before the VOP can be decoded.

In order to avoid these situations, the Video Buffering Verifier mechanism imposes strict times for starting and ending any VOP decoding—a *constraint imposed by the VCV model.*

The Video Buffering Verifier models provide the mechanisms allowing any encoder to produce bitstreams that are decodable by any decoder compliant with the selected profile@level. This mechanism allows limiting the amount of decoding resources needed at the receiving terminals while simultaneously ensuring the timely reconstruction of the encoded information.

It is important to highlight that it is a major task of the encoder to simulate each of the Video Buffering Verifier models in order to produce bitstreams compliant with the intended profile and level. If any of these models tends to be violated, the encoder has to take appropriate countermeasures to avoid it. Although the Video Buffering Verifier is defined for the decoders, it is in fact a major module of any encoder generating compliant sets of bitstreams.

A.2 DEFINITION OF LEVELS FOR VIDEO PROFILES

Table A.1 describes the MPEG-4 Visual levels for the Version 1 and Version 2 profiles, including natural visual (or video) data (the so-called MPEG-4 video profiles). Note that Level 0 for the `simple` profile has been defined in the 2d amendment to the 2d Edition of the MPEG-4 Visual standard [N3904].

Table A.2 describes the MPEG-4 Visual levels for the `studio` profiles defined in the 1st amendment to the 2d Edition of the MPEG-4 Visual standard [N3898].

Table A.3 describes the MPEG-4 Visual levels for the `Advanced Simple` and `Fine Granularity Scalability` profiles defined in the 2d amendment to the 2d Edition of the MPEG-4 Visual standard [N3904].

A.3 DEFINITION OF LEVELS FOR SYNTHETIC PROFILES

This section describes the MPEG-4 Visual levels for the profiles including only synthetic visual data. Note that the profiles including only texture object types are considered synthetic profiles here because the video texture coding tool was developed by the MPEG SNHC subgroup with the initial target of coding textures to map over 2D and 3D meshes.

Table A.1 Levels for the video profiles

Visual profile	Level	Typical visual session size	Max. number of objects[1]	Max. number objects per type	Max. unique quant. tables	Max. VMV buffer size (MB units)[2]	Max. VCV buffer size (MB)[3]	VCV decoder rate (MB/s)[4]	VCV boundary MB decoder rate (MB/s)[5]	Max. total VBV buffer size (units of 16,384 bits)	Max. VOL VBV buffer size (units of 16,384 bits)[6]	Max. video packet length (bits)[7]	Max. sprite size (MB units)	Wavelet restrictions	Max. bit rate (kbit/s)	Max. enhancement layers/object
Simple[8]	L0	QCIF	1	1 × Simple	1	198	99	1,485	N.A.[9]	10	10	2,048	N.A.	N.A.	64	N.A.
Simple	L1	QCIF	4	4 × Simple	1	198	99	1,485	N.A.	10	10	2,048	N.A.	N.A.	64	N.A.
Simple	L2	CIF	4	4 × Simple	1	792	396	5,940	N.A.	40	40	4,096	N.A.	N.A.	128	N.A.
Simple	L3	CIF	4	4 × Simple	1	792	396	11,880	N.A.	40	40	8,192	N.A.	N.A.	384	N.A.
Advanced Real-Time Simple	L1	QCIF	4	4 × Simple or Adv. Real Time Simple	1	198	99	1,485	N.A.	10	10	8,192	N.A.	N.A.	64	N.A.
Advanced Real-Time Simple	L2	CIF	4	4 × Simple or Adv. Real-Time Simple	1	792	396	5,940	N.A.	40	40	16,384	N.A.	N.A.	128	N.A.
Advanced Real-Time Simple	L3	CIF	4	4 × Simple or Adv. Real-Time Simple	1	792	396	11,880	N.A.	40	40	16,384	N.A.	N.A.	384	N.A.
Advanced Real-Time Simple	L4	CIF	16	16 × Simple or Adv. Real-Time Simple	1	792	396	11,880	N.A.	80	80	16,384	N.A.	N.A.	2000	N.A.
Simple Scalable	L1	CIF	4	4 × Simple or Simple Scalable	1	1782	495	7,425	N.A.	40	40	2,048	N.A.	N.A.	128	1 spatial or temporal enhancement layer
Simple Scalable[10]	L2	CIF	4	4 × Simple or Simple Scalable	1	3168	792	23,760	N.A.	40	40	4,096	N.A.	N.A.	256	1 spatial or temporal enhancement layer
Core	L1	QCIF	4	4 × Core or Simple	4	594	198	5,940	2,970	16	16	4,096	N.A.	N.A.	384	1
Core	L2	CIF	16	16 × Core or Simple	4	2,376	792	23,760	11,880	80	80	8,192	N.A.	N.A.	2,000	1

Table A.1 Levels for the video profiles (Continued)

Visual profile	Level	Typical visual session size	Max. number of objects[1]	Max. number objects per type	Max. unique quant. tables	Max. VMV buffer size (MB units)[2]	Max. VCV buffer size (MB)[3]	VCV decoder rate (MB/s)[4]	VCV boundary MB decoder rate (MB/s)[5]	Max. total VBV buffer size (units of 16,384 bits)	Max. VOL VBV buffer size (units of 16,384 bits)[6]	Max. video packet length (bits)[7]	Max. sprite size (MB units)	Wavelet restrictions	Max. bit rate (kbit/s)	Max. enhancement layers/object
Advanced Core	L1	QCIF	4	4 × Core or Simple or Adv. Scalable Texture	4	594	198	5940	2970	16	8	4096	N.A.	See Table A.5	384	1
Advanced Core	L2	CIF	16	16 × Core or Simple or Adv. Scalable Texture	4	2376	792	23,760	11,880	80	40	8,192	N.A.	See Table A.5	2,000	1
Core Scalable	L1	CIF	4	4 × Core or Simple or Core Scalable or Simple Scalable	4	2,376	792	14,850	7,425	64	64	4,096	N.A.	N.A.	768	1
Core Scalable	L2	CIF	8	8 × Core or Simple or Core Scalable or Simple	4	2,970	990	29,700	14,850	80	80	4,096	N.A.	N.A.	1,500	1
Core Scalable	L3	CCIR601	16	16 × Core or Simple or Core Scalable or Simple Scalable	4	12,906	4,032	120,960	60,480	80	80	16,384	N.A.	N.A.	4,000	2
Main	L2	CIF	16	16 × Main or Core or Simple	4	3,960	1,188	23,760	11,880	80	80	8,192	1,584	Scalable Texture @L1	2,000	1

Table A.1 Levels for the video profiles (Continued)

Visual profile	Level	Typical visual session size	Max. number of objects[1]	Max. number objects per type	Max. unique quant. tables	Max. VMV buffer size (MB units)[2]	Max. VCV buffer size (MB)[3]	VCV decoder rate (MB/s)[4]	VCV boundary MB decoder rate (MB/s)[5]	Max. total VBV buffer size (units of 16,384 bits)	Max. VOL VBV buffer size (units of 16,384 bits)[6]	Max. video packet length (bits)[7]	Max. sprite size (MB units)	Wavelet restrictions	Max. bit rate (kbit/s)	Max. enhancement layers/object
Main	L3	CCIR 601	32	32 × Main or Core or Simple	4	11,304	3,240	97,200	48,600	320	320	16,384	6,480	Scalable Texture @L1	15,000	1
Main	L4	1920 × 1088	32	32 × Main or Core or Simple	4	65,344	16,320	489,600	244,800	760	760	16,384	65,280	Scalable Texture @L2	38,400	1
Advanced Coding Efficiency	L1	CIF	4	4 × Adv. Coding Efficiency or Core or Simple	4	1,188	792	11,880	5,940	40	40	8,192	N.A.	N.A.	384	1
Advanced Coding Efficiency	L2	CIF	16	16 × Adv. Coding Efficiency or Core or Simple	4	2,376	1,188	23,760	11,880	80	80	8,192	N.A.	N.A.	2,000	1
Advanced Coding Efficiency	L3	CCIR 601	32	32 × Adv. Coding Efficiency or Core or Simple	4	9,720	3,240	97,200	48,600	320	320	16,384	N.A.	N.A.	15,000	1
Advanced Coding Efficiency	L4	1920 × 1088	32	32 × Adv. Coding Efficiency or Core or Simple	4	48,960	16,320	489,600	244,800	760	760	16,384	N.A.	N.A.	38,400	1
N-Bit	L2	CIF	16	16 × Core or Simple or N-Bit	4	2,376	792	23,760	11,880	80	80	8,192	N.A.	N.A.	2,000	1

1. Enhancement layers are not counted as separate objects.

2. The maximum VMV buffer size is the bound on the memory (in macroblock units) which can be used by the VMV algorithm. This algorithm (see [MPEG4-2, subclause D.5]) models the pixel memory needed by the entire visual decoding process. This includes the memory needed for reference VOPs in the prediction of P-, B-, and S(GMC)-VOPs and the storage of the reconstructed VOPs until they are released by the decoder, plus the memory required to queue B-VOPs until composition occurs. For the profiles that contain more than one layer, the memory requirements include all base and enhancement layers. When belonging to different, overlapping objects, some of these macroblocks may overlay on the display; separate memory is required (prior to composition) in the VMV.

3. The maximum VCV buffer size (cumulative over all layers of all VOs) is twice the maximum number of macroblocks per VOP in the profile and level combination except for the Simple, Simple Scalable (Level 1), and Advanced Real-Time Simple profiles. For the Simple and the Advanced Real-Time Simple profiles, this value is the maximum number of macroblocks per VOP. For the Simple Scalable profile (Level 1), it is 1.25 times the maximum number of macroblocks per VOP. The limit applies both to the VCV buffer and the boundary MB VCV buffer.

4. The VCV decoder rate is the vcv_decoder_rate (H) referred to in [MPEG4-2, subclause D.4]; this parameter is the number of MB/s based on the typical spatial and temporal resolutions, as follows:

   • 1,485 MB/s corresponds to QCIF at 15 Hz
   • 5,940 MB/s corresponds to CIF at 15 Hz and also twice QCIF at 30 Hz
   • 11,880 MB/s corresponds to CIF at 30 Hz
   • 7,425 MB/s corresponds to 1.25 times CIF at 15 Hz
   • 23,760 MB/s corresponds to twice CIF at 30 Hz
   • 97,200 MB/s corresponds to twice ITU-R 601 at 30 Hz
   • 489,600 MB/s corresponds to twice 1920×1088 at 30 Hz

5. The VCV boundary MB decoder rate column bounds the number of macroblocks containing nontrivial shape information (boundary, not transparent or opaque). The VCV boundary MB decoder rate constrains the total number of boundary MBs in all VOLs, concurrently. Note that the boundary macroblocks are added to both the VCV and boundary MB VCV buffers.

6. The total (aggregated) vbv_buffer_size is the sum of the individual VBV buffer occupancies at any given time (in units of 16,384 bits) for all VOLs of all VOs. This total VBV size is limited by the profile and level.

7. The maximum video packet length is defined as the maximum number of bits of data_partitioned_motion_shape_texture() in one video packet. The constraint applies only when the data-partitioning tool is enabled in the bitstream. When data partitioning is disabled, there is no limit on the video packet length.

8. For Simple@Level 0, the following restrictions apply:

   • The maximum frame rate is 15 frames/s.
   • The maximum f_code is 1.
   • The intra_dc_vlc_threshold is 0.
   • The maximum horizontal luminance pixel resolution is 176 pels/line.
   • The maximum vertical luminance pixel resolution is 144 pels/VOP.
   • If AC/DC prediction is used, the following restriction applies: QP value will not be changed within a VOP (or within a video packet if video packets are used in a VOP). If AC/DC prediction is not used, there are no restrictions to changing QP value.

9. *Not Applicable.*

10. The conformance point for the base layer of the Simple Scalable profile is the Simple@L1 when Simple Scalable@L1 is used and Simple@L2 when Simple Scalable@L2 is used.

Table A.2 Levels for the Studio profiles

Visual profile	Level	Typical visual session formats[1]	Max. pixel depth	Max. number of objects	Max. number per type	Max. VMV buffer size (sample)[2]	Max. VCV buffer size (sample)[3]	VCV decoder rate (sample/s)	VCV boundary MB decoder rate (sample/s)	Max. total VBV buffer size	Max. VOL VBV buffer size	Max. video packet length (bits)	Max. sprite size (sample)[4]	Wavelet restrictions	Max. bit rate (Mbit/s)	Max. enhancement layers/object
Simple Studio	L1	ITU-R601:4224 ITU-R601:444	10	1	1 × Simple Studio	1313280	1313280	33177600	33177600	576	576	N.A.	N.A.	N.A.	180	N.A.
Simple Studio	L2	ITU-R709.601:422 ITU-R601:444444	10	1	1 × Simple Studio	4194304	4194304	125,829120	125,829120	1920	1920	N.A.	N.A.	N.A.	600	N.A.
Simple Studio	L3	ITU-R709.601:444 ITU-R709.601:4224	12	1	1 × Simple Studio	6291456	6291456	188,743680	188,743680	2880	2880	N.A.	N.A.	N.A.	900	N.A.
Simple Studio	L4	ITU-R709.60P:444 ITU-R709.601:444444 2K×2K×30P:444	12	1	1 × Simple Studio	12582912	12582912	377487360	377487360	4320	4320	N.A.	N.A.	N.A.	1800	N.A.
Core Studio	L1	ITU-R601:4224 ITU-R601:444	10	4	4 × Core Studio or Simple Studio	5253120	2626560	66355200	66355200	576	576	N.A.	8294400	N.A.	90	N.A.
Core Studio	L2	ITU-R709.601:422 ITU-R601:444444	10	4	4 × Core Studio or Simple Studio	16777216	8388608	251658240	251658240	1920	1920	N.A.	50135040	N.A.	300	N.A.

Table A.2 Levels for the Studio profiles (Continued)

Visual profile	Level	Typical visual session formats[1]	Max. pixel depth	Max. number of objects	Max. number per type	Max. VMV buffer size (sample)[2]	Max. VCV buffer size (sample)[3]	VCV decoder rate (sample/s)	VCV boundary MB decoder rate (sample/s)	Max. total VBV buffer size	Max. VOL VBV buffer size	Max. video packet length (bits)	Max. sprite size (sample)[4]	Wavelet restrictions	Max. bit rate (Mbit/s)	Max. enhancement layers/object
Core Studio	L3	ITU-R709. 601:444 ITU-R709. 601:4224	10	8	8 × Core Studio or Simple Studio	25165824	12582912	377487360	377487360	2880	2880	N.A.	75202560	N.A.	450	N.A.
Core Studio	L4	ITU-R709. 60P:444 ITU-R709. 60I:444444 2K×2K×30P:444	10	16	16 × Core Studio or Simple Studio	50331648	25165824	754974720	754974720	4320	4320	N.A.	150994944	N.A.	900	N.A.

1. ITU-R 709 is ITU-R BT. 709 and ITU-R 601 is ITU-R BT. 601; 444444 means 444(RGB) + 3 auxiliary channels; 4224 means 422(YUV)+ 1 auxiliary channel.
2. VMV is defined by the number of samples which belong to the bounding box of texture regardless of shape information. VMV also includes auxiliary channel samples.
3. VCV is defined by the number of samples which belong to the bounding box of texture regardless of shape information. VCV also includes auxiliary channel samples.
4. Maximum sprite size is defined by the number of samples for sprite memory.

Table A.3 Levels for the Advanced Simple and FGS profiles[1, 2, 3, 4]

Visual profile	Level	Typical visual session size	Max. number of objects	Max. number per type	Max. unique quant. tables	Max. VMV buffer size (MB units)	Max. VCV buffer size (MB)	VCV decoder rate (MB/s)	Max. percentage of intra MBs with AC/DC prediction in VCV buffer	Max. total VBV buffer size (units of 16384 bits)	Max. VOL VBV buffer size (units of 16384 bits)	Max. video packet length (bits)	Max. bit rate (kbit/s)[5]	Max. number of coded VOP-bit/s[6]
Adv.Sim.	L0	176×144	1	1× AS or Simple	1	297	99	2,970	100	10	10	2,048	128	N.A.
Adv.Sim.	L1	176×144	4	4× AS or Simple	1	297	99	2,970	100	10	10	2,048	128	N.A.
Adv.Sim.	L2	352×288	4	4× AS or Simple	1	1,188	396	5,940	100	40	40	4,096	384	N.A.
Adv.Sim.	L3	352×288	4	4× AS or Simple	1	1,188	396	11,880	100	40	40	4,096	768	N.A.
Adv.Sim.	L4	352×576	4	4× AS or Simple	1	2,376	792	23,760	50	80	80	8,192	3,000	N.A.
Adv.Sim.	L5	720×576	4	4× AS or Simple	1	4,860	1,620	48,600	25	112	112	16,384	8,000	4
FGS	L0	176×144	1	1× AS or FGS or Simple	1	297	99	2,970	100	10	10	2,048	128	4
FGS	L1	176×144	4	4× AS or FGS or Simple	1	297	99	2,970	100	10	10	2,048	128	4
FGS	L2	352×288	4	4× AS or Simple	1	1,188	396	5,940	100	40	40	4,096	384	4
FGS	L3	352×288	4	4× AS or FGS or Simple	1	1,188	396	11,880	100	40	40	4,096	768	4
FGS	L4	352×576	4	4× AS or FGS or Simple	1	2,376	792	23,760	50	80	80	8,192	3,000	4
FGS	L5	720×576	4	4× AS or FGS or Simple	1	4,860	1,620	48,600	25	112	112	16,384	8,000	4

1. The following restriction applies to Level 0 of Advanced Simple profile and FGS profile: If AC/DC prediction is used, the QP value will not be changed within a VOP (or within a video packet if video packets are used in a VOP). If AC/DC prediction is not used, there are no restrictions to changing the QP value.
2. The number of FGS, FGST, or FGS-FGST layers is always one. If the FGS layer and the FGST layer are separated, the number of total enhancement layers is two.
3. The interlace tools are not used for Levels L0, L1, L2, and L3 of the Advanced Simple and FGS profiles.
4. It is inherent in the FGS profile that the base and enhancement layers are tightly coupled to each other. To avoid unnecessary memory storage, the following constraints apply to the decoding time relationship of the enhancement layer and the base layer:
   • Decoding and composition (or presentation in a decoder without compositor) of each FGS or FGST VOP will be performed in the same time unit.
   • Decoding of each FGS and FGST VOP shall be performed immediately after the reference base layer VOP(s) are decoded without violating the preceding constraint.
5. For the FGS profile, this column is the maximum base-layer bit rate.
6. The maximum number of coded VOP-bit/s takes into consideration the shifted bits after applying frequency weighting and/or selective enhancement.

A.3.1 Scalable Texture Profile

This profile includes three levels defined in Table A.4.

A.3.2 Simple Face Animation Profile

All MPEG-4 facial animation decoders (for all object types) are required to generate as their output a facial model including all the feature points defined in MPEG-4 Visual, even if some of the feature points will not be affected by any information received from the encoder.

The Simple Face Animation object type is not required to implement the viseme_def/expression_def functionality [MPEG4-2]. The Simple Face Animation profile has two levels.

Level 1

1. Number of objects: 1
2. The total FAP decoding frame rate in the bitstream will not exceed 72 Hz
3. The decoder will be capable of a face model rendering update of at least 15 Hz
4. Maximum bit rate: 16 kbit/s

Level 2

1. Maximum number of objects: 4
2. The FAP decoding frame rate in the bitstream will not exceed 72 Hz (this means that the FAP decoding frame rate is to be shared among the objects)
3. The decoder will be capable of rendering the face models with the update rate of at least 60 Hz, sharable between faces, with the constraint that the update rate for each individual face is not required to exceed 30 Hz
4. Maximum bit rate: 32 kbit/s

A.3.3 Simple FBA Profile

All MPEG-4 Visual FBA decoders (for all object types) are required to generate at their output a humanoid model including all the feature points and joints defined in MPEG-4 Visual [MPEG4-2], even if some of the feature points and joints will not be affected by any information received from the encoder. The Simple FBA profile has two levels.

Level 1

1. Number of objects: 1
2. The total FBA decoding frame rate in the bitstream will not exceed 72 Hz
3. The decoder will be capable of a humanoid model rendering update of at least 15 Hz

4. Maximum bit rate: 32 kbit/s

5. The decoder is not required to animate Spine3, Spine4, and Spine5 BAP groups [MPEG4-2]

Level 2

1. Maximum number of objects: 4

2. The FBA decoding frame rate in the bitstream will not exceed 72 Hz (this means that the FBA decoding frame rate is to be shared among the objects)

3. The decoder will be capable of rendering the humanoid models with the update rate of at least 60 Hz, sharable among humanoids, with the constraint that the update rate for each individual humanoid is not required to exceed 30 Hz

4. Maximum bit rate: 64 kbit/s

A.3.4 Advanced Core and Advanced Scalable Texture Profiles

The levels for the Advanced Core and Advanced Scalable Texture profiles are defined in Table A.5. Notice that Advanced Core includes video as well as texture object types and thus has two types of level constraints (see Table A.1 and Table A.5).

A.4 DEFINITION OF LEVELS FOR SYNTHETIC AND NATURAL HYBRID PROFILES

The levels for the profiles supporting both video object types and synthetic visual object types are specified by giving bounds for the natural objects and for the synthetic objects. Parameters such as bit rate can be combined across natural and synthetic objects.

A.4.1 Basic Animated Texture Profile

The Basic Animated Texture profile has two levels.

Level 1 Equivalent to Simple Facial Animation@L1 + Scalable Texture @L1 + the following restrictions on Basic Animated Texture objects

1. Maximum number of mesh objects (with uniform topology): 4

2. Maximum total number of nodes (vertices) in mesh objects: 480 (equivalent to 4 times the number of nodes of a uniform mesh covering a QCIF image with 16×16 pixel elements)

3. Maximum frame rate of a mesh object: 30 Hz

4. Maximum bit rate of mesh objects: 64 kbit/s

Table A.4 Levels for the Scalable Texture profile

Profile	Level	Default wavelet filter	Max. download filter length	Max. no. of decomposition levels	Typical visual session size[1]	Max. Qp value (bits)	Max. number of pixels/session	Max. number of bit planes for DC values	VCV decoder rate (equivalent MB/s)[2]	Max. VCV buffer size (equivalent MB)[2]
Scalable Texture	L1	Integer	OFF	5	704×576	8	405,504	13	1,584	1,584
Scalable Texture	L2	Integer	ON, 15	8	2048×2048	10	4,194,304	16	16,384	16,384
Scalable Texture	L3	Float, Integer	ON, 15	10	8192×8192	12	67,108,864	18	262,144	262,144

1. This column is for informative use only. It provides an example configuration of the maximum number of pixels per session.
2. This still texture VCV model is separate from the global video VCV model. An equivalent MB corresponds to 256 pixels.

Table A.5 Levels for the Advanced Core and Advanced Scalable Texture profiles

Visual profile	Level	Default wavelet filter	Max. download filter length	Max. number of decomposition levels	Typical visual session size[1]	Max. Qp value (bits)	Max. number of pixels/session[2]	VCV decoder rate (equivalent MB/s)[3]	Max. number of bit planes for DC values	Max. VCV buffer size (equivalent MB)	Max. STO[4] packet length (bits)	Max. number of pixels/tile	Max. number of tiles
Advanced Core	L2	Integer	ON, 15	8	8192×8192	10	67,108,864	262,144	16	262,144	8,192	262,144	2,048
Advanced Core	L1	Integer	OFF	5	2048×2048	8	4,194,304	16,384	13	16,384	4,096	65,536	1,024
Advanced Scalable Texture	L3	Float, Integer	ON, 15	10	8192×8192	12	67,108,864	262,144	18	262,144	8,192	67,108,864	4,096
Advanced Scalable Texture	L2	Integer	ON, 15	8	2048×2048	10	4,194,304	16,384	16	16,384	4,096	4,194,304	2,048
Advanced Scalable Texture	L1	Integer	OFF	5	704×576	8	405,504	1,584	13	1,584	2,048	405,504 (4×CIF)	1,024

1. This column is for informative use only. It provides an example configuration of the maximum number of pixels per session.
2. When the number of pixels per session is larger than the maximum number of pixels per tile, tiling_disable shall be 0.
3. This still texture VCV model is separate from the global video VCV model. An equivalent MB corresponds to 256 pixels.
4. Still texture object.

Level 2   Equivalent to `Simple Facial Animation@L2` + `Scalable Texture @L2` + the following restrictions on `Basic Animated Texture` objects

1. Maximum number of mesh objects (with uniform topology): 8
2. Maximum total number of nodes (vertices) in mesh objects: 1,748 (equivalent to 4 times the number of nodes of a uniform mesh covering a CIF image with 16×16 pixel elements)
3. Maximum frame rate of a mesh object: 60 Hz
4. Maximum bit rate of mesh objects: 128 kbit/s

A.4.2 `Hybrid` Profile

The `Hybrid` profile has two levels.

Level 1   Equivalent to `Core@L1` + `Basic Animated Texture@L1` + the following restrictions on `Animated 2D Mesh` objects

1. Maximum number of mesh objects (with uniform or Delaunay topology): 4 (the same as the maximum number of objects in visual session)
2. Maximum total number of nodes (vertices) in mesh objects: 480 (equivalent to 4 times the number of nodes of a uniform mesh covering a QCIF image with 16×16 pixel elements)
3. Maximum frame rate of a mesh object: 30 Hz (the same as the maximum frame rate for video objects)
4. Maximum bit rate of mesh objects: 64 kbit/s

Level 2   Equivalent to `Core@L2` + `Basic Animated Texture@L2` + the following restrictions on `Animated 2D Mesh` objects

1. Maximum number of mesh objects(with uniform or Delaunay topology): 8 (the same as the maximum number of objects in visual session)
2. Maximum total number of nodes (vertices) in mesh objects: 1,748 (equivalent to 4 times the number of nodes of a uniform mesh covering a CIF image with 16×16 pixel elements)
3. Maximum frame rate of a mesh object: 60 Hz (2 × the maximum frame rate for video objects)
4. Maximum bit rate of mesh objects: 128 kbit/s

A.5 REFERENCES

[MPEG1-2] ISO/IEC 11172-2:1993. *Coding of Moving Pictures and Associated Audio for Digital Storage Media at Up to About 1.5 Mbit/s—Part 2: Video*, 1993.

[MPEG2-2] ISO/IEC 13818-2:2000. *Generic Coding of Moving Pictures and Associated Audio Information—Part 2: Video*, 2000.

[MPEG4-1] ISO/IEC 14496-1:2001. *Coding of Audio-Visual Objects—Part 1: Systems*, 2d Edition, 2001.

[MPEG4-2] ISO/IEC 14496-2:2001. *Coding of Audio-Visual Objects—Part 2: Visual*, 2d Edition, 2001.

[N3898] MPEG Final Draft Amendment. *Studio Profiles*. Doc. ISO/MPEG N3898, Pisa MPEG Meeting, January 2001.

[N3904] MPEG Final Draft Amendment. *Streaming Video Profiles*. Doc. ISO/MPEG N3904, Pisa MPEG Meeting, January 2001.

[Nunes] Nunes, P., *Rate Control for Object-Based Video Coding Architectures*, Ph.D. thesis, Instituto Superior Técnico, Lisboa, Portugal, to be submitted.

Levels for Audio Profiles

by Fernando Pereira

Keywords: audio profile, audio level, conformance

This appendix includes the profiles and levels specified in MPEG-4 Audio both in Version 1 and in Version 2 [MPEG4-3]. The MPEG-4 Audio standard defines (as of October 2001) 23 object types and 8 profiles as explained in Chapter 13.

Four audio profiles have been defined in MPEG-4 Audio Version 1: Main, Scalable, Speech, and Synthetic. Four additional profiles were specified in MPEG-4 Audio Version 2: High-Quality Audio, Low-delay Audio, Natural Audio, and Mobile Audio Internetworking (MAUI).

Tables 13.3 and 13.4 show which audio coding tools include each audio object type and which audio object types include each audio profile.

B.1 COMPLEXITY UNITS

Complexity units are defined to give an approximation of the decoder complexity in terms of processing power and RAM usage required for processing MPEG-4 Audio bitstreams in dependence of specific parameters.

The approximate processing power is given in *Processor Complexity Units* (PCU), specified in integer numbers of MOPS (Millions of Operations Per Second). The approximated RAM usage is given in *RAM Complexity Units*

(RCU), specified in (mostly) integer numbers of kWords (1,000 Words) [MPEG4-4]. The RCU numbers do not include working buffers that can be shared among different objects and/or channels.

If a level for a profile is specified by the maximum number of complexity units, then a flexible configuration of the decoder handling different types of objects is allowed under the constraint that both values for the total decoding complexity and sampling rate conversion (if needed) do not exceed this limit.

Table B.1 gives complexity estimates for the different audio object types.

Table B.1 Complexity of audio object types

Object type	Parameters[1]	PCU (MOPS per channel)	RCU (kWords per channel)
AAC Main[2]	$f_s = 48$ kHz	5	5
AAC LC[2]	$f_s = 48$ kHz	3	3
AAC SSR[2]	$f_s = 48$ kHz	4	3
AAC LTP[2]	$f_s = 48$ kHz	4	4
AAC Scalable[2,3]	$f_s = 48$ kHz	5	4
TwinVQ[2]	$f_s = 24$ kHz	2	3
CELP	$f_s = 8$ kHz	1	1
CELP	$f_s = 16$ kHz	2	1
CELP	$f_s = 8/16$ kHz (bandwidth scalable)	3	1
HVXC	$f_s = 8$ kHz	2	1
General MIDI		4	1
TTSI[4]			
Wavetable Synthesis	$f_s = 22.05$ kHz	Depends on bitstreams[5]	Depends on bitstreams[5]
Main Synthetic		Depends on bitstreams[5]	Depends on bitstreams[5]
Algorithmic Synthesis and AudioFX		Depends on bitstreams[5]	Depends on bitstreams[5]
ER AAC LC[2]	$f_s = 48$ kHz	3	3
ER AAC SSR[2]	$f_s = 48$ kHz	4	3
ER AAC LTP[2]	$f_s = 48$ kHz	4	4
ER AAC Scalable[2,3]	$f_s = 48$ kHz	5	4
ER TwinVQ[2]	$f_s = 24$ kHz	2	3

Table B.1 Complexity of audio object types (Continued)

Object type	Parameters[1]	PCU (MOPS per channel)	RCU (kWords per channel)
ER BSAC[2]	f_s = 48 kHz (input buffer size = 26,000bits)	4	4
	f_s = 48 kHz (input buffer size = 106,000 bits)	4	8
ER AAC LD[2]	f_s = 48 kHz	3	2
ER CELP	f_s = 8 kHz	2	1
	f_s = 16 kHz	3	1
ER HVXC	f_s = 8 kHz	2	1
ER HILN[3]	f_s = 16 kHz, n_s = 93	15	2
	f_s = 16 kHz, n_s = 47	8	2
ER Parametric[6,7]	f_s = 8 kHz, n_s = 47	4	2

1. f_s—sampling frequency; r_f—ratio of sampling rates; n_s—maximum number of sinusoids being synthesized.
2. PCU proportional to sampling frequency.
3. Includes core decoder.
4. The complexity for speech synthesis is not taken into account.
5. See MPEG-4 Part 4: Conformance Testing [MPEG4-4].
6. Parametric code in HILN mode; for HVXC mode, see ER HVXC.
7. PCU depends on f_s and n_s (see next paragraph).

The computational complexity of HILN depends on the sampling frequency, f_s, and the maximum number of sinusoids, n_s, to be synthesized simultaneously. The value of n_s for a frame is the total number of harmonic and individual lines synthesized in that frame (i.e., the number of starting + continued + ending lines). For f_s in kHz, the PCU in MOPS is calculated as follows:

$$PCU = (1 + 0.15 \times n_s) \times f_s / 16$$

The typical maximum values of n_s are 47 for 6 kbit/s and 93 for 16 kbit/s HILN bitstreams.

B.2 DEFINITION OF LEVELS FOR AUDIO PROFILES

The levels for the audio profiles are specified using the complexity units previously defined. A number of zero stages of interleaving for the error-protection (EP) tool indicates that error protection is not used for that particular level.

The notation used to specify the number of audio channels indicates the number of full-bandwidth channels and the number of low-frequency enhancement channels. For example, *5.1* indicates 5 full-bandwidth channels and one low-frequency enhancement channel.

B.2.1 Main Profile

Since the Main profile includes natural and synthetic object types in MPEG-4 Audio Version 1, levels are defined as a combination of the two different types of levels using the two different metrics defined for natural audio tools (computation-based metrics) and synthetic audio tools (macro-oriented metrics).

Four levels are defined for the Main profile as a combination of constraints to the set of natural objects (corresponding to the object types not included in the Synthetic profile) and synthetic objects (corresponding to the object types in the Synthetic profile) in the scene.

For the set of natural audio objects in the scene, the following constraints are defined using the complexity units (PCU, RCU):

- ☞ **Natural Audio 1:** PCU < 40, RCU < 20
- ☞ **Natural Audio 2:** PCU < 80, RCU < 64
- ☞ **Natural Audio 3:** PCU < 160, RCU < 128
- ☞ **Natural Audio 4:** PCU < 320, RCU < 256

For the set of synthetic audio objects in the scene, the constraints correspond to the three levels defined in Section B.2.4 for the Synthetic profile.

In conclusion, the four levels for the Main profile are defined by the combination of the constraints on the natural and synthetic objects:

- ☞ **Level 1:** Natural Audio 1 + Synthetic Audio Level 1
- ☞ **Level 2:** Natural Audio 2+ Synthetic Audio Level 1
- ☞ **Level 3:** Natural Audio 3: + Synthetic Audio Level 2
- ☞ **Level 4:** Natural Audio 4 + Synthetic Audio Level 3

B.2.2 Scalable Profile

Three levels are defined by configuration; complexity units define the fourth level:

- ☞ **Level 1:** Maximum sampling rate of 24 kHz, one mono object (all object types)
- ☞ **Level 2:** Maximum sampling rate of 24 kHz, one stereo object or two mono objects (all object types)
- ☞ **Level 3:** Maximum sampling rate of 48 kHz, one stereo object or two mono objects (all object types)

☞ **Level 4:** Maximum sampling rate of 48 kHz, one 5.1 channels object or multiple objects with, at maximum, one integer factor sampling rate conversion for a maximum of two channels; flexible configuration is allowed with PCU < 30 and RCU < 19

B.2.3 Speech Profile

Two levels are defined in terms of the number of objects:[1]

☞ **Level 1**: One speech object
☞ **Level 2**: Up to 20 speech objects

B.2.4 Synthetic Profile

Three levels are defined for this profile:

Level 1

1. Low processing (exact numbers found in MPEG-4 Part 4: Conformance Testing [MPEG4-4])
2. Only core sample rates may be used
3. No more than one TTSI object

Level 2

1. Medium processing (exact numbers in MPEG-4 Part 4: Conformance Testing [MPEG4-4])
2. Only core sample rates may be used
3. No more than four TTSI objects

Level 3

1. High processing (exact numbers in MPEG-4 Part 4: Conformance Testing [MPEG4-4])
2. No more than twelve TTSI objects

For the case of scalable coding schemes, only the first instantiation of each object type will be counted to determine the number of objects relevant to the level definition and complexity metric. For example, in a scalable coder consisting of a CELP core coder and two enhancement layers implemented by means of AAC LC scalable objects, one CELP object and one AAC LC scalable object and their associated complexity metrics are counted because almost no overhead is associated with the second (and any further) generic audio enhancement layer.

---

1. See Table B.1 for the complexity of the various types of speech objects.

B.2.5 High-Quality Audio **Profile**

Eight levels are defined for this profile as shown in Table B.2.

B.2.6 Low-delay Audio **Profile**

Eight levels are defined for this profile as shown in Table B.3.

B.2.7 Natural Audio **Profile**

Four levels are defined for this profile as shown in Table B.4. Note that no RCU limitations are specified for this profile.

B.2.8 Mobile Audio Internetworking **Profile**

Six levels are defined for this profile as shown in Table B.5.

Table B.2 Levels for the High-Quality Audio profile

Level	Max. channels/ object	Max. sampling rate [kHz]	Max. PCU[1]	Max. RCU[1]	EP-Tool: Max. redundancy by class FEC[2](%)	EP-Tool: Max. no. of stages of inter-leaving /object
1	2	22.05	5	8	0	0
2	2	48	10	8	0	0
3	5.1	48	25	12^3	0	0
4	5.1	48	100	42^3	0	0
5	2	22.05	5	8	20	9
6	2	48	10	8	20	9
7	5.1	48	25	12^3	20	22
8	5.1	48	100	42^3	20	22

1. Levels 5 to 8 do not include RAM and computational complexity for the EP tool.
2. This number does not cover Forward Error Correction (FEC) for the EP header (i.e., FEC for the EP header is always permitted). In case of several audio objects, the limit is valid independently for each audio object. This value is the maximum redundancy for the audio object, which has the longest frame length, for each profile@level.
3. Sharing of work buffers among multiple objects or channel-pair elements is assumed.

Table B.3 Levels for the Low-delay Audio profile

Level	Max. channels/ object	Max. sampling rate [kHz]	Max. PCU[1]	Max. RCU[1]	EP-Tool: Max. redundancy by class FEC[2](%)	EP-Tool: Max. no. of stages of inter-leaving/ object
1	1	8	2	1	0	0
2	1	16	3	1	0	0
3	1	48	3	2	0	0
4	2	48	24	12[3]	0	0
5	1	8	2	1	100	5
6	1	16	3	1	100	5
7	1	48	3	2	20	5
8	2	48	24	12[3]	20	9

1. Levels 5 to 8 do not include RAM and computational complexity for the EP tool.
2. This number does not cover FEC for the EP header (i.e., FEC for the EP header is always permitted). In case of several audio objects, the limit is valid independently for each audio object. This value is the maximum redundancy for the audio object, which has the longest frame length, for each profile@level.
3. Sharing of work buffers among multiple objects or channel-pair elements is assumed.

Table B.4 Levels for the Natural Audio profile

Level	Max. sampling rate [kHz]	Max. PCU[1]	EP-Tool: Max. redundancy by class FEC[2](%)	EP-Tool: Max. no. of stages of interleaving / object
1	48	20	0	0
2	96	100	0	0
3	48	20	20	9
4	96	100	20	22

1. Levels 3 and 4 do not include computational complexity for the EP tool.
2. This number does not cover FEC for the EP header (i.e., FEC for the EP header is always permitted). In case of several audio objects, the limit is valid independently for each audio object. This value is the maximum redundancy for the audio object, which has the longest frame length, for each profile@level.

Table B.5 Levels for the `Mobile Audio Internetworking` profile

Level	Max. channels/ object	Max. sampling rate [kHz]	Max. PCU[1]	Max. RCU[1,2]	Max. no. of audio objects	EP-Tool: Max. redundancy by class FEC[3](%)	EP-Tool: Max. no. of stages of inter-leaving/ object
1	1	24	2.5	4	1	0	0
2	2	48	10	8	2	0	0
3	5.1	48	25	12[4]		0	0
4	1	24	2.5	4	1	20	5
5	2	48	10	8	2	20	9
6	5.1	48	25	12[4]		20	22

1. Levels 4 to 6 do not include RAM and computational complexity for the EP tool.
2. The maximum RCU for one channel in any object in this profile is four. For the ER BSAC, this limits the input buffer size. The maximum possible input buffer size in bits is given in [MPEG4-3].
3. This number does not cover FEC for the EP header (i.e., FEC for the EP header is always permitted). In case of several audio objects, the limit is valid independently for each audio object. This value is the maximum redundancy for the audio object, which has the longest frame length, for each profile@level.
4. Sharing of work buffers among multiple objects or channel-pair elements is assumed.

B.3 REFERENCES

[MPEG4-3] ISO/IEC 14496-3:2001. *Coding of Audio-Visual Objects—Part 3: Audio*, 2d Edition, 2001.

[MPEG4-4] ISO/IEC 14496-4:2001. *Coding of Audio-Visual Objects—Part 4: Conformance Testing*, 2d Edition, 2001.

Levels for Graphics Profiles

by Fernando Pereira

Keywords: graphics profile, graphics level, conformance

*A*s explained in Chapter 13, the MPEG-4 Systems standard defines (as of December 2001) four graphics profiles: Simple 2D, Complete 2D, Complete, and 3D Audio.

Moreover, four additional graphics profiles were under study or already included in amendments under ballot: X3D Interactive was under study [N4506] and Simple 2D+Text, Core 2D, and Advanced 2D were under ballot [N4408].

By December 2001, no levels had been approved for the graphics profiles but the levels presented in the following were already included in amendments under ballot [N4408]. The following sections present the levels defined or proposed for the MPEG-4 graphics profiles already approved or under ballot.

C.1 SIMPLE 2D PROFILE

The Simple 2D graphics profile provides only those BIFS graphics elements that are necessary to place one or more visual objects in a scene.

Level 1 The restrictions listed in Table C.1 apply for the Simple 2D graphics profile at Level 1 [N4408].

Table C.1 Restrictions for Level 1 of the `Simple 2D` profile

Nodes	Restrictions
Appearance	`material` not supported `textureTransform` not supported
Bitmap	No restriction
Shape	No restriction

C.2 SIMPLE 2D + TEXT PROFILE

The `Simple 2D + Text` graphics profile is designed for applications in which the only graphics to be used are text elements (perhaps colored or transparent, maybe in addition to audio and visual objects).

Level 1 The restrictions listed in Table C.2 apply for the `Simple 2D + Text` graphics profile at Level 1 [N4408].

C.3 CORE 2D PROFILE

The `Core 2D` graphics profile is designed for applications using some simple graphics elements (perhaps in addition to audio and visual objects).

Table C.2 Restrictions for Level 1 of the `Simple 2D + Text` profile

Nodes	Restrictions
Appearance	`textureTransform` not supported
Background2D	Only 1 `Background2D` node allowed in a scene for color-only background `url, set_bind` not supported
Bitmap	No restriction
FontStyle	No restriction
Material2D	`lineProps` not supported Used for colored and/or transparent text and transparent visual objects
Rectangle	No restriction
Shape	No restriction
Text	`maxExtend` not supported No texture mapping allowed 1,200 characters maximum in the scene at once

Level 1 The restrictions listed in Table C.3 apply for the `Core` 2D graphics profile at Level 1 [N4408].

Table C.3 Restrictions for Level 1 of the `Core` 2D profile

Nodes	Restrictions
`Appearance`	`textureTransform` not supported
`Background2D`	Only 1 `Background2D` node allowed in a scene for color and image background only `set_bind` not supported
`Bitmap`	No restriction
`Circle`	No texture mapping allowed
`Color`	No restriction (not used at this level)
`Coordinate2D`	4 points maximum
`FontStyle`	No restriction
`IndexedFaceSet2D`	15 `IndexedFaceSet2D` nodes maximum in a scene `set_colorIndex, set_coordIndex, set_texCoordIndex` not supported `EventIns` are ignored; the only field that can be modified is `coord` `color` not supported `colorIndex, colorPerVertex, texCoordIndex` not supported The number of points is restricted to be equal to 4 (quadrilateral) `texCoord` field is always considered to be (00 10 11 01) `coordIndex` field is always considered to be (0 1 2 3 –1) `convex` is always considered to be TRUE Face list shall be defined as follows: • Each face contains at least three noncoincident vertices • A given `coordIndex` is not repeated in a face • The vertices of a face shall define a planar polygon • The vertices of a face shall not define a self-intersecting polygon
`Material2D`	`lineProps` not supported
`PixelTexture`	32×32 maximum image size 8 `PixelTexture` nodes maximum in a scene at once
`Rectangle`	No restriction
`Shape`	No restriction
`Text`	`maxExtend` not supported No texture mapping allowed 6,480 characters maximum in the scene at once

Level 2   The restrictions listed in Table C.4 apply for the `Core 2D` graphics profile at Level 2 [N4408].

Table C.4 Restrictions for Level 2 of the `Core 2D` profile

Nodes	Restrictions
Appearance	`textureTransform` not supported
Background2D	Only 1 `Background2D` node allowed in a scene for color and image background only `set_bind` not supported
Bitmap	No restriction
Circle	No texture mapping allowed
Color	255 colors maximum in the scene at a time
Coordinate2D	255 points maximum in the scene at a time
FontStyle	No restriction
IndexedFaceSet2D	31 `IndexedFaceSet2D` nodes maximum in a scene `set_colorIndex, set_coordIndex, set_texCoordIndex` not supported `EventIns` are ignored `colorIndex, colorPerVertex, texCoordIndex` not supported `convex` is always considered to be TRUE 255 total indices maximum in all index fields in the scene Face list shall be defined as follows: • Each face is terminated with −1, including the last face in the array • Each face contains at least three noncoincident vertices • A given `coordIndex` is not repeated in a face • The vertices of a face shall define a planar polygon • The vertices of a face shall not define a self-intersecting polygon
Material2D	`lineProps` not supported
PixelTexture	32×32 maximum image size 8 `PixelTexture` nodes maximum in a scene at a time
Rectangle	No restriction
Shape	No restriction
Text	`maxExtend` not supported No texture mapping allowed 6,480 characters maximum in the scene at a time

C.4 ADVANCED 2D PROFILE

The Advanced 2D graphics profile is designed for applications using advanced graphics elements (possibly in addition to audio and visual objects).

Level 1 The restrictions listed in Table C.5 apply for the Advanced 2D graphics profile at Level 1 [N4408]:

Table C.5 Restrictions for Level 1 of the Advanced 2D profile

Nodes	Restrictions
Appearance	No restriction
Background2D	No restriction
Bitmap	No restriction
Circle	No restriction
Color	65,535 colors maximum in the scene at once
Coordinate2D	65,535 points maximum in the scene at once
Curve2D	255 Curve2D nodes maximum in a scene 65,535 total elements in all type fields in the scene
FontStyle	No restriction
IndexedFaceSet2D	255 IndexedFaceSet2D nodes maximum in a scene 65,535 total indices maximum in all index fields in the scene- Face list shall be defined as follows: • Each face is terminated with –1, including the last face in the array • Each face contains at least three noncoincident vertices • A given coordIndex is not repeated in a face • The vertices of a face shall define a planar polygon • The vertices of a face shall not define a self-intersecting polygon
IndexedLineSet2D	255 IndexedLineSet2D nodes maximum in a scene 65,535 total indices maximum in all index fields in the scene
LineProperties	No restriction
Material2D	No restriction
MaterialKey	No restriction
MatteTexture	Only one MatteTexture node allowed No BLUR function for s>2 Total pixel area of MatteTexture less than or equal to twice CIF

Table C.5 Restrictions for Level 1 of the Advanced 2D profile (Continued)

Nodes	Restrictions
PixelTexture	32×32 maximum image size 8 PixelTexture nodes maximum in a scene at once
Rectangle	No restriction
Shape	No restriction
Text	12,288 characters maximum in the scene at once
TextureCoordinate	65,535 coordinates maximum in the scene at once
TextureTransform	No restriction

Level 2   The restrictions listed in Table C.6 apply for the Advanced 2D graphics profile at Level 2 [N4408].

Table C.6 Restrictions for Level 2 of the Advanced 2D profile

Nodes	Restrictions
Appearance	No restriction
Background2D	No restriction
Bitmap	No restriction
Circle	No restriction
Color	65,535 colors maximum in the scene at once
Coordinate2D	65,535 points maximum in the scene at once
Curve2D	32,767 Curve2D nodes maximum in a scene 1,048,575 total elements in all type fields in the scene
FontStyle	No restriction
IndexedFaceSet2D	32,767 IndexedFaceSet2D nodes maximum in a scene 1,048,575 total indices maximum in all index fields in the scene
IndexedLineSet2D	32,767 IndexedLineSet2D nodes maximum in a scene 1,048,575 total indices maximum in all index fields in the scene
LineProperties	No restriction
Material2D	No restriction
MaterialKey	No restriction

Table C.6 Restrictions for Level 2 of the `Advanced 2D` profile (Continued)

Nodes	Restrictions
`MatteTexture`	No BLUR function for s>2 Total pixel area of `MatteTexture` less than or equal to twice CIF
`PixelTexture`	32×32 maximum image size 8 `PixelTexture` nodes max in a scene at once
`Rectangle`	No restriction
`Shape`	No restriction
`Text`	1,048,575 characters maximum in the scene at once
`TextureCoordinate`	65,535 coordinates maximum in the scene at once
`TextureTransform`	No restriction

C.5 REFERENCES

[N4408] MPEG Systems. *Text of ISO / IEC 14496-1:2001 / FDAM2—Extensible MPEG-4 Textual Format (XMT), Systems Extensions and Additional Profiles.* Doc. ISO/MPEG N4408, Pattaya MPEG Meeting, December 2001.

[N4506] MPEG Requirements. *MPEG-4 Profiles Under Consideration.* Doc. ISO/MPEG N4506, Pattaya MPEG Meeting, December 2001.

Levels for Scene Graph Profiles

by Fernando Pereira

Keywords: scene graph profile, scene graph level, conformance

As explained in Chapter 13, the MPEG-4 Systems standard defines (as of December 2001) five scene graph profiles: Simple 2D, Complete 2D, Complete, Audio, and 3D Audio [N3850, N4264]. From these profiles, only Complete 2D and Complete had no levels defined.

Moreover, six additional scene graph profiles were under study or already included in amendments under ballot in December 2001: X3D Interactive and Advanced Main 2D[1] were under study [N4506] and Basic 2D, Core 2D, Advanced 2D, and Main 2D were under ballot [N4408].

The following sections present the levels defined or proposed for the MPEG-4 scene graph profiles already approved or under ballot.

D.1 SIMPLE 2D PROFILE

The Simple 2D scene graph profile provides only those BIFS scene graph elements necessary to place one or more audiovisual objects in a scene (e.g., for broadcast television). It allows the presentation of audiovisual content with

---

1. Advanced Main 2D was removed from the profiles under consideration list in March 2002.

potential update of the complete scene but no interaction capabilities. There is one level (Level 1) already approved for this profile and another one is under ballot (Level 2).

Level 1   This level defines a scene that includes only audio and video objects; there are no capabilities to transform or manipulate the objects in the scene. It is intended for very simple, low-complexity applications with image/video composition in 2D.

The restrictions listed in Table D.1 apply for the Simple 2D scene graph profile at Level 1 [N3850].

The metric shall be the pixel metrics and BIFSConfig.isPixel=1.

A cascade of Transform2D nodes is not allowed. Children nodes of a Transform2D node shall not be Transform2D nodes. Only one initial update to convey the complete scene graph is allowed.

Level 2   The restrictions listed in Table D.2 apply for the Simple 2D scene graph profile at Level 2 [N4408].

For Simple 2D@L2, the maximum number of nodes in a scene is limited to 64, including all instances of these nodes through the DEF/USE mechanism.

D.2 AUDIO PROFILE

The Audio scene graph profile provides for a set of BIFS scene graph elements for usage in audio-only applications (e.g., radio broadcasting).

Levels 1 to 4   Levels for the Audio scene graph impose the limitations on the BIFS tools using the parameters presented in Table D.3 [N4264].

Table D.1 Restrictions on the Transform2D node for Level 1 of the Simple 2D profile

Field Name	Restrictions
addChildren	Ignored
removeChildren	Ignored
children	X
center	Ignored
rotationAngle	0
scale	1, 1
scaleOrientation	0
translation	X

X = allowed; else: default value

Table D.2 Restrictions for the `Simple 2D` profile at Level 2

Nodes	Restrictions
OrderedGroup	addChildren and removeChildren not supported order not supported 31 children maximum allowed
Sound2D	intensity, spatialize, location not supported
Transform2D	addChildren and removeChildren not supported center, rotationAngle, ScaleOrientation not supported (only translations are allowed) 31 children maximum allowed
Scene Update	No restriction
AudioClip	pitch, description not supported
AudioSource	addChildren, removeChildren, children, pitch, speed, numChan, phaseGroup not supported
ImageTexture	repeatS, repeatT not supported
MovieTexture	speed ignored (no rewind nor fast forward) repeatS, repeatT not supported

Table D.3 BIFS complexity restriction parameters

BIFS Tool	Restriction Parameters
BIFS Field Update	Maximum reaction time until a BIFS field update is achieved
AudioMix, AudioSwitch, AudioSource	Maximum width, maximum depth of the subtree, click-free switching
AudioDelay, AudioClip, AudioBuffer	Total buffer memory, click-free delay
Sample Rate Conversion	Total conversion-processing power, sample-rate conversion ratios
AudioFX	According to the restrictions of SA approved by the Audio group (SAOL-level definition based on the complexity metrics)
Sound, Sound2D	#spatialized

For the definition of the levels presented in Table D.4, the following parameter definitions are relevant:

☞ **Reaction time of a BIFS field update:** Maximum time in ms until the changes are audible

☞ **Audio subtree depth:** Maximum number of consecutive nodes from the output of an `AudioSource` or `AudioClip` node to the input of a `Sound/Sound2D` node

☞ **Audio subtree width:** Maximum number of parallel channels from the output of an `AudioSource` or `AudioClip` node to the input of a `Sound/Sound2D` node

☞ **Total memory buffer:** Amount of memory needed to store samples shared between the different `AudioDelay`, `AudioClip`, and `AudioBuffer` nodes present in a scene according to the formula:

Total Memory = SUM[NbChannels(j) × NbBufferedSamples(j)]
where
    j is the considered node
    NbChannels(j) is the number of channels for this node
    NbBufferedSamples(j) = Delay(j) × SamplingFrequency(j)

☞ **Total conversion processing power:** Amount of PCU shared among the different sampling conversions present in a scene according to the complexity units defined for the audio levels (see Appendix B)

☞ **Spatialization:** Number of possible spatialized channels

Levels for the `Audio` scene graph profile are defined in Table D.4.

For the `AudioFX` node, the complexity limitations presented in Table D.5 apply.

Table D.4 Restrictions for the `Audio` profile at Levels 1 to 4

Audio Parameter	Level 1	Level 2	Level 3	Level 4
Reaction Time (ms)	64	32	32	16
Width	8	32	64	128
Depth	1	4	6	8
Click-Free Fadings	Not required	Yes[1]	Yes[1]	Yes—HQ[2]
Total Memory Buffer	256 ksamples	512 ksamples	2 Msamples	6 Msamples (2 s for 64 channels at 48 kHz)
Sampling Ratio Conversion Factor	1	Integer	Any allowed ratio	Any allowed ratio

Table D.4 Restrictions for the `Audio` profile at Levels 1 to 4 (Continued)

Audio Parameter	Level 1	Level 2	Level 3	Level 4
Total Conversion Processing Power	0 (sampling rate conversion is forbidden)	16 PCU	64 PCU	128 PCU
`AudioFX` Complexity	Very Low Complexity (see Table D.5)	Low Complexity (see Table D.5)	Medium Complexity (see Table D.5)	High Complexity (see Table D.5)
Spatialization	0	4	16	32

[1]Indicates that a simple click-free fading technique (e.g., linear interpolation) is required.
[2]Indicates that a high-quality click-free technique must be supported.

Table D.5 Complexity restrictions for `AudioFX` node levels

Parameter	Very Low Complexity (Level 1)	Low Complexity (Level 2)	Medium Complexity (Level 3)	High Complexity (Level 4)
Total opcode Calls	1 M	1 M	4 M	8 M
Floating-Point Operations	0	4 M	12 M	20 M
Multiplications	0	2 M	8 M	16 M
Tests	0	1 M	4 M	8 M
Math Methods	0	2 M	6 M	12 M
Noise Generators	0	0.05 M	0.2 M	0.5 M
Interpolations	0	0.3 M	1.2 M	2 M
Multiply-and-Add	2 M	2 M	4 M	8 M
Filters	0.2 M	0.2 M	1 M	4 M
Effects	96 k	96 k	0.4 M	2 M
Allocated Memory	96 k	96 k	1 M	16 M

D.3 3D Audio Profile

The 3D Audio scene graph profile provides for a set of BIFS scene graph elements for usage in audio-only applications.

Levels 1 to 4 The levels specified in Table D.6 are based on a sampling rate of 44,100 Hz at 16-bit resolution [N4264].

Table D.6 Restrictions for the 3D Audio profile at Levels 1 to 4

	Level 1	Level 2	Level 3	Level 4
Maximum number of spatialized sources per scene[1]	8	32	64	128
Number of temporal sections whose levels and time limits can be controlled individually for each source	1	1	2	3
Maximum number of independent late reverberation processes per scene	1	1	2	4
Maximum number of control frequencies in reverberation process filters, source directivity filters, and material filters	2	2	3	3

[1]These spatialized sources can include discrete reflections that are perceptually equivalent to individual sound sources.

D.4 BASIC 2D PROFILE

The Basic 2D scene graph profile is designed for very simple scenes that may handle only few (possibly only one) audio and visual elements. This profile includes basic 2D composition and audio and video node interfaces. The application area for the Basic 2D scene graph profile is related to audio-video-only scene description.

The only allowed BIFS node for audio objects is AudioSource. The allowed BIFS nodes for visual objects are ImageTexture and MovieTexture.

Level 1 Level 1 of the Basic 2D scene graph profile is used to describe one audio and/or one visual object only. The restrictions listed in Table D.7 apply for the Basic 2D scene graph profile at Level 1 [N4408].

The usage of repeated OrderedGroup nodes to build bigger scenes is forbidden.

D.5 CORE 2D PROFILE

The Core 2D scene graph profile includes basic 2D composition, 2D texturing, local interaction, local animation, BIFS updates, quantization, and access to Web links and subscenes, in addition to audio and visual elements. It also introduces tools such as back channel (ServerCommand) and VoD features (MediaControl and MediaSensor).

Level 1 The restrictions listed in Table D.8 apply for the Core 2D scene graph profile at Level 1 [N4408].

Table D.7 Restrictions for Level 1 of the `Basic` 2D profile

Nodes	Restrictions
OrderedGroup	addChildren and removeChildren not supported 2 children maximum allowed Only 1 OrderedGroup node per scene used as top node
Sound2D	intensity, spatialize, location not supported
AudioSource	addChildren, removeChildren, children, pitch, speed, numChan, phaseGroup not supported
ImageTexture	repeatS, repeatT not supported
MovieTexture	speed ignored (no rewind nor fast forward) repeatS, repeatT not supported

Table D.8 Restrictions for Level 1 of the `Core` 2D profile

Nodes	Restrictions
Anchor	addChildren and removeChildren not supported 31 children maximum allowed
ColorInterpolator	255 key-value pairs
Conditional	No restriction
CoordinateInterpolator2D	31 coordinates per keyValue 255 key-value pairs
Inline	No restriction
InputSensor	Restricted to mice, keyboards, remote controls
PositionInterpolator2D	255 key-value pairs
ROUTE	No restriction
ScalarInterpolator	255 key-value pairs
TimeSensor	Ignored if cycleInterval < 0.03 s
TouchSensor	No restriction
MediaControl	mediaSpeed not supported (no rewind nor fast forward) 31 url maximum
MediaSensor	info ignored
OrderedGroup	addChildren and removeChildren not supported 31 children maximum allowed

Table D.8 Restrictions for Level 1 of the `Core 2D` profile (Continued)

Nodes	Restrictions
`QuantizationParameter`	`isLocal, useEfficientCoding` not supported
`ServerCommand`	No restriction
`Sound2D`	`intensity, spatialize, location` not supported
`Switch`	No restriction
`Transform2D`	`addChildren` and `removeChildren` not supported `center, rotationAngle, scaleOrientation` not supported (only translations are allowed) `31 children` maximum allowed
`Valuator`	No restriction
`Node Update`	Add and remove commands for `children` fields are not allowed
`ROUTE Update`	No restriction
`Scene Update`	No restriction
`AudioClip`	`pitch, description` ignored
`AudioSource`	`addChildren, removeChildren, children, pitch, speed, numChan, phaseGroup` not supported (no rewind nor fast forward)
`ImageTexture`	`repeatS, repeatT` not supported
`MovieTexture`	`speed` ignored (no rewind nor fast forward) `repeatS, repeatT` not supported

The maximum number of nodes that is allowed in a scene compliant with `Core 2D@L1` is 8,191, including all instances of these nodes through the DEF/USE mechanism.

Level 2   The restrictions listed in Table D.9 apply for the `Core 2D` scene graph profile at Level 2 [N4408].

Table D.9 Restrictions for Level 2 of the `Core 2D` profile

Nodes	Restrictions
`Anchor`	`127 children` maximum allowed
`ColorInterpolator`	255 key-value pairs
`Conditional`	No restriction

Table D.9 Restrictions for Level 2 of the Core 2D profile (Continued)

Nodes	Restrictions
CoordinateInterpolator2D	127 coordinates per keyValue 255 key-value pairs
Inline	No restriction
InputSensor	restricted to mice, keyboards, remote controls
MediaControl	mediaSpeed not supported (no rewind nor fast forward) 127 url maximum
MediaSensor	info ignored
OrderedGroup	127 children maximum allowed
PositionInterpolator2D	255 key-value pairs
QuantizationParameter	isLocal, useEfficientCoding not supported
ROUTE	No restriction
ScalarInterpolator	255 key-value pairs
Sound2D	intensity, spatialize, location not supported
ServerCommand	No restriction
Switch	No restriction
TimeSensor	Ignored if cycleInterval < 0.03 s
TouchSensor	No restriction
Transform2D	addChildren and removeChildren not supported center, rotationAngle, scaleOrientation not supported (only translations are allowed) 127 children maximum allowed
Node Update	No restriction
ROUTE Update	No restriction
Scene Update	No restriction
Valuator	No restriction
AudioClip	pitch, description ignored
AudioSource	addChildren, removeChildren, children, pitch, speed, numChan, phaseGroup not supported (no rewind nor fast forward)
ImageTexture	repeatS, repeatT not supported
MovieTexture	speed ignored repeatS, repeatT not supported

The maximum number of nodes that is allowed in a scene compliant with `Core 2D@L2` is 32,767, including all instances of these nodes through the `DEF`/`USE` mechanism.

D.6 ADVANCED 2D PROFILE

The `Advanced 2D` scene graph profile comprises all `Basic 2D` and `Core 2D` scene graph functionalities. The `Advanced 2D` scene graph profile also allows advanced 2D composition, advanced local interaction, streamed animation (`BIFS-Anim`), scripting, advanced audio, and `PROTO`s.

Level 1 The restrictions listed in Table D.10 apply for the `Advanced 2D` scene graph profile at Level 1 [N4408].

The maximum number of nodes that is allowed in a scene compliant with `Advanced 2D@L1` is 32,767, including all instances of these nodes through the `DEF`/`USE` or `PROTO` mechanisms.

Table D.10 Restrictions for Level 1 of the `Advanced 2D` profile

Nodes	Restrictions
Anchor	511 `children` maximum allowed
AnimationStream	No restriction
AudioBuffer	No restriction
AudioClip	`pitch` not supported
AudioSource	63 `children` maximum allowed `pitch`, `speed`, `numChan`, `phaseGroup` not supported (no rewind nor fast forward)
AudioSwitch	No restriction
ColorInterpolator	1,023 key-value pairs
Conditional	No restriction
CoordinateInterpolator2D	128 coordinates per `keyValue` 1,023 key-value pairs
DiscSensor	No restriction
Group	511 `children` maximum allowed
ImageTexture	No restriction
Inline	No restriction
InputSensor	No restriction
Layer2D	511 `children` maximum allowed

Table D.10 Restrictions for Level 1 of the Advanced 2D profile (Continued)

Nodes	Restrictions
MediaBuffer	No restriction
MediaControl	63 url maximum
MediaSensor	No restriction
MovieTexture	speed ignored (no rewind nor fast forward)
Node Update	No restriction
OrderedGroup	511 children maximum allowed
PlaneSensor2D	No restriction
PositionInterpolator2D	1,023 key-value pairs
PredictiveMF coding	No restriction
PROTO	31 fields 31 eventIns 31 eventOuts 31 exposedFields 7 levels
ProximitySensor2D	No restriction
QuantizationParameter	No restriction
ROUTE	No restriction
ROUTE Update	No restriction
ScalarInterpolator	1,023 key-value pairs
Scene Update	No restriction
Script	31 eventIns 31 eventOuts 31 fields
ServerCommand	No restriction
Sound2D	spatialize, location not supported
Switch	No restriction
TermCap	No restriction
TimeSensor	Ignored if cycleInterval < 0.03 s
TouchSensor	No restriction
Transform2D	511 children maximum allowed
Valuator	No restriction
WorldInfo	No restriction

Level 2   The restrictions listed in Table D.11 apply for the `Advanced` 2D scene
graph profile at Level 2 [N4408].

The maximum number of nodes that is allowed in a scene compliant with
`Advanced` 2D@L2 is 131,071, including all instances of these nodes through the
`DEF`/`USE` or `PROTO` mechanisms.

Table D.11 Restrictions for Level 2 of the `Advanced` 2D profile

Nodes	Restrictions
Anchor	16,383 `children` maximum allowed
AnimationStream	No restriction
AudioBuffer	No restriction
AudioClip	`pitch` not supported
AudioSource	255 `children` maximum allowed `pitch`, `speed`, `numChan`, `phaseGroup` not supported (no rewind nor fast forward)
AudioSwitch	No restriction
ColorInterpolator	16,383 key-value pairs
Conditional	No restriction
CoordinateInterpolator2D	1,023 coordinates per `keyValue` 16,383 key-value pairs
DiscSensor	No restriction
Group	16,383 `children` maximum allowed
ImageTexture	No restriction
Inline	No restriction
InputSensor	No restriction
Layer2D	16,383 `children` maximum allowed
MediaBuffer	No restriction
MediaControl	63 `url` maximum
MediaSensor	No restriction
MovieTexture	`speed` ignored (no rewind nor fast forward)
Node Update	No restriction
OrderedGroup	16,383 `children` maximum allowed
PlaneSensor2D	No restriction
PositionInterpolator2D	16,383 key-value pairs

Table D.11 Restrictions for Level 2 of the Advanced 2D profile (Continued)

Nodes	Restrictions
PredictiveMF coding	No restriction
PROTO	255 fields 255 eventIns 255 eventOuts 255 exposedFields 7 levels
ProximitySensor2D	No restriction
QuantizationParameter	No restriction
ROUTE	No restriction
ROUTE Update	No restriction
ScalarInterpolator	16,383 key-value pairs
Scene Update	No restriction
Script	255 eventIns 255 eventOuts 255 fields
ServerCommand	No restriction
Sound2D	spatialize, location not supported
Switch	No restriction
TermCap	No restriction
TimeSensor	Ignored if cycleInterval < 0.03 s
TouchSensor	No restriction
Transform2D	16,383 children maximum allowed
Valuator	No restriction
WorldInfo	No restriction

D.7 MAIN 2D PROFILE

The Main 2D scene graph profile includes basic 2D composition, 2D texturing, local interaction, local animation, BIFS updates, and access to Web links and subscenes, in addition to audio and visual elements. It also introduces tools such as back channel (ServerCommand), VoD features (MediaControl, MediaSensor) and Flextime nodes (TemporalGroup, TemporalTransform) for advanced, flexible synchronization.

Level 1   The restrictions listed in Table D.12 apply for the `Main 2D` scene graph profile at Level 1 [N4408]:

The maximum number of nodes that is allowed in a scene compliant with `Main 2D@L1` is 8,191, including all instances of these nodes through the `DEF`/ `USE` mechanism.

Where `addChildren` and `removeChildren` fields are not supported, this only refers to in-scene routing. Node update is permitted to add and remove children from all `MFNode` fields.

Table D.12 Restrictions for Level 1 of the `Main 2D` profile

Nodes	Restrictions
Anchor	addChildren and removeChildren not supported 31 children maximum allowed
AudioClip	pitch, description ignored
AudioSource	addChildren, removeChildren, children, pitch, speed, numChan, phaseGroup not supported (no rewind nor fast forward)
ColorInterpolator	255 key-value pairs
Conditional	No restriction
CoordinateInterpolator2D	31 coordinates per keyValue 255 key-value pairs
DiscSensor	No restriction
Inline	TBD
ImageTexture	repeatS, repeatT not supported; always treated as TRUE
Group	TBD
InputSensor	restricted to mice, keyboards, remote controls
Layer2D	addChildren and removeChildren not supported 31 children maximum allowed
MediaControl	mediaSpeed not supported (no rewind nor fast forward) 31 url maximum
MediaSensor	info ignored
MovieTexture	speed ignored (no rewind nor fast forward) repeatS, repeatT not supported; always treated as TRUE
Node Update	No restriction
OrderedGroup	addChildren and removeChildren not supported 31 children maximum allowed

Table D.12 Restrictions for Level 1 of the Main 2D profile (Continued)

Nodes	Restrictions
PlaneSensor2D	No restriction
PositionInterpolator2D	255 key-value pairs
ProximitySensor2D	No restriction
QuantizationParameter	TBD
ROUTE	No restriction
ROUTE Update	No restriction
ScalarInterpolator	255 key-value pairs
Scene Update	No restriction
ServerCommand	No restriction
Sound2D	intensity, spatialize, location not supported
Switch	31 choices maximum allowed
TemporalGroup	addChildren and removeChildren not supported 7 children maximum allowed
TemporalTransform	addChildren and removeChildren not supported 31 children maximum allowed speed not supported for media referenced via url (speed only affects in scene elements from BIFS) stretchMode values linear and repeat not supported shrinkMode value linear not supported
TimeSensor	Ignored if cycleInterval < 0.03 s
TouchSensor	No restriction
Transform2D	addChildren and removeChildren not supported center, rotationAngle, scaleOrientation not supported (only translations and scalings are allowed) 31 children maximum allowed
Valuator	No restriction
WorldInfo	TBD

Level 2 The restrictions listed in Table D.13 apply for the Main 2D scene graph profile at Level 2 [N4408]:

The maximum number of nodes that is allowed in a scene compliant with Main 2D@L2 is 65,535, including all instances of these nodes through the DEF/USE mechanism.

Where `addChildren` and `removeChildren` fields are not supported, this only refers to in-scene routing. Node update is permitted to add and remove children from all `MFNode` fields.

Table D.13 Restrictions for Level 2 of the `Main 2D` profile

Nodes	Restrictions
Anchor	addChildren and removeChildren not supported 255 children maximum allowed
AudioClip	pitch, description ignored
AudioSource	addChildren, removeChildren, children, pitch, speed, numChan, phaseGroup not supported (no rewind nor fast forward)
ColorInterpolator	511 key-value pairs
Conditional	No restriction
CoordinateInterpolator2D	511 coordinates per keyValue 511 key-value pairs
DiscSensor	No restriction
Inline	TBD
ImageTexture	repeatS, repeatT not supported; always treated as TRUE
Group	TBD
InputSensor	restricted to mice, keyboards, remote controls
Layer2D	addChildren and removeChildren not supported 255 children maximum allowed
MediaControl	255 url maximum
MediaSensor	info ignored
MovieTexture	speed ignored repeatS, repeatT not supported; always treated as TRUE
Node Update	No restriction
OrderedGroup	addChildren and removeChildren not supported 255 children maximum allowed
PlaneSensor2D	No restriction
PositionInterpolator2D	1,023 key-value pairs
ProximitySensor2D	No restriction

Table D.13 Restrictions for Level 2 of the Main 2D profile (Continued)

Nodes	Restrictions
QuantizationParameter	TBD
ROUTE	No restriction
ROUTE Update	No restriction
ScalarInterpolator	1,023 key-value pairs
Scene Update	No restriction
ServerCommand	No restriction
Sound2D	spatialize, location not supported
Switch	255 choices maximum allowed
TemporalGroup	addChildren and removeChildren not supported 15 children maximum allowed
TemporalTransform	addChildren and removeChildren not supported 255 children maximum allowed
TimeSensor	Ignored if cycleInterval < 0.03 s
TouchSensor	No restriction
Transform2D	addChildren and removeChildren not supported 255 children maximum allowed
Valuator	No restriction
WorldInfo	TBD

D.8 REFERENCES

[N3850] MPEG Systems. *MPEG-4 Systems, Version 2*. Doc. ISO/MPEG N3850, Pisa MPEG Meeting, January 2001.

[N4264] MPEG Systems. *Text of ISO/IEC 14496-1:2001/COR1*. Doc. ISO/MPEG N4264, Sydney MPEG Meeting, July 2001.

[N4408] MPEG Systems. *Text of ISO/IEC 14496-1:2001/FDAM2—Extensible MPEG-4 Textual Format (XMT), Systems Extensions and Additional Profiles*. Doc. ISO/ MPEG N4408, Pattaya MPEG Meeting, December 2001.

[N4506] MPEG Requirements. *New MPEG-4 Profiles Under Consideration*. Doc. ISO/ MPEG N4506, Pattaya MPEG Meeting, December 2001.

MPEG-J Code Samples

by Viswanathan Swaminathan, Alex MacAulay,
and Gianluca De Petris

Keywords: MPEG-J, MPEGlets, Java examples, MPEG-J examples,
Resource APIs, Network APIs, Scene APIs, Section Filtering and
Service Information APIs, application engine

This appendix presents some detailed examples of MPEGlets to give you an understanding of the steps needed to create an MPEG-J application. It is intended for readers already introduced to the Java programming language.

E.1 SCENE APIS

The following example shows a complete MPEGlet[1] using the Scene API to invert the color of a shape node when the user clicks on it. In this example, SceneExampleMpeglet implements the MPEGlet interface, SceneListener interface, and the EventOutListener interface. Initially, the SceneManager object is obtained from the MpegjTerminal and the MPEGlet adds itself as the SceneListener. Then the MPEGlet retrieves a reference to the TouchSensorNode (defined with the name TOUCH) and registers itself to be notified when the user clicks on the TouchSensorNode. Finally, the MPEGlet inverts the color of

---

1. MPEG-J applications are called MPEGlets. All MPEG-J applications should implement the MPEGlet Java interface defined in MPEG-J.

the shape node (defined with the name "MATERIAL") when the MPEGlet
receives the touch event.

```java
import org.iso.mpeg.mpegj.*;
import org.iso.mpeg.mpegj.scene.*;

public class SceneExampleMpeglet implements
      MPEGlet, SceneListener, EventOutListener {

    Scene m_scene = null;
    boolean m_isTouched = false;

    public void run() {
        try {
            // Get the scene manager and register to be
            // notified when the scene is ready.
            MpegjTerminal terminal =
                new MpegjTerminal(this);
            SceneManager manager =
                terminal.getSceneManager();
            manager.addSceneListener(this);

            // Wait until the scene is ready.
            synchronized (this) {
                while (m_scene == null) {
                    wait();
                }
            }

            // Get the TouchSensor node.
            Node touch = m_scene.getNode("TOUCH");

            // Register to be notified when it's touched.
            touch.addEventOutListener(
                EventOut.TouchSensor.touchTime, this);

            // Wait until it's touched.
            synchronized (this) {
                while (!m_isTouched) {
                    wait();
                }
            }
            // Invert the color of a shape as feedback.
            Node material2d = m_scene.getNode("MATERIAL");
            FieldValue val = material2d.getEventOut(
                EventOut.Material2D.emissiveColor);
            final float[] color =
                ((SFColorFieldValue) val).getSFColorValue();
            color[0] = 1 - color[0];
            color[1] = 1 - color[1];
            color[2] = 1 - color[2];
            material2d.sendEventIn(
                    EventIn.Material2D.emissiveColor,
                    new SFColorFieldValue() {
```

```
                              public float[] getSFColorValue() {
                                  return color;
                              }
                          }
                      );
              } catch (MPEGJException e) {
                  e.printStackTrace();
              } catch (InterruptedException e) {
                  return;
              }
          }

          // Called by the scene manager when the scene is ready.
          public synchronized void notify(int msg, Scene scene) {
              if (msg == SceneListener.Message.SCENE_READY) {
                  m_scene = scene;
                  notify();
              }
          }

          // Called when the touchTime eventOut is triggered.
          public synchronized void notify(int id, FieldValue v) {
              m_isTouched = true;
              notify();
          }

          public void init() {}
          public void stop() {}
          public void destroy() {}
      }
```

E.2 RESOURCE AND DECODER APIS

Examples in this section show how the Resource and Decoder APIs can be used to monitor resources and adapt to time varying resource conditions. Initially, an MPEGlet using both the Resource and Decoder APIs is presented. In the next two sections examples of event handlers are presented.

Here the MPEGlet RM_Example obtains a handle to the ResourceManager through the MpegjTerminal. Two event handlers, rendererEH and decoderEH, are defined for handling renderer and decoder events. The MPEGlet obtains a handler to the Renderer and registers the rendererEH. The MPDecoder object associated with the node m_node is obtained. The defined event handler decoderEH is registered with the MPDecoder object to handle the events that occur in the associated decoder. This example also shows how to stop and restart the decoder. Further, it shows how to replace the obtained MPDecoder with another decoder object of the same type.

```
import org.iso.mpeg.mpegj.resourceManager.*;
import org.iso.mpeg.mpegj.decoder.*;
import org.iso.mpeg.mpegj.scene.*;
```

```
import org.iso.mpeg.mpegj.*;
import org.iso.mpeg.mpegj.net.*;

public class RM_Example implements MPEGlet{
   MpegjTerminal mpegjTerminal;
   private ResourceManager resourceManager;
   RendererEventHandler rendererEH;
   DecoderEventHandler decoderEH;
   // other declarations

   Node m_node;
   Renderer m_renderer;
   java.util.Vector m_decoders;

   public void RM_Example() {
      m_decoders = new java.util.Vector( 3, 3);
   }

   public void init(){

      // intialize the mpegjterminal
      mpegjTerminal = new MpegjTerminal(MPEGlet);

      // other initializations go here

      // get resource manager from the mpegjterminal
      try
      {
         resourceManager = mpegjTerminal.getResourceManager();
      }catch(MPEGJException ex){ }

      // create event handlers for renderer and decoders
      rendererEH = new EventHandler();
      decoderEH = new EventHandler();

      // getting the renderer from the Resource Manager
      try {
         m_renderer = resourceManager.getRenderer();
      } catch( RendererNotFoundException rnfe) { }

      // to add event listener to the Renderer
      if( m_renderer != null )
         m_renderer.addMPRendererMediaListener(rendererEH);

      // to get the required decoder from the resource manager
      try {


         MPdecoderImp dec1
            = resourceManager.getDecoder(m_node);
         // registering the decoder as listener to the events
         dec1.addMPDecoderMediaListener(decoderEH);

      } catch( DecoderNotFoundException  dnfe) { }
```

```
                catch (BadNodeException bne) { }

            // to change a decoder associated with a node
            // get a decoder from the available decoders list
            MPdecoder decoder =
                resourceManager.getAvailableDecoder(decoder_type);


            try {
                resourceManager.changeDecoder( node, decoder)
            } catch ( DecoderNotFoundException dnfe ) { }
                catch( BadNodeException bne) { }

            // to stop a decoder and restarting again
            decoder.stop();
            decoder.start();


            // to retrieve all capabilities(static, dynamic, profile)
            // of the terminal
            try {
                CapabilityManager cm =
                    resourceManager.getCapabilityManager();
            }catch (CapabilityManagerNotFoundException cmnfe ) { }
            // other code goes here
        }
        public void stop(){}
        public void destroy(){}
    }
```

E.2.1 Listener Class for Decoder Events

Here is a simple example to illustrate how to implement an event handler to
receive and handle the decoder events. DecoderEventHandler implements
MPDecoderMediaListener. The event handler obtains the decoder object that
is the source of these events; it also prints out the event condition for
STREAM_START, for example.

```
        import org.iso.mpeg.mpegj.resourceManager.*;
        import org.iso.mpeg.mpegj.decoder.*;

        public class DecoderEventHandler implements
            MPDecoderMediaListener {
            public void mPDecoderMediaHandler(MPDecoderMediaEvents
                event){
                MPDecoderImp dec = ((MPDecoderImp)event.getSource());

                int condition = event.getCondition();
                System.out.println("Event in Decoder with condition "
                    + condition );
                // we can stop the decoder and restart it again
                dec.stop();
                // can change the decoder if we want
```

```
         // restart it again
         dec.start();
      }
      public DecoderEventHandler( ) {
         super();
      }
   }
```

E.2.2 Listener Class for Renderer Events

Like the example in the previous section, this example implements a class
that handles events, but now the events are sent by the Renderer. Here the
RendererEventHandler implements MPRendererMediaListener and prints out
the condition associated with the event MISSED_FRAME, for example.

```
import org.iso.mpeg.mpegj.resourceManager.*;
 public class RendererEventHandler implements
   MPRendererMediaListener
{
   public void mPCompositeMediaHandler(MPRendererMediaEvents
      event)    {
      System.out.println("Renderer Event with condition  " +
         event.getCondition() );
      // other code goes here
   }
   public RendererEventHandler( ) {
      super();
   }
}
```

E.3 NETWORK APIS

This section illustrates how to use the Network APIs through a simple example that enables and disables channels. Here the MPEGlet Net_Example
obtains a handle to the NetworkManager through the MpegjTerminal. Using
the NetworkManager object, the MPEGlet obtains the ChannelController
object. The MPEGlet uses the ChannelController to enable and disable a
channel.

```
import org.iso.mpeg.mpegj.resourceManager.*;
import org.iso.mpeg.mpegj.scene.*;
import org.iso.mpeg.mpegj.*;
import org.iso.mpeg.mpegj.net.*;

public class Net_Example implements MPEGlet{
   MpegjTerminal mpegjTerminal;
   private NetworkManager netManager;
   ChannelController cc;
   public void Net_Example() {
   }
```

```
        public void init(){
           mpegjTerminal = new MpegjTerminal(MPEGlet);
           try {
              netManager = mpegjTerminal.getNetworkManager();
           } catch(NetworkManagerNotFoundException ex){ }
              catch(MPEGJException ex){ }

        // to get the channel controller used to enable/ disable
        // the channels
           cc = netManager.getChannelController();

           // to enable a channel
           cc.enable( serviceSessionID, channelID);
           // to disable a channel
           cc.disable( serviceSessionID, channelID);
        }
        public void stop(){}
        public void destroy(){}
     }
```

E.4 SECTION FILTERING APIS

This section presents a simple example of how the section information can be extracted from the MPEG-2 Transport Stream. Here a SectionFilterListener is registered with a SimpleSectionFilter, which filters out sections of a given program identifier (PID) defined in MPEG-2 Systems.

```
import org.iso.mpeg.mpegj.*;
import java.lang.Boolean;

public class SI_SF_Example implements MPEGlet {
     MpegjTerminal mpegjTerminal;
     SimpleSectionFilter ssFilter;
     SectionFilterListener sfListener;
     int milliSecs;

     // Class public methods
     public SI_SF_Example() {

        // initialize the mpegjterminal
        mpegjTerminal = new MpegjTerminal(this);
        // other initializations go here

        ssFilter = new SimpleSectionFilter();
        sfListener = new SectionFilterListener();

        // Specify an object to be notified of events relating to
        //this SectionFilter object.
        ssFilter.addSectionFilterListener( sfListener);

        //Set the time-out for this section filter
        ssFilter.setTimeOut( milliSecs);
```

```
        // create mask and value parameters
        byte[] posValue = new byte[12];
        byte[] posMask  = new byte[12];
        for ( int i=0; i<12; i++ )   {
            posValue[i] = 0;
            posMask[i]  = 0;
        }
        posMask[0] =  (byte)0xFF; // only check first byte
        posValue[0] = (byte)0;    // table_id PAT

        //sets the SectionFilter object as filtering only for
        //sections matching a specific PID and
        //table_id, and where contents of the section match the
        //specified filter pattern.
        ssFilter.startFilter (
            0          // index, the number of this section filter
            , 100        // id, uniquely identifying this filter
                       // action
            , ssFilter // the listener to receive the events
                       // indicating a new section has arrived
            , 0        // PID, in the case of the PAT 0
            , posMask  // mask, which bits to check
            , posValue // value, the value checked bits
                       // should have
            , null     // neg masking not done, always call this
                       // function in ssFilter, other startfilter
                       // methods are incorrect.
            , null
        );


        Section m_section = ssFilter.getSection()
        try {
            byte[] m_data = m_section.getData();
        } catch (NoDataAvailableException ndae) { }

        // sections matching this SectionFilter object will stop.
        ssFilter.stopFiltering();
    }
    public void init() {}
    public void stop(){}
    public void destroy(){}
}
```

Index

A

absolute category rating method, 713–14

accessUnitDuration, 96

accessUnitEndFlag element, 95

access units, 45, 92–93

accessUnitStartFlag element, 95

AC/DC prediction, 302, 312

    for intra macroblocks, 315–16

AcousticMaterial, 573–74

AcousticsScene, 572–73

activateEvent, 201

ACTS European Project TAPESTRIES, 678

adaptive arithmetic entropy coding, 434

adaptive differential pulse code modulation

    (ADPCM), 453, 454

adaptive intra refresh, 346, 354

adaptive window shape selection, 495

addSceneListener method, 166

ad-hoc groups (AHG), 16

Advanced **AudioBIFS,** 546

    for enhanced presentation of 3D sound

        scenes, 569–80

        AcousticMaterial, 573–74

        AcousticScene, 572–73

        DirectiveSound, 570–72

        PerceptualParameters, 574–77

    examples of use of nodes, 577–80

Advanced Audio Coding (AAC), 50

    coupling channels, 506–7

    intensity stereo coding, 506

    MPEG-4 and, 508–18

Advanced Coding Efficiency (ACE) profile,

    376, 599

Advanced Real-Time Simple (ARTS) profile,

    593

    error-robustness test, 696–98, 699–700,

        700–701, 702–4

    temporal resolution stability test, 699,

        700, 701, 704

advanced rendering techniques, 386

Advanced Simple profile, 376, 598

advanced synchronization mechanisms, 45

affine transform, 97

aggregated fields, 646

Algorithmic Synthesis and AudioFx, 553–

    54, 555

algorithmic synthesis and processing, 549–52

alpha coding, 319

alpha masks, 318

amendments, 14

amplitude scalability, 341

analysis by synthesis procedure, 453

analysis filterbanks, 488–89

Anchor node, 109

angle, 571

animation, 203–5. *See also* face and body animation (FBA)

animation definition
for **IndexedFaceSet** node, 399
for **Transform** node, 398

animation frames, 111–12

animation mask, 111

animation module, 197

animation rules using **bodyDefTables**, 401

AnimationStream node, 111

application engine, 152

application scenarios, 181
adaptive rich media content for wireless devices, 181
content personalization, 182
enhanced interactive electronic program guide, 181–82
enriched interactive digital television, 182

application-specific descriptors, 234

a priori standardization, 17

arbitrarily shaped texture, 362

arbitrarily shaped video objects, 593–94, 599–600
coding of, 294, 318
binary alpha masks, 318–23
boundary macroblocks, 324–30
gray-level alpha masks, 323–24

architecture, MPEG-J, 154–56

arithmetic encoder, 318

arithmetic encoding of binary alpha blocks, 321–22

aspect ratio, $112n$

assessment trial, 674

asymmetry of masking, 489

asynchronous callback, 232

Asynchronous Transfer Mode (ATM) networks, 43, 228, 350

atoms, 257–59, 262–65
inside movie, 262–63
inside sample table, 264–65
inside track, 263
top-level, 262

audio, synthetic-natural hybrid coding (SNHC) of, 546–48

AudioBIFS nodes, 545, 546, 551, 560–62
AudioBIFS scene graph, 566–67
AudioBuffer, 564–65
AudioDelay, 564
AudioFX, 563–64
AudioMix, 562–63
AudioSource, 562
AudioSwitch, 563
interface with structured audio coding and, 554
ListeningPoint, 565–66
Sound2D, 565

AudioBIFS scene graph, 566–67

AudioBuffer, 136, 564–65

AudioClip node, 559

audio coding, 46

audio composition, 557–80
advanced **AudioBIFS**, 569–80
AcousticMaterial, 573–74
AcousticScene, 572–73
DirectiveSound, 570–72
PerceptualParameters, 574–77
AudioBIFS nodes, 560–62
AudioBIFS scene graph, 566–67
AudioBuffer, 564–65
AudioDelay, 564
AudioFX, 563–64
AudioMix, 562–63
AudioSource, 562
AudioSwitch, 563
ListeningPoint, 565–66
Sound2D, 565
enhanced modeling of 3D audio scenes in MPEG-4, 567–69
VRML sound model in **BIFS**, 558–59

AudioDelay, 564

audio effects processing, 555

AudioFX, 553–54, 563–64

AudioMix, 562–63

audio object types, 602–7

audio-only objects, 39

audio on the Internet test, 710, 724
    assessment method and design, 726–
        27
    conditions, 724–25
    data analysis, 727–28
    material, 725–26
    results, 728–30
audio profiles, 602–9, 606–7
audio signals
    flow of, 547
    object-based representation of, 451
AudioSource node, 66, 562
audio streams, integration of synthetic and
        natural, 547–48
audio subtree, 545
AudioSwitch, 563
audio testing for validation, 709–49
    general aspects, 710
        laboratory setup, 712
        selection of test material, 710–11
        selection of test subjects, 711
        test plan, 712
        training phase, 712–13
    on Internet test, 724
        assessment method and design,
            726–27
        conditions, 724–25
        data analysis, 727–28
        material, 725–26
        results, 728–30
    methods, 713
        absolute category rating (ACR), 713–
            14
        multistimulus test with hidden
            reference and anchors
            (MUSHRA), 715–17
        paired comparison (PC), 714–15
    narrowband digital audio broadcasting
        test, 717–18
        assessment method and test design,
            720–21
        conditions, 718–19
        data analysis, 721
        material, 719–20
        results, 722–24

speech communication test, 730
    assessment method and design,
        733–34
    conditions, 730–32
    data analysis, 734
    material, 732–33
    results, 734–36
Version 2 coding efficiency, 736–37
    assessment method and test design,
        740
    data analysis, 740
    test material, 739–40
    test results, 740–44
    tests conditions, 737–39
Version 2 error-robustness test, 744
    conditions, 744–46
    data analysis, 747
    material, 747
    method and experimental design,
        747
    results, 747–49
audio tools, 50–51
audiovisual coding standard, 3
audiovisual objects, 39, 628
    data for, 44–45
AU_sequenceNumber, 95
authoring, 40
authoring system, 586
autoReverse attribute, 203
auxiliary subgroups, 16

B

background noise, 475
backward-adaptive prediction scheme, 502
backward compatibility, 359
backward prediction scheme, 502
backward reference, 304
backward shape, 338
bad frame masking, 480
Bark Scale, 489
Bark-Scale Envelope Coding, 514
base classes, 69
BaseCommand, 651
BaseDescriptor, 70, 647

basic animated texture profile, 446

bidirectionally predicted Video Object Planes (VOPs), 302

bidirectional motion compensation, 309–11

BIFS-Anim (animation), 46, 110, 140, 191

BIFS (binary format for scenes), 46, 48, 66, 103–46, 188, 587

    advanced features, 129

        encapsulation and reuse, 131–33

        scripting, 130–31

        text layout, 134–36

    basic features by example, 114

        icons and buttons, 119–21

        magnifying glass, 127–29

        movie with subtitles, 117–19

        simple 3D scene, 124–27

        slides and transitions, 121–24

        trivial scene, 114–17

    basics of, 104

        binary encoding, 113–14

        fields and **ROUTEs,** 106–8

        hyperlinks, 109

        node types, 108

        quantization, 114

        scene and nodes, 104–6

        scene changes, 109–12

        scene rendering, 112

        subscenes, 109

    carousel, 95

    flex time, 138–39

    new sensors, 137–38

    nodes, 140

        animation-related, 142–43

        audio, 145

        general use, 140–41

        geometry, 144

        interactivity-related, 143

        media, 136

        Script, 172

        SNHC visual-related, 145–46

        texture-related, 143–44

        visual, 141–42

    profiles, 139–40

BIFS (binary format for scenes) decoder, 154–56, 632

BIFS (binary format for scenes) streams, 82, 259

BIFS (binary format for scenes) update stream, 172

BIFS-Commands, 82, 110–11, 140, 215–19

BIFS_FIELD macros, 638

BifsFieldTable, 634

BIFS/OD decoders, 628

bilevel quantizer mode, 366

binary alpha blocks (BABs), 318

    arithmetic encoding of, 321–22

    context computation for, 319–21

binary alpha map, 49

binary alpha mask, 318–23

binary encoding, 113–14

binary-level APIs, 630

binary MP4 file format, 41–42

binary_only shape coding, 322–23

binary shapes

    coding for, 318–19

    spatial scalability for, 334–36

binaural masking level depression (BMLD), 505

bit, 67

bit allocation, 489

Bitmap node, 116

bit plane representation, 341

bit plane shift, 342

bit-rate controllability, 454

bit-rate scalability, 454, 488, 529

bit reservoir, 507

bit-sliced arithmetic coding, 523–25

bitstream conformance, 585

bitstream format, 535

bitstream multiplexing, 507–8

bitstream packetization, 363, 372–74

block boundary mirroring, 307

body animation parameters (BAPs), 391, 406

    coding schemes, 406–10

body definition parameters (BDPs), 391, 399–401

    coding schemes, 401–3

bodyDefTables, animation rules using, 401

boundary macroblocks, 324–30
  motion compensation for, 324–25
  texture coding for, 327–30
boundary vertices, 415
bounding, 428
box in transition, 123–24
Box node, 126
buffer and time management, 628
buffer API, 631
A Bug's Life (movie), 385
B-VOPs, 334
byte code, 151

C

Callback suffix, 232
CapabilityManager, 174–75
CCETT, 25
Channel Association Tag (CAT), 243–44
channel control, 179
channel coupling, 505
channel primitives, 233–35, 243–45
channels, 229, 652
Chiariglione, Leonardo, 15
child element, 197*n*
chrominance sampling, 360
chunk, 257–58
Circle node, 121
C++ language object, 628
classes, 151
class loader, 155
class overloading, 638
ClockReference object, 644–46
clock reference stream, 97
cloning mechanism, 640
codebook, 453
Code Excited Linear Prediction (CELP), 453,
  454
  coder, 50, 455–56
  decoder, 456–57
coding efficiency, 6
  for low and medium bit-rates test, 691
    conditions, 691–92
    data analysis, 693
    materials, 692

  method and design, 692–93
  results, 693–94
Coding of Moving Picture and Audio,
  13. *See also* Moving Picture
  Experts Group; Working Group
  11
coefficient scanning (CS), 366
CoEnd, 45
collaborative phase, 20
colocated macroblock, 310
ColorInterpolator node, 124
comfort noise, 461
Command Frames, 113
competitive phase, 20
completely opaque pixels, 318
completely transparent pixels, 318
compliance, 585
component selection, perceptual models for,
  528
CompositeTexture2D node, 112
CompositeTexture3D node, 112
compositeUnitDuration, 96
composition buffer, 644
composition time, 93, 98
Composition Time Stamp (CTS), 158, 643
composition units, 100
compositor, 629
compositor API, 630
compression, 293
compression efficiency, 6–7
computer graphics, 385
computer vision, 385
Conditional BIFS node, 172
conditional expressions, 646
Conditional node, 108, 110, 120, 213
Cone node, 126
conformance, 624
conformance points, 52, 585, 587, 624
conformance testing, 584
connectivity coding, 429–33
connectivity information, 422
constant end-to-end delay assumption,
  100
constraints field, 134
content authors, 40

content-based coding test, 688
    conditions, 688–89
    data analysis, 690
    material, 689
    method and design, 689–90
    results, 690–91
content-based interactivity, 5–6
content-based manipulation and bitstream
        editing, 6
content-based multimedia data access tools,
        5–6
content-based scalability, 7
content complexity description, 77–79
content filtering, 47
ContentIdentificationDescriptor, 89
content storage, 228
context-based arithmetic encoding (CAE),
        318
context computation for binary alpha blocks,
        319–21
continue flags, 535
control data, 45
control plane, 229
control points, 89
convergence, 12–13
CoordinateInterpolator(2D) node,
        124
core experiments (CE), 19, 21
corrigenda, 14–15
CoStart, 45
critical bands, 489
critical error conditions, 702
cropping criterion, 444
cross-standard interoperability, 189–90
CSound computer music programming
        language, 549
CullOrRender method, 663
current triangle, 419
Curve2D node, 122
cyclic intra refresh, 354
cyclic redundancy codes (CRC),
        476
Cylinder node, 126
CylinderSensor node, 121

D

DataChannel :: OnPacketReceived func-
        tions, 643
data partitioning, 350–52
    syntax for, 351
data primitives, 237–38
data references, 261
    information for, 257
decay time, early, 576
Decoder :: SetFormat, 655
decoder API, 162, 176–77, 631
DecoderConfigDescriptor, 74, 644
decoder conformance, 585
decoder plug-ins, 654–56
DecoderSpecificInfo descriptor, 74, 111
decoding buffer, 644
DecodingBufferSize value, 644
decoding time, 93
Decoding Time Stamp (DTS), 158, 643
DEF and **USE** handling, 657–58
default facial expression and body posture,
        392–93
deformation rules, using **faceDefTables,** 398
Delaunay triangulation, 415–17, 418–19
Delete command, 216
deleteRoute method, $167n$
DeleteValue, 635
        Delivery Application Interface (DAI),
        228
delta motion vector (MV), 310
delta vector, 309
demographically focused programming, 59
DeMUX, 296
depacketization, 99
derived classes, 69
destroy ( ) method, 156
deterministic mapping of XMT-A, 222–23
differential coding, 433–34
digital audio broadcasting (DAB), 3, 43
digital audio effects, 548
digital radio broadcasting, 710
digital still cameras, 2
digital television, 3, 12
Digital Video Broadcasting-Multimedia
        Home Platform (DVB-MHP), 54–55

dimension converter, 465–66
directed graph, 106*n*
DirectiveSound, 570–72, 579
directivity, 571
direct mode
  in bidirectional motion compensation, 309–11
  in bidirectional prediction, 303
direct sound, filtering, 576–77
discrete cosine transform (DCT), 302
  frame/field, 344
  shape-adaptive, 328–30
discrete cosine transform (DCT)-based coding mode, 408–10
discrete cosine transform (DCT) transform coefficients, quantization of, 312–15
discrete wavelet transform (DWT), 365–66
DiscSensor node, 121
display order of Video Object Planes (VOPs), 304
DMIF (Delivery Multimedia Integration Framework), 42, 228, 628
  Application Interface, 42, 51, 86, 228, 629, 631
  ClientFilter, 653
  Default Signaling Protocol, 247
  Filter, 653
  instance, 100, 229
  layer, 229
  Network Interface (DNI), 238–40
  Plug-In Interface (DPI), 653–54
  plug-ins, 632, 652–54
  Signaling Protocol, 51
  stacks, 629
  users, 229
double, 67
double-blind-triple-stimulus with hidden reference method, 714
double stimulus continuous quality scale (DSCQS) method, 676–77
double stimulus impairment scale (DSIS) method, 675–76
drawable context, 658
dual graph representation, 426
DVD, 3

DynamicCapability interface, 174
dynamic compliance, 585
dynamic crosstalk, 505
Dynamic Link Libraries (DLLs), 628
dynamic range control, 508
dynamic resolution conversion (DRC), 355

E

early room response, energy of, 576
ECMAScript, 46, 131
edit list, 255, 257
efficient, object-based coding, of sounds, 548
elementary streams (ESs), 42, 44–45, 72, 221–22, 586
  transport of, over Internet Protocol (IP), 283–85
element_name, 67
ElevationGrid node, 126
embedded zero-tree wavelet (EZW), 365
embedding coding, 454
empty time, 255
enable/disable signaling, 503
encapsulation and reuse, 131–33
EncDec function, 648, 650
encoder, 586
  aspects of optimization, 534–35
encoding of scalefactors, 497
encoding process, 468
enhanced modeling of 3D audio scenes in MPEG-4, 567–69
entry points to MPEG-4 content, object descriptor (OD) as, 65–86
  content complexity description, 77–79
  linking scene to its media streams, 82–85
  stream description, 72–75
  streaming, 79–82
  stream relationship description, 76–77
  syntactic description language, 66–70
envelopment, 576
error concealment, 477, 480–81, 517
error-prone environments, robustness in, 7, 49
error-protection (EP) class, 479

error-protection (EP) tool, 476, 517
 setting for, 479–80
  error-resilience coding tools, 295, 476
error-resilience mode, 435–37
error-resilience test, 429, 508–9, 680–81
 conditions, 681–84
 data analysis, 686
 material, 684
 method and design, 684–86
 results, 686–88
error-resilient bitstream payload syntax, 744
error-resilient bitstream reordering, 476
error-resilient CELP, 481–82
error-resilient coding, 345–55
error-resilient HVXC, 477–81
error-resilient syntax, 477
error robustness, 475–77, 517–18, 694
error robustness bitstream payload syntax,
 518
error-sensitivity categories (ESCs), 476, 477–
 78, 518
Escape Huffman codebook, 517
ES_DescriptorImp :: SetDecoder, 655
ES descriptors (ESDs), 219
ES_IDs, assigning, to streams, 277–79
ETSI Digital Video Broadcasting-Multimedia Home Platform (DVB-MHP), 54–56
European Computer Manufacturers' Association (ECMA), 108*n*
European Telecommunication Standards
 Institute (ETSI), 730
eventIn, 110–11
event processing, 46
EventTiming, 200
exceptional BAB, 336
excitation signal, 481
execution engine, 152
execution of OD objects, 651–52
executive API, 630
executive object, 632
existent signaling protocols, 246–48
eXperimentation Model (XM), 20*n*
exposedField, 106, 110
eXtended boundary prediction mode, 437
Extensible 3D (X3D), 387

eXtensible Markup Language (XML), 53–54
extensible media (xMedia) objects, 195, 197–
 98
Extensible MPEG-4 Textual (XMT) format,
 48, 187–224, 205–10
 cross-standard interoperability, 189–90
 objectives, 188–89
 two-tier architecture, 190–93
 XMT-A format, 210
  deterministic mapping of, 222–23
  document structure, 210–11
  interoperability with X3D, 223–24
  object descriptor framework, 219–22
  scene description, 212–19
  timing, 211–12
 XMT-Ω format, 193–94
  animation, 203–5
  examples, 205–10
  objects, 195, 197–98
  reusing Synchronized Multimedia
   Integration Language (SMIL)
   in, 194–95
  spatial layout, 205
  time manipulations, 202–3
  timing and synchronization, 198–
   202
extension profile level descriptor, 77
ExtensionProfileLevelDescriptor, 78
Extrusion node, 126

F

face and body animation (FBA), 49, 389–92
 body animation parameters, 406
 body definition parameters, 399–401
 decoding, 296
 default facial expression and body posture, 392–93
 FAB and BAP coding schemes, 406–10
 FDP and BDP coding schemes, 401–3
 interface, 410
 objects, 389, 393–96
 text-to-speech interface and, 410
face animation-related parameters, 555
face animation tables (FAT), 390

face definition parameters (FDPs), 391, 396–99
  coding schemes for, 401–3
faceDefTables, deformation rules using, 398
facial animation parameter (FAP), 389–90, 403–6
  bookmarks for, 51
  coding schemes for, 406–10
  interpolation table for, 399
fade transition, 124
fast harmonic synthesis, by inverse fast Fourier transform (IFFT), 471–72
Fetch functions, 645–46
FetchNext functions, 645–46
field IDs, 168
field padding, 344
fields, characteristics of, 106–8
FieldValue interfaces, 165
Field values, 168–70
File Transfer Protocol (FTP), 230
filterbank, 493–95
fine grain scalability (FGS), 49, 509
fine pitch search, 464
fine rate control, 458
first-in-first-out (FIFO) buffers, 631
first quantization method (MPEG quantization), 313–15
fixed end-to-end delay, 100
flexBehavior attribute, 203
flexBehaviorDefault attribute, 203
FlexMux, 42, 44, 48, 51, 228, 229, 248–51, 654
  Clock Reference (FCR), 251
  layer, 681
  streams
    conveying, in sections, 277
    over Internet Protocol (IP) and timing models, 288–89
    timing of, 251–53
FlexMuxTimingDescriptor, 251
FlexTime, 138–39, 203
float values, 113
fly transition, 123, 124
FontStyle node, 118–19, 134
forest split, 437
formal standardization process, 13–15

forward compatibility, 359
forward-error-correction codes (FEC), 476
forward reference, 303–4
forward shape, 338
frame application, 629
frame-based coding mode, 408
frame events, 176
frame/field DCT, 344
frame/field motion compensation and interlaced direct mode, 344
frequency, 571
frequency domain division of response, 575
frequency selective switch (FSS), 512, 521
full enhancement, 340

G

gain control, 491–93
Gaussian noise, 470
general audio, 607
general audio coding, 50, 487–540, 602–3
  MPEG-2 advanced, 490
    bitstream multiplexing, 507–8
    filterbank, 493–95
    gain control, 491–93
    joint stereo, 504–7
    noiseless, 497–99
    prediction, 502–4
    quantization, 495–97
    temporal noise shaping, 499–502
  MPEG-4 additions to AAC, 508–18
    error robustness, 517–18
    long-term prediction, 511–13
    low-delay, 515–17
    perceptual noise substitution, 509–11
    twinVQ, 513–15
  MPEG-4 HILN, 50, 530, 538–39
    bitstream formal, 535
    parameter quantization and subdivision coding, 536–37
    spectral model parameter coding, 537

MPEG-4 scalable, 518–19
   bit-sliced arithmetic, 523–25
   large-step, 519–23
  parametric, 525–26
   encoding and decoding concepts,
    528–29
   source and perpetual models,
    526–28
  time/frequency, 488–90
generalized audio coding, 555
general **MIDI**, 553
geometry, 422, 433–34
geometry motion, 417
geometry node, 116
getNode method, 167
global gain, 537
global motion compensation (GMC), 303,
  307–9, 357
glue definitions, 228
graphics authoring tools, 383
graphics profiles, 139–40, 587, 609–13
gray-level shape coding, 323–24
gray-scale alpha values, 311
gray-scale map, 49
Grimstadt MPEG meeting, 5, 37
Group node, 126
grouping, as harmonic tone, 534
grouping/interleaving, 499
grouping nodes, 105
group of blocks (GOBs), 348
group of VOPs (GOV), 298
groups field, 134
groupsIndex field, 134

H

half-pel motion compensation, 305
handler information, 257
hardware, decoding, 586
harmonic coding for voiced segments,
  462
harmonic magnitudes
  estimation of, 464
  vector quantization of, 464
harmonic models, 527

harmonic spectral magnitudes
  scalable vector quantization scheme for,
   466–68
  vector quantization (VQ) of, 465–68
harmonic tone, 530
  detection of, 531
  grouping as, 534
harmonic vector excitation coding (HVXC),
  50, 454
  coder, 50, 462–69
  decoder, 470–75
harpsichord or pitch pipe test signals, 504
header atom, 257
header extension code, 353–54
heuristic delivery-timing model of Real-Time
  Transport Protocol (RTP), 288
hierarchical embedded coding, 519
hierarchical transmission system, in scalable
  video coding, 294–95
hinters, 261
hinting, 260–62
horizontal padding, 324
host language, 191
Huffman code-based entropy coding, 491
Huffman codeword reordering (HCR), 476,
  518
Huffman coding, 321–22, 489, 497–99
Hughes Aircraft Company, 25
hybrid natural and synthetic data coding, 6
hybrid natural/synthetic audio, 608
hyperlink, 2, 109
Hypertext Markup Language (HTML), 384
  browsers, 47

I

idleFlag, 95
image source method, 573
ImageTexture node, 108, 119
I-MOP coding, 417–19
implicit skipped mode, 311
improved error robustness, 696
in-band descriptor delivery, 253
incremental rendering, 429
IndexedFaceSet2D node, 122, 413

`IndexedFaceSet` node, 126, 573–74, 640
    animation definition for, 399
`IndexedLineSet2D` node, 121
`IndexedLineSet` node, 126
individual sinusoids, 530
    extraction of, 532–33
information, prosodic, 556
informative tools, 670
initial object descriptor (IOD), 77, 233
initial scene, 109–10, 113
initial triangle, 419
`init ( )` method, 156
`Inline` node, 77n, 82, 84, 109
`InputSensor` node, 137
`Insert` command, 216
`insertRoute` method, 167n
`InsertValue`, 635
instantaneous decoding, 100
`int`, 67
integer values, 113
integrated parametric coder, 539
intellectual property identification (IPI) data
    sets, 89
intellectual property management and pro-
    tection (IPMP), 89–90, 188
    API, 631
    descriptors, 89, 656
    filter plug-ins, 629
    hooks, 48
    information, 47
    plug-ins, 656
    streams, 86, 89, 656
    tools, 160
intellectual property rights, 47
intensity stereo coding, 504–5, 505–7
interactive broadcasting, 57–59
interactivity, 383
interface between structured audio coding
    and `AudioBIFS,` 554
interframe prediction, 457
interlaced coding, 344
interlaced video, 49
interleaved VQ, 515
interleaving of spectral coefficients,
    514

inter mode, 334–35
International Organization for Standardiza-
    tion (ISO), 5
international standards, 13–14
International Telecommunications Union-
    Telecommunications Standardization
    Sector (ITU-T) Low Bit-rate Coding
    (LBC) group, 5
Internet Engineering Task Force (IETF), 228
Internet Protocol (IP)
    transporting MPEG over, 280–89
    transport of elementary streams over,
        283–85
    transport of SL-packetized streams over,
        285–88
Internet Protocol (IP) and timing models,
    FlexMux streams over, 288–89
Internet Protocol (IP) networks, carriage of
    MPEG-4 on, 52
interoperability, 3, 189, 584
    cross-standard, 189–90
    with X3D, 223–24
interpolator nodes, 108, 110
intra coding mode, 408
intra DC coefficient quantization, 313
intra macroblocks, AC/DC prediction for,
    315–16
intra mode, 334
Inverse Fast Fourier Transform (IFFT), fast
    harmonic synthesis by, 462, 471–72
Inverse Modified Discrete Cosine Transform
    (IMDCT) stages, 493
inverse TNS filtering, 500–501
Ishihara's tables, 671–72
ISO/IEC 21000, 4
ISO/IEC MPEG-1 Systems, 53
ISO/IEC MPEG-1 Video, 53
ISO/IEC standard, 384
ISO/MPEG standardization activities, princi-
    ples in, 295
`ISPostDecoding` function, 656
ITU-T, 730
    H.223 videoconferencing multiplex, 228
I-VOPs, 332

J

Java, 53, 150–51
Java application engine, 151
Java Archive (JAR) tools, 159
Java interfaces, 155n
Java language API, 46, 48
Java objects, 155n
Java programs, 48
JavaScript, 46, 131
Java virtual machine (JVM), 151
Joint Model (JM), 24n
joint stereo coding, 504–7
JPEG format, 2
just-in-time, 260
just noticeable differences (JNDs), 528

K

Kaiser-Bessel-derived (KBD) window, 495

L

language code, 555
large-step scalability, 519
large-step scalable audio coding, 519–23
Layer2D node, 123–24
Layout node, 134
leaf nodes, 105
least significant bit (LSB) plane, 341
levels, defined, 52, 587
life-cycle file format, 253–54
linear approach, 37
linear predictive coding (LPC), 452
    analysis, 464
    residual signals, 452
    spectral estimation, 513
    synthesis filter, 465
    vocoder, 452–53
linear predictive (LP) analysis, 452
LineProperties node, 122
line spectral pairs (LSP), 457, 480
line spectrum frequencies (LSF), 457
list description, 113
ListeningPoint, 565–66
Load By Time Stamp, 158

long-term prediction (LTP), 456, 508,
    511–13
long windows, 494
loop field, 134
lossless compression, 362n
lossy compression, 362n
low bit-rate coding (LBC) standard, 5
low-complexity (LC)
    encoding and decoding, 362
    profile, 491
low-delay AAC, 509, 515–17
L/R correlation, 510–11
luminance and chrominance (YUV) bitmaps,
    624

M

macroblocks (MBs), 298
    boundary
        motion compensation for, 324–25
        texture coding for, 327–30
    colocated, 310
main profile, 491, 620
manifold model, 425n
marching pattern, 432–33
marker bit stuffing, 322
Material 2D node, 122, 213
Material 3D node, 123
mean frequency, 534
MediaBuffer node, 136
MediaControl node, 136
media data, 258
media decoder plug-ins, 629
media events, 176
Media node, 136, 632, 651
MediaObjects, 628, 632
media objects, 66. *Also see* objects
media profiles, 588
MediaSensor node, 136
MediaStream objects, 632, 644
media streams, 72
    linking scene to its, 82–85
media time, 98
Meet, 45
memory-constrained applications, 374

mesh connectivity, 415–17
mesh decoding, 296
mesh geometry, 417
mesh motion coding, 419–20
mesh traversal, 419
metadata, 42, 45, 86, 254, 257
MFNode fields, 634–36
MIDI Manufacturers Association (MMA), 552
mime-type, 197*n*
Mitsubishi, 25
mobile connections, 2
mock testing, 713
modal density, 575–76
modeling time, 90–92
Modified Discrete Cosine Transform (MDCT), 489
modulated noise reference unit (MNRU), 713
mono/stereo scalability, 522
Monoyer Optometric Table, 671
moov atom, 257
most significant bit (MSB) plane, 341
motion, 419
motion compensation (MC)
    for boundary MBs, 324–25
    hybrid coding, 354
    prediction, 302
motion decoding and compensation, 296
motion estimation, 302
motion vector (MV), delta, 310
MovieTexture node, 108, 119
Moving Picture Experts Group (MPEG), launching of MPEG-4 project by, 2
MP3, 3
MP4, 48
MPDecoder interface, 176–77
MPEG Applications and Operational Environments (AOE) subgroup, 4–5
MPEG Component API, 180
MPEG-1, 3
MPEG-2, 227
    transporting MPEG-4 over, 268–80
MPEG-2 Section Filter API, 180
MPEG-2 section syntax, 272
MPEG-2 standards, 3

MPEG-2 STD extensions, 279–80
MPEG-2 Systems
    carriage of MPEG-4 on, 51–52
    specifications, 228
    transport of MPEG-4 elementary streams over, 274–75
    transport of MPEG-4 scenes over, 275–80
MPEG-2 transport streams, 43
MPEG-2 Video, 300
MPEG-4, 44, 227
    carriage of
        on IP networks, 52
        on MPEG-2 systems, 51
    other multimedia standards and, 52–56
    transporting
        over IP, 280–89
        over MPEG-2, 268–80
MPEG-4 additions to AAC, 508–18
    error robustness, 517–18
    long-term prediction, 511–13
    low-delay, 515–17
    perceptual noise substitution, 509–11
    twinVQ, 513–15
MPEG-4 Animation Framework eXtension, 384
MPEG-4 applications, 12–13, 56–60
    interactive broadcasting, 57–59
    multimedia conferencing and communities, 59–60
    multimedia portals, 56–57
MPEG-4 Audio, 46, 451
MPEG-4 audiovisual session, 152
MPEG-4 CELP coding, 455
    decoder, 456–57
    encoder, 455–56
    multipulse excitation, 458–59
    parameter decoding, 457–58
    regular pulse excitation, 459–60
    scalability, 460–61
    silence compression, 461
MPEG-4 content, 227–89
    delivery framework, 229–31
        DMIF-application interface, 231–38
            channel primitives, 233–35
            data primitives, 237–38

MPEG-4 content (continued)
    QoS monitoring primitives, 235–36
    service primitives, 232
    user command primitives, 236–37
   DMIF network interface, 238–40
    channel primitives, 243–45
    service primitives, 241–42
    session primitives, 240–41
    TransMux primitives, 242–43
    user command primitives, 245–46
   existent signaling protocols, 246–48
  file format, 253–54
   atoms, 262–65
    inside movie, 262–63
    inside sample table, 264–65
    inside track, 263
    top-level, 262
   hinting, 260–62
   MP4 example, 266–68
   physical structure, atoms and containers, offsets and pointers, 257–59
   random access, 265–66
   systems concepts in MP4, 259
   temporal structure, tracks and streams, time and durations, 254–56
   track types and storage, 259–60
  FlexMux tool, 248–51
   timing of stream, 251–53
  object descriptors as entry points to, 65–86
  transporting MPEG-4 elementary streams over MPEG-2 systems, 274–75
  transporting MPEG-4 over IP, 280–89
  transporting MPEG-4 over MPEG-2, 268–80
  transporting MPEG-4 scenes over MPEG-2 systems, 275–80
MPEG-4 content access procedure, 85–86
MPEG-4 delivery, 228
MPEG-4 encoders, 625

MPEG-4 eXtensible MPEG-4 Textual format (XMT), 41–42
MPEG-4 files, 43
MPEG-4 HILN audio coding, 530
  bitstream formal, 535
  parameter quantization and subdivision coding, 536–37
  spectral model parameter coding, 537
MPEG-4 HVXC coding, 461–62
  coder, 462–69
  decoder, 470–75
MPEG-4 Industry Forum (M4IF), 4, 30–31
  Web site (*www.m4if.org/join.html*), 31
MPEG-4 licenses, 31
MPEG-4 objectives, 4–13
  functionalities, 5
   compression efficiency, 6–7
   content-based interactivity, 5–6
   universal access, 7
  requirements, 7–10
  tools, 10–12
MPEG-4 player architecture, 628–31
  structure, 631–32
MPEG-4 presentations, 65
  association of, 83–85
MPEG-4 project, 2, 4, 37–61
  design goals, 38–40
  end-to-end walkthrough, 40–43
  terminal architecture, 43–47
  tools, 47
   audio, 50–51
   Delivery Multimedia Integration Framework (DMIF), 51
   levels, 52
   other MPEG-4, 51–52
   profiles, 52
   systems, 47–48
   visual, 48–50
MPEG-4 Proposal Package Description (PPD), 5, 25
MPEG-4 scalable audio coding, 518–19
  bit-sliced arithmetic, 523–25
  large-step, 519–23
MPEG-4 schedule, 25–30
MPEG-4 speech coders, 454

MPEG-4 standard, 586
    adoption of, 31
    scope of, 37–38
MPEG-4 standard organization, 21–25
    Part 1, systems, 21–22
    Part 2, visual, 22
    Part 3, audio, 22
    Part 4, conformance testing, 23
    Part 5, reference software, 23
    Part 6, delivery multimedia integration
        framework (DMIF), 23
    Part 7, optimized visual reference soft-
        ware, 24
    Part 8, carriage of MPEG-4 contents over
        Internet Protocol (IP) networks, 24
    Part 9, reference hardware description,
        24
    Part 10, advanced video coding, 24
MPEG-7, 3–4
MPEG-7 stream, 87–88
MPEG-21 project, 4
MPEG-21 vision, 4
MPEG-J, 46, 48, 53, 149–84, 188, 195
    architecture, 154–56
    delivery
        class dependency, 158–59
        elementary stream, 159
        time stamp semantics, 158
        MPEGlets, 156
            life cycle, 156–57
            name scope, 157
        profiles, 180–81
        security, 159–60
MPEG-J APIs, 160–62
    decoder, 176–77
    network, 178–79
        channel control, 179
        query, 179
    resource, 173–76
        CapabilityManager, 174–75
        events, 175–76
        ResourceManager, 173–74
    scene, 164–65
        field IDs, 168
        field values, 168–70

        node interface, 168
        node types, 168
        node values, 170–72
        scene interface, 167–68
        SceneListener interface, 165–67
        SceneManager interface, 165
    service information and section
        filtering
        MPEG-2 Section Filter, 180
        MPEG Component, 180
        Resource Notification, 180
        Service Information API, 180
    terminal, 163–64
MPEG-J applications, 154, 181–82
MPEG-J application engine, 154
MPEG-J architecture, 153
MPEG-J decoder, 155
MPEG-J dimension, 587
MPEG-J elementary stream, 152
MPEG-J profiles, 180–81, 588
MPEG-J profiling, 619–20
MPEG-J reference software, 182–83
MPEGlets, 46, 48, 53, 56, 152, 153, 154, 180
    life cycle, 156–57
    name scope, 157
MPEG *modus operandi,* 15–21
    mission, 16–17
    principles, 17–18
    standards development approach, 18–20
        core experiments, 21
        verification models, 20–21
M/S stereo coding, 505
multimedia conferencing and communities,
    59–60
Multimedia Content Description Interface, 3
multimedia portals, 56–57
multiphase excitation, 458–59
multiple concurrent data streams, coding of,
    6–7
multiple quantizer mode, 366
multipulse excitation (MPE), 454, 456
multiscale zero-tree wavelet entropy coder
    (MZTE), 365
MUSHRA method, 715–17

Musical Instrument Digital Interface (MIDI),
        549, 551
    Downloadable Sounds 2 format, 552–53
    format, 50
    general, 553
Music-N languages, 549
MuxCode mode, 248–49
MuxCodeTable, 249

N

narrowband digital audio broadcasting test,
        717–18
    assessment method and test design, 720–
        21
    conditions, 718–19
    data analysis, 721
    material, 719–20
    results, 722–24
narrowband mode, 455
natural audio, 452
    coding for, 545
    signals in, 709
natural objects, 451
natural speech, 451
natural video coding, 293–376, 385
    of arbitrarily shaped video objects,
        318
        arithmetic encoding of binary alpha
            blocks, 321–22
        binary only shape coding, 322–23
        binary shape coding, 318–19
        context computation for binary
            alpha blocks, 319–21
        gray-level shape coding, 323–24
        motion compensation for boundary
            MBs, 324–25
        predictive motion vector coding,
            325–27
        texture coding for boundary MBs,
            327–30
    basic principles, 295–300
    functionalities and application scenarios,
        294–95
    of rectangular video objects, 300–303

AC/DC prediction for intra macrob-
            locks, 315–16
        alternative scan modes, 316–18
        direct mode in bidirectional motion
            compensation, 309–11
        global motion compensation, 307–9
        new motion-compensation tools, 303
        new texture tools, 311–12
        quantization of DCT transform coef-
            ficients, 312–15
        quarter-pel motion compensation,
            304–7
        temporal prediction structure, 303–4
    scalable, 330–32
        SNR fine granularity scalability,
            340–43
        spatial scalability, 332–36
        temporal scalability, 337–40
    special tools, 343
        error-resilient, 345–55
        interlaced, 344
        reduced resolution, 355–56
        short video header mode, 359
        sprite, 356–59
        texture coding for high-quality appli-
            cations, 359–61
    visual texture, 361–62
        bitstream packetization, 372–74
        shape-adaptive wavelet, 368
        spatial and quality scalability, 369–
            72
        tiling, 374–75
        tools, 363–64
        wavelet, 364–68
natural video objects, 46
N-bit tool, 311
Network APIs, 162, 178–79
network query, 179
Network Time Protocol (NTP), 282
neutral sets, 416
new motion-compensation tools, 303
NewNode interface, 171
NEWPRED, 354
new sensors, 137–38
new texture coding tools, 311–12

node coding table, 635
Node Data Type tag, 113
NodeField, 632–33
node ID, 109*n*, 167
Node interface, 168
node reference, 170
node types, 108, 168, 635
node values, 170–72
noise, 530
noise allocation, 489
noise component modeling, 534
noise definition, 531
noise energies, transmission of, 510
noiseless coding, 497–99
noise models, 527
noise shaping
    by predictive coding, 500
    by using scalefactors, 496
noise update frame, 475
Non-Backward Compatible (NBC) coding
        scheme, 490
nonframing, 254
nonlinear, object-based interactive approach,
        37
nonmanifold model, 425*n*
non-normative areas, 12
non-normative reference software modules,
        625
nonuniform quantization, 495–96
N-order extensions, 321
NormalInterpolator node, 124
normal play time, 98
normative software module, 624
normative tools, 670
NTT, 25

O

object-based video coding, 293
object-based representation of audio signals,
        451
object-based scalability, 330
object-based scene, 11
object clock reference (OCR), 644
    time stamp for, 97

object content information (OCI), 45, 86–88,
        188
ObjectDescriptor, 647
ObjectDescriptorImp :: SetStream, 655
object descriptor (OD), 39, 45, 46, 70–71, 646
    as entry points to MPEG-4 content, 65–
        86
        content complexity description, 77–
            79
        linking scene to its media streams,
            82–85
        stream description, 72–75
        streaming, 79–82
        stream relationship description, 76–
            77
        syntactic description language, 66–
            70
    execution of objects, 651–52
    parsing stream, 647–51
    streaming, 79–82
    streaming configuration information in,
        81–82
    syntactic description language, 646–47
object descriptor (OD) dimension, 587
object descriptor (OD) framework, 47, 104,
        188, 192, 219–22
object descriptor (OD) management, 627
object descriptor (OD) objects, execution of,
        651–52
object descriptor (OD) profiles, 588, 619
object descriptor (OD) stream, parsing, 647–
        51
ObjectDescriptorUpdateImp :: Commit
        function, 651–52
objects, 151
    metadata, 86–88
    scalability of, 49
object time bases, 97–98
object type, 586–87
occlusion effect, 577
OCI streams, 86
off-line creation, 40
omniDirectivity, 577
one functionality, one tool principle, 17
OnPacketReceived, 643

`OpenGLVisualRenderer :: Paint` method, 662–63

open-loop pitch search, 464

open-source project, 624

orchestra file, 550

`OrientationInterpolator` node, 124

overlapped block motion compensation (*OBMC*), *303*

P

packet-based resynchronization, 347–50

Packet IDentifier (PID), 271

Packetized Elementary Stream (PES), 269–70

    conveying FlexMux streams in, 276–77

    conveying-SL-packetized streams in, 276

packetizing streams, sync layer, 93–96

`PacketSequenceNumber,` 95, 643

padding of transparent pels, 328

padding process, 324

paired comparison method, 714–15

parameter decoding, 457–58, 538–39

    signal synthesis and, 529

parameter estimation, 528–29, 531–34

parameter quantization

    coding and, 529

    perceptual models for, 528

    subdivision coding and, 536–37

parametric audio coding, 525–26

    encoding and decoding concepts, 528–29

    source and perceptual models, 526–28

parametric coders, 50, 454

parametric encoding and decoding concepts, 528–29

Paris MPEG meeting, 4

parsing code, 633

parsing macros, 637

parsing OD stream, 647–51

parsing table, 634

partial enhancement, 338

patent pools, 31n

Peak Signal-to-Noise ratio (PSNR), 409–10

perceived room response, effect of source-listener distance in, 576

perceptual models, 489, 569

    for component selection, 528

    for parameter quantization, 528

perceptual noise substitution (PNS), 508, 509–11

    signaling in, 510

`PerceptualParameters` node, 574–77, 579

perceptual weighting filter, 465

periodic component coding, 514

periodic spatial resynchronization, 347

`Personal` profile, 619

photometry, 422

    coding for, 433–34

    information in, 422

physical models, 526, 569

piece mode, 358

pitch lag, 511

pitch modification, 474

pitch-shifting, 529

`PlaneSensor2D` node, 121

`PlaneSensor` node, 121

plug-ins, 57, 630, 652

    decoder, 654–56

    DMIF, 652–84

    IPMP, 656

pointers, 46

`PointSet` node, 126

polyphase filterbanks, 489

`PositionInterpolator2D` node, 124

`PositionInterpolator` node, 124

power spectral density (PSD), 500

prediction, 502–4

prediction error, 302

    coding of, 365–66

predictive coding, 302

    noise shaping by, 500

    techniques in, 502

predictive embedded zero-tree wavelet coder (PEZW), 365

predictive motion vector coding, 324–27

predictor control, 457

predictor reset, 503

pre-echo effect, 494

presentation engine, 150, 152

primitives, 653

processor complexity units (PCU), 606
profile@level combinations, 585–86
profileLevelIndicationIndexDescriptor, 74
profiles, defined, 52, 587
profiling and conformance, 583–620
    audio, 602–9
    goals and principles, 584–88
    graphics, 609–13
    MPEG-J, 619–20
        Main profile, 620
        Personal profile, 619
    object descriptor, 619
    policy and version management, 588–92
    scene graph, 613–19
    visual, 592–601
Program Association Table (PAT), 272
Program Map Table (PMT), 272
Program Specific Information (PSI), 272
program stream (PS), 268, 270
Program Stream Map (PSM), 270
progressive transmission mode, 437
progressive video, 49
prosodic information, 556
prosody, 555
PROTO, 108, 110, 132, 639–43
proxy objects, 657
publishing, 40
P-VOPs, 334

Q

qosReport field, 235
QP node, 114
Q_0 quantization error, 367
Quality of Service (QoS) descriptor, 74–75, 234
Quality of Service (QoS) management, 230–31
Quality of Service (QoS) monitoring, 51
Quality of Service (QoS) monitoring primitives, 235–36
quality scalability, 49, 341
quality update mode, 358

quantization, 114, 495–97
    of DCT transform coefficients, 312–15
quantization scaling factor, 408
quarter-pel motion compensation, 303, 304–7
quick error recovery, 696
QuickTime, 53, 228
    file format for, 254

R

RAM complexity units (RCU), 606
random access, 265–66, 293
randomAccessPointFlag, 95
Real-Time Streaming Protocol (RTSP), 230, 247
Real-Time Transfer Protocol (RTTP), 52
Real-Time Transport Protocol (RTP), 229
    heuristic delivery-timing model of, 288
    payload format in, 282
receiver modeling, 568
recovery mechanisms, 47
Rectangle node, 121
rectangular video, 593, 597–99
rectangular video objects, coding of, 300–303
    AC/DC prediction for intra macroblocks, 315–16
    alternative scan modes, 316–18
    direct mode in bidirectional motion compensation, 309–11
    global motion compensation, 307–9
    new motion-compensation tools, 303
    new texture tools, 311–12
    quantization of DCT transform coefficients, 312–15
    quarter-pel motion compensation, 304–7
    temporal prediction structure, 303–4
reduced resolution coding, 355–56
reduced resolution VOPs, 355
reference block, 307
reference image, 302
reference points, 357
reference software, 623–65
    MPEG-4 player architecture, 628–31
        structure, 631–32
    object descriptors, 646

reference software (continued)
    execution of OD objects, 651–52
    parsing OD stream, 647–51
    syntactic description language, 646–47
    plug-ins, 652
        decoder, 654–56
        DMIF, 652–84
        IPMP, 656
    PROTOs, 639–43
    scene graph, 632–33
        parsing, 633–38
        rendering, 638–39
    synchronization, 643
        layer, 643–44
        MediaStream objects, 644–46
    systems, 625–28
    3D compositor, 660–61
        differences between 2D and, 661–62
        overview of key classes, 662
        rendering process, 662–63
        support for 2D nodes, 664
        user navigation, 664–65
        using core framework, 662
    2D compositor, 656–57
        DEF and **USE** handling, 657–58
        rendering optimization, 658–59
        synchronization, 659–60
        using core framework, 657
reference video encoder, 625
reflection coefficients, 527
region-based video coding, 293
region of sound, rendering, 572
registration descriptor, 75
regular pulse excitation (RPE), 454, 456, 459–60
rendering, 112*n,* 154
rendering optimization, 658–59
RenderMan, 383
Replace command, 216
Replace Scene command, 113
ReplaceValue, 635
resolution coding, reduced, 355–56
Resource APIs, 161–62, 173–76
ResourceManager, 173–74

Resource Notification API, 180
restricted boundary prediction mode, 437
resynchronization, 347
resync markers, 347
reverbDelay, 573
reverbFreq, 573
reverbLevel, 573
reverbTime, 573
reversible variable-length coding (RVLC), 352–53
    tool for, 476, 517–18
robustness in error-prone environments, 7, 49
room effect, filtering of, 576–77
root triangle, 434
Route class, 633
Route object, 633
ROUTEs, 107–8, 110, 140, 215
    modifying, 167–68
RTP/UDP over IP, 43
run ( ) method, 156
run-time subroutines, 637

S

scalability, 293, 460–61
    amplitude, 341
    bit rate, 488
    large-step, 519
    mono/stereo, 522
    quality or SNR, 341
    SNR fine granularity, 340–43
    spatial, 332–36, 334–36
    spatial and quality, 369–72
    temporal, 337–40
    view-dependent, 440–42
scalable audio coding, 509
scalable coding schemes, 330
scalable sampling rate (SSR) profile, 491
scalable vector graphics (SVG), 55, 197
scalable vector quantization scheme, for harmonic spectral magnitudes, 466–68
scalable video coding, 294–95, 330–32
    spatial, 332–36
    temporal, 337–40
ScalarInterpolator node, 124

scalefactor bands, 496

scalefactors

encoding of, 497

noise shaping by using, 496

scaling index, 408

scan interleaving (SI) method, 335–36

scanning process, 316–17

scene, 104–6

scene APIs, 161, 164–65

scene changes, 109–12

scene description, 104, 212–19

data, 45, 48

information, 45–46

language, 39, 42

profiles, 587

scene description tools, 545

scene graph, 139, 632–33

API, 631

handling, 627

nodes, 587

parsing, 633–38

profiling, 613–19

rendering, 638–39

Scene interface, 167–68

SceneListener interface, 165–67

SceneManager interface, 165

scene tree, 105

scene updates, 55, 109

scheduler description, 549

score and control languages, 549

screen, 103n

scripting, 129, 130–31, 195

Script node, 108

scrollRate field, 134

scrollVertical field, 134

searching applications, 47

second quantization method (H.263 quantization), 315

sectioning, 497

SegmentDescriptor, 136

segments, 408

semantic description and access management, 86–88

ServerCommand node, 139

service information API, 180

service primitives, 232–33, 241–42

Session Description Protocol (SDP), 281

session primitives, 240–41

Session Service Access Point, 231

SetDecoder calls Decoder :: Setup, 655

set partitioning in hierarchical threes (SPHIT), 365

Set-top box, 55–56

SFNote fields, 634–36

SF2DNode, 635

SF3DNode, 635

SFGeometryNode, 635

shadow sync samples, 255–56

shape-adaptive discrete cosine transform (DCT), 328–30

shape-adaptive wavelet coding, 363, 368

Shape node, 108, 117

short filterbank windows, handling of, 503–4

short video header mode, 359

short windows, 494

signal decomposition, 528–29, 531–34

signal modification, 529

signal synthesis, 538–39

Signal-to-Noise (SNR) fine granularity scalability, 340–43

significance map, 365

silence compression, 455, 461

simple face animation profile, 446

simple FBA profile, 446

simple mode, 248

Simulation Model (SM), 20

simultaneous double stimulus for continuous evaluation method, 677–80

single quantizer mode, 366

single stimulus method, 674–75

sinusoidal extraction, 531

sinusoidal magnitudes and frequencies, 454

sinusoidal models, 526–27

size field, 134

skipped mode, 311

implicit, 311

SLConfigDescriptor elements, 93

SL-packetized streams, 93
conveying, in PES, 276
transport of, over Internet Protocol (IP),
285–88
smoothScroll field, 134
SnakeWipe transition, 189
software, decoding, 586
Sony, 25
Sound2D, 565
sounds
composition, 103–4
efficient, object-based coding of, 548
spatial presentation of, 548
sound synthesis, controlling, in structured
audio, 551–52
source-level APIs, 630
source-listener distance, 576
source modeling, 567
spatial and quality scalability, 369–72
spatial composition information, 103
spatial layout, 205
spatial presentation of sound, 548
spatial scalability, 49, 332–36, 334–36
for binary shape, 334–36
speaker-related information, 555
special video coding tools, 343
error-resilient, 345–55
interlaced, 344
reduced resolution, 355–56
short video header mode, 359
sprite, 356–59
texture, for high-quality applications,
359–61
specification development subgroups, 16
spectral coefficients, interleaving of, 514
spectral model parameter coding, 537
spectral normalization, 513–14
spectrum normalization, 515
speech and general audio coding, 603–4,
607
speech coding, 46, 50, 451–83, 603
error robustness, 475–77
error-resilient CELP, 481–82
error-resilient HVXC, 477–81
HVXC, 461–62

MPEG-4 CELP, 455
decoder, 456–57
encoder, 455–56
multipulse excitation, 458–59
parameter decoding, 457–58
regular pulse excitation, 459–60
scalability, 460–61
silence compression, 461
MPEG-4 coders, 454
speech communication, 710
speech communication test, 730
assessment method and design, 733–34
conditions, 730–32
data analysis, 734
material, 732–33
results, 734–36
Sphere node, 126
SphereSensor node, 121
SpotLight, 126
sprite, 356
sprite coding, 356–59
sprite motion trajectories, 357
Sprite VOP (S-VOP), 357
square brackets, 67
start code emulation, 322
Start Loading Time Stamp, 158
StaticCapability interface, 174
static compliance, 585
still visual objects, 594, 600
stitching, 425
stitching mode, 434–35
stochastic codebook, 456
StreamConsumer, 651
StreamConsumer :: SetStream, 652
stream description, 72–75
streamed graphics, 58
streaming object descriptor (OD), 79–82
streamPriority element, 73
stream relationship description, 76–77
stream segments, 136
structured audio
applications in, 554–55
coding for, 548–55
algorithmic synthesis and processing, 549–52

applications, 554–55

interface between **AudioBIFS** and, 554

interface with **AudioBIFS** and, 554

wavetable synthesis, 552–54

controlling sound synthesis in, 551–52

object types, 553–54

structured audio orchestra language (SAOL), 384, 549–50

structured audio sample bank format (SASBF), 549

structured audio score language (SASL), 549

subband-by-subband scanning order, 367

subframes, 456

subimages, 375

subscenes, 109

substructure, 250

sum/difference coding, 504–5

SupplementaryContentIdentificationDe-scriptor, 89

sweep rate, 534

synchronization, 627, 643, 659–60

layer, 643–44

MediaStream objects, 644–46

synchronization layer (SL), 45, 643–44

management of, 629

syntax, 229

tools, 588

synchronization modules, 200

Synchronized Multimedia Integration Language (SMIL), 48, 55, 187, 191

reusing, in XMT-Ω, 194–95

Syntactic Description Language (SDL), 66–70, 248, 635, 646–47

synthesis engine of SA decoder, 550

synthesized audio coding, 50

synthesized speech coding, 50–51

synthetic and hybrid natural/synthetic visual, 600–601

synthetic and natural hybrid coding (SNHC), 384, 546

of audio, 546–48

audio and audio composition, 545–80

need for, 385–88

synthetic audio, 452, 604, 607

coding of, 545

tools for, 384–85

synthetic music coding, 46

synthetic objects, 451

synthetic speech, 452, 604–7

synthetic visual objects, 46, 594

system decoder model (SDM), 47, 98–100, 252

sYstems Model (YM), 20n

Systems reference software, 625–28

T

talking heads, 385

technical reports, 15

television paradigm, 1–2

temporal composition information, 104

temporal envelope, estimation of, 532

TemporalGroup, 139

temporal noise shaping (TNS), 499–502

filtering, 500

inverse, 500–501

temporal prediction, 332

structure in, 303–4

temporal random access, 6

temporal resolution stability, 694, 696

temporal scalability, 49, 337–40

TemporalTransform, 139

terminal APIs, 161, 163–64

Test Model (TM), 20

text layout, 134–36

Text node, 117, 126

text-to-speech (TTS), 384–85

text-to-speech (TTS) sequence, 556

text-to-speech (TTS) synthesis, 46

text-to-speech (TTS) tools, 50

text-to-speech interface (TTSI), 410, 546, 555–57, 604–5

texture coding

for boundary MBs, 327–30

for high-quality applications, 359–61

texture decoding, 296

texture mapping, 388

Texture node, 108, 123

TextureTransform node, 123

T/F duality, 499–500

3D audio scenes, enhanced modeling of, 567–69

3D compositor, 660–61
  differences between 2D and, 661–62
  overview of key classes, 662
  rendering process, 662–63
  support for 2D nodes, 664
  user navigation, 664–65
  using core framework, 662

3D mesh coding, 49, 389, 421–22
  examples, 437–40
  objects, 422–25
  scheme, 425–37

3D rendering, 112

3D shopping portal, 57

3rd Generation Partnership Project, 598

tiling, 374–75

tiling scheme, 363

time base, 96

time containers, 199

Time Domain Aliasing Cancellation (TDAC), 493

time-domain division of room impulse response, 575

time/frequency (T/F) coders, 487

time/frequency audio coping, 488–90

Time Manipulations, 202–3

time scale modification, 473–74

time-scaling, 529

TimeSensors, 107, 110

time stamps, 91–92, 92–93, 95–96, 110, 158

time values, 113

timing-identification synchronization, 47

timing model, 273–74

timing model and synchronization of streams, 90
  distributed content, time bases, and OCR streams, 96–98
  media time, 98
  modeling time, 90–92
  packetizing streams, sync layer, 93–96
  system decoder model, 98–100
  time stamps and access units, 92–93

timing modules, 198–202

top node, 126

topological analysis, 426–29

TouchSensor node, 120, 121

track identifier, 254–55

track references, 256

training listening, 713

trak atom, 257–59

transform coding, 500

transform-domain weighted interleave vector quantization (TwinVQ), 508–9, 513–15

Transform node, 105
  animation definition for, 398

transient models, 527

transitional BAB, 336

transmission buffering delay, 696

transmission order, 304

TransMux Association Tag (TAT), 243

TransMux channels, 229

TransMuxes, 229

TransMux primitives, 242–43

transport stream (TS), 268, 270–73

transport tools, 48

tree-depth scanning order, 367

triangle data, 433

triangle tree, 426
  coding, 432–33

triangular mesh, 426

triangulation, Delaunay, 418–19

triangulation, uniform, 417–18

2D compositor, 656–57
  DEF and **USE** handling, 657–58
  differences between 3D compositors and, 661–62
  rendering optimization, 658–59
  synchronization, 659–60
  using core framework, 657

2D mesh coding, 49, 389, 411
  object, 411–12
  scheme, 413–21

2D mesh map, 426

2D mesh object, 411–12

2D rendering, 112

2D scene, 104
2D segmented video, 57

U

unequal error protection (UEP), 476, 518
unflatness, 488
Unified Modeling Language (UML) representations, 162
uniform triangulation, 415, 417–18
universal access, 7, 39
unsigned int, 67
unvoiced segments, vector excitation coding for, 468–69
unvoiced signals, vector excitation coding (VXC) of, 464–65
upStream flag, 74
USE clause, 105–6
user command primitives, 236–37, 245–46
User Datagram Protocol (UDP), 281
User Interaction Decoder, 137
User Interaction Stream, 137
user plane, 229

V

validation. *See* audio testing for validation; video testing for validation
vantage points, 664
variable bit rate coding, 474–75
variable length method, 435
vector description, 113
vector excitation coding (VXC), 454
    for unvoiced segments, 462, 468–69
    for unvoiced signals, 464–65
vector quantization (VQ), 454, 514–15
    of harmonic magnitudes, 464
    of harmonic spectral magnitudes, 465–68
verification models, 20–21
verification tests, 19, 669–70
Version 2 coding efficiency, 736–37
    assessment method and test design, 740
    data analysis, 740
    test conditions, 737–39
    test material, 739–40
    test results, 740–44

Version 2 error-robustness test, 744
    conditions, 744–46
    data analysis, 747
    test material, 747
    method and experimental design, 747
    test results, 747–49
vertex graph, 426, 430n
    coding of, 429–32
vertical padding, 324
very critical error conditions, 698, 702
Very High Compression Efficiency, 37
Very Low Bit Rate Audiovisual Coding, 37
Very Low Bit Rates, 37
Video-CD, 3
video coding
    standards in, 1
    tools for, 682
video compression tools, 48–49
Video frames, 659–60
video object, 297
video object layers (VOLs), 297
Video Object Planes (VOPs), 26, 45, 92, 297–98
    bidirectionally predicted, 302
    bounding box, 298–99
video-only objects, 39
video packets, 316n, 327
video testing for validation, 669–704
    **Advanced Real-Time Simple
        (ARTS)** profile test, 694–96
        conditions, 696–99
        data analysis, 701–2
        material, 699–700
        method and design, 700–701
        results, 702–4
    coding efficiency for low and medium
        bit-rates test, 691
        conditions, 691–92
        data analysis, 693
        materials, 692
        method and design, 692–93
        results, 693–94
    content-based coding test, 688
        conditions, 688–89
        data analysis, 690

video testing for validation (continued)
material, 689
method and design, 689–90
results, 690–91
error-resilience test, 680–81
conditions, 681–84
data analysis, 686
material, 684
method and design, 684–86
results, 686–88
general aspects, 670–71
laboratory setup, 672
selection of test material, 671
selection of test subjects, 671–72
test plan, 672–73
training phase, 673
test methods, 673–74
double stimulus continuous quality
scale method, 676–77
double stimulus impairment scale
method, 675–76
simultaneous double stimulus
for continuous evaluation
method, 677–80
single stimulus method, 674–75
view-dependent mask, 445–46
view-dependent object, 442–44
view-dependent scalability, 440–42
coding scheme, 444–46
object, 442–44
parameters, 443–44
Viewpoint node, 126
virtual codebooks tool, 476, 517
virtual **DescEncDec** function, 650
Virtual Reality Model Language (VRML),
104, 187, 384
virtual samples, 305
visual objects (VOs), 297, 593
visual profiles, 597–99
visual profiling, 592–601
visual sequence (VS), 297
visual SNHC tools, 383–447
2D mesh coding, 411
object, 411–12
scheme, 413–21

3D mesh coding, 421–22
examples, 437–40
object, 422–25
scheme, 425–37
face and body animation, 389–92
body animation parameters, 406
body definition parameters, 399–401
default facial expression and body
posture, 392–93
FAB and BAP coding schemes, 406–
10
facial animation parameters, 403–6
facial definition parameters, 396–99
FBA and text-to-speech interface,
410
FBA objects, 393–96
FDP and BDP coding schemes, 401–
3
view-dependent scalability, 440–42
coding scheme, 444–46
object, 442–44
profiles and levels, 446–47
visual texture coding (VTC), 294, 346, 361–
62, 375, 385
bitstream packetization, 372–74
shape-adaptive wavelet, 368
spatial and quality scalability, 369–72
tiling, 374–75
tools, 363–64
wavelet, 364–68
vocoder principle, 509
voiced/unvoiced (V/UV) decision, 452, 464
VRML sound model in **BIFS,** 558–59
V/UV mode indicator, 458

W

warping process, 308
wavelet and zero-tree-based compression
algorithm, 363
wavelet transform coding, 294, 295, 362, 364–
68
wavetable synthesis, 552–54, 553, 555
W3C Scalable Vector Graphics (SVG)
format, 54

W3C Synchronized Multimedia Integration Language (SMIL), 45, 54
W3C XML, 53–54
W3C XML Schema language, 188
W3C XML standard, 187
Web browsers, 47
Web3D consortium, 48, 387, 388
Web3D Virtual Reality Modeling Language, 53
Web-like content viewing, 59
Web paradigm, 2
weighted vector quantization, 514–15
weighting matrices, 313
wideband mode, 455
windowing, 531–32
Working Group 11 (WG11) of Subcommittee 29 (Coding of Audio, Picture, Multimedia and Hypermedia Information) of the ISO/IEC (International Organization for Standardization/International Electrotechnical Commission) Joint Technical Committee 1 (JTC1), 13
working model, 19
wrap field, 134

X

X3D, 55
interoperability with, 223–24
XMT, 55, 187

XMT-A format, 54, 189, 191–92, 210
deterministic mapping of, 222–23
document structure, 210–11
interoperability with X3D, 223–24
object descriptor framework, 219–22
scene description, 212–19
timing, 211–12
XMT two-tier architecture, 190–93
XMT-Ω events, 201–2
XMT-Ω examples, 205–10
XMT-Ω format, 54, 189, 190–91, 192, 193–94
animation, 203–5
examples, 205–10
objects, 195, 197–98
reusing Synchronized Multimedia Integration Language (SMIL) in, 194–95
spatial layout, 205
time manipulations, 202–3
timing and synchronization, 198–202

Z

zero-padding, 328
zero-tree coding (ZTC), 294, 295, 362, 366
zip tools, 159
ZLIB compression format, 159